1993

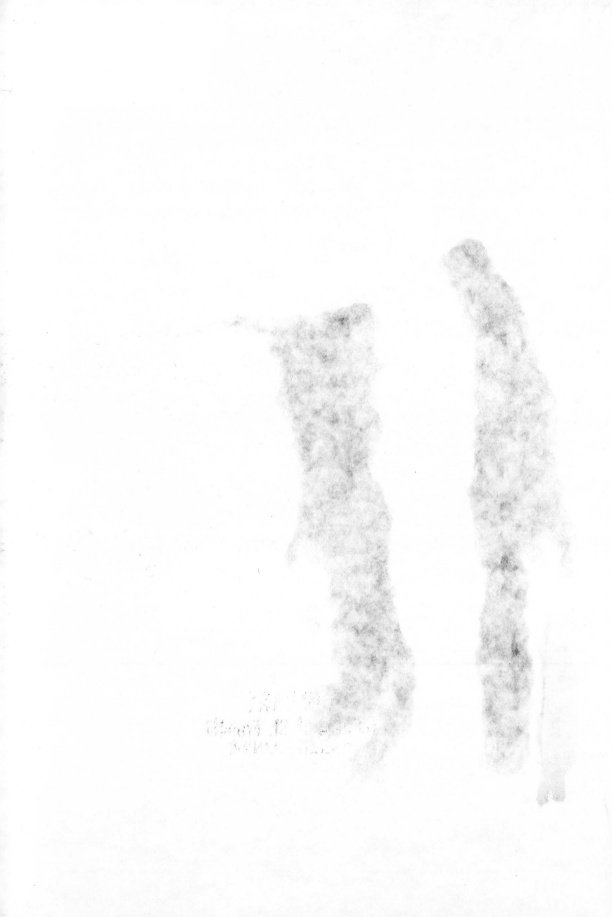

A Study of Enzymes

Volume I
Enzyme Catalysis, Kinetics, and Substrate Binding

Author
Stephen A. Kuby
Professor of Biochemistry
Research Professor of Medicine
Laboratory for Study of Hereditary and Metabolic Disorders
Departments of Biochemistry and Medicine
University of Utah
Salt Lake City, Utah

CRC Press
Boca Raton Ann Arbor Boston

Library of Congress Cataloging-in-Publication Data

Enzyme catalysis, kinetics, and substrate binding.

(A Study of enzymes; v. 1)
Includes bibliographical references and index.
1. Enzyme kinetics. 2. Enzymes. I. Kuby, Stephen Allen, 1925—
II. Series.
QP601.S677 vol. 1 [QP601.3] 574.19′25s 90-2276
[574.19′25]
ISBN 0-8493-6987-8

Direct all inquiries to CRC Press, Inc., 2000 Corporate Blvd., Boca Raton, Florida 33431.

Library of Congress Number 90-2276
International Standard Book Number 0-8493-6987-8
Printed in the United States

Vere scire, esse per causas scire
(To know truly, is to know by causes)

Francis Bacon (1561—1626)

Dedicated to my loyal, loving wife, Josette, without whose persistence this task would never have been completed; and also to my students and colleagues, whose sometimes lively discussions, requests, and whose very insistence forced me to bring to a close these writings of several years. My gratitude to my brother George, for his mathematical critique, to Gerald for many of the figures and schemes drawn by computer, and to Rose and Lori for much of the typing.

PREFACE

Volume I has been completed first. It is largely a summary of almost 4 decades of our own work, and accordingly many of the examples and illustrations are drawn from our own published works. Although not every topic in Enzyme Kinetics or in Equilibrium Substrate Binding will be covered herein, there is enough breadth and scope to cover much that has been of interest to us and enzymologists, in general. It has been used as a text in Biochemistry 707 (Topics in Biophysical Chemistry, with Special Emphasis on Enzyme Kinetics), and might form the nucleus of a similar course, but hopefully it should be of some interest to graduate students and to teachers and researchers alike, if they find enzymology the fascinating and absorbing subject that it has been to this author over these past years.

Interestingly, the subject still appears to be in a remarkable state of flux as witnessed by the very recent questioning of the generality of a very important concept, viz., that of the "rate-limiting step" in certain steady-state enzymatic reactions by D. Northrop (606,607) and the possibility of a more consistent redefinition by W. Ray (608). In Chapter 6 (on Kinetic Isotope Effects) this interesting problem will be briefly touched upon in relation to the recent attempts to utilize the potential mechanistic applications of "kinetic isotope effects".

Even the subject of "equilibrium ligand binding", which superficially might be considered to be in a current state of almost static development, has been the very recent subject of a vigorous controversy in regard to the correct estimation of molecular receptor capacity by Scatchard plots in certain binding studies (Klotz, 620; and Feldman, 621). The subject of "equilibrium ligand binding" will be developed in Chapter 8, and Klotz's arguments (620) will be very briefly presented there.

Volume II on Enzyme Mechanisms will follow, with the aid of my colleagues and associates, and will present selected studies which are currently being vigorously pursued by a number of approaches and techniques which are at the "cutting edge" of biochemistry.

SELECTED BIBLIOGRAPHY

A number of excellent reviews and monographs have been written on the subject of "Enzyme Kinetics" and "Equilibrium Ligand Binding Measurements"; some of these are presented below, as well as a few specific references which will prove to be valuable in our discussions. In addition, a few general references to mathematics, and chemical kinetics which have been found useful by the author, also listed below. Finally, some references are cited to the "Mechanism of Action of Enzymes", and which will have a bearing on these writings and those to follow.

Thus, refer to the following references:

1. Enzyme kinetics
   A. Reviews: References 7, 9, 10, 22, 40, 51, 52, 82, 96, 103, 164, 169, 184, 202, 489, 494, 554, 555—557, 568, 569, 574, 602, 693, 773, 774, 935, 963.
   B. Some specific references: References 4, 8, 11, 14, 15, 17—20, 24, 26, 31, 38, 39, 42—47, 49, 50, 53—58, 60—62, 64—66, 68, 69, 71, 84, 157, 158, 165—168, 180, 181, 203—227, 490, 508, 581, 584, 585, 588—591, 601, 606—608, 610, 617—619, 625, 629, 630, 639, 653, 657, 658, 680, 681—685, 689, 690, 692, 812, 813, 841, 842, 936.
2. Mathematical references: References 170—179, 228—230, 382, 621.
3. Chemical kinetics: References 231—236, 577, 587, 592, 595, 609, 631, 634, 640—642, 644, 659, 684.
4. Equilibrium ligand binding measurements: References 58, 59, 74, 113, 115, 144—148, 237—249, 573, 596, 599, 600, 602, 603, 611, 620—624, 632, 633, 638, 646—648, 650, 652, 655, 687, 722, 938.
5. Mechanism of action of enzymes: References 103, 190, 191, 202, 250—270, 297, 494, 518, 555—557, 571, 572, 575, 576, 578—580, 582, 583, 588, 593, 598, 602—605, 612—616, 626—628, 635—637, 645, 649, 651, 655, 656, 660, 683, 686, 688, 807, 954, 966.

ACKNOWLEDGMENT

Much of the work by S. K. and co-workers cited in this volume was supported, in part, by grants-in-aid from the National Institutes of Health, U.S. Public Health Service.

THE AUTHOR

Stephen A. Kuby, Ph.D., is a Professor of Biochemistry and Research Professor of Medicine at the University of Utah School of Medicine, Salt Lake City, Utah. He is also the head of the Biochemical Division of the Laboratory for the Study of Hereditary and Metabolic Disorders, University of Utah Research Park, Salt Lake City, Utah 84108. He graduated *Summa cum Laude* from New York University in 1948 with an A.B. in Chemistry. He obtained his M.S. in Biochemistry in 1951, and his Ph.D. in Biochemistry in 1953 at the University of Wisconsin under Professor H. A. Lardy. As a recipient of the U.S. Public Health Service Fellowship, he spent his post-doctorate period with Professor B. Chance at the University of Pennsylvania (Johnson Foundation for Medical Physics) and with Professor H. Theorell at the Karolinska Institute (Medical Nobel Institute, Stockholm, Sweden). Following a period at the Enzyme Institute (University of Wisconsin) as an Assistant Professor, in 1963 he joined the faculty at the University of Utah, and has held his present positions since 1969. His research interests have dealt with many aspects of enzyme and protein chemistry and certain aspects of medicinal chemistry and inherited disorders (including the muscular dystrophies). Mechanistic enzymology is his current interest, and "state of the art" approaches are being applied to study of enzyme action, e.g., of the kinases.

He is a member of the American Society of Biochemistry and Molecular Biology, the American Chemical Society Division of Biochemistry (Chemistry), Sigma Xi, Phi Beta Kappa, American Association for the Advancement of Science, and the New York Academy of Sciences. He has been a member of the Subcommittee on Enzymes, Committee on Biological Chemistry, Division of Chemistry and Chemical Technology, National Academy of Sciences and National Research Council; and also a member of the Physiological Chemistry Study Section, Division of Research Grants, National Institutes of Health, Publich Health Service, Department of Health, Education and Welfare. His long list of publications reflect these research interests.

TABLE OF CONTENTS

Volume I

TABLE OF CONTENTS

Volume II

Section XI: Hydrolases — Mechanisms

Section I
Enzyme Catalysis and Steady-State Kinetics

The aim of chemical kinetics is to predict the rates of chemical reactions and to describe the course of the reactions. It is a more difficult field than thermodynamics and the prediction of chemical equilibria, because the latter are concerned only with the initial and final states and not with time nor with mechanisms, nor with intermediate states. Farrington Daniels, Chemical kinetics, in *Outlines of Physical Chemistry*, John Wiley & Sons, New York, chapter XIV, p. 342.

Chapter 1

HISTORICAL INTRODUCTION, THEORIES OF ENZYME CATALYSIS, AND SOME ELEMENTARY CONSIDERATIONS OF ENZYME KINETICS

I. INTRODUCTION

Any understanding of other branches of biochemistry ultimately depends, to a large degree, on some knowledge of the nature, properties (both physical and chemical), and actions of the enzymes, which in turn are responsible for the rate of chemical changes which occur in all living organisms. The myriad of reactions and chemical pathways involved in the intermediary metabolism of protein, fat, and carbohydrate and the intricate chemical reactions concerned with the energetics of the cell are mediated through protein catalysts. Many physiological processes such as muscular contraction, blood clotting, excretion, detoxication, digestion, etc. are intimately linked with the chemistry of the cell, which invariably hinge on enzymatically catalyzed reactions. To quote Dixon,[9] "Life depends on a complex network of chemical reactions brought about by specific enzymes, and any modification of the enzyme pattern may have far-reaching consequences for the living organism . . . "

Enzymes are biocatalysts and as such their function is to influence the rates of chemical reactions. If they act as true catalysts, not the chemical equilibria; they may increase the rate of a particular chemical reaction, but ideally they do not influence the final equilibrium concentrations of reactants and products. However, in several notable cases where investigations were made under conditions where the enzyme concentration was not catalytic, but on the contrary, stoichiometric with respect to the substrate, the final state of the system was shown to be influenced by the enzyme protein concentration. Thus, the enzyme molecule exerts its effective catalytic function by participating directly in the reaction as if it were a reactant. In general, however, where concentrations of enzymes are employed which are several orders of magnitude below that of the substrate(s), the enzyme approaches a true catalyst in nature, and, under these conditions, influences only the rates of chemical reactions and not the final equilibrium state.

Also, enzymes catalyze the rates of reaction in both directions, e.g., substrate(s) $\overset{\text{enzyme}}{\rightleftharpoons}$ product(s). Many enzyme-catalyzed reactions are readily reversible; thermodynamically, all reactions are theoretically reversible. What is meant by the case of an enzymatically catalyzed irreversible reaction is one in which, within experimental error, the reverse reaction cannot be measured with ordinary chemical procedures; however, the reverse reaction might be measured, e.g., with radioactive tracer techniques (see Chapter 6). In any case, the equilibrium for these so-called irreversible cases is considered to be far to the right, making the reverse reaction negligible for these cases.

Until recently, almost all enzymes which have been isolated were considered to be protein in nature (See References 717 to 721.) Consequently, in studying an enzyme per se one must take all the necessary precautions ordinarily taken in dealing with the very labile complex protein molecule, i.e., he must consider the effect of temperature, pH, ionic strength, specific ion effects, solvent medium, etc., as far as the stability of the enzyme protein molecule is concerned.

A quantitative description of enzyme kinetics incorporates all the known chemical laws of catalytic reactions; the rate of the enzymatic reaction can often be adequately expressed in terms of reaction-rate mathematics. The difficulty lies in the complexity of the kinetics, for the rates of enzymatically catalyzed reactions are influenced by almost every physical and chemical variable known, e.g., (1) the enzyme concentration, (2) the substrate(s) concentration (where the substrate is the molecule with which the enzyme interacts to give the products of the chemical reaction), (3) competitive, noncompetitive, and uncompetitive inhibitors, (4) coenzymes, activators, cofactors, and in some cases, "allosteric effectors" — the so-called moderators or modulators, (5) pH, (6) temperature, (7) the ionic environment, and (8) dielectric constant of the medium. One may go on to list several other possible variables, e.g., pressure and even viscosity, where the latter is described in the recent kinetics study on adenosine deaminase by Kurz et al.[959]

II. HISTORICAL DEVELOPMENT OF ENZYME KINETICS

During the late 19th and early 20th century, many investigators sought to explain the catalytic activity and the progress of reactions involving enzymes in precise mathematical terms of known principles of equilibrium and mass action.

In 1902 Henri[11] and Brown[12] independently suggested that an enzyme-substrate complex was an obligate intermediate in the catalytic reaction. In fact, Henri derived a mathematical equation to account for the effect of substrate concentration on the velocity, which he naively proceeded to integrate, not realizing all the possible reasons (which we are aware of now) that a (P) vs. t or (S) vs. t plot might increase or decrease more rapidly than expected. We will return to this very shortly. The effect of pH on the enzyme activity was pointed out by Sorenson in 1909.[13] In 1913, Michaelis and Menten[14] rediscovered the equation of Henri and precisely tested it with a firm control of the pH. Moreover, they showed that the Henri-Michaelis-Menten equation could be derived on simple chemical equilibrium principles or a quasi-equilibrium concept. In 1925, Briggs and Haldane[4] introduced the steady-state concept to enzyme kinetics. Interestingly, the quasi- or rapid-equilibrium approach and the steady-state approach are both used today to explain the kinetic properties of enzymes.

Until the mid-1950s, most studies on the kinetics of enzyme-catalyzed reactions were based on the Henri-Michaelis-Menten or Briggs-Haldane equations especially for unireactant enzymes. From the mid-1950s to the early 1960s, attempts were made to analyze the kinetics of two-substrate and even three-substrate enzyme reactions, using largely the rapid or quasi-equilibrium assumptions of Michaelis and Menten. Some attempts, however, were made to incorporate equations based on steady-state concepts of Briggs and Haldane, especially by Dalziel, Alberty, Hearon, and others. Finally, when King and Altman[15,16] demonstrated a convenient schematic procedure for the derivation of complex steady-state reactions, the way was now open for the expression of many complex reactions by convenient shorthand procedures, developed by a number of people, including Cleland.[34] In 1965, Monod, Wyman, and Changeux[17] (see also References 18 and 19) presented a kinetic model for *allosteric* enzymes (regulatory enzymes which displayed sigmoidal rather than hyperbolic velocity curves). A year later, Koshland, Nemethy, and Filmer[20] presented an alternate model based on the flexible enzyme-induced fit model of Koshland.[21]

Coincident with the development of steady-state enzyme kinetics (pre-steady-state enzyme kinetics is an additional subject developed largely by Chance, Eigen, Roughton, and others), the measurement of the thermodynamic equilibrium-substrate binding properties of enzyme proteins became a sophisticated tool for the elucidation and confirmation of enzyme mechanisms largely through the fundamental efforts of Edsall, Wyman, Scatchard, and Klotz who were responsible for defining, in precise thermodynamic terms, the ligand binding properties of proteins.

III. SOME THEORETICAL ASPECTS OF ENZYME CATALYSIS

A number of theories have been proposed to explain the extraordinarily high effectiveness of enzyme catalysis and the unusual degree of substrate specificity which they display. Several of these theories will be discussed or listed below.

A. REACTIONS IN SEQUENCE AND COVALENT CATALYSIS

It is generally found that catalysts often facilitate the reactions between reactants by permitting the overall reaction to occur by a sequence of intermediate reactions, each one of which is more rapid than the overall reaction. Thus, a reaction between A and B, which may occur in the absence of an enzyme catalyst by passage over a comparatively high free-energy barrier, may in the presence of an enzyme E first react to form EA, to be followed by a reaction of EA with B. If each of the barriers are lower for the enzyme mechanism, its rate will be faster than that for the uncatalyzed reaction. As an example, consider the enzyme rhodanese which catalyzes the overall reaction

$$SSO_3^{2-} + CN^- \longrightarrow SO_3^{2-} + SCN^- \tag{1}$$

In the absence of the enzyme, the reaction is relatively slow as a result of electrostatic repulsion between ionic reactants which bear charges of like sign. In the presence of the enzyme, the mechanism has been demonstrated by Westley[521,522] to be

$$\begin{aligned} E + SSO_3^{2-} &\longrightarrow E\text{-}S + SO_3^{2-} \\ E\text{-}S + CN^- &\longrightarrow SCN^- + E \end{aligned} \tag{2}$$

Since there is no electrostatic hindrance for either of these two intermediate steps, each of which is more rapid than the uncatalyzed reaction, the enzyme-catalyzed reaction is very much accelerated. For a review of those enzymic reactions involving covalent intermediates, see References 523 and 524 and refer to a discussion of the recently deduced X-ray crystal structure of rhodanese[909] and that of the flexible protein in solution.[910]

B. PROXIMITY AND ORIENTATION EFFECTS

Apparently, as early as 1825, Michael Faraday suggested that the main function of the catalyst was to cause reaction molecules to come into close contact. The idea, therefore, that the catalytic role of the enzyme was to act as a site(s) for simultaneous binding of the substrates is one of the oldest theories on the mechanism of action of enzymes. This idea would have the surface of the enzyme (or its active site) serve to increase the local concentration of reactants and thereby make collision between reactant molecules more probable.[516] However, any quantitative contribution of this proximity effect to the enormous catalytic power of enzymes, in general, is very likely of a limited nature because of the small number of active sites involved. The matter has been discussed in detail by Koshland[525,526] and by Jencks.[527,528] A very simple argument based on free energy changes demonstrates that the proximity effect *alone* cannot have a favorable effect on enzyme-catalyzed rates, especially at low substrate concentrations.

However, if the active site acts as an *orienting* influence on the substrate molecules when they are in close proximity, so that their reactive groups are placed in juxtaposition in a highly specific manner, it is conceivable that a considerable acceleration in velocity may take place and that the stereochemistry of the enzymic reaction may be maintained. The magnitude of this effect would depend on the number of catalytic groups involved as well as the restraints imposed by the three-dimensional angles of contact. However, experimental quantitative evidence for this theory is lacking (see References 529 and 530). Page and Jencks[528] have adopted the view that orientation effects are likely to be of secondary importance in enzyme reactions, and they have suggested that translational and rotational motions provide an important entropic driving force for reactions.

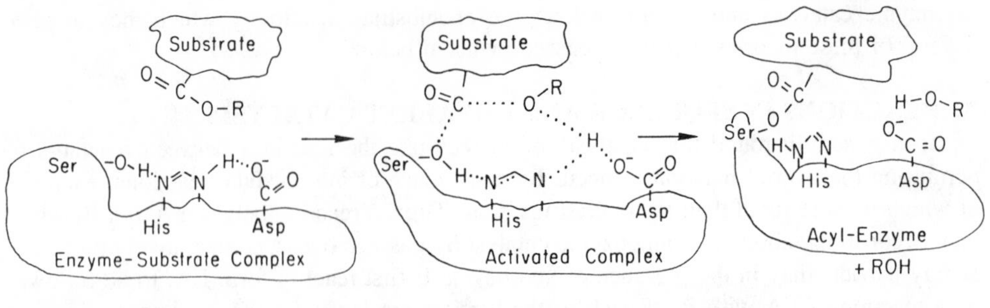

FIGURE 1. Postulated proton-transfer processes during acylation of trypsin and chymotrypsin.[541-543,549]

C. BIFUNCTIONAL AND ACID-BASE CATALYSIS AND OTHER ELECTRONIC EFFECTS, E.G., PUSH-PULL MECHANISM, FACILITATED PROTON TRANSFER, AND NEIGHBORING-GROUP EFFECTS

Since enzymatic reactions occur largely within the polar solvent water and accordingly are influenced greatly by changes in pH, it is only logical that a contribution by Brönsted acid or the conjugate bases[531] might be involved in a number of cases. A large body of literature has dealt with nonenzymatic models of this *acid-base catalytic effect* (for reviews, see References 523, 533, and 534). A frequently cited example invokes the known presence of a histidine residue in the active site of several esterases, peptidases, and proteinases (see References 535 to 537). Because the imidazole group is a fair catalyst in model ester hydrolytic reactions, it is tempting to postulate that the imidazole group acts either as a general base catalyst or as a nucleophile. However, if the imidazole group participated in the nucleophilic attack on an ester, presumably an acyl intermediate would form with the imidazole group. However, the acyl intermediate, which is actually found in chymotrypsin and other similar enzymes, is actually with a serine hydroxyl group positioned close to the active imidazole group. Consequently, it has been speculated that the imidazole group acts as a general base to accept a proton from the serine hydroxyl group and thereby to form a strong nucleophile in the ionized serine hydroxyl group. The imidazolium ion could then act as a general acid catalyst which might protonate the base which is released. Another version of this mechanism is the *push-pull* arrangement of an acid and a base acting in concert (e.g., see References 538 to 540) which is more likely to exist in the highly structured enzyme protein catalyst than in more general nonenzymic models.

Wang[541,549] put forth the suggestion that *proton transfers* may be *facilitated* by a hydrogen-bonded network in the enzyme, with special reference to trypsin, chymotrypsin, the subtilisins, and other hydrolytic enzymes. Some evidence for Wang's viewpoint is provided by studies in D_2O, where, for example, the rate of deacylation of the acyl subtilisin in D_2O is about one third that in H_2O. Another example of *facilitated proton transfer* is to be found in the suggested mechanism[542,543] of action of trypsin and chymotrypsin during acylation as shown in Figure 1, which makes use of the properties of the imidazole group as proton-transfer agent. Eigen et al.[544,545] have shown that the tautomeric character of the unprotonated imidazole ring permits a proton to be accepted and donated simultaneously. According to Figure 1, therefore, the efficiency of acylation in these enzymes is due to the fact that the imidazole group is at the same time assisting in the removal of the proton from the hydroxyl group of the seryl residue and interacting with the carboxyl group of the aspartyl residue. The transfer of the proton of the imidazole ring to the alcoholic oxygen atom of the substrate is aided by the carboxyl group of the aspartyl residue. (Compare also the excellent review by Kanamori and Roberts[567] on applications of ^{15}N NMR studies to such problems. Also, see the recent studies on the three-dimensional structure of the Asn[102] mutant of trypsin and the proposed role of Asp[102] in serine proteases catalysis.[974,975] The high efficiencies of certain enzymes have been explained by *neighboring-group effects*. It is well known that the rates of proton transfer can be strongly influenced by neighboring groups. In the case of the ribonuclease-catalyzed hydrolysis of polyribonucleotides, a cyclic phosphodiester intermediate is formed. The presence of a carbonyl group is found to be essential for catalytic activity, and it has been postulated[546,547] that the oxygen atom of the group forms a hydrogen bond with a hydroxyl group of the sugar ring; as a result, the oxygen atom assists in the polarization of the -OH group which aids in the formation of the cyclic ester.

Warshel and Levitt,[613] in a theoretical study, have drawn attention to a dielectric, electrostatic, and steric stabilization of the carbonium ion in the reaction of lysozyme, a study which will be further elaborated in the next section.

D. ELECTROSTATIC EFFECTS

Perutz, in a critical discussion of the *electrostatic effects* in proteins and enzymes,[694] has drawn attention to what he has considered to be the "first successful attempt at predicting even the approximate structure of a protein from its sequence on the basis of Kauzmann's principles."[695] Thus, Levitt and Warshel[696] attempted to fold up the chain of pancreatic trypsin inhibitor by energy minimization with the aid of a computer analysis. They made use of simplified side chains and assigned to each of them a hydrophobic energy calculated from the energy of transfer between water and ethanol, which was taken as the difference between the energy of the side chain when isolated in water and when surrounded by other residues. With the use of this principle plus that of closest packing, the chain could be folded up to something which approached the correct conformation with a root-mean-square deviation of 3.7 Å from the true mode. Perutz has stated[694] that "The result is impressive as a first step, but it also shows that hydrophobic bonding and close-packing alone are not sufficient. To get the structure right, electrostatic effects due to hydrogen bonding would also have had to be taken into account."

Pauling[697] had predicted that the enzymic active site would be complementary to the transition state of the substrate (see also Lipscomb[807] and this prediction appeared to have been confirmed when the X-ray structure of the first enzyme, lysozyme, was solved, and also later when "catalytically active antibodies" were discovered which mimicked the transition state of certain reactions (see References 992, 993, 997, and 1008). Thus, Phillips and colleagues[698,699] and Vernon[700] have suggested the following mechanism:

1. The substrate becomes attached to the enzyme and is held in position by hydrogen bonds and other forces. In the process, the sugar ring residue D of the substrate becomes distorted and takes up a conformation favorable for the formation of a carbonium ion.
2. A proton is transferred to the glycosidic oxygen atom from Glu-35.
3. Heterolysis of the C_1-O bond gives a carbonium ion which is stabilized by interaction with Asp-52.
4. The disaccharide EF diffuses away and a water molecule attacks the carbonium ion, thus completing the hydrolysis.

These proposals have acted as a stimulus for research in this area. Synthetic compounds with attached carboxyl groups to mimic those of the active site of the enzymes were found unfortunately to hydrolyze in water at a much slower rate than that of the substrate by the enzyme. Perutz[694] wondered whether the distortion of the substrate itself had been neglected. Warshel and Levitt[613] and then Warshel[701] made a structural and thermodynamic study of those factors they felt might be responsible for the catalytic mechanism of lysozyme. They found that distortion of the substrate can be neglected, eventually because the enzyme is too "soft" to bring it about. The decisive factor, they believed, consists of the much stronger electrostatic effects that the charged groups can exert on the substrate when attached to the enzyme as compared to the effect of the same charged groups in water. This is the result because, in bulk water, solvation by water dipoles makes the stability of oppositely charged ions almost independent of distance, whereas, in the active site of the enzyme, permanent and induced dipoles can generate a minimum of free energy at a separation of 3 to 5 Å between the charges, because the stabilization of charges is larger than in bulk water. As a consequence, the calculated free energy of the transition state in the cleft of lyoszyme is lower by 6 kcal/mol than that of an equivalent system with fixed anionic charges in water, compared to an experimentally estimated difference of approximately 7 kcal/mol. Therefore, as Perutz[694] had commented, "The model accounts quantitatively for the observed differences in reaction rates in the enzyme and in water."

In a recent paper, Hwang and Warshel[949] conducted semiquantitative calculations of the "catalytic free energies" in genetically modified subtilisin enzymes. By "site-directed mutagenesis", three mutant enzymes could be studied theoretically: the Asn-155 → Thr, Asn-155 → Leu, and Asn-155 → Ala mutants of subtilisin.[950,951] The *catalytic* free energy and *binding* free energies of the native enzyme and the above three mutants of subtilisin were calculated by the "empirical valence bond method"[952] and the free energy perturbation method.[952,953] The calculated changes in free energies between the mutant and native enzymes proved to be within only approximately 1 kcal/mol of the observed values. Hwang and Warshel[949] attributed the calculated changes in catalytic free energies almost entirely due to the electrostatic interaction between the enzyme-water system and the charges of the reacting system, and they felt that this evaluation supports the idea that, "The electrostatic free energy associated with the changes of charges of the reacting system is the key factor in enzyme catalysis."[949]

E. INDUCED-FIT MECHANISM OF THE ENZYME AND STRAIN OR DISTORTION THEORY OF THE SUBSTRATE BOUND TO THE ENZYME (RELATION OF ENERGY DERIVED FROM SUBSTRATE BINDING AND CATALYSIS)

Koshland[532,548,550] developed his *induced-fit* theory, based, to some degree, on suggestions made earlier by Laidler[551] which had been derived on estimates for entropies and volumes of activation for several enzyme reactions. The theory presupposes that the catalytic groups at the active site of the "free enzyme" molecule are not precisely in the correct positions to exert their effective catalytic function. Following the binding of the substrate, however, the binding forces between the enzyme and the substrate force the enzyme into a conformation which is more active catalytically. In the case of a relatively poor substrate or an inhibitor (e.g., a substrate analogue) which may also bind to the active site, these compounds will not possess the necessary structural features to induce the appropriate conformational changes. Bender et al.[552] have noted that the induced-fit theory can account for those cases with relatively large positive and negative entropics of activation which are not readily interpreted otherwise. The theory also can account for the observation that substrate specificity is often found to be more important at high rather than low substrate concentrations. Hexokinase is an enzyme to which the induced-fit hypothesis had been applied. As first noted by Colowick,[553] hexokinase not only catalyzes the phosphoryl group transfer from ATP to the 6-hydroxyl group of glucose, but also to water, i.e., the enzyme also possesses ATP-ase activity. However, the rate of reaction with water is only 5×10^{-6} that with glucose as the phosphoryl group acceptor; the difference in rate calculates to be equivalent to approximately 7 kcal of $\Delta G^{\ddagger}$. Thus, the suggestion has been made that a water molecule lacks those chemical side groups necessary to induce a distortion of the enzyme, which glucose possesses, and which will form a more reactive complex with the enzyme.

It is of interest that J. B. S. Haldane[554] put forth an alternative theory which was later developed by Jencks.[555] The theory postulates that the binding forces between the substrate and the enzyme are directly utilized to bring about the strain or distortion in the substrate molecule. From the simplest viewpoint, the enzyme is assumed to be rigid with the substrate becoming distorted during the binding process. Jencks[555] (see also References 528, 556, and 557) discusses various lines of evidence in support of the strain theory, e.g., strained cyclic compounds are often observed to hydrolyze more rapidly than less strained cyclic compounds or straight-chain compounds. Additional evidence was derived from X-ray studies, e.g., of lysozyme.[558,559] These studies on lysozyme had demonstrated that the sugar ring of the inhibitor (the trisaccharide of *N*-acetylglucosamine) adjacent to the bond to be broken is in a strained, planar configuration, whereas all other sugar rings are in the unstrained chair configuration. It is also obvious that the *induced-fit* and *strain theories* may be combined into a single theory in which both the enzyme and the substrate become distorted during the formation of the enzyme-substrate complex with accompanying enhancement of rates at relatively high substrate concentrations. Wishnia[560,732] has proposed the alternative theory, i.e., that the free enzyme is strained, with the release of the strain upon the binding of the substrate.

Finally, rate enhancements from this type of catalysis may be large, especially if rotational entropy is also reduced as a result of reversible binding interactions.[556,807,808]

F. PRODUCTIVE AND NONPRODUCTIVE BINDING OF SUBSTRATES

This hypothesis was developed by Bernhard and Gutfreund largely to explain certain substrate specificity results[561] (see also References 562 to 566). Their idea is that a substrate may also become bound to an enzyme in a catalytically unfavorable position. Thus, a good substrate may have several ligand sites for binding which are complementary to binding sites on the enzyme. A poor substrate which lacks one of these ligand binding sites, or possesses them in an incorrect position or alignment, may bind correctly to the enzyme and give rise to products; however, it may also bind incorrectly to the enzyme to give rise to a catalytically inactive complex. Thus, such a substrate will be a less effective substrate, and its V_{max} will be less than that for the productively bound substrate by the fraction of the substrate molecules that are bound productively, apart from other factors. This theory has been successfully applied by Niemann and co-workers[564-566] to chymotrypsin-catalyzed reactions, and it has provided a satisfactory interpretation of the reactivities of a number of substrates of chymotrypsin.

This hypothesis is capable of providing explanations for a number of results, but other cases require a combination of several of the hypotheses enumerated above. The fact that hexokinase catalyzes the transphosphorylation of glucose with ATP, 2×10^5 more rapidly than its ATP-ase activity with water, cannot logically be explained as an unproductive binding of water; rather, the induced fit or strain hypothesis would better provide an explanation of this case.

Before leaving this section on enzyme catalysis it behooves us to comment on the basic aspect, the enormous acceleration of chemical reactions by some enzymes, i.e., by factors of 10^{14} or larger (note that these factors are dependent on the particular choice of the standard state selected, e.g., see Reference 702). In an interesting review, Lipscomb[807] succinctly discusses from a chemical standpoint several known contributions to these large rate-enhancement factors in enzymatic catalysis, and then he critically examines the possible mechanisms of carboxypeptidease A, as a specific case.

G. ELECTRO-MECHANO-CHEMICAL MODEL AND ENERGY FUNNEL MODEL

Finally, several models developed largely from a physical standpoint have been proposed, e.g., the *electro-mechano-chemical model* developed by Green and Ji[615] which views the enzyme specifically as a thermally activated inducer of bond polarization. The oscillatory motion of the protein supposedly generates, transiently, a stabilized conformational state in which the substrate molecule (bound at the active center) is subjected to a highly localized electric field which effectively polarizes a key bond(s) whose labilization (or cleavage) is critical to catalysis. Supposedly, the bombardment of the protein by solvent/solute particles is the driving force for this oscillatory motion. This model assumes that a critical internal coordinate (e.g., the separation of two charged moieties at the active center) is coupled specifically to low-frequency collective vibrational modes of the protein structure. As the protein oscillates (''breathes'') from an expanded state to a contracted state, kinetic energy and potential energy would alternately interchange. According to Green[703] '' . . . as the protein undergoes its *fluctuating motions,* charges and dipoles are periodically separated (generating electric energy), the relative positions of atoms and groups undergo periodic change (generating mechanical energy), and chemical bonds such as hydrogen bonds are made and broken (generating chemical energy).'' Therefore, we see the possibility of interconverting three forms of energy.

Jencks[534] has pointed out that such an oscillating model, unfortunately, would demand relatively strong coupling among various parts of the protein; in general, the catalytic process would involve the coordination of a number of molecular events.

Careri and co-workers,[704-709] however, have given redress to this ''oscillating enzyme model'' by statistical considerations and suggested that particular classes of fluctuations in the protein conformation provide a mechanism for coordinating the thermal energy to produce high free-energy events at the active center. Accordingly, the ''secret of an enzyme''[704] might lie in its ability to let the relevant conformational events occur with a well-defined time correlation. This does not mean a rigid coupling between conformational variables, rather a statistical correlation in their rate of change around the equilibrium position. To establish the theoretical connection between fluctuations and enzyme function Careri[704] employs a statistical mechanical treatment.

Another approach in the continuing development of the ''mechano-chemical'' models assumes that the enzymatic function is governed by ''transient strains'' which are developed during transitions between conformations. The theory was constructed by Gavish[710-712] and amplified by Frauenfelder's group (see Reference 713). This model probes more deeply in an exploration of the manner in which the rate of conformational transitions depends on structural parameters and solvent properties. Most importantly, the Gavish-Frauenfelder model yielded an explicit formulation of the enzymic catalytic turnover number (described as ''k_{cat}'') as a function of dynamic protein-solvent interactions.

Lastly, in an excellent review of these proposed models, Welch, Somogyi, and Damjanovich[614] develop their own version which permits an actual formulation of all the enzymic catalytic constants based on the dynamics of protein-solvent interactions and which they term the ''energy-funnel model''. This theoretical model has been developed over the past decade by Somogyi and Damjanovich (see References cited in Reference 614) and applied to organized multienzyme systems by Welch.[714-716]

An incisive and theoretical discourse on the thermodynamic aspects of enzyme mechanisms and the ''fluctuation theory'' will be dealt with in detail by R. Lumry in Volume II, Chapter 1, Section I of this work (see also Reference 955). R. J. P. Williams also will provide new insights in the relationship between ''Complexation and Catalysis'' in Volume II, Chapter 1, Section II of this work (see also Reference 956).

IV. SOME ELEMENTARY CONSIDERATIONS OF ENZYME KINETICS

A. ONE-SUBSTRATE REACTIONS HAVING A SINGLE INTERMEDIATE

The simplest reaction mechanism which may apply to some one-substrate systems, uncomplicated by the reverse reaction or by the effects of modifiers (activators or inhibitors) is the one first proposed by Brown[12] for the action of invertase on sucrose:

$$E + S \xrightleftharpoons[k_{-1}]{k_{+1}} ES$$

$$ES \xrightarrow{k_{+2}} E + P_1 + P_2$$

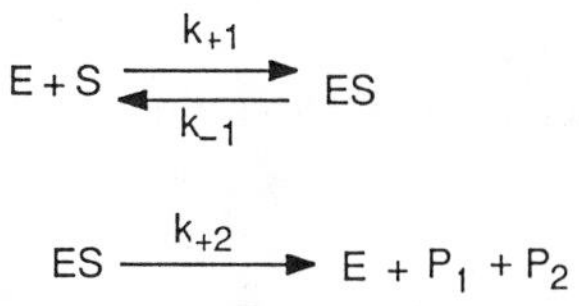

SCHEME I.

In 1913, Michaelis and Menten[14] formulated a rate expression based on this mechanism of Brown but made the assumption that the second reaction does not disturb the first equilibrium (i.e., $k_{-1} \gg k_{+2}$), an approximation which is currently called the rapid or quasi-equilibrium hypothesis. Briggs and Haldane[4] gave a more general formulation on the basis of the steady-state hypothesis, i.e., that the net rate of change (ES) of the central *complex* is zero. Thus,

$$\frac{d(ES)}{dt} = k_{+1} (E) (S) - k_{-1} (ES) - k_{+2} (ES) = 0 \tag{3}$$

In most "initial velocity" enzyme kinetics studies, the initial concentration of substrate is far greater than the total enzyme concentration, i.e., $(S_0) \gg (E_t)$, otherwise, if this condition is not satisfied, the steady-state may never be attained during the course of the reaction.

Since $(E_t) = (E) + (ES)$, Equation 3 may be rewritten as

$$k_{+1} [E_t - (ES)] (S) - k_{-1} (ES) - k_{+2} (ES) = 0 \tag{4}$$

which leads to the solution for (ES):

$$(ES) = \frac{k_{+1} (E_t) (S)}{k_{-1} + k_{+2} + k_{+1} (S)} \tag{5}$$

The rate of reaction is therefore

$$v = k_{+2} (ES) \tag{6}$$

$$= \frac{k_{+1} k_{+2} (E_t) (S)}{k_{-1} + k_{+2} + k_{+1} (S)} \tag{7}$$

$$= \frac{k_{+2} (E_t) (S)}{\dfrac{(k_{-1} + k_{+2})}{k_{+1}} + (S)} \tag{8}$$

Or, setting $k_{+2} E_t = V_{max}$, the maximal velocity at high (S), and

$$\left(\frac{k_{-1} + k_{+2}}{k_{+1}} \right) = K_m \tag{9}$$

the Michaelis constant; and $v = v_0$, the initial velocity corresponding to an initial substrate concentration (S_0):

$$\therefore \quad v_0 = \frac{V_{max} (S_0)}{K_m + (S_0)} \tag{10}$$

the familiar Michaelis-Menten equation.

In the original Michaelis-Menten derivation, the assumption was made that $k_{-1} \gg k_{+2}$, so that the constant, K_m, approached

$$K_s \simeq \frac{k_{-1}}{k_{+1}} \tag{11}$$

an equilibrium constant.

It is of interest that van Slyke and Cullen[30] in 1914 formulated this rate equation on the assumption that the reaction described by Scheme I is essentially irreversible, i.e., $k_{-1} \ll k_{+2}$. Their treatment, therefore, proved to be another special case of the treatment of Briggs and Haldane, with K_m reducing to $\frac{k_{+2}}{k_{+1}}$. Of course, initial rate studies on the variation of the rate with initial substrate concentration cannot provide information on the relative magnitude of k_{-1} and k_{+2}; only by studies on the kinetics conducted during the pre-steady-state or transient region can these estimations be made.

It is readily seen that Equation 10 predicts the extreme types of behavior observed for enzymatic reactions behaving according to the Michaelis-Menten equation, that is, (1) at low concentrations of substrate, where $(S_0) \ll K_m$ and (S_0) may be neglected in the denominator of Equation 10, then

$$v_0 \simeq \frac{V_{max}\,(S_0)}{K_m} \tag{12}$$

or the kinetics are first order with respect to (S) and (2) at high concentrations of substrate, where $(S_0) \gg K_m$, and K_m may be neglected in comparison with (S_0), the rate now approaches

$$v_0 \simeq V_{max} = k_{+2}\,(E_t) \tag{13}$$

and the kinetics are now zero order in (S).

Note that $k_{+2} = \frac{V_{max}}{(E_t)}$ and k_{+2} is sometimes called k_{cat}, especially in proteolytic enzyme studies. Also, the ratio $\frac{k_{cat}}{K_m}$ has been termed the "catalytic efficiency", and possesses the units of an apparent second order velocity constant, i.e., $M^{-1}\ s^{-1}$. Both k_{cat} and $\frac{k_{cat}}{K_m}$ are now often used as comparative kinetic parameters, especially in enzyme-engineering studies where the wild-type enzyme is to be compared against a mutant enzyme in which one or more amino acid residues have been replaced, e.g., by "site-directed mutagenesis" means (see References 954 and 987).

Thus, Equation 10 is the equation of *rectangular hyperbola* with an asymptote at V_{max} (see Figure 2) and a value of K_m = (S) when $v_o = \frac{V_{max}}{2}$.

A number of linear plots have been suggested to enable one to more readily estimate the kinetic parameters, V_{max} and K_m. The one most commonly employed today is the one first suggested by Lineweaver and Burk.[31] Equation 10 may be placed in reciprocal form:

$$\frac{1}{v_0} = \frac{K_m}{V_{max}}\frac{1}{(S_0)} + \frac{1}{V_{max}} \tag{14}$$

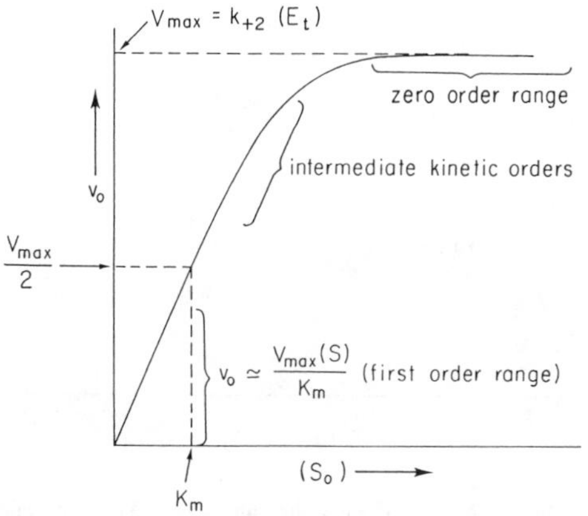

FIGURE 2. Equation 10 plotted as a rectangular hyperbola with $v_0 \simeq f(S_o)$ and V_{max} as the asymptote.

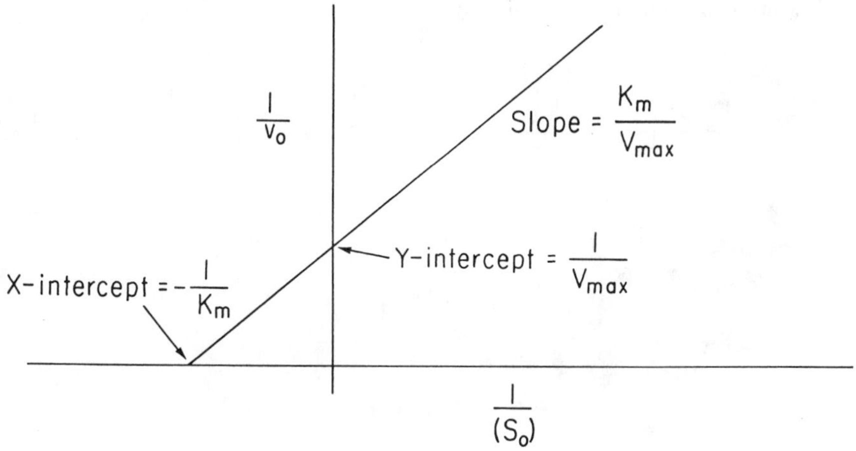

FIGURE 3. Double-reciprocal plot according to Reference 31.

and a plot of $\frac{1}{v_o}$ vs. $\frac{1}{(S_0)}$ will yield a straight line of slope $\frac{K_m}{V_{max}}$ and Y-intercept $= \frac{1}{V_{max}}$.

In the double-reciprocal plot, a simple least-square plot is usually not adequate for statistically evaluating the kinetic parameters and appropriate weighting of the points may be necessary. Various discussions of this problem have been given[32,33] and there are several computer programs available[34-36] including a recent one by R. J. Leatherbarrow termed Enzfitter.[1006] Moreover, other plots of the data (see p. 535 to 537) which give a more even distribution of points have been suggested.[37-39] See p. 540 to 542 for the "direct linear plot" of Eisenthal and Cornish-Bowden.[772]

B. INHIBITION BY HIGH SUBSTRATE CONCENTRATION

It is not uncommon to find in single substrate reactions that the hyperbolic law (or the Michaelis equation) is not obeyed at high substrate concentrations, but that the rate passes through a maximum as the substrate concentration is increased and then falls. The simplest mechanism that explains this behavior was theoretically treated by Haldane,[40] and involves the formation of an (ES_2) complex which may or may not break down to products. If (ES_2)

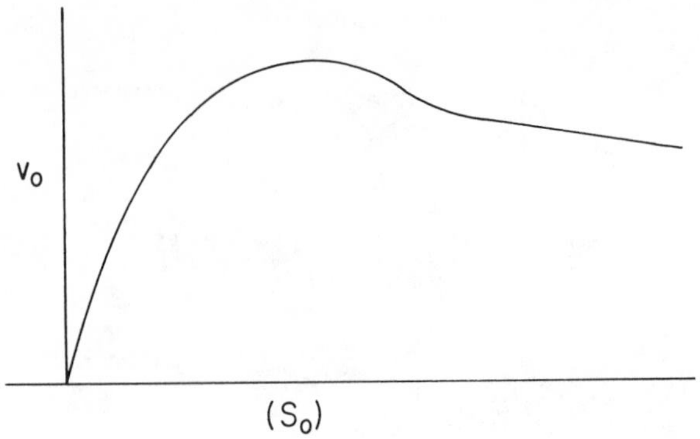

FIGURE 4. A case of substrate inhibition. (See Reference 41 for the hydrolysis of Cbz-gly-L-tyr catalyzed by carboxypeptidase A.)

does not react further, the rate approaches zero. If it reacts less rapidly than (ES), the rate approaches a limiting value that is lower than the maximum value, as shown in Figure 4.

Such a mechanism may be formulated as

$$
\begin{array}{c}
(K_S) \\
E + S \underset{k_{-1}}{\overset{k_{+1}}{\rightleftharpoons}} (ES) \overset{k_{+2}}{\longrightarrow} E + P \\
+ \\
S \\
(K_a) \; k_{+a} \updownarrow k_{-a} \quad k'_{+2} \\
(ES_2) \overset{k'_{+2}}{\longrightarrow} ES + P
\end{array}
$$

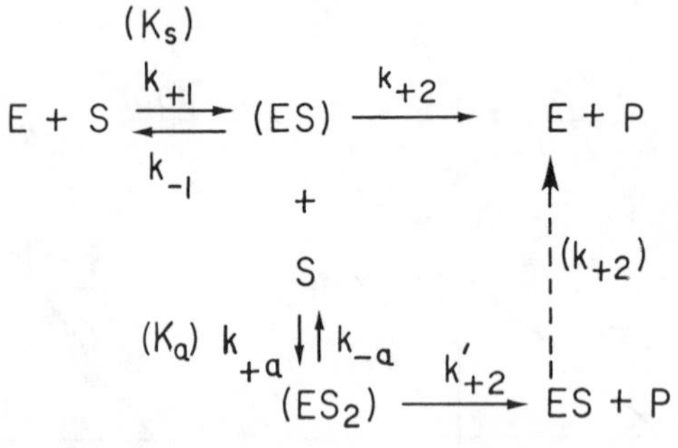

SCHEME II.

The total concentration of enzyme is

$$(E_t) = (E) + (ES) + (ES_2) \tag{15}$$

and the steady-state equations for (ES) and (ES$_2$) are

$$k_{+1} (E) (S) - [k_{-1} + k_{+2} + k_a (S)] (ES) + k_{-a} (ES_2) = 0 \tag{16}$$

and

$$k_a (ES) (S) - (k_{-a} + k'_{+2}) (ES_2) = 0 \tag{17}$$

Equations 16 and 17 permit (E) and (ES$_2$) to be expressed in terms of (ES) and (S), and insertion of the expressions into Equation 15 leads to

$$(E_t) = (ES) \left\{ \frac{k_{-1} + k_{+2} + k_a (S)}{k_{+1} (S)} - \frac{k_a k_{-a}}{k_{+1}(k_{-a} + k'_{+2})} + 1 + \frac{k_a(S)}{k_{-a} + k'_{+2}} \right\} \tag{18}$$

Two cases result:

Case 1: If $k'_{+2} = O$, $V_o = k_{+2}(ES)_1$ and reduces to

$$v_0 = \frac{k_{+2}(E_t)(S)}{\left\{\frac{(k_{-1} + k_{+2})}{k_{+1}} + (S) + \frac{k_{+a}}{k_{-a}}(S)^2\right\}} \tag{19}$$

Define

$$K_m = \frac{k_{-1} + k_{+2}}{k_{+1}};\ K_a = \frac{k_{+a}}{k_{-a}};\qquad V_{max} = k_{+2}E_t$$

and Equation 19 may be rewritten as

$$v_0 = \frac{V_{max}(S)}{K_m + (S) + K_a(S)^2} = \frac{V_{max}}{\frac{K_m}{(S)} + 1 + K_a(S)} \tag{20}$$

(See Appendix 1 for the derivation of Equations 20 and 24.[41])

Thus, at high (S) concentratons, the $(S)^2$ term dominates in the denominator and the rate becomes inversely proportional to (S) and approaches zero at high concentrations of substrate. A maximum rate is obtained at a substrate concentration given by

$$(S)_{max} = \left(\frac{K_m}{K_a}\right)^{1/2} \tag{21}$$

Case 2: If $k_{+2}' \neq O$, the overall rate of product formation is given by

$$v_{overall} = k_{+2}(ES) + k_{+2}'(ES_2) \tag{22}$$

$$= \left\{k_{+2} + k_{+2}'\frac{k_a(S)}{(k_{-a} + k_{+2}')}\right\}(ES) \tag{23}$$

If we define

$$K_m' = \frac{k_{-a} + k_{+2}'}{k_{+a}};\ K_m = \frac{k_{-1} + k_{+2}}{k_{+1}}$$

it reduces to

$$v_{overall} = \frac{\left(k_{+2} + \frac{k_{+2}'(S)}{K_m'}\right)E_t(S)}{K_m + \left(\frac{k_{+a}}{k_{+1}} - \frac{k_{-a}}{k_{+1}K_m'} + 1\right)(S) + \frac{(S)^2}{K_m'}} \tag{24}$$

At sufficiently high substrate concentrations, the rate approaches a new limiting rate, v_{lim}, given by

$$v_{lim} = k_{+2}'E_t \text{ or } V_{max,2} \tag{25}$$

Three different kinds of behavior are special cases of Equation 24 and are illustrated in Figure 5 where curve a is the situation where normal hyperbolic kinetics are obeyed, curve

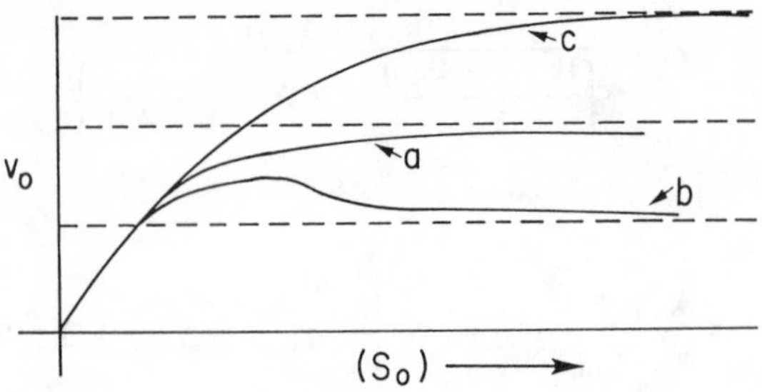

FIGURE 5. Plots showing (a) simple hyperbolic kinetics; (b) inhibition by high substrate concentration; and (c) activation by high substrate concentration.

b shows the behavior when the rate of breakdown of ES_2 is very small compared to that of ES, and curve c is an unusual case which corresponds to a high rate of reaction through ES_2, a case which could be referred to as "substrate activation".

C. INTEGRATED FORM OF THE MICHAELIS-MENTEN EQUATION AND APPLICATIONS

Most steady-state kinetics studies are carried out by initial-rate methods, and the usual Michaelis-Menten equation for a single-substrate reaction is a differential velocity equation, where

$$v_0 = \frac{d(P)}{dt} = \frac{-d(S)}{dt}$$

To find the relation between $v_0 = f(S_o)$, in order to deduce a value for K_m and V_{max} by such initial-rate measurements, the reaction should proceed to a negligible extent during the course of each measurement. Henri, on the other hand, integrated his expression relating the initial velocity to substrate disappearance or product formation. Recently, there has been a renewed interest in integrated forms of Henri-Michaelis-Menten expressions,[22,23] especially in those cases where it may be difficult to determine low product concentrations, and it becomes necessary to allow the reaction to proceed until a substantial fraction of the initial (S_o) is converted to product. Theoretically, the use of integrated expressions provides an infinite number of points during the progress curve for calculation of the kinetic parameters. Thus, consider the single substrate case where a decrease in velocity with time results only from a decrease in substrate saturation of the enzyme, and not from product inhibition nor in the approach to equilibrium, i.e., assume $K_{m,s} \ll K_{m,p}$ and that K_{eq} is very large: let

$$v = \frac{-d(S)}{dt} = \frac{V_{max}\,(S)}{K_m + (S)} \tag{26}$$

the usual Michaelis equation. Rearranging,

$$V_{max}\,dt = -\left(\frac{K_m + (S)}{(S)}\right) d\,(S) \tag{27}$$

Integrating between t_o, S_o, and t, S:

$$V_{max}\int_{t_0}^{t} dt = -\int_{(S_0)}^{(S)} \left(\frac{K_m + (S)}{(S)}\right) d\,(S) \tag{28}$$

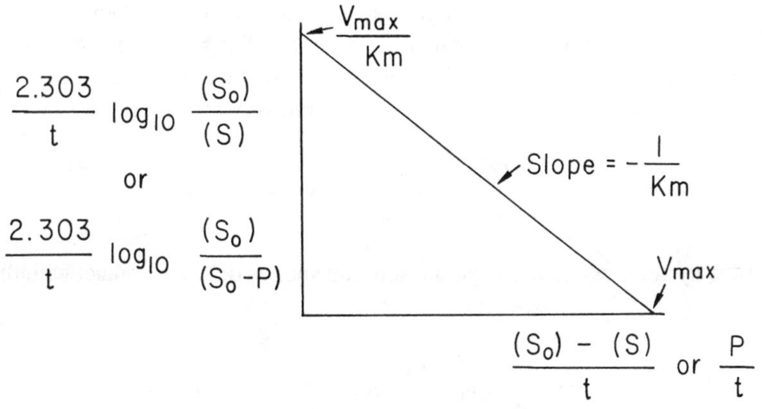

FIGURE 6. Plot of Henri's Integrated Expression.

Separating terms

$$V_{max}\int_{t_0}^{t} dt = -K_m\int_{(S_0)}^{(S)} \frac{ds}{(S)} - \int_{(S_0)}^{(S)} d\,(S) \qquad (29)$$

$$V_{max}\,t = -K_m \ln \frac{(S)}{(S_0)} - ((S) - (S)_0)$$

$$\text{or } V_{max}\,t = 2.303\ K_m \log_{10} \frac{(S_0)}{(S)} + ((S)_0 - (S)) \qquad (30)$$

or Henri's equation. where $(S)_0 - (S) = (P)$, concentration of product at time t. Rearranging:

$$\frac{2.303}{t}\log_{10}\frac{(S_0)}{(S)} = -\frac{1}{K_m}\frac{(S_0 - S)}{t} + \frac{V_{max}}{K_m} \text{ or}$$

$$\frac{2.303}{t}\log_{10}\frac{(S_0)}{(S)} = -\frac{1}{K_m}\frac{(P)}{t} + \frac{V_{max}}{K_m} = \frac{2.303}{t}\log\left(\frac{S_0}{(S_0 - P)}\right) \qquad (31)$$

$\therefore$ plot of left-hand side vs. $\left(\frac{S_0 - S}{t}\right)$ or $\frac{(P)}{t}$ yields a value for K_m and V_{max}, as shown in Figure 6. This plot was first used by Walker and Schmidt.[24]

Kuby in 1953[25] demonstrated for β-D galactosidase from *E. coli* (K-12) acting on *o*-nitrophenyl β-D galactoside as substrate that Henri's integrated expression (integrated by a series approximation) was followed up to almost 95% hydrolysis.

An interesting engineering application of the integrated Michaelis equation is the case where an enzyme is attached to a water-insoluble polymer and the substrate is allowed to flow through it, e.g., Lilly et al.[26] and O'Neill et al.,[27] considered the case of CM-cellulose-ficin (which is similar to trypsin in its specificity). Thus, expressing the Michaelis equation by

$$\frac{-d(A)}{dt} = \frac{k_c(E_0)(A)}{K_A + (A)} \qquad (32)$$

where k_c = limiting rate constant at high concentrations of (A) (*N*-benzoylarginine ethyl ester) or $k_d \cdot (E_0) = V_{max}$ (in moles per liter per minute). (E_0) = total enzyme concentration in moles per liter, and K_A = Michaelis constant for A. After integration of this equation which is, as before, subject to the boundary condition: (A) = (A_o) when t = o, gives

$$K_A \ln \frac{(A_0)}{(A)} + (A_0) - (A) = k_c(E_0)t \tag{33}$$

In the flow system, where the substrate solution spends time t in contact with the enzyme catalyst, t is given by

$$t = \frac{\nu_f}{Q} \text{ where } \nu_f = \text{``void'' or ``free'' volume}$$

$$\text{and } Q = \text{volume rate of flow} \tag{34}$$

The total enzyme concentration, (E_o), equals the total number of moles of enzyme, $(E)_t$, divided by the total volume, ν_t, of the system:

$$(E_0) = \frac{(E)_t}{\nu_t} \tag{35}$$

The fraction, f, of substrate converted is

$$f = \frac{(A_0) - (A)}{(A_0)} \tag{36}$$

Substitution of

$$t = \frac{\nu_f}{Q}, (E_0) = \frac{(E)_t}{\nu_t}, \text{ and } f = \left(\frac{(A_0)-(A)}{(A_0)}\right)$$

into Equation 33 leads to

$$K_A \ln\left(\frac{1}{1-f}\right) + (A_0)f = \frac{k_c(E)_t}{\nu_t} \text{ and } -K_A \ln(1-f) + (A_0)f = \frac{k_c(E_t)}{\nu_t}\frac{\nu_f}{Q} \tag{37}$$

If one lets $\beta = \frac{\nu_f}{\nu_t}$ = "voidage" of the column and $C = k_c\,(E)_t\beta$, referred to as the reaction capacity of the packed bed reactor, then

$$f(A_0) - K_A \ln(1-f) = \frac{k_c\,(E)_t\beta}{Q} = \frac{C}{Q}, \text{ i.e.,}$$

$$f(A_0) - K_A \ln(1-f) = \frac{C}{Q} \tag{38}$$

Equation 38 relates the fraction f of substrate converted during the passage through the column containing the immobilized enzyme to the flow rate Q, in terms of the initial substrate concentration (A_0), the Michaelis constant K_A, and the reaction capacity of the column C. The way in which f varies with (Q/C) is shown in Figure 7. For zero flow rate [infinite

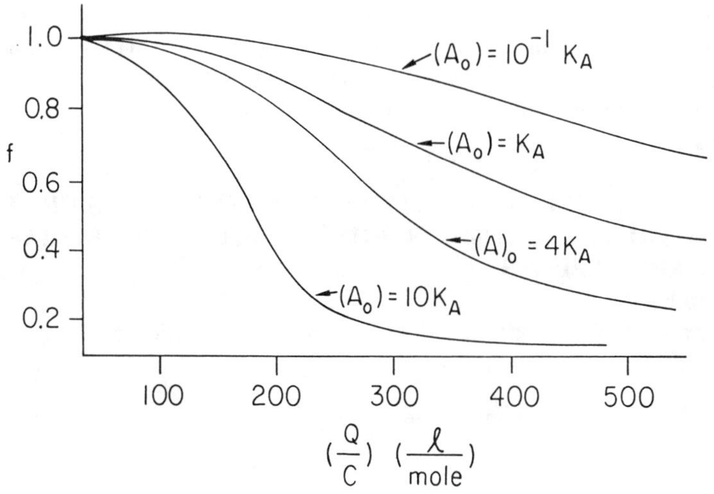

FIGURE 7. Fraction (f) of substrate converted with $\left(\frac{Q}{C}\right)$ for various values of $\frac{(A_o)}{K_A}$.

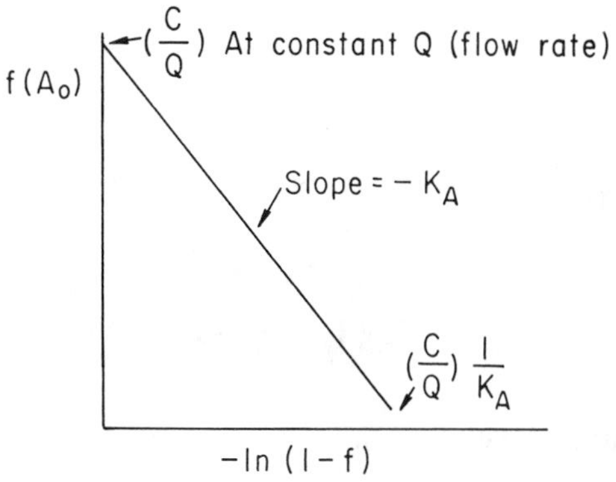

FIGURE 8. Plot of $f(A_o)$ vs. $-\ln(1-f)$.

residence time ($Q = O$)], there is complete conversion; for infinite flow rate ($Q = \infty$), there is no conversion.

In order to determine K_A and $\frac{C}{Q}$ from the experimental data, $f(A_o)$ may be plotted against $-\ln(1-f)$ as shown in Figure 8 where $f(A_o) = \frac{C}{Q} - [-K_A \ln(1-f)]$.

From $\left(\frac{C}{Q}\right)$, if Q, β, and $(E)_t$ are known, k_c, the limiting kinetic rate constant at high substrate concentrations, can be calculated, since

$$C = k_c\,(E)_t\beta$$

Lilly et al.,[26] showed that $-K_A$, the slope, was not quite independent of flow rate for the hydrolysis of *N*-benzoylarginine ethyl ester by a column of CM-cellulose ficin, but some variation was observed. Therefore, the flow is more complex than envisaged; each cellulose-ficin particle may be surrounded by a diffusion layer and the reaction rate may be controlled to some extent by the rate of diffusion of the substrate through this layer as described in a detailed theoretical treatment by Kobayashi and Laidler[754] (see below, Section VII).

For a recent review of commercial applications of immobilized enzyme systems in fixed-bed reactions see Reference 939.

D. CLASSICAL SINGLE-SUBSTRATE INHIBITION SYSTEMS OF LINEAR RECIPROCAL FORM IN THEIR PRIMARY PLOTS AND LINEAR IN THEIR SECONDARY REPLOTS

1. Competitive Inhibition

Consider the simplest model describing competitive inhibition of a single substrate (S) by inhibitor, I:

$$
\begin{array}{c}
E + S \underset{k_{-1}}{\overset{k_{+1}}{\rightleftharpoons}} ES \xrightarrow{k_{+2}} E + P \\
+ \\
I \\
K_i \updownarrow \\
EI
\end{array}
$$

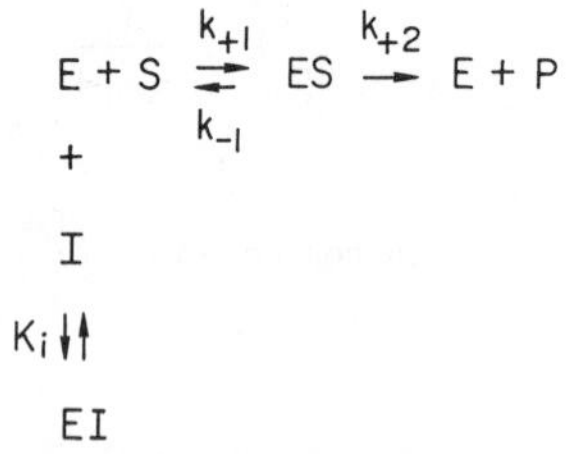

SCHEME III.

Derivation of the following expression, Equation 39, follows readily:

$$
v_f = \frac{V_{max}(S_0)}{K_m \left(1 + \dfrac{(I)}{K_i}\right) + (S_0)}
\tag{39}
$$

and which may be compared with Equation 10 without (I), that is

$$
v_f = \frac{V_{max,f}(S_0)}{K_m + (S_0)}
\tag{10a}
$$

where

$$
K_i = \frac{(E)(I)}{(EI)}; \quad K_m = \frac{k_{-1} + k_{+2}}{k_{+1}}; \quad V_{max} = k_{+2} E_t
\tag{40}
$$

Or, in reciprocal form:

$$
\frac{1}{v_f} = \frac{K_m}{V_{max,f}} \left(1 + \frac{(I)}{K_i}\right) \frac{1}{(S_0)} + \frac{1}{V_{max,f}}
\tag{41}
$$

Refer to

$$
\frac{1}{v_f} = \frac{K_m}{V_{max}} \frac{1}{(S_0)} + \frac{1}{V_{max,f}}
\tag{42}
$$

for the case without an inhibitor.

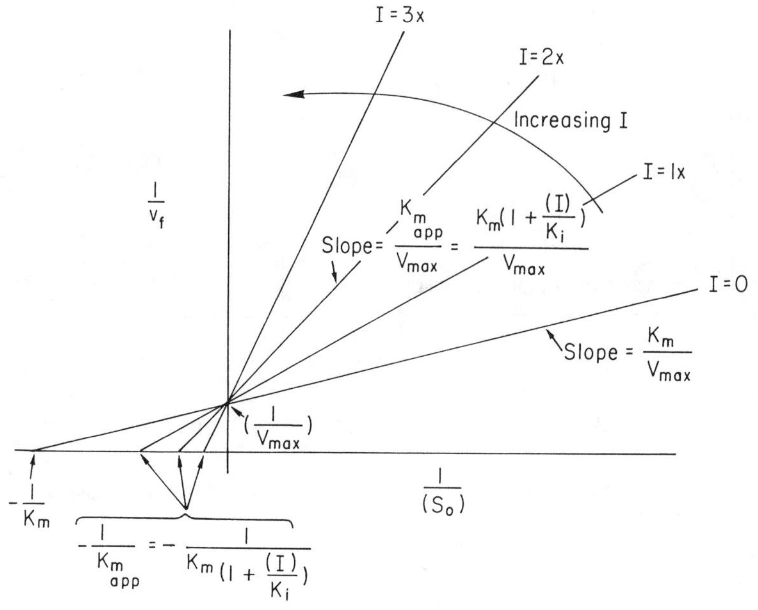

FIGURE 9. $\frac{1}{v}$ vs. $\frac{1}{S}$ plots in the presence of several fixed values of competitive inhibitor, I.

It is evident that the experimental value for K_m will be increased by the term $(1 + \frac{I}{K_i})$ on the addition of the competitive inhibitor, I. A $\frac{1}{v_f}$ vs. $\frac{1}{(S_o)}$ plot in the presence of several fixed concentrations of a competitive inhibitor is given in Figure 9.

To evaluate K_i, a replot may be made of either the slopes vs. (I) or $K_{m_{app}}$ vs. I as given in Figure 10A and B.

Thus,

$$\text{slope} = \frac{K_m}{V_{max,f}}\left(1 + \frac{(I)}{K_i}\right) = \frac{K_m}{V_{max,f}\,K_i}(I) + \frac{K_m}{V_{max,f}}$$

$$\left(\text{or slope} = \frac{K_{m\ app}}{V_{max,f}}\right) \tag{43}$$

or

$$K_{m_{app}} = K_m\left(1 + \frac{(I)}{K_i}\right) = \frac{K_m(I)}{K_i} + K_m \tag{44}$$

Figure 10A and B provides the solutions for the secondary slopes and secondary intercepts for both plots.

If the reaction has a large numerical value for K_{eq}, and if the product(s) does not have appreciable affinity for the enzyme, then the integrated Henri-Michaelis-Menten equation in the presence of a competitive inhibitor may be written as

$$\frac{2.303}{t}\log_{10}\frac{(S_0)}{(S)} = -\frac{1}{K_m\left(1 + \frac{(I)}{K_i}\right)}\frac{(S_0 - S)}{t} + \frac{V_{max}}{K_m\left(1 + \frac{(I)}{K_i}\right)} \tag{45}$$

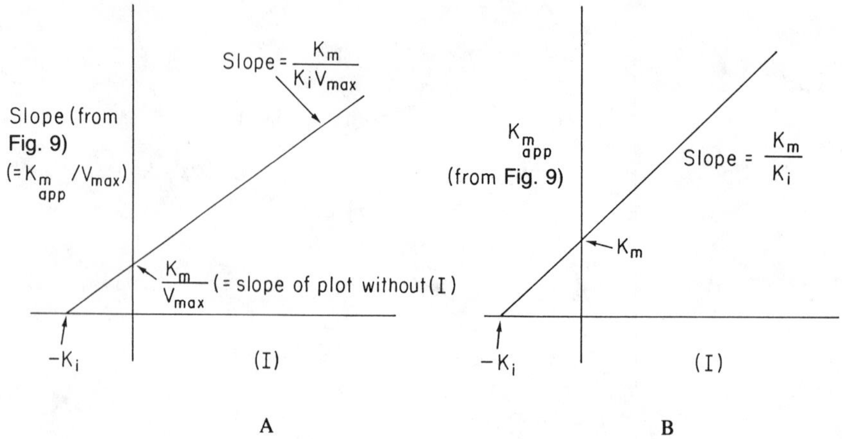

FIGURE 10. Evaluation of K_i by (A): replot of slopes vs. (I) (Equation 43); or (B): replot of $K_{M_{app}}$ vs. (I) (Equation 44).

or, since (P) = $(S_o) - (S)$,

$$\frac{2.303}{t}\log_{10}\left(\frac{S_0}{S}\right) = -\frac{1}{K_m\left(1+\frac{(I)}{K_i}\right)}\frac{(P)}{t} + \frac{V_{max}}{K_m\left(1+\frac{(I)}{K_i}\right)} \tag{46}$$

Refer to Equation 31.

Equations 45 and 46 assume that (I) remains constant as (S) decreases; therefore, (I) *cannot* be an "alternative" substrate. (For a discussion of alternative substrates and an interesting application of their use to study enzyme-catalyzed chemical modifications, see Reference 830). The determination of (P) at various times during the progress of the reaction permits $K_{m_{app}}$ and V_{max} to be determined; a family of curves can be obtained for several fixed-inhibitor concentrations (see Figure 11) where the values for K_m and K_i can be estimated from appropriate replots of slopes or ordinate-intercepts.

2. Noncompetitive Inhibition

The simplest model describing noncompetitive inhibition by I, of a single substrate reaction is given by Scheme IV.

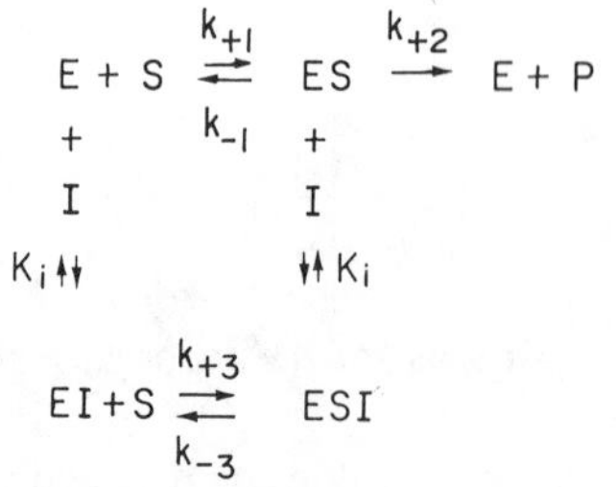

SCHEME IV.

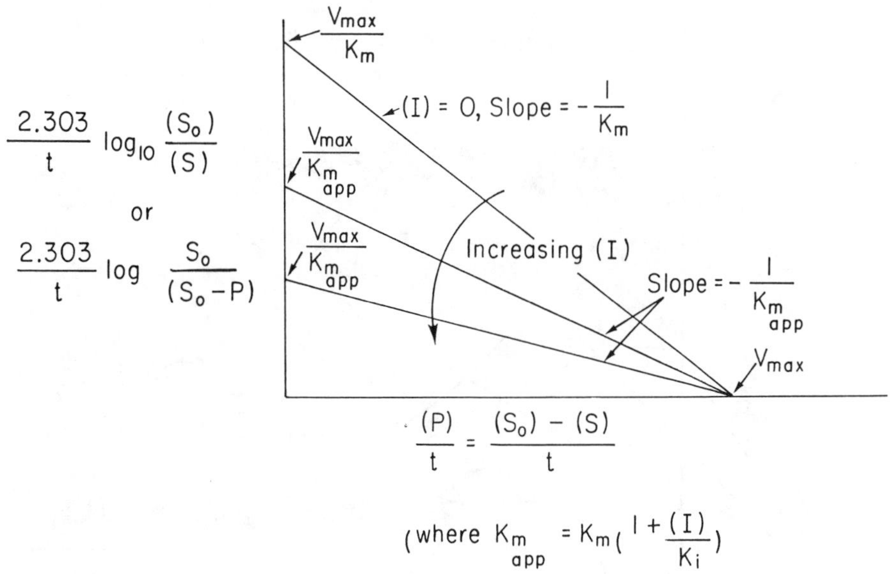

FIGURE 11. Plot of the integrated Henri-Michaelis-Menten equation in the presence of a competitive inhibitor (Equation 46) where $K_{m_{app}} = Km\ (1 + \frac{(I)}{K_i})$.

where

$$K_i = \frac{(E)(I)}{(EI)} = \frac{(ES)(I)}{(ESI)}$$

$$K_s = \frac{k_{-1}}{k_{+1}} = \frac{(E)(S)}{(ES)} = \frac{k_{-3}}{k_{+3}} = \frac{(EI)(S)}{(ESI)};$$

i.e., independent binding of I and S to the enzyme is assumed.

$$K_m = \left(\frac{k_{-1} + k_{+2}}{k_{+1}}\right) \text{ and } V_{max} = k_{+2}\, E_t \text{ which leads to}$$

$$v_f = \frac{V_{max,f}\,(S_0)}{K_m\left(1 + \frac{(I)}{K_i}\right) + (S_0)\left(1 + \frac{(I)}{K_i}\right)} \tag{47}$$

or in reciprocal form, to

$$\frac{1}{v_f} = \frac{K_m}{V_{max}}\left(1 + \frac{(I)}{K_i}\right)\frac{1}{(S_0)} + \frac{1}{V_{max}}\left(1 + \frac{(I)}{K_i}\right) \tag{48}$$

This equation indicates that both slope and Y-intercept of the reciprocal plot would be increased by the factor $(1 + \frac{(I)}{K_i})$ compared to the case where (I) = O; whereas, the X-intercept will remain constant and equal to $-\frac{1}{K_m}$. Thus, one may define

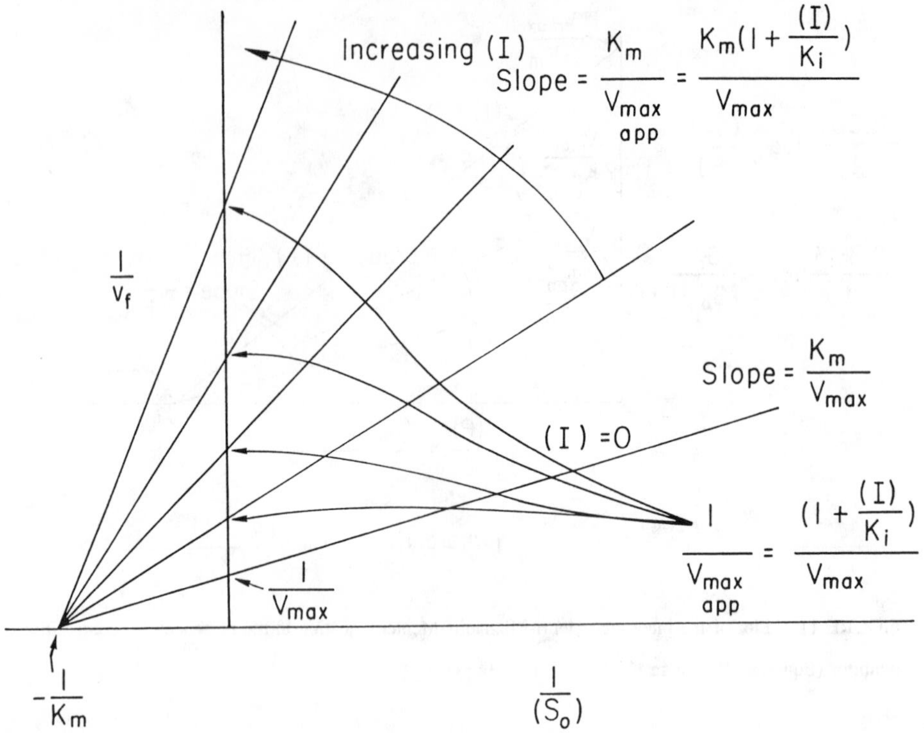

FIGURE 12. A plot of $\frac{1}{v}$ vs. $\frac{1}{(S_o)}$ with several fixed concentrations of a noncompetitive inhibitor.

$$\frac{1}{V_{\substack{max\\app}}} = \frac{1}{V_{max}}\left(1 + \frac{(I)}{K_i}\right) \tag{49}$$

In Figure 12 is shown a plot of $\frac{1}{v_f}$ vs. $\frac{1}{(S)_0}$ with several fixed concentrations of a noncompetitive inhibitor.

Replots of the slope and Y-intercepts from the reciprocal plot of Figure 12 are based on

$$\text{Slope} = \frac{K_m}{V_{\substack{max\\app}}} = \frac{K_m}{V_{max}} + \frac{K_m}{V_{max}\,K_i}\,(I)$$

$$\text{Y-intercept} = \frac{1}{V_{\substack{max\\app}}} = \frac{1}{V_{max}} + \frac{(I)}{V_{max}\,K_i} \tag{50}$$

and are shown in Figure 13A and B.

The integrated rate expression of Equation 47 for the case of a single-substrate reaction in the presence of a fixed concentration of a noncompetitive inhibitor and for the case of a large K_{eq} with the product(s) showing only slight affinity for the enzyme is

$$\frac{2.303}{t}\log_{10}\frac{(S_0)}{(S)} = \frac{-1}{K_m}\frac{(P)}{t} + \frac{V_{max}}{K_m\left(1 + \frac{(I)}{K_i}\right)} \tag{50a}$$

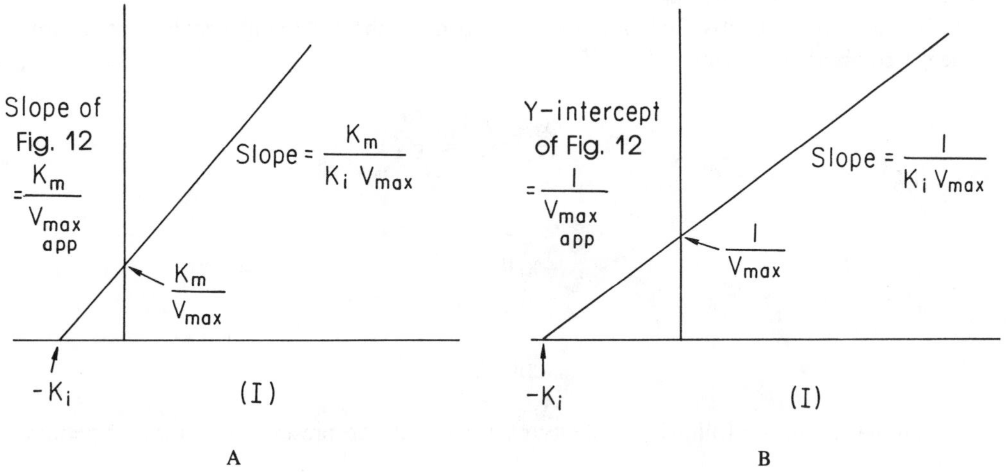

FIGURE 13. Secondary replots of data from Figure 12 primary plots. (A) Slope vs. (I) and (B) Y-intercept vs. (I).

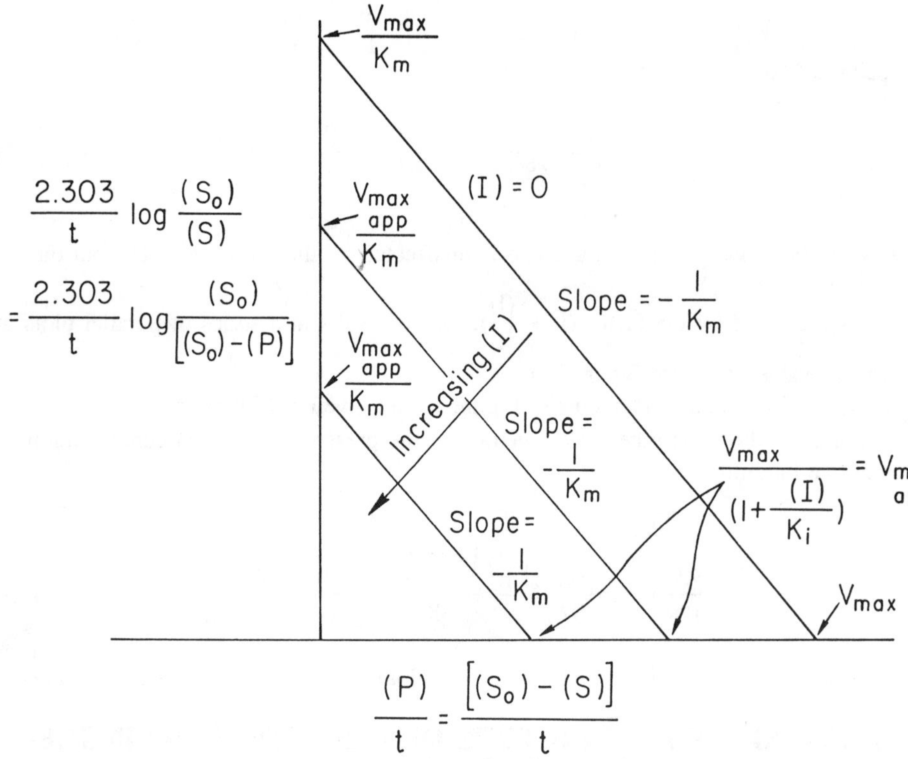

FIGURE 14. Plot of left-hand side of Equation 50 vs. $\frac{(p)}{t}$ for several fixed concentrations of I, a noncompetitive inhibitor.

where P = $((S_o) - S)$ with plots of the left-hand side of Equation 50a vs. $\frac{P}{t} = \frac{((S_o)-(S))}{t}$ at several fixed concentrations of (I) given in Figure 14.

3. Uncompetitive Inhibition

Classical uncompetitive inhibition is described by the following simple mechanism, where I combines only with (ES).

$$
E + S \underset{k_{-2}}{\overset{k_{+1}}{\rightleftharpoons}} (ES) \overset{k_{+2}}{\longrightarrow} E + P
$$

$$
+
$$

$$
I
$$

$$
K_i \updownarrow
$$

$$
(ESI)
$$

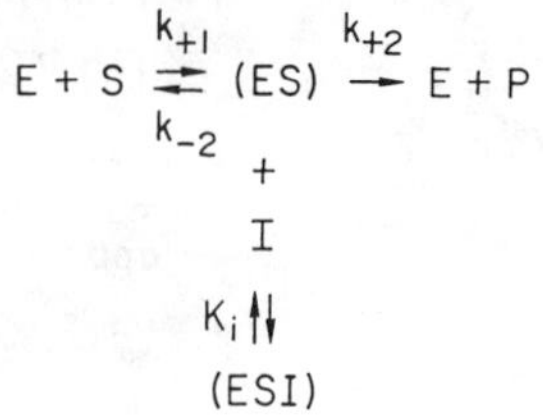

SCHEME V.

This results in the following velocity expression in the presence of an uncompetitive inhibitor:

$$
v_f = \frac{V_{max,f} (S_0)}{K_m + (S_0) \left(1 + \dfrac{(I)}{K_i}\right)} \tag{51}
$$

or in reciprocal form

$$
\frac{1}{v_f} = \frac{K_m}{V_{max}} \frac{1}{(S_0)} + \frac{1}{V_{max}} \left(1 + \frac{(I)}{K_i}\right) \tag{52}
$$

Thus, in the $\dfrac{1}{v_f}$ vs. $\dfrac{1}{(S_0)}$ plot, the slope is unaffected by the presence of (I), but the Y-intercept is increased by the factor $(1 + \dfrac{(I)}{K_i})$, which results in a series of parallel plots at several fixed values of I (see Figure 15).

Replots of the y- and x-intercepts in Figure 15 are given in Figure 16.

The corresponding integrated rate equation in the presence of a fixed concentration of uncompetitive inhibitor is

$$
\frac{2.303}{t} \log_{10} \frac{(S)_0}{(S)} = - \frac{\left(1 + \dfrac{(I)}{K_i}\right)}{K_m} \frac{(P)}{t} + \frac{V_{max}}{K_m} \tag{53}
$$

where $(P) = (S_o) - (S)$. Figure 17 provides the plot of this integrated velocity equation.

V. PARTIAL AND MIXED TYPE OF INHIBITION IN SINGLE-SUBSTRATE SYSTEMS

A. PARTIAL COMPETITIVE INHIBITION

A simple rapid (or quasi-) equilibrium mechanism describing this case is given by Scheme VI.

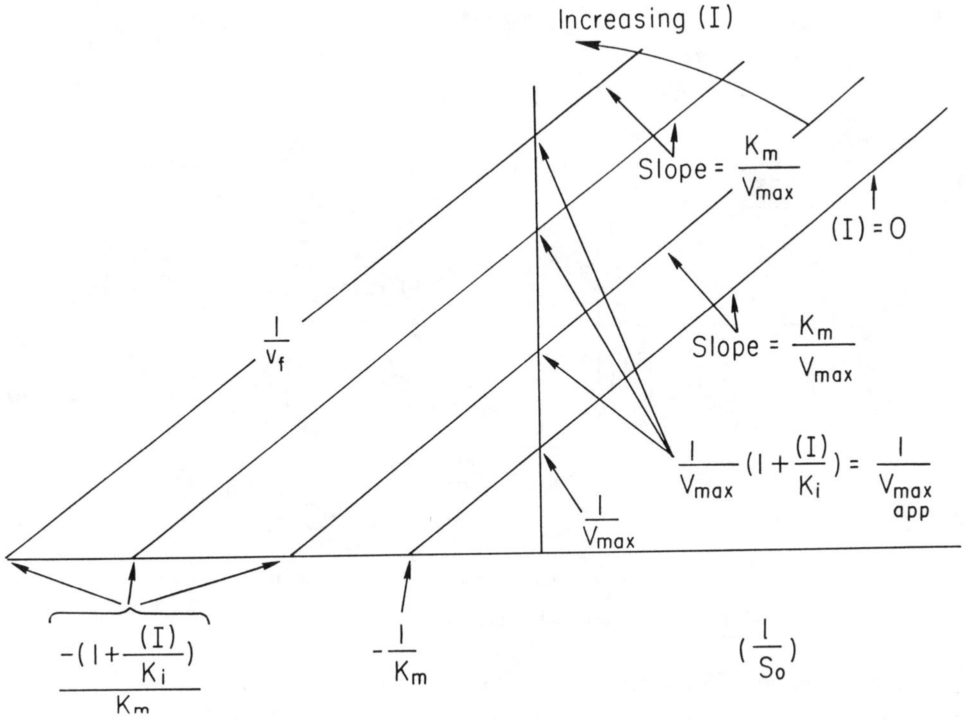

FIGURE 15. $\frac{1}{v_f}$ vs. $\frac{1}{(S_o)}$ plots in the presence of fixed concentrations of I, an uncompetitive inhibitor.

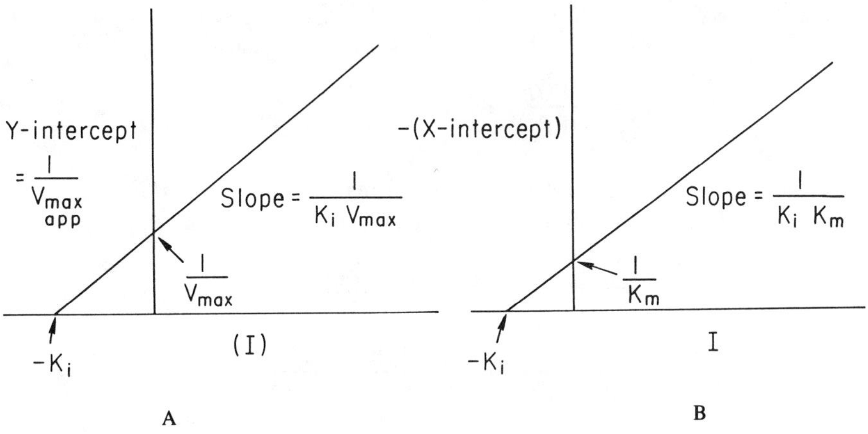

FIGURE 16. Replots of Y- and X-intercepts of Figure 15.

$$E + S \underset{}{\overset{K_s}{\rightleftharpoons}} ES \xrightarrow{k_{+2}} E + P$$

$$+ \qquad +$$

$$I \qquad I$$

$$K_i \updownarrow\downarrow \qquad \alpha K_i \updownarrow\downarrow$$

$$EI + S \underset{}{\overset{\alpha K_s}{\rightleftharpoons}} ESI \xrightarrow{k_{+4} = k_{+2}} EI + P$$

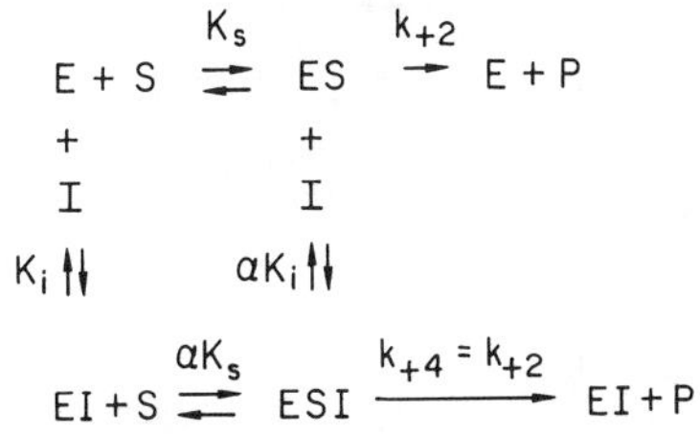

SCHEME VI.

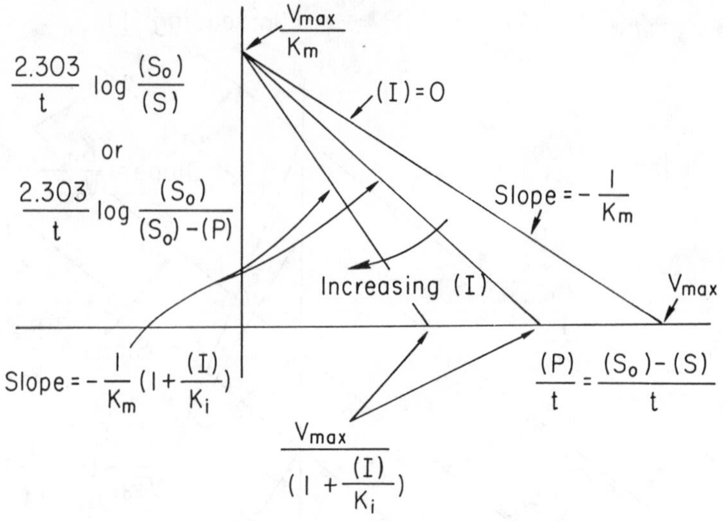

FIGURE 17. Plot of left-hand side of Equation 53 vs. $\frac{(P)}{t}$ in the presence of a fixed concentration of (I), an uncompetitive inhibitor.

In this case, both ES and ESI can break down to product(s), and therefore the reaction velocity can never be driven to zero by an excess concentration of I, hence the name partial competitive inhibitor.

Define the several equilibrium constants as follows:

$$K_s = \frac{(E)(S)}{(ES)}; \ \alpha K_s = \frac{(EI)(S)}{(ESI)} \text{ where } \alpha > 1$$

$$K_i = \frac{(E)(I)}{(EI)}; \ \alpha K_i = \frac{(ES)(I)}{(ESI)} \tag{54}$$

and k_{+2} and k_{+4} = rate constants for product formation from ES or ESI.

Assume rapid or quasi-equilibrium conditions and that $k_{+2} = k_{+4}$.

$$v_f = k_{+2}\,(ES) + k_{+4}\,(ESI) \tag{55}$$

$$\frac{v_f}{E_t} = \frac{k_{+2}\,(ES) + k_{+4}\,(ESI)}{(E) + (ES) + (EI) + (ESI)} \tag{56}$$

$$\frac{v_f}{k_{+2}\,E_t} = \frac{\frac{(S)}{K_s} + \frac{(S)(I)}{\alpha K_s\,K_i}}{1 + \frac{(S)}{K_s} + \frac{(I)}{K_i} + \frac{(I)(S)}{\alpha K_s\,K_i}} = \frac{v_f}{V_{max,f}} \tag{57}$$

where

$$V_{max,f} = k_{+2}\,E_t \tag{58}$$

Therefore

$$\frac{v_f}{V_{max,f}} = \frac{(S)}{K_s \dfrac{\left(1 + \dfrac{(I)}{K_i}\right)}{\left(1 + \dfrac{(I)}{\alpha K_i}\right)} + (S)} \tag{59}$$

or

$$\frac{v_f}{V_{max,f}} = \frac{(S)}{K_{s_{app}} + (S)} \tag{60}$$

where

$$K_{s_{app}} = K_s \frac{\left(1 + \dfrac{(I)}{K_i}\right)}{\left(1 + \dfrac{(I)}{\alpha K_i}\right)} \tag{61}$$

Note that at infinitely high or saturating inhibitor concentration:

$$\lim_{I \to \infty} \left(K_{s_{app}}\right) \longrightarrow \frac{K_s \left(\dfrac{I}{K_i}\right) = \alpha K_s}{\dfrac{1}{\alpha}\left(\dfrac{I}{K_i}\right)} = \alpha K_s \tag{62}$$

Therefore

$$\frac{v_f}{V_{max,f}} = \frac{(S)}{\alpha K_s + (S)} \tag{63}$$

A plot of v_f vs. $\left(\frac{I}{K_i}\right)$ for a partial and a pure competitive inhibitor is given below in Figure 18.

In reciprocal form, the corresponding velocity equation for partial competitive inhibition is

$$\frac{1}{v_f} = \frac{K_s}{V_{max,f}} \frac{\left(1 + \dfrac{(I)}{K_i}\right)}{\left(1 + \dfrac{(I)}{\alpha K_i}\right)} \frac{1}{(S)} + \frac{1}{V_{max,f}} \tag{64}$$

and its double-reciprocal plot is given in Figure 19.

As I increases, the slope of the above double-reciprocal plot increases by the factor $K_{s_{app}}/V_{max}$. Unlike a pure competitive inhibitor, this slope approaches a finite limit as $(I) \rightarrow \infty$, that is, $\frac{\alpha K_s}{V_{max}}$, with a corresponding x-intercept $= -\frac{1}{\alpha K_s}$.

B. PARTIAL NONCOMPETITIVE INHIBITION

Consider a system where both the substrate and inhibitor combine reversibly and *in-*

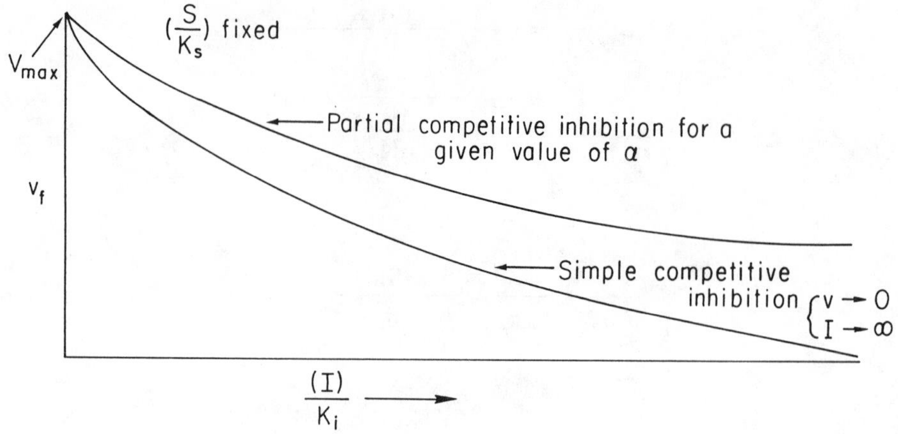

FIGURE 18. Plot of v_f vs. $\left(\frac{I}{K_i}\right)$ for a partial and a pure competitive inhibitor.

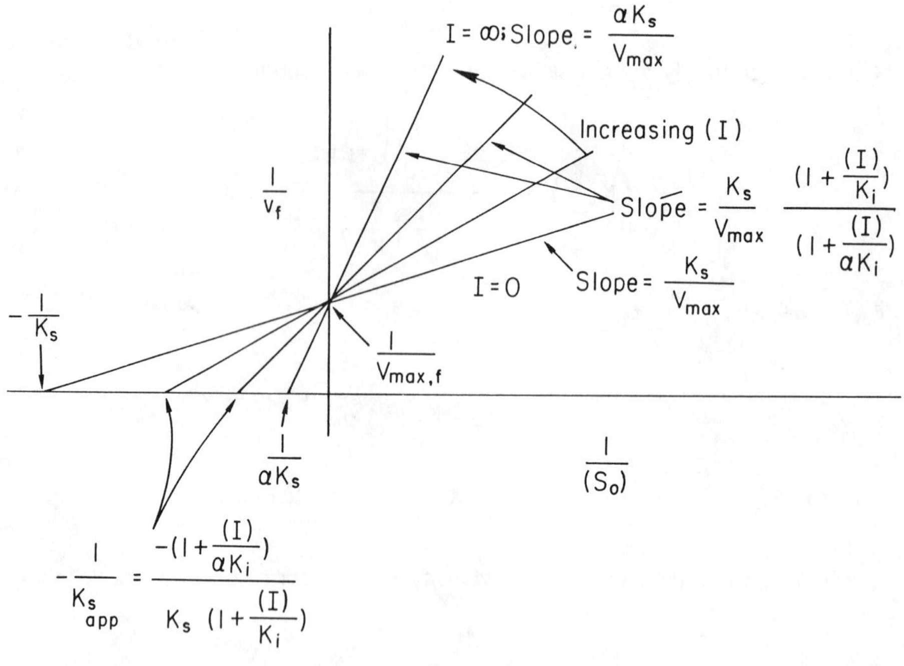

FIGURE 19. $\frac{1}{v_f}$ vs. $\frac{1}{(S_0)}$ plots in the presence of fixed concentrations of a partial competitive inhibitor. Plots with I made with a constant value for α.

dependently with the enzyme, as shown in Scheme VII, to produce ES, EI, and ESI complexes.

$$E + S \underset{}{\overset{K_s}{\rightleftharpoons}} ES \xrightarrow{k+2} E + P$$

$$EI + S \underset{}{\overset{K_s}{\rightleftharpoons}} ESI \xrightarrow{\beta k+2} E + P$$

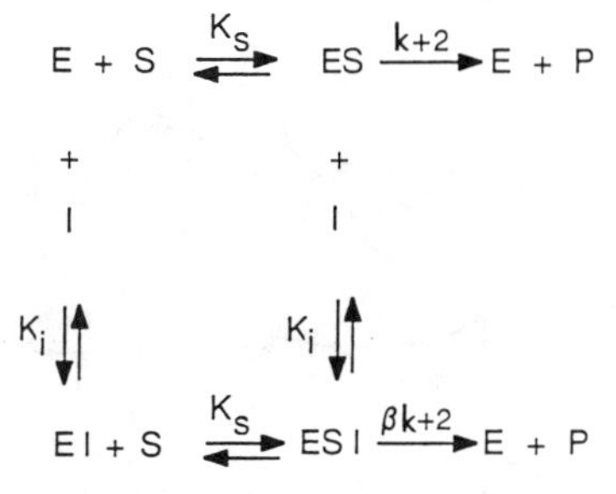

SCHEME VII.

In this case, in contrast to Scheme VI, the rates of breakdown of ES and ESI to product are not considered equal, and if ES produces a product more effectively than ESI, I will act only as a partial noncompetitive inhibitor.

Thus, with rapid (quasi-) equilibrium conditions considered to hold

$$v_f = k_{+2} (ES) + \beta k_{+2} (ESI); \text{ where } \beta < 1 \tag{65}$$

$$\frac{v_f}{E_t} = \frac{k_{+2} \dfrac{(S)}{K_s} + \beta k_{+2} \dfrac{(S)(I)}{K_s K_i}}{1 + \dfrac{(S)}{K_s} + \dfrac{(I)}{K_i} + \dfrac{(S)(I)}{K_s K_i}} \tag{66}$$

Setting $k_{+2}(E_t) = V_{max,f}$

$$\frac{v_f}{V_{max,f}} = \frac{(S)}{K_s \dfrac{\left(1 + \dfrac{(I)}{K_i}\right)}{\left(1 + \dfrac{\beta(I)}{K_i}\right)} + (S) \dfrac{\left(1 + \dfrac{(I)}{K_i}\right)}{\left(1 + \dfrac{\beta(I)}{K_i}\right)}} \tag{67}$$

If one defines

$$V_{max,f \atop app} = V_{max,f} \frac{\left(1 + \dfrac{\beta(I)}{K_i}\right)}{\left(1 + \dfrac{(I)}{K_i}\right)} \tag{68}$$

therefore

$$\frac{v_f}{V_{max,f \atop app}} = \frac{(S)}{K_s + (S)} \tag{69}$$

or, in reciprocal form

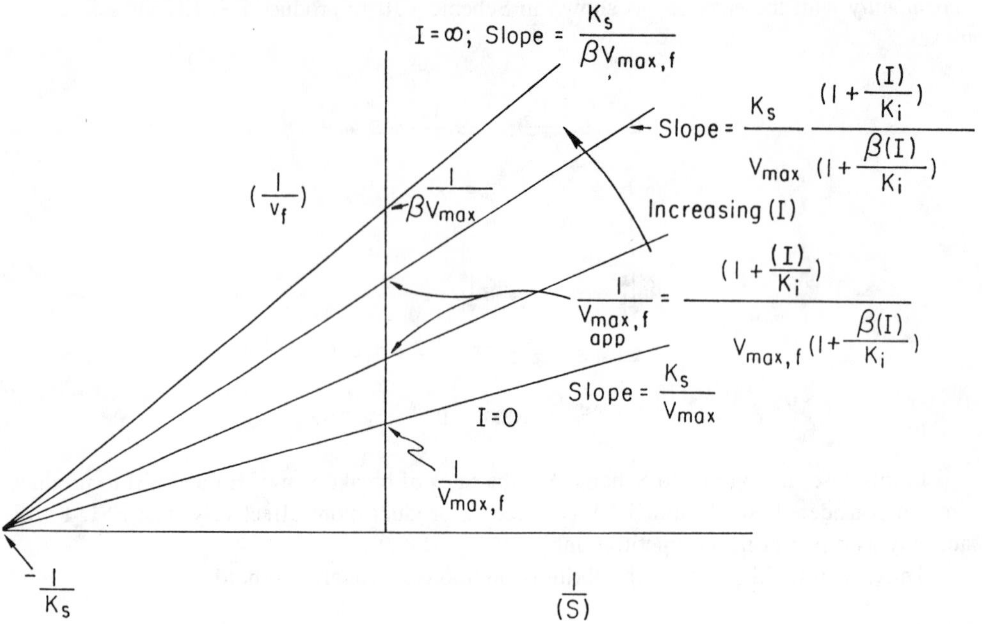

FIGURE 20. $\frac{1}{v_f}$ vs. $\frac{1}{(S)}$ plots in the presence of several fixed concentrations of partial noncompetitive inhibitor.

$$\frac{1}{v_f} = \frac{K_s}{V_{max,f}} \frac{\left(1 + \frac{(I)}{K_i}\right)}{\left(1 + \frac{\beta(I)}{K_i}\right)} \frac{1}{(S)} + \frac{1}{V_{max,f}} \frac{\left(1 + \frac{(I)}{K_i}\right)}{\left(1 + \frac{\beta(I)}{K_i}\right)} \tag{70}$$

which is then plotted in Figure 20.
(Note: at high (I), i.e., as (I) → ∞, Equation 70 reduces to

$$\frac{v_f}{\beta V_{max,f}} = \frac{(S)}{Ks + (S)} \tag{71}$$

C. MIXED TYPE OF INHIBITION

A mixed type of inhibitor will affect both K_m and V_{max}. It can arise as the result of several mechanisms, only one of which will be described here, that is, a case yielding linear secondary replots.

This system may be considered a mixture of partial competitive and pure noncompetitive inhibition. In the simplest system ESI is nonproductive (or $\beta = O$ in the mechanism described in Scheme VII), whereas, $\alpha > 1$ in the mechanism described in Scheme VI.

Therefore:

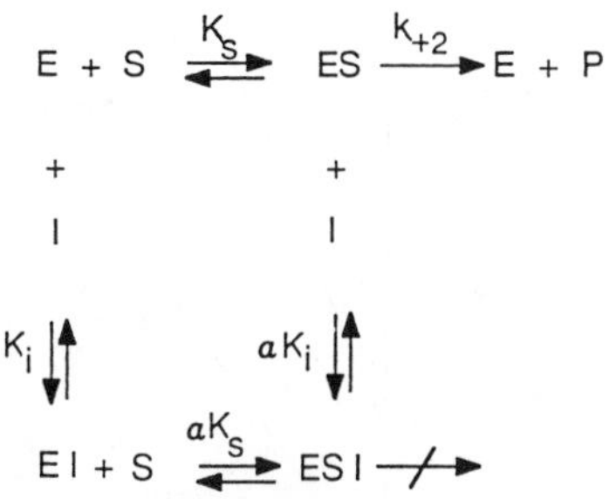

SCHEME VIII.

Deriving the velocity equation from rapid (quasi-) equilibrium assumptions:

$$v_f = k_{+2}(ES); \frac{v_f}{(E_t)} = \frac{k_{+2}(ES)}{(E) + (ES) + (EI) + (ESI)} \tag{72}$$

Therefore,

$$\frac{v_f}{V_{max,f}} = \frac{\frac{(S)}{K_s}}{1 + \frac{(S)}{K_s} + \frac{(I)}{K_i} + \frac{(S)(I)}{\alpha K_s K_i}} \tag{73}$$

Rearranging,

$$\frac{v_f}{V_{max,f}} = \frac{(S)}{K_s\left(1 + \frac{(I)}{K_i}\right) + (S)\left(1 + \frac{(I)}{\alpha K_i}\right)} \tag{74}$$

Defining

$$V_{\underset{app}{max,f}} = \frac{V_{max,f}}{\left(1 + \frac{(I)}{\alpha K_i}\right)} \tag{75}$$

and

$$K_{\underset{app}{s}} = K_s \frac{\left(1 + \frac{(I)}{K_i}\right)}{\left(1 + \frac{(I)}{\alpha K_i}\right)}; \text{ which leads to } \frac{v_f}{V_{\underset{app}{max}}} = \frac{(S)}{K_{\underset{app}{s}} + (S)} \tag{76}$$

In double reciprocal form, the Lineweaver-Burk equation is

$$\frac{1}{v_f} = \frac{K_s}{V_{max,f}}\left(1 + \frac{(I)}{K_i}\right)\frac{1}{(S)} + \frac{1}{V_{max,f}}\left(1 + \frac{I}{\alpha K_i}\right) \tag{77}$$

Because the ESI complex does not break down to products, the velocity can be reduced to zero as (I) $\rightarrow \infty$, and the limiting slope in a $\frac{1}{v}$ vs. $\frac{1}{(S)}$ plot, at I = ∞, will be a vertical

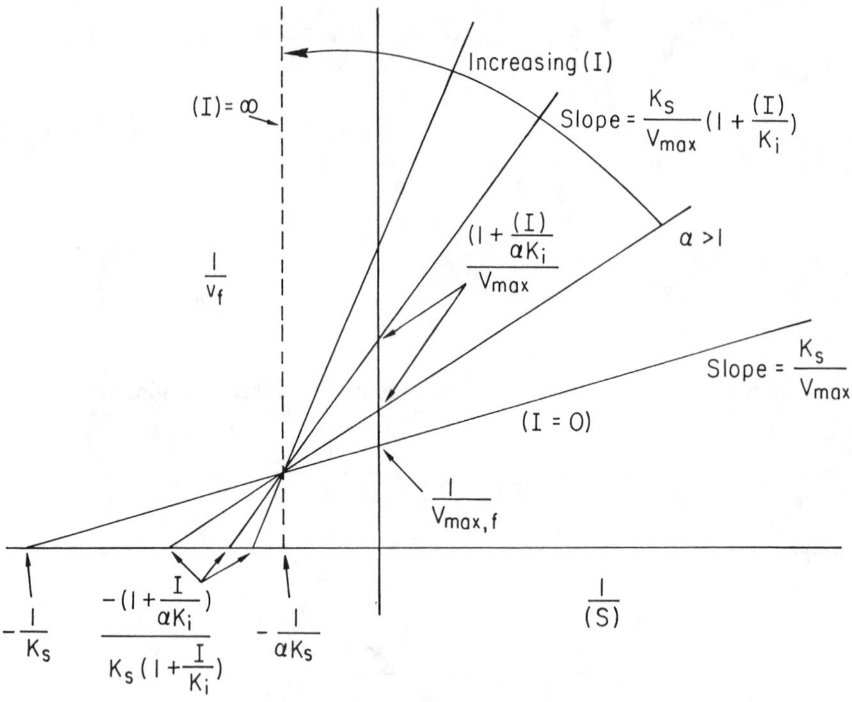

FIGURE 21. Primary plot of $\frac{1}{v_f}$ vs. $\frac{1}{(S)}$ (Equation 76) for several values of a mixed type of inhibitor; with $\alpha > 1$.

line through the common intersection point in the left quadrant and parallel to the $\frac{1}{v}$ axis. With $\alpha > 1$ and $\beta = O$, this common intersection point is above the $\left(\frac{1}{S}\right)$-axis in the left-hand quadrant. If $\alpha < 1$ and $\beta = O$, the double-reciprocal plots will intersect below the $\left(\frac{1}{S}\right)$ axis.

Thus, the primary plots of the reciprocal plot data are given in Figure 21.

In this case of a mixed type of inhibitor, the slope factor $\left(1 + \frac{(I)}{K_i}\right)$ and Y-intercept factor $(1 + \frac{(I)}{\alpha K_i})$ are not identical; however, secondary plots can be constructed from

$$\begin{cases} \text{Slope} = \dfrac{K_s}{V_{max,f}} + \dfrac{K_s(I)}{V_{max}K_i} \\ \text{and Y-intercept} = \dfrac{1}{V_{\substack{max\\app}}} = \dfrac{1}{V_{max,f}} + \dfrac{1}{\alpha K_i V_{max,f}}(I) \end{cases} \tag{78}$$

to yield values for α, K_i, K_s, and $V_{max,f}$ (see Figure 22).

D. TIGHTLY BOUND INHIBITORS

In cases where K_i is very low, or the enzyme has a very high affinity for the inhibitor, the value of (I) in the velocity equations for inhibition cannot be assumed to be equal to

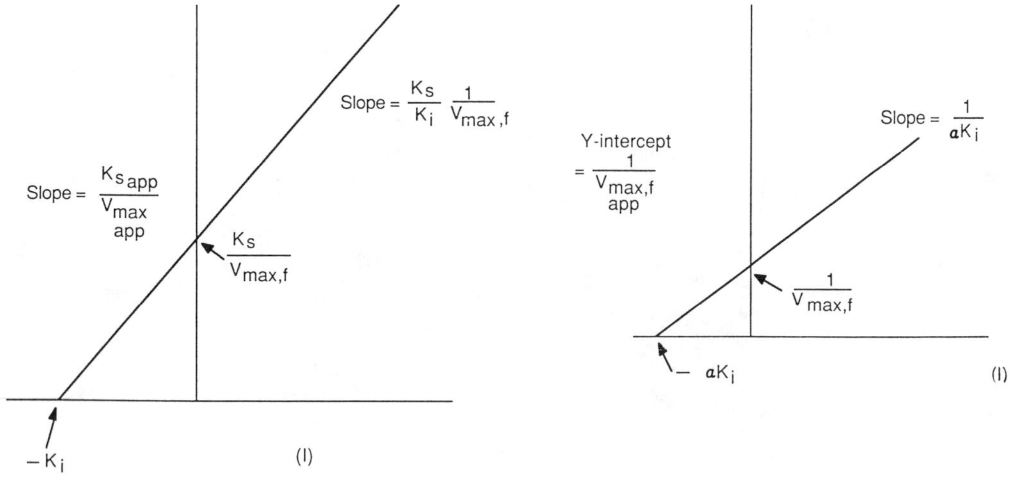

FIGURE 22. Secondary replots of slopes and intercepts from primary reciprocal plots of Figure 21 for the case of a mixed-type inhibitor.

(I_t), the total concentration of added I, since a significant fraction of (I) will be bound to the enzyme. Such cases can be derived for classical competitive and noncompetitive inhibitors where the conservation expression for $(I_t) = (I) + (EI) + (ESI) + \ldots$ must now be considered in addition to that for the enzyme, i.e., $(E_t) = (E) + (ES) + (EI) + (ESI) + \ldots$ Whereas, if $(S_t) \gg (E_t)$, it is assumed that $(S_t) \simeq (S)$.[154,155,843] Also see the discussion by Morrison on "slow-binding inhibition" by potent reversible enzyme inhibitors[989] and its relationship with analogues of intermediates of enzymic reactions.[990,991]

An interesting situation arises in the cases of the unique antienzyme globulins which may specifically inhibit an enzyme or isoenzyme.

Thus, consider the case of rabbit anti-calf muscle-type myokinase,[29] which specifically inhibits the calf muscle-type myokinase (molecular weight = 21,200) but not the calf liver adenylate kinase. The highly purified antibody was estimated to have a molecular weight of 134,000 and to consist of two heavy (42,000 mol wt) and two light chains (25,000 mol wt). The possibility of several kinds of stoichiometric binding had to be considered: for $E_2 \cdot I \rightleftharpoons 2E + I$ (or 2 Ag to 1 Ab) where $K'_d = E_2 \cdot I \rightleftharpoons 2E + I$. I = Ab, the antibody, and E = Ag, the antigen; or a stoichiometry of 1 Ag to 1 Ab, i.e., for $E \cdot I \rightleftharpoons E + I$, and $K'_d = (E)\,(I)/(EI)$.

In the former case (2 Ag to 1 Ab), on the basis of mass action, and defining $i = 1 - \left(\frac{v_i}{v_o}\right)$ = fraction inhibition, the following equation may be derived (see Appendix 2):

$$(I_t)\,\frac{(2-i)}{(i)} = \frac{(2-i)^2}{(2-2i)^2}\,\frac{K'_d}{(E_t)} + (E_t) \qquad (79)$$

A plot of the left-hand side vs. $\left(\frac{2-i}{2-2i}\right)^2$ should then be linear with E_t = the ordinate-intercept and the slope = K'_d/E_t.

However, for the latter case (1 Ag to 1 Ab), the equation

$$\frac{(I_t)}{(i)} = K'_d\left(\frac{1}{1-i}\right) + (E_t) \qquad (80)$$

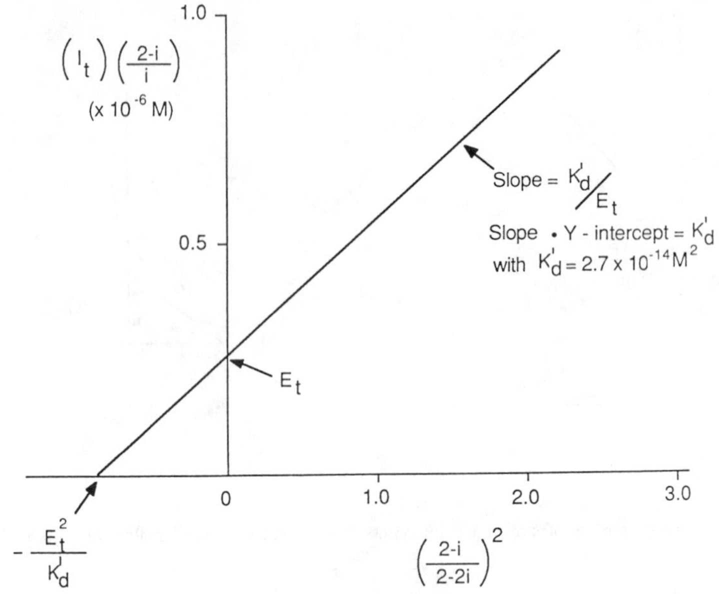

FIGURE 23. A kinetic determination of the stoichiometry and the dissociation constant for the reaction of calf muscle myokinase with anticalf muscle myokinase.[29]

results, and a plot of $\frac{I_t}{(i)}$ vs. $\frac{(1)}{(1-i)}$ should be linear with a Y-intercept $=$ E_t and a slope $=$ K'_d.

As shown in Figure 23,[29] a linear plot results for the first case and threfore a 2:1 stoichiometry can be assumed (2 Ag to 1 Ab), with a calculated $K'_d = 2.7 \times 10^{-14}$ M^2.

E. TIGHTLY BOUND COFACTORS OR SUBSTRATES

Similar to the case where there is a tightly bound inhibitor (see Section D above), in those cases where the K_m or K'_D is very low, or the enzyme has a very high affinity for a cofactor or substrate, a significant fraction of the cofactor will remain bound to the enzyme. An interesting case may be found among the flavin enzymes; thus, consider the *binding* of FAD to apo-NADPH-cytochrome c reductase, as estimated by a kinetic approach.[283]

On the basis of mass action, for the case of a stoichiometry of 1 FAD to 1 polypeptide chain of 34,000 g/mol, i.e., assuming each subunit binds independently,

$$E{\cdot}FAD \rightleftarrows E + FAD; \quad K'_D = \frac{(E)(FAD)}{(E{\cdot}FAD)} \tag{81}$$

Let $a = \frac{v_0}{V_{max}}$ = fraction activation, and the following equation may be readily derived:

$$\frac{(FAD)_t}{a} = \frac{K'_D}{(1-a)} + E_t \tag{82}$$

Thus, a plot of $\frac{(FAD)_t}{a}$ vs. $\frac{1}{(1-a)}$ should be linear with a slope $=$ K'_D (compare Reference 155). Similarly, if a stoichiometry of 2 FAD per subunit were to hold, i.e.,

$$E \cdot (FAD)_2 \rightleftarrows 2FAD + E; \; K'_D = \frac{(E)(FAD)^2}{E \cdot (FAD)_2} \tag{83}$$

which leads to

$$\frac{(FAD)_t}{(a)} = \frac{K^{1/2}}{(1-a)^{1/2}(a)^{1/2}} + 2\,E_t$$

and a plot of

$$\frac{(FAD)_t}{(a)} \text{ vs. } \frac{1}{(1-a)^{1/2}} \frac{1}{(a)^{1/2}}$$

would be linear.

In Figure 24 (taken from Reference 283 on a study of flavin interactions in the ale yeast NADPH-cytochrome c reductase), the lowest plot shows that the apoenzyme still retained a small amount of activity in the absence of added FAD (approximately 90% of the flavin had been removed in the preparation of the apoenzyme, and the added FAD appeared to titrate the remaining "apoenzyme"). The upper right-hand plot permitted an estimate for V_{max} for these conditions (and $K_{m_{app}}$), and the upper left-hand plot then allowed the final estimate of $K'_D \simeq 4.7 \times 10^{-8}M$. In addition, the linear plot of $\frac{(FAD)_t}{a}$ vs. $\frac{1}{(1-a)}$ supported the idea that each monomeric unit binds 1 mol of FAD in an equivalent and independent fashion. Similarly, the K'_D for FMN binding to the apoenzyme (the data are not shown here) was estimated to be approximately $4.4 \times 10^{-8}M$ (and $K_{m_{app}} \simeq 2 \times 10^{-8}$) and essentially identical with the K'_D for FAD. Either FMN or FAD may activate the enzymatic activity of NADPH-cytochrome c reductase yielding very similar turnover numbers (FMN/FAD $\simeq$ 1.07).

VI. EFFECT OF TEMPERATURE

The effect of temperature on enzyme-catalyzed reactions is in fact still a problem which is not *completely* solved because of the complexity involved. Our understanding of temperature effects is largely based on those aspects better understood in classical chemical kinetics. For that reason, a quick review will be presented of the effect of temperature on chemical reaction rates. Additional discussion will be given in Chapter 4.

The van't Hoff equation,

$$\left(\frac{\partial \ell n K_{eq}}{\partial T_p}\right) = \frac{\Delta H^\circ}{RT^2} \tag{84}$$

provided a relation between the equilibrium constant, K_{eq}, of a reversible reaction at thermodynamic equilibrium where all the reactants were in their respective standard states and ΔH° is the standard enthalpy change and T, the absolute temperature. At constant volume, i.e., for solutions or for ideal gas reactions, Equation 84 may be replaced by

$$\frac{d\ell nK}{dT} = \frac{\Delta E^\circ}{RT^2} \tag{85}$$

where ΔE° is the change in internal energy of the system. If $K = \frac{k_{+1}}{k_{-1}}$, it follows that

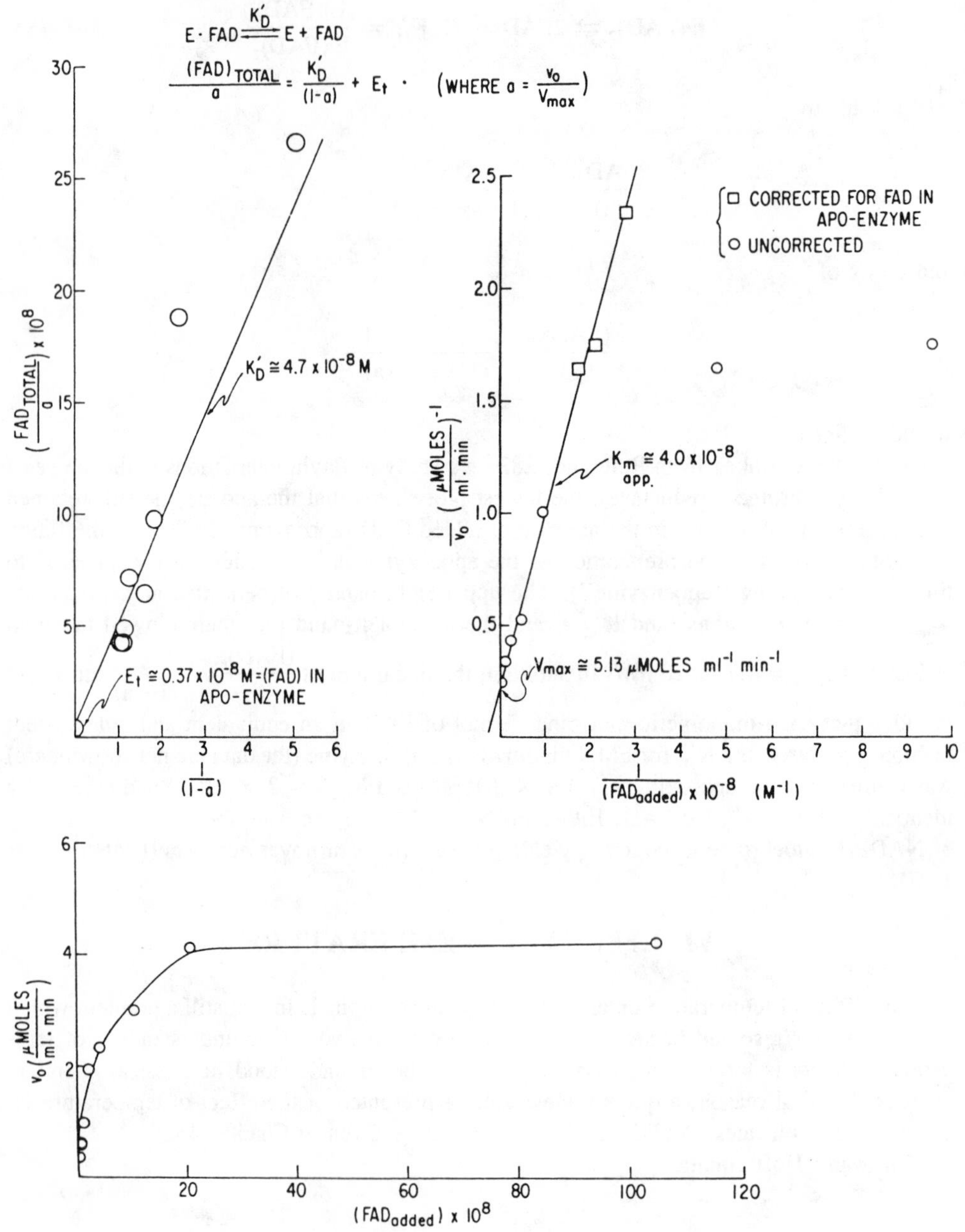

FIGURE 24. Kinetic evaluation of the dissociation constant for FAD and apo-NADPH-cytochrome c reductase. Upper left-hand plot: $\left(\frac{FAD_{total}}{a}\right) \times 10^{-8}$ is plotted against $1/(1 - a)$; where $a - v_o/V_{max}$ = fraction activation. Upper right-hand plot: $1/v_o$ in $(\mu mol)^{-1}$ ml min vs. $(FAD_{added})^{-1} \times 10^{-8}$ (in M^{-1}): (□) corrected for FAD added in the "apoenzyme", (○) uncorrected. Bottom plot: v_o (in μmol min^{-1} ml^{-1}) vs. $(FAD_{added}) \times 10^{-8}$ (in M).[283]

$$\frac{d\ell nk_{+1}}{dT} - \frac{d\ell nk_{-1}}{dT} = \frac{\Delta E^\circ}{RT^2} \tag{86}$$

At van't Hoff's suggestion, Equation 84 may be split up into

$$\frac{d\ell nk_{+1}}{dT} = \frac{\Delta E_{+1}}{RT^2} + B \text{ and } \frac{d\ell nk_{-1}}{dT^2} = \frac{\Delta E_{-1}}{RT^2} + B \tag{87}$$

where

$$\Delta E_{+1} - \Delta E_{-1} = \Delta E^\circ \tag{88}$$

is a constant.

Experimentally, it was found that the temperature variation of the specific rate constant was best expressed by assuming B = O, so that the Arrhenius equation, i.e., Equation 89, resulted:

$$\frac{d\ell nk}{dT} = \frac{\Delta E_{act}}{RT^2} \tag{89}$$

where ΔE_{act} is called the Arrhenius "energy of activation". ΔE_{act} is of great theoretical importance as we shall see shortly. The Arrhenius equation in its integrated form:

$$\ell nk = \frac{-\Delta E_{act}}{RT} + \text{constant} \tag{90}$$

predicts a linear plot of ℓnk vs. $\frac{1}{T}$, whose slope $= \frac{-\Delta E_{act}}{R}$, a relation found to hold for a large number of reactions, provided ΔE_{act} is independent of temperature. One may write the integrated Arrhenius expression as follows:

$$k = Ae^{-\Delta E_{act}/RT} \tag{91}$$

(note: by taking the natural logs of both sides, Equation 90 results). A complete theory of reaction rates involves an interpretation of the two quantities, the preexponential factor, A (called the "frequency factor"), and the ΔE_{act} (the Arrhenius energy of activation). It is believed that when two reactant molecules, which possess the necessary energy of activation, come together, they first form an *activated complex* or *transition state,* and this decomposes at a definite rate to yield the products of the reaction. The formation of this intermediate state is considered to be characteristic of all chemical changes and even of certain physical processes taking place at a definite rate. Consider, for example, the single displacement type of reaction: A + B-C $\leftrightarrows$ A-B + C, where the atom A is gradually brought up to the diatomic molecule B-C, in which the atoms are vibrating at their normal distance apart. When the reactants are relatively far apart, the potential energy of the system is unaffected, but when A approaches B-C, the nuclei of the latter are to some degree forced apart, and the potential energy increases. This process continues until a configuration, which may be represented as A-B-C, is attained where it is equally possible for A to unite with B, forming A-B + C, as it is for B to remain attached to C. This state is, in fact, the activated complex, or transition state, through which the system A + B-C must pass before it can be converted into A-B + C, and vice versa; thus,

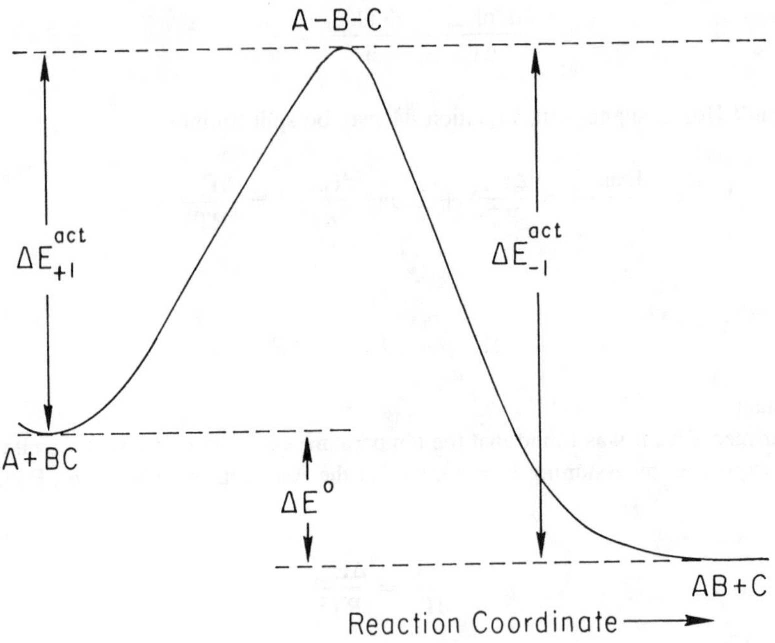

FIGURE 25. Change of potential energy in a hypothetical exothermic reaction: A + B-C ⇌ A-B-C ⇌ A-B + C.

A + B-C	⇄	A-B-C	⇄	A-B + C	
initial state		activated complex		final state	(91A)

In the condition of the "activated complex" the potential energy is at a maximum. The change of potential energy may thus be represented diagrammatically as in Figure 25 for an exothermic reaction where the energy level of the product is ΔE° below that of the reactants. The difference in energy between A + B-C and the activated complex is equal to ΔE^{act}_{+1}, the energy of activation of the forward reaction, and that between the complex and A-B + C is the activation energy, ΔE^{act}_{-1}, for the reverse reaction. The difference between ΔE^{act}_{+1} and ΔE^{act}_{-1} is thus equal to ΔE°, the total energy change in the reaction under consideration. Another important point to be derived from Figure 25 is that the energy of activation of an endothermic reaction must be at least equal to the heat absorbed in the reaction; highly endothermic changes are, therefore, likely to be very slow except at relatively high temperatures.

In this connection, a brief discussion about *potential-energy surfaces* is appropriate and we will continue with our discussion of the single-displacement reaction described by Equation 91A (which incidentally is characteristic of a number of reactions, including some enzyme reactions). For energetic reasons, it is most likely for A to approach B-C along the line of centers, forming a linear A . . . B . . . C complex. In the initial state of the reaction the A-B distance is large, while in the final state B-C is large; and all intermediate states of the linear complex are represented by specifying the distances A-B and B-C. If one plots the potential energy of the system vs. the A-B and B-C distances, the result, as illustrated in Figure 26 is known as a *potential-energy surface*.

On the left-hand face of Figure 26 appears a dissociation-energy curve for the B-C molecule, the group A being so far removed that it does not influence the shape of this curve. Similarly, on the right-hand face, where C is far removed, is the dissociation energy

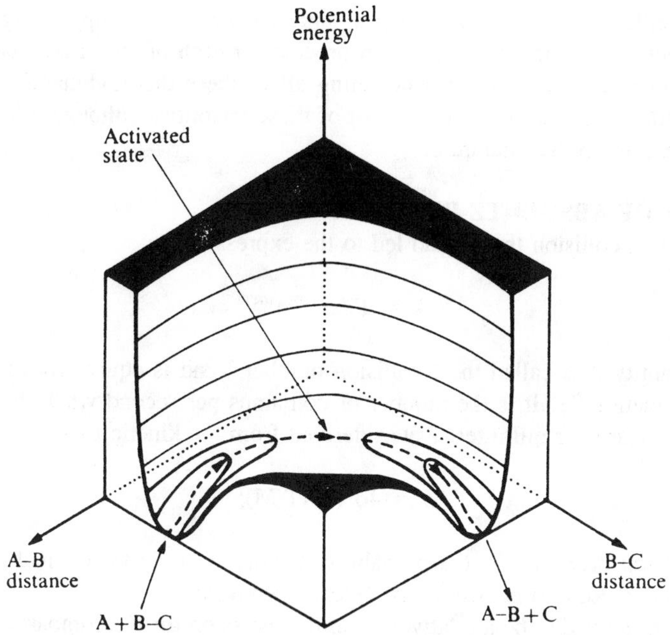

FIGURE 26. A potential-energy surface for the system A + B-C → A-B + C, where potential energy is plotted vs. the A-B and B-C distances. (From Laidler, K. J. and Bunting, P. S., *The Chemical Kinetics of Enzyme Action,* 2nd ed., Clarendon Press, Oxford, 1973, 44. With permission.)

curve for the molecule A-B. The initial and final states correspond to points towards the bottoms of these two curves, as shown in the diagram. Potential-energy surfaces are often represented as contour diagrams, and we will see another example when we apply it later to a discussion of kinetic-isotope effects (see Chapter 6). The path by which the system passes from the initial to the final state is determined by the shape of the potential-energy surface. In Figure 26 may be seen two valleys meeting at a *col* or "saddle point" in the interior of the diagram. The system at the *col* is referred to as the activated complex or transition state. The path of least resistance, which the reaction will consequently follow, will correspond to motion up the left-hand valley, over the *col*, and into the other valley. The rate of the reaction is determined primarily by the height of the barrier at the *col*, compared to the initial energy. If the barrier is high, a large proportion of collisons between A and B-C will not have sufficient energy to surmount the barrier; whereas, a low barrier will be surmounted by a larger proportion of colliding systems. The *col* represents the point of no return; once the system has reached the *col*, it almost inevitably passes down into the second valley and gives rise to the products of the reaction. A complete understanding of the mechanism of an elementary process involves knowing the potential-energy surface for the system and calculating the rate of passage of the system over the barrier. Unfortunately, it is still not possible to calculate reliable potential-energy surfaces except for the simplest reactions, e.g., H + H_2; it is therefore impossible to estimate activation energies for complex systems. Nevertheless, it is very useful to regard chemical reactions in terms of potential-energy surfaces, which provide valuable insights into mechanisms and will facilitate our later discussion of kinetic-isotope effect (see Chapter 6).

In the case of the simplest single substrate, single product enzymatically catalyzed reversible reaction, involving an E·S complex and an E·P complex, the effect of temperature on this reaction will be the resultant of the separate effects on these individual stages. Thus, there are at least 18 thermodynamic parameters for the forward and reverse reaction, each proceeding in at least three successive stages, and each with its enthalpy change, free-energy change, and entropy change of the reaction process for each of the three separate changes. Thus far, no one has succeeded in calculating all of these thermodynamic quantities and admittedly, little is known today about most of these quantities although a few such values have been presented in the literature.

A. THEORY OF ABSOLUTE REACTION RATES

The modified collision theory had led to the expression,

$$k = PZe^{-\Delta E_{act}/RT} \quad (92)$$

where the quantity Z is called the "collision number" and is equivalent to the frequency factor A in Equation 91. It is the number of collisions per second when there is only one molecule of reactant per milliliter. Z was derived from the kinetic theory of gases,

$$Z = 4\sigma^2(\pi RT/M)^{1/2} \quad (93)$$

and σ = collison diameter. P, the probability factor, or steric factor made allowance for any effects causing deviations from any "ideal" behavior.

Equation 92 proved to be a relatively useful expression for the comparison of a number of similar chemical reactions, or for the comparison of the effects of a number of substrates or substrate analogues for a given enzyme.

Thus, consider Figure 27 and Table 1, taken from a study on *Escherichia coli* β-D galactosidase by Kuby and Lardy.[78] The six substrates were employed at sufficiently high concentrations so that their zero-order velocity constants (k_0) approached their respective limiting maximal velocities. Figure 27 shows the result of plotting in the conventional Arrhenius manner, $-\log k_0$ vs. $\frac{1}{T}$ (T = absolute temperature). Only in the case of the methyl derivative, and for all practical purposes in the case of the *n*-butyl derivative, are the plots linear over the entire temperature range studied (0 to approximately 37°C). The others deviated from linearity, and the greatest deviation is observed in the case of the *O*-nitrophenyl derivative. For purposes of calculating ΔE_{act} values, the curves have been fitted to two straight lines (note that Kistiakowsky and Lumry[296] considered any sharp change in slope unlikely). In Table 1, a summary is given of the data in terms of the average Arrhenius energies of activation (ΔE_{act}) and log $PZ_{average}$ values (calculated from $k = PZ \cdot e^{-\Delta E_{act}/RT}$ for the designated temperature ranges. An increase in temperature is associated with a decrease in ΔE_{act} and a parallel decrease in log PZ; thus, apparently both ΔH_{act} and ΔS_{act} vary concomitantly with temperature with ΔG_{act} remaining constant.

As an alternative point of view, Eyring,[483,485] in particular, provided a much more complete interpretation of the significance of the "frequency factor" in the reaction rate equation (Equation 91) which has been called the "Theory of Absolute Reaction Rates" since it made use of the statistical mechanical approach. By this theory, before two (or more) molecules which possess the requisite energy can react they must first collide and form the activated complex for the reaction which then decomposes; the problem is to calculate from statistical mechanical theory, and making use of potential-energy surfaces, the frequency with which this event happens. The *activated complex* is treated by statistical mechanical methods just like the normal molecule, except that in addition to having three translational degrees of freedom, it has a fourth degree of freedom of movement along the reaction coordinate. It is in the direction of this coordinate that the activated complex approaches the top of the energy barrier, crosses it, and then falls to pieces. Since a molecule can have only *3n* energy variables in all, if there are *n* atoms in the molecule, and, since

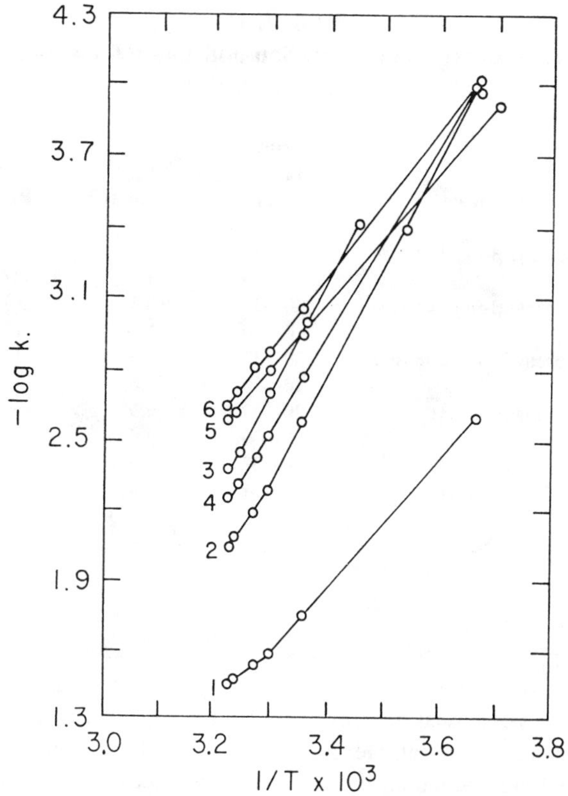

FIGURE 27. Arrhenius plots for several substrates: 1, *O*-nitrophenyl β-D-galactoside (initial concentration 1.00×10^{-3} *M*); 2, *p*-nitrophenyl β-D-galactoside (initial concentration 1.00×10^{-3} *M*); 3, *o*-nitrophenyl α-L-arabinoside (initial concentration 6.00×10^{-3} *M*); 4, phenyl β-D-galactoside (initial concentration 2.60×10^{-3} *M*); 5, methyl β-D-galactoside (initial concentration 4.80×10^{-3} *M*); 6, *n*-butyl β-D-galactoside (initial concentration 5.00×10^{-3} *M*); $[Na^+] = 0.14$ *M* (phosphate buffer), pH 7.25; k_0, the initial zero order velocity constant.[78]

the activated complex is assumed to have an additional degree of translational freedom, it must have one vibrational degree of freedom less than a normal molecule with the same number of atoms. The activated complex is thus treated as a molecule in which one of the vibrations is very stiff, so that the particular frequency is extremely large.

If we consider a process involving the reacting substances A, B, etc., which form the activated complex $X^{\ddagger}$ as an intermediate stage in the reaction,

$$A + B + \ldots \longrightarrow X^{\ddagger} \longrightarrow \text{products} \tag{94}$$

the rate of the reaction is equal to the activity of the activated complexes at the top of the barrier multiplied by the frequency in crossing the barrier. If a length δ in the reaction coordinate at the top of the barrier is taken as representing the activated state, and $\bar{v}$ is the mean velocity of the activated complexes in the same coordinate, the frequency of crossing the barrier is given by $(\bar{v}/\delta)$. If the number of activated complexes per unit volume lying in the length δ is $a'_{\ddagger}$, then it follows that

$$\text{Rate of reaction} = a'_{\ddagger}(\bar{v}/\delta) \tag{95}$$

TABLE 1
Average Energies of Activation and Log PZ Factors for Several Substrates

Substrate	Temp. range, (°C)	$\Delta E^{act}_{av.}$ (kcal/mol)	log $(PZ_{av.})^a$
o-Nitrophenyl β-D-galactoside	30—37	7.3	3.7
	0—30	13.1	7.7
p-Nitrophenyl β-D-galactoside	30—37	15.5	8.9
	0—30	21.0	12.8
o-Nitrophenyl α-L-arabinoside	30—37	19.3	11.2
	16—30	21.6	12.9
Phenyl β-D-galactoside	30—37	16.9	10.7
	0—30	19.6	12.6
Methyl β-D-galactoside	0—37	12.6	7.3
n-Butyl β-D-galactoside	25—37	13.4	6.8
	0—25	14.3	7.5

[a] Calculated from $k_0 = PZe^{-\Delta E^{act}_{av}/RT}$.

Taken from Reference 78.

These activated complexes are regarded as differing from normal molecules in the one respect that one of the vibrational degrees of freedom is replaced by translational motion in the reaction coordinate. These complexes, however, can be treated as normal molecules with one very stiff vibrational mode, by expressing their activity $a_\ddagger$ at the top of the energy barrier in the form

$$a'_\ddagger = a_\ddagger \, (2\pi m_\ddagger kT)^{1/2} \, (\delta/h^3) \tag{96}$$

where $m_\ddagger$ is the effective mass of the activated complex in the reaction coordinate k is the Boltzmann constant, and h is the Planck constant. The factor $(2\pi m_\ddagger kT)^{1/2}$ (δ/h) is the translational partition function (Q_{tr}) of the activated complex in the reaction path; it may be taken as a measure of the probability of occurrence of the activated complex at the top of the barrier. The mean velocity $\bar{v}$ of motion in one coordinate can be obtained by a method based on the Maxwell distribution law and the value is found to be $(kT/2\pi m_\ddagger)^{1/2}$. If this value, together with Equation 96, is introduced into Equation 95, the final result takes the surprisingly simple form

$$\text{Rate of reaction} = a_\ddagger \, (kT/h) \tag{97}$$

It follows, therefore, that the effective rate of crossing the energy barrier by the activated complexes is equal to kT/h, which is a universal frequency, having the dimensions of time^{-1}; its value is dependent only on the temperature and is independent of the nature of the reactants and the type of reaction.

If the specific reaction rate constant is k_r,* the reaction rate can be expressed in the familiar manner as

* In this section only, to distinguish the specific reaction rate constant from the Boltzmann constant (k), the symbol "k_r" is used to denote the specific reaction rate constant.

$$\text{Rate of reaction} = k_r\, a_A\, a_B \ldots \quad (98)$$

(where a_A and a_B are the activities of A and B, respectively).

Therefore, from Equation 98 and Equation 97,

$$k_r = \frac{kT}{h} \frac{a_\ddagger}{a_A a_B \ldots} = \frac{kT}{h} K^\ddagger \quad (99)$$

where $K^\ddagger$ is the constant for the equilibrium, that is,

$$A + B + \ldots \rightleftarrows X^\ddagger \quad (100)$$

which is supposed to exist between the reactants and the activated complex. If the system is assumed to behave ideally, the activities may be replaced by the usual concentrations.

In Equation 99, if all the molecules passing through the activated state do not proceed to decomposition, the "transmission coefficient," κ, is introduced with a value less than one so that the frequency of passage of activated complexes over the barrier is less than kT/h, or

$$k_r = \kappa \frac{kT}{h} K^\ddagger \quad (101)$$

For practically all enzymatic reactions considered here (except perhaps those involving the hemoproteins), κ is taken as 1.0.

From a thermodynamic standpoint, the equations of the "absolute rate reaction theory" may be placed in a form which is most useful for chemically catalyzed and enzyme-catalyzed reactions. Thus, the equilibrium constant $K^\ddagger$ may be expressed in terms of the free energy of activation in the *standard state*, $\Delta G^\ddagger$, i.e., analogous to the familiar thermodynamic equations:

$$\Delta G^\circ = -RT\,\ell n\, K_{eq} \text{ where } \Delta G^\circ = \Delta H^\circ - T\Delta S^\circ \text{ and}$$

$$\Delta E^\circ = \Delta H^\circ - RT \quad (102)$$

(ΔG°, ΔH°, ΔS°, and ΔE° are the standard free energy change, enthalpy change, entropy change, and internal energy change, respectively), we may write

$$-\Delta G^\ddagger = RT\ell n\, K^\ddagger;\ \Delta G^\ddagger = \Delta H^\ddagger - T\Delta S \quad (103)$$

and substitute into Equation 99

$$k_r = \left(\frac{kT}{h}\right) K^\ddagger$$

Thus

$$k_r = \left(\frac{kT}{h}\right) e^{-\Delta G^\ddagger/RT}$$

$$k_r = \left(\frac{kT}{h}\right) e^{-\Delta H^\ddagger/RT}\, e^{\Delta S^\ddagger/R} \quad (104)$$

where $\Delta E^{\ddagger} = \Delta H^{\ddagger} - RT$. We can assume $\Delta E^{\ddagger}$ to be the Arrhenius energy of activation, ΔE_{act}, for all practical purposes. It should be noted that strictly speaking, $\Delta G^{\ddagger}$, $\Delta S^{\ddagger}$, and $\Delta H^{\ddagger}$ represent the difference in free energy, entropy, and heat content, respectively, between the activated complex and the reactants, all referred to their standard states.

The equations of the absolute rate reaction theory may also be expressed in terms of partition functions, expressions which will be useful when kinetic-isotope effects are discussed.

Thus, if all products and reactants are in their standard states and relative to an arbitrary energy zero, the thermodynamic equilibrium constant may be written, in terms of the respective partition functions, as

$$K_{eq} = \frac{(Q^{\circ})^{\nu}_{products}}{(Q^{\circ})^{\nu}_{reactants}} e^{-\Delta E^{\circ}_{0}/RT} \tag{105}$$

where Q is the complete partition function, including translational and internal contributions. Consequently, $K^{\ddagger}$, the equilibrium constant for the reaction given by Equation 100 may be represented by

$$K^{\ddagger} = \frac{Q^{\ddagger}}{Q_A Q_B \ldots} e^{-\Delta E^{\ddagger}/RT} \tag{106}$$

and therefore,

$$k_r = \left(\frac{kT}{h}\right) \frac{Q^{\ddagger}}{Q_A Q_B \ldots} e^{-\Delta E^{\ddagger}/RT} \tag{107}$$

Equation 107 provides a basis for the calculation of the specific rate of a reaction solely from a knowledge of the physical properties of the reacting species. The values of Q_A and Q_B can be derived from statistical methods and spectral, physical, and thermodynamic data,[484] and, in principle, $Q^{\ddagger}$ may be evaluated from the dimensions and vibration frequencies of the activated complex from the potential energy surface. It is for these reasons the theory is called the "absolute rate theory". However, in only a very few simple cases can the potential energy surface be obtained with sufficient accuracy to make the results reliable. Enzymatic reactions, today, are still far too complex for quantitative evaluation by this method. Qualitatively, however, useful information is derivable (especially in a discussion of mechanisms).

One final point before leaving the discussion of activation energies is the relation between the *activation energy* and *zero-point energy* as applied to kinetic-isotope effects.

Consider the difference in reactivity of two isotopes of hydrogen, H_2 and D_2. Assume these hydrogen molecules react with a molecule AB, thus

$$H_2 + AB \rightleftarrows HA + HB; \; D_2 + AB \rightleftarrows DA + DB \tag{108}$$

and it may be assumed that the activated complexes HHAB or DDAB will be formed. There is reason for believing that the potential energies of the two complexes are very similar, and hence the difference in activation energies of the two reactions will depend on the difference in energies of the two isotopic forms of molecular hydrogen; this, in essence, is the difference in their "zero-point" energies. If E^* is the energy of the activated complex, and ϵ_1 and ϵ_2 are the zero-point energies of molecular hydrogen and deuterium, respectively, then the activation energies ΔE_1^{act} and ΔE_2^{act} will be

$$\Delta E_1^{act} = E^* - \epsilon_1 \text{ and } \Delta E_2^{act} = E^* - \epsilon_2 \tag{109}$$

We can express the specific rate of reaction in the Arrhenius form:

$$k = Ae^{-\Delta E_{act}/RT} \tag{110}$$

and if it may be assumed as approximately true, that the frequency factor A is the same for the two isotopic reactions, it follows that:

$$\frac{k_1}{k_2} = \frac{Ae^{-\Delta E_1^{act}/RT}}{Ae^{-\Delta E_2^{act}/RT}} = \frac{e^{-(E^*-\epsilon_1)/RT}}{e^{-(E^*-\epsilon_2)/RT}} \tag{111}$$

Therefore,

$$\frac{k_1}{k_2} = e^{(\epsilon_1-\epsilon_2)/RT} \tag{112}$$

where $\epsilon_1 - \epsilon_2$ is the difference in the zero-point energies, which is about 1790 cal.

In the reaction of hydrogen and bromine, the rate-determining steps are probably: $H_2 + Br \rightleftarrows H\,Br + H$; and $D_2 + Br \rightleftarrows D\,Br + D$ for the two isotopes. The ratio of the specific rates found experimentally is given by

$$\frac{k_1}{k_2} = e^{2130/RT}$$

which is in fair agreement with the theory.

VII. KINETICS OF IMMOBILIZED ENZYMES

There has been a recent resurgence of interest in the study of insoluble enzyme derivatives (e.g., see References 731, 738, 750 to 752, 760, and 766). This subject was pioneered by Katchalski and co-workers (see, e.g., References 728 to 731). Besides engineering applications briefly alluded to above (Section IV.C), those enzymes fixed within or on artificial membranes are now attracting attention in a variety of applications, e.g., in analytical devices,[732-734,760] as suggested therapeutic agents,[735,760] and even as models of biological membranes.[728,736,737] For these reasons, the kinetic behavior (both theoretical and experimental) of these membrane-immobilized enzyme systems has received and is receiving considerable attention.[760,766,960]

In the case of cellular metabolism, in view of the intricate nature of the cell structure and the relatively large concentrations of certain enzymes *in vivo* compared to their substrates,[961,962] one may not readily assume the conditions of a steady-state and a homogeneous distribution of freely diffusible intermediates and products within the cell.

Thus, Lilly and co-workers had discovered early[739] that the rate of hydrolysis of an *O*-nitrophenyl β-D galactoside solution, which had been passed through a porous sheet of DEAE-cellulose containing β-D galactosidase, was dependent on the flow. They had also derived an equation which accounted for the diffusion and electrical potential gradients within a Nernst type of diffusion layer, and they showed that the equation was of a form which agreed experimentally with the behavior for ficin (see Section IV.C) and for creatine kinase attached to carboxymethyl cellulose.[739]

Katchalski and co-workers published[740] a mathematical treatment of substrate and product distribution in membranes containing enzymes. This treatment was then modified to account for diffusion through the unstirred liquid film which was adjacent to the membrane, since it was found experimentally that liquid film diffusion did affect the apparent rate constants of the fixed enzyme in the case of alkaline phosphatase-collodion membranes.[741] This was followed by a theoretical analysis by Goldman and Katchalski[742] of the kinetic behavior of a two-enzyme-containing membrane capable of carrying out two consecutive reactions.

Laidler and co-workers[743] derived equations which described the kinetics of reaction in an enzyme-containing membrane immersed in a substrate solution; they considered the partitioning of the substrate between the liquid and membrane phases. Later, they examined various methods for evaluating the true kinetic parameters of an enzyme within a support[744] and experimentally attacked this problem.[745]

Broun and co-workers[746] studied the transport of glucose (substrate) from one solution to another through a glucose oxidase membrane and concluded that the Michaelis constant for the fixed enzyme was approximately the same as for the soluble enzyme. In addition, they described regulation and transport models which were based on membranes containing bound enzymes.[737]

Kasche and co-workers have derived equations for the steady-state catalysis by an enzyme immobilized in spherical gel particles[747,748] and showed that catalysis by the bound enzyme at low substrate concentrations differs from that by the free, unbound enzyme.

The main difficulties involved in the measurement and use of immobilized enzyme rate constants are (1) experimental measurements of the Michaelis-Menten overall rate constants (K_m and V_{max}, which will be denoted as $\bar{K}$ and $\bar{V}$ for membrane-bound parameters, see Equation 10 above) are complicated by transport properties (transport and diffusion coefficients of the substrate and product in both the aqueous and the membrane media), and also by the distribution of substrate and product between the aqueous and membrane phases and (2) the interpretation of $\bar{V}$ is difficult to make in terms of its more fundamental rate constant ($\bar{k}_2$, for the rate of product formation from the enzyme-substrate complex; see Scheme I above for the assumed mechanism), which then requires some knowledge of the enzyme activity in the phase of the membrane. The approach we will describe here is based on that of Boguslaski et al.,[749] which will be limited to enzymes which are immobilized in membranes.

Their approach gives estimates of $\bar{K}$ and $\bar{V}$ independent of transport properties.

Assume that the immobilized enzyme follows simple Michaelis-Menten kinetics* (and

* Sundaram et al.,[743] had early come to the conclusion that the rate equation of the enzyme within a solid support (e.g., in the form of a cylindrical disc) was also approximately of the Michaelis-Menten form, i.e., in their nomenclature:

$$v = \frac{k_c' [E]_s [S]}{Km(app) + [S]} \tag{115}$$

where $[E]_s$ is the enzyme concentration within the solid support. This equation may be compared to the comparable one for the case of the enzyme and substrate both in free solution, that is,

$$v = \frac{k_c [E]_o [S]}{Km + [S]} \tag{116}$$

where k_c is the catalytic constant, K_m is the Michaelis constant and both constants are functions of the individual rate constants for the particular mechanism, E_o is the total enzyme concentration in solution, and [S] is the initial substrate concentration in free solution. In Equation 115, the apparent Michaelis constant K_m (app) is related to the Km' value for the immobilized enzyme (which does not involve partitioning and diffusional effects) by:

$$Km\ (app) = Km'/PF \tag{117}$$

follows Equation 10); note the conditions described earlier (in section IV.A) which apply to the Michaelis equation, i.e., (1) at low concentrations of substrate, $v_o \cong \frac{V_{max}}{K_m}(S_o)$, where $(S_o) \ll Km$ (Equation 12), and for the reaction in the membrane phase:

$$\bar{v} \cong \frac{\bar{V}(\bar{S})}{\bar{K}} \tag{113}$$

and (2) at high concentrations of substrate; $v_o \cong V_{max} = k_{+2}\, E_t$, where $(S_o) \gg K_m$, (Equation 13), and for the reaction in the membrane phase:

$$\bar{v} \cong \bar{V} \tag{114}$$

They[749] describe three kinds of membrane-solution systems which can be used with the membrane: (1) covering a sensor immersed in a solution (see Figure 28A), (2) separating two solutions (see Figure 28B), and (3) immersed in a solution. In Figure 28, the transport of substrate S is considered to occur from solution 1 (the donor solution) towards the sensor, or solution 2 (the acceptor solution). As before, those quantities which have a bar or without a bar represent concentrations in the membrane and solution phases, respectively. Subscripts "1" and "2" represent concentrations in the bulk donor or acceptor solutions, and when accompanied by the additional subscript "m" they represent concentrations at the interfaces. $\bar{D}_s$ and $\bar{D}_p$ represent diffusion coefficients of S and P in the membrane phase. Only one-dimensional diffusion is assumed through the membrane film where $\bar{x}$ is the distance from the solution-1 membrane interface and $\bar{X}$ is the (uniform) membrane thickness. For the case of (3), the immersed membrane system (not shown in Figure 28) is a special case of (2), the two-solution system, in which both solutions are identical.

Their derivations for the two-solution system, case (2), will be limited to the condition where $(\bar{S}) \gg \bar{K}$. The fluxes of substrate and product through the membrane are the result of diffusion and the enzyme-catalyzed reaction. Diffusion and the enzymatic reaction in the membrane are related through Equations 119 and 120.

$$\bar{D}_s \frac{d^2\bar{S}}{d\,\bar{x}^2} = \bar{V} \tag{119}$$

where P is the partition coefficient (or the ratio of the surface concentration of the substrate in the support of thickness, ℓ, to that in the solution) and F is a function given by

$$F = \frac{\tanh}{\gamma\ell}\,\gamma\ell = \text{"Thiele function"} \tag{118}$$

which commonly appears in catalytic problems where diffusion is important and

$$\gamma = \frac{1}{2}\left\{\frac{k\acute{c}\,[E]_s}{D\,K\acute{m}}\right\}^{1/2}$$

where D = diffusion coefficient for the substrate in the solid support. The above relationships (derived and applied by Laidler and associates[731,743,750]) which apply to an enzyme immobilized in a cylindrical support may also apply with some modification to other particle shapes. Hinberg et al.,[753] for example, have developed a treatment for an enzyme immobilized in spherical particles, In those systems there is a similar parameter, F, related to the kinetic and diffusional parameters and to the radius of the sphere.

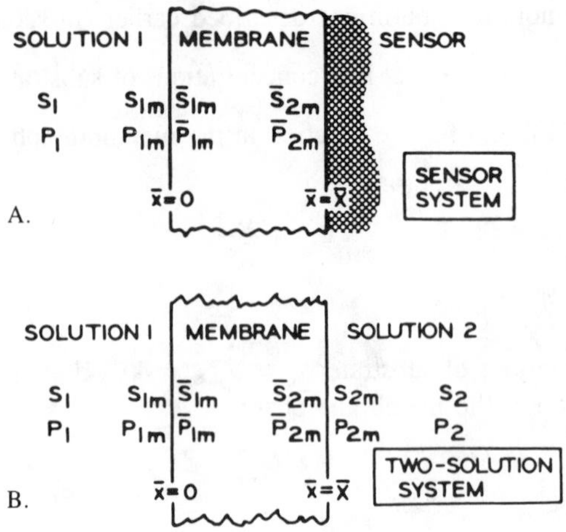

FIGURE 28. Steady-state conditions in solution membrane systems. (From Boguslaski, R. C., Blaedel, W. J., and Kissel, T. R., *Insolubilized Enzymes,* Salmona, M., Saronio, C., and Garattini, S., Eds., Raven Press, New York, 1974, 87. With permission.)

$$\bar{D}_p \frac{d^2\bar{P}}{d\,\bar{x}^2} = -n\bar{V} \qquad (120)$$

(for a stoichiometric enzymatic reaction of $\bar{S} \xrightarrow{Eng} n\,\bar{P}$). These equations may be integrated with the boundary conditions provided in Figure 28 to give Equations 121 and 122.

$$\bar{S} = \frac{\bar{V}\bar{X}^2}{2\bar{D}_s}\left(\frac{\bar{x}}{\bar{X}}\right)^2 + \left(\bar{S}_{2m} - \bar{S}_{1m} - \frac{\bar{V}\,\bar{X}^2}{2\,\bar{D}_s}\right)\left(\frac{\bar{x}}{\bar{X}}\right) + \bar{S}_{1m} \qquad (121)$$

$$\bar{P} = \frac{-n\bar{V}\,\bar{X}^2}{2\,\bar{D}_p}\left(\frac{\bar{x}}{\bar{X}}\right)^2 + \left(\bar{P}_{2m} - \bar{P}_{1m} + \frac{n\bar{V}\bar{X}^2}{2\,\bar{D}_p}\right)\left(\frac{\bar{x}}{\bar{X}}\right) + \bar{P}_{1m} \qquad (122)$$

These last two equations describe the distribution of substrate and product concentrations within the membrane. For the situation where product formation is used to follow the reaction, the following derivation is presented for the flux equations. Thus, differentiation of Equation 122 with respect to distance yields the product gradient throughout the membrane

$$\frac{d\,\bar{P}}{d\,\bar{x}} = \frac{-n\,\bar{V}\,\bar{x}}{\bar{D}_p} + \left(\bar{P}_{2m} - \bar{P}_{1m} + \frac{n\bar{V}\bar{X}^2}{2\,\bar{D}_p}\right)\left(\frac{1}{\bar{X}}\right) \qquad (123)$$

The evaluation of the gradient at the membrane surface provides expressions for the product fluxes ($\bar{J}_p$) at the membrane surfaces.

$$\bar{J}_{P1m} = -\,\bar{D}_P\left(\frac{d\,\bar{P}}{d\,\bar{x}}\right)_{\bar{x}=0} = \frac{-\bar{D}_P}{\bar{X}}\left(\bar{P}_{2m} - \bar{P}_{1m} + \frac{n\bar{V}\bar{X}^2}{2\,\bar{D}_p}\right) \qquad (124)$$

$$\bar{J}_{p2m} = -\bar{D}_P\left(\frac{d\bar{P}}{d\bar{x}}\right)_{\bar{x}=\bar{X}} = -\frac{\bar{D}_P}{\bar{X}}\left(\bar{P}_{2m} - \bar{P}_{1m} - \frac{n\bar{V}\bar{X}^2}{2\,\bar{D}_p}\right) \qquad (125)$$

These last two equations must be expressed in terms of measurable solution concentrations if the product fluxes ($\bar{J}_{plm} + \bar{J}_{p2m}$) are to be related to the measured bulk solution concentrations; this is accomplished with the aid of Equations 126 to 128,

$$\bar{J}_{p1m} = J_{p1m}$$

$$\bar{J}_{p2m} = J_{p2m} \qquad (126)$$

$$\frac{\bar{P}1m}{P1m} = \frac{\bar{P}2m}{P2m} = \delta_p \text{ (the distribution ratio)} \qquad (127)$$

$$J_{p1m} = -t_p\,(P_{1m} - P_1)$$

$$J_{p2m} = -t_p\,(P_2 - P_{2m}) \qquad (128)$$

where t_p is the mass transport coefficient. Note that Equation 126 describes the equality of the membrane and solution fluxes in the steady-state at the membrane-solution interface; Equation 127 provides the relationship between the solution and membrane surface-product concentrations through the distribution ratio, δ_p; Equation 128 is the product flux through the boundary layer between the bulk solution and the membrane. With the above relationships, Equations 124 to 128, the surface concentrations (which are not measurable) may be eliminated to give Equations 129 and 130.

$$J_{p1m} = \frac{P_1 - P_2}{\dfrac{2}{t_p} + \dfrac{\bar{x}}{\delta_p\,\bar{D}_p}} - \frac{n\bar{V}\bar{X}}{2} \qquad (129)$$

$$J_{p2m} = \frac{P_1 - P_2}{\dfrac{2}{t_p} + \dfrac{\bar{x}}{\delta_p\,\bar{D}_p}} + \frac{n\bar{V}\bar{X}}{2} \qquad (130)$$

These two equations describe the product flux through the membrane-solution interfaces and similar equations may be set up for the substrate fluxes.

The experimental method for measuring the steady-state fluxes is to pass the solutions 1 and 2 by the membrane surfaces at constant flow rates of F_1 and F_2, respectively. The sole source of substrate is solution 1 which contains the substrate at an entering or initial concentration of S_i. The steady-state fluxes may then be calculated from the steady-state concentrations of substrate and product.

For the case where $\bar{S} \gg \bar{K}$ (or at excess substrate concentration), the following equations describe the steady-state concentrations of substrate and product in the two solutions separated by an immobilized-enzyme membrane of area $\bar{A}$ and thickness $\bar{X}$

$$S_2\,\frac{F_2}{\bar{A}} = \frac{S_1 - S_2}{\dfrac{2}{t_s} + \dfrac{\bar{X}}{\delta_s\,\bar{D}_s}} - \frac{\bar{V}\bar{X}}{2} \qquad (131)$$

$$P_1 \frac{F_1}{\bar{A}} = \frac{P_1 - P_2}{\dfrac{2}{t_p} + \dfrac{\bar{X}}{\delta_p \bar{D}_p}} + \frac{n\bar{V}\bar{X}}{2} \tag{132}$$

$$P_2 \frac{F_2}{\bar{A}} = \frac{P_1 - P_2}{\dfrac{2}{t_p} + \dfrac{\bar{X}}{\delta_p \bar{D}_p}} + \frac{n\bar{V}\bar{X}}{2} \tag{133}$$

The above three equations therefore provide a steady-state solution for the case where an enzyme is imbedded in a membrane and separates the solutions, and, for example, when $\bar{S} \gg \bar{K}$ at high substrate concentration.[755] Under conditions where these equations are valid, a plot of S_2 vs. $S_1 - S_2$ should yield a straight line with a negative intercept equal to $\bar{V}\bar{X}/2$. Also, if $F_1 = F_2$, then P_1 should equal P_2, since the membrane would be at a steady state and saturated with respect to the substrate.

Then the first term on the right-hand side of Equations 132 and 133 become zero, and yields, therefore, a value of $\bar{V}$ from measurements of solution product concentrations (P_1 and P_2), donor and acceptor solution flow rates (F_1 and F_2), membrane area ($\bar{A}$), and membrane thickness ($\bar{X}$), all of which are measurable quantities.

Before leaving this section on kinetics of immobilized enzymes, it is important to note that there have been only a few detailed investigations made on the flow kinetics of enzymes involving two substrates. One such example is that of Daka and Laidler[756] who attached rabbit muscle lactate dehydrogenase to the inner surface of nylon tubing and made a study of the flow kinetics for the reaction between pyruvate and NADH. Their studies showed that there was a considerable degree of diffusion control under all conditions employed. They indicated that there is still a need for a general theoretical treatment of two-substrate systems for both membranes and tubes.

Illustrative of the recent interest in the kinetics of immobilized enzymes, Aflalo and DeLuca[960] described an interesting case for continuous monitoring of ATP in the microenvironment of immobilized enzymes producing or consuming ATP (e.g., hexokinase or pyruvate kinase). They made use of immobilized firefly luciferase (on Sepharose® 4B beads) as a localized probe for ATP and outlined the basic principles for its quantitative analysis. Coimmobilized or separately immobilized firefly luciferase coupled to the other immobilized kinases monitored the ATP concentration in the microenvironment and the bulk medium, respectively. Their results demonstrated the relative accumulation and depletion of ATP in the microenvironment of coimmobilized pyruvate kinase and hexokinase, respectively, whereas, such phenomena were not detected in the bulk medium. This method enabled them to directly assess the concentration gradients of ATP between both compartments.[960]

VIII. SOME PRACTICAL CONSIDERATIONS IN ENZYME ASSAYS AND ENZYME KINETIC MEASUREMENTS

A recent review on the subject by Allison and Purich[757] will be found to be very helpful. There are usually two separate sets of objectives in the practical design of an enzyme assay: (1) enzyme assays for the distinct purpose of following the purification and isolation of a specific enzyme or to estimate quantitatively a particular enzyme in physiological fluids or in tissues in order to assist in a clinical test or diagnosis or (2) enzyme kinetic measurements, and, in particular, to develop a set of reliable and accurate initial-rate measurements for the purpose of deducing a plausible kinetic mechanism for a particular enzyme or enzyme system.

For the former case, the enzyme assay is often concerned with an estimate of the particular enzyme (e.g., by a defined unit of activity) in a variety of samples. Thus, of prime consideration is a reliable, routine, reproducible, and highly specific assay. Often the sensitivity is of consideration, where the enzyme content might be low, and, since a large number of determinations is to be anticipated, the convenience of the assay will often play some role. If possible, an attempt is made to optimize enzyme assay conditions by making measurements under conditions of optimal concentrations of substrate(s), cofactors, activators, and pH and buffer species. Optimal conditions may not always be at saturation for the substrate. Also, to minimize other interfering reactions, often conditions have to found, e.g., pH, buffer species, ionic strength, or specific inhibitors, which will render these reactions small or at least permit their estimation, in turn. In the case of a number of ATP-requiring enzymes, i.e., the kinases (e.g., creatine kinase), it is often found that optimal concentrations of ATP_{total} and Mg_{total} are achieved when they are in a 1:1 ratio,[761] as a result of the relatively large complexation constant of $MgATP^{2-}$ (see discussion in Chapter 5, Section III.B and Table 11 for a list of complexation constants pertinent to this discussion). In the case of the most commonly used assay today for creatine kinase, that is, a coupled-enzyme assay which involves the following reactions catalyzed by hexokinase and glucose 6-phosphate dehydrogenase:

$$ADP_t + Cr{\sim}P_t \xrightarrow[Mg_t]{\text{Creatine Kinase}} ATP_t + Cr$$

$$ATP_t + Glc \xrightarrow[Mg_t]{\text{Hexokinase}} ADP_t + \text{Glc-6-P}$$

$$\text{Glc-6-P} + NADP^+ \xrightarrow{\text{G-6-P-dehydrogenase}} \text{6-P-gluconate} + NADPH + H^+ \quad (134)$$

where NADPH, the indicator product, is most conveniently determined spectrophotometrically at 340 nm (the wavelength maximum of NADPH as determined by P. L. Biochemicals, Inc. Circular OR-18, Fourth Printing, 1968).

However, the final product of the glucose 6-phosphate dehydrogenase reaction is actually the δ-lactone of 6-phosphogluconate, which has a lifetime of minutes at pH values below pH 6, and seconds or milliseconds above pH 7. Thus, this coupled-enzyme assay for ATP, below pH 6, would run into difficulties quantitatively; NADPH, the indicator product itself, is unstable below pH 6 making this a highly unsatisfactory assay under acidic conditions. Finally, Cr ~ P, one of the substrates, is one of the most acid-labile compounds known.

All these considerations would dictate that any clinical assay should be run at an alkaline pH value, even though the pH optimum for the so-called "reverse reaction"[48] is at pH 6 to 6.5.[761] Unfortunately, it is believed that most clinical assays today are run at a pH 6 to 6.5 in order to increase its sensitivity. There are, however, other means of increasing the sensitivity of the reaction, e.g., NADPH may be estimated fluorometrically at 10 to 100 times the sensitivity of the spectrophotometric assay ($E_M = 6.22 \times 10^3$ at 340 nm[765]) or adenosine triphosphate (ATP), by coupling it with the luminescent luciferase reaction[762] may be estimated in the nanomole to picomole range. Finally, in certain crude tissues or fluids, the adenylate kinase reaction is a potent competitor of the creatine kinase reaction:

$$2ADP_t \xrightarrow[Mg_t]{\text{Adenylate Kinase}} ATP_t + AMP_t \quad (135)$$

and may result in false positive reactions.

One way to eliminate it is to work at higher ionic strengths where the adenylate kinase reaction is inhibited,[767] or to add 5′-AMP, the product of the adenylate kinase reaction in an effort to partially inhibit it. However, at concentrations of AMP which will inhibit 70 to 90% of the adenylate kinase (1 to 3 m*M* adenosine monophosphate (AMP)), almost half of the creatine kinase activity is inhibited. AP_5A [P^1P^5-di(adenosine-5′)pentaphosphate] was found by Lienhard and Secemski[763] to be a potent inhibitor of muscle-type or cytoplasmic adenylate kinase (myokinase) with a $K_i \cong 10^{-8}$ *M*, and accordingly would be a far superior choice (0.1 m*M* AP_5A would inhibit almost quantitatively). Hamada and Kuby,[50] who confirmed Lienhard and Secemski's observation, unfortunately found that AP_5A was a far less potent inhibitor of the mitochondrial type of adenylate kinase ($K_i \cong 10^{-7}$ to 10^{-6} *M*) and, if its presence were suspected, then another approach would be dictated. One control, for example, would be to assay with and without $Cr{\sim}P_t$ and hopefully to correct the creatine kinase reaction for all those extraneous ADP-utilizing reactions which might be present. However, this necessitates two measurements and, in an age of rapid, routine, and pre-packaged reagents, often the clinical laboratory is unwilling to sacrifice the time and convenience for accuracy and reliability.

During the isolation of the ATP-creatine transphosphorylase (creatine kinase) from human muscle,[286] it was found prudent to follow the isolation of the enzyme by three procedures, as shown in Table 2, which is also illustrative of a typical protocol used for following the isolation of an enzyme.

A few words more are applicable largely to the enzymologist interested in following the purification of an enzyme.

As pointed out earlier (Section IV.A) the Michaelis equation (Equation 10) shows that at relatively low substrate concentrations (where $S_0) \ll K_m$) then the velocity follows first order kinetics:

$$v_0 \cong \frac{V_{max}}{K_m}(S_0) = \frac{k_{+2}\,E_t}{K_m}(S_0) \qquad (136)$$

In a number of reactions (e.g., some peptidases) the solubility of the substrate limits the reaction to initially low substrate concentrations or in the case of catalase, for example, high concentrations of H_2O_2 inactivate the enzyme; in those cases it is more judicious to select a first-order velocity coefficient (of 1.0 min^{-1}) as a unit of activity, even though the I.U.B. had recommended[768] that, if possible, activities be expressed in a zero-order fashion (e.g., 1.0 mol min^{-1} equal to 1.0 unit).

The primary consideration under all conditions and degrees of purification is that the enzyme assay be a quantitative measure of the enzyme content, and any defined unit of activity that leads to a reproducible, quantitative estimation is to be considered satisfactory.

For the enzyme kineticist, his problem for a given enzymic reaction is to develop quantitative determinations of initial rates under a large set of conditions which involve variation in initial substrate(s) concentrations.

For single-substrate reactions, it is commonly stated that a 25-fold variation in substrate concentration (from $^1/_5$ K_m to 5 × K_m) is often a satisfactory range to yield an excellent spread in a double-reciprocal plot for accurate estimations of K_m and V_{max} (i.e., by a suitable statistical analysis, see Chapter 10, Section IV). However, one should take care that substrate inhibition does not set in; also, if any metals or cofactors are required, they may have to be considered in the protocol, i.e., several families or sets of plots will have to be made at several fixed values of cofactor. This problem is then magnified in two- or three-substrate reactions, where the protocol will demand a large number of determinations, with one substrate at a time systematically varied, and with each of the others accordingly systematically fixed, as discussed, for example, by Fromm.[758]

One situation commonly encountered (e.g., with spectrophotometric assays) when the K_m is approached for one of the substrates, especially if the K_m is rather low compared to

TABLE 2
Fractionation of Human Muscle ATP-Creatine Transphosphorylase (Preparation 5) Initially at 925 g of Tissue (Wet)[286]

Fraction no.	Vol (ml)	Protein conc (mg/ml)	Total activity (units × 10^{-4})	Sp act (U/mg of protein) pH-Stat	Colori-metric[a]	Sphotophoto metric[b]	Purification Overall	Over preceding step	Recovery of act (%) Overall	Over preceding step
1 Homogenate in 0.01 *M* KCl	1560	16.67	34.6[c]	13.31[c]	5.25				100[c]	100[c]
			35.6[d]	13.69[d]	5.75	11.85[d]			100[d]	100[d]
2 60% (95%) Ethanol; 0.03 *M* $MgSO_4$; 10°C, followed by extraction reprecipitation and dialysis	192	17.4	26.8[c]	80.3[c]	22.92		6.0[c]	6.0[c]	77.5[c]	77.5[c]
			28.4[d]	85.7[d]	25.14	82.53	6.3[d]	6.3[d]	79.8[d]	79.8[d]
3 Phosphocellulose chromatography	70	23.4	17.0[c]	103.6[c]	34.9		7.7[c]	1.3[c]	49.1[c]	63.4[c]
			17.3[d]	105.5[d]	40.8	91.5	7.7[d]	1.2[d]	48.6[d]	60.9[d]
4 DEAE-Sephadex® A-50 chromatography, conc with phosphocellulose, 0.90 saturation $(NH_4)_2SO_4$	62	16.12	13.0[c]	130.11[c]	49.33		9.8[c]	1.3[c]	37.6[c]	76.5[c]
			14.7[d]	147.13[d]	53.625	171.7	10.7[d]	1.4[d]	41.3[d]	85.0[d]
5 0.45—0.77 saturation $(NH_4)_2SO_4$	27.7	35.51	12.0[c]	122.0[c]	29.76	147.2	9.2[c]	0.94[c]	34.7[c]	92.3[c]
			14.8[d]	152.5[d]	56.788	176.6	11.1[d]	1.04[d]	41.6[d]	100.7[d]
6 Crystallization										
1 Crystals	21.5	39.8	11.59[c]	135.4[c]	45.0	172.7	10.2[c]	1.11[c]	33.5[c]	96.6[c]
			11.98[d]	140[d,f]	58.1[a,e]	198.5	10.2[d]	0.92[d]	33.6[d]	80.9[d]
2 Crystals	20.0	28.7	10.07[c]	130.1[c]	53.3	145.7	9.8[c]	0.96[c]	29.1[c]	86.9[c]
			10.92[d]	141.5[d]	65.5[a,e]	165.5	10.3[d]	1.01[d]	30.7[d]	91.2[d]

Note: For this preparation no. 5, a portion of the skeletal muscle from the upper part of the leg and a portion of the psoas muscle, total weight of 925 g, had been obtained from two autopsy cases, a male, age 28, death as a result of congenital heart disease and a male, age 35, death as a result of electric shock.

[a] Colorimetric method, see text, 1 unit = k^1 = 1 ml/μmol/min. Aliquots diluted as in c and d, respectively.

[b] Coupled-enzyme spectrometric procedure, 1 unit = 1 μmol/min; aliquots diluted as in c and d.

[c] Refers to results obtained from aliquots of respective fractions (see text, 1 unit = 1 μeq/min by pH-stat procedure) diluted initially in 0.001 *M* glycine (pH 8.8) at 5°C and corrected for slow ATPase activity; following fraction 2, the ATPase could be neglected.

[d] Refers to cases where aliquots of respective fraction were diluted initially in 0.01 *M* β-mercaptoethanol, 0.001 *M* EDTA (pH 7.5); purification and recoveries are calculated from pH-stat data in fifth and sixth columns.

[e] For preparation no. 1, first crystals had a specific activity of 89.36 U/mg and second of 72.91 U/mg by the colorimetric assay; for preparation no. 2, first crystals had a specific activity of 82.0 U/mg by colorimetric assay.

[f] Preparation no. 8: first crystals had a specific activity of 180 U/mg by pH-stat assay.

the molar extinction coefficient of the indicating reaction, is that the extent of reaction is rather small for accurate estimation of the initial velocity. In these cases, it is often prudent, where possible, to introduce a regenerating system for the limiting substrate and so to extend the initial rate and to maintain the initial substrate concentration (incidentally, it also may eliminate product inhibition, if present). This situation was encountered by Johnson and Kuby[872] in a kinetics study of the FMN-containing ale yeast-NADH(NADPH)-cytochrome c reductase, where the K_m for NADPH was of the order of 5×10^{-8} *M*. At low values of NADPH, 10^{-7} to 10^{-8} *M*, a glucose 6-phosphate dehydrogenase-regenerating system for NADPH proved very effective in maintaining the fixed concentrations of NADPH, with ferro-cytochrome c^{2+} monitored either at its Soret or Alpha band (415 or 550 nm, respectively).

Another important consideration when the effect of pH is under investigation is to evaluate the effect of the buffer species themselves. This problem is not always an easy one and will require preliminary tests with a variety of buffer species over the range in pH values to be explored. Thus, in the case of brewers' yeast glucose 6-phosphate dehydrogenase, phosphate buffers were competitive inhibitors as determined very early by Theorell.[769] Kuby and Roy[71] after exploring several buffers under conditions where the two-subunit species of enzyme could be stabilized in the absence of superimposed association-dissociation reactions which might affect the kinetic results, they came to the conclusion that tris-acetate-EDTA buffers, at constant ionic strength, yielded either a uniform degree of interaction or provided a minimal degree of buffer species interaction over the pH range explored.

In the case of those reactions whose substrates may involve their magnesium complexes, e.g., $MgATP^{2-}$ or $MgADP^{-}$, the calculations of the actual substrate concentrations require that a set of conservation equations be employed (see Chapter 5, Equations 95 and 96) with the individual ionization or complexation constants involved being assigned or measured under the particular conditions of temperature, pH, ionic strength employed (see Chapter 5, Table 5). Calculations by this method will be illustrated later (Chapter 6, Section I) in the case of a study of the kinetics of the adenylate kinase reaction.[50]

$$MgATP^{2-} + AMP^{2-} \rightleftarrows MgADP^{-} + ADP^{3-} \quad (137)$$

Finally, although a great deal has been discussed about the theoretical aspects of setting up an adequate coupled-enzyme assay,[764] each assay should be investigated experimentally to ascertain whether it is indeed satisfactory regarding the various components required for the individual study (including coupled enzymes, their substrates, and their cofactors); the investigator should not rely solely on the suggested values of the literature. An interesting situation developed when Hamada and Kuby[50] studied the initial velocity of the forward reaction ($MgATP^{2-} + AMP^{2-} \rightarrow$) of both the liver-type and muscle-type adenylate kinases. They measured the reaction by a spectrophotometric coupled-enzyme procedure for ADP_0 (adenosine diphosphate) employing pyruvate kinase and lactate dehydrogenase; consequently, their reaction mixture also contained phosphoenolpyruvate (PEP), the substrate for the pyruvate-kinase reaction. However, it was found that phosphoenolpyruvate acts as an inhibitor of the calf liver enzyme, but apparently not of the calf or rabbit muscle enzyme. PEP was shown to be a competitive inhibitor of $MgATP^{2-}$ ($K_i \cong 1.2$ m*M*), and a noncompetitive inhibitor (of the mixed type) with respect to AMP^{2-}, $MgADP^{-}$, and ADP^{3-} ($K_i \cong$ 5 to 6 m*M*). Thus, at the 4-m*M* concentrations of PEP employed in the assay system for the forward direction, the liver enzyme would be severely inhibited (but not the muscle enzyme). For this reason, the PEP concentration was decreased to 0.25 m*M*, where the inhibition would be insignificant. Thus, an initial exploration of the interactions of all components of a coupled-enzyme assay on the particular enzyme being studied is a wise and helpful hint.

Chapter 2

A DESCRIPTION OF STEADY-STATE KINETICS AND QUASI- OR RAPID EQUILIBRIUM KINETICS BY A DEVELOPMENT OF THE RATE EXPRESSIONS FOR SEVERAL SELECTED MECHANISMS AND THEIR CHARACTERISTICS

I. PROCEDURES FOR THE DERIVATION OF RATE EXPRESSIONS

A. STRAIGHT ELIMINATION PROCEDURE FOR STEADY-STATE KINETICS

A simple case has been dealt with in Chapter 1 for the case of

$$E + S \underset{k_{-1}}{\overset{k_{+1}}{\rightleftharpoons}} ES \overset{k_{+2}}{\longrightarrow} E + \text{Products},$$

SCHEME I.

to obtain

$$v_{0,f} = \frac{V_{max,f}}{[K_{m,s}/(S_0)] + 1}; \text{ where } V_{max,f} = k_{+2}E_t;\ K_{m,s} = \left(\frac{k_{-1} + k_{+2}}{k_{+1}}\right) \quad (1)$$

B. THE DETERMINANT PROCEDURE FOR STEADY-STATE KINETICS

Let us consider the following reversible mechanism for a one-substrate reaction and derive it by either method #1 or #2.

Method #1

$$E + S \underset{k_{-1}}{\overset{k_{+1}}{\rightleftharpoons}} (ES) \underset{k_{-2}}{\overset{k_{+2}}{\rightleftharpoons}} (EP) \underset{k_{-3}}{\overset{k_{+3}}{\rightleftharpoons}} E + P$$

SCHEME II.

This actually is the only logical mechanism which does not violate the principle of microscopic reversibility and incorporates the minimum number of intermediates.

Set up the following differential equations:

$$\frac{d(ES)}{dt} = k_{+1}\,(E)(S) - (k_{-1} + k_{+2})(ES) + k_{-2}(EP)$$

$$\frac{d(EP)}{dt} = k_{-3}\,(E)(P) + k_{+2}\,(ES) - (k_{-2} + k_{+3})(EP)$$

$$\frac{-d(S)}{dt} = k_{+1}\,(E)(S) - k_{-1}(ES);\ \frac{dP}{dt} = k_{+3}\,(EP) - k_{-3}\,(E)(P) \quad (2)$$

Solve the set of differential equations by assuming a steady-state for both complexes (ES) and (EP), simultaneously adjusted, and thus, set

$$\frac{d(ES)}{dt} = 0 \text{ and } \frac{d(EP)}{dt} = 0 \tag{3}$$

Our conservation equation for enzyme species is, of course,

$$E_t = (E) + (ES) + (EP) \tag{4}$$

where E_t = total enzyme concentration.

The total substrate in all forms (including the derived product) is a constant; therefore, the time derivative of the total substrate is zero. That is

$$S_t = (S) + (ES) + (EP) + (P) \tag{5}$$

and the derivative is:

$$\frac{d(S)}{dt} + \frac{d(ES)}{dt} + \frac{d(EP)}{dt} + \frac{d(P)}{dt} = 0 \tag{6}$$

In the steady-state, the two middle terms drop out (i.e., $\frac{d(ES)}{dt} = 0; \frac{d(EP)}{dt} = 0$). Therefore,

$$\frac{d(S)}{dt} + \frac{d(P)}{dt} = 0 \text{ or } \frac{-d(S)}{dt} = \frac{d(P)}{dt} \tag{7}$$

Thus, either expression $\left(\frac{-dS}{dt} \text{ or } \frac{dP}{dt}\right)$ can be used to solve for the rate, v; i.e., $v = \frac{-d(S)}{dt}$ or $v = \frac{d(P)}{dt}$ in the hypothetical steady state.

We must now solve for the unknowns (E), (ES), and (EP). We can use the equation $\frac{d(ES)}{dt}$ and $\frac{d(EP)}{dt}$ with both set equal to zero, and the conservation equation for E_t = (E) + (ES) + (EP); substitute the results into either equation, $\frac{-d(S)}{dt}$ or $\frac{d(P)}{dt}$ to obtain v, and we find by this process of elimination that the solution is

$$v = \frac{E_t\,(k_{+1}k_{+2}k_{+3}\,(S) - k_{-1}k_{-2}k_{-3}\,(P))}{[k_{+1}(S)(k_{+2}+k_{-2}+k_{+3}) + k_{-3}(P)(k_{-1}+k_{+2}+k_{-2}) + k_{-1}k_{-2} + k_{-1}k_{+3} + k_{+2}k_{+3}]} \tag{8}$$

However, even for this simple case, the method of elimination results in very tedious algebra.

Method #2: Cramer's Rule

To apply the method of determinants to solving simultaneous linear equations, take as an example a set of three equations in three unknowns, that is, x, y, and z (the three variables) or

$$
\begin{aligned}
a_1\, x + a_2\, y + a_3\, z &= A \\
b_1\, x + b_2\, y + b_3\, z &= B \\
c_1\, x + c_2\, y + c_3\, z &= C
\end{aligned}
\qquad (9)
$$

where *A, B,* and *C* are constants (independent of x, y, and z). To solve, substitute the column vector $\begin{vmatrix} A \\ B \\ C \end{vmatrix}$ for one of the columns of the three by three determinant, Δ, to give Δ_1, Δ_2, Δ_3. The solution to this system of equations (Cramer's rule) is simply

$$
x = \frac{\Delta_1}{\Delta};\ y = \frac{\Delta_2}{\Delta};\ \text{and } z = \frac{\Delta_3}{\Delta}
\qquad (10)
$$

$$
\text{where } \Delta_1 = \begin{vmatrix} A & a_2 & a_3 \\ B & b_2 & b_3 \\ C & c_2 & c_3 \end{vmatrix};\ \Delta_2 = \begin{vmatrix} a_1 & A & a_3 \\ b_1 & B & b_3 \\ c_1 & C & c_3 \end{vmatrix};\ \text{and } \Delta_3 = \begin{vmatrix} a_1 & a_2 & A \\ b_1 & b_2 & B \\ c_1 & c_2 & C \end{vmatrix}
$$

where

$$
\Delta = \begin{vmatrix} a_1 & a_2 & a_3 \\ b_1 & b_2 & b_3 \\ c_1 & c_2 & c_3 \end{vmatrix}
\qquad (11)
$$

To illustrate, solve Scheme II again:

$$
E + S \underset{k_{-1}}{\overset{k_{+1}}{\rightleftharpoons}} (ES) \underset{k_{-2}}{\overset{k_{+2}}{\rightleftharpoons}} (EP) \underset{k_{-3}}{\overset{k_{+3}}{\rightleftharpoons}} E + P
$$

SCHEME II.

Solve, by setting up the following three equations for the three unknowns, (E), (ES), and (EP) and in the steady-state, putting $\frac{d(ES)}{dt} = \frac{d(EP)}{dt}$ to zero.

$$
\begin{aligned}
& (E) + (ES) + (EP) = E_t \\
\frac{d(ES)}{dt} &= k_1\, (S)(E) - (k_{-1} + k_2)ES + k_{-2}\, (EP) = 0 \\
\frac{d(EP)}{dt} &= k_{-3}\, (P)(E) + k_2\, (ES) - (k_{-2} + k_3)(EP) = 0
\end{aligned}
\qquad (12)
$$

Therefore, by the method of determinants:

$$
\Delta = \begin{vmatrix} 1 & 1 & 1 \\ k_1 S & -(k_{-1} + k_2) & k_{-2} \\ k_{-3} P & k_2 & -(k_{-2} + k_3) \end{vmatrix}
\qquad (13)
$$

Here, the column vector *A, B, C* is actually

$$\begin{vmatrix} E_t \\ 0 \\ 0 \end{vmatrix}$$

Accordingly,

$$\Delta_1 = \begin{vmatrix} E_t & 1 & 1 \\ 0 & -(k_{-1} + k_2) & k_{-2} \\ 0 & k_2 & -(k_{-2} + k_3 \end{vmatrix} \tag{14}$$

and

$$E = \frac{\Delta_1}{\Delta} \tag{15}$$

Expansion of the determinants give

$$\begin{aligned} \Delta = {} & 1\,[(k_{-1} + k_2)\,(k_{-2} + k_3) - k_2\,k_{-2}] \\ & -1\,[-k_1(S)\,(k_{-2} + k_3) - k_{-2}\,k_{-3}\,(P)] \\ & +1\,[k_1\,k_2\,(S) + k_{-3}\,(P)\,(k_{-1} + k_2] \end{aligned} \tag{16}$$

$$\Delta_1 = E_t\,[(k_{-1} + k_2)\,(k_{-2} + k_3) - k_2\,k_{-2}] \tag{17}$$

$$\text{and } E = \frac{Et\,[(k_{-1} + k_2)(k_{-2} + k_3) - k_2\,k_{-2}]}{[(k_{-1} + k_2)(k_{-2} + k_3) - k_2\,k_{-2} + k_1(S)(k_{-2} + k_3) + k_{-2}\,k_{-3}(P) + k_1\,k_2(S) + k_{-3}(P)(k_{-1} + k_2)]} \tag{18}$$

Similarly, $(ES) = \dfrac{\Delta_2}{\Delta}$

$$\Delta_2 = \begin{vmatrix} 1 & E_t & 1 \\ k_1S & 0 & k_{-2} \\ k_{-3}P & 0 & -(k_{-2} + k_3) \end{vmatrix} = E_t\,[k_1S(k_{-2} + k_{+3}) + k_{-2}\,k_3\,(P)] \tag{19}$$

Therefore,

$$(ES) = \frac{E_t\,[k_1\,(S)(k_{-2} + k_{+3}) + k_{-2}\,k_{-3}\,(P)]}{[(k_{-1} + k_2)(k_{-2} + k_3) - k_2\,k_{-2} + k_1\,(S)(k_{-2} + k_3) + k_{-2}\,k_{-3}\,(P) + k_1\,k_2\,(S) + k_{-3}\,(P)(k_{-1} + k_2)]} \tag{20}$$

$$\text{and } (EP) = \frac{\Delta_3}{\Delta} \tag{21}$$

$$\Delta_3 = \begin{vmatrix} 1 & 1 & E_t \\ k_1\,S & -(k_{-1} + k_2) & 0 \\ k_{-3}\,(P) & k_2 & 0 \end{vmatrix} = E_t\,[k_1\,k_2\,(S) + k_{-3}(P)(k_{-1} + k_2)] \tag{22}$$

Therefore

$$(EP) = \frac{E_t\,[k_1\,k_2\,(S) + k_{-3}\,(k_{-1} + k_2)\,P]}{[(k_{-1} + k_2)(k_{-2} + k_3) - k_2\,k_{-2} + k_1(S)(k_{-2} + k_3) + k_{-2}\,k_{-3}\,(P) + k_1k_2(S) + k_{-3}\,(P)(k_{-1} + k_2)]} \tag{23}$$

As a check, add the terms for (E) + (ES) + EP and we should arrive at our original conservation equation for E_t = (E) + (ES) + (EP) =

$$\frac{\Delta_1}{\Delta} + \frac{\Delta_2}{\Delta} + \frac{\Delta_3}{\Delta} = \frac{\Delta_1 + \Delta_2 + \Delta_3}{\Delta} \tag{24}$$

or

$$(E) + (ES) + (EP) = \frac{E_t\,[(k_{-1} + k_2)(k_{-2} + k_3) - k_{+2}k_{-2} + k_1(S)(k_{-2} + k_3) + k_{-2}k_{-3}(P) + k_1k_2(S) + k_{-3}\,(k_{-1} + k_2)\,(P)]}{\Delta} = E_t\left(\frac{\Delta}{\Delta}\right) = E_t \tag{25}$$

Before continuing with the deriviation, note that in the case of

$$E = \frac{\Delta_1}{\Delta} = \frac{E_t\,[(k_{-1} + k_2)(k_{-2} + k_3) - k_2\,k_{-2}]}{\Delta} \tag{26}$$

or

$$\frac{E}{E_t} = \frac{[(k_{-1} + k_2)(k_{-2} + k_3) - k_2\,k_{-2}]}{\Delta} = \frac{\left(\frac{\Delta_1}{E_t}\right)}{\Delta} \tag{27}$$

that the fraction

$$\frac{E}{E_t} \propto [(k_{-1} + k_2)(k_{-2} + k_3) - k_2\,k_{-2}] \tag{28}$$

with the proportionality factor equal to $\left(\frac{1}{\Delta}\right)$, or that $\left(\frac{E}{E_t}\right) \propto \left(\frac{\Delta_1}{E_t}\right)$ with $\left(\frac{1}{\Delta}\right)$ as the proportionality factor.

Similarly, for $\frac{(ES)}{E_t}$ and $\frac{(EP)}{E_t}$, they are proportional to $\frac{\Delta_2}{E_t}$ and $\frac{\Delta_3}{E_t}$, respectively, with proportionality factors which are all identical and equal to $\left(\frac{1}{\Delta}\right)$.

This generality, that

$$\frac{E_n}{E_t} \propto \frac{\Delta_n}{E_t} \tag{29}$$

(i.e., E_n, n^{th} species, and Δ_n, the corresponding n^{th} determinant) with proportionality factor equal to $\left(\frac{1}{\Delta}\right)$ will become useful when we turn next to the symbolic matrix or algorithm procedure of King and Altman, since it is really based on this observation.

To continue the derivation, which now follows simply, we can make use of either equation to complete the derivation:

$$\begin{cases} \dfrac{-dS}{dt} = k_1(S)(E) - k_{-1}(ES) \\ \\ \dfrac{+dP}{dt} = k_3(EP) - k_{-3}(P)(E) \end{cases} \tag{30}$$

since $\dfrac{-dS}{dt} = \dfrac{dP}{dt} = v$

Substitution into either equation for the terms E and ES or EP and E, respectively, and after collection of terms in (S) and (P) as before, it leads again to Equation 31 (or Equation 8):

$$v = \frac{E_t\,(k_1\,k_2\,k_3(S) - k_{-1}\,k_{-2}\,k_{-3}\,(P))}{k_1(S)(k_2 + k_{-2} + k_3) + k_{-3}\,(P)\,(k_{-1} + k_2 + k_{-2}) + (k_{-1}\,k_{-2} + k_{-1}\,k_{+3} + k_2\,k_3)} \tag{31}$$

(but this time with much less labor than by the process of elimination; also, note that in most cases, for simplicity, we omitted the subscripts "t").

2. Overall Velocity Expression and the Haldane Relation

Before going on to the schematic rules of King and Altman[15] let us consider Equation 31 in some more detail.

Thus, divide numerator and denominator by the constant term found in the denominator, i.e., by $(k_{-1}\,k_{-2} + k_{-1}\,k_{+3} + k_2\,k_3)$.
Therefore,

$$v = E_t\,\frac{\left\{\dfrac{k_1\,k_2\,k_3}{k_{-1}\,k_{-2} + k_{-1}\,k_3 + k_2\,k_3}(S) - \dfrac{k_{-1}\,k_{-2}\,k_{-3}\,(P)}{k_{-1}\,k_{-2} + k_{-1}\,k_3 + k_2\,k_3}\right\}}{1 + \dfrac{k_1\,(S)\,(k_2 + k_{-2} + k_3)}{(k_{-1}\,k_{-2} + k_{-1}\,k_3 + k_2\,k_3)} + \dfrac{k_{-3}\,(P)\,(k_{-1} + k_2 + k_{-2})}{k_{-1}\,k_{-2} + k_{-1}\,k_3 + k_2\,k_3}} \tag{32}$$

Define

$$\frac{k_1\,(k_2 + k_{-2} + k_3)}{k_{-1}\,k_{-2} + k_{-1}\,k_3 + k_2\,k_3} = \frac{1}{K_{s,f}} \tag{33}$$

(or the reciprocal form of the Briggs-Haldane constant for the forward reaction involving S, the substrate) and

$$\frac{V_{max,f}}{K_{s,f}} = \frac{k_1\, k_2\, k_3\, E_t}{k_{-1}\, k_2 + k_{-1}\, k_3 + k_2\, k_3} \tag{34}$$

($V_{max,f}$ = maximal velocity for forward reaction).
Therefore,

$$V_{max,f} = \frac{k_2\, k_3\, E_t}{k_2 + k_{-2} + k_3} \tag{35}$$

(after substitution of $K_{s,f}$).
Similarly, define for the reverse direction:

$$\frac{k_{-3}\,(k_{-1} + k_2 + k_{-2})}{k_{-1}\, k_{-2} + k_{-1}\, k_3 + k_2\, k_3} = \frac{1}{K_{p,b}} \tag{36}$$

or reciprocal of Briggs-Haldane constant for back or reverse reaction, involving P, the product and

$$\frac{V_{max,b}}{K_{p,b}} = \frac{k_{-1}\, k_{-2}\, k_{-3}\, E_t}{k_{-1}\, k_{-2} + k_{-1}\, k_3 + k_2\, k_3} \tag{37}$$

($V_{max,b}$ = maximal velocity for reverse or back reaction), and therefore,

$$V_{max,b} = \frac{k_{-1}\, k_{-2}\, E_t}{k_{-1} + k_2 + k_{-2}} \tag{38}$$

(after substitution of $K_{p,b}$).
Therefore, substitute

$$v = \frac{\left\{\frac{V_{max,f}}{K_{s,f}}(S) - \frac{V_{max,b}}{K_{p,b}}(P)\right\}}{\left\{1 + \frac{(S)}{K_{s,f}} + \frac{(P)}{K_{p,b}}\right\}} \tag{39}$$

Note: If the reverse reaction is negligible, one can neglect the rate of the back reaction, i.e., neglect:

$$\frac{\left\{\frac{V_{max,b}\,(P)}{K_{p,b}}\right\}}{\text{Denominator}}$$

Therefore,

$$v_f = \frac{\frac{V_{max,f}(S)}{K_{s,f}}}{1 + \frac{(S)}{K_{s,f}} + \frac{(P)}{K_{p,b}}} \tag{40}$$

where v_f = velocity of forward reaction, or

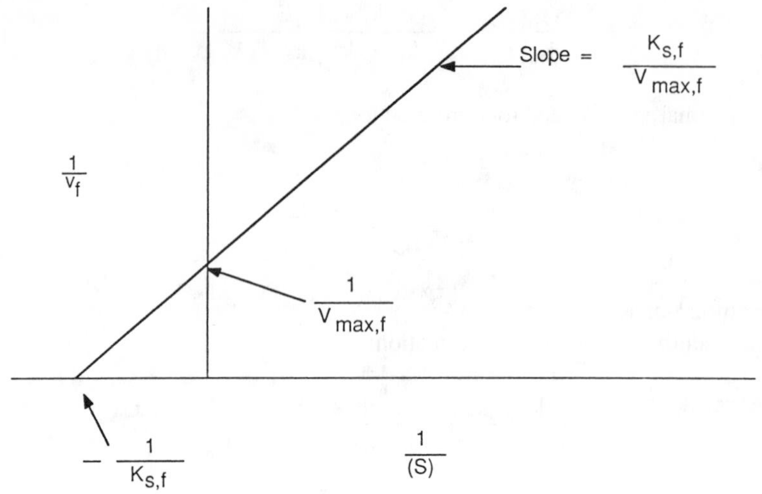

FIGURE 1. Plot of $\frac{1}{v_f}$ vs. $\frac{1}{(S)}$.

$$\frac{v_f}{V_{max}} = \frac{1}{1 + \frac{K_{s,f}}{(S)}\left(1 + \frac{P}{K_{P,b}}\right)} \tag{41}$$

which is the familiar equation for competitive inhibition, i.e., competitive inhibition of the forward reaction by the product (see Chapter 1, Section IV. D.1).

If (P) = 0 initially, then the equation reduces to the familiar Michaelis-Menten equation, i.e.,

$$v_f = \frac{\frac{V_{max,f}\,(S)}{K_{s,f}}}{\left(1 + \frac{(S)}{K_{s,f}}\right)} = \frac{V_{max,f}\,(S)}{(K_{s,f} + (S))} = \frac{V_{max,f}}{\left(\frac{K_{s,f}}{(S)}\right) + 1} \tag{42}$$

which may be rearranged in the familiar double-reciprocal form: $\frac{1}{v_f} = \frac{1}{V_{max,f}} + \frac{1}{V_{max,f}}\frac{K_{s,f}}{(S)}$, and a plot of $\frac{1}{v_f}$ (vs.) $\frac{1}{(S)}$ yields values for the kinetic prameters, $K_{s,f}$ and $V_{max,f}$ (see Figure 1).

Similarly, for the reverse reaction,

$$\frac{v_b}{V_{max,b}} = \frac{1}{\left(\frac{K_{p,b}}{(P)} + 1\right)} \tag{43}$$

and one may obtain, therefore, estimates for all kinetic parameters: $K_{s,f}$, $V_{max,f}$, $K_{P,b}$, and $V_{max,b}$.

When (S) and (P) $\neq$ 0, the overall rate equation is Equation 39. At equilibrium, when $v = 0$,

$$\left(\frac{V_{max,f}}{K_{s,f}}\right)(S)_{eq} = \left(\frac{V_{max,b}}{K_{p,b}}\right)(P)_{eq}$$

or

$$\frac{V_{max,f}\,K_{p,b}}{V_{max,b}\,K_{s,f}} = \frac{(P)_{eq}}{(S)_{eq}} = K_{eq} \tag{44}$$

where $K_{eq} = \dfrac{(P)_{eq}}{(S)_{eq}}$ for the overall equilibrium of $S \rightleftarrows P$.

This type of relationship between the overall equilibrium constant, K_{eq}, and the kinetically obtained parameters, i.e., the maximal velocities and the Michaelis constants for the forward and reverse directions, was first pointed out by Haldane[40] and is usually called the *Haldane relation*. For some bisubstrate reactions, as we will see shortly, there may be several Haldane relations which must be derived for each mechanism.

The agreement between the directly measured thermodynamic equilibrium constant and that estimated kinetically from the Haldane relation, or relations, provides additional confidence in the assigned mechanism.

If we substitute our definitions for $\dfrac{V_{max}}{K_{s,f}}$ (Equation 34) and $\dfrac{V_{max}}{K_{p,b}}$ (Equation 37) into the particular Haldane relation (Equation 44) for the mechanism under consideration (Scheme II) we get

$$k_{eq} = \frac{(P)_{eq}}{(S)_{eq}} = \left(\frac{V_{max,f}}{K_{s,f}}\right)\left(\frac{K_{p,b}}{V_{max,b}}\right) = \frac{k_1k_2k_3}{k_{-1}k_{-2}k_{-3}} = K_1 \cdot K_2 \cdot K_3 \tag{45}$$

where the equilibrium constants are defined according to the mechanism given, that is,

$$K_1 = \frac{(ES)}{(E)(S)} = \frac{k_1}{k_{-1}};\ K_2 = \frac{(EP)}{(ES)} = \frac{k_2}{k_{-2}};\ K_3 = \frac{(E)(P)}{(EP)} = \frac{k_3}{k_{-3}} \tag{46}$$

Thus, the system of definitions is self-consistent, in that the overall equilibrium constant, K_{eq}, can be described in terms of the product of the ratios of kinetic constants for each discrete step.

C. KING-ALTMAN SCHEMATIC METHOD (SYMBOLIC-MATRIX METHOD) FOR DERIVING STEADY-STATE RATE EQUATIONS FOR ENZYME-CATALYZED REACTIONS

First, to quote from King and Altman,[15] "If the mechanism for an enzyme-catalyzed reaction involves *n* different enzyme-containing species, the usual steady-state assumption [i.e., recall $\dfrac{d(ES)}{dt} = \dfrac{d(EP)}{dt} = 0$ in the case we just discussed] leads, after first eliminating the concentration of one of the enzyme-containing species by the relationship $E_t = \Sigma\,(EX_i)$ [that is, the conservation equation for total enzyme], to a system of $n-1$ non-homogeneous linear equations [i.e., not all the right-hand sides are zero; at least one right-hand side term is not zero]." Our present discussion will also be limited to those systems in which E_t is very small compared to the concentrations of the substrate (or substrates), product (or products), etc., their true concentrations (of S, P, etc.) being approximately equal to their stoichiometric concentrations.

Now, the solution of these equations for the concentrations of the several EX_is is, as we have seen, a straightforward problem; however, for cases more complicated than the one we discussed, i.e., for two- or three-substrate reactions, or for the effect of inhibitors, etc., the solutions can be tedious even when the determinant method (i.e., Cramer's rule) is employed.

King and Altman, after making use of the determinant method for solving these sets of simultaneous equations, deduced from their results a set of *schematic rules* which enable one to express the rate equation for a given reaction mechanism without the tedium of expanding the individual determinants. It is a method which may be readily learned without demanding too much in the way of a mathematical background.

Let us quote a brief description of their method, as given by King and Altman[15,16] which will provide us with the schematic rules which we will attempt to translate into practical use. "The concentration of each EX_i relative to the total concentration of enzyme, $\frac{(EX_i)}{E_t}$, is a quotient of two summations of terms ([i.e., $\frac{\Sigma(\text{numerator})}{\Sigma(\text{denominator})} = \frac{EX_i}{E_t}$]), each term in ($\frac{EX_i}{E_t}$) being the product of n − 1 different rate constants and the appropriate concentrations. Each term in the numerator of the expression for $\frac{(EX_m)}{(E_t)}$ [or the mth enzyme containing species/E_t] involves the rate constants (and appropriate concentrations) associated with reaction steps which *individually or in sequence* lead to EX_m, the enzyme-containing species in question. The n − 1 rate constants in each term are associated with n − 1 reaction steps in which *each* of the [i.e., the other] n − 1 different enzyme-containing species, EX_i (where $i \neq m$) is a reactant. All of the possible combinations of n − 1 rate constants which conform to this requirement are present as numerator terms in the expression for $\frac{(EX_m)}{(E_t)}$. The denominator of the expression for $\frac{(EX_m)}{E_t}$ is the sum of the several different numerators."

We recall once again, that from Cramer's rule each $(EX_i) = \frac{\Delta i}{\Delta}$ and the applicauion of this point will become clearer if we take the same example as before to illustrate this method, that is, Scheme II.

This mechanism involves three enzyme-containing species, or n = 3; therefore, n − 1 = 2 rate constants (and associated concentration factors) are present as individual terms in the expression for each $\frac{(EX_i)}{(E_t)}$; thus, each term in each of the expressions for the relative concentration of E, ES, or EP contains two rate constants and the appropriate concentration factors for this particular mechanism (Scheme II).

We will represent the above mechanism (Scheme II) in triangular form, as King and Altman did, in order to illustrate most simply how the schematic method of King and Altman may yield the numerator terms for each EX_i:

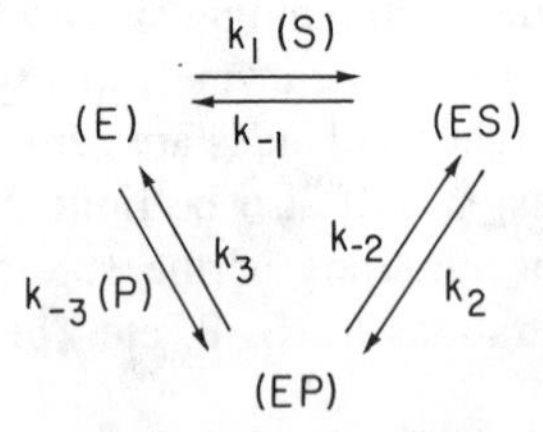

SCHEME III.

There will be nine individual denominator terms in the final rate expression (refer to Equation 8) and this may be predicted from probability theory. Thus, if the mechanism involves *m* reversible steps interconverting the *n* different EX_is, the total number of combinations of the m steps taken n − 1 at a time is ${}_mC_{n-1}$.

$$
\begin{aligned}
{}_mC_{n-1} &= \text{number of interconversion patterns with } (n-1) \text{ lines} \\
&= \frac{m!}{(n-1)!\,(m-n+1)!}
\end{aligned}
\qquad (47)
$$

where n = the number of enzyme species, i.e., the number of corners in the basic figure and m = the number of lines in the basic figure, i.e., number of reversible steps interconverting the n different EX_is.

Thus, for the mechanism given in Scheme III, n = 3 and m = 3.

Therefore the number of patterns with n − 1 = 3 − 1 = 2 lines =

$$
\frac{3!}{(3-1)!\,(3-3+1)!} = \frac{3!}{2! \times 1!} = \frac{3 \times 2 \times 1}{2 \times 1} = 3 \text{ patterns}
$$

Therefore, in the denominator three *two-line* figures for each of n or three enzyme species or nine terms (i.e., 3 × 3).

Thus, for the nine individual denominator terms in the final rate expression, they schematically can be represented in Scheme IV where each of the terms in the rate equation will be shown below the corresponding schematic arrangement.

In the following nine denominator terms, note that the patterns are the same in all cases, but differ only in the direction of the arrows.

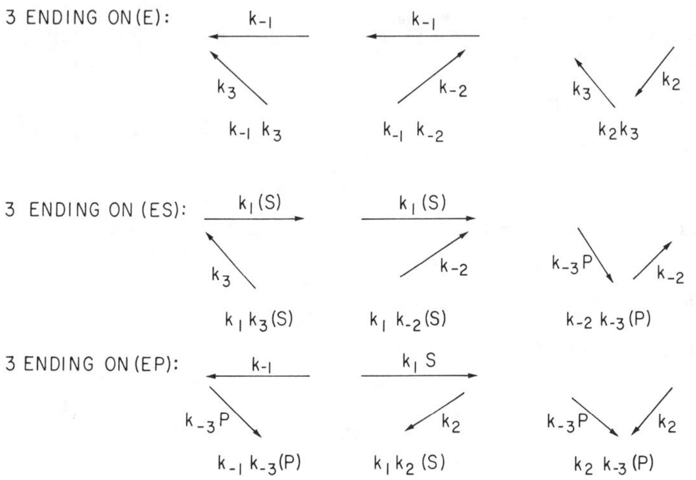

SCHEME IV.

The patterns containing one less line than the basic figure represent all the ways any one enzyme species can be formed from the others by paths *not involving closed loops*. The total number of interconversion patterns is given above by Equation 47. These, then, are the total possible arrangements of line segments for the three enzyme-containing species (E, ES, EP), or the total *number* of *combinations* of (n − 1) rate constants and associated concentration factors where n is the number of enzyme-containing species (three in this case) which are present in the expression for the *relative concentration* of each enzyme-containing species (EX_i). For the simple mechanism given above (Scheme II) none of the (n − 1)-lined interconversion patterns calculated by Equation 47 contain closed loops. However, Equation 47 gives the maximum number of (n − 1)-lined patterns, which may include closed loops or cycles for more complex systems. These closed loops or cycles must be subtracted from the total. These closed loops may be calculated in the following manner.

The number of patterns with (n − 1) lines containing a given closed loop or cycle of *r* sides = ${}_{m-r}C_{n-1-r}$ = r combinations which involve this cycle.

$$ {}_{m-r}C_{n-1-r} = \frac{(m-r)!}{(n-1-r)!\,(m-n+1)!} \tag{48} $$

We will make use of Equation 48 later in more complex situations, if necessary.

The *relative concentration* of each enzyme-containing species is proportional to the summation of three terms (the three individual products of rate constants and associated concentration factors) which are schematically depicted by the combination of line segments (in this case, two segments per term, and three terms for each enzyme-containing species). The proportionality constant in each case is

$$ \frac{1}{(E) + (ES) + (EP)} = \frac{1}{E_t} $$

Thus,

$$ \frac{(E)}{E_t} = k_{-1}k_3 + k_{-1}k_{-2} + k_2k_3 $$

$$ \frac{(ES)}{E_t} = k_1k_3(S) + k_1k_{-2}(S) + k_{-2}k_{-3}(P) $$

$$ \frac{(EP)}{E_t} = k_{-1}k_{-3}(P) + k_1k_2(S) + k_2k_{-3}(P) \tag{49} $$

where we had noted from the determinant method that $\frac{1}{\Delta}$ *was the proportionality factor* which in turn could be taken as proportional to $\frac{1}{E_t}$.

Since the net rate of production of the product is

$$ v = \frac{d(p)}{dt} = k_3(EP) - k_{-3}(E)(P) \tag{50} $$

which, as you recall from the determinant method is

$$ \frac{dP}{dt} = k_3\frac{\Delta_3}{\Delta} - k_{-3}(P)\frac{\Delta_1}{\Delta} = \frac{k_3\Delta_3 - k_{-3}(P)\Delta_1}{\Delta} $$

$\left(\text{Multiply Equation 50 by } \frac{E_t}{E_t} \text{ or by } \frac{E_t}{(E) + (ES) + EP}\right)$

Therefore,

$$v = \frac{[k_3(EP) - k_{-3}(E)(P)]\,E_t}{[(E) + (ES) + (EP)]} \tag{51}$$

Since the proportionality factor for the numerator terms and the denominator terms is the same (namely, (E) + (ES) + (EP)), the proportionality factor cancels, or

$$v = \frac{\left\{k_3\frac{(EP)}{E_t} - k_{-3}\frac{(E)(P)}{E_t}\right\}E_t}{\left\{\frac{(E)}{E_t} + \frac{(ES)}{E_t} + \frac{(EP)}{E_t}\right\}} \tag{52}$$

Now, substitute the sums of terms for each expression in Equation 52 (i.e., substitute the proportional values for $\frac{(EP)}{E_t}$, $\frac{(E)}{E_t}$, and $\frac{(ES)}{E_t}$ in both numerator and denominator as given in Equation 49).

Therefore,

$$v = E_t\,\frac{\begin{array}{l} k_3[k_{-1}k_{-3}(P) + k_1k_2(S) + k_2k_{-3}(P)] \\ -k_{-3}(P)[k_{-1}k_{+3} + k_{-1}k_{-2} + k_2k_3] \end{array}}{\begin{array}{l} k_{-1}k_3 + k_{-1}k_{-2} + k_2k_3 + k_1k_3(S) \\ + k_1k_{-2}(S) + k_{-2}k_{-3}(P) + k_{-1}k_{-3}(P) + k_1k_2(S) + k_2k_{-3}(P) \end{array}} \tag{53}$$

By collecting the terms, we get

$$v = \frac{\{k_1k_2k_3\,(S) - k_{-1}k_{-2}k_{-3}\,(P)\}\,E_t}{\begin{array}{l} k_{-1}k_3 + k_{-1}k_{-2} + k_2k_3 + (S)(k_1k_3 + k_1k_{-2} + k_1k_2) \\ +(P)(k_{-2}k_{-3} + k_{-1}k_{-3} + k_2k_{-3}) \end{array}} \tag{54}$$

which is the same equation we derived before, but this time with less effort.

Note that the *patterns* of line segments (Scheme IV) which are the schematic representations of the three numerator terms in the expressions for each $\frac{(EX_i)}{E_t}$ (or denominator terms in the final velocity equation) are the *same* for *each* of the *enzyme-containing species;* these patterns differ only with the identification of the processes by the introduction of the arrowheads, i.e., the *patterns are the same in each case, but the direction of the arrows differ.*

Also, the expression for the relative concentration of each enzyme-containing species in a reversible reaction involves the same number of numerator terms (or denominator terms in the final velocity expression), that is, the number of combinations of arrows (or combinations of rate constants with associated concentration factors, and the total number of arrows in each combination) is the same for each enzyme-containing species.

Thus, one can empirically determine how many combinations of arrows one can construct for the individual terms, remembering the following rules.

1. Only *one* arrow of each reversible step can be used.
2. The arrows should start from *each* of the enzyme-containing species, *except* the one in question, and should include *all* of the enzyme-containing species except the one in question (i.e., except the one for whose term a derivation is attempted).
3. At least *one* (or more) of the arrows should end on the enzyme-containing species in question.

Finally, one may use the probability expression derived by King and Altman, Equation 47, to compute the number of interconversion patterns with (n − 1) lines for each enzyme species, where the number of closed loops, calculated by Equation 48, must be subtracted from the total number of patterns.

The basis for this schematic method of King and Altman rests upon the determinant solution which we have previously discussed and which King and Altman discuss in a more rigorous fashion. Also, Wong and Hanes[8] discuss a number of bisubstrate examples, solved by means of the method of King and Altman, and the schematic rules for their solution. More recently, Chou and Forsén[156] have reinforced mathematically these graphical rules for deriving steady-state rate expressions for enzyme-catalyzed reactions and which are, in principle, based on the determinant solutions.

To recapitulate, the method consists of setting out diagrammatically arrangements of arrows in accordance with certain rules; these arrangements define the terms in the overall rate equation. For the *numerator* one notes all the possible combinations of arrows which fulfill the following requirements: (1) one arrow (not more) starts from each enzyme-containing species (including the free enzyme itself), (2) the arrows must not form a closed ring, and (3) not more than one direction of any step must be shown. It must be remembered that in some mechanisms depicted diagrammatically according to these rules, the first and last species (e.g., the free enzyme, E) are the same.

1. A Random Steady-State Bisubstrate Reaction to Yield Reversibly a Single Product

To further illustrate the method, consider the mechanism of an enzyme catalyzing a reaction between two substrates, A and B, with the formation of a single product, P. This mechanism might be applicable to an aldolase-type reaction. We will presume that the reaction sequence involves the formation of a single ternary complex EXY, by a combination of A and B, or B and A, successively with the enzyme in a random order of addition with the enzyme, and that the breakdown of the single ternary complex leads to the formation of the product and free enzyme. Such a *bi uni system* in the nomenclature of Cleland[43] has been solved by Wong and Hanes[8] using the King and Altman[15] procedure.

The mechanism may be written as

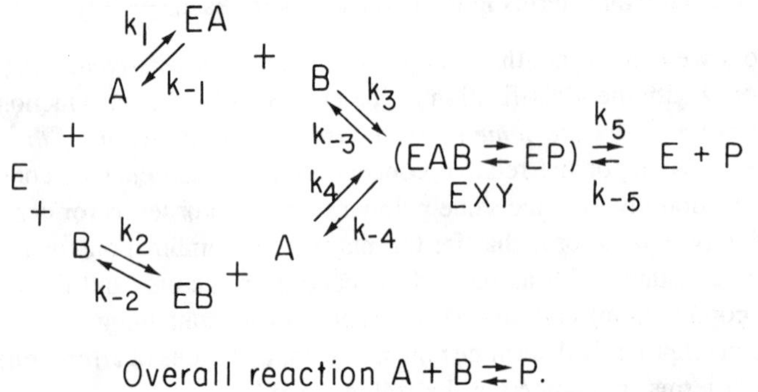

SCHEME V.

Or, the basic King-Altman figure and the eight interconversion patterns are

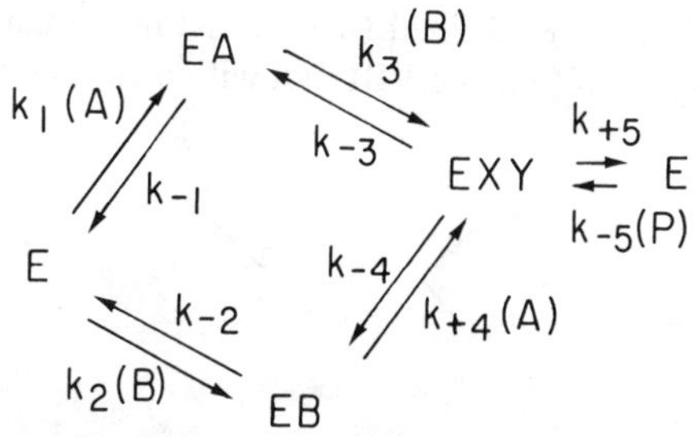

SCHEME VI.

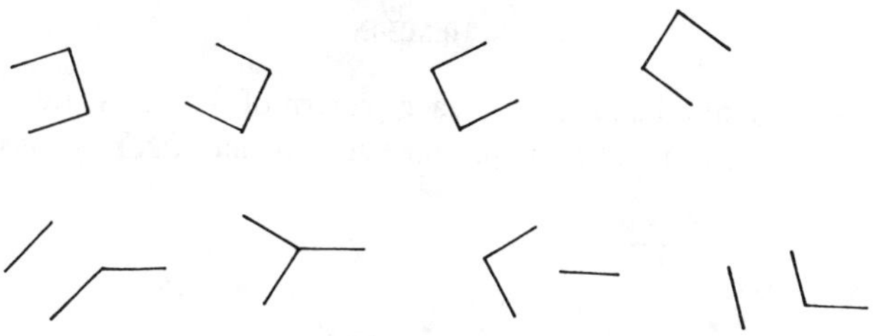

SCHEME VII.

However, it is also possible to write the King-Altman figure in terms of two loops in order to demonstrate use of the probability equations (Equations 47 and 48).

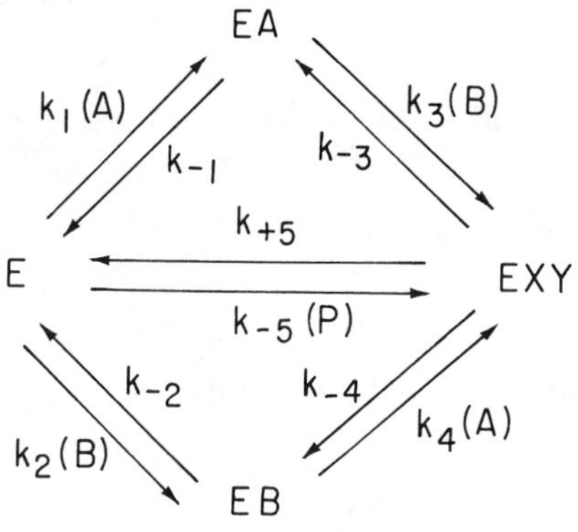

SCHEME VIII.

From Equation 47, the number of patterns, m, with n − 1 = 4 − 1 = 3 lines is

$$ {}_{m}C_{n-1} = \frac{m!}{(n-1)!(m-n+1)!} = \frac{5!}{(4-1)!(5-4+1)!} = \frac{5!}{(3!)(2!)} = 10 \qquad (55) $$

However, there are two closed loops, each of three sides, and from Equation 48 where r = 3 and m = 5 as before.

$$ {}_{m-r}C_{n-1-r} = \frac{(m-r)!}{(n-1-r)!(m-n+1)!} = \frac{(5-3)!}{(4-1-3)!(5-4+1)!} = \frac{2!}{0!2!} = 1 $$

Note that 0! = 1 or, there is one pattern for each closed loop of which there are two.

Therefore, there are 10 − 2 = 8, 3-lined (n − 1) King-Altman interconversion patterns for the basic 4-cornered figure of Scheme VIII, but with no closed loops, as given in Scheme IX.

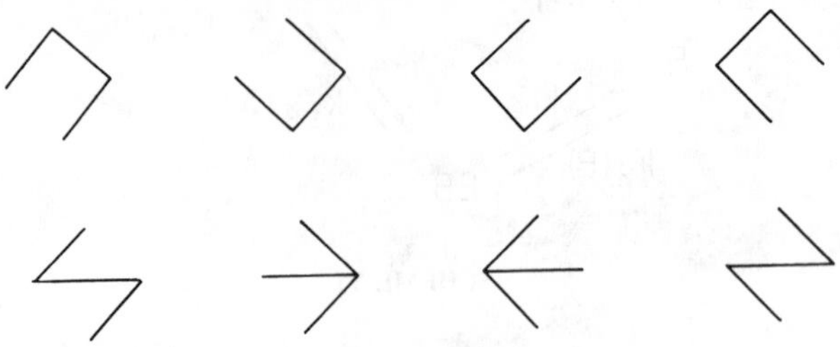

SCHEME IX.

Thus, there are 32 denominator terms (each pattern of 3 arrows, i.e., n = 4 and (n − 1) = 3) with 8 ending on (E) and 8 ending on EA, EB, and EXY, respectively.

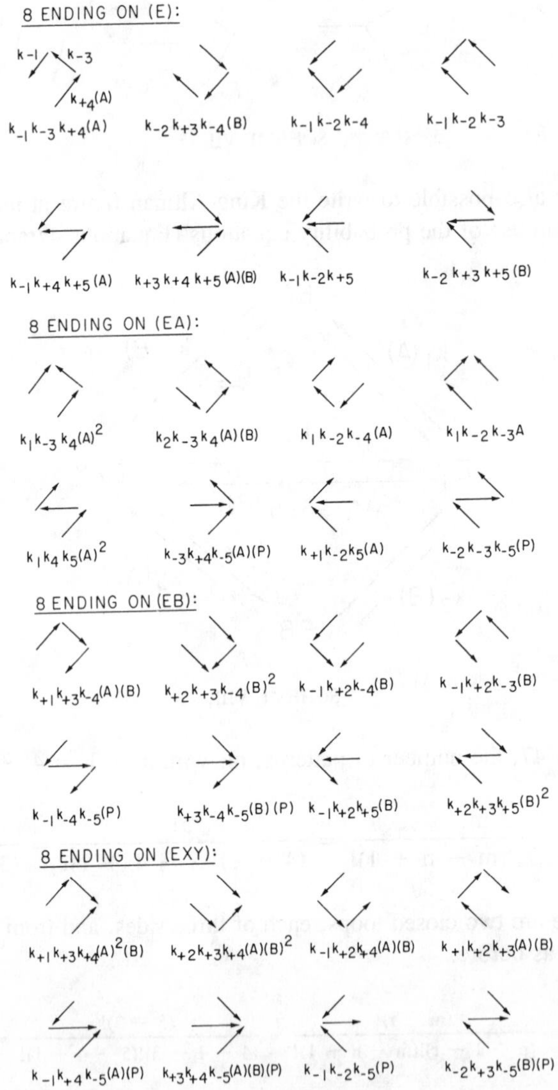

SCHEME X.

Substitution in

$$\frac{dp}{dt} = k_{+5}(EXY) - k_{-5}(E)(P) = v \tag{56}$$

i.e., after multiplying Equation 56 $\frac{E_t}{E_t}$ and substitution into Equation 57.

$$v = \frac{(k_{+5}(EXY) - k_{-5}(E)(P))E_t}{(E) + (EA) + (EB) + (EXY)} \tag{57}$$

Collect terms and one obtains Equation 58.

$$\frac{v}{E_t} = \frac{(K_1(A)(B) + K_2(A)^2(B) + K_3(A)(B)^2 - K_4(P) - K_5(A)(P) - K_6(B)(P))}{K_7 + K_8(A) + K_9(B) + K_{10}(A)(B) + K_{11}(A)^2 + K_{12}(B)^2 + K_{13}(A)^2(B) + K_{14}(A)(B)^2 + K_{15}(P) + K_{16}(A)(P) + K_{17}(B)(P) + K_{18}(A)(B)(P)} \tag{58}$$

where the 18 constants are combinations of rate constants which cannot be simplified further into convenient kinetic constants:

$$\begin{aligned}
&K_1 = k_{+1}k_{-2}k_{+3}k_{+5} + k_{-1}k_{+2}k_{+4}k_{+5};\ K_2 = k_{+1}k_{+3}k_{+4}k_{+5};\ K_3 = k_{+2}k_{+3}k_{+4}k_{+5};\\
&K_4 = k_{-1}k_{-2}k_{-3}k_{-5} + k_{-1}k_{-2}k_{-4}k_{-5};\ K_5 = k_{-1}k_{-3}k_{+4}k_{-5};\ K_6 = k_{-2}k_{+3}k_{-4}k_{-5};\\
&K_7 = k_{-1}k_{-2}k_{-3} + k_{-1}k_{-2}k_{-4} + k_{-1}k_{-2}k_{+5};\ K_8 = k_{+1}k_{-2}k_{-3} + k_{+1}k_{-2}k_{-4} + k_{+1}k_{-2}k_{+5}\\
&+ k_{-1}k_{-3}k_{+4} + k_{-1}k_{+4}k_{+5};\ K_9 = k_{-1}k_{+2}k_{-3} + k_{-1}k_{+2}k_{-4} + k_{-1}k_{+2}k_{+5}\\
&+ k_{-2}k_{+3}k_{-4} + k_{-2}k_{+3}k_{+5};\ K_{10} = k_{+1}k_{-2}k_{+3} + k_{-1}k_{+2}k_{+4} + k_{+1}k_{+3}k_{-4}\\
&+ k_{+2}k_{-3}k_{+4} + k_{+3}k_{+4}k_{+5};\ K_{11} = k_{+1}k_{-3}k_{+4} + k_{+1}k_{+4}k_{+5};\ K_{12} = k_{+2}k_{+3}k_{-4}\\
&+ k_{+2}k_{+3}k_{+5};\ K_{13} = k_{+1}k_{+3}k_{+4};\ K_{14} = k_{+2}k_{+3}k_{+4};\ K_{15} = k_{-1}k_{-2}k_{-5}\\
&+ k_{-1}k_{-4}k_{-5} + k_{-2}k_{-3}k_{-5};\ K_{16} = k_{-1}k_{+4}k_{-5} + k_{-3}k_{+4}k_{-5};\ K_{17} = k_{-2}k_{+3}k_{-5}\\
&+ k_{+3}k_{-4}k_{-5};\ K_{18} = k_{+3}k_{+4}k_{-5}
\end{aligned} \tag{59}$$

However, this very formidable equation may be reduced to a workable equation if either P = 0 or A = B = 0, i.e., for either the forward or reverse reaction with no product or products added initially, and for measurements of initial velocity, so that the product (or products) accumulation is small.

Therefore, for the first case, for the initial forward reaction, setting P = 0 in both the numerator and denominator:

$$\frac{V_{o,f}}{E_t} = \frac{K_1(A)(B) + K_2(A)^2(B) + K_3(A)(B)^2}{[K_7 + K_8(A) + K_9(B) + K_{10}(A)(B) + K_{11}A^2 + K_{12}(B)^2 + K_{13}(A)^2(B) + K_{14}(B)^2]} \tag{60}$$

and in the second case for the initial reverse reaction, setting (A) = (B) = 0 and reversing the sign:

$$\frac{V_{0,b}}{E_t} = \frac{K_4(P)}{K_7 + K_{15}(P)} \tag{61}$$

The equation for v_b is of the same form as the Briggs-Haldane equation (for a one-substrate reaction), but with a more complex expression for V_{max} and K_m, that is, divide numerator and denominator by K_{15}

$$\frac{v_b}{E_t} = \frac{\left(\frac{K_4}{K_{15}}\right)(P)}{\left(\frac{K_7}{K_{15}}\right) + (P)} \tag{62}$$

and then by (P). Therefore:

$$\frac{v_b}{E_t} = \frac{\left(\frac{K_4}{K_{15}}\right)}{\frac{(K_7/K_{15})}{(P)} + 1} \tag{63}$$

which may be compared with

$$v = \frac{V_{max}}{\frac{K_m}{(S)} + 1} \tag{63a}$$

Define

$$V_{max,b} = \frac{K_4}{K_{15}} E_t = \frac{(k_{-1}k_{-2}k_{-3}k_{-5} + k_{-1}k_{-2}k_{-4}k_{-5})}{k_{-1}k_{-2}k_{-5} + k_{-1}k_{-4}k_{-5} + k_{-2}k_{-3}k_{-5}} E_t$$

$$K_{m,p} = \frac{K_7}{K_{15}} = \frac{k_{-1}k_{-2}k_{-3} + k_{-1}k_{-2}k_{-4} + k_{-1}k_{-2}k_{-5}}{k_{-1}k_{-2}k_{-5} + k_{-1}k_{-4}k_{-5} + k_{-2}k_{-3}k_{-5}} \tag{64}$$

Therefore:

$$v_b = \frac{V_{max,b}}{\frac{K_{m,p}}{(P_0)} + 1} \tag{64a}$$

and the kinetic parameters are the same as the Briggs-Haldane expression except the definitions in terms of kinetic or rate constants differ from the one-substrate reactions we discussed.

Both $V_{max,b}$ and $K_{m,p}$ may be evaluated by a double-reciprocal plot (Figure 2). On the other hand, the equation for $v_{0,f}$ cannot be reduced to the form of the simple Michaelis equation for either substrate, because one still has terms involving $(B)^2$ or $(A)^2$. Thus, if (B) is held fixed and (A) is systematically varied, then the equation for v_f becomes simpler in form:

$$\frac{v_f}{E_t} = \frac{i[A]^2 + j[A]}{k + \ell[A]^2 + m(A)} \tag{65}$$

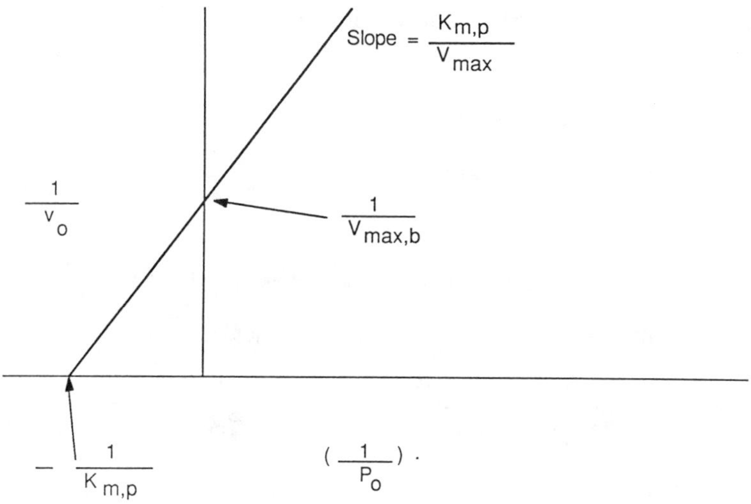

FIGURE 2. Plot of $\frac{1}{v}$ vs. $\frac{1}{(B)}$ from Equation 64a.

where i = K_2(B), j = K_1(B) + $K_3(B)^2$, k = K_7 + K_9(B) + $K_{12}(B)^2$, ℓ = K_{11} + K_{13}(B), and m = K_8 + K_{10}(B) + $K_{14}(B)^2$.

Wong and Hanes[8] have described such a system of *second degree* in (A) (i.e., A is found to the second power in the rate expression). This situation arises from the fact that A combines with two different enzyme-containing species (either E or EB) in a random manner, and in this case the system cannot be treated in terms of a "simple" Michaelis constant for A, and the double-reciprocal plots are nonlinear, i.e., nonlinear plots for $\frac{1}{v_o}$ vs. $\frac{1}{B}$, at fixed A; since, in the latter case, if A is held fixed and B is varied

$$\frac{v_f}{E_t} = \frac{i'(B^2) + j'(B)}{k' + \ell'(B)^2 + m'(B)} \tag{66}$$

where i′ = K_3A, j′ = K_1(A) + $K_2(A)^2$, k′ = K_7 + K_8(A) + $K_{11}(A)^2$, ℓ' = K_{12} + K_{14}(A), and m′ = K_9 + K_{10}(A) + $K_{13}(A)^2$.

However, although the reciprocal plots may be nonlinear, the departure from linearity may be impossible to detect if both routes to EAB are about equally favorable. The curvature of the reciprocal plot may occur close to the $\frac{1}{v}$ axis. The system may then appear to be rapid equilibrium random (i.e., a quasi-equilibrium random mechanism of two basic types — with or without independent binding of the substrates) which we will discuss shortly; however, in this case the K_m values may not equal the $\bar{K}_s$ values (dissociation constants) determined from binding studies. Equations 65 and 66 are also known in the Cleland nomenclature[43] as "2/1" functions, because in reciprocal form (after dividing numerator and denominator by S^2), $1/(S)^2$ appears in the numerator, while 1/(S) appears in the denominator.

However, if one of the pathways for the mechanism we discussed (Scheme V) was eliminated, the coefficients involving the $(A)^2$ terms become zero and normal Michaelis-type kinetics may appear to result. In one case, it appears to reduce, in part, (if A is varied and B is fixed) to the so-called "ordered bisubstrate reaction" where A must first combine with E and then B to yield (EXY), as modified by the effect of B, as we will see.

Thus, let $k_{+4} \rightarrow 0$ or $k_{+4}(A) \rightarrow 0$ in Equation 60.
Therefore,

$$K_1 \rightarrow k_{+1}k_{-2}k_{+3}k_{+5} = K_1'$$

$$K_2 \rightarrow 0$$

$$K_3 \rightarrow 0$$

$$K_8 \rightarrow k_{+1}k_{-2}k_{-3} + k_{+1}k_{-2}k_{-4} + k_{+1}k_{-2}k_{+5} = K_8'$$

$$K_{10} \rightarrow k_{+1}k_{-2}k_{+3} + k_{+1}k_{+3}k_{-4} = K_{10}$$

$$K_{11} \rightarrow 0$$

$$K_{13} \rightarrow 0$$

$$K_{14} \rightarrow 0$$

Therefore:

$$\frac{v_f}{E_t} = \frac{K_1'(A)(B)}{K_7 + K_8'(A) + K_9(B) + K_{10}'(A)(B) + K_{12}(B)^2} \tag{67}$$

If (B) is held fixed and (A) is varied, then

$$\frac{v_f}{E_t} = \frac{\delta_1(A)}{\delta_2 + \delta_3(A)} \tag{68}$$

where $\delta_1 = K'_1(B)$, $\delta_2 = K_7 + K_9(B) + K_{12}(B)^2$, and $\delta_3 = K'_8 + K'_{10}(B)$.
Rearrange Equation 68

$$\frac{v_f}{E_t} = \frac{\delta_1}{\dfrac{\delta_2}{(A)} + \delta_3} = \frac{\dfrac{\delta_1}{\delta_3}}{\dfrac{\delta_2}{\delta_3}\dfrac{1}{(A)} + 1} \tag{69}$$

$$v_f = \frac{V_{max,f}^{app}}{\dfrac{K_{m_{app}}^{(A)}}{(A)} + 1} \tag{70}$$

where we define

$$V_{max,f}^{app} = \left(\frac{\delta_1}{\delta_3}E_t\right) = \frac{K_1'(B)E_t}{K_8' + K_{10}'(B)} = \frac{k_{+1}k_{-2}k_{+3}k_{+5}(B)E_t}{(k_{+1}k_{-2}k_{-3} + k_{+1}k_{-2}k_{-4} + k_{+1}k_{-2}k_{+5}) + (B)[k_{+1}k_{-2}k_{+3} + k_{+1}k_{+3}k_{-4}]}$$

$$K_{m_{app}}^{(A)} = \left(\frac{\delta_2}{\delta_3}\right) = \frac{K_7 + K_9(B) + K_{12}(B)^2}{K_8' + K_{10}'(B)} \tag{71}$$

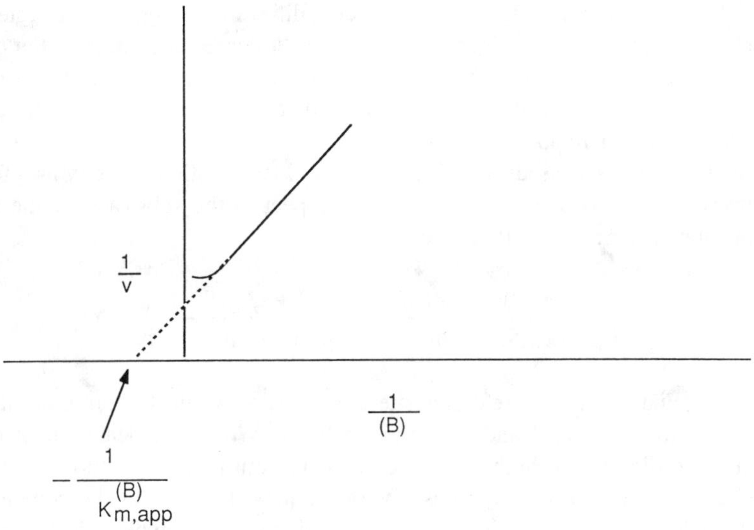

FIGURE 3. Plot of $\frac{1}{v}$ vs. $\frac{1}{(B)}$ from Equation 74.

and linear double-reciprocal plots of $\frac{1}{v}$ vs. $\frac{1}{(A)}$ (at fixed B) results; however, $k_{mapp}^{(A)}$ has no simple meaning in this case.

On the other hand, if now A is fixed and B is varied,

$$\frac{v_f}{E_t} = \frac{\delta_4(B)}{\delta_5 + \delta_6(B) + K_{12}(B)^2} \tag{72}$$

where $\partial_4 = K_1'(A)$; $\partial_5 = K_7 + K_8'(A)$; $\partial_6 = K_9 + K_{10}'(A)$; however, it still retains a $(B)^2$ term and therefore $\frac{1}{v_0}$ vs. $\frac{1}{(B)}$ plot at fixed (A) will still be nonlinear; divide through by $\delta_6(B)$:

$$\frac{v_f}{E_t} = \frac{(\delta_4/\delta_6)}{1 + \left(\frac{\delta_5}{\delta_6}\right)\frac{1}{(B)} + \frac{K_{12}(B)}{\delta_6}} \tag{73}$$

$$(\delta_4/\delta_6)\frac{E_t}{v_f} = 1 + \frac{\delta_5}{\delta_6}\frac{1}{(B)} + \frac{K_{12}(B)}{\delta_6} \tag{74}$$

A $\frac{1}{v}$ vs. $\frac{1}{(B)}$ plot at fixed (A) would be nonlinear as shown in Figure 3. However, a $k_{mapp}^{(B)}$ value might be obtained by extrapolation using small values of B, but large values for $\frac{1}{(B)}$, so that $\frac{K_{12}}{\delta_6}(B) \rightarrow 0$.

This case clearly illustrates the extreme possibilities to be encountered in two-substrate steady-state reactions which, in general, are quite complex; however, experimental conditions may often be found which will allow simplifying conditions to prevail. In fact, in cases of higher-degree reactions, it is difficult to evaluate steady-state kinetics unless experimental conditions are found to reduce the velocity expressions to one of first degree.

This leads us now to a discussion of some generalities involving two-substrate reactions whose mechanisms result in first degree equations (first degree according to the definitions of Wong and Hanes,[8] i.e., to those cases where there are no second-degree terms or squared terms in the rate expressions. Much later we will briefly discuss some three-substrate reactions which also lead to linear reciprocal plots.

Wong and Hanes[8] expressed as a general equation the reaction velocity as a function of a given substrate in terms of a polynomial with respect to the substrate for the numerator and denominator of the rate equation, i.e.,

$$\frac{v}{E_t} = \frac{a_0(S)^z + a_1(S)^{z-1} + \ldots + a_{z-1}(S) + a_z}{b_0(S)^z + b_1(S)^{z-1} + \ldots + b_{z-1}(S) + b_z} \tag{75}$$

where $a_0 \ldots a_z$ and $b_o \ldots b_z$ are composite constants, as we have seen in our illustration, where one substrate is varied and the other is fixed. Wong and Hanes then designated mechanisms according to the highest degree in (S) in both numerator and denominator.

In the case of two-substrate systems, they described the form of the equation for 2nd degree systems for approximately 6 different mechanisms, and then described about 24 cases for the first degree system which they showed would, in general, conform to the velocity equation for the forward reaction only

$$\frac{v_f}{E_t} = \frac{a_0(S_1)(S_2)}{b_0(S_1)(S_2) + b_1(S_1) + b_2(S_2) + b_3} \tag{76}$$

where a_0 and $b_o - b_3$ are composite constants and S_1 and S_2 are the two substrates involved in the forward reaction.

D. TWO-SUBSTRATE SYSTEMS OF THE FIRST DEGREE

We will restrict our next discussion to those two-substrate systems of the first degree. Contributions by Alberty, King, Dalziel, Frieden, Boyer, Dixon, Laidler, Cleland, and Wong, to mention a few names, and a number of others have made solutions of enzyme kinetic mechanisms, which conform to the first-degree expression, easier to interpret and to analyze (as well as for three-substrate reactions).

We have mentioned one particular random mechanism, a so-called random bi uni system, in the Cleland nomenclature, that was second degree in A and B for the forward reaction, but first degree in P for the reverse reaction.

There is one important special case of a random mechanism, however, which is the case where a quasi-equilibrium prevails. Thus, either substrate may add to the enzyme without any specific order, i.e., randomly, but if all the steps leading to the formation of the ternary complex are rapid and approach equilibrium conditions except, for example, in the breakdown of the ternary complex, EAB, the rate-limiting step is,

$$\ldots \rightleftarrows EAB \xrightarrow[\text{step}]{\text{slowest}} \ldots \text{products} \tag{77}$$

then first degree velocity expressions and linear reciprocal plots will result.

Dalziel has analyzed the yeast alcohol dehydrogenase reaction in this fashion[44-47] and Kuby and Noltmann[48] have treated the rabbit muscle ATP-creatine transphosphorylase kinetics in a similar fashion, except that in this latter case, specific conditions, i.e., Mg-complexation with the nucleotide substrates, required a further elaboration; in the case of the rabbit muscle creatine kinase, the additional condition was invoked that the binding of the individual substrates to the enzyme were considered to be independent of one another. This condition, however, could not be invoked in the case of the calf brain ATP-creatine transphosphorylase,[49] and it was necessary to consider two intrinsic dissociation constants for each substrate, one from the binary enzyme complex (e.g., $\bar{K}_{s,1}$) and one from the ternary enzyme complex (e.g., $K_{s,1}$) with a rate-limiting step at the interconversion of the ternary complexes. (See also References 66, 83, 84, 338, 831 to 840, and 849 to 851.)

A similar mechanism was discussed by Hamada and Kuby[50] for the adenylate kinases of muscle and liver. We will return to a discussion of these mechanisms, but for the present, we will attempt to deal with and discuss some generalities which Dalziel first presented for two-substrate reactions which yield first-degree expressions, i.e., linear double-reciprocal plots, especially the ordered reactions (i.e., for an ordered reaction, A must add to the enzyme first followed by B (or vice versa) and there is a definite order as to which of the products come off the enzyme first, etc.). It is also called a compulsory-binding-ordered mechanism in contrast to a random-binding mechanism (or random addition of substrates).

It was shown by Alberty[51,52] and others, that for those cases where the rate expressions contain no squared terms, a general rate equation could be given, e.g., for the forward reaction only (v_f):

$$v_f = \frac{V_{max,f}}{1 + \frac{K_A}{(A)} + \frac{K_B}{(B)} + \frac{K_{AB}}{(A)(B)}} \tag{78}$$

where (A) and (B) are the two substrates, K_A and K_B are Michaelis constants, K_{AB} is a composite constant, and $V_{max,f}$ = forward maximum velocity.

1. The Theorell-Chance Mechanism for Liver Alcohol Dehydrogenase

Dalziel[44] studied this type of equation and arrived at a graphical method of determination of the three constants, similar to that of Florini and Vestling.[53] We will return to this shortly, but first let us take as an example used by Dalziel, the simplest case of an *ordered reaction,* involving only *two binary enzyme complexes,* that is, the so-called Theorell-Chance[54] mechanism originally proposed as the mechanism for the liver alcohol dehydrogenase-catalyzed reaction. (Dalziel later showed that this is only a limiting mechanism for high substrate concentrations.[46])

Theorell and Chance[54] proposed that in this reaction the enzyme reacts with DPN to form an E·DPN complex which then reacts with ethanol to yield an E·DPNH complex and acetaldehyde; then, this binary complex D·DPNH breaks down finally to E + DPNH. They, therefore, did not consider it necessary to invoke a ternary complex, but only two binary complexes (E·DPN or E·DPNH) and that E·alcohol or E·acetaldehyde were kinetically negligible.

Thus, if the overall reaction is

$$S_1 + S_2 \underset{\text{alc.dehyd.}}{\overset{\text{liver}}{\rightleftarrows}} P_1 + P_2 \tag{79}$$

the Theorell-Chance mechanism may be written as

$$E + S_1 \underset{k_{-1}}{\overset{k_{+1}}{\rightleftharpoons}} ES_1$$

$$ES_1 + S_2 \underset{k_{-2}}{\overset{k_{+2}}{\rightleftharpoons}} EP_1 + P_2$$

$$EP_1 \underset{k_{-3}}{\overset{k_{+3}}{\rightleftharpoons}} E + P_1$$

SCHEME XI.

or in the short-hand nomenclature of Cleland:

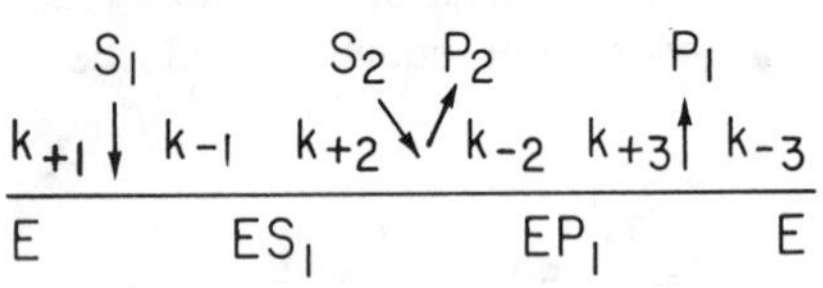

SCHEME XII.

Schematically, the basic King-Altman figure may be written as

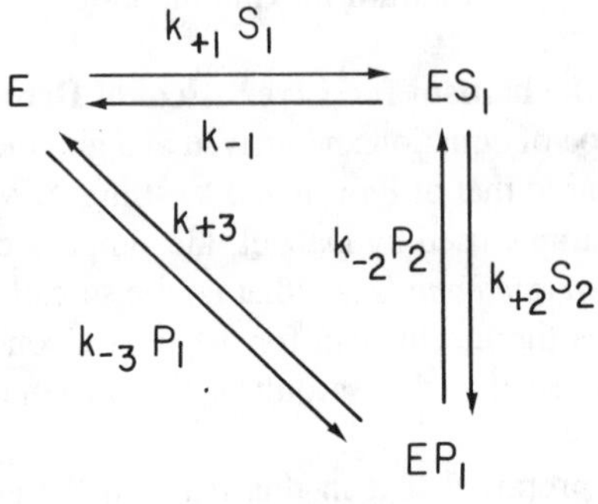

SCHEME XIII.

The distribution of enzyme forms is obtained from three two-lined ($n - 1 = 3 - 1 = 2$) interconversion patterns:

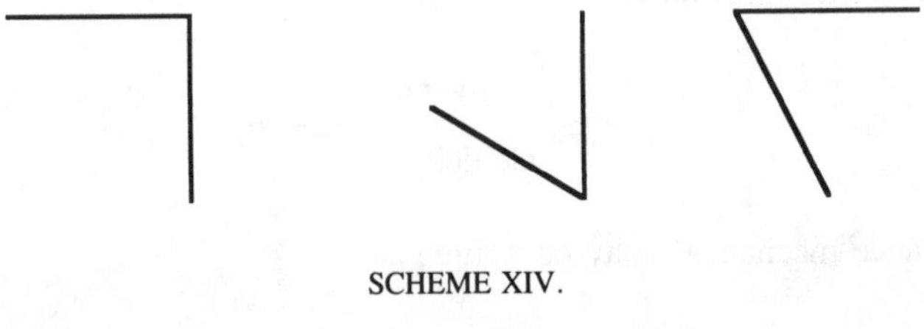

SCHEME XIV.

$$\frac{(E)}{E_t} = k_{-1}k_{-2}(P_2) + k_{+2}k_{+3}(S_2) + k_{-1}k_{+3}$$

$$\frac{(ES_1)}{E_t} = k_{+1}k_{-2}(S_1)(P_2) + k_{-2}k_{-3}(P_1)(P_2) + k_{+1}k_{+3}(S_1)$$

$$\frac{(EP_1)}{E_t} = k_{+1}k_{+2}(S_1)(S_2) + k_{+2}k_{-3}(S_2)(P_1) + k_{-1}k_{-3}(P_1) \quad (80)$$

Setting $v = k_{+1}(E)(S_1) - k_{-1}(ES_1)$ or

$$v = \frac{[k_{+1}(E)(S_1) - k_{-1}(ES_1)]}{(E) + (ES_1) + (EP_1)} E_t \quad (81)$$

Substituting and grouping denominator terms:

$$\frac{v}{E_t} = \frac{k_{+1}k_{+2}k_{+3}(S_1)(S_2) - k_{-1}k_{-2}k_{-3}(P_1)(P_2)}{\begin{gathered}[k_{-1}k_{+3} + k_{+1}k_{+3}(S_1) + k_{+2}k_{+3}(S_2) + k_{+1}k_{+2}(S_1)(S_2) + k_{+1}k_{-2}(S_1)(P_2) \\ + k_{+2}k_{-3}(S_2)(P_1) + k_{-2}k_{-3}(P_1)(P_2) + k_{-1}k_{-3}(P_1) + k_{-1}k_{-2}(P_2)]\end{gathered}} \quad (82)$$

We can at this point define certain coefficient ratios as described by Cleland[43] and convert the overall expression into one containing the measured kinetic parameters.

Thus, multiply Equation 82 through by E_t and then rewrite it as

$$v = \frac{(num_1)(S_1)(S_2) - (num_2)(P_1)(P_2)}{\begin{gathered}[constant + coef_{s_1}(S_1) + coef_{s_2}(S_2) + coef_{s_1s_2}(S_1)(S_2) + coef_{s_1p_2}(S_1)(P_2) \\ + coef_{s_2p_1}(S_2)(P_1) + coef_{p_1p_2}(P_1)(P_2) + coef_{p_1}(P_1) + coef_{p_2}(P_2)]\end{gathered}} \quad (83)$$

where num_1 = coefficient of the positive numerator term and contains E_t, num_2 = coefficient of the negative numerator term and contains E_t, const = the constant term in the denominator, $coef_{s1}$ = coefficient of (S_1) in the denominator, $coef_{s2}$ = coefficient of (S_2) in the denominator, $coef_{s1s2}$ = coefficient of (S_1) (S_2) in the denominator, $coef_{s1p2}$ = coefficient of (S_1) (P_2) in the denominator, $coef_{s2p1}$ = coefficient of (S_2) (P_1) in the denominator, $coef_{p1p2}$ = coefficient of $(P_1)(P_2)$ in the denominator, $coef_{p1}$ = coefficient of (P_1) in the denominator, and $coef_{p2}$ = coefficient of (P_2) in the denominator.

The final velocity equation is obtained by redefining the coefficients of Equation 83 as kinetic parameters that can be experimentally determined.

The following definitions will permit us to simplify the velocity equation into the usual Michaelis form. These definitions can be verified by substitution of their definitions in terms of kinetic rate constants (see above).

Thus,

$$V_{max,f} = \frac{num_1}{coef_{s_1s_2}}; \quad V_{max,r} = \frac{num_2}{coef_{p_1p_2}};$$

$$K_{eq} = \frac{num_1}{num_2} = \frac{k_{+1}k_{+2}k_{+3}}{k_{-1}k_{-2}k_{-3}} \frac{E_t}{E_t} = \frac{k_{+1}k_{+2}k_{+3}}{k_{-1}k_{-2}k_{-3}} = \frac{(P_1)(P_2)}{(S_1)(S_2)} \quad (84)$$

Also, define:

$$K_{s_1} = \frac{coef_{s_2}}{coef_{s_1s_2}}; \quad K_{s_2} = \frac{coef_{s_1}}{coef_{s_1s_2}}$$

$$\bar{K}_{s_1} = \frac{coef_{p_2}}{coef_{s_1p_1}} = \frac{constant}{coef_{s_1}} = \frac{k_{-1}k_{+3}}{k_{+1}k_{+3}} = \frac{k_{-1}}{k_{+1}}$$

(Note $\bar{K}_{s1}$, as well as $\bar{K}_{p1}$ below, is a true thermodynamic binary dissociation constant for the dissociation of S_1 from ES_1; however, there is *no* physical meaning for $\bar{K}_{s2}$ or $\bar{K}_{p2}$ as a result of the mechanism (see Scheme 13).

$$K_{p_2} = \frac{coef_{p_1}}{coef_{p_2p_1}}; \quad K_{p_1} = \frac{coef_{p_2}}{coef_{p_2p_1}}$$

$$\bar{K}_{p_1} = \frac{coef_{s_2}}{coef_{s_2p_1}} = \frac{constant}{coef_{p_1}} = \frac{k_{-1}k_{+3}}{k_{-1}k_{-3}} = \frac{k_{+3}}{k_{-3}}$$

Now, if the numerator and denominator of Equation 83 are multiplied by

$$\frac{num_2}{coef_{s_1s_2}coef_{p_1p_2}} \left\{ = \frac{k_{-1}k_{-2}k_{-3}}{(k_{+1}k_{+2})(k_{-2}k_{-3})} = \frac{k_{-1}}{k_{+1}k_{+2}} \right\}$$

then the (S_1) (S_2) numerator term and the (S_1), (S_2), and (S_1) (S_2) terms can be defined in terms of kinetic parameters. If we also multiply the numerator (P_1) (P_2) term and the denominator terms (P_1), (P_2), (P_1) (P_2), and (S_1) (P_2) by $\left(\frac{num\ 1}{num_1}\right)$, then they can be reduced to kinetic parameters (and K_{eq}).

The $(S_1)(P_2)$ and $(S_2)(P_1)$ terms simplify after substitution of:

$$coef_{s_1p_2} = coef_{p_2}/\bar{K}_{s_1} \text{ and } coef_{s_2p_1} = \frac{coef_{s_2}}{\bar{K}_{p_1}}$$

and the constant term in the denominator will simplify after the substitution of const $= \bar{K}_{s1} \cdot coef_{s1}$.

Therefore,

$$v = \frac{\left[\frac{num_1(S_1)(S_2)num_2}{coef_{s_1s_2}coef_{p_1p_2}} - \frac{num_2(P_1)(P_2)num_2num_1}{coef_{s_1s_2}coef_{p_1p_2}num_1}\right]}{\begin{aligned}&\frac{constant \cdot num_2}{coef_{s_1s_2}coef_{p_1p_2}} + \frac{coef_{s_1}(S_1)num_2}{coef_{s_1s_2}coef_{p_1p_2}} + \frac{coef_{s_2}(S_2)num_2}{coef_{s_1s_2}coef_{p_1p_2}} \\ &+ \frac{coef_{s_1s_2}(S_1)(S_2)num_2}{coef_{s_1s_2}coef_{p_1p_2}} + \frac{coef_{s_1p_2}(S_1)(P_2)num_2num_1}{coef_{s_1s_2}coef_{p_1p_2}num_1} \\ &+ \frac{coef_{s_2p_1}(S_2)(P_1)num_2}{coef_{s_1s_2}coef_{p_1p_2}} + \frac{coef_{p_1p_2}(P_1)(P_2)num_2num_1}{coef_{s_1s_2}coef_{p_1p_2}num_1} \\ &+ \frac{coef_{p_1}(P_1)num_2num_1}{coef_{s_1}s_2coef_{p_1p_2}num_1} + \frac{coef_{p_2}(P_2)num_2num_1}{coef_{s_1s_2}coef_{p_1p_2}num_1}\end{aligned}} \tag{85}$$

and

$$v = \frac{V_{max,f}V_{max,r}\left[(S_1)(S_2) - \frac{(P_1)(P_2)}{K_{eq}}\right]}{\begin{aligned}&V_{max,r}\bar{K}_{s_1}K_{s_2} + V_{max,r}K_{s_2}(S_1) + V_{max,r}K_{s_1}(S_2) \\ &+ V_{max,r}(S_1)(S_2) + \frac{V_{max,f}K_{p_1}(S_1)(S_2)(P_2)}{K_{eq}\bar{K}_{s_1}} + \frac{V_{max,r}K_{s_1}(S_2)(P_1)}{\bar{K}_{p_1}} \\ &+ \frac{V_{max,f}(P_1)(P_2)}{K_{eq}} + \frac{V_{max,f}K_{p_2}(P_1)}{K_{eq}} + \frac{V_{max,f}K_{p_1}(P_2)}{K_{eq}}\end{aligned}} \tag{86}$$

If we limit our discussion to the initial forward velocity, without products, the initial forward velocity becomes

$$\frac{v_f}{V_{max,f}} = \frac{(S_1)(S_2)}{\bar{K}_{s_1}K_{s_2} + K_{s_2}(S_1) + K_{s_1}(S_2) + (S_1)(S_2)} \tag{87}$$

or

$$v_f = \frac{V_{max,f}}{1 + \frac{K_{s_1}}{(S_1)} + \frac{K_{s_2}}{(S_2)} + \frac{\bar{K}_{s_1}K_{s_2}}{(S_1)(S_2)}} \tag{88}$$

which is of the same form as the Alberty generalized equation (Equation 78) with $\bar{K}_{s1}$ K_{s2} equal to the composite constant K_{AB}.

To estimate the constants, e.g., when (S_1) is varied:

$$\frac{v_f}{V_{max,f}} = \frac{1}{\left(1 + \frac{K_{s_2}}{(S_2)}\right) + \frac{K_{s_1}}{(S_1)}\left(1 + \frac{\bar{K}_{s_1}K_{s_2}}{K_{s_1}(S_2)}\right)} \tag{89}$$

or the reciprocal:

$$\frac{1}{v_f} = \frac{1}{V_{max,f}}\left(1 + \frac{K_{s_2}}{(S_2)}\right) + \frac{K_{s_1}}{V_{max,f}}\frac{1}{(S_1)}\left(1 + \frac{\bar{K}_{s_1}K_{s_2}}{K_{s_1}(S_2)}\right) \tag{90}$$

and a primary plot is made of $\frac{1}{v_f}$ vs. $\frac{1}{S_1}$ (see Figure 4).

(Note that since

$$\bar{K}_{s_1} \cdot K_{s_2} = K_{s_1} \cdot \bar{K}_{s_2}; \quad \frac{\bar{K}_{s_1}K_{s_2}}{K_{s_1}} = \bar{K}_{s_2} \tag{91}$$

where one requires an arbitrary definition for $\bar{K}_{s2}$ from product inhibition data.)

The reciprocal plots may intersect above, on, or below the horizontal axis depending on the ratio of $\frac{\bar{K}_{s1}}{K_{s1}}$.

If $\bar{K}_{s1}$ is very small compared to K_{s1}, the slopes of the plots become insensitive to changes in the concentration of the paired substrate and the family of plots will be essentially parallel, and may actually appear to be "ping-pong"[43] in nature.

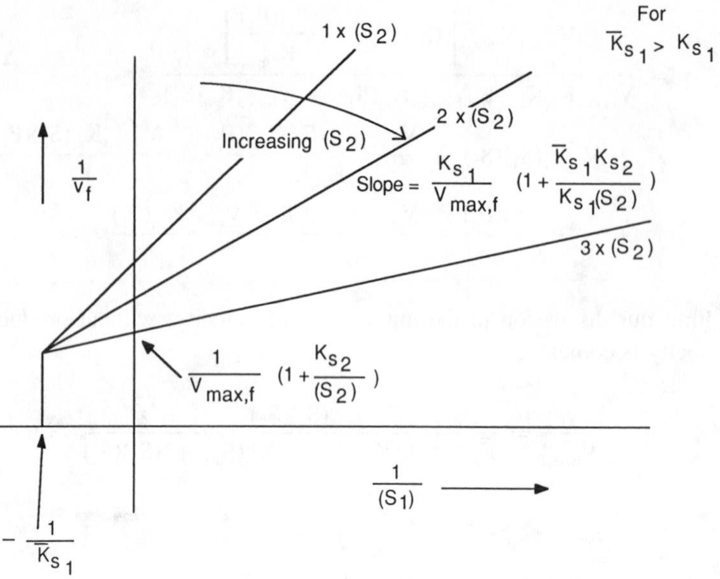

FIGURE 4. Plots of $\frac{1}{v_f}$ vs. $\frac{1}{(S_1)}$ at fixed values of (S_2) for the Theorell-Chance mechanism.

Secondary plots, based on Equations 92 and 93 ((A) for the Y-intercepts and (B) for the slopes of Figure 4, vs. $\left(\frac{1}{(S_2)}\right)$) provide a means of estimating the kinetic parameters and are shown in Figures 5 and 6, respectively.

$$\text{(A) Y-intercept vs. } \frac{1}{(S_2)}; \qquad \text{Y-intercept} = \frac{1}{V_{max,f}} + \frac{K_{s_2}}{V_{max,f}}\frac{1}{(S_2)} \tag{92}$$

$$\text{(B) Slope vs. } \frac{1}{(S_2)}: \qquad \text{Slope} = \frac{K_{s_1}}{V_{max,f}} + \frac{\bar{K}_{s_1}K_{s_2}}{V_{max,f}}\frac{1}{(S_2)} \tag{93}$$

Thus, from both plots, all four constants, K_{s1}, $\bar{K}_{s1}$, K_{s2}, and $V_{max,f}$ may be evaluated. Similarly, when (S_2) is varied and (S_1) is fixed:

$$\frac{v_f}{V_{max,f}} = \frac{1}{\left(1 + \frac{K_{s_1}}{(S_1)}\right) + \frac{K_{s_2}}{(S_2)}\left(1 + \frac{\bar{K}_{s_1}}{(S_1)}\right)} \tag{94}$$

or

$$\frac{1}{v_f} = \frac{1}{V_{max,f}}\left(1 + \frac{K_{s_1}}{(S_1)}\right) + \frac{K_{s_2}}{(S_2)}\frac{1}{V_{max,f}}\left(1 + \frac{\bar{K}_{s_1}}{(S_1)}\right) \tag{95}$$

and a plot of $\frac{1}{v_f}$ vs. $\frac{1}{(S_2)}$ is given in Figure 7. The secondary plots from Figure 7 are based on Equations 96 and 97.

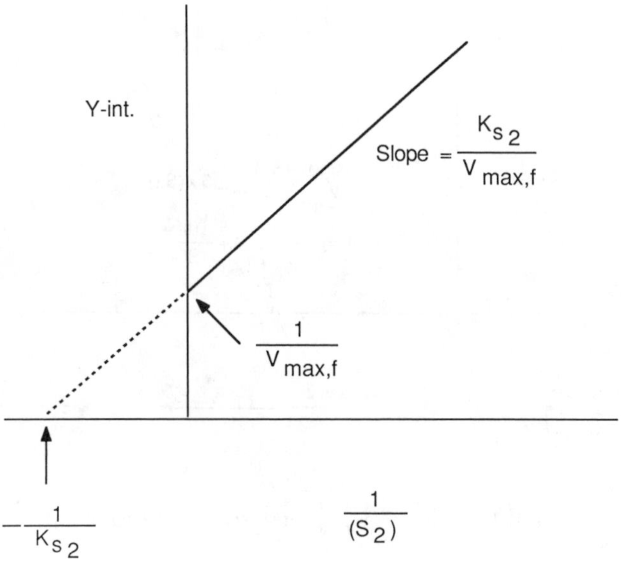

FIGURE 5. Secondary plot (from Figure 4) of Y-intercept vs $\frac{1}{(S_2)}$; based on Y-intercept $= \frac{1}{V_{max,f}} + \frac{K_{S_2}}{V_{max,f}} \frac{1}{(S_2)}$ (Equation 92).

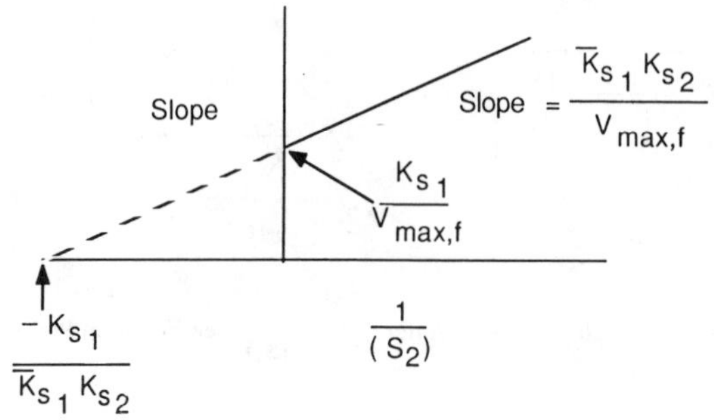

FIGURE 6. Secondary plot of slope vs. $\frac{1}{(S_2)}$ (from Figure 4); based on slope $\frac{K_{S_1}}{V_{max,f}} + \frac{\overline{K}_{s_1}K_{s_2}}{V_{max,f}} \frac{1}{(S_2)}$ (Equaton 93).

$$\text{Y-intercept} = \frac{1}{V_{max,f}} + \frac{1}{V_{max,f}} \frac{K_{s_1}}{(S_1)} \tag{96}$$

and

$$\text{slope} = \frac{K_{s_2}}{V_{max,f}} + \frac{K_{s_2}\overline{K}_{s_1}}{V_{max,f}} \frac{1}{(S_1)} \tag{97}$$

From Figure 8A and B $\left(\text{i.e., Y-int. vs. } \perp \text{ and slope vs. } \frac{1}{(S_1)}\right)$ again, all four values, K_{s1}, $\bar{K}_{s1}$, K_{s2}, and $V_{max,f}$, may be obtained.

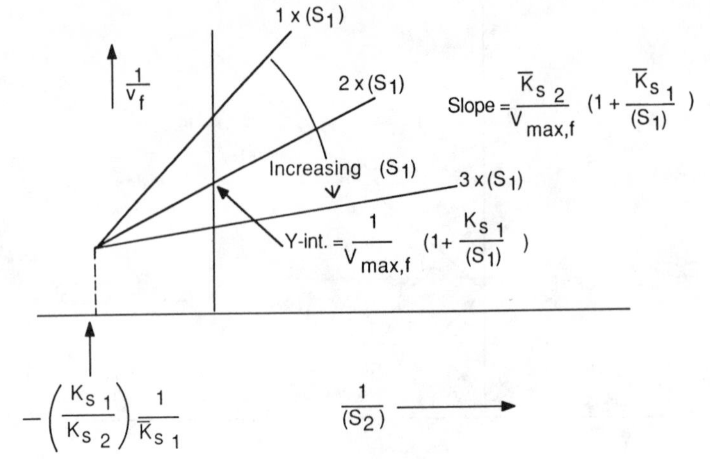

FIGURE 7. Plots of $\frac{1}{v_f}$ vs. $\frac{1}{(S_2)}$ at several fixed values of (S_1).

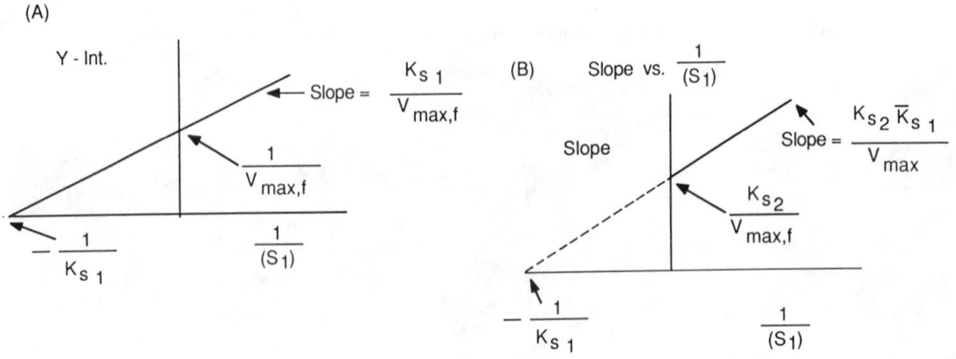

FIGURE 8. Secondary plots of Figure 7. (A) Y-intercept vs. $\frac{1}{(S_1)}$ (see Equation 96) and (B) slope vs. $\frac{1}{(S_1)}$ where

$$\text{slope} = \frac{K_{s2}}{V_{max,f}} + \frac{K_{s_2}\bar{K}_{s_1}\frac{1}{(S_1)}}{V_{max,f}}$$ (Equation 97).

Similarly, equations for the initial reverse reaction may be set up. If one divides Equation 86 by $\frac{V_{max,f}}{K_{eq}}$, one can place it in an overall velocity expression for the reverse reaction.

$$v = \frac{V_{max,r}K_{eq}\left[(S_1)(S_2) - \frac{(P_1)(P_2)}{K_{eq}}\right]}{\begin{array}{l} \frac{V_{max,r}}{V_{max,f}}K_{eq}\bar{K}_{s_1}K_{s_2} + \frac{V_{max,r}}{V_{max,f}}K_{eq}K_{s_2}(S_1) \\ + \frac{V_{max,r}}{V_{max,f}}K_{eq}K_{s_1}(S_2) + K_{p_1}(P_2) + K_{p_2}(P_1) \\ + \frac{V_{max,r}}{V_{max,r}}K_{eq}(S_1)(S_2) + \frac{K_{P_1}}{\bar{\bar{K}}_{s_1}}(S_1)(P_2) \\ + \frac{V_{max,r}}{V_{max,f}}K_{eq}\frac{K_{s_1}}{\bar{\bar{K}}_{P_1}}(S_2)(P_1) + (P_2)(P_1) \end{array}} \qquad (98)$$

To find the equivalence of, e.g., the constant term, write it out in ''coefficient'' form. Thus,

$$\frac{V_{max,r}}{V_{max,f}} K_{eq}\bar{K}_{s_1}K_{s_2} = \frac{\left(\dfrac{num_2}{coef_{p_1p_2}}\right)}{\left(\dfrac{num_1}{coef_{s_1s_2}}\right)} \left(\frac{num_1}{num_2}\right)\left(\frac{constant}{coef_{s_1}}\right)\left(\frac{coef_{s_1}}{coef_{s_1s_2}}\right)$$

$$= \left(\frac{constant}{coef_{p_1p_2}}\right) = \frac{constant}{coef_{p_1}} K_{p_2} = \bar{K}_{p_1}K_{p_2}$$

Therefore,

$$K_{eq} = \frac{\bar{K}_{p_1}K_{p_2}}{\bar{K}_{s_1}K_{s_2}}\left(\frac{V_{max,f}}{V_{max,r}}\right) \tag{99}$$

which is a Haldane relation.

If one adds new definitions, which will be required for product inhibition studies and which are purely arbitrary, then other Haldane relations will be forthcoming:

$$\bar{K}_{s_2} = \frac{coef_{p_1}}{coef_{s_2p_1}} = \frac{constant}{coef_{s_2}} = \frac{k_{-1}}{k_{+2}}$$

$$\bar{K}_{p_2} = \frac{coef_{s_1}}{coef_{s_1p_2}} = \frac{constant}{coef_{p_2}} = \frac{k_{+3}}{k_{-2}} \tag{100}$$

(Note that neither $\bar{K}_{s2}$ nor $\bar{K}_{p2}$ are true thermodynamic dissociation constants, since the true dissociation constants for the ES_2 and EP_2 complexes are assumed to approach infinity by this mechanism).

There are actually 16 Haldane relations or equations, including the following four equations (where $\left(\dfrac{V_{max,f}}{V_{max,r}}\right)$ is raised to the first power and which requires these arbitrary definitions of $\bar{K}_{s2}$ and $\bar{K}_{p2}$):

$$K_{eq} = \frac{V_{max,f}}{V_{max,r}}\frac{K_{p_2}}{K_{s_1}}\frac{\bar{K}_{p_1}}{\bar{K}_{s_2}} = \frac{V_{max,f}}{V_{max,r}}\frac{K_{p_2}\bar{K}_{p_1}}{K_{s_1}\bar{K}_{s_2}} = \frac{V_{max,f}}{V_{max,r}}\frac{\bar{K}_{p_2}K_{p_1}}{K_{s_1}\bar{K}_{s_2}} = \frac{V_{max,f}}{V_{max,r}}\frac{\bar{K}_{p_2}K_{p_1}}{\bar{K}_{s_1}K_{s_2}} \tag{101}$$

From the Haldane relations, the following equalities are evident (see also Figure 4):

$$\bar{K}_{s_1}K_{s_2} = K_{s_1}\bar{K}_{s_2}$$

$$K_{p_2}\bar{K}_{p_1} = \bar{K}_{p_2}K_{p_1} \tag{102}$$

We will return to this topic when we discuss the overall velocity equation for the so-called ''Ordered Bi Bi Reaction''. Initial velocity studies will *not* distinguish the Theorell-Chance mechanism from the ordered Bi Bi reaction as we will see later; however, the overall velocity expression for the Theorell-Chance mechanism lacks any $(S_1S_2P_2)$ and $(S_2P_1P_2)$ terms in the denominator of Equation 82, leading, therefore, to different product inhibition patterns. Also, we will see later, product inhibition data in this case (i.e., the Theorell-Chance mechanism) will identify only S_1-P_1 and S_2-P_2 pairs and will not disclose the order of substrate addition and product release. Equilibrium binding studies will permit identification of S_1 and P_1, since only these ligands bind to the free (E) species.

A recent study of a Theorell-Chance Kinetic mechanism may be found in the reports by Gates and Northrop.[994-996]

2. Dalziel's Empirical Approach

Dalziel[44] preferred a more empirical approach to that which we have discussed and which has dealt with the overall velocity expression (which has really been Cleland's approach[43]).

Thus, we return to Equation 82.

For the forward reaction v_f (i.e., with $P_1 = P_2 = 0$), Dalziel expressed it in terms of reciprocal velocities:

$$\frac{E_t}{v_f} = \frac{1}{k_{+3}} + \frac{1}{k_{+1}(S_1)} + \frac{1}{k_{+2}(S_2)} + \frac{k_{-1}}{k_{+1}k_{+2}(S_1)(S_2)} \tag{103}$$

and for the reverse reaction (i.e., $S_1 = S_2 = 0$):

$$\frac{E_t}{v_r} = \frac{1}{k_{-1}} + \frac{1}{k_{-3}(P_1)} + \frac{1}{k_{-2}(P_2)} + \frac{k_{+3}}{k_{-2}k_{-3}(P_1)(P_2)} \tag{104}$$

Because of the symmetry of the mechanism, the *form* of the rate expressions (written in reciprocal fashion as one would use for Lineweaver-Burk plots) is the same for forward and reverse reactions. If either substrate concentration is varied, with the other held constant and $\frac{1}{v}$ vs. $\frac{1}{(S)}$ plotted, linear plots are to be expected. Dalziel then proposed that a *general expression* could be written for two-substrate reactions (where the expressions did not involve squared terms):

$$\frac{E_t}{v_{initial}} = \phi_0 + \frac{\phi_1}{(S_1)} + \frac{\phi_2}{(S_2)} + \frac{\phi_{12}}{(S_1)(S_2)} \tag{105}$$

for either forward or reverse directions, where the Φs are considered to be kinetic coefficients and are functions of particular velocity constants.

Designate Φ_0, Φ_1, Φ_2, and Φ_{12} for the forward reaction and Φ'_0, Φ'_1, Φ'_2, and Φ'_{12} for the reverse reaction.

This is, of course, analogous to Alberty's general equation[51,52] for first-degree reactions, which in reciprocal form would be

$$\frac{V_{max,f}}{v_f} = 1 + \frac{K_1}{(S_1)} + \frac{K_2}{(S_2)} + \frac{K_{12}}{(S_1)(S_2)} \tag{106}$$

and we may derive the relations between the Φs and Alberty's "Michaelis constants" and maximal velocities for forward and reverse direction as Dalziel[44] did for several mechanisms.

In the case of the Theorell-Chance mechanism *only*:

$$\phi_0 = \frac{1}{k_{+3}};\ \phi_1 = \frac{1}{k_{+1}};\ \phi_2 = \frac{1}{k_{+2}};\ \text{and } \phi_{12} = \frac{k_{-1}}{k_{+1}k_{+2}}$$

$$\phi'_0 = \frac{1}{k_{-1}};\ \phi'_1 = \frac{1}{k_{-3}};\ \phi'_2 = \frac{1}{k_{-2}};\ \text{and } \phi'_{12} = \frac{k_{+3}}{k_{-2}k_{-3}} \tag{107}$$

There are complementary relationships between the kinetic coefficients of the forward and reverse direction and for this case:

$$\phi_0 = \frac{\phi_1'\phi_2'}{\phi_{12}'} = \frac{1}{k_{+3}}$$

$$\phi_0' = \frac{\phi_1\phi_2}{\phi_{12}} = \frac{1}{k_{-1}} \tag{108}$$

Substitute definitions for Equation 108:

$$\frac{\dfrac{1}{k_{-3}} \cdot \dfrac{1}{k_{-2}}}{\dfrac{k_{+3}}{k_{-2}k_{-3}}} = \frac{1}{k_{+3}} = \phi_0 \tag{108a}$$

Dalziel tabulated these complementary relationships for several mechanisms and used these relations as distinguishing features of various mechanisms. Also, he derived the Haldane relations and for the case of the Theorell-Chance mechanism:

$$K_{eq} = \frac{\phi_{12}'}{\phi_{12}} = \frac{\phi_0'\phi_1'\phi_2'}{\phi_0\phi_1\phi_2} \tag{109}$$

Proof for Equation 109:

$$K_{eq} = \frac{\dfrac{k_{+3}}{k_{-2}k_{-3}}}{\dfrac{k_{-1}}{k_{+1}k_{+2}}} = \frac{k_{+1}k_{+2}k_{+3}}{k_{-1}k_{-2}k_{-3}} \tag{109a}$$

Another set of requirements are laid down for this particular mechanism — that it must fulfill these Haldane relations. Similarly, Dalziel tabulated the Haldane relations and the relationships between the kinetic coefficients for several different mechanisms.

To evaluate the kinetic coefficients, Dalziel simply rearranged his general equation for a two-substrate reaction of the first degree in (S_1) and (S_2), i.e.,:

$$\frac{E_t}{v_f} = \phi_0 + \frac{\phi_1}{(S_1)} + \frac{\phi_2}{(S_2)} + \frac{\phi_{12}}{(S_1)(S_2)} \tag{110}$$

Factor out $\dfrac{1}{(S_1)}$:

$$\frac{E_t}{v_f} = \left(\phi_0 + \frac{\phi_2}{(S_2)}\right) + \left(\phi_1 + \frac{\phi_{12}}{(S_2)}\right)\frac{1}{(S_1)} \tag{111}$$

and he plotted the so-called primary plots of $\dfrac{E_t}{v_f}$ vs. $\dfrac{1}{(S_1)}$ at several fixed values of (S_2) as we did before in the case of the Cleland-type plots (see Figure 9). The secondary plots were conducted of the Y-intercepts and slopes vs. $\dfrac{1}{(S_2)}$.

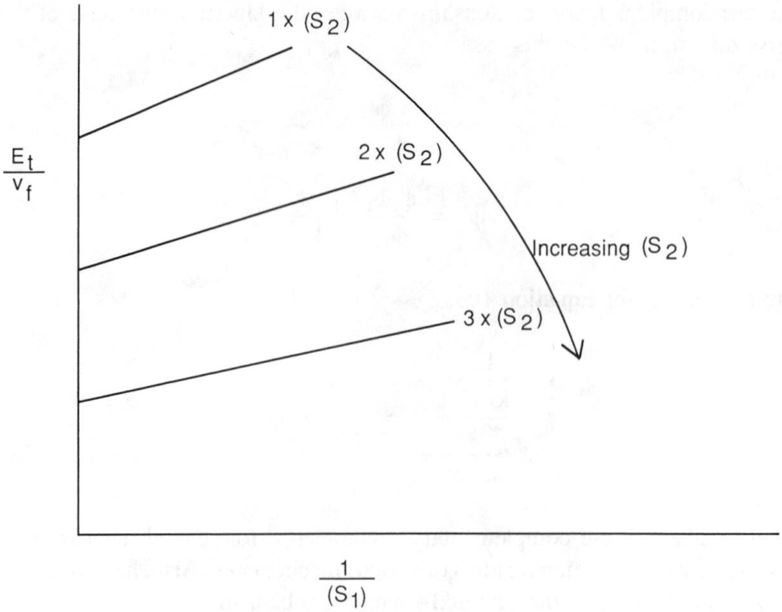

FIGURE 9. Dalziel primary plots of $\frac{E_t}{v_f}$ vs. $\frac{1}{(S_1)}$ at several fixed values of (S_2).

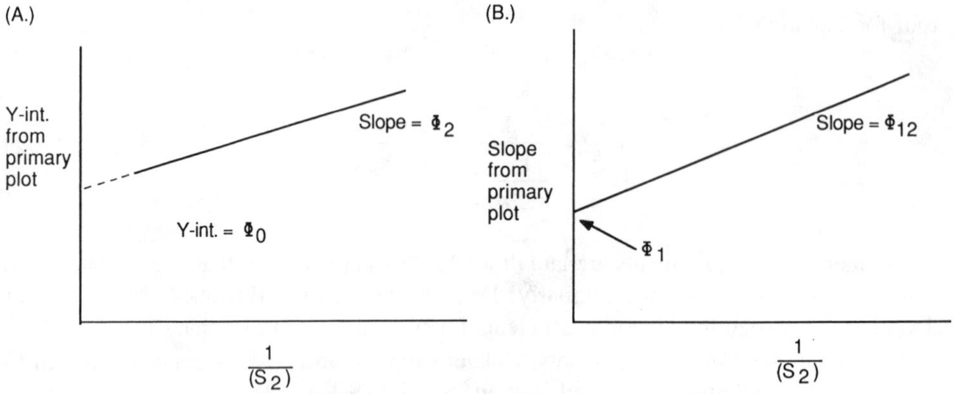

FIGURE 10. Secondary plots from Figure 9.

Thus,

$$\text{Y-intercept} = \phi_0 + \frac{\phi_2}{(S_2)}; \quad \text{Slope} = \phi_1 + \frac{\phi_{12}}{(S_2)} \tag{112}$$

Similarly, if $\frac{1}{v_f}$ vs. $\frac{1}{(S_2)}$ were plotted at fixed values for (S_1)

$$\frac{E_t}{v_f} = \left(\phi_0 + \frac{\phi_1}{(S_1)}\right) + \left(\phi_2 + \frac{\phi_{12}}{(S_1)}\right)\frac{1}{(S_2)} \tag{113}$$

in the primary plots, then the secondary plots would yield Figure 11.

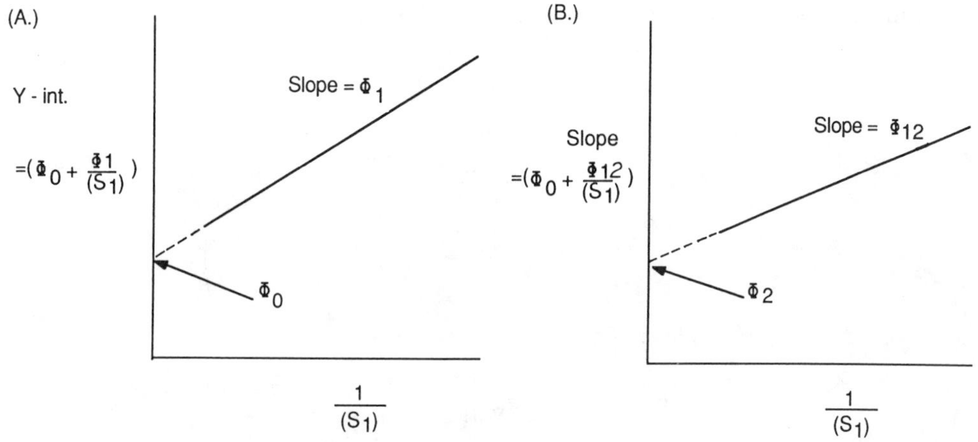

FIGURE 11. Secondary plots of Y-intercepts and slopes from Equation 113.

By repeating the procedure for the reverse reaction, all *eight* coefficients can be obtained, and for the case of the Theorell-Chance mechanism, all *six* velocity constants can be calculated.

Since

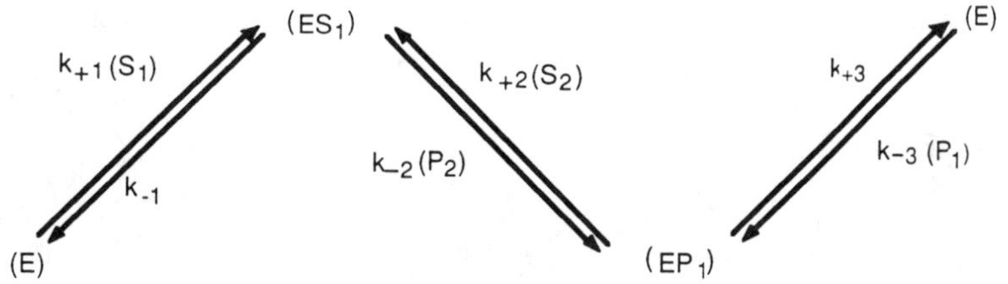

SCHEME XV.

then

$$\phi_0 = \frac{1}{k_{+3}};\quad \phi_1 = \frac{1}{k_{+1}};\quad \phi_2 = \frac{1}{k_{+4}};\quad \text{and } \phi_{12} = \frac{k_{-1}}{k_{+1}k_{+2}}$$

$$\phi_0' = \frac{1}{k_{-1}};\quad \phi_1' = \frac{1}{k_{-3}};\quad \phi_2' = \frac{1}{k_{-2}};\quad \text{and } \phi_{12}' = \frac{k_{+3}}{k_{-2}k_{-3}} \qquad (114)$$

If one could measure $\frac{k_{-1}}{k_{+1}}$ directly or $\frac{k_{+3}}{k_{-3}}$ directly by an alternate technique, i.e., the dissociation constants for the E·DPN complex and the E·DPNH complex, then one could confirm the mechanism by comparing the calculated values (from kinetics) with the measured thermodynamic values.

Theorell attempted this by making use of the spectrophotometric shift which occurred when liver alcohol dehydrogenase and DPNH combined (i.e., E·DPNH)[55] and calculated its dissociation constant; later he made use of fluorometric techniques. With this approach of Theorell in mind, and utilizing their kinetic treatment as he had done for the Theorell-Chance mechanism, Dalziel attempted to subdivide the steady-state treatments for the cases which conform to his general equation (for $S_1 + S_2 \rightleftarrows P_1 + P_2$) into four types of kinetic mechanisms, and summarized the following information:

or in the Cleland shorthand notation commonly employed today:

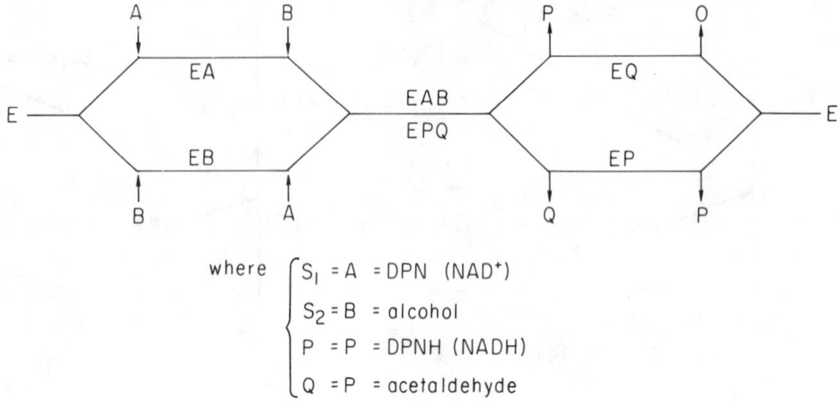

SCHEME XVIII.

This mechanism is similar to the one we discussed earlier, for the aldolase-type mechanism, except that in this case there are two ternary complexes, ES_1S_2 and EP_1P_2, and the forward and reverse directions are equivalent with two substrates and two products.

The nomenclature of Kuby et al.[58] is probably easier to use for rapid or quasi-equilibrium systems and was employed for the glucose 6-phosphate dehydrogenase case.

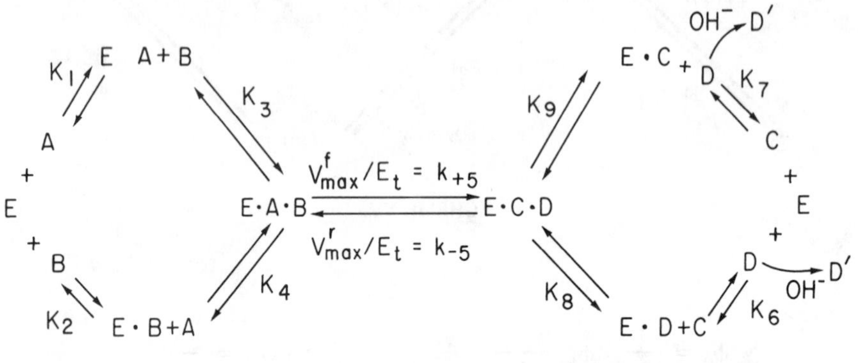

SCHEME XIX.

where A = $NADP^+$, C = NADPH, B = β-D-glucose 6-phosphate, D = 6-phosphoglucono-δ-lactone, i.e., product of A = C and product of B = D and vice versa. (Note: D′ is the nonenzymatic hydrolysis product of D, i.e., 6-phosphogluconic acid.)

To derive the expressions for a rapid or quasi-equilibrium mechanism is relatively simple, for example consider only the forward reaction with k_{+5} (EAB) as the rate-limiting step.

The conservation equation for the enzyme is

$$E_t = E + EA + EB + EAB \tag{115}$$

and to express the equivalent equation for E_t in terms of only (EAB), substitute from the various expressions for equilibria; i.e.,

$$\frac{(E)(A)}{(EA)} = K_1;\quad K_2 = \frac{(E)(B)}{(EB)};\quad K_3 = \frac{(EA)(B)}{(EAB)};\quad \text{and } K_4 = \frac{(EB)(A)}{(EAB)}$$

1. The meanings of the kinetic coefficients (i.e., the Φs) in terms of their velocity constants (the ks)
2. The relations between the coefficients, i.e. the Φs
3. The Haldane relations

Points 2 and 3 were used in an effort to demonstrate how one could distinguish kinetically between the four types of mechanisms he described.[44]

Let us turn to a particular case which interested Dalziel, that is, the case of the yeast alcohol dehydrogenase-catalyzed reaction.

Nygaard and Theorell[56] had applied the Haldane relations as derived by Alberty[57] to initial rate data in order to calculate K_{eq} and compared the results with the values directly measured. This enabled them to reach a decision that a "rapid-equilibrium" mechanism appeared to be a satisfactory hypothesis, and that the Theorell-Chance mechanism could be rejected for the yeast enzyme, in contrast to the liver enzyme.

3. Quasi- or Rapid Equilibrium Kinetics for a Random Two-Substrate Two-Product Reaction

The quasi- or rapid equilibrium random mechanism would be as follows (Scheme 16):

$$E + S_1 \overset{K_1}{\rightleftarrows} ES_1 \qquad EP_2 \overset{K_9}{\rightleftarrows} E + P_2$$

$$E + S_2 \overset{K_2}{\rightleftarrows} ES_2 \qquad EP_1 \overset{K_8}{\rightleftarrows} E + P_1$$

$$ES_1 + S_2 \overset{K_3}{\rightleftarrows} ES_1S_2 \qquad EP_1P_2 \overset{K_7}{\rightleftarrows} EP_2 + P_1$$

$$ES_2 + S_1 \overset{K_4}{\rightleftarrows} ES_1S_2 \qquad EP_1P_2 \overset{K_6}{\rightleftarrows} EP_1 + P_2$$

$$ES_1S_2 \underset{k_{-5}}{\overset{k_{+5}}{\rightleftarrows}} EP_1P_2$$

SCHEME XVI.

Where S_1 and P_1 would represent DPN^+ (NAD^+) and DPNH (NADH), respectively, and S_2 and P_2 representa alcohol and acetaldehyde, respectively, where all the capital Ks are dissociation constants and all equilibria, except reaction number 5 are considered to be adjusted so rapidly that only k_{+5} and k_{-5} are considered to be rate-limiting steps for the forward and reverse reactions, respectively, and quasi-equilibria exist for all other steps.

Symbolically, the mechanism can be written:

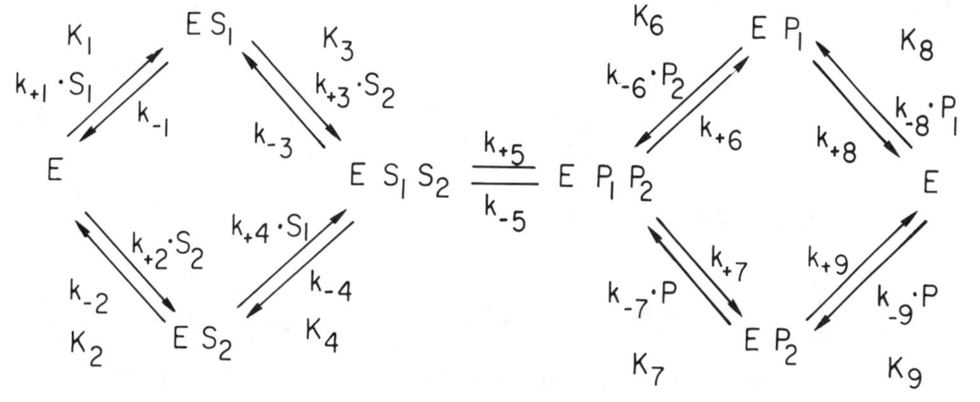

SCHEME XVII.

Note that the symmetry requires that $K_1 \cdot K_3 = K_2 \cdot K_4$ and all four equilibrium constants are not independent.

The equation for E_t therefore, can be expressed in terms of either K_1 or K_2 (and the other constants minus K_1 or K_2).

Therefore,

$$E_t = \left[1 + \frac{K_4}{(A)} + \frac{K_3}{(B)} + \frac{K_1K_3}{(A)(B)}\right](EAB) \tag{116}$$

Since

$$v_0^f = k_{+5}(EAB) \tag{117}$$

and substituting:

$$v_0^f = \frac{V_{max}^f}{1 + \frac{K_4}{(A)} + \frac{K_3}{(B)} + \frac{K_1K_3}{(A)(B)}}; \quad \begin{matrix} K_1K_3 = K_2K_4 \\ V_{max} = k_{+5}E_t \end{matrix} \tag{118}$$

Similarly, for the reverse reaction, the initial velocity, v_0^r is

$$v_0^r = \frac{V_{max}^r = k_{-5}E_t}{1 + \frac{K_8}{(C)} + \frac{K_9}{(D)} + \frac{K_7K_9}{(C)(D)}} \tag{119}$$

and $K_7K_9 = K_6\,K_8$.

4. Overall Velocity Expressions for Random Two-Substrate Two-Product Reactions

The overall velocity expression is relatively easy to derive,[49] e.g., for the case of the mechanism detailed in Scheme 19 that is *without dead end complexes* (which we will discuss shortly) in terms of the forward velocity

$$v^f = \frac{V_{max}^f\left[(A)(B) - \frac{(C)(D)}{K'_{eq}}\right]}{K_1K_3 + K_3(A) + (A)(B) + K_4(B) + \frac{K_1K_3(C)(D)}{K_7K_9} + \frac{K_1K_3(C)}{K_7} + \frac{K_1K_3K_8(D)}{K_7K_9}} \tag{120}$$

where

$$K'_{eq} = \frac{(C)(D)}{(A)(B)}; \quad \begin{Bmatrix} K_1K_3 = K_2K_4 \\ K_6K_8 = K_7K_9 \end{Bmatrix} \tag{121}$$

and there are four Haldane expressions:

$$K'_{eq} = \frac{V_{max}^f}{V_{max}^r}\frac{K_7K_9}{K_1K_3} = \frac{V_{max}^f}{V_{max}^r}\frac{K_6K_8}{K_2K_4} = \frac{V_{max}^f}{V_{max}^r}\frac{K_7K_9}{K_2K_4} = \frac{V_{max}^f}{V_{max}^r}\frac{K_6K_8}{K_1K_3} \tag{122}$$

We can also write Equation 120 in terms of both V_{max} terms (V_{max}^f and V_{max}^r). Thus, multiply numerator and denominator by V_{max}^r:

$$v = \frac{V^f_{max}V^r_{max}\left[(A)(B) - \frac{(C)(D)}{K'_{eq}}\right]}{K_1K_3V^r_{max} + V^r_{max}K_3(A) + V^r_{max}K_4(B) + V^r_{max}(A)(B) + \frac{K_1K_3}{K_7K_9}V^r_{max}(C)(D) + \frac{K_1K_3}{K_7}V^r_{max}(C) + \frac{K_1K_3K_8}{K_7K_9}V^r_{max}(D)} \tag{123}$$

Note: $\frac{K_1K_3}{K_7K_9}V^r_{max} = \frac{V^f_{max}}{K'_{eq}}$; $\frac{K_1K_3}{K_7}V^r_{max} = \frac{K_9V^f_{max}}{K'_{eq}}$; and $\frac{K_1K_3V^r_{max}}{K_7K_3} = \frac{V^f_{max}}{K'_{eq}}$

Therefore,

$$v = \frac{V^f_{max}V^r_{max}\left[(A)(B) - \frac{(C)(D)}{K'_{eq}}\right]}{K_1K_3V^r_{max} + K_3V^r_{max}(A) + V^r_{max}K_4(B) + V^r_{max}(A)(B) + \frac{V^f_{max}}{K'_{eq}}(C)(D) + \frac{K_9V^f_{max}}{K'_{eq}}(C) + \frac{K_8V^f_{max}}{K'_{eq}}(D)} \tag{124}$$

If either or both dead end or *abortive* complexes, E·B·C or E·A·D, could form as they had been deduced for the rabbit muscle ATP-creatine transphosphorylase,[48] then the following equations must be added:

$$(E \cdot B) + (C) \overset{K\alpha}{\rightleftarrows} (E \cdot B \cdot C) \overset{K\beta}{\rightleftarrows} (E \cdot C) + (B)$$

$$(E \cdot A) + (D) \overset{K\gamma}{\rightleftarrows} (E \cdot A \cdot D) \overset{K\delta}{\rightleftarrows} (ED) + (A).$$

SCHEME XX.

and equalities for Equation 121 must be extended to include

$$\begin{cases} K_\alpha K_2 = K_\beta K_7 \\ K_\gamma K_1 = K_\delta K_6 \end{cases} \tag{125}$$

which leads to the overall velocity expression:

$$v^f = \frac{V^f_{max}\left\{(A)(B) - \frac{(C)(D)}{K'_{eq}}\right\}}{K_1K_3 + K_3(A) + K_4(B) + (A)(B) + \frac{K_1K_3(C)(D)}{K_7K_9} + \frac{K_1K_3(C)}{K_7} + \frac{K_1K_3K_8(D)}{K_7K_9} + \frac{K_1K_3(C)(B)}{K_2K\alpha} + \frac{K_3(D)(A)}{K_\gamma}} \tag{126}$$

One notes that *both* overall velocity expressions (Equations 120 and 126), in the case of (C) = D → 0, reduce to the same limiting forward velocity expression we have already presented (and thus, will not be distinguished by initial velocity studies alone) that is,

$$v^f_0 = \frac{V^f_{max}}{1 + \frac{K_4}{(A)} + \frac{K_3}{(B)} + \frac{K_1K_3}{(A)(B)}} \tag{127}$$

or for the reverse reaction, setting the concentrations of products to zero, i.e., (A) = (B) → 0,

$$v_0^r = \frac{V_{max}^r}{1 + \frac{K_8}{(C)} + \frac{K_9}{(D)} + \frac{K_7K_9}{(C)(D)}} \tag{128}$$

However, the product inhibition patterns will be altered and these will distinguish the two cases.[58]

5. Product Inhibition Patterns in the Case of Dead-End Complexes

Thus, for the case of product (C), inhibition with the dead-end complexes of Scheme XX and with D = 0, Equation 126 reduces to

$$\frac{V_{max}^f}{v_0^f} = 1 + \frac{K_3}{(B)} + \frac{K_3}{(B)}\left[\frac{K_1}{(A)} + \frac{K_1(C)}{K_7(A)}\right] + \frac{K_4}{(A)}\left[1 + \frac{(C)}{K_\alpha}\right] \tag{129}$$

and for fixed (B):

$$\frac{1}{v_0^f} = \frac{1}{V_{max}^f}\left[1 + \frac{K_3}{(B)}\right] + \frac{K_4}{(A)}\frac{1}{V_{max}^f}\left[1 + \frac{K_2}{(B)}\left(1 + \frac{(C)}{K_7}\right) + \frac{(C)}{K_\alpha}\right] \tag{130}$$

Thus, (C) would affect only the slope in a $\frac{1}{v_0^f}$ vs. $\frac{1}{(A)}$ plot and would therefore be a "competitive" inhibitor with respect to (A).

Whereas, for fixed (A):

$$\frac{1}{v_0^f} = \frac{1}{V_{max}^f}\left[1 + \frac{K_4}{(A)}\left(1 + \frac{(C)}{K_\alpha}\right)\right] + \frac{K_3}{(B)}\frac{1}{V_{max}^f}\left[1 + \frac{K_1}{(A)}\left(1 + \frac{(C)}{K_7}\right)\right] \tag{131}$$

(C) would affect both slope and intercept and therefore (C) would act as a "noncompetitive" inhibitor with respect to (B).

Now consider the case without dead-end complexes, i.e., Equation 120, the case for inhibition by the product (C) (with D = O); the resulting equation is

$$v_0^f = \frac{V_{max}^f}{\frac{K_1K_3}{(A)(B)} + \frac{K_3}{(B)} + \frac{K_4}{(A)} + \frac{K_1K_3(C)}{K_7(A)(B)} + 1} \tag{132}$$

For fixed (B) and varied (A):

$$\frac{V_{max}^f}{v_0^f} = 1 + \frac{K_3}{(B)} + \frac{K_4}{(A)}\left[1 + \frac{K_3K_1}{(B)K_4} + \frac{K_1K_3(C)}{K_7K_4(B)}\right]$$

$$\frac{1}{v_0^f} = \frac{1}{V_{max}^f}\left(1 + \frac{K_3}{(B)}\right) + \frac{K_4}{V_{max}^f}\frac{1}{(A)}\left[1 + \frac{K_1K_3}{(B)K_4}\left(1 + \frac{(C)}{K_7}\right]\right. \tag{133}$$

which predicts that (C) would be "competitive" vs. (A).

For fixed (A) and varied (B):

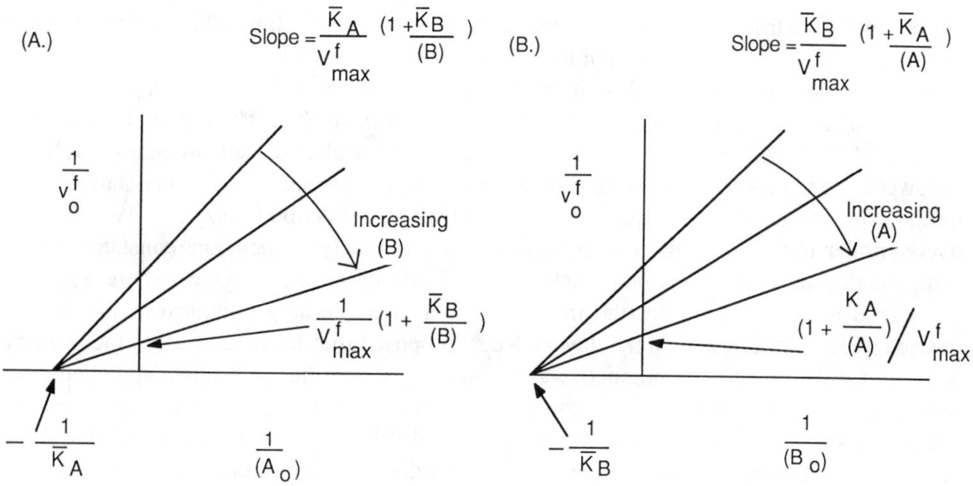

FIGURE 12. Plots of $\frac{1}{v_0}$ vs. $\frac{1}{(A_o)}$ or $\frac{1}{(B_o)}$ with fixed concentrations of complementary substrates for mechanism following Equation 136.

$$\frac{V^f_{max}}{v^f_0} = 1 + \frac{K_4}{(A)} + \frac{K_3}{(B)}\left[1 + \frac{K_1}{(A)} + \frac{K_1}{K_7}\frac{(C)}{(A)}\right]$$

$$\frac{1}{v^f_0} = \frac{1}{V^f_{max}}\left(1 + \frac{K_4}{(A)}\right) + \frac{1}{V_{max}}\frac{K_3}{(B)}\left[1 + \frac{K_1}{(A)}\left(1 + \frac{(C)}{K_7}\right)\right] \tag{134}$$

and again (C) would be "competitive" vs. (B).

6. Special Case of Independent Binding of Two Substrates

In the special case of *independent binding* of A and B, i.e., when

$$K_1 \simeq K_4 = \overline{K}_A$$

$$K_2 \simeq K_3 = \overline{K}_B \tag{135}$$

(see Scheme XIX), the equation for the forward velocity (refer to Equation 127) reduces to

$$v^f_0 = V^f_{max}\left[\frac{\overline{K}_A\overline{K}_B}{(A)(B)} + \frac{\overline{K}_B}{(B)} + \frac{\overline{K}_A}{(A)} + 1\right]^{-1} \tag{136}$$

as found for the glucose 6-phosphate dehydrogenase by Kuby et al.[58]

In this special case of independent binding of either substrate, in the double-reciprocal plots a constant ratio of slope/ordinate-intercept results as shown in Figure 12 for either substrate where

$$\frac{1}{v^f_0} = \frac{1}{V^f_{max}}\left[1 + \frac{\overline{K}_B}{(B)}\right] + \frac{\overline{K}_A}{V^f_{max}}\frac{1}{(A_0)}\left[1 + \frac{\overline{K}_B}{(B)}\right];$$

$$\frac{1}{v^f_0} = \frac{1}{V^f_{max}} = \left[1 + \frac{\overline{K}_A}{(A)}\right] + \frac{\overline{K}_B}{V^f_{max}}\frac{1}{(B_0)}\left[1 + \frac{\overline{K}_A}{(A)}\right]$$

and $\bar{K}_A$ or $\bar{K}_B$ are true thermodynamic dissociation constants which may be verified by direct equilibrium ligand-binding studies of the substrates (as measured by Kuby et al.[58] for glucose 6-phosphate dehydrogenase and by Kuby et al.[59] for rabbit muscle creatine kinase).

It is interesting that in 1962, Dalziel[46] and just prior to this, Theorell and McKinley-McKee,[60,61,813,841,842] came to the conclusion that the liver alcohol dehydrogenase probably followed a similar mechanism as the yeast alcohol dehydrogenase, i.e., a steady-state random mechanism, but not a quasi-equilibrium mechanism, and thus both enzymes were qualitatively similar in their mechanism, but differed quantitatively in their rate constants. Thus, compare the mechanism given in Scheme XVII for the yeast enzyme, where k_{+5} was considered to be rate limiting (the interconversion of the ternary complexes); but for the liver enzyme, Theorell and McKinley-McKee[60] proposed that for certain cases the primary pathway for dissociation of products was the upper one and the mechanism approached an ordered mechanism (we will discuss this shortly) with $k_{+6} \gg k_{+7}$. Also, they suggested that if k_{+5} and k_{-5} were large and not rate limiting, and if $k_{+1} = k_{-7}$, they claimed that these conditions would then reduce the complex second-degree expression for the initial steady-state rate equation[45] to one which approximated the Theorell-Chance equation, i.e., to the simple equation we discussed before for the case where the ternary complexes were neglected. In addition, Dalziel felt that additional qualifications and conditions were necessary to make their assumptions valid, among which conditions, the following was necessary: $k_{+6} \gg k_{+8}$ (see Scheme XVII), that is, in essence that the last step $(EP_1) \xrightarrow{K_{+8}} (E) + (P_1)$, was slow, if not rate limiting. Refer to the original papers for a clearer description of their arguments. However, we will see later a relatively simple way to eliminate the ternary complexes to reduce an ordered mechanism to the Theorell-Chance mechanism.

7. Steady-State Mechanism for a Compulsory Order of Addition of the Two Substrates and Dissociation of the Two Products that is, an Ordered Bi Bi System and Illustration of Cleland's Symbolism

We will turn now to a more complete discussion of this steady-state mechanism, i.e., an ordered Bi Bi system, and an illustration of Cleland's symbolism. Frieden[62] apparently provided the first general mechanism for this type of reaction.

However, we will derive the overall velocity expression using largely Cleland's nomenclature to illustrate his symbolism[43] (except that we will use, e.g., K_{m_a} as the Michaelis constant for A, as recommended by the I.U.B.[63]

Thus, for an ordered Bi Bi system, that is, an ordered reaction that is bireactant in both directions (and one in which the ternary complex EAB can only be formed from EA, or in the reverse reaction where EPQ is formed only from EQ).

$$(E) + (A) \underset{k}{\overset{k}{\rightleftharpoons}} (EA) \qquad (EQ) \underset{k_{-4}}{\overset{k_{+4}}{\rightleftharpoons}} (E) + (Q)$$

$$+ \qquad\qquad +$$

$$B \qquad\qquad P$$

$$k_{-2} \updownarrow k_{+2} \qquad k_{-3} \updownarrow k_{+3}$$

$$(EAB) \underset{k_{-p}}{\overset{k_{+p}}{\rightleftharpoons}} (EPQ)$$

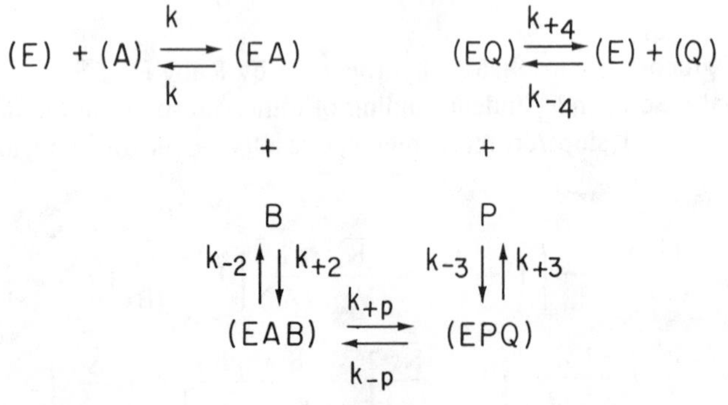

SCHEME XXI.

Assume that the product of A = Q, the product of B = P, and vice versa. It is not always easy to establish the relationships; therefore, product inhibition data may be necessary.

If $k_{+3} \gg k_{-2}$ and $k_{-4} \gg k_{+3}$, then (EAB) → 0 and we have the Theorell-Chance mechanism, as we will see later.)

We can derive the rate expression in two ways: one in which the interconversion of the so-called transitory or central ternary complexes $(EAB) \underset{k_{-p}}{\overset{k_{+p}}{\rightleftarrows}} (EPQ)$ are neglected and in essence replaced only by (EAB), i.e., only one ternary complex. This is the usual method of Cleland. The same form of the velocity expression will result, however, if we incorporate this additional reaction, but the meaning of the kinetic parameters in terms of their kinetic rate constants will differ, as discussed by Plapp.[64]

The Cleland schematic form for the mechanism is

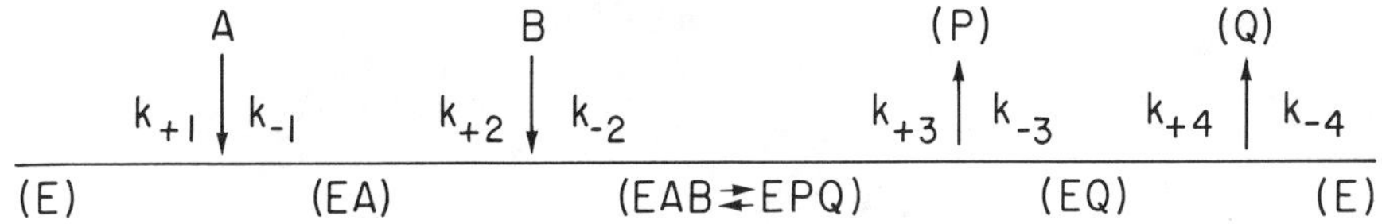

SCHEME XXII.

The basic King-Altman figure has four corners (or n = 4) if we neglect the interconversion of the ternary complexes:

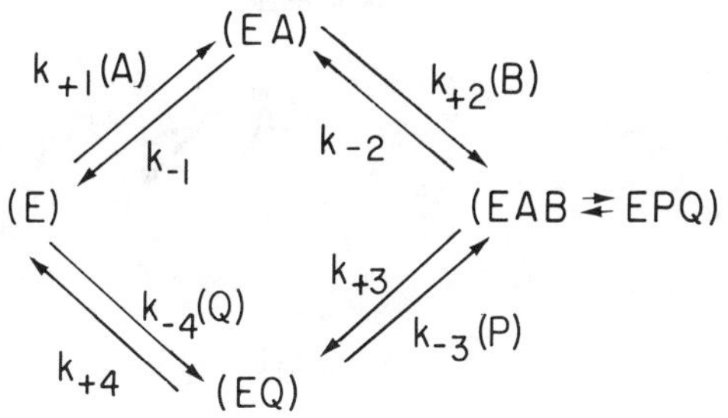

SCHEME XXIII.

Or, it can be written as

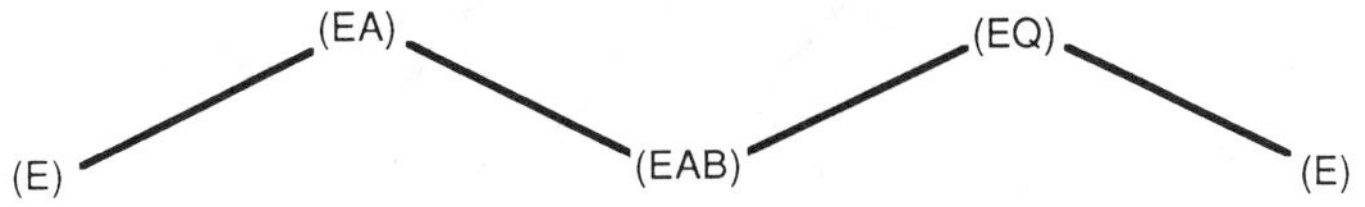

SCHEME XXIIIA.

and therefore, no loops. Note also that (B) cannot form the binary complex (EB) in this mechanism and that (P) does not form an (EP) binary complex in the reverse reaction. Presumably, (A) and (Q) are the related substrate-product pair and (B)-(P) is the second substrate-product pair.

The distribution of enzyme species is obtained from four three-lined interconversion patterns, where n = 4, n − 1 = 3, m = 4.

$$ {}_{m}C_{n-1} = \frac{m!}{(n-1)!(m-n+1)!} = \frac{4!}{3!(4-4+1)} = 4 $$

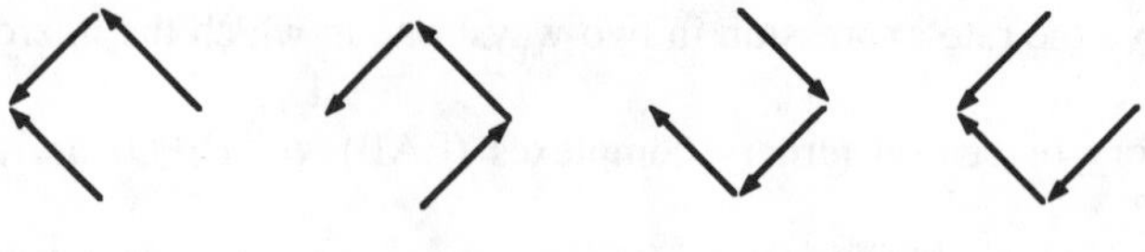

SCHEME XXIV.

where,

$$ \frac{(E)}{E_t} = \frac{k_{+4}k_{-1}k_{-2} + k_{-3}(P)k_{-2}k_{-1} + k_{+2}(B)k_{+3}k_{+4} + k_{-1}k_{+3}k_{+4}}{\text{denominator}} \tag{137} $$

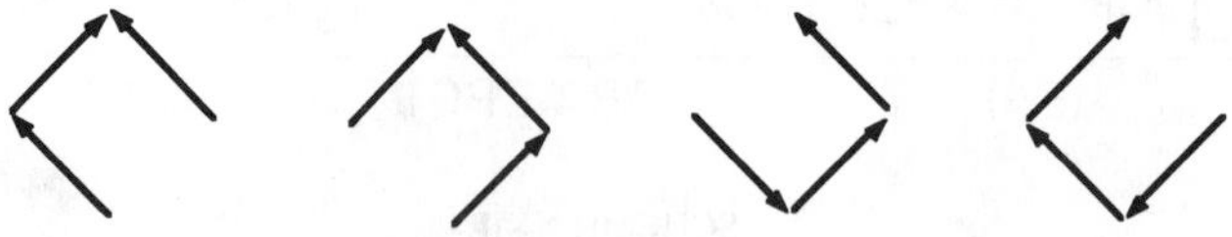

SCHEME XXV.

$$ \frac{(EA)}{E_t} = \frac{k_{+4}k_{+1}(A)k_{-2} + k_{+1}(A)k_{-3}(P)k_{-2} + k_{-4}(Q)k_{-3}(P)k_{-2} + k_{+3}k_{+4}k_{+1}(A)}{\text{denominator}} \tag{138} $$

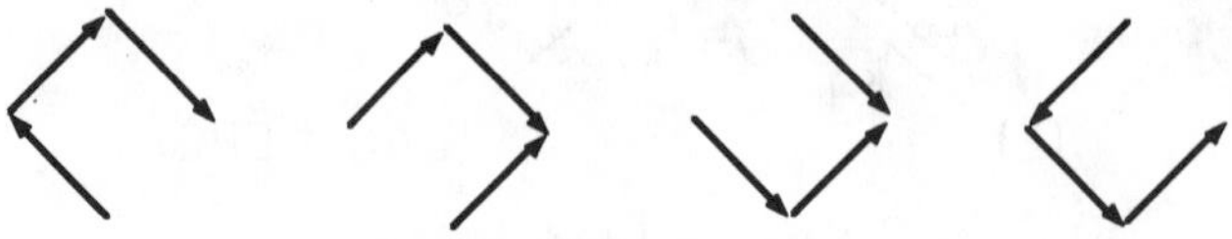

SCHEME XXVI.

$$ \frac{(EAB + EPQ)}{E_t} = \frac{k_{+4}k_{+1}(A)k_{+2}(B) + k_{+1}(A)k_{+2}(B)k_{-3}(P) + k_{-4}(Q)k_{-3}(P)k_{+2}(B) + k_{-1}k_{-4}(Q)k_{-3}(P)}{\text{denominator}} \tag{139} $$

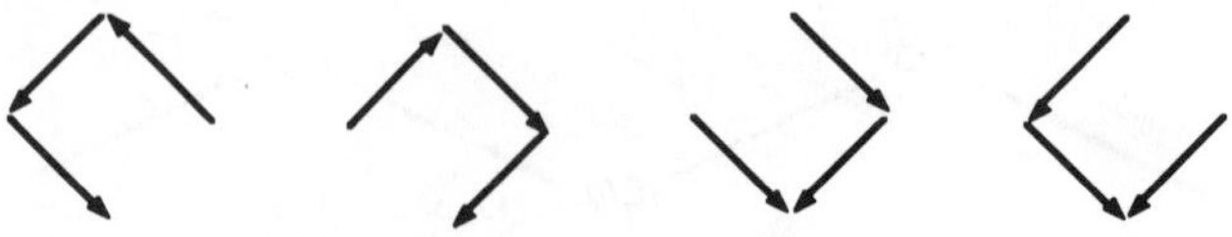

SCHEME XXVII.

$$ \frac{(EQ)}{E_t} = \frac{k_{-4}(Q)k_{-1}k_{-2} + k_{+1}(A)k_{+2}(B)k_{+3} + k_{-4}(Q)k_{+3}k_{+2}(B) + k_{-1}k_{-4}(Q)k_{+3}}{\text{denominator}} \tag{140} $$

Letting $v = k_{+1}\,(E)\,(A) - k_{-1}\,(EA)$ (or for that matter, any other difference between the forward and reverse velocities of a given step), we get, after grouping denominator terms:

$$v = \frac{[k_{+1}(E)(A) - k_{-1}(EA)]}{[(E) + (EA) + (EAB + EPQ) + (EQ)]} E_t \tag{141}$$

and substituting

$$\frac{v}{E_t} = \frac{[k_{+1}k_{+2}k_{+3}k_{+4}(A)(B) - k_{-1}k_{-2}k_{-3}k_{-4}(P)(Q)]}{\begin{array}{l} k_{-1}k_{+4}(k_{-2} + k_{+3}) + k_{+1}k_{+4}(k_{-2} + k_{+3})(A) + k_{+2}k_{+3}k_{+4}(B) + k_{-1}k_{-2}k_{-3}(P) \\ + k_{-1}k_{-4}(k_{-2} + k_{+3})(Q) + k_{+1}k_{+2}(k_{+3} + k_{+4})(A)(B) + k_{+1}k_{-2}k_{-3}(A)(P) \\ + k_{-3}k_{-4}(k_{-1} + k_{-2})(P)(Q) + k_{+2}k_{+3}k_{-4}(B)(Q) + k_{+1}k_{+2}k_{-3}(A)(B)(P) \\ + k_{+2}k_{-3}k_{-4}(B)(P)(Q) \end{array}} \tag{142}$$

First, express the equation in terms of the shorthand nomenclature, or coefficients, in the manner of Cleland.[43] (Incidentally, the (ABP) and (BPQ) terms are absent in the Theorell-Chance mechanisms because the ternary complex → 0.)

Thus,

$$v = \frac{[num_1(A)(B) - num_2(P)(Q)]}{\begin{array}{l} constant + coef_A(A) + coef_B(B) + coef_P(P) + coef_Q(Q) + coef_{AB}(A)(B) \\ + coef_{AP}(A)(P) + coef_{PQ}(P)(Q) + coef_{BQ}(B)(Q) + coef_{ABP}(A)(B)(P) \\ + coef_{BPQ}(B)(P)(Q) \end{array}} \tag{143}$$

Now, if the numerator and denominator of Equation 142 are multiplied by

$$\frac{(num_2)}{(coef_{AB})(coef_{PQ})} = \frac{k_{-1}k_{-2}k_{-3}k_{-4}}{[k_{+1}k_{+2}(k_{+3} + k_{+4})][(k_{-3}k_{+4}(k_{-1} + k_{-2})]}$$

and multiply also the (PQ) numerator term and the (P), (Q), (PQ) (A) (P) and (BPQ) denominator terms by $\left(\frac{num_1}{num_1}\right)$:

$$v = \frac{\left(\frac{num_1 num_2(A)(B)}{coef_{AB}coef_{PQ}} - \frac{num_2 num_2}{coef_{AB}coef_{PQ}}(P)(Q)\left(\frac{num_1}{num_1}\right)\right)}{\begin{array}{l} constant \times \frac{num_2}{coef_{AB}coef_{PQ}} + coef_A(A) \times \frac{num_2}{coef_{AB}coef_{PQ}} + coef_B(B)\frac{num_2}{coef_{AB}coef_{PQ}} \\ + coef_P(P)\frac{num_2}{coef_{AB}coef_{PQ}}\left(\frac{num_1}{num_1}\right) + coef_Q(Q)\frac{num_2}{coef_{AB}coef_{PQ}}\frac{num_1}{num_1} \\ + coef_{AB}(A)(B)\frac{num_2}{coef_{AB}coef_{PQ}} + coef_{AP}(A)(P)\frac{num_2}{coef_{AB}coef_{PQ}}\frac{num_1}{num_1} \\ + coef_{PQ}(P)(Q)\frac{num_2}{coef_{AB}coef_{PQ}}\frac{num_1}{num_1} + coef_{BQ}(B)(Q)\frac{num_2}{coef_{AB}coef_{PQ}} \\ + coef_{ABP}(A)(B)(P)\frac{num_2}{coef_{AB}coef_{PQ}} + coef_{BPQ}(B)(P)(Q)\frac{num_2}{coef_{AB}coef_{PQ}}\frac{num_1}{num_1} \end{array}} \tag{144}$$

Now, the (A) (B) numerator term and the (A), (B), (AB), and (ABP) denominator terms can be defined directly in terms of kinetic parameters or constants. The (P) (Q) numerator term and the (P), (Q), (PQ), and (BPQ) denominator terms are now also reduced to kinetic parameters or constants (and K_{eq}) after the additional multiplication of $\frac{num_1}{num_1}$ which has already been done.

The (A) (P) term simplifies after multiplication by $\frac{num_1}{num_1}$ and substitution of $\frac{coef_P}{K_{ia}}$ for

$coef_{AP}$. (We have used $\bar{K}_A$ for k_{ia} before.) The (BQ) term simplifies after substitution of $coef_B/K_{mA}$ for $coef_{AB}$. The constant term in the denominator is simplified by substituting $coef_A/K_{mB}$ for $coef_{AB}$ and K_{ib} is set equal to $\frac{coef_{PQ}}{coef_{BPQ}}$, although there is no E·B complex (nor E·P complex, but one can define a $K_{ip} = \frac{coef_{AB}}{coef_{ABP}}$).

Now the equation can be written in terms of the kinetic constants of Cleland (the kinetic parameters): V^f_{max} and V^r_{max}, the maximal velocities; K_{mA}, K_{mB}, K_{mP}, and K_{mQ}, the Michaelis constants; K_{ia}, K_{ib}, K_{ip}, and K_{iq}, the product inhibition constants (we used $\bar{K}_A$, $\bar{K}_B$, $\bar{K}_P$, and $\bar{K}_Q$ for these terms before); where only K_{ia} and K_{iq} are the binary dissociation constants for EA and EQ, respectively, but not K_{ib} or K_{ip} for this mechanism. We can also define a dissociation constant from the ternary complex shortly.[65]

Thus,

$$\frac{num_1}{num_2} = K_{eq};\quad V^f_{max} = \frac{num_1}{coef_{AB}};\quad V^r_{max} = \frac{num_2}{coef_{PQ}}$$

$$K_{mB} = \frac{coef_A}{coef_{AB}};\quad K_{mA} = \frac{coef_B}{coef_{AB}};\quad K_{ia} = \frac{coef_P}{coef_{AB}} = \frac{k_{-1}}{k_{+1}} = \frac{constant}{coef_A} \tag{145}$$

$$v = \frac{\left[V^f_{max}V^r_{max}(A)(B) - V^f_{max}V^r_{max}\frac{1}{K_{eq}}(P)(Q)\right]}{\begin{aligned}&K_{mB}K_{ia}V^r_{max} + K_{mB}V^r_{max}(A) + K_{mA}V^r_{max}(B) + V^f_{max}\frac{K_{mQ}}{K_{eq}}(P) + V^f_{max}\frac{K_{mp}(Q)}{K_{eq}}\\&+ V^r_{max}(A)(B) + \frac{V^f_{max}K_{mQ}(A)(P)}{K_{eq}K_{ia}} + \frac{V^f_{max}(P)(Q)}{K_{eq}} + \frac{V^r_{max}K_{mA}(B)(Q)}{K_{iq}}\\&+ \frac{V^r_{max}(A)(B)(P)}{K_{ip}} + \frac{V^f_{max}(B)(P)(Q)}{K_{ib}K_{eq}}\end{aligned}} \tag{146}$$

Therefore, rewriting the equation for an ordered Bi Bi system in Cleland's nomenclature

$$v = \frac{V^f_{max}V^r_{max}\left((A)(B) - \frac{(P)(Q)}{K_{eq}}\right)}{\begin{aligned}&V^r_{max}K_{ia}K_{mB} + V^r_{max}K_{mB}(A) + V^r_{max}K_{mA}(B) + V^f_{max}\frac{K_{mQ}(P)}{K_{eq}} + \frac{V^f_{max}K_{mP}(Q)}{K_{eq}}\\&+ V^r_{max}(A)(B) + \frac{V^f_{max}K_{mQ}(A)(P)}{K_{eq}K_{ia}} + \frac{V^f_{max}(P)(Q)}{K_{eq}} + \frac{V^r_{max}K_{mA}(B)(Q)}{K_{iq}}\\&+ \frac{V^r_{max}(A)(B)(P)}{K_{ip}} + \frac{V^f_{max}(B)(P)(Q)}{K_{ib}K_{eq}}\end{aligned}} \tag{147}$$

where a list of the definitions we have employed is

$$K_{m_A} = \frac{coef_B}{coef_{AB}};\quad K_{m_B} = \frac{coef_A}{coef_{AB}};\quad K_{ia} = \frac{coef_P}{coef_{AP}} = \frac{constant}{coef_A} = \frac{k_{-1}}{k_{+1}}$$

$$K_{ib} = \frac{coef_{PQ}}{coef_{BPQ}} = \frac{k_{-1} + k_{-2}}{k_{+2}};\quad K_{m_P} = \frac{coef_Q}{coef_{PQ}};\quad K_{m_Q} = \frac{coef_Q}{coef_{PQ}}$$

$$K_{ip} = \frac{coef_{AB}}{coef_{ABP}} = \frac{k_{+3} + k_{+4}}{k_{-3}};\quad K_{iq} = \frac{coef_B}{coef_{BQ}} = \frac{constant}{coef_Q} = \frac{k_{+4}}{k_{-4}}$$

$$V^f_{max} = \frac{num_1}{coef_{AB}};\quad V^r_{max} = \frac{num_2}{coef_{PQ}};\quad K_{eq} = \frac{num_1}{num_2} \tag{148}$$

(K_{ia} and K_{iq} are true dissociation constants, but K_{ip} and K_{ib} are not.)

We have seen that K_{ib} and K_{ip} cannot be the binary dissociation constants, since EB and EP do not exist for this mechanism, and they are not the ternary dissociation constants for B and P from the ternary complexes; however, these ternary dissociation constants can be expressed in terms of the other kinetic parameters (see Scheme XXI):[65]

$$K_{\underline{\underline{B}}} = \frac{k_{-2}}{k_{+2}} = \frac{K_{ib}K_{m_A}V^r_{max}}{K_{ia}V^f_{max}};\quad K_{\underline{\underline{P}}} = \frac{k_{+3}}{k_{-3}} = \frac{K_{ip}K_{m_Q}}{K_{iq}}\frac{V^f_{max}}{V^r_{max}} \tag{149}$$

Proof of

$$K_{\underline{\underline{B}}} = \frac{k_{-2}}{k_{+2}} = \left(\frac{\left(\frac{k_{-1}k_{-2}}{k_{+2}}\;\frac{k_{+2}k_{+3}k_{+4}}{(k_{+1}k_{+2})(k_{+3} + k_{+4})}\;\frac{k_{-1}k_{-2}k_{-3}k_{-4}}{(k_{-3}k_{-4})(k_{-1} + k_{-2})}\right)}{\left(\frac{k_{-1}}{k_{+1}}\right)\left(\frac{k_{+1}k_{+2}k_{+3}k_{+4}}{k_{+1}k_{+2}(k_{+3} + k_{+4})}\right)}\right)$$

In 1973, Plapp[64] redefined the kinetic constants for the case where the interconversion of the ternary complexes $EAB \underset{k_{-p}}{\overset{k_{+p}}{\rightleftarrows}} EPQ$ is taken into account.

$$K_{ia} = \frac{k_{-1}}{k_{+1}} \qquad K_{iq} = \frac{k_{+4}}{k_{-4}}$$

$$K_{m_A} = \frac{k_{+3}k_{+4}k_{+p}}{k_{+1}(k_{+3}k_{+4} + k_{+3}k_{+p} + k_{+4}k_{+p} + k_{+4}k_{-p}} \qquad K_{m_Q} = \frac{k_{-1}k_{-2}k_{-p}}{k_{-4}(k_{-1}k_{-2} + k_{-1}k_{+p} + k_{-1}k_{-p} + k_{-2}k_{-p}}$$

$$K_{ib} = \frac{k_{-1}k_{-2} + k_{-1}k_{+p} + k_{-1}k_{-p} + k_{-2}k_{-p}}{k_{+2}(k_{+p} + k_{-p})} \qquad K_{ip} = \frac{k_{+3}k_{+4} + k_{+3}k_{+p} + k_{+4}k_{+p} + k_{+4}k_{-p}}{k_{-3}(k_{+p} + k_{-p})}$$

$$K_{m_B} = \frac{k_{+4}(k_{-2}k_{+3} + k_{-2}k_{-p} + k_{+3}k_{+p})}{k_{+2}(k_{+3}k_{+4} + k_{+3}k_{+p} + k_{+4}k_{+p} + k_{+4}k_{-p})} \qquad K_{m_P} = \frac{k_{-1}(k_{-2}k_{+3} + k_{-2}k_{-p} + k_{+3}k_{+p})}{k_{-3}(k_{-1}k_{-2} + k_{-1}k_{+p} + k_{-1}k_{-p} + k_{-2}k_{-p})}$$

$$\frac{V^f_{max}}{E_t} = \frac{k_{+3}k_{+4}k_{+p}}{k_{+3}k_{+4} + k_{+3}k_{+p} + k_{+4}k_{+p} + k_{+4}k_{-p}} \qquad \frac{V^r_{max}}{E_t} = \frac{k_{-1}k_{-1}k_{-p}}{k_{-1}k_{-2} + k_{-1}k_{+p} + k_{-1}k_{-p} + k_{-2}k_{-p}} \tag{150}$$

K_{ia} and K_{iq} are the only true dissociation constants obtainable from initial velocity data.

Equation 147 may be rewritten in terms of V^f_{max} after dividing through by V^r_{max} and using one of the definitions of K_{eq};

$$K_{eq} = \frac{V^f_{max}}{V^r_{max}} \frac{K_{mP}}{V_{mB}} \frac{K_{iq}}{K_{ia}} \tag{151}$$

(from one of the Haldane equations), then,

$$v_f = \frac{V^f_{max}[(A)(B) - (P)(Q)/K_{eq}]}{\begin{aligned} &K_{ia}K_{mB} + K_{mB}(A) + K_{mA}(B) + (A)(B) + \frac{K_{ia}K_{mB}}{K_{iq}}(Q) \\ &+ \frac{K_{mQ}K_{ia}K_{mB}}{K_{iq}K_{mp}}(P) + \frac{K_{mB}K_{mQ}}{K_{mp}K_{iq}}(A)(P) + \frac{K_{ia}K_{mB}}{K_{mp}K_{iq}}(P)(Q) \\ &+ \frac{K_{mA}}{K_{iq}}(B)(Q) + \frac{(A)(B)(P)}{K_{ip}} + \frac{K_{mB}K_{ia}(B)(P)(Q)}{K_{mp}K_{iq}K_{ib}} \end{aligned}} \tag{152}$$

or the net steady-state velocity equation for the forward direction. Another Haldane equation for the ordered Bi Bi case is

$$K_{eq} = \left(\frac{V^f_{max}}{V^r_{max}}\right)^2 \frac{K_{ip}K_{mQ}}{K_{ib}K_{mA}} \tag{153}$$

which can be proved after substitution of the coefficients.

Thus:

$$K_{eq} = \frac{\left(\frac{num_1}{coef_{AB}}\right)^2 \frac{coef_{AB}}{coef_{ABP}} \frac{coef_P}{coef_{PQ}}}{\left(\frac{num_2}{coef_{PQ}}\right)^2 \frac{coef_{PQ}}{coef_{BPQ}} \frac{coef_B}{coef_{AB}}} = \left(\frac{num_1}{num_2}\right)^2 \frac{coef_P}{coef_{ABP}} \frac{coef_{BPQ}}{coef_B} \tag{154}$$

and

$$K_{eq} = \frac{(k_{+1}k_{+2}k_{+3}k_{+4})^2}{(k_{-1}k_{-2}k_{-3}k_{-4})^2} \cdot \frac{k_{-1}k_{-2}k_{-3}}{k_{+1}k_{+2}k_{+3}} \cdot \frac{k_{+2}k_{-3}k_{-4}}{k_{+2}k_{+3}k_{+4}} = \frac{k_{+1}k_{+2}k_{+3}k_{+4}}{k_{-1}k_{-2}k_{-3}k_{-4}}$$

Every kinetic mechanism has at least one Haldane equation relating K_{eq} to the kinetic parameters. These Haldane relations have the form

$$K_{eq} = \left(\frac{V^f_{max}}{V^r_{max}}\right)^n \frac{K_{(P)}K_{(q)}K_{(r)}\cdots}{K_{(a)}K_{(b)}K_{(c)}\cdots} \tag{155}$$

The Ks will be either inhibition or Michaelis constants. The choice is not arbitrary but rather is set for any given mechanism. There will always be at least one Haldane equation where n = 1, which can be used to eliminate K_{eq} from the velocity equation. If the velocity equation contains a constant term in the denominator, as in the case we have seen for the ordered Bi Bi reaction, we can conveniently obtain a Haldane equation with n = 1 by rearranging the equation to express the net velocity of the reverse reaction as follows: Equation 147 is divided by $\frac{V^f_{max}}{K_{eq}}$ throughout, to obtain a form giving the net steady-state velocity in the reverse direction.

$$v = \frac{V^r_{max}K_{eq}\left[(A)(B) - \frac{(P)(Q)}{K_{eq}}\right]}{\frac{V^r_{max}}{V^f_{max}}K_{ia}K_{mB}K_{eq} + \frac{V^r_{max}}{V^f_{max}}K_{mB}K_{eq}(A) + \frac{V^r_{max}}{V^f_{max}}K_{mA}K_{eq}(B) + K_{mQ}(P) + K_{mp}(Q) + \frac{V^r_{max}}{V^f_{max}}K_{eq}(A)(B) + \frac{K_{mQ}}{K_{ia}}(A)(P) + (P)(Q) + \frac{V^r_{max}}{V^f_{max}}\frac{K_{mA}K_{eq}}{K_{iq}}(B)(Q) + \frac{V^r_{max}}{V^f_{max}}K_{eq}(A)(B)(P) + \frac{(B)(P)(Q)}{K_{ib}}} \qquad (156)$$

Multiply both sides by (-1):

$$-v = \frac{V^r_{max}K_{eq}\left[\frac{(P)(Q)}{K_{eq}} - (A)(B)\right]}{(\text{denominator})} \qquad (157)$$

and defining $(-v) = v^{reverse} = v^r$.

The denominator of the equation contains a number of terms that have K_{eq} as a factor; e.g., the constant term, from which a Haldane relation can be derived for the system. Thus, the constant denominator term containing K_{eq} is written out first in coefficient form:

$$\frac{K_{ia}K_{mB}V^r_{max}K_{eq}}{V^f_{max}} = \frac{\left(\frac{\text{constant}}{\text{coef}_A}\right)\left(\frac{\text{coef}_A}{\text{coef}_{AB}}\right)\left(\frac{\text{num}_2}{\text{coef}_{PQ}}\right)\left(\frac{\text{num}_1}{\text{num}_2}\right)}{\frac{\text{num}_1}{\text{coef}_{AB}}} = \frac{\text{constant}}{\text{coef}_{PQ}} = \frac{\text{constant}}{\frac{(\text{coef}_Q)}{K_{mp}}} = K_{iq}K_{mp} \qquad (158)$$

since

$$K_{mp} = \frac{\text{coef}_Q}{\text{coef}_{PQ}} \text{ or } \text{coef}_{PQ} = \frac{\text{coef}_Q}{K_{mp}} \text{ and } K_{iq} = \frac{\text{constant}}{\text{coef}_Q}$$

Therefore,

$$K_{ia}K_{mB}\frac{V^r_{max}}{V^f_{max}}K_{eq} = K_{iq}K_{mp}$$

and

$$K_{eq} = \frac{V^f_{max}K_{iq}K_{mp}}{V^r_{max}K_{ia}K_{mB}} \qquad (159)$$

Other Haldane equations for this system might be found by recombining coefficients into new definitions; the same procedure can be followed using one of the other denominator terms containing K_{eq}, or examine the definitions of the various constants to see which combinations reduce to

$$K_{eq} = \frac{k_{+1}k_{+2}k_{+3}k_{+4}}{k_{-1}k_{-2}k_{-3}k_{-4}}.$$

Note that the numerator of the Haldane equation must contain a constant for each product and the denominator one for each substrate.

The initial forward velocity in the absence of products is, if $(P) = (Q) = 0$ in Equation 152:

$$v_0^f = \frac{V_{max}^f(A)(B)}{K_{ia}K_{mB} + K_{mB}(A) + K_{mA}(B) + (A)(B)} \tag{160}$$

for a steady-state mechanism, which is identical to that for the rapid equilibrium random Bi Bi system we have already discussed, but in terms of the "Kuby" nomenclature.

$$v_0^f = \frac{V_{max}^f(A)(B)}{\bar{K}_A K_B + K_B(A) + K_A(B) + (A)(B)} \tag{161}$$

where $K_{ia} = \bar{K}_A$; $K_A = K_{mA}$; $K_B = K_{mB}$.
For, if we divide by (A) (B):

$$v_0^f = \frac{V_{max}^f}{1 + \dfrac{K_A}{(A)} + \dfrac{K_B}{(B)} + \dfrac{\bar{K}_A K_B}{(A)(B)}}$$

or

$$v_0^f = \frac{V_{max}^f}{1 + \dfrac{K_4}{(A)} + \dfrac{K_3}{(B)} + \dfrac{K_1 K_3}{(A)(B)}} \tag{127}$$

Therefore, one cannot distinguish in form, from initial velocity data alone, the ordered Bi Bi system from the quasi-equilibrium random system.

8. Rapid Equilibrium Ordered Bi Bi System

Interestingly, for a rapid equilibrium system, (A) must combine with (E) first and (B) can add only to the (EA) complex, i.e.:

$$E + A \underset{k_{-1}}{\overset{\bar{K}_A/k_{+1}}{\rightleftarrows}} EA$$

$$+$$

$$B$$

$$K_B \quad \underset{k_{-2}}{\overset{k_{+2}}{\rightleftarrows}}$$

$$EAB \xrightarrow{k_p} E + Products$$

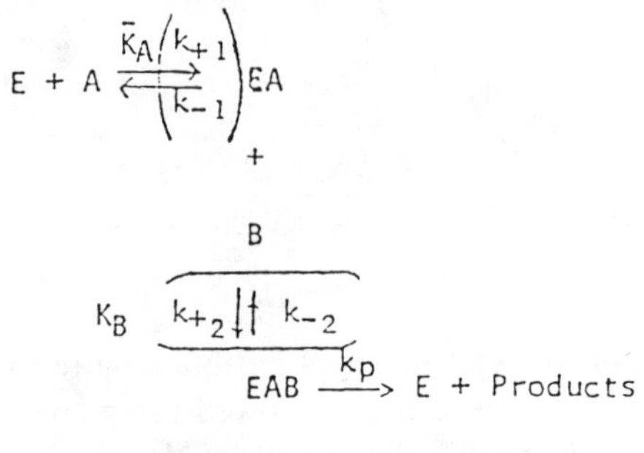

SCHEME XXVIII.

Or, in a rapid equilibrium ordered Bi Bi system, the K_{mA} (B) term in Equation 160 is absent, or the K_A (B) term in Equation 161 is absent. They both reduce to the form

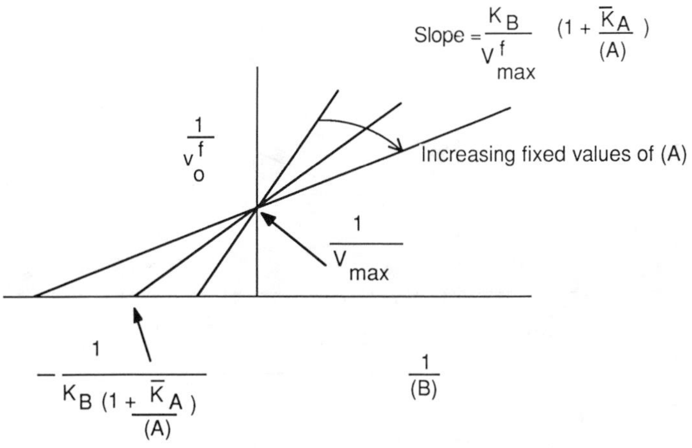

FIGURE 13. Plots of $\frac{1}{v_0^f}$ vs. $\frac{1}{(B)}$ for a rapid equilibrium ordered Bi Bi system.

$$v_0^f = \frac{V_{max}^f(A)(B)}{\bar{K}_A K_B + K_B(A) + (A)(B)} = \frac{V_{max}^f}{\frac{\bar{K}_A K_B}{(A)(B)} + \frac{K_B}{(B)} + 1} \tag{162}$$

or,

$$v_0^f = \frac{V_{max}^f}{\frac{K_B}{(B)}\left(1 + \frac{\bar{K}_A}{(A)}\right) + 1} \tag{163}$$

(refer to Chapter 9, Equation 28), and it appears in a form as if (A) is a competitive inhibitor of (B) and therefore a common intersection point in the double-reciprocal plot occurs right on the Y-axis.

Therefore, when a plot of $\frac{1}{v_0}$ vs. $\frac{1}{(B)}$ is made with fixed (A) and varied (B), Figure 13 results. Whereas, when (A) is the variable substrate and (B) is fixed

$$\frac{1}{v_0^f} = \frac{\bar{K}_A}{V_{max}^f} \frac{(K_B)}{(B)} \frac{1}{(A)} + \frac{1}{V_{max}^f}\left(1 + \frac{K_B}{(B)}\right) \tag{164}$$

and Figure 14 results.

Schimerlik and Cleland[66] found that the creatine kinase mechanism went from a rapid equilibrium random Bi Bi above approximately pH 7 to an ordered Bi Bi rapid equilibrium mechanism at pH 7.*

Now return to Equation 160:

$$v_0^f = \frac{V_{max}^f(A)(B)}{K_{ia}K_{m_B} + K_{m_B}(A) + K_{m_A}(B) + (A)(B)} \tag{160}$$

* Huang (in Appendix of Reference 1009) recently provided a theoretical basis and diagnostic aid for bireactant mechanisms that can give rise to typical equilibrium-ordered kinetic patterns, including creatine kinase. He discussed various sequential cases and applied his evaluation to a study of the kinetic mechanism of Type II Calmodulin-Dependent Protein Kinase.

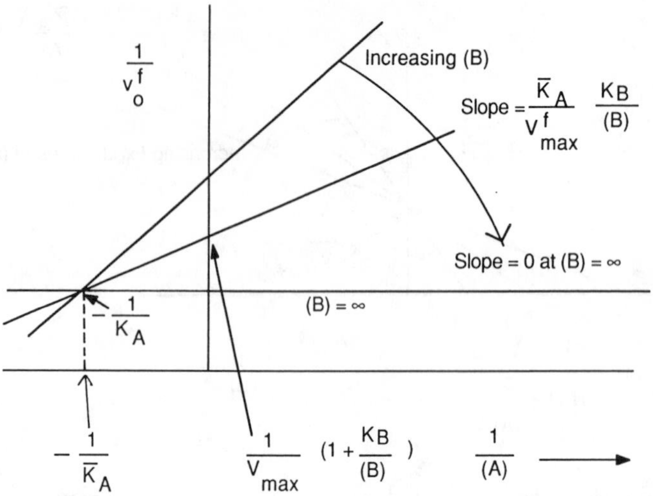

FIGURE 14.   Plots of $\frac{1}{v_0^f}$ vs. $\frac{1}{(A)}$ for a rapid equilibrium ordered Bi Bi system.

for the initial velocity equation for a steady-state ordered Bi Bi system. When (A) is varied

$$\frac{v}{V_{max}^f} = \frac{1}{\dfrac{K_{ia}K_{m_B}}{(A)(B)} + \dfrac{K_{m_B}}{(B)} + \dfrac{K_{m_A}}{(A)} + 1} \tag{160a}$$

or in reciprocal form

$$\frac{1}{v_0^f} = \frac{1}{V_{max}^f}\left(1 + \frac{K_{m_B}}{(B)}\right) + \frac{1}{V_{max}^f}\frac{K_{m_A}}{(A)}\left(1 + \frac{K_{ia}}{K_{m_A}}\frac{K_{m_B}}{(B)}\right) \tag{161a}$$

or when (B) is varied

$$\frac{1}{v_0^f} = \frac{1}{V_{max}^f}\left(1 + \frac{K_{m_A}}{(A)}\right) + \frac{1}{V_{max}^f}\frac{K_{m_B}}{(B)}\left(1 + \frac{K_{ia}}{(A)}\right) \tag{162}$$

The reciprocal plots may intercept above, on, or below the horizontal axis, depending on the ratio of K_{m_A}/K_{ia}, and if K_{ia} is very small compared to K_{m_A}, the slopes of the plots become insensitive to changes in the concentration of the paired substrate and the family of plots will be essentially parallel, as in a "ping-pong" mechanism, which we will see later.

Typical curves for the case of $K_{ia} > K_{m_A}$ follow.

Primary plot: $\frac{1}{v_0^f}$ vs. $\frac{1}{(A)}$ is shown in Figure 15. Secondary plots of Y-int vs. $\frac{1}{(B)}$ which follow the equations:

$$\text{Secondary Plots: Y-int.} = \frac{1}{V_{max}^f} + \frac{K_{m_B}}{V_{max}^f}\frac{1}{(B)}$$

$$\text{Secondary Plots: Slope} = \frac{K_{m_A}}{V_{max}^f} + \frac{K_{ia}}{V_{max}^f}\frac{K_{m_B}}{(B)}$$

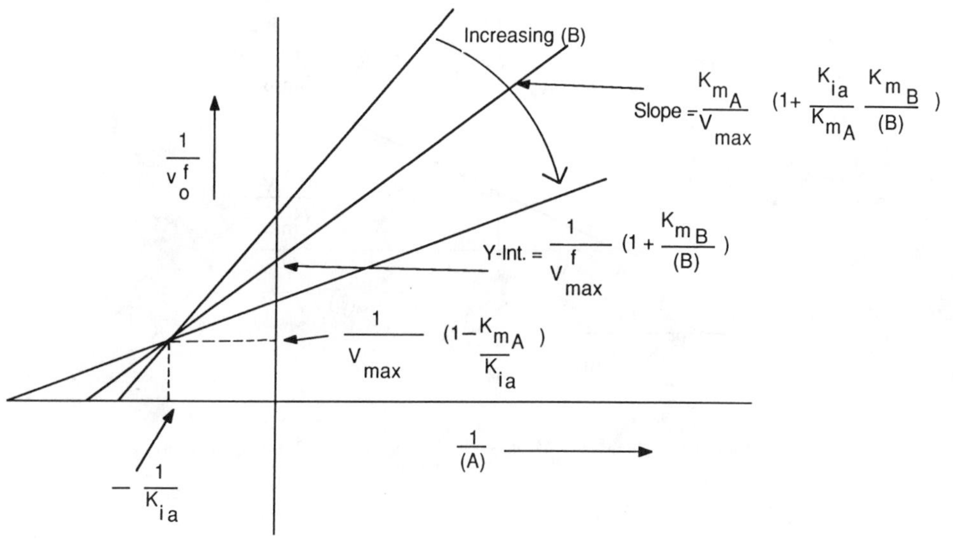

FIGURE 15. Primary plot: $\frac{1}{v_0^f}$ vs. $\frac{1}{(A)}$, for case where $K_{ia} > K_{MA}$.

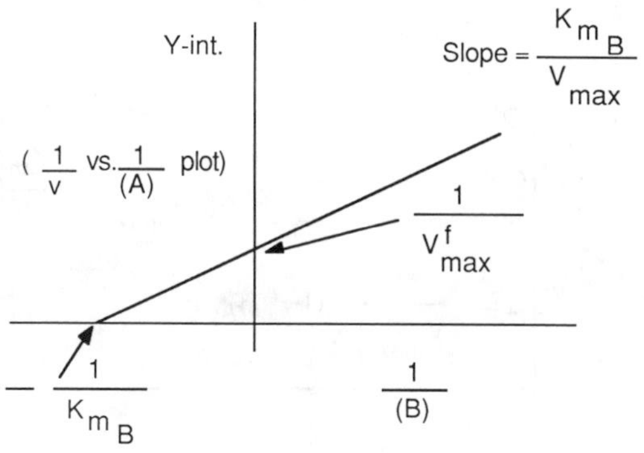

FIGURE 16. Y-intercept $= \frac{1}{V_{max}^f} + \frac{K_{mB}}{V_{max}^f}\frac{1}{(B)}$; secondary plot of data from Figure 15: Y-int. vs $\frac{1}{(B)}$.

are shown in Figures 16 and 17, respectively. Primary plot for $\frac{1}{v_0^f}$ vs. $\frac{1}{(B)}$ for $K_{ia} > K_{m_A}$ as shown in Figure 18.

The common intersection point is not K_{ib} since there is no K_{ib} in the velocity equation, according to the mechanism.

Secondary plots of Y-int vs. $\frac{1}{(A)}$ and slope vs. $\frac{1}{(A)}$ (from Figure 18) are given in Figure 19 and 20, respectively.

All four kinetic parameters for the forward reaction, V_{max}^f, K_{ia}, K_{m_B}, and K_{m_A} are capable of being estimated. However, it is not always easy to determine which substrate corresponds to "A" and which to "B" by initial velocity data alone and product inhibition data are often necessary, as we shall see.

Similarly, by appropriate plots, all four parameters of the reverse reaction may be calculated.

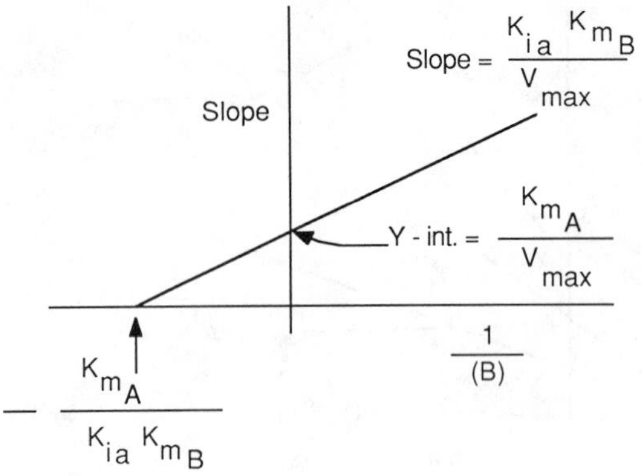

FIGURE 17. Slope $= \frac{K_{m_A}}{V^f_{max}} + \frac{K_{ia}}{V^f_{max}} \frac{K_{m_B}}{(B)}$; secondary plot of data from Figure 15: slope vs. $\frac{1}{(B)}$.

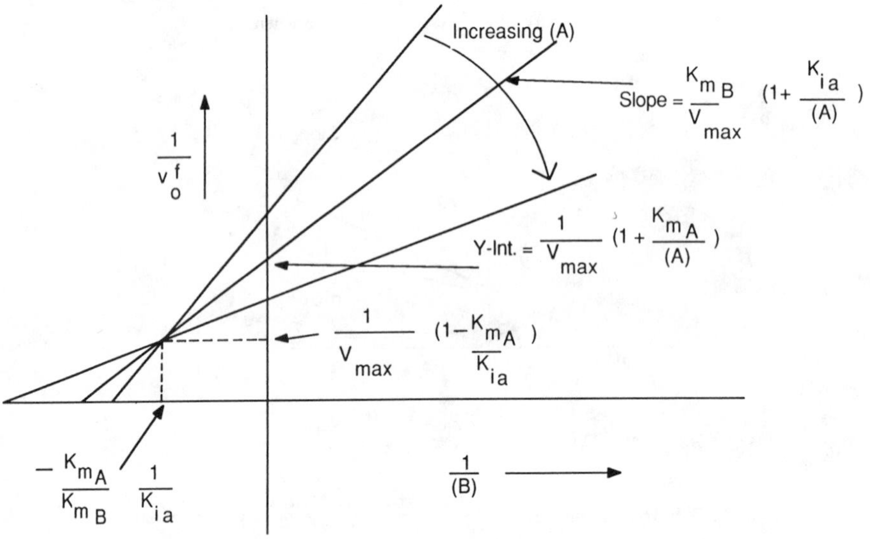

FIGURE 18. Primary plot for $\frac{1}{v^f_0}$ vs. $\frac{1}{(B)}$, for the case where $K_{ia} > K_{M_A}$.

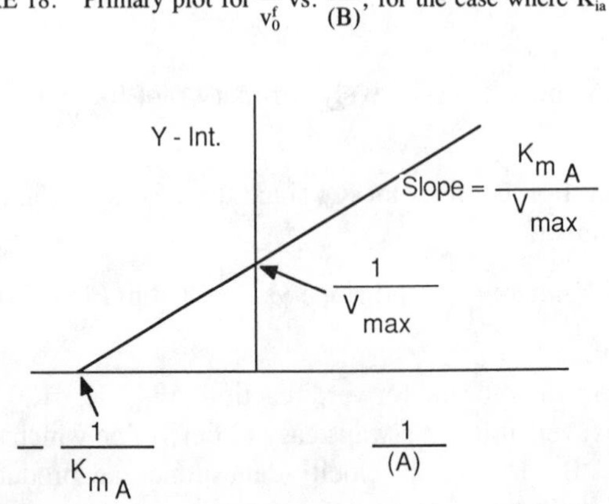

FIGURE 19. Secondary plot from Figure 18 of Y-int. vs. $\frac{1}{(A)}$.

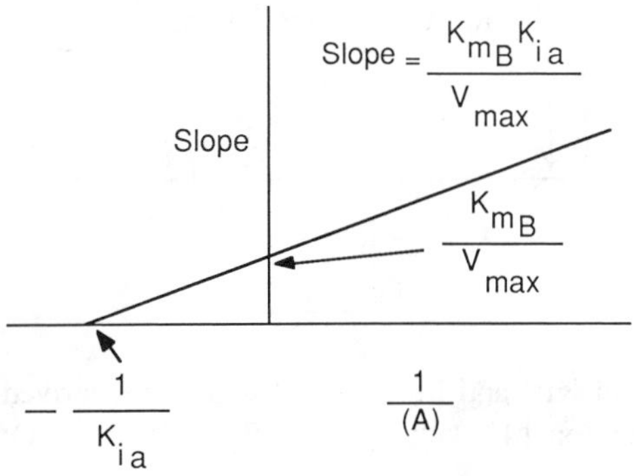

FIGURE 20. Secondary plot from Figure 18 of slope vs $\frac{1}{(A)}$.

Assuming the mechanism holds, as described in Scheme XXI, without including the interconversion of the central complexes in the reaction sequence, then all the reaction-rate constants may be calculated, i.e., k_{+1}, k_{-1}, k_{+2}, k_{-2}, k_{+3}, k_{-3}, k_{+4}, and k_{-4}, as follows:

$$k_{+1} = \frac{V^f_{max}}{K_{m_A}E_t}; \quad k_{-1} = \frac{V^f_{max}K_{ia}}{K_{m_A}E_t}; \quad k_{+2} = \frac{V^f_{max}}{K_{m_B}E_t}\left(1 + \frac{k_{-2}}{k_{+3}}\right);$$

$$\frac{1}{k_{-2}} = \frac{E_t}{V^r_{max}} - \frac{1}{k_{-1}}; \quad \frac{1}{k_{+3}} = \frac{E_t}{V^f_{max}} - \frac{1}{k_{+4}}; \quad k_{-3} = \frac{V^r_{max}}{K_{m_P}E_t}\left(1 + \frac{k_{+3}}{k_{-2}}\right);$$

$$k_{+4} = \frac{V^r_{max}K_{iq}}{K_{m_Q}E_t}; \quad k_{-4} = \frac{V^r_{max}}{K_{m_Q}E_t} \tag{163}$$

If the interconversions of the central complexes are included, only the calculations of k_{+1}, k_{-1}, k_{+4}, and k_{-4} are still valid because the additional rate constants (k_{+p} and k_{-p}) cancel out (e.g., they appear in the same way in $\frac{V^f_{max}}{E_t}$ and K_{m_A} and in $\frac{V^r_{max}}{E_t}$ and K_{m_Q}).

Finally, the distribution equations in the Cleland nomenclature may be listed for the ordered Bi Bi system:

$$\frac{(E)}{E_t} = \frac{K_{ia}K_{m_B}V^r_{max} + \frac{K_{m_Q}}{K_{eq}} V_f(P) + K_{m_A}V^r_{max}(B)}{\text{denominator of velocity Equation 147}} \tag{164}$$

$$\frac{(EA)}{E_t} = \frac{K_{m_B}V^r_{max}(A) + \frac{K_{m_Q}V^f_{max}}{K_{ia}K_{eq}}(A)(P) + \frac{K_{m_A}V^r_{max}}{K_{ia}K_{eq}}(P)(Q)}{\text{denominator of velocity Equation 147}} \tag{165}$$

$$\frac{(EAB + EPQ)}{E_t} = \frac{\left(\frac{V^r_{max} - V^f_{max}}{K_{iq}} K_{m_Q}\right)(A)(B) + \left(\frac{V^f_{max}}{K_{eq}} - \frac{V^r_{max}K_{m_A}}{K_{eq}K_{ia}}\right)(P)(Q) + \frac{V^r_{max}(A)(B)(P)}{K_{ip}} + \frac{V^f_{max}(B)(P)(Q)}{K_{ip}K_{eq}}}{\text{denominator of velocity equation}} \tag{166}$$

$$\frac{(EQ)}{E_t} = \frac{\frac{K_{m_P}V^f_{max}(Q)}{K_{eq}} + \frac{K_{m_Q}V^f_{max}(A)(B)}{K_{iq}} + \frac{K_{m_A}V^r_{max}(B)(Q)}{K_{iq}}}{\text{denominator of velocity equation}} \tag{167}$$

If now we return to the Theorell-Chance mechanism:

$$
\begin{array}{cccc}
& A & B \quad P & Q \\
& k_{+1} \downarrow\, k_{-1} & k_{+2} \searrow \nearrow k_{-2} & k_{+3} \uparrow\, k_{-3} \\
\hline
E & EA & EQ & E
\end{array}
$$

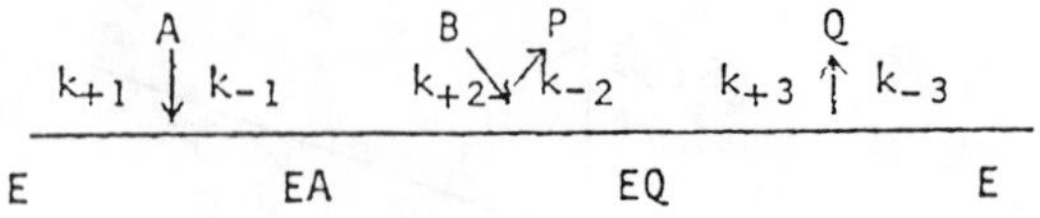

SCHEME XXIX.

see Schemes XI, XII, and XIII and Equation 82 which we derived and which may be put in the same form as Equation 142 with $S_1 = A$; $S_2 = B$; $P_1 = Q$; and $P_2 = P$.

$$
\frac{v}{E_t} = \frac{[k_{+1}k_{+2}k_{+3}(A)(B) - k_{-1}k_{-2}k_{-3}(P)(Q)]}{\begin{array}{c} k_{-1}k_{+3} + k_{+1}k_{+3}(A) + k_{+2}k_{+3}(B) + k_{-1}k_{-2}(P) + k_{-1}k_{-3}(Q) \\ + k_{+1}k_{+2}(A)(B) + k_{+1}k_{-2}(A)(P) + k_{+2}k_{-3}(B)(Q) + k_{-2}k_{-3}(P)(Q) \end{array}} \tag{168}
$$

Compared to our generalized equation for ordered Bi Bi (Equation 142) we note that the Theorell-Chance mechanism lacks the (A) (B) (P) and (B) (P) (Q) terms in the denominator.

Therefore, the Theorell-Chance mechanism is really a specific type of ordered reaction and, in fact, derivable from the generalized equation for an ordered Bi Bi reaction. After substituting and defining coefficient ratios as described for the ordered Bi Bi system, we obtain the same equation (Equation 86) we already derived and now we can rewrite it in Cleland's nomenclature (i.e., for the Theorell-Chance mechanism).

$$
v = \frac{V_{max}^f V_{max}^r \left((A)(B) - \dfrac{(P)(Q)}{K_{eq}} \right)}{\begin{array}{c} V_{max}^r K_{ia} K_{mB} + V_{max}^r K_{mB}(A) + V_{max}^r K_{mA}(B) + \dfrac{V_{max}^f K_{mQ}}{K_{eq}}(P) + \dfrac{V_{max}^f K_{mP}}{K_{eq}}(Q) \\[4mm] + V_{max}^r(A)(B) + \dfrac{V_{max}^f K_{mQ}}{K_{eq}K_{ia}}(A)(P) + \dfrac{V_{max}^r K_{mA}}{K_{iq}}(B)(Q) + \dfrac{V_{max}^f}{K_{eq}}(P)(Q) \end{array}} \tag{169}
$$

In the absence of products, in the case of the Theorell-Chance mechanism, the velocity equation for the net steady-state forward velocity is the same as that for the usual ordered Bi Bi reaction, as we have seen. Let $(P) = (Q) = 0$

$$
\frac{v_0^f}{V_{max}^f} = \frac{(A)(B)}{K_{ia}K_{mB} + K_{mB}(A) + K_{mA}(B) + (A)(B)} \tag{170}
$$

(i.e., for Theorell-Chance or for ordered Bi Bi reaction), or if we return to the equation in terms of reaction-rate constants (Equation 168) and set $(P) = (Q) = 0$:

$$
\frac{v_0^f}{E_t} = \frac{k_{+1}k_{+2}k_{+3}(A)(B)}{k_{-1}k_{+3} + k_{+1}k_{+3}(A) + k_{+2}k_{+3}(B) + k_{+1}k_{+2}(A)(B)} \tag{171}
$$

If we return to Equation 142 for the generalized ordered Bi Bi reaction and if we also set $(P) = (Q) = 0$:

$$
\frac{v_0^f}{E_t} = \frac{k_{+1}k_{+2}k_{+3}k_{+4}(A)(B)}{k_{-1}k_{+4}(k_{-2} + k_{+3}) + k_{+1}k_{+4}(k_{-2} + k_{+3})(A) + k_{+2}k_{+3}k_{+4}(B) + k_{+1}k_{+2}(k_{+3} + k_{+4})(A)(B)} \tag{172}
$$

Now, if we set

$$k_{+3} >> k_{-2}$$
$$k_{+4} >> k_{+3}$$

$$
\begin{array}{ccccc}
& A & B & P & Q \\
& k_{+1}\downarrow k_{-1} & k_{+2}\downarrow k_{-2} & k_{+3}\uparrow k_{-3} & k_{+4}\uparrow k_{-4} \\
\hline
E & EA & (EAB \rightleftarrows EPQ) & EQ & E
\end{array}
$$

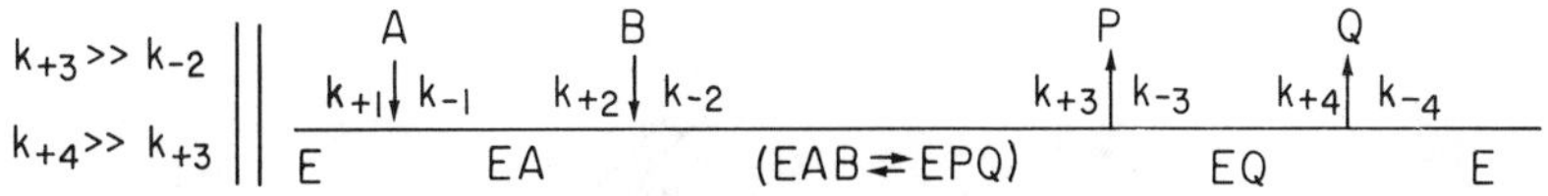

SCHEME XXX.

This reduces to

$$
\begin{array}{cccc}
& A & B\ P & Q \\
& k_{+1}\downarrow k_{-1} & k_{+2}\downarrow\ \uparrow k_{-2} & k_{+3}\uparrow k_{-3} \\
\hline
E & EA & EQ & E
\end{array}
$$

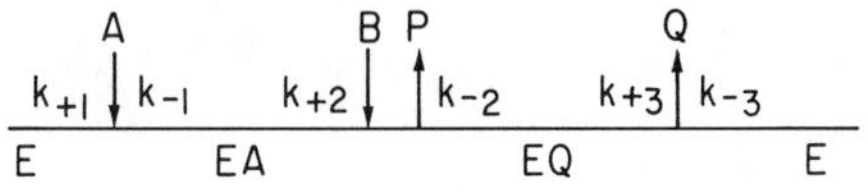

SCHEME XXXI.

(when the ternary complexes are reduced to zero).

Thus, Equation 172 reduces to Equation 171 after dividing through by k_{+4}; and therefore the Theorell-Chance mechanism is only a limiting case of the ordered Bi Bi mechanism. It does not follow that the ternary complexes cannot form; it is just that they become negligible in the steady state under certain conditions.

9. A Case Where a Product is Released between the Additions of the Two Substrates

We will now turn to the case in which a product is released between the additions of the two substrates, a reaction mechanism which was described initially by Alberty[57] and by Koshland[67] but which has now been popularized as ping-pong kinetics by Cleland.[43]

These reaction mechanisms are common in group transfer or covalently substituted enzyme reactions, e.g., in transaminase reactions, in reactions which yield a phosphoryl enzyme, as in nucleoside diphosphokinase, or in the flavin enzyme mechanisms, where the enzyme, for all practical purposes, exists in two forms.

We will consider the case for a mechanism which is referred to as "ping-pong Bi Bi", i.e., where there are two substrates and two products. In Cleland's shorthand notation the reaction can be written as

$$
\begin{array}{ccccc}
& A & P & B & Q \\
& k_{+1}\downarrow k_{-1} & k_{+2}\uparrow k_{-2} & k_{+3}\downarrow k_{-3} & k_{+4}\uparrow k_{-4} \\
\hline
E & (EA \rightleftarrows FP) & & (FB \rightleftarrows EQ) & E
\end{array}
$$

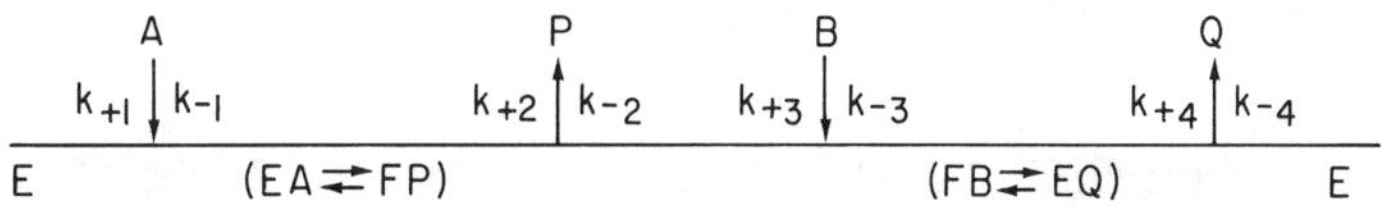

SCHEME XXXII.

The mechanism involves two different stable enzyme forms, (E) and (F), and two sets of central complexes. We shall ignore the interconversion of the central complexes in our derivation:

$$
E + A \underset{k_{-1}}{\overset{k_{+1}}{\rightleftarrows}} (EA \rightleftarrows FP) \underset{k_{-2}}{\overset{k_{+2}}{\rightleftarrows}} F + P
$$

$$
F + B \underset{k_{-3}}{\overset{k_{+3}}{\rightleftarrows}} (FB \rightleftarrows EQ) \underset{k_{-4}}{\overset{k_{+4}}{\rightleftarrows}} E + Q
$$

SCHEME XXXIII.

Note that there are no ternary complexes, EB, or FA, but there are EA and FB and FP and EQ binary complexes. The basic King-Altman figure is Scheme XXXIV or XXXV.

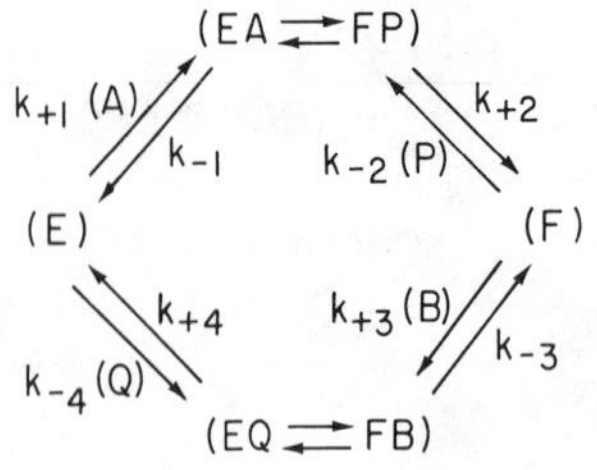

SCHEME XXXIV.

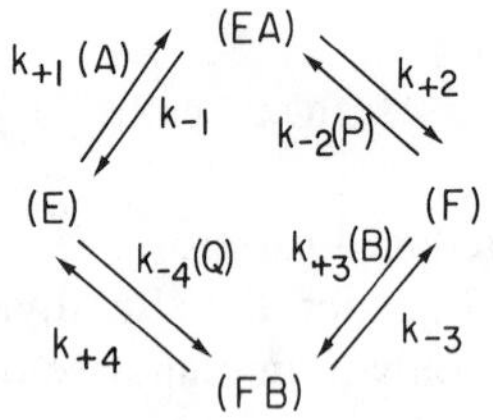

SCHEME XXXV.

The four three-lined interconversion patterns are

SCHEME XXXVI.

and yield the following distribution equations:

$$\frac{(E)}{E_t} = \frac{k_{+4}k_{-1}k_{-2}(P) + k_{+2}k_{+3}(B)k_{+4} + k_{+3}(B)\,k_{+4}k_{-1} + k_{-3}k_{-2}(P)k_{-1}}{\text{denominator}} \quad (173)$$

$$\frac{(EA + FP)}{E_t} = \frac{k_{+4}k_{+1}(A)k_{-2}(P) + k_{-4}(Q)k_{-3}k_{-2}(P) + k_{+3}(B)k_{+4}k_{+1}(A) + k_{-3}k_{-2}(P)k_{+1}(A)}{\text{denominator}} \quad (173a)$$

$$\frac{(F)}{E_t} = \frac{k_{+4}k_{+1}(A)k_{+2} + k_{-4}(Q)k_{-3}k_{+2} + k_{-1}k_{-4}(Q)k_{-3} + k_{-3}k_{+1}(A)(k_{+2})}{\text{denominator}} \quad (174)$$

$$\frac{(FB + EQ)}{E_t} = \frac{k_{-2}(P)k_{-1}k_{-4}(Q) + k_{-4}(Q)k_{+2}k_{+3}(B) + k_{-1}k_{-4}(Q)k_{+3}(B) + k_{+1}(A)k_{+2}k_{+3}(B)}{\text{denominator}} \quad (175)$$

Letting $v = k_{+1}(E)\,(A) - k_{-1}(EA + FP)$, and after substitution and grouping terms, we obtain:

$$v = \frac{[k_{+1}k_{+2}k_{+3}k_{+4}E_t(A)(B) - k_{-1}k_{-2}k_{-3}k_{-4}E_t(P)(Q)]}{\begin{array}{l} k_{+1}k_{+2}(k_{-3} + k_{+4})(A) + k_{+3}k_{+4}(k_{+2} + k_{-1})(B) + k_{-1}k_{-2}(k_{-3} + k_{+4})(P) \\ + k_{-3}k_{-4}(k_{-1} + k_{+2})(Q) + k_{+1}k_{+3}(k_{+2} + k_{+4})(A)(B) + k_{+1}k_{-2}(k_{-3} + k_{+4})(A)(P) \\ + k_{-2}k_{-4}(k_{-1} + k_{-3})(P)(Q) + k_{+3}k_{-4}(k_{-1} + k_{+2})(B)(Q) \end{array}} \quad (176)$$

We note that Equation 176 has no constant term in the denominator, and neither an (A) (B) (P) nor a (B) (P) (Q) term in the denominator (refer to Equation 142) for ordered Bi Bi.

Multiplying the numerator and denominator by the same factor we used for the ordered Bi Bi reaction, that is, by $(num_2)/[(coef_{AB})(coef_{PQ})]$ and the (P), (Q), (A) (P), and (P) (Q) terms by $\frac{num_1}{num_1}$, as well, we obtain

$$v = \frac{V^f_{max}V^r_{max}\left[(A)(B) - \frac{(P)(Q)}{K_{eq}}\right]}{\begin{array}{l} V^r_{max}K_{m_B}(A) + V^r_{max}K_{m_A}(B) + \frac{V^f_{max}K_{mQ}}{K_{eq}}(P) + \frac{V^f_{max}K_{mP}}{K_{eq}}(Q) \\ + V^r_{max}(A)(B) + \frac{V^f_{max}K_{mQ}}{K_{eq}K_{ia}}(A)(P) + \frac{V^f_{max}}{K_{eq}}(P)(Q) + \frac{V^r_{max}K_{m_A}}{K_{iq}}(B)(Q) \end{array}} \quad (177)$$

where all of the K_ms, K_{ia}, and K_{iq} are the same ratios of terms we had given for the ordered Bi Bi system, except now that

$$K_{ib} = \frac{coef_Q}{coef_{BQ}} = \frac{k_{-3}}{k_{+3}}; \quad \frac{coef_A}{coef_{AP}} = \frac{k_{+2}}{k_{-2}} = K_{ip} \quad (178)$$

(refer to Equation 148) and the following will be used later in deriving the Haldane equations:

$$V^f_{max} = \frac{num_1}{coef_{AB}}; \quad V^r_{max} = \frac{num_2}{coef_{PQ}}; \quad K_{m_A} = \frac{coef_B}{coef_{AB}}; \quad K_{m_B} = \frac{coef_A}{coef_{AB}}$$

$$K_{mP} = \frac{coef_{1Q}}{coef_{PQ}}; \quad K_{mQ} = \frac{coef_P}{coef_{PQ}}; \quad K_{eq} = \frac{num_1}{num_2}; \quad K_{ia} = \frac{k_{-1}}{k_{+1}} = \frac{coef_P}{coef_{AP}};$$

$$K_{iq} = \frac{coef_B}{coef_{BQ}} = \frac{k_{+4}}{k_{-4}} \quad (179)$$

Setting (P) = (Q) = O, the forward velocity is given by

$$\frac{v^f_0}{V^f_{max}} = \frac{(A)(B)}{K_{m_B}(A) + K_{m_A}(B) + (A)(B)} \quad (180)$$

where there is no constant term, or

$$\frac{v^f_0}{V^f_{max}} = \frac{1}{\frac{K_{m_B}}{(B)} + \frac{K_{m_A}}{(A)} + 1} \quad (181)$$

when compared to the general equation for the initial forward velocity for the ordered Bi Bi mechanism:

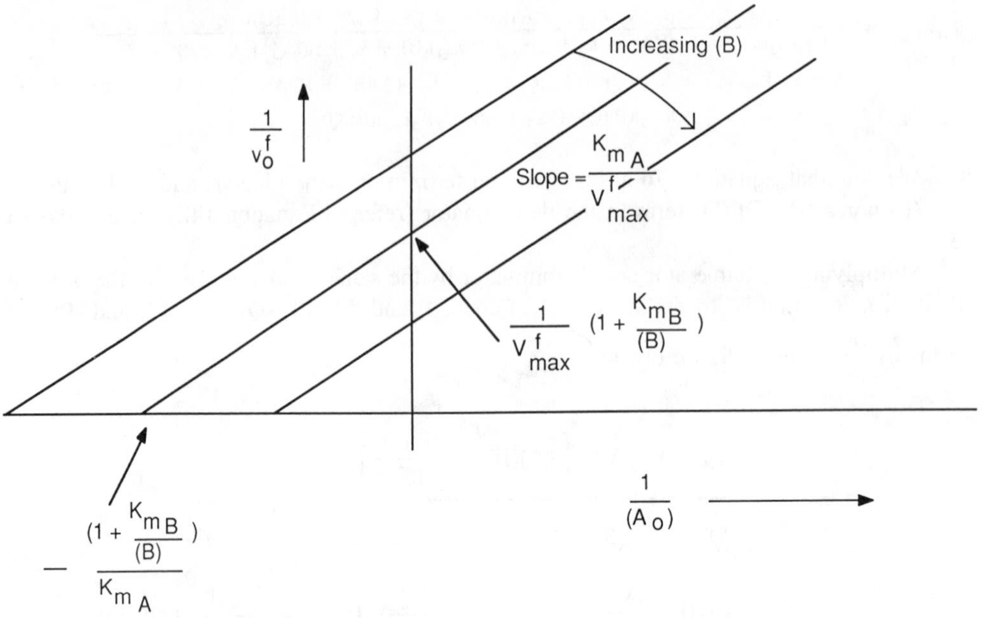

FIGURE 21. Primary plots of $\frac{1}{v_0^f}$ vs. $\frac{1}{(A_0)}$ (cf. Equation 184).

$$\frac{v_0^f}{V_{max}^f} = \frac{1}{\frac{K_{ia}K_{m_B}}{(A)(B)} + \frac{K_{m_B}}{(B)} + \frac{K_{m_A}}{(A)} + 1} \tag{182}$$

or

$$\frac{v_0^f}{V_{max}^f} = \frac{(A)(B)}{K_{ia}K_{m_B} + K_{m_B}(A) + K_{m_A}(B) + (A)(B)} \tag{183}$$

(Refer to Equation 160.) The constant term has now dropped out of Equation 180 and in reciprocal form for either varied (A) or (B) and fixed paired substrate

$$\frac{1}{v_0^f} = \frac{K_{m_A}}{V_{max}^f}\frac{1}{(A)} + \frac{1}{V_{max}^f}\left(1 + \frac{K_{m_B}}{(B)}\right) \tag{184}$$

$$\frac{1}{v_0^f} = \frac{K_{m_B}}{V_{max}^f}\frac{1}{(B)} + \frac{1}{V_{max}^f}\left(1 + \frac{K_{m_B}}{(A)}\right) \tag{185}$$

and which lead to the primary plots of $\frac{1}{v_0^f}$ vs. $\frac{1}{A_0}$ given in Figure 21 for Equation 184. The secondary plots of the data from Figure 21 are derived from Equation 186, i.e.,

$$\text{Y-int.} = \frac{1}{V_{max}^f} + \frac{K_{m_B}}{V_{max}^f}\frac{1}{(B)};\ \text{X-int.} = (-)\frac{1}{K_{m_A}}(-)\frac{K_{m_B}}{K_{mA}}\frac{1}{(B)},\ \text{or}$$

$$(-)\ \text{X-int.} = \frac{1}{K_{m_A}} + \frac{K_{m_B}}{K_{m_A}}\frac{1}{(B)} \tag{186}$$

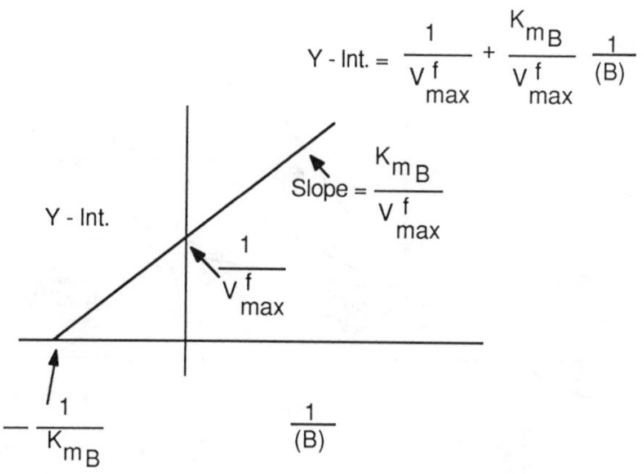

FIGURE 22. Secondary plot of data from Figure 21, of Y-int. vs. $\frac{1}{(B)}$ and based on Y-intercept $= \frac{1}{V^f_{max}} + \frac{K_{m_B}}{V^f_{max}} \frac{1}{(B)}$.

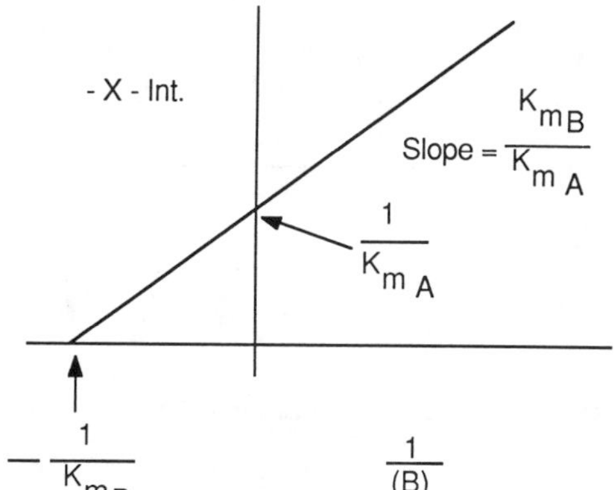

FIGURE 23. Secondary plot of data from Figure 21 of −X-int. vs. $\frac{1}{(B)}$ (see Equation 186).

These secondary plots of Y-int res $\left(\frac{1}{B}\right)$ and H, -X-int vs. $\left(\frac{1}{B}\right)$, are shown in Figure 22 and 23, respectively.

The primary plots and secondary plots for (B) as the varied substrate are symmetrical to those shown for varied (A). While the system is symmetrical with respect to (A) and (B), there is usually no difficulty in identifying which substrate is (A) and which is (B), since (A) must be the donor substrate with the group to be transferred and (B) the acceptor. Similarly, (P) must be the product of (A) while (Q) is the product of (B) plus the group transferred from (A).

A convenient way to analyze such data is to vary (A) and (B) together, while maintaining their concentrations at a fixed ratio, i.e., (B) = X(A), then the reciprocal equation becomes

$$\frac{1}{v^f_0} = \frac{K_{m_A}}{V^f_{max}} \left(1 + \frac{K_{m_B}}{XK_{m_A}}\right) \frac{1}{(A)} + \frac{1}{V^f_{max}} \tag{187}$$

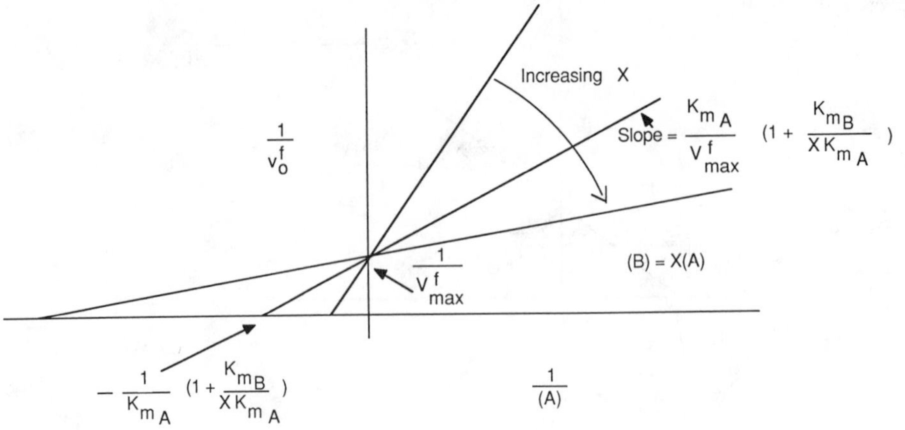

FIGURE 24. Plot of $\frac{1}{v_0^f}$ vs. $\frac{1}{(A)}$ according to Equation 187.

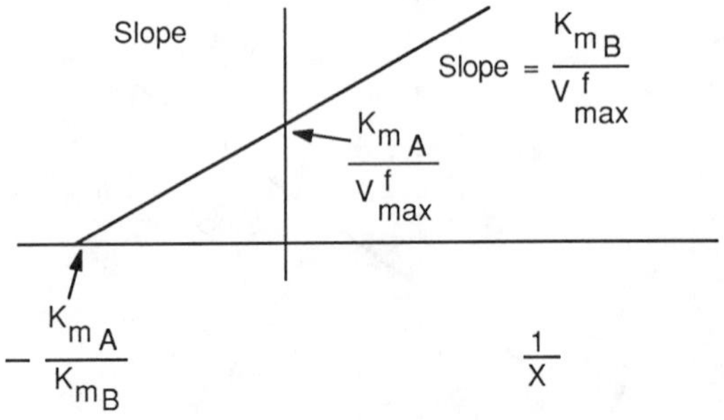

FIGURE 25. Secondary plot of data from Figure 24; slope vs. $\frac{1}{X}$.

and then plot $\frac{1}{v_0^f}$ vs. $\frac{1}{(A)}$, as shown in Figure 24. The secondary plots based on Equation 187a,

$$\text{Slope} = \frac{K_{m_A}}{V^f_{max}} + \frac{K_{m_B}}{X}\frac{1}{V^f_{max}} \tag{187a}$$

and on Equation 188 are shown in Figure 25 and 26.

$$-\left(\frac{1}{\text{X-intercept}}\right) = K_{m_A}\left(1 + \frac{K_{m_B}}{XK_{m_A}}\right); \quad -\left(\frac{1}{\text{X-intercept}}\right) = K_{m_A} + \frac{K_{m_B}}{X} \tag{188}$$

Usually, the family of parallel lines (Figure 21) and the linear reciprocal plot, when (A) + (B) are varied together (Figure 24), are sufficient to identify a ping-pong Bi Bi system. However, the ordered Bi Bi system will yield the same results if $K_{ia} \ll K_{m_A}$, as we have seen (Equation 162), and from Equation 182 for varied (A) in reciprocal form, i.e., for an ordered Bi Bi initial forward reaction with (P) = (Q) = O.

$$\frac{V^f_{max}}{v^f_0} = \frac{K_{m_A}}{(A)}\left(1 + \frac{K_{ia}K_{m_B}}{K_{m_A}}\frac{1}{(B)}\right) + 1 + \frac{K_{m_B}}{(B)} \tag{189}$$

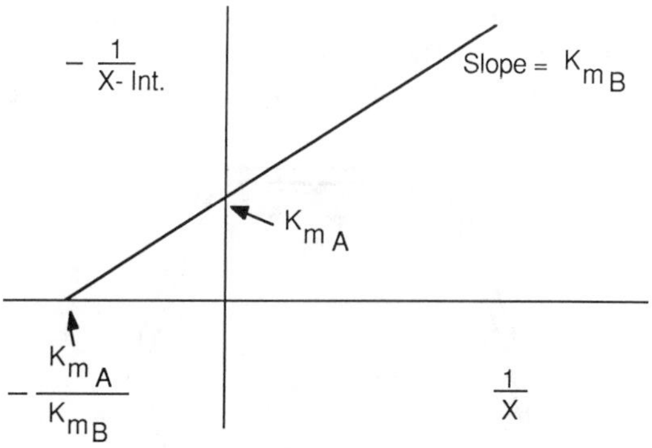

FIGURE 26. Secondary plot of data from Figure 24; $\frac{1}{-\text{X-int.}}$ vs. $\frac{1}{X}$.

If

$$K_{ia} \ll K_{m_A}, \text{ then } \frac{K_{ia}K_{m_B}}{K_{m_A}} \rightarrow 0 \tag{190}$$

and

$$\frac{1}{v_0^f} = \frac{1}{V_{max}^f}\frac{K_{m_A}}{(A)} + \frac{1}{V_{max}^f}\left(1 + \frac{K_{m_B}}{(B)}\right) \tag{191}$$

which is identical to Equation 184 for a ping-pong Bi Bi reaction.

Therefore, product inhibition studies are necessary to distinguish the two mechanisms, as we shall see later.

E. AN ORDERED TER TER SEQUENCE — A CASE OF A THREE-SUBSTRATE SYSTEM OF THE FIRST DEGREE

There is, in this context, another possibility to consider, that is, the case of an ordered ter ter system, i.e., an ordered three-reactant system which may actually degenerate into ping-pong kinetics.

Frieden[68] apparently was one of the first to attempt a steady-state solution for this complex mechanism. He expressed the initial forward velocity in the following form for a three-substrate obligatory-ordered system:

$$v_f = \frac{V_{max}^f}{1 + \frac{K_a}{(A)} + \frac{K_B}{(B)} + \frac{K_C}{(C)} + \frac{K_{AB}}{(A)(B)} + \frac{K_{BC}}{(B)(C)} + \frac{K_{ABC}}{(A)(B)(C)}} \tag{192}$$

The shorthand schematic in the Cleland nomenclature for an ordered ter ter sequence:

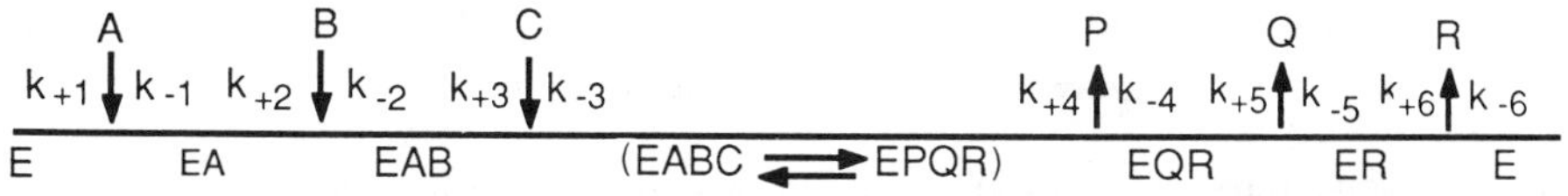

SCHEME XXXVII.

(One example of this type of reaction might be a flavoenzyme, e.g., NADPH-cytochrome c reductase, where (B) would be FAD).

The King-Altman figure would take the form, for the six enzyme species (n = 6),

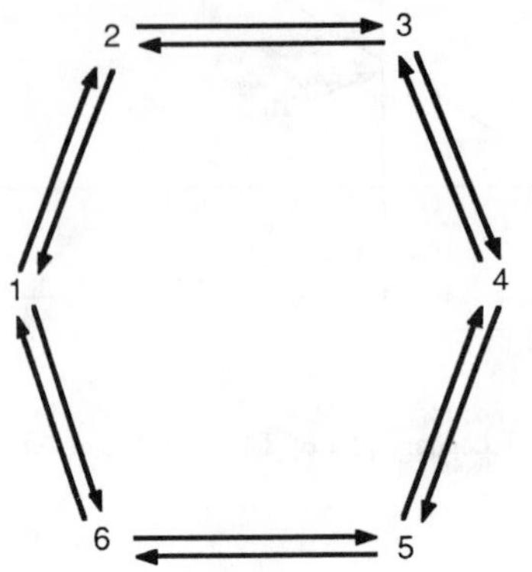

SCHEME XXXVIII.

and the six five-sided King-Altman interconversion patterns for each enzyme species/E_t are (since ${}_mC_{n-1} = \frac{m!}{(n-1)!(m-n+1)!} = \frac{(6)!}{(5)!(1)!} = 6$).

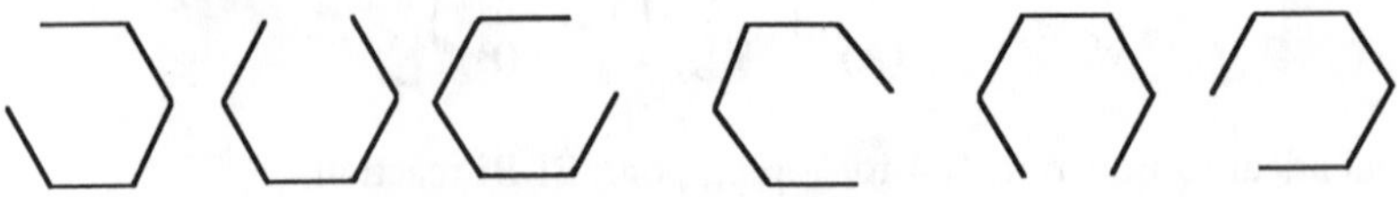

SCHEME XXXIX.

After substitution and collection of terms in the velocity expression, the overall steady-state rate equation contains 27 terms in the denominator:[164,169]

$$\begin{aligned}\frac{v}{E_t} = \frac{[k_{+1}k_{+2}k_{+3}k_{+4}k_{+5}k_{+6}(A)(B)(C) - k_{-1}k_{-2}k_{-3}k_{-4}k_{-5}k_{-6}(P)(Q)(R)]}{\begin{aligned}&k_{-1}k_{-2}k_{+5}k_{+6}(k_{-3} + k_{+4}) + k_{+1}k_{-2}k_{+5}k_{+6}(k_{-3} + k_{+4})(A) + k_{-1}k_{+3}k_{+4}k_{+5}k_{+6}(C)\\ &+ k_{+1}k_{+2}k_{+5}k_{+6}(k_{-3} + k_{+4})(A)(B) + k_{+1}k_{+3}k_{+4}k_{+5}k_{+6}(A)(C)\\ &+ k_{+2}k_{+3}k_{+4}k_{+5}k_{+6}(B)(C) + k_{+1}k_{+2}k_{+3}(k_{+4}k_{+5} + k_{+4}k_{+6} + k_{+5}k_{+6})(A)(B)(C)\\ &+ k_{-1}k_{-2}k_{-3}k_{-4}k_{+6}(P) + k_{-1}k_{-2}k_{+5}k_{-6}(k_{-3} + k_{+4})(R) + k_{-1}k_{-2}k_{-3}k_{-4}k_{-5}(P)(Q)\\ &+ k_{-1}k_{-2}k_{-3}k_{-4}k_{-6}(P)(R) + k_{-1}k_{-2}k_{-5}k_{-6}(k_{-3} + k_{+4})(Q)(R)\\ &+ k_{-4}k_{-5}k_{-6}(k_{-1}k_{-2} + k_{-1}k_{-3} + k_{-2}k_{-3})(P)(Q)(R) + k_{+1}k_{-2}k_{-3}k_{-4}k_{+6}(A)(P)\\ &+ k_{-1}k_{+3}k_{+4}k_{+5}k_{-6}(C)(R) + k_{+1}k_{+2}k_{-3}k_{-4}k_{+6}(A)(B)(P) + k_{+2}k_{+3}k_{+4}k_{+5}k_{-6}(B)(C)(R)\\ &+ k_{+1}k_{-2}k_{-3}k_{-4}k_{-5}(A)(P)(Q) + k_{-1}k_{+3}k_{+4}k_{-5}k_{-6}(C)(Q)(R)\\ &+ k_{+1}k_{+2}k_{+3}k_{-4}k_{+6}(A)(B)(C)(P) + k_{+1}k_{+2}k_{+3}k_{+4}k_{-5}(A)(B)(C)(Q)\\ &+ k_{+1}k_{+2}k_{-3}k_{-4}k_{-5}(A)(B)(P)(Q) + k_{+2}k_{+3}k_{+4}k_{-5}k_{-6}(B)(C)(Q)(R)\\ &+ k_{+2}k_{-3}k_{-4}k_{-5}k_{-6}(B)(P)(Q)(R) + k_{-1}k_{+3}k_{-4}k_{-5}k_{-6}(C)(P)(Q)(R)\\ &+ k_{+1}k_{+2}k_{+3}k_{-4}k_{-5}(A)(B)(C)(P)(Q) + k_{+2}k_{+3}k_{-4}k_{-5}k_{-6}(B)(C)(P)(Q)(R)\end{aligned}}\end{aligned} \tag{193}$$

In terms of the following defined kinetic parameters (or coefficient forms) this transforms into[164,169]

$$
\begin{aligned}
v = {} & \frac{V^fV^r\left((A)(B)(C) - \dfrac{(P)(Q)(R)}{K_{eq}}\right)}{V^rK_{ia}K_{ib}K_{m_C} + V^rK_{ib}K_{m_C}(A) + V^rK_{ia}K_{m_B}(C) + V^rK_{m_C}(A)(B)} \\
& + V^rK_{m_B}(A)(C) + V^rK_{m_A}(B)(C) + V^r(A)(B)(C) + \frac{V^fK_{ir}K_{m_Q}(P)}{K_{eq}} \\
& + \frac{V^fK_{iq}K_{mp}(R)}{K_{eq}} + \frac{V^fK_{m_R}(P)(Q)}{K_{eq}} + \frac{V^fK_{m_Q}(P)(R)}{K_{eq}} + \frac{V^fK_{mp}(Q)(R)}{K_{eq}} \\
& + \frac{V^f(P)(Q)(R)}{K_{eq}} + \frac{V^fK_{m_Q}(A)(P)}{K_{ia}K_{eq}} + \frac{V^rK_{ia}K_{m_B}(C)(R)}{K_{ir}} \\
& + \frac{V^fK_{m_Q}K_{ir}(A)(B)(P)}{K_{ia}K_{ib}K_{eq}} + \frac{V^rK_{m_A}(B)(C)(R)}{K_{ir}} + \frac{V^fK_{m_R}(A)(P)(Q)}{K_{ia}K_{eq}} \\
& + \frac{V^rK_{ia}K_{m_B}(C)(Q)(R)}{K_{iq}K_{ir}} + \frac{V^fK_{ir}K_{m_Q}(A)(B)(C)(P)}{K_{ia}K_{ib}K_{ic}K_{eq}} + \frac{V^fK_{ip}K_{m_R}(A)(B)(C)(Q)}{K_{ia}K_{ib}K_{ic}K_{eq}} \\
& + \frac{V^fK_{m_R}(A)(B)(P)(Q)}{K_{ia}K_{ib}K_{eq}} + \frac{V^rK_{m_A}(B)(C)(Q)(R)}{K_{iq}K_{ir}} + \frac{V^rK_{m_A}K_{ic}(B)(P)(Q)(R)}{K_{ip}K_{iq}K_{ir}} \\
& + \frac{V^rK_{ia}K_{m_B}(C)(P)(Q)(R)}{K_{ip}K_{iq}K_{ir}} + \frac{V^fK_{m_R}(A)(B)(C)(P)(Q)}{K_{ia}K_{ib}K_{ic}K_{eq}} \\
& + \frac{V^rK_{m_A}(B)(C)(P)(Q)(R)}{K_{ip}K_{iq}K_{ir}}
\end{aligned}
\qquad (194)
$$

Where the coefficients are:

The Michaelis constants:

$$
\begin{aligned}
& K_{m_A} = \frac{coef_{BC}}{coef_{ABC}}; \quad K_{m_B} \frac{coef_{AC}}{coef_{ABC}}; \quad K_{m_C} \frac{coef_{AB}}{coef_{ABC}} \\
& K_{mp} = \frac{coef_{QR}}{coef_{PQR}}; \quad K_{m_Q} = \frac{coef_{PR}}{coef_{PQR}}; \quad K_{m_R} \frac{coef_{PQ}}{coef_{PQR}}
\end{aligned}
\qquad (195)
$$

The inhibition constants:

$$
\begin{aligned}
K_{ia} &= \frac{constant}{coef_A} = \frac{coef_C}{coef_{AC}} = \frac{coef_P}{coef_{AP}} = \frac{coef_{PQ}}{coef_{APQ}} = \frac{k_{-1}}{k_{+1}} \\
K_{ib} &= \frac{coef_A}{coef_{AB}} = \frac{coef_{AP}}{coef_{ABP}} = \frac{coef_{APQ}}{coef_{ABPQ}} = \frac{k_{-2}}{k_{+2}} \\
K_{ic} &= \frac{coef_{ABP}}{coef_{ABCP}} = \frac{coef_{ABPQ}}{coef_{ABCPQ}} = \frac{coef_{BPQR}}{coef_{BCPQR}} = \frac{k_{-3}}{k_{+3}} \\
K_{ip} &= \frac{coef_{CQR}}{coef_{CPQR}} = \frac{coef_{ABCQ}}{coef_{ABCPQ}} = \frac{coef_{BCQR}}{coef_{BCPQR}} = \frac{k_{+4}}{k_{-4}} \\
K_{iq} &= \frac{coef_R}{coef_{QR}} = \frac{coef_{CR}}{coef_{CQR}} = \frac{coef_{BCR}}{coef_{BCQR}} = \frac{k_{+5}}{k_{-5}} \\
K_{ir} &= \frac{constant}{coef_R} = \frac{coef_P}{coef_{PR}} = \frac{coef_C}{coef_{CR}} = \frac{coef_{BC}}{coef_{BCR}} = \frac{k_{+6}}{k_{-6}}
\end{aligned}
\qquad (196)
$$

The Haldane equations are

$$K_{eq} = \frac{V^f_{max}K_{mP}K_{iq}K_{ir}}{V^r_{max}K_{ia}K_{ib}K_{mC}} = \frac{K_{ip}K_{iq}K_{ir}}{K_{ia}K_{ib}K_{ic}} \tag{197}$$

The rate constants are

$$k_{+1} = \frac{V^f_{max}}{K_{mA}E_t}; \quad k_{-1} = \frac{V^f_{max}K_{ia}}{K_{mA}E_t}; \quad k_{+2} = \frac{V^f_{max}}{K_{mB}E_t}; \quad k_{-2} = \frac{V^f_{max}K_{ib}}{K_{mB}E_t};$$

$$k_{+3} = \left(1 + \frac{k_{-3}}{k_{+4}}\right)\frac{V^f_{max}}{K_{mQ}E_t}; \quad \frac{1}{k_{-3}} = \frac{E_t}{V^r_{max}} - \frac{1}{k_{-1}} - \frac{1}{k_{-2}}; \quad k_{-4} = \left(1 + \frac{k_{+4}}{k_{-3}}\right)\frac{V^r_{max}}{K_{mp}E_t};$$

$$k_{+5} = \frac{V^r_{max}K_{iq}}{K_{mQ}E_t}; \quad k_{-5} = \frac{V^r_{max}}{K_{mQ}E_t}; \quad k_{+4} = \frac{E_t}{V^f_{max}} - \frac{1}{k_{-5}} - \frac{1}{k_{+6}};$$

$$k_{+6} = \frac{V^r_{max}K_{ir}}{E_tK_{mR}}; \quad k_{-6} = \frac{V^r_{max}}{E_tK_{mR}} \tag{198}$$

The initial forward velocity in the absence of products is

$$\frac{v^f_0}{V^f_{max}} = \frac{(A)(B)(C)}{K_{ia}K_{ib}K_{mC} + K_{ib}K_{mC}(A) + K_{ia}K_{mB}(C) + K_{mC}(A)(B) + K_{mB}(A)(C) + K_{mA}(B)(C) + (A)(B)(C)} \tag{199}$$

Dividing through by (A) (B) (C), or

$$\frac{v^f_0}{V^f_{max}} = \frac{1}{1 + \frac{K_{mA}}{(A)} + \frac{K_{mB}}{(B)} + \frac{K_{mC}}{(C)} + \frac{K_{ia}K_{mB}}{(A)(B)} + \frac{K_{ib}K_{mC}}{(B)(C)} + \frac{K_{ia}K_{ib}K_{mC}}{(A)(B)(C)}} \tag{200}$$

and compare with Frieden's equation (Equation 192). (See also Chapter 9, Section I and Equation 51 for an ordered three-substrate system with a rapid equilibrium random sequence in ligands A and B and in R and Q.)

If we take the reciprocal of Equation 200 and rearrange terms so that (A) is the variable, (C) is the varying fixed substrate, and (B) is the constant fixed substrate, then:

$$\frac{1}{v^f_0} = \frac{1}{V^f_{max}}\left[1 + \frac{K_{mB}}{(B)} + \frac{K_{mC}}{(C)}\left(1 + \frac{K_{ib}}{(B)}\right)\right] + \frac{K_{mA}}{V^f_{max}}\frac{1}{(A)}\left[1 + \frac{K_{ia}K_{mB}}{K_{mA}(B)} + \frac{K_{ia}K_{ib}K_{mC}}{K_{mA}(B)(C)}\right] \tag{201}$$

$$\frac{1}{v^f_0} = \frac{1}{V^f_{max}}\left[1 + \frac{K_{mB}}{(B)} + \frac{K_{mC}}{(C)}\left(1 + \frac{K_{ib}}{(B)}\right)\right] + \frac{K_{mA}}{V^f_{max}}\frac{1}{(A)}\left[1 + \frac{K_{ia}K_{mB}}{K_{mA}(B)}\left(1 + \frac{K_{ib}K_{mC}}{K_{mB}(C)}\right)\right] \tag{202}$$

Now, if one plots $\frac{1}{v_0}$ vs. $\left(\frac{1}{A_0}\right)$ at constant $(B)_1$ (Figure 27A), or at a constant $(B)_2$ (Figure 27B). Since

$$\underset{(1/v\ \text{vs.}\ 1/A)}{\text{Slope}} = \frac{K_{mA}}{V^f_{max}}\left[1 + \frac{K_{ia}K_{mB}}{K_{mA}}\frac{1}{(B)}\left(1 + \frac{K_{ib}K_{mC}}{K_{mB}}\frac{1}{(C)}\right)\right] \tag{203}$$

$$\underset{(1/v\ \text{vs.}\ 1/A)}{\text{Slope}} = \frac{K_{mA}}{V^f_{max}} + \frac{K_{ia}K_{mB}}{V^f_{max}}\frac{1}{(B)} + \frac{K_{ia}K_{ib}K_{mC}}{V^f_{max}}\frac{1}{(B)}\frac{1}{(C)} \tag{204}$$

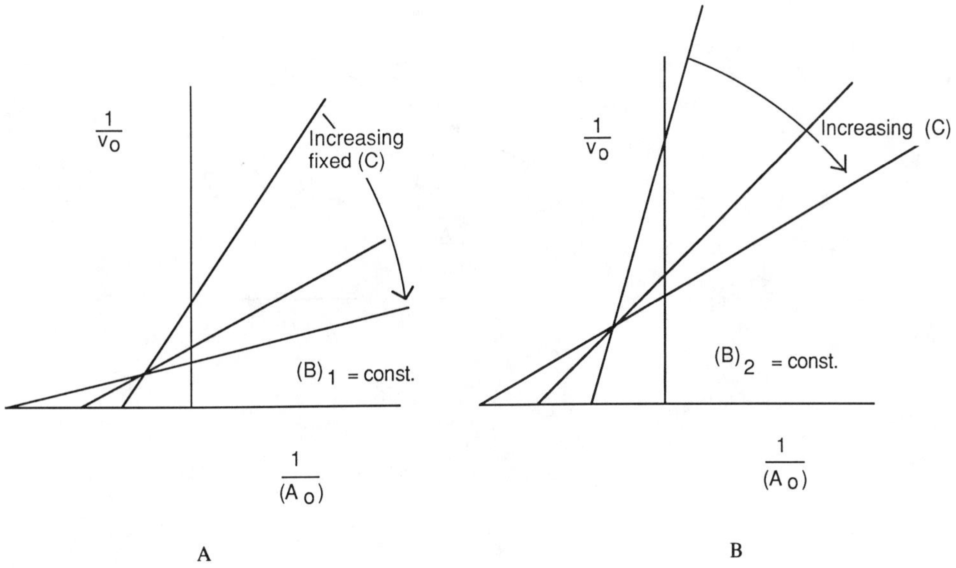

FIGURE 27. Plot of $\frac{1}{v_0^f}$ vs. $\frac{1}{(A_0)}$ at constant $(B)_1$ (Figure 27A) or at constant $(B)_2$ (Figure 27B) according to Equation 205.

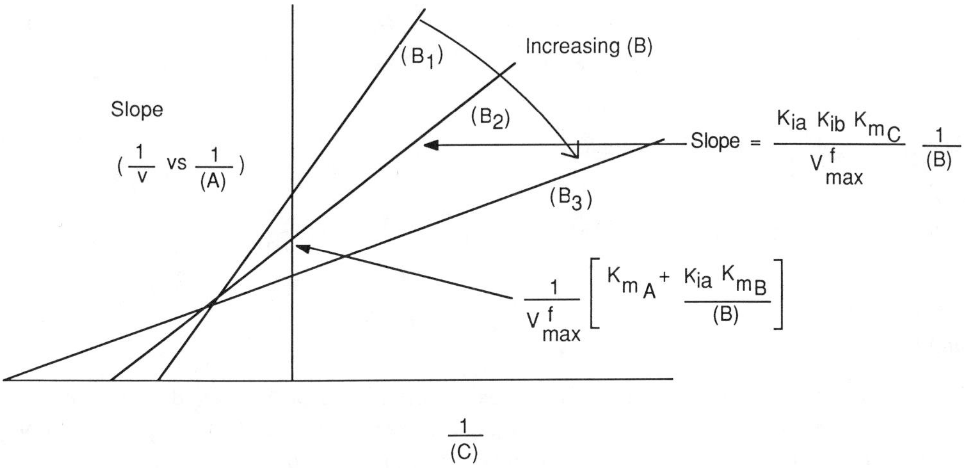

FIGURE 28. Secondary plot of slopes vs. $\frac{1}{(C)}$, according to Equation 205. Data from Figure 27.

or

$$\underset{(1/v \text{ vs. } 1/A)}{\text{Slope}} = \frac{1}{V^f_{max}} \left[K_{m_A} + K_{ia} \frac{K_{m_B}}{(B)} \right] + \frac{K_{ia} K_{ib} K_{m_C}}{V^f_{max}} \frac{1}{(B)} \frac{1}{(C)} \tag{205}$$

The secondary plots of the slopes are therefore plotted in Figure 28, vs. $\frac{1}{(c)}$.

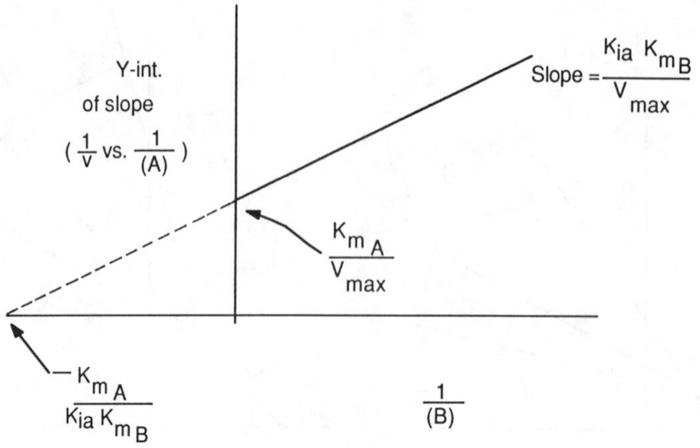

FIGURE 29. Tertiary plot of Y-int (from Figure 28) vs. $\frac{1}{(B)}$ to yield a value of K_{MA}/K_{max}.

$$\left(\text{When slope}_{\text{sec (or x-intercept)}} = 0;\ \frac{1}{(C)} = -\frac{\left[K_{m_A} + \frac{K_{ia}K_{m_B}}{(B)}\right]}{K_{ia}K_{ib}\frac{K_{m_C}}{(B)}} = \text{x-intercept of } \underset{(1/v \text{ vs. } 1/A)}{\text{slope}} \text{ vs. } \frac{1}{(C)} \text{ plot.}\right.$$

$$\left. = \frac{-K_{m_A}}{K_{ia}K_{ib}\frac{K_{m_C}}{(B)}} - \frac{K_{m_B}}{K_{ib}K_{m_C}}\right) \qquad (206)$$

The tertiary plot of the Y-intercept is

$$\text{Y-int. of } \underset{(1/v \text{ vs. } 1/A)}{\text{Slope}} = \left(\frac{K_{m_A}}{V^f_{max}} + \frac{1}{V^f_{max}}\frac{K_{ia}K_{m_B}}{(B)}\right)$$

and plotted as shown in Figure 29 to yield $\frac{K_{m_A}}{V_{max}}$.

By repeating this for the other three cases (i.e., Figure 30A to D) the other kinetic parameters are eventually obtained. Therefore, treat each substrate as the varied substrate at different fixed concentrations of a second substrate, maintaining a constant concentration of the third substrate (which can be changed for a different family of plots) and from the secondary plots and tertiary plots we can evaluate most, if not all, of the kinetic parameters. The kinetics of three-substrate reactions have been studied by a number of workers;[844] Dalziel[152] has carried out a systematic study of the possible pathways that may be involved. In the case of the rare four-substrate reaction, Elliott and Tipton[153] have made a detailed study of such systems.

If now we return to Equation 202 (refer to Equation 200) for the v_0^f, for the case of a three-substrate reaction, with (A) as the variable substrate, (B) as a constant fixed substrate, and (C) as varying, but fixed substrate:

$$\frac{1}{v_0^f} = \frac{1}{V^f_{max}}\left[1 + \frac{K_{m_B}}{(B)} + \frac{K_{m_C}}{(C)}\left(1 + \frac{K_{ib}}{(B)}\right)\right] + \frac{K_{m_A}}{V^f_{max}}\frac{1}{(A)}\left[1 + \frac{K_{ia}K_{m_B}}{K_{m_A}(B)}\left(1 + \frac{K_{ib}K_{m_C}}{K_{m_B}(C)}\right)\right] \qquad (202)$$

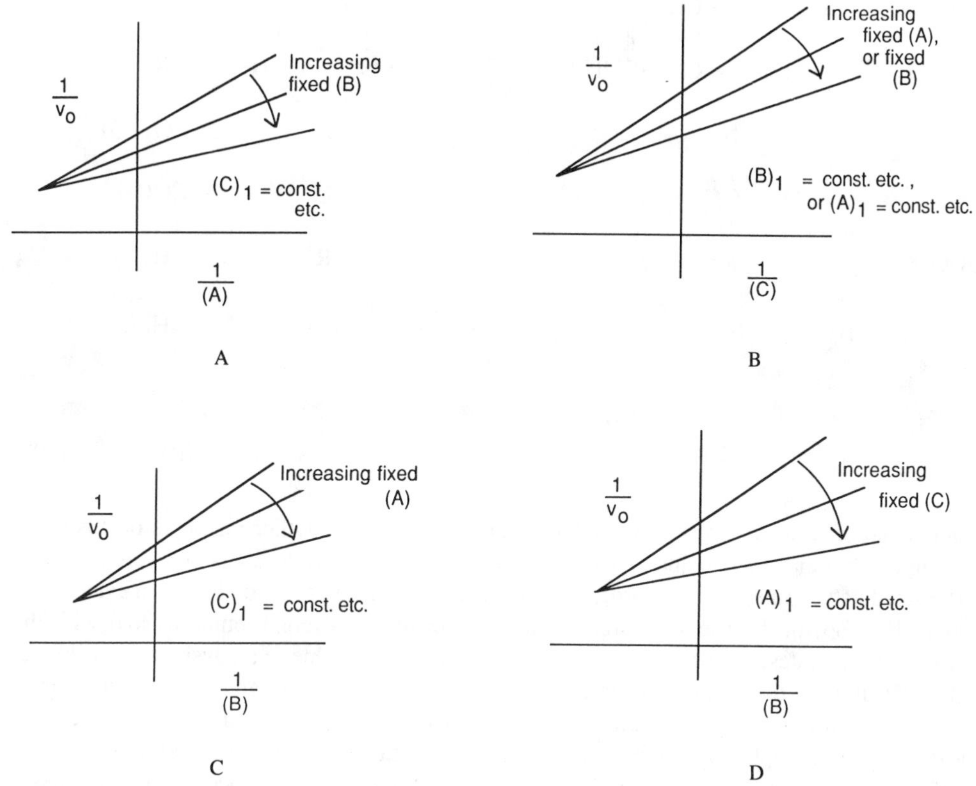

FIGURE 30.

Now

$$\left.\begin{array}{l}\text{if constant (B)} >> K_{ib} \\ \text{or constant (B)} >> K_{m_B}\end{array}\right\} \quad \text{and (B)} >> \frac{K_{ia}K_{m_B}}{K_{m_A}}$$

These conditions are readily obtainable if $K_{ib} = 10^{-8} \simeq K_{m_B}$ as, for example, the dissociation constant of FAD from the NADPH cytochrome c reductase and if (B) is fixed at approximately 10^{-5} to 10^{-6} *M*.[829]

Therefore,

$$\frac{1}{v_0^f} \rightarrow \frac{1}{V_{max}^f}\left[1 + \frac{K_{m_C}}{(C)}\right] + \frac{K_{m_A}}{V_{max}^f}\frac{1}{(A)} \tag{207}$$

which is identical in form to Equation 185 with (C) now replacing (B).

1. A Steady-State Kinetic Analysis of the Prolyl-4-Hydroxylase Reaction

This analysis was recently attempted by Soskel and Kuby.[271] Prolyl-4-hydroxylase in the presence of α-ketoglutarate, O_2, Fe^{2+}, and ascorbate catalyzes the hydroxylation of prolyl residues in collagen (or in synthetic substrates) in the sequence x-pro-gly where x is any amino acid except glycine,[272] with the liberation of CO_2 and succinate as products. Thus, this enzyme catalyzes a ter ter reaction:

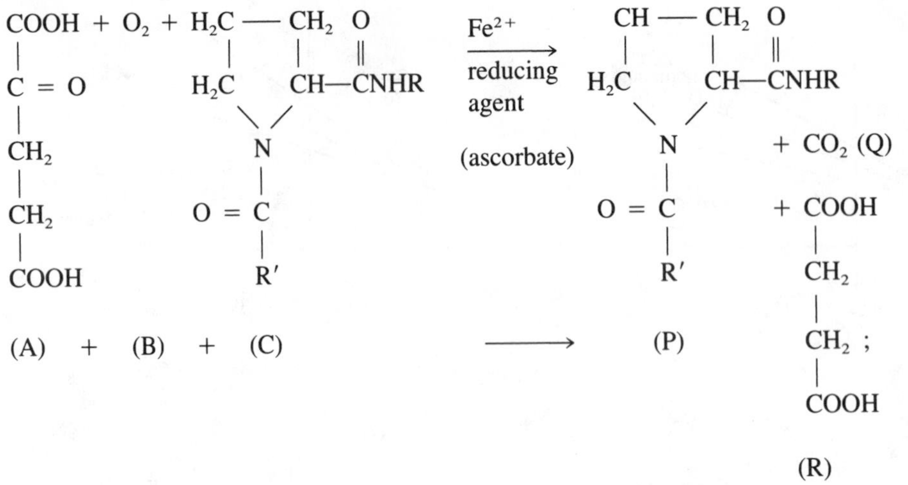

$$
\begin{array}{l}
\text{COOH} + \text{O}_2 + \text{H}_2\text{C}\!\!-\!\!\text{CH}_2 \; \text{O} \\
| \qquad\qquad\quad | \qquad | \quad \| \\
\text{C} = \text{O} \qquad\quad \text{H}_2\text{C} \quad \text{CH}\!\!-\!\!\text{CNHR} \\
| \qquad\qquad\qquad\quad \diagdown\!\diagup \\
\text{CH}_2 \qquad\qquad\qquad \text{N} \\
| \qquad\qquad\qquad\qquad | \\
\text{CH}_2 \qquad\qquad\quad \text{O} = \text{C} \\
| \qquad\qquad\qquad\qquad | \\
\text{COOH} \qquad\qquad\quad \text{R}'
\end{array}
\xrightarrow[\substack{\text{reducing} \\ \text{agent} \\ \text{(ascorbate)}}]{\text{Fe}^{2+}}
\begin{array}{l}
\text{CH}\!\!-\!\!\text{CH}_2 \; \text{O} \\
| \qquad | \quad \| \\
\text{H}_2\text{C} \quad \text{CH}\!\!-\!\!\text{CNHR} \qquad + \text{CO}_2 \; (\text{Q}) \\
\diagdown\!\diagup \\
\text{N} \qquad\qquad\quad + \text{COOH} \\
| \qquad\qquad\qquad\qquad | \\
\text{O} = \text{C} \qquad\qquad\quad \text{CH}_2 \\
| \qquad\qquad\qquad\qquad | \\
\text{R}' \qquad\qquad\qquad \text{CH}_2 \; ;
\end{array}
$$

$$(\text{A}) \;+\; (\text{B}) \;+\; (\text{C}) \qquad\longrightarrow\qquad (\text{P}) \qquad\qquad \begin{array}{l} | \\ \text{COOH} \end{array}$$

$$(\text{R}) \qquad (208)$$

and is classified as an iron-α-ketoglutarate dioxygenase.[273,274] Other members of this interesting class of enzymes include lysyl hydroxylase[275,276] and γ-butyrobetaine hydroxylase.[277] The steady-state kinetics of the prolyl-4-hydroxylase-catalyzed reaction has been extensively studied[278-280] with the use of a highly purified enzyme from chick embryos and with the synthetic substrates, (pro-pro-gly)$_5$ and (pro-pro-gly)$_{10}$. The kinetic analysis of Kivirikko et al.[278-280] utilized the rules and tables of Cleland[82,195] and Plowman[282] for multisubstrate reactions. Since the reaction utilizes five reactants, three of which are substrates, Soskel and Kuby[271] attempted an analysis of the published kinetic data of Kivirikko et al.,[278,279] with the use of the overall velocity equation for an ordered ter ter mechanism (see above Equation 194).

For the forward direction (i.e., assuming $P = Q = R = 0$), Equation 194 reduced to

$$\frac{V_{max}^f}{v_0^f} = 1 + \frac{K_{mA}}{(A)} + \frac{K_{mB}}{(B)} + \frac{K_{mC}}{(C)} + \frac{K_{ia}K_{mB}}{(A)(B)} + \frac{K_{ib}K_{mC}}{(B)(C)} + \frac{K_{ia}K_{ib}K_{mC}}{(A)(B)(C)} \qquad (209)$$

To analyze the data graphically, three conditions were selected (and these are summarized in Table 1). Thus, for condition 1: (A), or α-ketoglutarate, is varied, (C), or (pro-pro-gly)$_5$ or (pro-pro-gly)$_{10}$, is fixed at different concentrations, and (B), or O_2, is held constant. Solving for the varied substrate, we obtain

$$\frac{V_{max}^f}{v_0^f} = \left(\frac{1}{A}\right) \left\{ K_{mA} + \frac{K_{ia}K_{mB}}{(B)} + \frac{K_{ia}K_{ib}K_{mC}}{(B)(C)} \right\} + 1 + \frac{K_{mB}}{(B)} + \frac{K_{mC}}{(C)} + \frac{K_{ib}K_{mC}}{(B)(C)} \qquad (210)$$

A *primary* plot in double-reciprocal form would yield a family of lines which intersect to the left of the $\frac{1}{v}$ - axis, with the slopes and Y- and X-intercepts given in Table 1. A secondary plot of the slopes of the primary plot vs. $\frac{1}{(C)}$ at constant (B) yields a similar pattern whose slopes and Y- and X-intercepts are also given in Table I. Finally, a tertiary plot of the Y-intercept of the secondary plot vs. $\frac{1}{(B)}$ yields a straight line of

$$\text{slope} = \frac{K_{ia}K_{m_B}}{V^f_{max}} \text{ and Y-intercept} = \frac{K_{m_A}}{V^f_{max}}$$

(see Table 1). Similarly, these manipulations can be carried out for conditions 2 and 3, i.e., with (C) varied, (A) fixed, and at different concentrations with (B) held constant — or with (C) varied, (B) fixed, and (A) held constant. (See Table I for a summary of pertinent solutions of the primary, secondary, and tertiary plots under consideration.) Lastly, in condition 4, if (B) is held at saturation (while (C) is varied at fixed values of (A)), i.e., by allowing (B) to approach infinity, the $\frac{1}{(B)}$ terms approach zero in the $\frac{1}{v^f_0}$ vs. $\frac{1}{(C)}$ equation (see above Equation 210) which reduces to

$$\frac{V^f_{max}}{v} = \frac{1}{(A)} K_{m_A} + \frac{K_{m_C}}{(C)} + 1 \tag{211}$$

Thus, in condition 4, parallel lines are produced in the primary plot; however, the secondary plot allows direct determination of K_{m_A} and V^f_{max} and completes the information needed to calculate all the kinetic parameters (V^f_{max}, K_{ia}, K_{ib}, K_{m_A}, K_{m_B}, and K_{m_C}).

The patterns from Myllylä et al.[279] can now be compared to those predicted above. Condition 1 (α-ketoglutarate, (A) varied, peptide (pro-pro-gly)$_5$ (C) fixed at different concentrations, oxygen (B) held constant) yields lines intersecting to the left of the $\frac{1}{v_0}$ axis, a pattern identical to that predicted for the ordered ter ter mechanism. Since the reaction was carried out at only one concentration of oxygen, the secondary plot yields only a single line, and the tertiary plot yields a single value. Myllylä et al.[279] further describe the primary plot of condition 2 (peptide (C) varied, α-ketoglutarate (A) fixed at different concentrations, oxygen (B) held constant) as one which consists of lines intersecting to the left of the $\frac{1}{v_o}$ axis. A similar pattern is described for the primary plot of condition 3 (peptide varied, oxygen fixed at different concentrations, and α-ketoglutarate held constant). No information is given in regard to the secondary or tertiary plots of conditions 2 and 3. Condition 4 (oxygen saturated, α-ketoglutarate varied, and peptide fixed at different concentrations) is shown by Myllylä et al.[279] for the peptide (pro-pro-gly)$_{10}$, but not for the peptide (pro-pro-gly)$_5$. This pattern, which consists of parallel lines, is consistent with the pattern predicted for the ordered ter ter mechanism (*vide supra*), but a comparison of the kinetic patterns of this longer peptide with those of the shorter peptide cannot be made directly. Thus, for conditions 1, 2, 3, and possibly 4, the prolyl-4-hydroxylase reaction appears to fit the primary plots predicted for the initial forward velocity expression for an ordered ter ter mechanism (Equation 209). A further discussion of the mechanism of the prolyl-4-hydroxylase reaction is postponed until the *product* inhibition data[279] are discussed and compared against the predicted patterns for an ordered ter ter reaction (see Chapter 3, Table 2).

2. Alternate Nomenclature

Although the Cleland nomenclature apparently is now in general use, there are other methods of defining groups of reaction-rate constants which have been used, or are still in use, especially for specific cases and for which they are more applicable. We will illustrate these several sets of nomenclature for the two-substrate compulsory binding mechanism (or ordered Bi Bi reaction) which can be written in general form for the forward initial velocity and in the absence of products as

TABLE 1
Results of Double-Reciprocal Plots of Initial Forward Velocity Expressions for an Ordered Ter Ter Kinetic Mechanism[271]

Condition 1[a]

	Primary plot 1/v vs. 1/(A)	Secondary plot slope vs. 1/(C)	Tertiary plot Y-intercept vs. 1/(B)
Y-intercept	$\left(\frac{1}{V^f}\right)\left(1 + \frac{K_{mB}}{(B)} + \frac{K_{mC}}{(C)} + \frac{K_{ib}K_{mC}}{(B)(C)}\right)$	$\left(\frac{1}{V^f}\right)\left(K_{mA} + \frac{K_{ia}K_{mB}}{(B)}\right)$	$\frac{K_{mA}}{V^f}$
Slope	$\left(\frac{1}{V^f}\right)\left(K_{mA} + \frac{K_{ia}K_{mB}}{(B)} + \frac{K_{ia}K_{ib}K_{mC}}{(B)(C)}\right)$	$\left(\frac{1}{V^f}\right)\frac{K_{ia}K_{ib}K_{mC}}{(B)(C)}$	$\frac{K_{ia}K_{mB}}{V^f}$
X-intercept	$-\frac{((B)(C) + (C)K_{mB} + (B)\,K_{mC} + K_{ib}K_{mC})}{((B)(C)K_{mA} + (C)K_{ia}K_{mB} + K_{ia}K_{ib}K_{mC})}$	$\frac{-(B)K_{mA} - K_{ia}K_{mB}}{K_{ia}K_{ib}K_{mC}}$	$-\frac{K_{mA}}{K_{ib}K_{mB}}$

Condition 2[b]

	Primary plot 1/v vs. 1/(C)	Secondary plot slope vs. 1/(A)	Tertiary plot Y-intercept vs. 1/(B)
Y-intercept	$\left(\frac{1}{V^f}\right)\left(1 + \frac{K_{mA}}{(A)} + \frac{K_{mB}}{(B)} + \frac{K_{ia}K_{mB}}{(A)(B)}\right)$	$\left(\frac{1}{V^f}\right)\left(K_{mC} + \frac{K_{ia}K_{mC}}{(B)}\right)$	$\frac{K_{mC}}{V^f}$
Slope	$\left(\frac{1}{V^f}\right)\left(K_{mC} + \frac{K_{ib}K_{mC}}{(B)}\frac{K_{ia}K_{ib}K_{mC}}{(A)(B)}\right)$	$\left(\frac{1}{V^f}\right)\left(\frac{K_{ia}K_{ib}K_{mC}}{(B)}\right)$	$\frac{K_{ib}K_{mC}}{V^f}$
X-intercept	$-\frac{((A)(B) + (B)K_{mA} + (A)K_{mB} + K_{ia}K_{mB}}{((A)(B)K_{mC} + (A)\,K_{ib}K_{mC} + K_{ia}K_{ib}K_{mC})}$	$-\frac{((B) + K_{ib})}{K_{ia}K_{ib}}$	$-\frac{1}{K_{ib}}$

Condition 3[c]

	Primary plot 1/v vs. 1/(C)	Secondary plot slope vs. 1/(B)	Tertiary plot slope vs. 1/(A)
Y-intercept	Same as condition 2	$\frac{K_{mC}}{V^f}$	$\frac{K_{ib}K_{mC}}{V^f}$
Slope	Same as condition 2	$\left(\frac{1}{V^f}\right)\left(K_{ib}K_{mC} + \frac{K_{ia}K_{ib}K_{mC}}{(A)}\right)$	$\frac{K_{ia}K_{ib}K_{mC}}{V^f}$
X-intercept	Same as condition 2	$-\frac{(A)}{(A)K_{ib} + K_{ia}K_{ib}}$	$-\frac{1}{K_{ia}}$

Condition 4[d]

	Primary plot 1/v vs. 1/(C)	Secondary plot Y-intercept vs. 1/(A)
Y-intercept	$\left(\frac{1}{V^f}\right)\left(+ \frac{K_{mA}}{(A)}\right)$	$\frac{1}{V^f}$
Slope	$\frac{K_{mC}}{V^f}$	$\frac{K_{mA}}{V^f}$
X-intercept	$-\frac{((A) + K_{mA})}{(A)K_{mA}}$	$-\frac{1}{K_{mA}}$

Note: (A) = α-ketoglutarate, (B) = O_2, and (C) = peptide substrate—$(\text{pro-pro-gly})_5$ or $(\text{pro-pro-gly})_{10}$.

[a] (A) is varied, (B) is held constant, and (C) is fixed at different concentrations.

[b] (A) is fixed at different concentrations, (B) is held constant, and (C) is varied.

[c] (A) is held constant, (B) is fixed at different concentrations, and (C) is varied.

[d] (A) is fixed at different concentrations, (B) is saturated, and (C) is varied.

$$v_0^f = \frac{\delta_1 E_t(A)(B)}{\delta_2 + \delta_3(A) + \delta_4(B) + \delta_5(A)(B)} \tag{212}$$

where δ_1, δ_2, δ_3, δ_4, and δ_5 are different groups of rate constants.

According to Alberty,[57] divide numerator and denominator by δ_5 and define the resulting constant terms as follows:

$$v = \frac{\delta_1/\delta_5 E_t}{\frac{\delta_2}{\delta_5}\frac{1}{(A)}\frac{1}{(B)} + \frac{\delta_3}{\delta_5}\frac{1}{(B)} + \frac{\delta_4}{\delta_5}\frac{1}{(A)} + 1}$$

$$v_f = \frac{V_f(A)(B)}{K_{AB} + K_B(A) + K_A(B) + (A)(B)} \tag{213}$$

and written in reciprocal form:

$$\frac{V_f}{v_f} = + \frac{K_A}{(A)} + \frac{K_B}{(B)} + \frac{K_{AB}}{(A)(B)} \tag{214}$$

where

$$V_f = \frac{\delta_1}{\delta_5} E_t; \quad K_{AB} = \frac{\delta_2}{\delta_5}; \quad K_B = \frac{\delta_3}{\delta_5}; \quad K_A = \frac{\delta_4}{\delta_5}$$

K_A and K_B are limiting Michaelis constants, i.e., when its paired substrate concentration approaches infinity (K_A when $(B) \rightarrow \infty$; K_B when $(A) \rightarrow \infty$). V_f is the maximal velocity of the forward reaction, while K_{AB} is a complex constant with the dimensions of M^2.

The method proposed by Dalziel[44] follows: divide numerator and denominator of Equation 212 by δ_1:

$$\left(v_f = \frac{E_t(S_1)(S_2)}{(\delta_2/\delta_1) + (\delta_3/\delta_1)(S_1) + (\delta_4/\delta_1)(S_2) + (\delta_5/\delta_1)(S_1)(S_2)}\right)$$

$$v_f = \frac{E_t(S_1)(S_2)}{\phi_{12} + \phi_2(S_1) + \phi_1(S_2) + \phi_0(S_1)(S_2)} \tag{215}$$

or, in reciprocal form

$$\frac{E_t}{v_f} = \phi_0 + \frac{\phi_1}{(S_1)} + \frac{\phi_2}{(S_2)} + \frac{\phi_{12}}{(S_1)(S_2)} \tag{216}$$

where the two substrates, as we have seen, are indicated by (S_1) and (S_2), and

$$\phi_{12} = \delta_2/\delta_1; \quad \phi_2 = \delta_3/\delta_1; \quad \phi_1 = \delta_4/\delta_1; \quad \phi_0 = \delta_5/\delta_1$$

Similar to the Alberty system, the dimensions of the different constants are not all the same, thus, ϕ_o = minutes, ϕ_1 and ϕ_2 = M × minutes, and ϕ_{12} = M^2 × minutes.

To show a comparison with Alberty's system, divide by ϕ_o:

$$\frac{E_t}{v}\frac{1}{\phi_0} = 1 + (\phi_1/\phi_0)\frac{1}{(S_1)} + (\phi_2/\phi_0)\left(\frac{1}{S_2}\right) + (\phi_{12}/\phi_0)\frac{1}{(S_1)}\frac{1}{(S_2)}$$

where

$$V_f = E_t/\phi_0;\ \ K_A = \phi_1/\phi_0;\ \ K_B = \phi_2/\phi_0;\ \ K_{AB} = \phi_{12}/\phi_0$$

Bloomfield et al.[69] have proposed the following method.

Divide numerator and denominator of Equation 212 by δ_2 and define $V_f = V_{AB}$

$$v = \frac{\dfrac{V_{AB}}{K_{AB}}(A)(B)}{1 + \dfrac{(A)}{K_A} + \dfrac{(B)}{K_B} + \dfrac{(A)(B)}{K_{AB}}} \text{ or } v = \frac{V_{AB}(A)(B)}{K_{AB} + \dfrac{K_{AB}(A)}{K_A} + \dfrac{K_{AB}(B)}{K_B} + (A)(B)} \tag{217}$$

and

$$\frac{V_{AB}}{v_f} = 1 + \frac{K_{AB}}{K_B}\frac{1}{(A)} + \frac{K_{AB}}{K_A}\frac{1}{(B)} + K_{AB}\frac{1}{(A)}\frac{1}{(B)} \tag{218}$$

After division of Equation 212 by δ_2, divide through by δ_5/δ_2 and take the reciprocal:

$$\frac{\delta_1}{\delta_5}\frac{E_t}{v} = 1 + \frac{\delta_4}{\delta_5}\frac{1}{(A)} + \frac{\delta_3}{\delta_5}\frac{1}{(B)} + \frac{\delta_2}{\delta_5}\frac{1}{(A)(B)}\bigg]$$

$$\therefore V_{AB} = \frac{\delta_1}{\delta_5}E_t,\ K_A = \frac{\delta_2}{\delta_3},\ K_B\ \frac{\delta_2}{\delta_4},\ \text{and } K_{AB} = \frac{\delta_2}{\delta_5}$$

$$\left(\text{Since } \frac{K_{AB}}{K_A} = \frac{\delta_3}{\delta_5} \text{ and } K_{AB} = \frac{\delta_2}{\delta_5}\text{; therefore, } K_A = \frac{\delta_2}{\delta_3}\text{;}\right.$$

$$\left.\frac{K_{AB}}{K_B} = \frac{\delta_4}{\delta_5}\text{; therefore, } K_B = \frac{\delta_2}{\delta_4}\right)$$

Cleland's method involves dividing numerator and denominator of Equation 212 by δ_5, as in the case of Alberty's equation, but define the complex term K_{AB} as $K_{ia}K_b$ (i.e., $K_{ia}\,K_b = (\delta_2/\delta_5)$; $K_{ia} = (\delta_2/\delta_5)\dfrac{1}{K_b} = (\delta_2/\delta_5)/(\delta_3/\delta_5) = \delta_2/\delta_3$).

$$v_f = \frac{V_1(A)(B)}{K_{ia}K_b + K_b(A) + K_a(B) + (A)(B)} \tag{219}$$

or in reciprocal form:

$$\frac{V_1}{v_f} = 1 + \frac{K_a}{(A)} + \frac{K_b}{(B)} + \frac{K_{ia}K_b}{(A)(B)} \tag{220}$$

where

$$V_1 = \left(\frac{\delta_1}{\delta_5}\right) E_t; \quad K_a = \frac{\delta_4}{\delta_5}; \quad K_b \frac{\delta_3}{\delta_5}; \quad \text{but } K_{ia} = \frac{\delta_2}{\delta_3}$$

Therefore,

$$K_{AB} = K_{ia}K_b = \delta_2/\delta_3 \times \delta_3/\delta_5 = \delta_2/\delta_5$$

These symbols in the different nomenclature systems are compared in Table 2 for a two-substrate compulsory binding mechanism (i.e., an ordered Bi Bi sequential reaction, as in the Cleland description).

Additional symbolism applicable to studies of kinetic isotope effects[490] will be presented later.

3. Random Three-Substrate (Terreactant) Rapid (Quasi-) Equilibrium System

The equilibria shown in Scheme XL is for a case where an enzyme requires the simultaneous presence of all three of its substrates for catalytic activity. If A, B, and C are the three substrates which yield products (e.g., P, Q, and R), and if there is a random addition of each of the substrates to yield (via the three binary complexes EA, EB, and EC), the three ternary complexes (EAB, EBC, and EAC) which in turn yield, by a random order of addition, the single quaternary complex, EABC, which finally breaks down to eventually yield the free enzyme (E) and the products (P, Q, and R), then the mechanism can be described or visualized in the form of a ''cubic'' equilibrium as shown in Scheme XL:

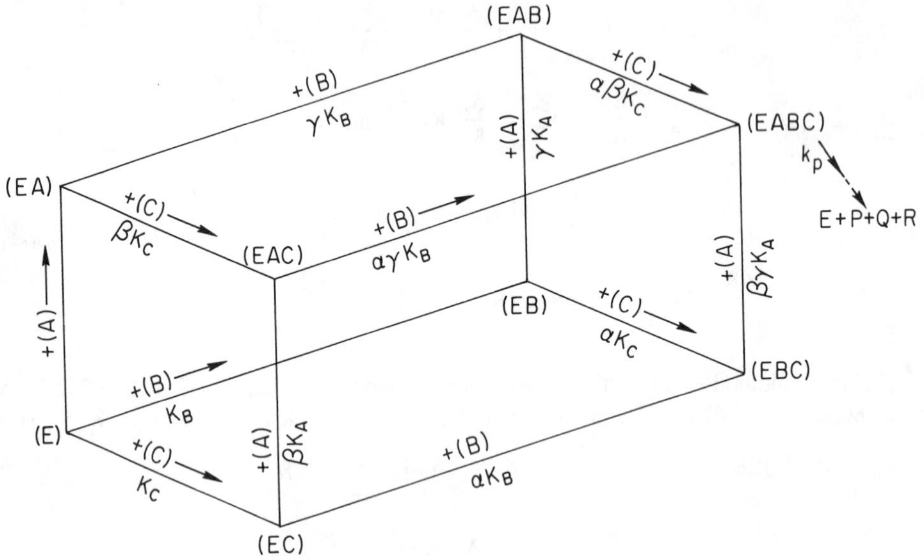

SCHEME XL. Equilibria between the various enzyme species in a random rapid (quasi-) equilibrium three-substrate reaction.

or, perhaps in clearer fashion in Scheme XLI.

TABLE 2
Different Symbols Used for Kinetic Parameters (Constants) of a Compulsory Ordered Two-Substrate Reaction without Products

Symbols used in this text								
Steady state	Quasi-equilibrium	Definition	Alberty[57]	Dalziel[44]	Bloomfield et al.[69]	Cleland[43]	Mahler and Cordes[187]	Enzyme Commission[63]
K_{mA}	K_A	Limiting Michaelis constant for (A)	K_A	ϕ_1/ϕ_0	K_{AB}/K_B	K_a	K_a	K_m^A
K_{mB}	K_B	Limiting Michaelis constant for (B)	K_B	ϕ_2/ϕ_0	K_{AB}/K_A	K_b	K_b	K_m^B
K_{ia}	$\bar{K}_A$	Dissociation constant for (A)	K_{AB}/K_B	ϕ_{12}/ϕ_2	K_A	K_{ia}	$\bar{K}_a$	K_s^A
V_{max}^f	V_{max}^f	Limiting maximum velocity	V_f	ℓ/ϕ_0	V_{AB}	V_1	V_1	V
V_{max}^f/E_t	V_{max}^f/E_t	Turnover number	V_f/E_0	$\frac{1}{\phi_0}$	V_{AB}/E_0	V_1/E_t	$V_1\ell_0$	None

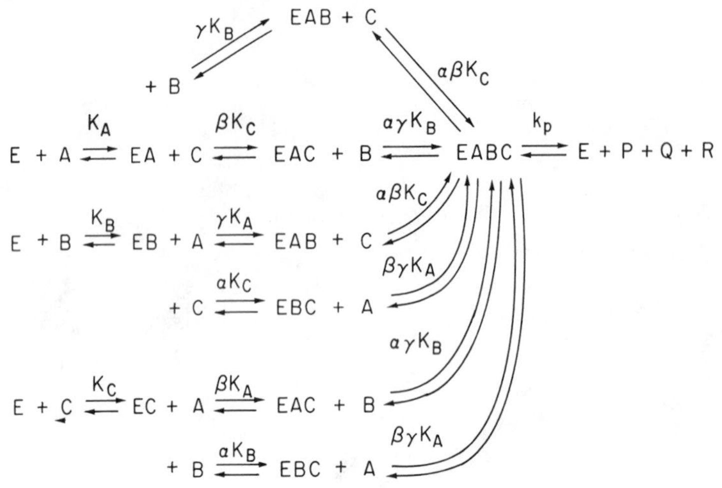

SCHEME XLI.

The interaction factors, α, β, and γ, indicate the effect on the dissociation constant for a given ligand when other ligands are, in turn, bound to their individual binding sites on the enzyme.

The resulting velocity equation for the forward reaction is

$$\frac{v_f}{V^f_{max}} = \frac{\dfrac{(A)(B)(C)}{\alpha\beta\gamma K_A K_B K_C}}{1 + \dfrac{(A)}{K_A} + \dfrac{(B)}{K_B} + \dfrac{(C)}{K_C} + \dfrac{(A)(B)}{\gamma K_A K_B} + \dfrac{(A)(C)}{\beta K_A K_C} + \dfrac{(B)(C)}{\alpha K_B K_C} + \dfrac{(A)(B)(C)}{\alpha\beta\gamma K_A K_B K_C}} \tag{221}$$

If the concentrations of the substrates (A) and (B) are fixed and the concentration of (C) is varied, after rearrangement:

$$\frac{v_f}{V^f_{max}} = \frac{(C)}{\alpha\beta K_C\left[1 + \dfrac{\gamma K_A}{(A)} + \dfrac{\gamma K_B}{(B)} + \dfrac{\gamma K_A K_B}{(A)(B)}\right] + (C)\left[1 + \dfrac{\beta\gamma K_A}{(A)} + \dfrac{\alpha\gamma K_B}{(B)} + \dfrac{\alpha\beta\gamma K_A K_B}{(A)(B)}\right]} \tag{222}$$

In a double reciprocal plot of $\dfrac{1}{v_f}$ vs. $\dfrac{1}{(C)}$ and several fixed concentrations of (B) (and constant A), (or constant (B) and several fixed concentrations of A), the family of intersecting lines will intersect to the left of the $\dfrac{1}{v}$-axis either on, above, or below the $\dfrac{1}{(C)}$ - axis, dependent on the values for the factors α, β, and γ. If $\alpha = \beta = \gamma = 1$, the double-reciprocal plots will intersect *on* the $\dfrac{1}{C}$ - axis.

If Equation 222 is again rearranged to express (C) as the varied substrate, (B) as the changing fixed substrate, and (A) as the constant substrate:

$$\frac{v_f}{V_{max}} = \frac{(C)}{\alpha\beta K_C\left(1 + \dfrac{\gamma K_A}{(A)}\right)\left[1 + \dfrac{\gamma K_B(1 + K_A/(A))}{(B)(1 + \gamma K_A/(A))}\right] + (C)\left(1 + \dfrac{\beta\gamma K_A}{(A)}\right)\left[1 + \dfrac{\alpha\gamma K_B(1 + \beta K_A/(A))}{(B)(1 + \beta\gamma K_A/(A))}\right]} \tag{223}$$

If (B) > K_B, i.e., (B) is saturating, the terms in the brackets approach unity, and the equation reduces to

$$\lim_{(K_B/(B))\to 0}\left(\frac{v_f}{V_{max}}\right) = \frac{(C)}{\alpha\beta K_C\left(1 + \frac{\gamma K_A}{(A)}\right) + (C)\left(1 + \frac{\beta\gamma K_A}{(A)}\right)} \tag{224}$$

or,

$$\frac{V_{max}}{v_f} = \frac{\alpha\beta K_C}{(C)}\left(1 + \frac{\gamma K_A}{(A)}\right) + \left(1 + \frac{\beta\gamma K_A}{(A)}\right) \tag{225}$$

Under these conditions, the limiting and apparent K_C and V_{max} values are still a function of the constant (A):

$$K_C^* = \alpha\beta K_C \frac{\left(1 + \frac{\gamma K_A}{(A)}\right)}{\left(1 + \frac{\beta\gamma K_A}{(A)}\right)}; \quad V_{max}^* = \frac{V_{max}}{\left(1 + \frac{\beta\gamma K_A}{(A)}\right)} \tag{226}$$

The relation for the slope of the $\frac{1}{v}$ vs. $\frac{1}{(C)}$ plot with changing fixed (A) and constant (B), held at saturation, designated the slope of $\frac{1}{C}$(B sat) and the corresponding Y-intercept of $(\frac{B\ sat}{C})$.

$$\underset{1/C(B\ sat)}{\text{Slope}} = \frac{1}{V_{max}}\alpha\beta K_C\left(1 + \frac{\gamma K_A}{(A)}\right) = \frac{\alpha\beta K_C}{V_{max}} + \frac{\alpha\beta\gamma K_A K_C}{V_{max}}\frac{1}{(A)} \tag{227}$$

A replot of the slope of $\frac{1}{C}$ (B sat) vs. $\frac{1}{(A)}$ yields values for $\frac{\alpha\beta K_C}{V_{max}}$ (secondary Y-intercept) and

$$\frac{\alpha\beta\gamma K_A K_C}{V_{max}} \text{ (secondary slope), and } \frac{\text{secondary slope}}{\text{secondary Y-intercept}} = \gamma K_A.$$

Also,

$$\underset{1/C(B\ sat)}{\text{Y-intercept}} = \frac{1}{V_{max}}\left(1 + \frac{\beta\gamma K_A}{(A)}\right) = \frac{1}{V_{max}} + \frac{1}{V_{max}}\frac{\beta\gamma K_A}{(A)} \tag{228}$$

and, in turn, a replot of the Y-intercept vs. $\frac{1}{A}$, yields as the secondary Y-intercept = $\frac{1}{V_{max}}$ and secondary slope = $\frac{\beta\gamma K_A}{V_{max}}$, or secondary slope/secondary intercept = $\beta\gamma K_A$. Between the two replots, a unique value for V_{max} and β are forthcoming, as well as γK_A, $\alpha\beta K_C$, and $\beta\gamma K_A$.

Similarly, when A is held at saturating concentrations, from Equation 221:

$$\lim_{(A/K_A)\to\infty}\left(\frac{v_f}{V^f_{max}}\right) = \frac{\dfrac{(A)(B)(C)}{\alpha\beta\gamma K_A K_B K_C}}{\dfrac{A}{K_A} + \dfrac{(A)(B)}{\gamma K_A K_B} + \dfrac{(A)(C)}{\beta K_A K_C} + \dfrac{(A)(B)(C)}{\alpha\beta\gamma K_A K_B K_C}} \tag{229}$$

$$\frac{v}{V_{max}} = \frac{\dfrac{(C)}{\alpha\beta\gamma K_B K_C}}{\dfrac{1}{B} + \dfrac{1}{\gamma K_B} + \dfrac{(C)}{\beta(B)K_c} + \dfrac{(C)}{\alpha\beta\gamma K_B K_C}} \tag{230}$$

$$\frac{v}{V_{max}} = \frac{(C)}{\dfrac{\alpha\beta\gamma K_B K_C}{(B)} + \alpha\beta K_C + \dfrac{\alpha\gamma(C)K_B}{(B)} + (C)} \tag{231}$$

$$\frac{v}{V_{max}} = \frac{(C)}{\alpha\beta K_C\left(1 + \dfrac{\gamma K_B}{(B)}\right) + (C)\left(1 + \dfrac{\alpha\gamma K_B}{(B)}\right)} \tag{232}$$

and

$$\frac{V_{max}}{v} = \frac{\alpha\beta K_C}{(C)}\left(1 + \frac{\gamma K_B}{(B)}\right) + \left(1 + \frac{\alpha\gamma K_B}{(B)}\right) \tag{233}$$

$\frac{1}{v}$ vs. $\frac{1}{C}$ (with A at saturation, and B at several fixed values) yields as

$$\text{slope} = \frac{\alpha\beta K_C}{V_{max}}\left(1 + \frac{\gamma K_B}{(B)}\right), \text{ or } \underset{1/C(A\ sat)}{\text{Slope}} = \frac{\alpha\beta K_C}{V_{max}} + \frac{\alpha\beta\gamma K_B K_C}{V_{max}(B)} \tag{234}$$

and

$$\text{Y-intercept} = \frac{1}{V_{max}}\left(1 + \frac{\alpha\gamma K_B}{(B)}\right), \text{ or } \underset{1/C(A\ sat)}{\text{Y-intercept}} = \frac{1}{V_{max}} + \frac{\alpha\gamma K_B}{V_{max}}\frac{1}{(B)} \tag{235}$$

Secondary plots (A = saturated) of $\frac{1}{v}$ vs. $\frac{1}{(C)}$ (at several fixed values of (B)) are shown in Figure 31 A and B (for slope vs. $\frac{1}{B}$ and Y-int. vs. $\frac{1}{B}$, respectively). Since

$$\frac{\text{x-intercept of sec slope plot}}{\text{x-intercept of sec Y-intercept plot}} = \alpha$$

and therefore, V_{max}, αBK_C, α, βK_C, γK_B, $\alpha\gamma K_B$, γK_B are determined; since β is estimated from Eq. II-228), K_c is now estimated. Similarly, if (C) is held at saturating concentrations, from Equation 221:

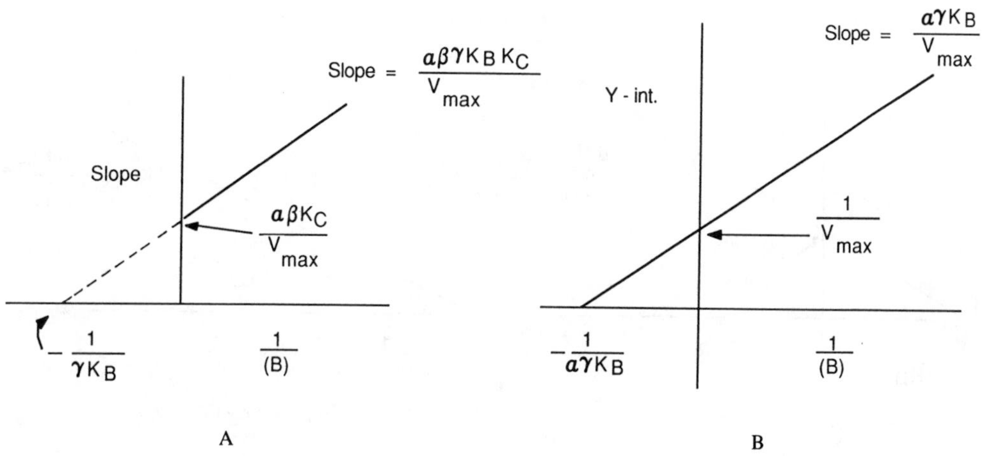

FIGURE 31. Secondary plots (data from Equation 233) of slope (A) and Y-int. (B) vs. $\frac{1}{(B)}$.

$$\underset{(C/K_C)\to\infty}{\text{LIM}}\left(\frac{v_f}{V^f_{max}}\right) = \frac{\dfrac{(A)(B)(C)}{\alpha\beta\gamma K_A K_B K_C}}{\dfrac{(C)}{K_C} + \dfrac{(A)(C)}{\beta K_A K_C} + \dfrac{(B)(C)}{\alpha K_B K_C} + \dfrac{(A)(B)(C)}{\alpha\beta\gamma K_A K_B K_C}} \tag{236}$$

$$\underset{(C/K_C)\to\infty}{\text{LIM}}\left(\frac{v_f}{V^f_{max}}\right) = \frac{\dfrac{(A)(B)}{\alpha\beta\gamma K_A K_B K_C}}{\dfrac{1}{K_C} + \dfrac{(A)}{\beta K_A K_C} + \dfrac{(B)}{\alpha K_B K_C} + \dfrac{(A)(B)}{\alpha\beta\gamma K_A K_B K_C}} \tag{237}$$

$$= \frac{(A)(B)}{\alpha\beta\gamma K_A K_B + \alpha\gamma(A)K_B + (\beta)(\gamma)(B)K_A + (A)(B)} \tag{238}$$

$$\frac{v_f}{V^f_{max}} = \frac{(A)}{\dfrac{\alpha\beta\gamma K_A K_B}{(B)} + \dfrac{\alpha\gamma(A)K_B}{(B)} + \beta\gamma K_A + (A)} \tag{239}$$

$$\text{and}\left(\frac{V^f_{max}}{v_f}\right) = \frac{\alpha\beta\gamma K_A K_B}{(A)(B)} + \frac{\alpha\gamma K_B}{(B)} + \frac{\beta\gamma K_A}{(A)} + 1 \tag{240}$$

$$\underset{(C/K_C)\to\infty}{\text{LIM}}\left(\frac{V^f_{max}}{v_f}\right) = \frac{\beta\gamma K_A}{(A)}\left(1 + \frac{\alpha K_A}{B}\right) + \left(1 + \frac{\alpha\gamma K_B}{(B)}\right) \tag{241}$$

$$\text{Slope}\left(\text{of } \frac{1}{v} \text{ vs. } \frac{1}{A} \text{ plot}\right) = \frac{\beta\gamma K_A}{V_{max}}\left(1 + \frac{\alpha K_B}{(B)}\right); \tag{242}$$

$$\text{Slope} = \frac{\beta\gamma K_A}{V_{max}} + \frac{\alpha\beta\gamma}{V_{max}}\frac{K_B}{(B)}$$

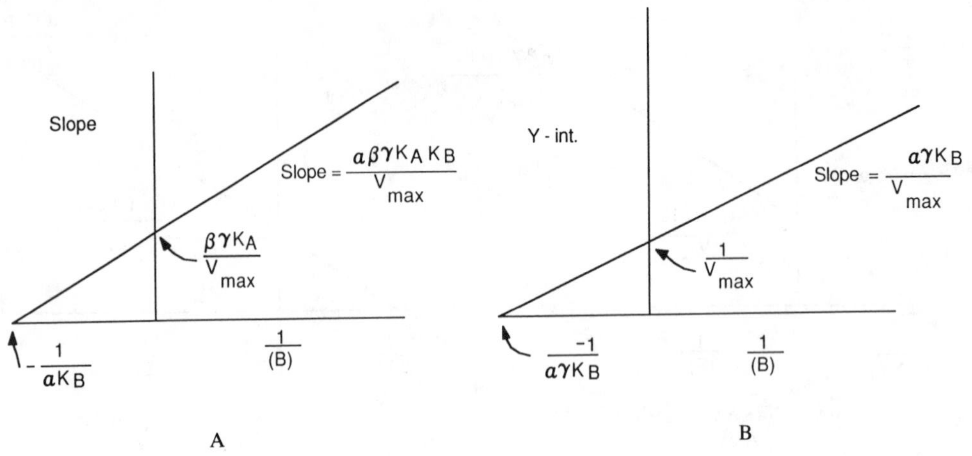

FIGURE 32. Secondary plots (data from Equation 241) of slope (A) and Y-int. (B) vs. $\frac{1}{(B)}$.

$$\text{Y-intercept (of } \frac{1}{v} \text{ vs. } \frac{1}{A} \text{ plot)} = \frac{1}{V_{max}} + \frac{\alpha\beta K_B}{(B)} \frac{1}{V_{max}} \tag{243}$$

Secondary plots are depicted in Figure 32A and B for slope and Y-int. vs. $\frac{1}{B}$. Therefore, since αK_B, $\alpha\gamma K_B$ are estimated, γ can be determined, as well as αK_B, $\beta\gamma K_A$.

With γ estimated here, β and α from Equation 235, $\therefore$ K_B is estimated here, as well as K_A and K_C from before, as well as V_{max}.

Therefore, all parameters are now estimated: α, β, γ, K_A, K_B, K_C, and V_{max}.

If it proves impractical to maintain one of the substrates at a saturating concentration, then a more tedious method can be developed with several families of reciprocal plots of, e.g., $\frac{1}{v}$ vs. $\frac{1}{C}$, at several fixed concentrations of (B) and at several constant values of (A).

Garfinkel et al.[957] have presented an interesting discussion of the recent application of microcomputers for the selection and determination of enzyme kinetic mechanisms. The steady-state kinetic measurements were fitted directly to programmed enzyme kinetic models, and the fitting was accomplished by a nonlinear regression analysis (see Chapter 9, Section II.B.2) operating in the computer language, Basic. They used the hexokinase system to test their computer analysis procedures.

Chapter 3

INHIBITION AND PRODUCT INHIBITION

The use of inhibitors, especially product inhibitors, are useful to distinguish between different mechanisms, especially for two-substrate reactions, as pointed out by Alberty in his classical paper.[70]

Different inhibition patterns may be found with different mechanisms, even when they lead to identical rate laws in the absence of inhibitors.

If product inhibition studies are carried out, together with an earlier suggestion of Alberty[57] for reversible systems, that is, a comparison of the apparent equilibrium constant which is directly estimated with that determined kinetically through the use of the Haldane relations (which we have already mentioned), then a self-consistent mechanism with some degree of confidence is likely to emerge.

An excellent review of the graphical properties of inhibitors, both linear and nonlinear cases, may be found in Dixon et al.[963] (see their Chapter 8, Table 2).

We have already indicated a few cases where specific inhibition patterns were characteristic for some of the mechanisms we previously considered; however, let us return to them and systematically develop their inhibition equations.

I. SINGLE-SUBSTRATE CASES — A SUMMARY

By way of a brief review, for single substrate reactions whose inhibition systems yield linear reciprocal primary and secondary plots (see Chapter 1.D), we can distinguish classically three main types of inhibitors. (1) competitive inhibition by (I) which competes for the substrate binding site and forms only an inactive EI complex (see Chapter 1, Scheme III, and Equations 42 and 41 and Figure 9).

$$\frac{1}{v_o^f} = \frac{1}{V_{max}^f} + \frac{1}{V_{max}^f}\frac{K_{ms}}{(S)}, \quad \text{in the absence of I;} \tag{1-42}$$

$$\frac{1}{v_o} = \frac{1}{V_{max}^f} + \frac{1}{V_{max}^f}\frac{K_{ms}}{(S)}\left(1 + \frac{(I)}{K_i}\right) \tag{1-41}$$

where only the slope, but not the Y-intercept is affected (see Figure 1). (2) Uncompetitive inhibition by (I), which reacts exclusively with the (ES) complex to form (EIS) (see Chapter 1, Scheme V, and Equation 52 and Figure 15).

$$\frac{1}{v_o} = \frac{1}{V_{max}^f}\left(1 + \frac{I}{K_i}\right) + \frac{K_{ms}}{V_{max}^f}\frac{1}{(S)} \tag{1-52}$$

where only the intercept is affected and not the slope, leading to a series of parallel lines in the double-reciprocal plot (as shown in Figure 2). (3) Noncompetitive inhibitors can combine with the enzyme at a site other than the substrate binding site, leading to both (EI) and (IES) complexes, and here both slope and Y-intercept are affected by the same degree $\left(1 + \frac{I}{K_i}\right)$.

In the case of "pure" noncompetitive inhibition, see Chapter 1, Scheme IV, Equation 48, and Figure 12.

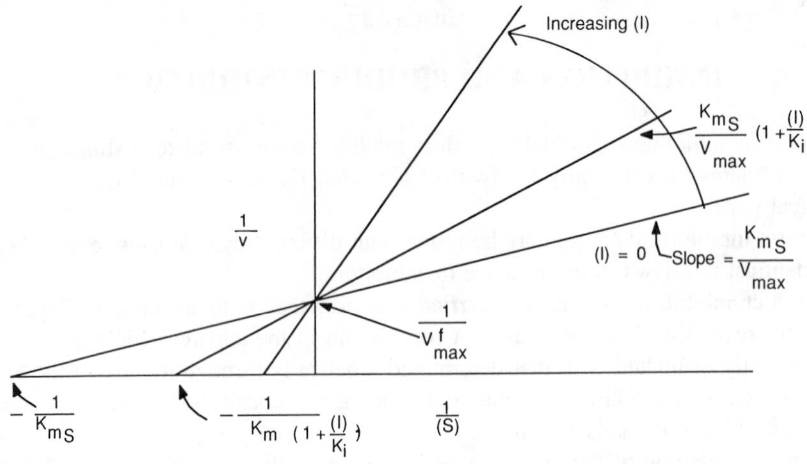

FIGURE 1.

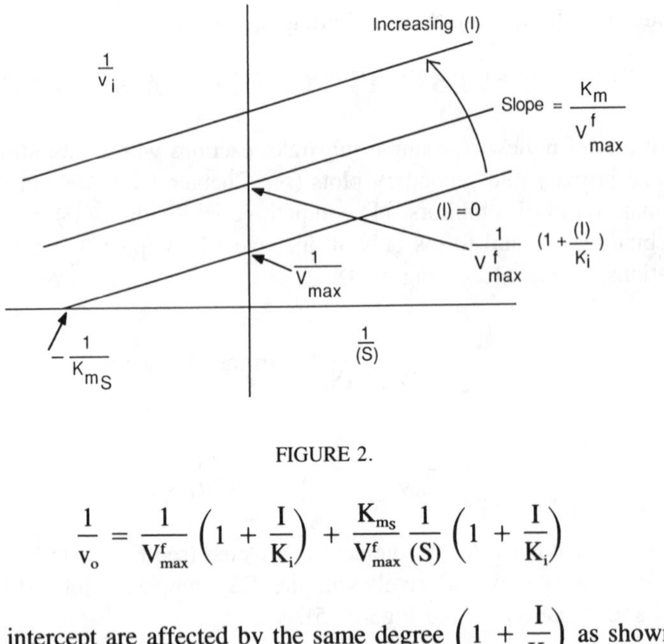

FIGURE 2.

$$\frac{1}{v_o} = \frac{1}{V^f_{max}}\left(1 + \frac{I}{K_i}\right) + \frac{K_{ms}}{V^f_{max}}\frac{1}{(S)}\left(1 + \frac{I}{K_i}\right) \tag{1-48}$$

both slope and intercept are affected by the same degree $\left(1 + \frac{I}{K_i}\right)$ as shown in Figure 3 and a common intersection point on the X-axis occurs $= \frac{-1}{K_{ms}}$.

In the case of the so-called "mixed-type" of inhibition which is a specific type of noncompetitive inhibition, the common intercept is not precisely on the X-axis (see Figure 4), and it is a case where both the slope and Y-intercept are affected, but not to the same degree (see Chapter 1, Scheme VIII, Equation 77, and Figure 21).

$$\frac{1}{v_o} = \frac{1}{V^f_{max}}\left(1 + \frac{I}{\alpha K_i}\right) + \frac{K_{ms}}{V^f_{max}}\frac{1}{(S_o)}\left(1 + \frac{I}{K_i}\right) \tag{1-77}$$

where (EI) and (ESI) complexes are inactive, but their K_is are not equal, and $\alpha K_i > K_i$: $K_{i_{ESI}}$ (for ESI $\overset{\alpha K_i}{\rightleftarrows}$ ES + I) and $K_{i_{EI}}$ (for EI $\overset{K_i}{\rightleftarrows}$ E + I).

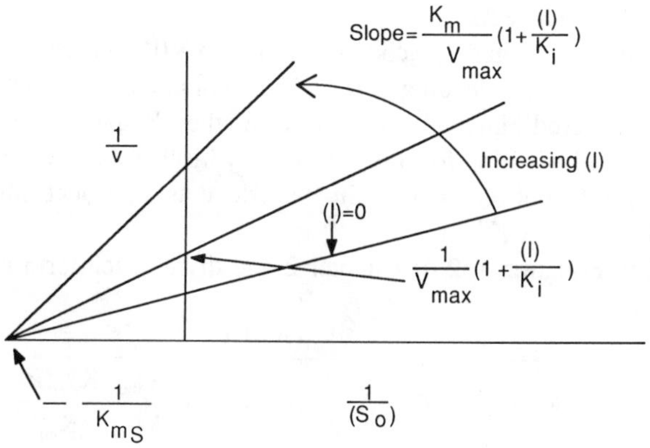

FIGURE 3.

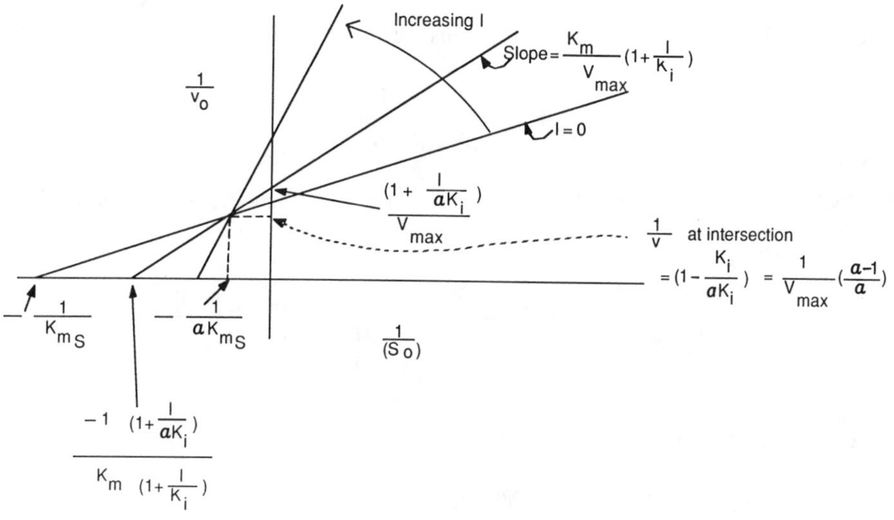

FIGURE 4.

II. MULTISUBSTRATE CASES

When more than one substrate participates in an enzymatic reaction, the kinetic effects exerted by inhibitors can become quite complex. Among other things, slopes and/or intercepts of the double-reciprocal plot may no longer be linear, but may become higher-order (usually second-order) functions of i in Cleland's nomenclature. Cleland[43] developed a set of rules for predicting by inspection the type of inhibition to be expected from an examination of the steady-state mechanisms containing no random sequences unless they were in rapid equilibrium; otherwise, the complete rate equation must be derived. In addition, different mechanisms generate quite different patterns for product inhibition and substrate inhibition, as well as for the effects observed on addition of dead-end inhibitors. To obtain a complete description, therefore, the complete rate equation should be derived, which in turn should take into account the complexes between all enzyme forms and inhibitors, and then the equation should be recast in a form to determine the effect on the initial rate which would be produced by varying any particular component.

We will consider, therefore, some of the mechanisms we have discussed and the effects of product inhibition as deduced from the complete velocity equation.

A. ORDERED BI BI SYSTEM

For this mechanism (Chapter 2, Schemes XXI to XXIII) where only the binary complexes, (EA) or (EQ) may form, leading to the ternary complexes; i.e., the binary complexes (EB) or (EP) are considered kinetically insignificant, then Equation 146 in Chapter 2 may be derived for the overall equation for this compulsory bisubstrate reaction (cf Equation 152 in Chapter 2). If (P) $\neq$ 0, but (Q) = 0, then P is added as a product inhibitor with respect to either (A) or (B).

Set (Q) = 0 into Equation 152 of Chapter 2 and divide each term by V^r_{max}:

$$v^f = \frac{V^f_{max}(A)(B)}{K_{ia}K_{m_B} + K_{m_B}(A) + K_{m_A}(B) + \frac{V^f_{max}}{V^r_{max}}\frac{K_{m_Q}(P)}{K_{eq}} + (A)(B) + \frac{V^f_{max}}{V^r_{max}}\frac{K_{mQ}}{K_{ia}}\frac{(A)(P)}{K_{eq}} + \frac{(A)(B)(P)}{K_{ip}}} \tag{1}$$

Recalling that one of the Haldane relations for

$$K_{eq} = \frac{V^f_{max}K_{iq}K_{mP}}{V^r_{max}K_{ia}K_{mB}} \tag{2-159}$$

and substituting K_{eq} into the (P) and the (A) (P) terms, then the denominator (P) term becomes

$$\frac{V^f_{max}K_{m_Q}(P)}{V^r_{max}K_{eq}} = \frac{V^f_{max}}{V^r_{max}}\frac{K_{m_Q}(P)}{\left[\frac{V^f_{max}K_{iq}K_{mP}}{V^r_{max}K_{ia}K_{mB}}\right]} = \frac{K_{m_Q}K_{m_B}K_{ia}(P)}{K_{iq}K_{mP}} \tag{2}$$

and the denominator (A) (P) term becomes

$$\frac{V^f_{max}K_{m_Q}(A)(P)}{V^r_{max}K_{ia}K_{eq}} = \frac{V^f_{max}K_{m_Q}(A)(P)}{V^r_{max}K_{ia}\left[\frac{V^f_{max}K_{iq}K_{mP}}{V^r_{max}K_{ia}K_{mB}}\right]} = \frac{K_{m_Q}K_{m_B}(A)(P)}{K_{iq}K_{mP}} \tag{3}$$

After substitution into and dividing by (A) (B), Equation 1 becomes

$$\frac{v^f}{V^f_{max}} = \frac{1}{\frac{K_{ia}K_{m_B}}{(A)(B)} + \frac{K_{m_B}}{(B)} + \frac{K_{m_A}}{(A)} + \frac{K_{m_Q}K_{m_B}K_{ia}(P)}{K_{iq}K_{mP}(A)(B)} + 1 + \frac{K_{m_Q}K_{m_B}(P)}{K_{iq}K_{mP}(B)} + \frac{(P)}{K_{ip}}} \tag{4}$$

For varied (A), fixed B, and fixed (P), the product:

$$\frac{v^f}{V^f_{max}} = \frac{1}{\left[1 + \frac{K_{m_B}}{(B)}\left(1 + \frac{K_{m_Q}(P)}{K_{iq}K_{mP}}\right) + \frac{P}{K_{iq}}\right] + \frac{K_{m_A}}{(A)}\left[1 + \frac{K_{ia}K_{m_B}}{K_{m_A}(B)}\left(1 + \frac{K_{m_Q}(P)}{K_{iq}K_{mP}}\right)\right]} \tag{5}$$

or

$$\frac{1}{v^f} = \frac{1}{V^f_{max}}\left[1 + \frac{K_{m_B}}{(B)}\left(1 + \frac{K_{m_Q}(P)}{K_{iq}K_{mp}}\right) + \frac{(P)}{K_{ip}}\right]$$

$$+ \frac{K_{m_A}}{(A)V^f_{max}}\left[1 + \frac{K_{ia}K_{m_B}}{K_{m_A}(B)}\left(1 + \frac{K_{m_Q}(P)}{K_{iq}K_{mp}}\right)\right] \tag{6}$$

It is evident that: (1) (P) will act as a noncompetitive inhibitor of the mixed type with respect to (A) when fixed (B) is not held at 100 × K_{m_B} or larger (i.e., at saturation of (B)); for both the slope and Y-intercept of a $\frac{1}{v}$ vs. $\frac{1}{(A)}$ plot will be affected by (P), but not to the same degree; and (2) if (B) is held at saturation, i.e., $\frac{K_{m_B}}{(B)} = \frac{1}{100}$, then Equation 6 reduces to

$$\frac{1}{v^f} \cong \frac{1}{V^f_{max}}\left[1 + \frac{(P)}{K_{ip}}\right] + \frac{K_{m_A}}{(A)V^f_{max}} \tag{7}$$

which still retains the $\frac{(P)}{K_{ip}}$ in the Y-intercept and thus, although at saturation, (B) reduces the slope to that which it would equal in the absence of (P), it cannot eliminate the Y-intercept inhibitor factor $\left(1 + \frac{(P)}{K_{ip}}\right)$; therefore, (P) acts as an uncompetitive inhibitor with respect to (A) at saturating (B).

Now, when (B) is varied with (A) held fixed and (P) fixed

$$\frac{v^f}{V^f_{max}} = \frac{1}{\left[1 + \frac{K_{m_A}}{(A)} + \frac{(P)}{K_{ip}}\right] + \frac{K_{m_B}}{(B)}\left[1 + \frac{K_{ia}}{(A)} + \frac{K_{m_Q}(P)}{K_{iq}K_{mp}} + \frac{K_{m_Q}}{K_{iq}}\frac{K_{ia}(P)}{K_{mp}(A)}\right]} \tag{8}$$

$$= \frac{1}{\left[1 + \frac{K_{m_A}}{(A)} + \frac{(P)}{K_{ip}}\right] + \frac{K_{m_B}}{(B)}\left[1 + \frac{K_{m_Q}(P)}{K_{iq}K_{mp}} + \frac{K_{ia}}{(A)}\left(1 + \frac{K_{m_Q}(P)}{K_{iq}K_{mp}}\right)\right]} \tag{9}$$

$$\frac{v^f}{V^f_{max}} = \frac{1}{\left[1 + \frac{K_{m_A}}{(A)} + \frac{(P)}{K_{ip}}\right] + \frac{K_{m_B}}{(B)}\left[\left(1 + \frac{K_{m_Q}(P)}{K_{iq}K_{mp}}\right)\left(1 + \frac{K_{ia}}{(A)}\right)\right]} \tag{10}$$

or in reciprocal form

$$\frac{1}{v^f_o} = \frac{1}{V^f_{max}}\left[1 + \frac{K_{m_A}}{(A)} + \frac{(P)}{K_{ip}}\right] + \frac{K_{m_B}}{(B)}\frac{1}{V^f_{max}}\left(1 + \frac{K_{m_Q}(P)}{K_{iq}K_{mp}}\right)\left(1 + \frac{K_{ia}}{(A)}\right) \tag{11}$$

(P) now appears to act as a noncompetitive inhibitor of the mixed-type again with respect to (B) as the varied substrate, at all concentrations of (A). Thus, with (A) held at saturation, or $\frac{K_{m_A}}{(A)} = \frac{1}{100} = \frac{K_{ia}}{(A)}$, Equation 11 reduces to

$$\frac{1}{v^f_o} \simeq \frac{1}{V^f_{max}}\left(1 + \frac{(P)}{K_{ip}}\right) + \frac{K_{m_B}}{(B)}\frac{1}{V^f_{max}}\left(1 + \frac{K_{m_Q}(P)}{K_{iq}K_{mp}}\right) \tag{12}$$

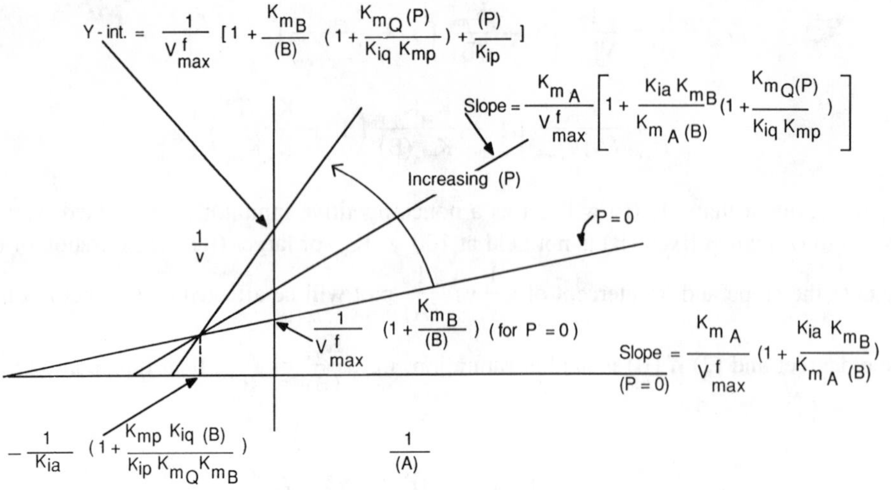

FIGURE 5A.

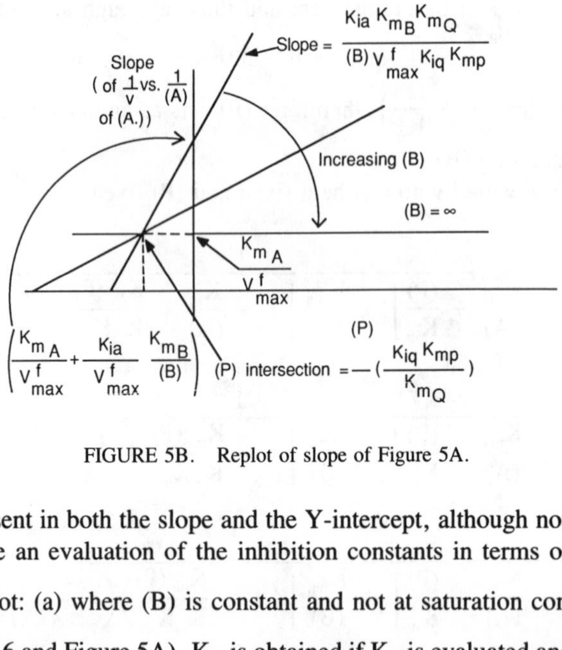

FIGURE 5B. Replot of slope of Figure 5A.

and (P) is still present in both the slope and the Y-intercept, although not in the same form.

We summarize an evaluation of the inhibition constants in terms of double-reciprocal plots. $\frac{1}{v}$ vs. $\frac{1}{(A)}$ plot: (a) where (B) is constant and not at saturation concentration and (Q) = 0 (see Equation 6 and Figure 5A). K_{iq} is obtained if K_{ia} is evaluated and with a knowledge of V^f_{max}, K_{m_Q}, K_{m_P}, and K_{m_B}. A replot of the data in Figure 5A is given in Figure 5B based on Equation 13:

$$\underset{(1/v\ \text{vs.}\ 1/(A))}{\text{Slope}} = \frac{K_{m_A}}{V^f_{max}} + \frac{K_{ia}K_{m_B}}{V^f_{max}(B)} + \frac{K_{ia}K_{m_B}K_{m_Q}(P)}{(B)V^f_{max}K_{iq}K_{mp}} \tag{13}$$

and this is a characteristic of a bisubstrate ordered reaction; at B = ∞, a zero slope is found in the secondary plot of Slope vs. (P). K_{ia} is obtained from K_{m_A}, K_{m_B}, V^f_{max}, (B).

(b) $\frac{1}{v}$ vs. $\frac{1}{(A)}$ plots where (B) is held constant at saturation, (Q) = 0, and (P) is fixed at several concentrations. (See Equation 7 and Figure 5.C).

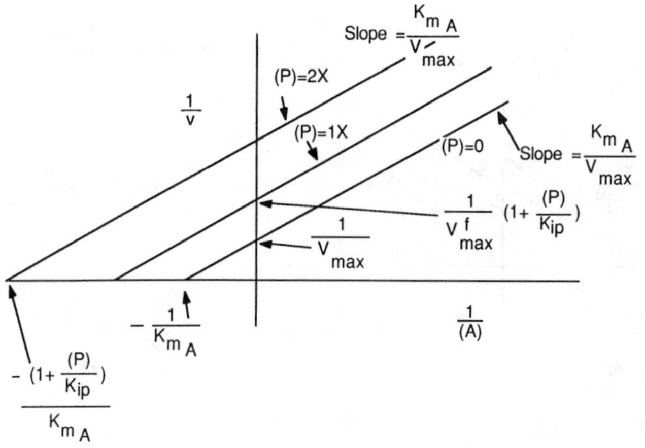

FIGURE 5C.

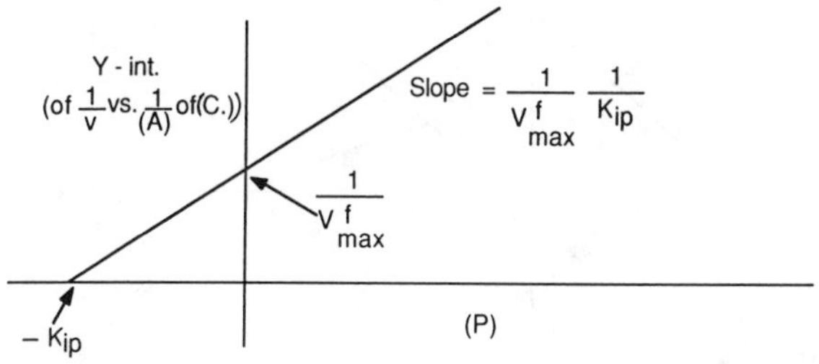

FIGURE 5D.

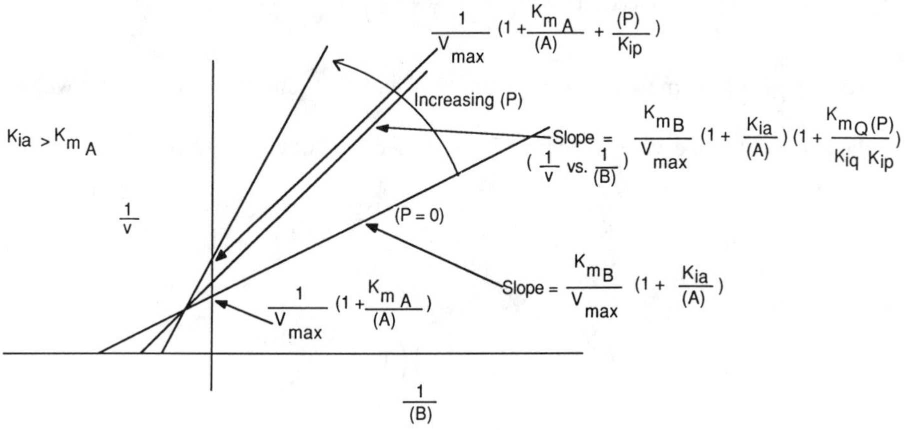

FIGURE 5E.

$$\text{Y-intercept} = \frac{1}{V^{f}_{max}} + \frac{(P)}{(V^{f}_{max})(K_{ip})} \tag{14}$$

(c) $\frac{1}{v}$ vs. $\frac{1}{(B)}$ plots (see Equation 11) where (B) is varied, (A) is constant and unsaturating, $P \neq 0$, and $Q = 0$, are shown in Figure 5E, and the secondary plots in Figure 5F. K_{ia} is

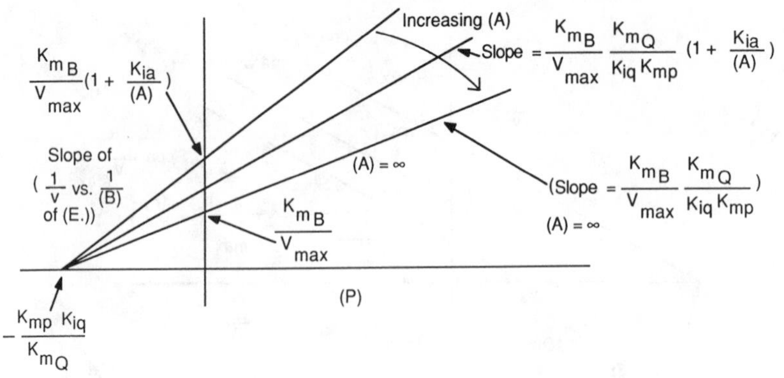

FIGURE 5F.

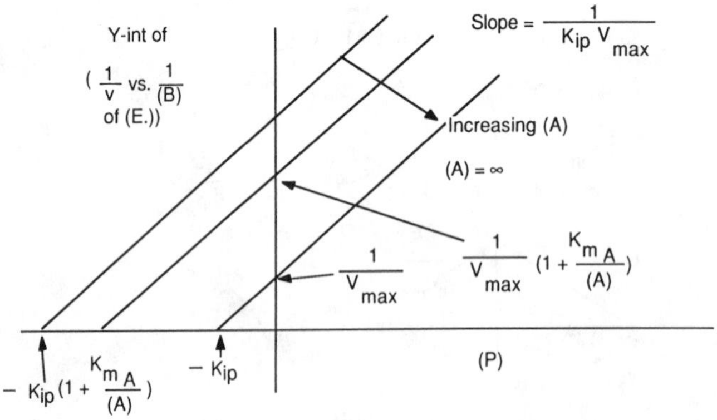

FIGURE 5G.

$$\text{Y-intercept} = \left(\frac{1}{V_{max}} + \frac{K_{mA}}{V_{max}}\frac{1}{(A)}\right) + \frac{(P)}{V_{max}K_{ip}}$$

obtained if $\left(\frac{K_{mB}}{V_{max}}\right)$ is known. K_{iq} is obtained if and K_{mp} and K_{mQ} are known, where the secondary plots of slope vs. $\frac{1}{P}$ (Figure 5F) are based on Equations 15 to 17.

$$\underset{(1/v\ \text{vs.}\ 1/(B))}{\text{Slope}} = \frac{K_{mB}}{V_{max}}\left(1 + \frac{K_{ia}}{(A)}\right)\left(1 + \frac{K_{mQ}(P)}{K_{iq}K_{mp}}\right) \tag{15}$$

$$= \left(\frac{K_{mB}}{V_{max}} + \frac{K_{mB}K_{ia}}{V_{max}(A)}\right)\left(1 + \frac{K_{mQ}(P)}{K_{iq}K_{mp}}\right) \tag{16}$$

$$\underset{(1/v\ \text{vs.}\ 1/(B))}{\text{Slope}} = \frac{K_{mB}}{V_{max}}\left(1 + \frac{K_{ia}}{(A)}\right) + \frac{K_{mB}}{V_{max}}\frac{K_{mQ}(P)}{K_{iq}K_{mp}}\left(1 + \frac{K_{ia}}{(A)}\right) \tag{17}$$

Figure 5G shows a secondary plot of Y-int vs. P, based on Equatin 17a:

$$\text{Y-int} = \left(\frac{1}{V_{max}} + \frac{K_{mA}}{V_{max}}\frac{1}{(A)}\right) + \frac{(P)}{V_{max}K_{ip}} \tag{17a}$$

and K_{ip} is obtained from this secondary plot.

Now, if (P) = 0, but Q ≠ 0 (Equation 147 or better 152 in Chapter 2) becomes, after dividing by V^r_{max} as before,

$$v = \frac{V^f_{max}(A)(B)}{K_{ia}K_{m_B} + K_{m_B}(A) + K_{m_A}(B) + \frac{V^f_{max}}{V^r_{max}}\frac{K_{m_P}(Q)}{K_{eq}} + (A)(B) + \frac{K_{m_A}(B)(Q)}{K_{ip}}} \qquad (18)$$

and K_{eq} is eliminated from the (Q) term by

$$K_{eq} = \frac{V^f_{max}}{V^r_{max}}\frac{K_{iq}K_{mP}}{K_{ia}K_{m_B}} \qquad (19)$$

as before. Therefore,

$$\frac{V^f_{max}}{V^r_{max}} \frac{K_{m_P}(Q)}{\left(\frac{V^f_{max}}{V^r_{max}}\frac{K_{iq}K_{mP}}{K_{ia}K_{m_B}}\right)} = \frac{K_{ia}K_{m_B}(Q)}{K_{iq}} \qquad (20)$$

Therefore,

$$v = \frac{V^f_{max}(A)(B)}{K_{ia}K_{m_B} + K_{m_B}(A) + K_{m_A}(B) + \frac{K_{ia}K_{m_B}(Q)}{K_{iq}} + (A)(B) + \frac{K_{m_A}(B)(Q)}{K_{iq}}} \qquad (21)$$

Now, if (A) is varied with (B) constant at several fixed values of (Q), the product, and P = 0,

$$\frac{v}{V^f_{max}} = \frac{1}{\frac{K_{ia}K_{m_B}}{(A)(B)} + \frac{K_{m_B}}{(B)} + \frac{K_{m_A}}{(A)} + \frac{K_{ia}K_{m_B}(Q)}{K_{iq}(A)(B)} + 1 + \frac{K_{m_A}(Q)}{K_{iq}(A)}} \qquad (22)$$

In double-reciprocal form:

$$\frac{1}{v^f} = \frac{1}{V^f_{max}}\left(+ \frac{K_{m_B}}{(B)}\right) + \frac{K_{m_A}}{V^f_{max}}\frac{1}{(A)}\left[+ \frac{(Q)}{K_{iq}} + \frac{K_{ia}K_{m_B}(Q)}{K_{m_A}K_{iq}(B)} + \frac{K_{ia}K_{m_B}}{K_{m_A}(B)}\right] \qquad (23)$$

$$\frac{1}{v^f} = \frac{1}{V^f_{max}}\left(+ \frac{K_{m_B}}{(B)}\right) + \frac{K_{m_A}}{V^f_{max}}\frac{1}{(A)}\left[1 + \frac{(Q)}{K_{iq}} + \frac{K_{ia}K_{m_B}}{K_{m_A}(B)} + \left(+ \frac{(Q)}{K_{iq}}\right)\right] \qquad (24)$$

$$\frac{1}{v^f} = \frac{1}{V^f_{max}}\left(+ \frac{K_{m_B}}{(B)}\right) + \frac{K_{m_A}}{V^f_{max}}\frac{1}{(A)}\left[1 + \frac{(Q)}{K_{iq}} + \left(1 + \frac{K_{ia}K_{m_B}}{K_{m_A}(B)}\right)\right] \qquad (25)$$

Therefore, (Q) affects only the slope and is a competitive inhibitor with respect to (A) at all (B) concentrations. (a) $\frac{1}{v}$ vs. $\frac{1}{(A)}$ plots where (B) is held constant, P = 0, and Q ≠ 0 are given in Figure 6A. A replot of Slope vs. (Q) is shown in Figure 6B. However, when (B) is varied, Equation 22 for (b) ($\frac{1}{v}$ vs. $\frac{1}{(B)}$) plots, where Q ≠ 0 and P = 0, now becomes:

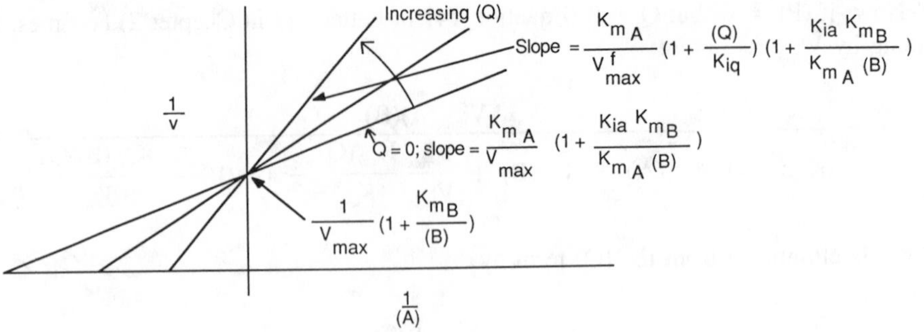

FIGURE 6A.

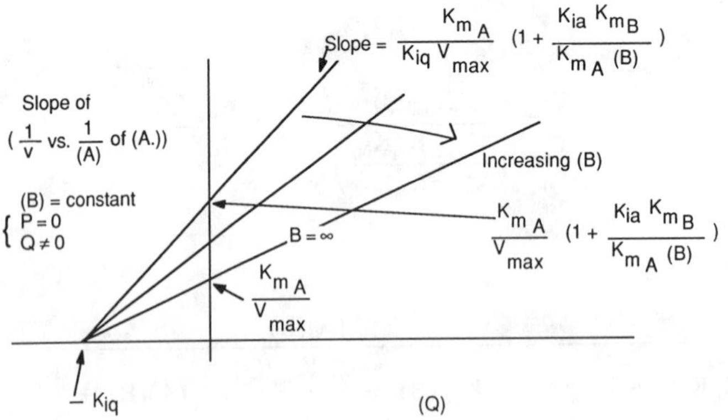

FIGURE 6B. Replot of slope vs. (Q).

$$\frac{1}{v_f} = \frac{1}{V^f_{max}}\left(+ \frac{K_{m_A}}{(A)} + \frac{K_{m_A}(Q)}{K_{iq}(A)}\right) + \frac{1}{V_{max}}\frac{K_{m_B}}{(B)}\left(+ \frac{K_{ia}}{(A)} + \frac{K_{ia}}{(A)}\frac{(Q)}{K_{iq}}\right) \tag{26}$$

$$\frac{1}{v_f} = \frac{1}{V^f_{max}}\left[+ \frac{K_{m_A}}{(A)}\left(+ \frac{(Q)}{K_{iq}}\right)\right] + \frac{1}{V_{max}}\frac{K_{m_B}}{(B)}\left[+ \frac{K_{ia}}{(A)}\left(+ \frac{(Q)}{K_{iq}}\right)\right] \tag{27}$$

Thus, (Q) affects both slope and Y-intercept, but not identically. Therefore, (Q) acts as a noncompetitive inhibitor with respect to (B) at concentrations of (A) less than saturation, but of the mixed type. If $\frac{K_{m_A}}{(A)} = \frac{1}{100} = \frac{K_{ia}}{(A)}$, or (A) is held at 100 × K_{m_A} or K_{ia} at saturation, Equation 27 reduces to

$$\frac{1}{v_f} = \frac{1}{V^f_{max}} + \frac{1}{V_{max}}\frac{K_{m_B}}{(B)} \tag{28}$$

and therefore, saturation with (A) completely overcomes the inhibition by (Q).

(c) $\frac{1}{v}$ vs. $\frac{1}{(B)}$ plots where Q ≠ 0, P = 0, and (A) is constant and below saturation, is given in Figure 7A and is identical to the case when $\frac{1}{v}$ vs. $\frac{1}{(B)}$ is plotted at fixed (A), (see Equation 160 in Chapter 2), $K_{ia} > K_{m_A}$, Q = 0, and P = 0 (see below).

A replot of Slope of Figure 7A plots are shown in Figure 7B. Replots of secondary plots of intercepts and slopes also yield K_{iq} (see Figure 7C, based on Equation 29)

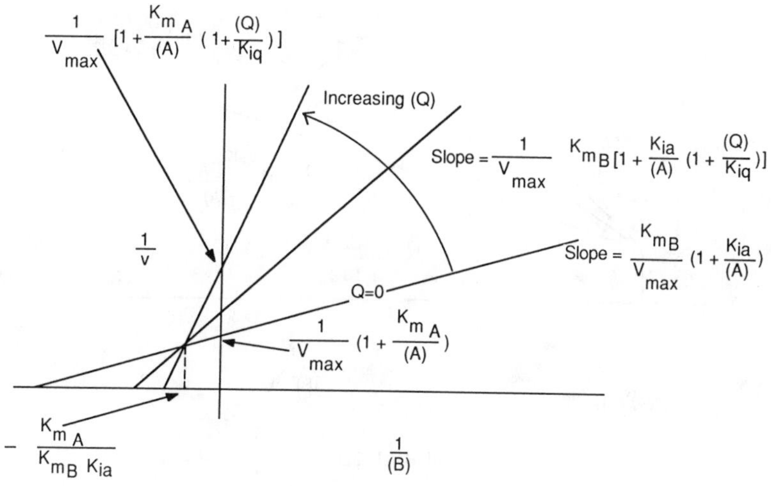

FIGURE 7A.

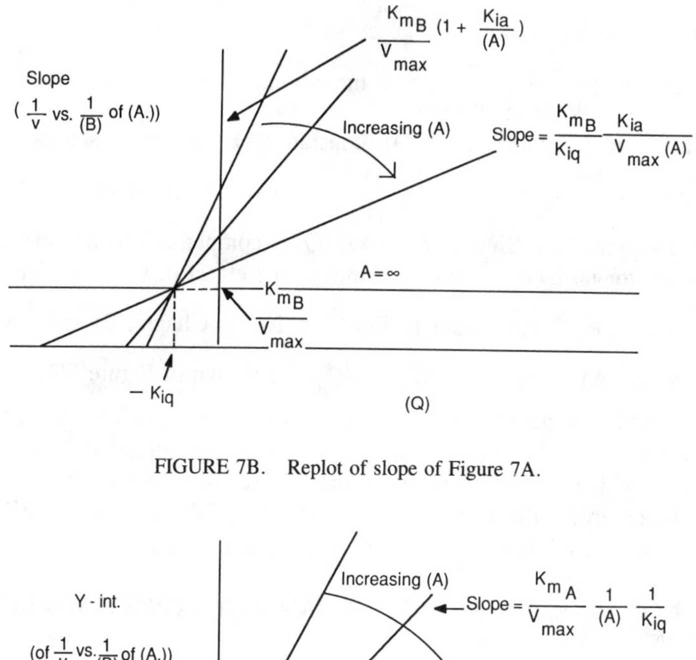

FIGURE 7B. Replot of slope of Figure 7A.

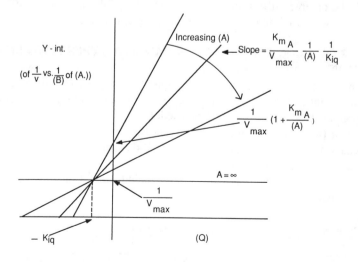

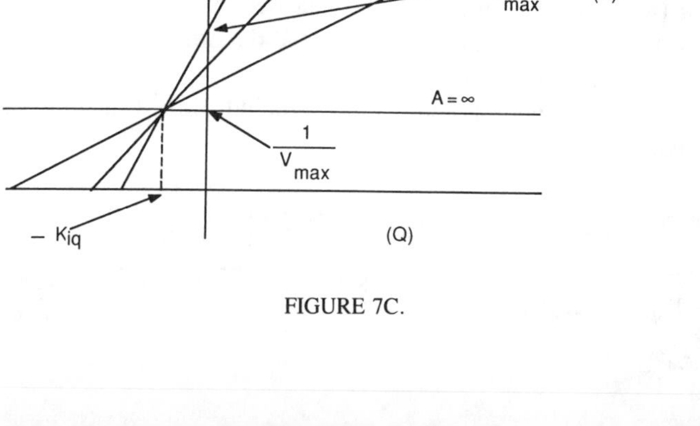

FIGURE 7C.

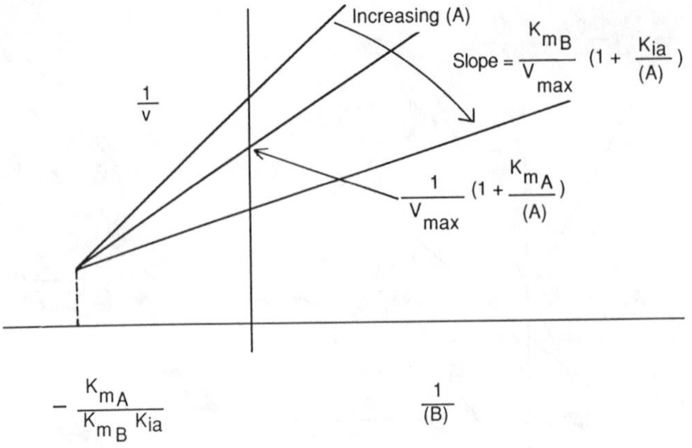

FIGURE 7D.

$$\text{Y-int} = \frac{1}{V_{max}} + \frac{K_{mA}}{(A)}\frac{1}{V_{max}} + \frac{K_{mA}}{V_{max}(A)}\frac{(Q)}{K_{iq}} \tag{29}$$

K_{iq} can be obtained from this plot.

It is significant that $\frac{1}{(B)_{int}} = \frac{-K_{mA}}{K_{mB}\,K_{ia}}$ at the common intersection of the $\frac{1}{v}$ vs. $\frac{1}{(B)}$ plot (Figure 7A) at several fixed values of (Q), constant (A), but below saturation; this same intersection point is obtained for the $\frac{1}{v}$ vs. $\frac{1}{(B)}$ plot at several fixed values of (A), when Q = 0 and P = 0 (Figure 7D). Since, of course, (Q) is competitive with respect to (A), this is not unexpected, for an increase in (A) decreases the effect of (Q) and vice versa. From Equation 160 in Chapter 2 (also refer to Equation 162 in Chapter 2) and $\frac{1}{v}$ vs. $\frac{1}{(B)}$ plots where Q = P = 0, (A) is fixed, and $K_{ia} > K_{mB}$, are shown in Figure 7D.

The product inhibition patterns allow one to decide as to the order of substrate addition and product release, i.e., (A) can be distinguished from (B) and (P) from (Q), since only (A) is competitive with respect to (Q). Also, the patterns of product inhibition permit one to distinguish the ordered BiBi system from the rapid equilibrium random BiBi system in the absence of dead-end inhibitors, which we will discuss shortly.

B. RAPID OR QUASI-EQUILIBRIUM OF RANDOM TYPES (TWO-SUBSTRATE REACTIONS)

Thus, consider the mechanism given in Chapter 2, Scheme XIX (to be compared with Cleland's nomenclature in Scheme XVIII, Chapter 2) for a rapid or quasi-equilibrium mechanism of the random type for a two-substrate reaction, where, to compare against the ordered mechanism for a two-substrate reaction, we must define that the product of (A) = (Q) (ordered mechanism) or C (random mechanism), and the product of B = P (ordered mechanism) or D (random mechanism), and vice versa.

Now, in the case where no dead-end complex could form, Equation 120 in Chapter 2 results for the random mechanism,

$$v^f = \frac{V^f_{max}\left[(A)(B) - \frac{(C)(D)}{K'_{equil}}\right]}{\left[K_1K_3 + K_3(A) + K_4(B) + (A)(B) + \frac{K_1K_3(C)(D)}{K_7K_9} + \frac{K_1K_3(C)}{K_7} + \frac{K_1K_3K_8(D)}{K_7K_9}\right]} \tag{2-120}$$

or the equivalent where $\overline{K}_A$ = dissociation of (A) from the binary complex and K_A = dissociation of (A) from the ternary complex

$$v^f = \frac{V^f_{max}\left[(A)(B) - \frac{(C)(D)}{K'_{equil}}\right]}{\overline{K}_A K_B + K_B(A) + K_A(B) + (A)(B) + \frac{\overline{K}_A K_B(C)(D)}{\overline{K}_C K_D} + \frac{\overline{K}_A K_B(C)}{\overline{K}_C} + \frac{\overline{K}_A K_B K_C(D)}{\overline{K}_C K_D}} \quad (30a)$$

which may be compared with the comparable equation for the steady-state ordered mechanism (see Equation 152 in Chapter 2)

$$v^f = \frac{V^f_{max}\left[(A)(B) - \frac{(P)(Q)}{K_{eq}}\right]}{K_{ia}K_{m_B} + K_{m_B}(A) + K_{m_A}(B) + (A)(B) + \frac{K_{ia}K_{m_B}(P)(Q)}{K_{m_P}K_{iq}}}$$

$$+ \frac{K_{ia}K_{m_B}(Q)}{K_{iq}} + \frac{K_{m_Q}K_{ia}K_{m_B}(P)}{K_{iq}K_{m_P}} + \frac{K_{m_B}K_{m_Q}(A)(P)}{K_{m_P}K_{iq}} + \frac{K_{m_A}(B)(Q)}{K_{iq}}$$

$$+ \frac{(A)(B)(P)}{K_{iq}} + \frac{K_{m_B}K_{ia}(B)(P)(Q)}{K_{m_P}K_{iq}K_{ib}} \quad (2\text{-}152)$$

where the last four terms are not present in the rapid equilibrium random mechanism.

It is worthwhile to repeat here that the quasi-equilibrium random BiBi mechanism contains all the terms in the denominator as the steady-state ordered BiBi, except the last four terms, that is, less the (A) (P), the (B) (Q), the (A) (B) (P), and the (B) (P) (Q) terms.

Also, in the Theorell-Chance equation, the (A) (B) (P) and (B) (P) (Q) terms could be missing from Equation 152, Chapter 2, and in the Ping Pong BiBi equation the constant term (first term) and the (A) (B) (P) and (B) (P) (Q) terms would be absent.

Now, to return to the case for inhibition by product C (i.e., Q) (with D = 0) (i.e., P = 0), Equation 132 in Chapter 2 results; which, for fixed but unsaturated (B), yields Equation 133 (See Scheme XIX in Chapter 2.)

$$\frac{1}{v^f_o} = \frac{1}{V^f_{max}}\left(1 + \frac{K_3}{(B)}\right) + \frac{K_4}{V^f_{max}}\frac{1}{(A)}\left[1 + \frac{K_1K_3}{(B)K_4}\left(1 + \frac{(C)}{K_7}\right)\right] \quad (2\text{-}133)$$

or

$$\frac{1}{v^f_o} = \frac{1}{V^f_{max}}\left(1 + \frac{K_B}{(B)}\right) + \frac{K_A}{V^f_{max}}\frac{1}{(A)}\left[1 + \frac{\overline{K}_A K_B}{(B)K_A}\left(1 + \frac{(C)}{\overline{K}_C}\right)\right] \quad (31)$$

which predicts that (C) or (Q) would be competitive with respect to (A).

If (B) is fixed at 100 K_3 (i.e., 100 × K_B), the inhibition is completely abolished.

For fixed (A), but not at saturation, Equation 134 in Chapter 2 results;

$$\frac{1}{v^f_o} = \frac{1}{V^f_{max}}\left(1 + \frac{K_4}{(A)}\right) + \frac{1}{V_{max}}\frac{K_3}{(B)}\left[1 + \frac{K_1}{(A)}\left(1 + \frac{(C)}{K_7}\right)\right] \quad (2\text{-}134)$$

or

$$\frac{1}{v_o^f} = \frac{1}{V_{max}^f}\left(1 + \frac{K_A}{(A)}\right) + \frac{1}{V_{max}}\frac{K_B}{(B)}\left[1 + \frac{\overline{K}_A}{(A)}\left(1 + \frac{(C)}{\overline{K}_C}\right)\right] \tag{32}$$

Therefore, (C) (or Q) would again be competitive with respect to (B); if (A) is fixed at least at 100 K_4 ($\overline{K}_A$), or 100 K_I (K_A), then again the inhibition is abolished.

In the case of the ordered BiBi, we find (Q) to be competitive with respect to (A), either at saturation or not at saturation. (Q) is noncompetitive (mixed type) with respect to (B) if not saturated with (A); however, the inhibition is abolished if saturated with (A). Thus, the two mechanisms could be distinguished.

However, we also found that for the case of both dead-end complexes E · B · C and E · A · D, as in Chapter 2, Scheme XX, (i.e., E · B · Q and E · A · P), the rapid equilibrium random mechanism yielded Equation 126 from Chapter 2,

$$v^f = \frac{V_{max}^f\left[(A)(B) - \dfrac{(C)(D)}{K'_{eq}}\right]}{K_1K_3 + K_3(A) + K_4(B) + (A)(B)\dfrac{K_1K_3}{K_7K_9}(C)(D) + \dfrac{K_1K_3(C)}{K_7} + \dfrac{K_1K_3K_8(D)}{K_7K_9} + \dfrac{K_1K_3}{K_2K_\alpha} + \dfrac{K_3(D)(A)}{K_\gamma}} \tag{2-126}$$

and for the case of product (C) (i.e., Q) inhibition with dead-end complexes and with (D) = 0, Equation 130, Chapter 2, resulted for fixed (B).

$$\frac{1}{v_o^f} = \frac{1}{V_{max}^f}\left[1 + \frac{K_3}{(B)}\right] + \frac{K_4}{(A)}\frac{1}{V_{max}^f}\left[1 + \frac{K_2}{(B)}\left(1 + \frac{(C)}{K_7}\right) + \frac{(C)}{K_\alpha}\right] \tag{2-130}$$

or

$$\frac{1}{v_o^f} = \frac{1}{V_{max}^f}\left[1 + \frac{K_B}{(B)}\right] + \frac{K_A}{(A)}\frac{1}{V_{max}^f}\left[1 + \frac{\overline{K}_B}{(B)}\left(1 + \frac{(C)}{\overline{K}_C}\right) + \frac{(C)}{K_\alpha}\right] \tag{33}$$

Thus, (C) (or Q) would be a competitive inhibitor with respect to (A) if (B) is not at saturation; with (B) = 100 K_3 (K_B) or 100 K_2 ($\overline{K}_B$), (C) would still be competitive.

Whereas, for fixed (A), Equation 131 in Chapter 2 results:

$$\frac{1}{v_o^f} = \frac{1}{V_{max}^f}\left[1 + \frac{K_4}{(A)}\left(+ \frac{(C)}{K_\alpha}\right)\right] + \frac{K_3}{(B)}\frac{1}{V_{max}^f}\left[1 + \frac{K_1}{(A)}\left(1 + \frac{(C)}{K_7}\right)\right] \tag{2-131}$$

or

$$\frac{1}{v_o^f} = \frac{1}{V_{max}^f}\left[1 + \frac{K_A}{(A)}\left(+ \frac{(C)}{K_\alpha}\right)\right] + \frac{K_B}{(B)}\frac{1}{V_{max}^f}\left[1 + \frac{\overline{K}_A}{(A)}\left(1 + \frac{(C)}{\overline{K}_C}\right)\right] \tag{34}$$

Therefore, (C) (or Q) is now noncompetitive (mixed type) with fixed (A) which is not saturated; if (A) is fixed at 100 K_A or $\overline{K}_A$ (i.e., 100 K_4 or K_1), the inhibition is abolished and identical with the ordered mechanism for product inhibition by (Q), but not by (P). However, direct equilibrium substrate binding measurements will readily rule out an ordered BiBi mechanism, e.g., if only EA or EQ binary complexes can form, to yield the ternary complexes, EAB ⇄ EPQ, whereas EB or EP are insignificant. Therefore, if by ligand

TABLE 1
Product Inhibition Patterns for Some Two-Substrate Mechanisms

Mechanism	Product inhibitor	Varied substrate: A, Not saturated[a] with B	A, Saturated with B	B, Not saturated with A	B, Saturated[a] with A
Ordered BiBi[b]	P	Non-compet. (M-T)	Uncompet.	Noncompet. (M-T)	Noncompet. (M-T)
	Q	Compet.	Compet.	Noncompet. (M-T)	No Inhib.
Theorell-Chance[c,d,e]	P	Noncompet. (M-T)	No inhib.	Compet.	Compet.
	Q	Compet.	Compet.	Noncompet. (M-T)	No inhib.
Ping Pong[e] BiBi	P	Noncompet. (M-T)	No inhib.	Compet.	Compet.
	Q	Compet.	Compet.	Noncompet. (M-T)	No inhib.
Rapid (quasi-)equilibrium, random BiBi (No dead-end complexes)	P (D)	Compet.	No inhib.	Compet.	No inhib.
	Q (C)	Compet.	No inhib.	Compet.	No inhib.
Rapid equilibrium ordered BiBi (no dead-end complexes)	P (D)	No inhib.	No inhib.	No inhib.	No inhib.
	Q (C)	Compet.	No inhib.	Compet.	No inhib.
Rapid equilibrium, random[d] BiBi with dead-end EAP and EBQ	P (D)	Noncompet. (M-T)	No inhib.	Compet.	Compet.
	Q (C)	Compet.	Compet.	Noncompet. (M-T)	No inhib.

Note: Compet. = competitive (only slope changes); Uncompet. = uncompetitive (only Y-intercept changes); Noncompet. = noncompetitive (both slope and Y-intercept change); Noncompet. (M-T) = noncompetitive (mixed type, common intersection point is above X-axis); and No inhib. = no inhibition.

[a] Saturation with the fixed paired substrate implies a concentration at approximated 100 × K_m.
[b] Product inhibition patterns are symmetrical.
[c] Product inhibition patterns identify substrate-product pairs, but do not disclose order of addition or release. Binding studies are necessary to identify A and Q.
[d] The product inhibition patterns are identical. The mechanisms can be distinguished only by binding studies. In a rapid equilibrium, random system, A, B, P, and Q all bind to free E. In a Theorell-Chance system, only A and Q bind to free E.
[e] Product inhibition patterns are identical; however, the mechanisms are easily distinguishable from initial velocity data.

binding measurements the intrinsic dissociation constants for EB or EP are found to be immeasurable (infinite), or measurable (finite), one may readily rule them in or out.

Finally, in Table 1, we summarize the product inhibition patterns for some of the two-substrate (or bireactant) mechanisms we have discussed.

Thymidine phosphorylase catalyzes the reversible phosphorolysis of thymidine (d Thd), deoxyuridine, and their analogues (but not deoxycytidine) to their respective bases and α-D-2-deoxyribose 1-phosphate (dRib-1-P): deoxynucleotide + Pi ⇌ base + dRib-1-P.

Thymidine phosphorylase also catalyzes a deoxyribosyl-transfer reaction of the deoxyribose moiety of one deoxynucleoside to a base to form a second deoxynucleoside, and there are apparently two mechanisms by which deoxyribosyl transfer can take place, that is, by the direct or indirect mechanism.[904] The direct transfer is as follows: $\text{deoxynucleoside}_1 + \text{base}_2 \rightleftharpoons \text{deoxynucleoside}_2 + \text{base}_1$. This mechanism is apparently followed by the thymidine phosphorylase from human leukocytes[905,906] The indirect transfer mechanism, on the other hand, proceeds as follows, with either free or enzyme-bound dRib-1-P as the intermediate: $\text{deoxynucleoside}_1 + P_i \rightleftharpoons \text{base}_1 + \text{dRib-1-P}$; $\text{dRib-1-P} + \text{base}_2 \rightleftharpoons \text{deoxynucleoside}_2 + P_i$.

Thymidine phosphorylases from *Escherichia coli*[907] and the rabbit small intestine[904] have been shown to catalyze deoxyribosyl transfer by the indirect mechanism. Recently, Iltzsch et al.[908] have proposed that the mouse liver enzyme also catalyzes the deoxyribosyl transferase reaction by an enzyme-bound 2-deoxyribose 1-phosphate intermediate. Their studies[908] on the thymidine phosphorylase from mouse liver included both initial velocity and product inhibition studies. The basic mechanism of this mouse liver enzyme was deduced to be a rapid equilibrium random BiBi mechanism with an enzyme-phosphate-thymine dead-end complex. Thymine displayed both substrate inhibition and nonlinear product inhibition, i.e., the slope and intercept replots vs. 1/[thymine] were nonlinear, indicative of more than one binding site on the enzyme for thymine and also that when thymine is bound to one of these sites, the enzyme is inhibited. Furthermore, both thymidine and phosphate showed cooperative effects (see Chapter 7) in the presence of thymine at concentrations above 60 μM, which suggested that the enzyme may have multiple interacting allosteric and/or catalytic sites. The deoxyribosyl transferase reaction catalyzed by this enzyme is phosphate dependent, requires nonstoichiometric amounts of phosphate, and can proceed by an enzyme-bound 2-deoxyribose 1-phosphate intermediate. These findings[908] were in accord with the rapid equilibrium random BiBi mechanism they proposed and also that the deoxyribosyl transfer by this mouse liver enzyme appeared to involve the indirect mechanism. Iltzsch[908] felt that their results suggested that phosphorolysis and deoxyribosyl transfer were catalyzed by the same site on thymidine phosphorylase.

C. PRODUCT INHIBITION PATTERNS FOR AN ORDERED TER TER MECHANISM

See Chapter 2, Scheme XXXVII for the mechanism, Equation 193 for the steady-state velocity equation, and Equation 194 for the overall velocity equation in terms of kinetic parameters.

For initial forward reaction (after dividing Equation 194 by V^r_{max}):

$$
\begin{aligned}
v^f_o = \; & \frac{V^f_{max}(A)(B)(C)}{K_{ia}K_{ib}K_{m_C} + K_{ib}K_{m_C}(A) + K_{ia}K_{m_B}(C) + K_{m_C}(A)(B)} \\
& + K_{m_B}(A)(C) + K_{m_A}(B)(C) + (A)(B)(C) + \frac{V^f_{max}K_{ir}K_{m_Q}(P)}{V^r_{max}K_{eq}} \\
& + \frac{V^f}{V^r}\frac{K_{iq}K_{m_P}(R)}{K_{eq}} + \frac{V^f}{V^r}\frac{K_{m_R}(P)(Q)}{K_{eq}} + \frac{V^f}{V^r}\frac{K_{m_Q}(P)(R)}{K_{eq}} \\
& + \frac{V^f}{V^r}\frac{K_{m_P}(Q)(R)}{K_{eq}} + \frac{V^f(P)(Q)(R)}{K_{eq}} + \frac{V^f K_{m_Q}K_{ir}(A)(P)}{V^r K_{ia}K_{eq}} \\
& + \frac{K_{ia}K_{m_B}(C)(R)}{K_{ir}} + \frac{V^f}{V^r}\frac{K_{m_Q}K_{ir}(A)(B)(P)}{K_{ia}K_{ib}K_{eq}} + \frac{K_{m_A}(B)(C)(R)}{K_{ir}} \\
& + \frac{V^f}{V^r}\frac{K_{m_R}(A)(P)(Q)}{K_{ia}K_{eq}} + \frac{K_{ia}K_{m_B}(C)(Q)(R)}{K_{iq}K_{ir}} + \frac{V^f}{V^r}\frac{K_{ir}K_{m_Q}(A)(B)(C)(P)}{K_{ia}K_{ib}K_{ic}K_{eq}} \\
& + \frac{V^f}{V^r}\frac{K_{ip}K_{m_R}(A)(B)(C)(Q)}{K_{ia}K_{ib}K_{ic}K_{eq}} + \frac{V^f}{V^r}\frac{K_{m_R}(A)(B)(P)(Q)}{K_{ia}K_{ib}K_{eq}} + \frac{K_{m_A}(B)(C)(Q)(R)}{K_{iq}K_{ir}} \\
& + \frac{K_{m_A}K_{ic}(B)(P)(Q)(R)}{K_{ip}K_{iq}K_{ir}} + \frac{K_{ia}K_{m_B}(C)(P)(Q)(R)}{K_{ip}K_{iq}K_{ir}} + \frac{V^f}{V^r}\frac{K_{m_R}(A)(B)(C)(P)(Q)}{K_{ia}K_{ib}K_{ic}K_{eq}} \\
& + \frac{K_{m_A}(B)(C)(P)(Q)(R)}{K_{ip}K_{iq}K_{ir}} \qquad (35)
\end{aligned}
$$

When (A) is varied in the presence of (P) (with Q = R = 0) at fixed (B) and (C) (after substitution of

$$K_{eq} = \frac{V_{max}^f}{V_{max}^r} \frac{K_{mP}K_{iq}K_{ir}}{K_{ia}K_{ib}K_{mC}}$$

in (P), (A) (P), and (A) (P) (B) terms).

$$\frac{v_o^f}{V_{max}^f} = \frac{(A)(B)(C)}{K_{ia}K_{ib}K_{mC} + K_{ib}K_{mC}(A) + K_{ia}K_{mB}(C) + K_{mC}(A)(B) + K_{mB}(A)(C)}$$

$$+ K_{mA}(B)(C) + (A)(B)(C) + \frac{K_{mC}K_{mQ}K_{ia}K_{ib}(P)}{K_{mP}K_{iq}}$$

$$+ \frac{K_{mC}K_{mQ}(A)(P)}{K_{mP}K_{iq}} + \frac{K_{mC}K_{mQ}(A)(B)(P)}{K_{mP}K_{iq}} + \frac{K_{mQ}K_{mC}(A)(B)(C)(P)}{K_{mP}K_{ic}K_{iq}} \quad (36)$$

Divide through by (B) (C):

$$\frac{v_o^f}{V_{max}^f} = \frac{(A)}{\dfrac{K_{ia}K_{ib}K_{mC}}{(B)(C)} + \dfrac{K_{ib}K_{mC}(A)}{(B)(C)} + \dfrac{K_{ia}K_{mB}}{(B)} + \dfrac{K_{mC}(A)}{(C)} + \dfrac{K_{mB}(A)}{(B)}}$$

$$+ K_{mA} + (A) + \frac{K_{mC}K_{mQ}K_{ia}K_{ib}(P)}{K_{mP}K_{iq}(B)(C)} + \frac{K_{mC}K_{mQ}(A)(P)}{K_{mP}K_{iq}(B)(C)}$$

$$+ \frac{K_{mC}K_{mQ}(A)(P)}{K_{mP}K_{iq}(C)} + \frac{K_{mQ}K_{mC}(A)(B)(C)(P)}{K_{mP}K_{ic}K_{iq}(B)(C)} \quad (37)$$

$$\frac{v_o^f}{V_{max}^f} = \frac{(A)}{(A)\left[1 + \dfrac{K_{mB}}{(B)} + \dfrac{K_{mC}}{(C)} + \dfrac{K_{ib}K_{mC}}{(B)(C)} + \dfrac{K_{mC}K_{mQ}}{K_{mP}K_{iq}}\dfrac{(P)}{(C)} + \dfrac{K_{mC}K_{mQ}K_{ib}}{K_{mP}K_{iq}}\dfrac{(P)}{(B)(C)}\right.}$$

$$\left. + \frac{K_{mC}K_{mQ}(P)}{K_{mP}K_{ic}K_{iq}}\right] + K_{mA}\left[1 + \frac{K_{ia}K_{ib}K_{mC}}{K_{mA}(B)(C)} + \frac{K_{ia}K_{mB}}{K_{mA}(B)} + \frac{K_{ia}K_{ib}K_{mC}K_{mQ}(P)}{K_{mA}K_{mP}K_{iq}(B)(C)}\right] \quad (38)$$

Divide through by (A):

$$\frac{v_o^f}{V_{max}^f} = \frac{1}{\left[1 + \dfrac{K_{mB}}{(B)} + \dfrac{K_{mC}}{(C)} + \dfrac{K_{ib}K_{mC}}{(B)(C)} + \dfrac{K_{mC}K_{mQ}}{K_{mP}K_{iq}}\dfrac{(P)}{(C)} + \dfrac{K_{mC}K_{mQ}K_{ib}(P)}{K_{mP}K_{iq}(B)(C)} + \dfrac{K_{mC}K_{mQ}(P)}{K_{iq}K_{ic}K_{iq}}\right]}$$

$$+ \frac{K_{mA}}{(A)}\left[1 + \frac{K_{ia}K_{ib}K_{mC}}{K_{mA}(B)(C)} + \frac{K_{ia}K_{mB}}{K_{mA}(B)} + \frac{K_{ia}K_{ib}K_{mC}K_{mQ}(P)}{K_{mA}K_{mP}K_{iq}(B)(C)}\right] \quad (39)$$

Take reciprocals:

$$\frac{1}{v_o^f} = \frac{1}{V_{max}^f}\left[1 + \frac{K_{mB}}{(B)} + \frac{K_{mC}}{(C)} + \frac{K_{ib}K_{mC}}{(B)(C)} + \frac{K_{mC}K_{mQ}}{K_{mP}K_{iq}}\frac{(P)}{(C)}\right.$$

$$+ \frac{K_{mC}K_{mQ}K_{ib}}{K_{mp}K_{iq}} \frac{(P)}{(B)(C)} + \frac{K_{mC}K_{mQ}(P)}{K_{mp}K_{ic}K_{iq}}\Bigg]$$

$$+ \frac{K_{mA}}{V^f_{max}} \frac{1}{(A)} \left[1 + \frac{K_{ia}K_{ib}K_{mC}}{K_{mA}(B)(C)} + \frac{K_{ia}K_{mB}}{K_{mA}(B)} + \frac{K_{ia}K_{ib}K_{mC}K_{mQ}(P)}{K_{mA}K_{mp}K_{iq}(B)(C)}\right] \quad (40)$$

Thus, (P) changes the slope and Y-intercept (but not to the same degree) in a $\frac{1}{v^f_0}$ vs. $\frac{1}{(A)}$ plot, with fixed (B) and (C) (both not at saturation), and therefore (P) is a noncompetitive inhibitor of the mixed type.

Saturation with either (B) or (C) eliminates the slope effect, but not Y-intercept, and therefore, (P) becomes an uncompetitive inhibitor. From above for varied (B) in the presence of (P) (Q = R = O) at fixed (A) and (C), divide Equation 36 through by (A) (B) (C).

$$\frac{v^f_o}{V^f_{max}} = \frac{1}{\left[\frac{K_{ia}K_{ib}K_{mC}}{(A)(B)(C)} + \frac{K_{ib}K_{mC}}{(B)(C)} + \frac{K_{ia}K_{mB}}{(A)(B)} + \frac{K_{mC}}{(C)} + \frac{K_{mB}}{(B)} + \frac{K_{mA}}{(A)} + 1 + \frac{K_{mC}K_{mQ}K_{ia}K_{ib}(P)}{K_{mp}K_{iq}(A)(B)(C)} + \frac{K_{mC}K_{mQ}K_{ib}(P)}{K_{mp}K_{iq}(B)(C)} + \frac{K_{mC}K_{mQ}}{K_{mp}K_{iq}}\frac{(P)}{(C)} + \frac{K_{mQ}K_{mC}(P)}{K_{mp}K_{ic}K_{iq}}\right]} \quad (41)$$

$$\frac{v^f_o}{V^f_{max}} = \frac{1}{\left[1 + \frac{K_{mA}}{(A)} + \frac{K_{mC}}{(C)} + \frac{K_{mC}K_{mQ}}{K_{mp}K_{iq}}\frac{(P)}{(C)} + \frac{K_{mQ}K_{mC}(P)}{K_{mp}K_{ic}K_{iq}}\right] + \frac{K_{mB}}{(B)}\left[1 + \frac{K_{ia}}{(A)} + \frac{K_{ib}K_{mC}}{K_{mB}(C)} + \frac{K_{ia}K_{ib}K_{mC}}{K_{mB}}\frac{1}{(A)}\frac{1}{(C)} + \frac{K_{mC}K_{mQ}K_{ib}}{K_{mB}K_{mp}K_{iq}}\frac{(P)}{(C)} + \frac{K_{mC}K_{mQ}K_{ia}K_{ib}(P)}{K_{mB}K_{mp}K_{iq}(A)(C)}\right]} \quad (42)$$

Therefore, the product inhibitor (P) alters both the slope and ordinate-intercept, but not to the same degree, and again, (P) is a noncompetitive inhibitor of the mixed type. Again, if (C) is held at saturating levels, the slope term is eliminated, but not the Y-intercept, and (P) becomes an uncompetitive inhibitor with respect to (B). However, if (A) is held at saturation, the slope term still remains and, together with the Y-intercept term, results in (P) remaining a noncompetitive inhibitor of the mixed type.

From Equation 41, for varied (C), in the presence of (P) (Q = R = O) at fixed (A) and (B):

$$\frac{v^f_o}{V^f_{max}} = \frac{1}{\left[1 + \frac{K_{mA}}{(A)} + \frac{K_{mB}}{(B)} + \frac{K_{ia}K_{mB}}{(A)(B)} + \frac{K_{mQ}K_{mC}(P)}{K_{mp}K_{ic}K_{iq}}\right] + \frac{K_{mC}}{(C)}\left[1 + \frac{K_{ib}}{(B)} + \frac{K_{ia}K_{ib}}{(A)(B)} + \frac{K_{mQ}(P)}{K_{mp}K_{iq}} + \frac{K_{mQ}K_{ib}}{K_{mp}K_{iq}}\frac{(P)}{(B)} + \frac{K_{mQ}K_{ia}K_{ib}(P)}{K_{mp}K_{iq}(A)(B)}\right]} \quad (43)$$

Therefore, product inhibitor (P) affects both Y-intercept and slope, but not to the same degree, and consequently (P) is a noncompetitive inhibitor of the varied substrate (C), but of the mixed type. If (B) is held at saturation, neither the slope term nor the Y-intercept term vanishes and (P) remains a noncompetitive inhibitor of the mixed type. Similarly, if (A) is held at saturation, (P) remains the same type of inhibitor (noncompetitive of the mixed type).

Now in the presence of (Q) (with P = R = O), from Equation 35, divide through by (A) (B) (C).

$$v_o^f = \frac{V_{max}^f}{\frac{K_{ia}K_{ib}K_{mC}}{(A)(B)(C)} + \frac{K_{ib}K_{mC}}{(B)(C)} + \frac{K_{ia}K_{mB}}{(A)(B)} + \frac{K_{mC}}{(C)} + \frac{K_{mB}}{(B)} + \frac{K_{mA}}{(A)} + 1 + \frac{V^f}{V^r}\frac{K_{ip}K_{mR}(Q)}{K_{ia}K_{ib}K_{ic}K_{eq}}} \tag{44}$$

After substitution of

$$K_{eq} = \frac{V_{max}^f}{V_{max}^r}\frac{K_{mp}K_{iq}K_{ir}}{K_{ia}K_{ib}K_{mC}} \qquad \text{(from Equation 2-197)}$$

$$\frac{v_o^f}{V_{max}^f} = \frac{1}{\left(\frac{K_{ia}K_{ib}K_{mC}}{(A)(B)(C)} + \frac{K_{ib}K_{mC}}{(B)(C)} + \frac{K_{ia}K_{mB}}{(A)(B)} + \frac{K_{mC}}{(C)} + \frac{K_{mB}}{(B)} + \frac{K_{mA}}{(A)} + 1 + \frac{K_{ip}K_{mC}K_{mR}(Q)}{K_{ic}K_{mp}K_{iq}K_{ir}}\right)} \tag{45}$$

Now, if (A) is varied at fixed (B) and (C) in the presence of (Q) as product inhibitor (with P = R = O):

$$\frac{1}{v_o^f} = \frac{1}{V_{max}^f}\left[1 + \frac{K_{mB}}{(B)} + \frac{K_{mC}}{(C)} + \frac{K_{ib}K_{mC}}{(B)(C)} + \frac{K_{ip}K_{mC}K_{mR}(Q)}{K_{ic}K_{mp}K_{iq}K_{ir}}\right] + \frac{K_{mA}}{(A)}\frac{1}{V_{max}^f}\left[1 + \frac{K_{ia}K_{mB}}{K_{mA}(B)} + \frac{K_{ia}K_{ib}K_{mC}}{K_{mA}(B)(C)}\right] \tag{46}$$

(Q), the product inhibitor, affects only the Y-intercept, with $\frac{1}{v}$ plotted against $\left(\frac{1}{(A)}\right)$, and acts as an uncompetitive inhibitor. Saturation with respect to (B) or (C), does not affect the inhibition.

Similarly, if (B) is varied at fixed (A) and (C) in the presence of (Q) as the product inhibitor (with P = R = O), then

$$\frac{1}{v_o^f} = \frac{1}{V_{max}^f}\left[1 + \frac{K_{mA}}{(A)} + \frac{K_{mC}}{(C)} + \frac{K_{ip}K_{mC}K_{mR}(Q)}{K_{ic}K_{mp}K_{iq}K_{ir}}\right] + \frac{K_{mB}}{(B)}\frac{1}{V_{max}^f}\left[1 + \frac{K_{ia}}{(A)} + \frac{K_{ib}K_{mC}}{K_{mB}(C)} + \frac{K_{ia}K_{ib}K_{mC}}{K_{mB}(A)(C)}\right] \tag{47}$$

(Q) still is an uncompetitive inhibitor affecting only the Y-intercept, whether or not (A) or (C) is held at saturation.

Finally, if (C) is varied at fixed (A) and (B) in the presence of (Q) as the product inhibitor (with P = R = O) then

$$\frac{1}{v_o^f} = \frac{1}{V_{max}^f}\left[1 + \frac{K_{mA}}{(A)} + \frac{K_{mB}}{(B)} + \frac{K_{ia}K_{mB}}{(A)(B)} + \frac{K_{ip}K_{mC}K_{mR}(Q)}{K_{ic}K_{mp}K_{iq}K_{ir}}\right] + \frac{K_{mC}}{(C)}\left[1 + \frac{K_{ib}}{(B)} + \frac{K_{ia}K_{ib}}{(A)(B)}\right] \tag{48}$$

Again, (Q) is an uncompetitive inhibitor affecting only the Y-intercept, whether or not (A) or (B) is at saturation.

Lastly, if (R) is the product inhibitor (with P = Q = O) from Equation 35, and after division through (A) (B) (C):

$$\frac{v_o^f}{V_{max}^f} = \frac{1}{\left(\frac{K_{ia}K_{ib}K_{mC}}{(A)(B)(C)} + \frac{K_{ib}K_{mC}}{(B)(C)} + \frac{K_{ia}K_{mB}}{(A)(B)} + \frac{K_{mC}}{(C)} + \frac{K_{mB}}{(B)} + \frac{K_{mA}}{(A)} + 1 + \frac{V_{max}^f}{V_{max}^r}\frac{K_{iq}K_{mP}(R)}{K_{eq}(A)(B)(C)} + \frac{K_{ia}K_{mB}(R)}{K_{ir}(A)(B)} + \frac{K_{mA}(R)}{K_{ir}(A)}\right)}$$

After substitution K_{eq} from of Equation 197 from Chapter 2 into the eighth denominator term, i.e.,

$$\frac{V_{max}^f}{V_{max}^r}\frac{K_{iq}K_{mP}(R)}{(A)(B)(C)\frac{V_{max}^f}{V_{max}^r}\frac{K_{mP}K_{iq}K_{ir}}{K_{ia}K_{ib}K_{mC}}} = \frac{K_{ia}K_{ib}K_{mC}(R)}{K_{ir}(A)(B)(C)} \tag{49}$$

for varied (A) with fixed (B) and (C):

$$\frac{v_o^f}{V_{max}^f} = \frac{1}{\left[1 + \frac{K_{mB}}{(B)} + \frac{K_{mC}}{(C)} + \frac{K_{ib}K_{mC}}{(B)(C)}\right] + \frac{K_{mA}}{(A)}\left[1 + \frac{K_{ia}K_{mB}}{K_{mA}(B)} + \frac{K_{ia}}{K_{mA}}\frac{K_{ib}K_{mC}}{(B)(C)} + \frac{(R)}{K_{ir}} + \frac{K_{ia}}{K_{mA}}\frac{K_{mB}}{K_{ir}}\frac{(R)}{(B)} + \frac{K_{ia}(R)}{K_{mA}(B)(C)}\frac{K_{ib}K_{mC}}{K_{ir}}\right]} \tag{50}$$

∴ R affects only the slope and not the Y-intercept and consequently is a competitive inhibitor with respect to (A) at fixed (B) and (C). Holding (B) or (C) at saturation does not relieve the inhibition entirely because of the $\frac{(R)}{K_{ir}}$ term and (R) remains a competitive inhibitor when (C) is held at saturation.

For varied (B) with fixed (A) and (C):

$$\frac{v_o^f}{V_{max}^f} = \frac{1}{\left[1 + \frac{K_{mA}}{(A)} + \frac{K_{mC}}{(C)} + \frac{K_{mA}}{K_{ir}}\frac{(R)}{(A)}\right] + \frac{K_{mB}}{(B)}\left[1 + \frac{K_{ia}}{(A)} + \frac{K_{ib}}{K_{mB}}\frac{K_{mC}}{(C)} + \frac{K_{ia}K_{ib}K_{mC}}{K_{mB}(A)(C)} + \frac{K_{ia}}{K_{ir}}\frac{(R)}{(A)} + \frac{K_{ia}K_{ib}K_{mC}(R)}{K_{mB}K_{ir}(A)(C)}\right]} \tag{51}$$

∴ R affects both slope and intercept in a $\frac{1}{v_0^f}$ vs. $\frac{1}{(B)}$ plot at fixed (A) and (C), neither of which are held at saturation, and is consequently a noncompetitive inhibitor of the mixed type (since both slope and Y-intercept are affected to different degrees). If (A) is held at saturation, the inhibition is relieved completely. If (C) is held at saturation, both slope and intercept are affected and, therefore, (R) remains a noncompetitive inhibitor of the mixed type; if, however, $K_{ia} \simeq K_{mA}$, (R) becomes a pure noncompetitive inhibitor when (C) is held at saturation.

For varied (C) with fixed (A) and (B):

$$\frac{v_o^f}{V_{max}^f} = \cfrac{1}{\left[1 + \frac{K_{m_A}}{(A)} + \frac{K_{m_B}}{(B)} + \frac{K_{ia}K_{m_B}}{(A)(B)} + \frac{K_{m_A}(R)}{K_{ir}(A)} + \frac{K_{ia}K_{m_B}(R)}{K_{ir}(A)(B)}\right] + \frac{K_{m_C}}{(C)}\left[1 + \frac{K_{ib}}{(B)} + \frac{K_{ia}K_{ib}}{(A)(B)} + \frac{K_{ia}K_{ib}(R)}{K_{ir}(A)(B)}\right]} \tag{52}$$

∴ (R) affects both slope and Y-intercept, but to different degrees, if neither (A) nor (B) is held at saturation, and (R) consequently is a noncompetitive inhibitor of the mixed type. If (A) is held at saturation, the inhibition disappears. If (B) is held at saturation, the Y-intercept term remains because of the $\frac{K_{m_A}(R)}{K_{ir}(A)}$ term and (R) becomes an uncompetitive inhibitor.

These product inhibition patterns for an ordered ter ter mechanism (i.e., an ordered three-substrate reaction, three-product reaction, Chapter 2, Scheme XXXVII) are summarized in Table 2.

1. Product Inhibition Patterns for Prolyl-4-Hydroxylase

At this point we may return to our discussion of the prolyl-4-hydroxylase reaction (Chapter 2). The initial velocity data of Myllylä et al.[279] had been shown qualitatively to agree with the data predicted from the velocity expressions derived for an ordered ter ter mechanism (Chapter 2, Equations 194 and 209). Product inhibition patterns for this particular three-substrate reaction are summarized above in Table 2. A similar table is given by Plowman[282] and was apparently made use of by Myllylä[279] in their product inhibition studies. Unfortunately, the only product inhibition data shown by Myllylä et al.[279] is for succinate; other data, however, are described and are summarized in Table 3.

These data are only superficially consistent with the inhibition patterns described in Table 2 for an ordered ter ter mechanism. Thus, in the case of succinate where the only quantitative data are provided, competitive inhibition with a α-ketoglutarate as the varied substrate would be predicted and was observed; whereas, noncompetitive inhibition of the mixed type would be predicted for the other substrates (below their saturation). However, pure, noncompetitive inhibition with the synthetic peptide (pro-pro-gly)$_5$ was observed with O_2 perhaps at or near saturation, a condition which might lead then to uncompetitive inhibition (see Table 2). However, the product inhibition patterns will also depend, to a minor degree, on the relative values of K_{ia} and K_{m_A} as shown for one case in Table 2. Unfortunately, quantitative product inhibition data on CO_2 are lacking. In a recent series of papers on the kinetics of the lysyl hydroxylase reaction by the same Finnish group,[275,276] it was shown that lysyl hydroxylase proved to have nearly identical initial velocity patterns as those of prolyl-4-hydroxylase and very similar product inhibitions data except for CO_2 which was studied in some detail in this case. For lysyl hydroxylase, CO_2 proved to be a noncompetitive inhibitor with respect to α-ketoglutarate,[276] whereas, uncompetitive inhibition was predicted by these authors from Plowman's table [282] (see also Table 2). They did, however, point out that a dead-end type of complex with CO_2 might lead to noncompetitive inhibition.[282]

In conclusion, the qualitative data on the prolyl-4-hydroxylase steady-state kinetics presented by Myllylä et al.[279] have been re-evaluated here in terms of an ordered ter ter kinetic mechanism, and the data appear to fit the predicted patterns in most respects both for initial velocity (Chapter 2, Table 1) and product inhibition studies (Tables 2 and 3).

When data are available for the reaction conditions which utilize different concentrations of oxygen and carbon dioxide, confirmation of these results may be possible. Also, when additional data become available, quantitation of the kinetic parameters will be permitted

TABLE 2
Product Inhibition Patterns for a Three-Substrate Reaction (Ordered Ter Ter)[b]

Product Inhibitor	Varied substrate A: Not saturated with B or C	A: Saturated with B or C	B: Not saturated with A or C	B: Saturated with A	B: Saturated with C	C: Not saturated with A or B	C: Saturated with A or B
P	Noncompet. (M-T)	Uncompet.	Noncompet. (M-T)	Noncompet. (M-T)	Uncompet.	Noncompet. (M-T)	Noncompet. (M-T)
Q	Uncompet.	Uncompet.	Uncompet.	Uncompet.	Uncompet.	Uncompet.	Uncompet.
R	Compet.	Compet.	Noncompet. (M-T)	No inhib.	Noncompet. (M-T)[a]	Noncompet. (M-T)	A at saturation, no inhib., B at saturation, uncompet.

Note: Noncompet. (M-T) = noncompetitive (mixed type, common intersection point is above X-axis); compet. = competitive (only Y-intercept changes); noncompet. = noncompetitive (both slope and Y-intercept changes); and no inhib. = no inhibition.

[a] If $K_{ia} \cong K_{mA}$, (R) is a pure noncompetitive inhibitor when (C) is held at saturation.

[b] See Equation 194 in Chapter 2.

TABLE 3
Product Inhibition Patterns for Prolyl-4-Hydroxylase

	(A) α-Ketoglutarate	(B) Oxygen	(C) (Pro-Pro-Gly)$_5$
(P) Collagen	Noncompetitive	Noncompetitive	Noncompetitive
(Q) Carbon dioxide	—	—	—
(R) Succinate	Competitive	Noncompetitive	Noncompetitive

by the graphical procedure outlined above (see Table 1 in Chapter 2). At that time an allowance for a dead-end type of inhibitor for CO_2 may have to be made. If the data are verified, i.e., with additional data for CO_2 and O_2, for example, it would imply that the order of substrate addition could be α-ketoglutarate, oxygen and peptide; the order of product release could be hydroxylated peptide (or collagen), carbon dioxide, and succinate, in agreement with the qualitative conclusions of Myllylä et al.[279] See also Reference 1003.

Chapter 4

EFFECTS OF pH AND TEMPERATURE

I. EFFECT OF pH ON THE REACTION VELOCITY

A. INTRODUCTION

Recently, the pH dependence of an enzyme-catalyzed reaction has become a popular approach to identifying substrate or activator moieties and enzymic active-site ionizable residues which play a role in substrate binding and in catalysis (see References 775 to 777). When properly applied, pH studies may yield a great deal of information concerning the mechanism of the enzyme-catalyzed reaction.

As will be seen below, the basic technique is simple and involves measuring the variation in the maximum velocity (V_{max}) and the ratio of V_{max}/Ks (which is an apparent second-order rate constant in units, e.g., of M^{-1} s^{-1}, if V_{max} is expressed in units of reciprocal seconds which appears to be the current usage) for the reaction of the substrate, and/or the Ki for the binding of an inhibitor or activator, as a function of the reaction pH. The kinetic parameter which is being measured, as we will see, will then begin to change as the reaction pH increases above or decreases below the pK_a, depending on whether the group must be protonated or deprontonated for proper function. The pH profile of the reaction is then simply constructed by plotting the log of the kinetic parameter vs. the pH of the reaction. The number of ionizing groups, as we will see, will be defined by the slope of the curve, while the apparent pK_a of the ionizing group(s) will be defined by the intersection of the asymptotes to the curve. It is common today to determine by computer fitting both the number of ionizing groups and their apparent pK_a values, i.e., by computer fitting the data to the rate equation which appropriately describes the type of ionization involved. It may be noted at this point that the pH profiles of functions which represent equilibrium binding (i.e., Ki for substrates or competitive inhibitors and Ki and Km for metal cofactors) will yield true pKa values. In the case of V_{max} or V_{max}/Ks vs. pH profiles, the apparent pKa value may actually be displaced from the true value if, under conditions that the enzyme activity measurement were made, the pH-dependent step was far from rate limiting.

Those enzymic catalytic groups which can be detected by pH studies include the ionizable groups derived from aspartate, glutamate, histidine, cysteine, lysine, arginine, and tyrosine.

Precise identification of the catalytic residue(s) is not a simple matter, however, since the pKa of an amino acid side chain in the environment of an active site can differ dramatically from that of an amino acid in aqueous solution. The enthalpies of ionization may assist in their identification, since (as will be described further below) these thermodynamic values are characteristic of the various ionizable groups; however, unfortunately, they also display a range in values.

Another technique which has been recently reintroduced was first proposed by Findlay et al.[778] It has been known for sometime that the range of pKa values of active-site carboxyl residues, for example, may overlap the range observed for histidine residues, thus making it difficult to differentiate between these two residues strictly on the basis of the measured pKa value (however, they may be distinguished on the basis of their heats of ionization, see below). In 1962, Findlay et al.[778] proposed a method which is currently called the ''solvent perturbation technique'', and which could, in principle, distinguish between the ionizations of neutral acid residues (carboxyl, sulfhydryl, and phenolic groups) and cationic acid residues (amine residues) on the basis of their difference in behavior in mixed-aqueous solvent systems.

Thus, when the loss of a proton generates an anionic conjugate base, as in the case of neutral acids, ionization results in a net increase in the number of ions in the solution

$$HA \rightleftarrows H^+ + A^- \quad (1)$$

Organic solvents shift the ionization equilibrium of neutral acids to the left, decreasing the number of protons in solution and increasing the pKa of the solute.[779-786] When the unionized acid is a cation, loss of a proton merely alters the size of the cations in the solution without changing the number of ions to be solvated:

$$HA^+ \rightleftarrows H^+ + A^0 \quad (2)$$

For these cationic acids, organic solvents, if they affect the pKa at all, they decrease it.[779,781,783,787]

The solvent perturbation method was used originally by Findlay et al.[778] to identify the catalytic histidine of bovine pancreatic ribonuclease. It is based on the assumption that the effect of the solvent on the ionization of the enzymic residue and on the ionization of the reaction buffer of the same acid type will be equivalent.

Experimentally, this technique involves a comparison of the pKa values measured in both neutral acid and cationic acid buffers in the presence and absence of the organic solvent system. The reaction pH is measured in the absence of the organic solvent and it is assumed in the calculation of the pKa from the rate data which are measured for reactions in the mixed solvent system. Thus, the pKa of a neutral acid enzymic group measured by using a cationic buffer would appear higher in the organic solvent system relative to the fully aqueous system while it would appear unchanged in the mixed solvent system buffered by a neutral acid. Similarly, a cationic enzymic group would show the same pKa in either a pure water or in a mixed solvent system when a cationic buffer is employed, while the pKa would appear lower in a mixed solvent system buffered by a neutral acid. In the original perturbation study by Findlay et al,[778] the precaution was taken, which has not been the case in most of the recent studies, to also measure the pH perturbation of the buffer systems employed.

Recently, Grace and Dunaway-Mariano[788] in an excellent study reevaluated the solvent perturbation technique as a method of identifying the catalytic groups of enzymes and extended their studies to a critical examination of the catalytic aspartate residue of yeast hexokinase. Their results indicate that the solvent perturbation technique must not be relied on indiscriminately, and, in their study, pointed out several precautions to be exercised in the application of this technique.

This solvent perturbation technique has now been applied to a number of cases in addition to the original study of Findlay et al.,[778] e.g., alkaline phosphatase of *Escherichia coli*,[789] glutamate dehydrogenase,[790] yeast inorganic pyrophosphatase,[791] yeast hexokinase,[792] arginine esterase,[793] dihydrofolate reductase,[794,795] rabbit muscle creatine kinase,[796] pigeon liver malic enzyme,[797] β-hydroxymethyl glutaryl-CoA reductase,[798] mitochondrial F-1 ATPase,[799] soybean β-amylase,[800] bovine brain hexokinase,[801] L-alanine dehydrogenase,[802] and the recent study on glucose 6-phosphate dehydrogenase from *Leuconostoc mesenteroides*.[846]

It should be remarked that whereas many enzymologists have accepted the solvent perturbation technique as almost routine, others (including the recent work of Grace and Dunaway-Mariano[788]) have cautioned against its use[804] or have actually rejected it.[805,806]

It has also become common, as part of the pH study, to include the pH dependence of isotopic effects (see Chapter 6, Section II) and solvent isotope effects (e.g., in D_2O). See also References 795, 845, and 980.

The following discussion will now deal quantitatively with some aspects of the pH dependence of enzymic reactions, in the steady state, and in aqueous solutions.

B. pH EFFECTS IN THE STEADY-STATE AND IN AQUEOUS SOLUTION

In the discussion that follows on the effects of pH on the initial velocity, it is assumed that effects due to enzyme stability or instability have been taken into account.

1. Single Substrate, Single Product Rapid Equilibrium Systems with Only One ES Complex

a. All Forms of E Bind S°, However Only HES Yields Product

The various ionic, i.e., protonated enzyme, species will be denoted simply by H_2E, HE, and E, where the net charges have been deleted for simplicity (i.e., they should be more rigorously listed as H_2E^{2+}, HE^+, E, or H_2E^+, HE^0, and E^-, for the fully protonated form, the form after its first dissociation, and after its second dissociation, etc.). The substrate is assumed to be uncharged, i.e., S^0, and the product, P^0.

For the forward reaction only

$$
\begin{array}{ccc}
E + S^0 & \xrightleftharpoons{\beta K_S} & ES \\
+ & & + \\
H^+ & & H^+ \\
\end{array}
$$

$K_{E2} \Big\updownarrow \qquad \Big\updownarrow K_{ES2} = K_{E2}/\beta \ (\text{or } \beta = \dfrac{K_{E2})}{K_{ES_2}}$

$$
HE + S^0 \xrightleftharpoons{K_S} HES \xrightarrow{k_P} HE + P^0
$$

$+ \qquad\qquad +$

$H^+ \qquad\qquad H^+$

$K_{E_I} \Big\updownarrow \qquad \Big\updownarrow K_{ES_1} = \alpha K_{EI} \ (\text{or } \alpha = \dfrac{K_{ES_1})}{K_{E_I}}$

$$
H_2E + S^0 \xrightleftharpoons{\alpha K_S} H_2ES
$$

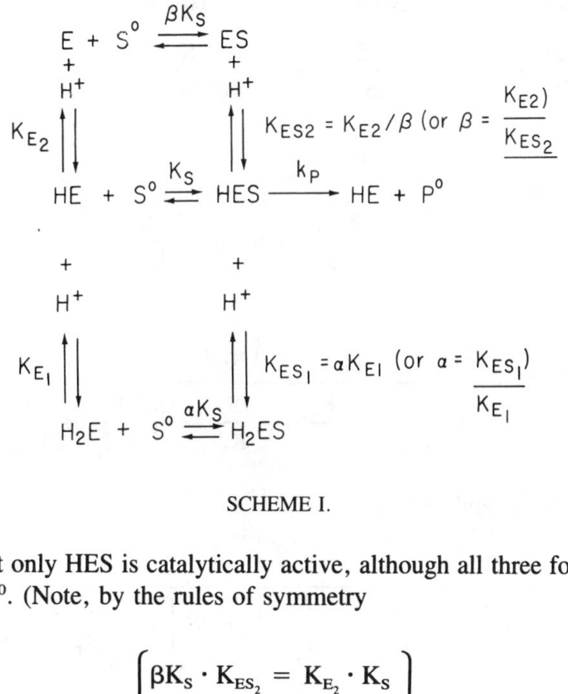

SCHEME I.

∴ It is assumed that only HES is catalytically active, although all three forms of E bind the neutral substrate, S^0. (Note, by the rules of symmetry

$$
\left\{
\begin{array}{l}
\beta K_S \cdot K_{ES_2} = K_{E_2} \cdot K_S \\
\alpha K_S \cdot K_{E_1} = K_{ES_1} \cdot K_S
\end{array}
\right\} \tag{3}
$$

Assume that the deprotonation, or the proton addition steps, are exceedingly fast compared to the catalytic step, and therefore that all the H^+ steps are at equilibrium. Assume that the remaining steps are in quasi-equilibrium and that the catalytic step is the rate-determining step. Therefore,

$$
v_f = k_p(HES) \tag{4}
$$

Divide both sides by E_t:

$$
\frac{v}{E_t} = \frac{k_p(HES)}{(H_2E) + (HE) + (E) + (H_2ES) + (HES) + (ES)} \tag{5}
$$

Substituting each term from the definitions of the various equilibria and transforming into the equation in terms of the HE form, dividing through by HE and setting

$$V^{f}_{max} = k_pE_t \tag{6}$$

$$\frac{v}{V_{max}} = \frac{\left(\dfrac{S}{K_S}\right)}{1 + \dfrac{(H^+)}{K_{E_1}} + \dfrac{K_{E_2}}{(H^+)} + \dfrac{(S)}{K_S} + \dfrac{(H^+)(S)}{K_{E_1}\alpha K_S} + \dfrac{K_{E_2}(S)}{(H^+)(\beta K_S)}} \tag{7}$$

where

$$K_{E_1} = \frac{(HE)(H^+)}{(H_2E)},\ K_{E_2} = \frac{(E)(H^+)}{(HE)},\ K_{ES_1} = \frac{(HES)(H^+)}{(H_2ES)} \text{ and } K_{ES_2} = \frac{(ES)(H^+)}{(HES)};$$

$$K_S = \frac{(HE)(S)}{(HES)},\ \alpha K_S = \frac{(H_2E)(S)}{(H_2ES)},\ \beta K_S = \frac{(E)(S)}{(ES)} \tag{8}$$

substituting for

$$\alpha = \frac{K_{ES_1}}{K_{E_1}} \text{ and } \beta = \frac{K_{E_2}}{K_{ES_2}} \tag{9}$$

$$\frac{v}{V_{max}} = \frac{\left(\dfrac{S}{K_S}\right)}{1 + \dfrac{(H^+)}{K_{E_1}} + \dfrac{K_{E_2}}{(H^+)} + \dfrac{(S)}{K_S} + \dfrac{(S)(H^+)}{K_SK_{ES_1}} + \dfrac{(S)K_{ES_2}}{K_S(H^+)}} \tag{10}$$

Multiply through by K_S and factoring

$$\frac{v}{V_{max}} = \frac{(S)}{K_S\left(1 + \dfrac{(H^+)}{K_{E_1}} + \dfrac{K_{E_2}}{(H^+)}\right) + (S)\left[1 + \dfrac{(H^+)}{K_{ES_1}} + \dfrac{K_{ES_2}}{(H^+)}\right]} \tag{11}$$

or

$$\frac{v}{V_{max}} = \frac{1}{\dfrac{K_S}{(S)}\left(1 + \dfrac{(H^+)}{K_{E_1}} + \dfrac{K_{E_2}}{(H^+)}\right) + \left[1 + \dfrac{(H^+)}{K_{ES_1}} + \dfrac{K_{ES_2}}{(H^+)}\right]} \tag{12}$$

$$\frac{1}{v} = \frac{1}{V_{max}}\left(1 + \frac{(H^+)}{K_{ES_1}} + \frac{K_{ES_2}}{(H^+)}\right) + \frac{1}{V_{max}}\frac{K_S}{(S)}\left(1 + \frac{(H^+)}{K_{E_1}} + \frac{K_{E_2}}{(H^+)}\right) \tag{13}$$

At any fixed H^+, the measured kinetic parameters are

$$V_{max,app} = \frac{V_{max}}{\left(1 + \dfrac{(H^+)}{K_{ES_1}} + \dfrac{K_{ES_2}}{(H^+)}\right)} \tag{14}$$

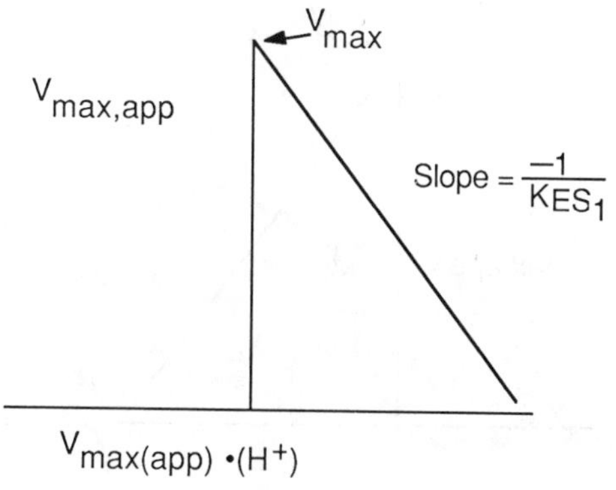

FIGURE 1.

$$K_{S,app} = \frac{K_S\left(1 + \frac{(H^+)}{K_{E_1}} + \frac{K_{E_2}}{(H^+)}\right)}{\left(1 + \frac{(H^+)}{K_{ES_1}} + \frac{K_{ES_2}}{(H^+)}\right)} \tag{15}$$

$$\left(\frac{K_{S,app}}{V_{max,app}}\right) = \frac{K_S}{V_{max}}\left(1 + \frac{(H^+)}{K_{E_1}} + \frac{K_{ES_2}}{(H^+)}\right) \tag{16}$$

and

$$\therefore \quad \frac{v}{V_{max,app}} = \frac{1}{\frac{K_{S,app}}{(S)} + 1} \tag{17}$$

To evaluate the acid dissociation constants, graphical replots of the $V_{max,app}$ and $K_{S,app}$ in the proper H^+ ranges will permit an evaluation of K_{ES_1}, K_{ES_2}, K_{E_1}, and K_{E_2}, as well as the pH-independent parameters, V_{max} and K_S.

Thus, when Equation 14 is applicable to in the acidic range, $\frac{K_{ES_2}}{(H^+)}$ may be neglected in the denominator; therefore,

$$V_{max,app}\left(1 + \frac{(H^+)}{K_{ES_1}}\right) \simeq V_{max}, \quad V_{max,app} + V_{max,app}\frac{(H^+)}{K_{ES_1}} = V_{max}, \quad \text{and}$$

$$V_{max,app} = V_{max} - V_{max,app}\frac{(H^+)}{K_{ES_1}} \tag{18}$$

and a plot of $V_{max,app}$ vs. $V_{max,app} \cdot (H^+)$ will yield V_{max} from the Y-intercept and $-\frac{1}{K_{ES_1}}$ from the slope (Figure 1), or, in the basic range, neglect $\frac{(H^+)}{K_{ES_1}}$ in the denominator of Equation 14. Therefore;

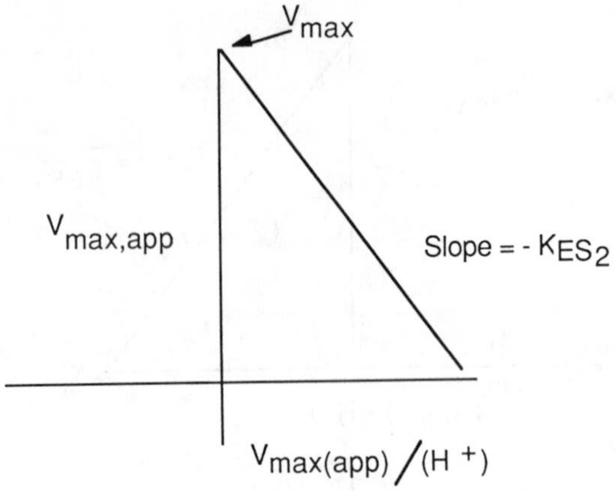

FIGURE 2.

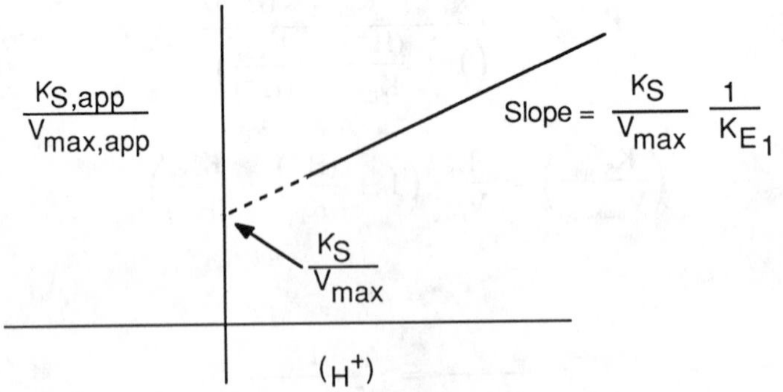

FIGURE 3.

$$V_{max,app} + V_{max,app}\frac{K_{ES_2}}{(H^+)} = V_{max}; \quad \text{or} \quad V_{max,app} = V_{max} - V_{max,app}\frac{K_{ES_2}}{(H^+)} \tag{19}$$

and a suitable plot yields K_{ES_2} and V_{max} (Figure 2).

Similarly, from Equation 16, the slope of the primary $\frac{1}{v_0}$ vs. $\frac{1}{S_0}$ plot is

$$\frac{K_{S,app}}{V_{max,app}} = \frac{K_S}{V_{max}} + \frac{K_S(H^+)}{V_{max}K_{E_1}} + \frac{K_S}{V_{max}}\frac{K_{E_2}}{(H^+)} \tag{20}$$

and under acidic conditions where $\frac{K_{E_2}}{(H^+)} \longrightarrow$ small,

$$\therefore \frac{K_{S,app}}{V_{max,app}} \cong \frac{K_S}{V_{max}} + \frac{K_S(H^+)}{V_{max}K_{E_1}} \tag{21}$$

and a plot of $\frac{K_{S,app}}{V_{max,app}}$ [= slope $\left(\text{of } \frac{1}{v} \text{ vs. } \frac{1}{S}\right)$ plot] vs. (H^+) yields as slope = $\frac{K_s}{V_{max}}\frac{1}{K_{E_1}}$ and Y-intercept = $\frac{K_s}{V_{max}}$. Therefore K_{E_1} can be obtained. Under basic conditions, where

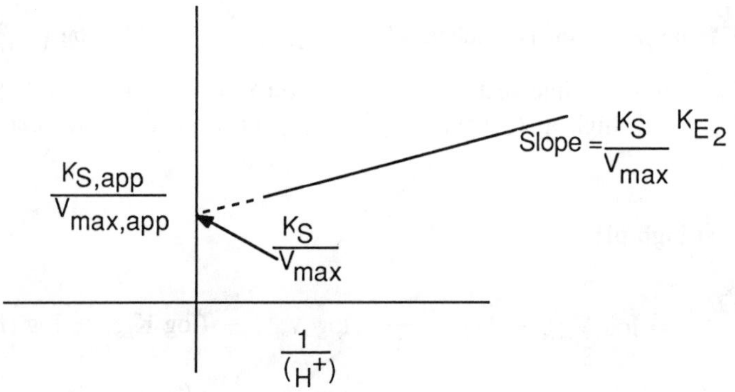

FIGURE 4.

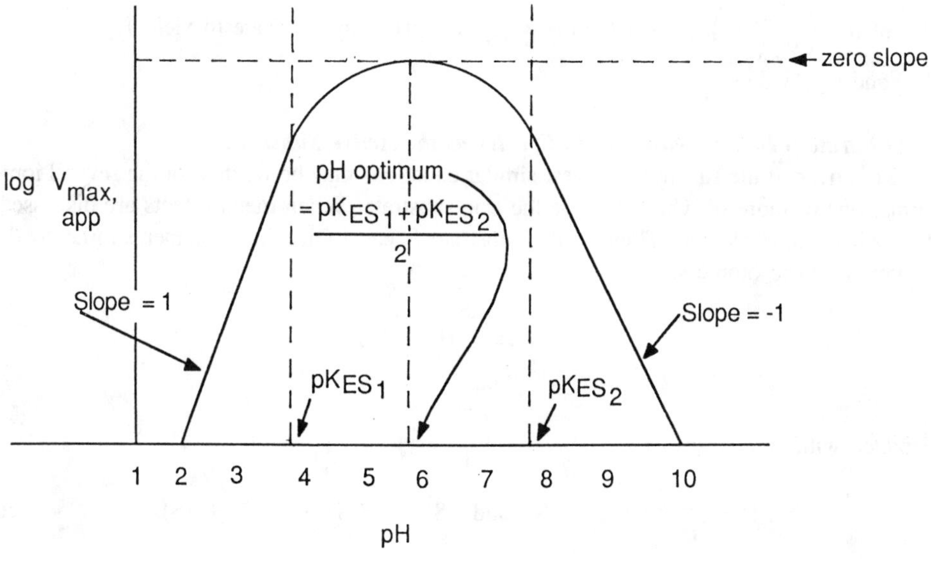

FIGURE 5.

$$\frac{(H^+)}{K_{E_1}} \rightarrow \text{small}, \quad \frac{K_{S,app}}{V_{max,app}} \simeq \frac{K_S}{V_{max}} + \frac{K_S}{V_{max}} K_{E_2} \frac{1}{(H^+)} \tag{22}$$

and a plot of $K_{S,app}/V_{max,app}$ vs. $\frac{1}{(H^+)}$ eventually yields K_{E_2} (Figure 4).

Dixon-Webb plots[10] are in essence log-log plots. Thus, for Equation 14 at low pH values, $\frac{K_{ES_2}}{(H^+)}$ can be neglected and also $\frac{(H^+)}{K_{ES_1}} >> 1$, so that "1" may also be neglected. Therefore,

$$\log V_{max,app} = \log V_{max} - \log\left(\frac{(H^+)}{K_{ES_1}}\right) = \log V_{max} + pH - pK_{ES_1} \tag{23}$$

or

$$\log\left(\frac{V_{max,app}}{V_{max}}\right) = pH - pK_{ES_1} \tag{24}$$

Therefore, in a plot of log $V_{max,app}$ vs. pH (Figure 5), the slope, at acidic pH values, has a value of 1, until a point is reached where $V_{max,app} = V_{max}$, the $\log\left(\frac{V_{max,app}}{V_{max}}\right) = 0$, and $pH = pK_{ES_1}$. Thus, the intersection point of the zero slope with the initial slope of 1.0, yields the pK_{ES_1}. At high pH, neglect $(H^+)/K_{ES_1}$ in Equation 14, then $V_{max,app} \cong \frac{V_{max}}{K_{ES_2}/(H^+)}$.

Therefore, at high pH,

$$\log V_{max,app} \simeq \log V_{max} - \log \frac{K_{ES_2}}{(H^+)} = \log V_{max} - (\log K_{ES_2} - \log (H^+))$$

$$= \log V_{max} + pK_{ES_2} - pH \tag{25}$$

and the intersection of the zero slope line with the slope $= -1$ line, yields p K_{ES_2}. A similar plot of $\log\left(\frac{K_{m,app}}{V_{max,app}}\right)$ vs. pH (or log $K_{m,app}$ vs. pH) may be made to yield K_{E_1}, K_{E_2}, and K_{ES_1} and K_{ES_2} values.

b. Substrate S Ionizes with the HS Species as the Active Substrate

Moreover, if the substrate ionizes, similar equations may be written for the several ionic forms, one or more of which may be the true substrate. When metal effects are discussed, this will become clearer. Thus, if the substrate S can ionize in a manner similar to the enzyme as a diprotonic species, i.e.,

$$\overset{+}{H_2S} \underset{K_{a_1}}{\rightleftharpoons} \overset{\circ}{HS} \underset{K_{a_2}}{\rightleftharpoons} \bar{S}$$

depicted without charges as

$$H_2S \underset{K_{a_1}}{\rightleftharpoons} HS \underset{K_{a_2}}{\rightleftharpoons} S \quad \text{and} \quad S_t = (H_2S) + (HS) + (S) \tag{26}$$

and only the HS species of substrate is assumed to be the active substrate species. Then, Equation 11 becomes

$$\frac{v_o^f}{V_{max}^f} = \frac{S_t\Big/\left(1 + \frac{(H^+)}{K_{a_1}} + \frac{K_{a2}}{(H^+)}\right)}{K_S\left(1 + \frac{(H^+)}{K_{E_1}} + \frac{K_{E2}}{(H^+)}\right) + S_t\left(1 + \frac{(H^+)}{K_{ES1}} + \frac{K_{ES2}}{(H^+)}\right)\Big/\left(1 + \frac{(H^+)}{K_{a_1}} + \frac{K_{a2}}{(H^+)}\right)} \tag{27}$$

c. All Forms of E Bind the Substrate, but Only HES and ES Yield the Product

Kuby and Roy[71] were able to evaluate the pKs of both the ternary and binary substrate complexes, as well as the $V_{max}^{acidic + neutral}$ and V_{max}^{basic}, which characterized the two-substrate system. For a single-substrate case, the mechanism would be

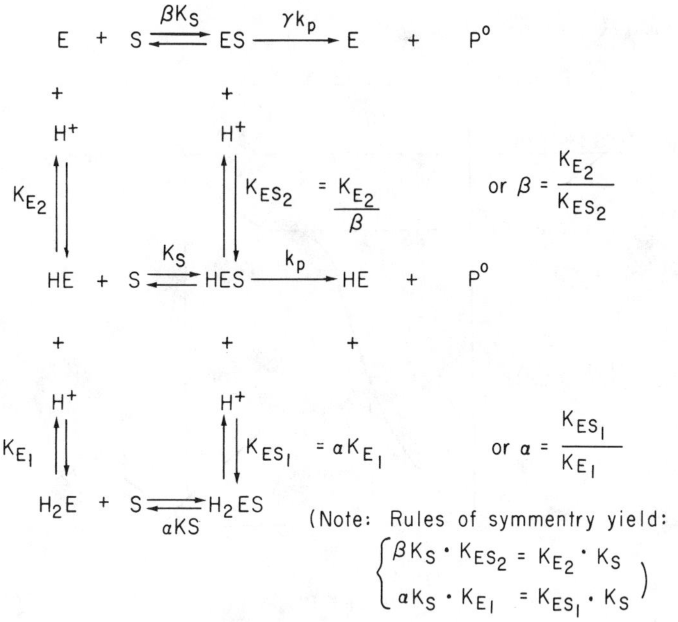

SCHEME II.

Note the rules of symmetry yield

$$\begin{cases} \beta K_S \cdot K_{ES_2} = K_{E_2} \cdot K_S \\ \alpha K_S \cdot K_{E_1} = K_{ES_1} \cdot K_S \end{cases} \tag{28}$$

The initial forward velocity equation becomes

$$\frac{v_f}{E_t} = \frac{k_p \dfrac{(S)}{K_S} + \gamma k_p \dfrac{(S)K_{ES_2}}{K_S(H^+)}}{1 + \dfrac{(S)}{K_S} + \dfrac{(H^+)}{K_{E_1}} + \dfrac{K_{E_2}}{(H^+)} + \dfrac{(S)(H^+)}{K_S K_{ES_1}} + \dfrac{(S)K_{ES_2}}{K_S(H^+)}} \tag{29}$$

If $k_p(E)_t = V_{max}$, factoring yields

$$\frac{v_f}{V_{max}} = \frac{(S)}{K_S\left(1 + \dfrac{(H^+)}{K_{E_1}} + \dfrac{K_{E_2}}{(H^+)}\right)\Big/\left(1 + \dfrac{\gamma K_{ES_2}}{(H^+)}\right) + (S)\left(1 + \dfrac{(H^+)}{K_{ES_1}} + \dfrac{K_{ES_2}}{(H^+)}\right)\Big/\left(1 + \dfrac{\gamma K_{ES_2}}{(H^+)}\right)} \tag{30}$$

Define

$$V_{max,app} = \frac{V_{max}\left(1 + \dfrac{\gamma K_{ES_2}}{(H^+)}\right)}{\left(1 + \dfrac{(H^+)}{K_{ES_1}} + \dfrac{K_{ES_2}}{(H^+)}\right)} \tag{31}$$

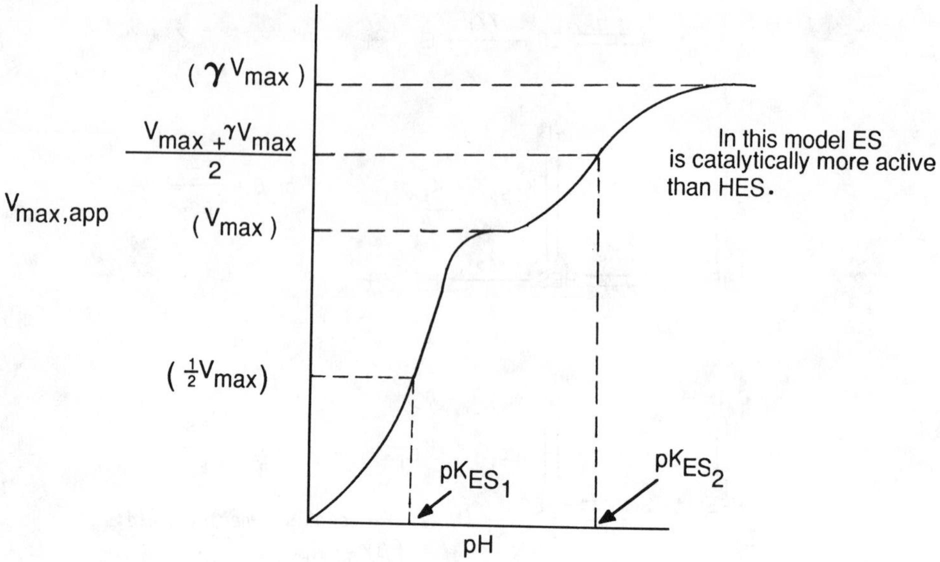

FIGURE 6A.

Or, if γ <1:

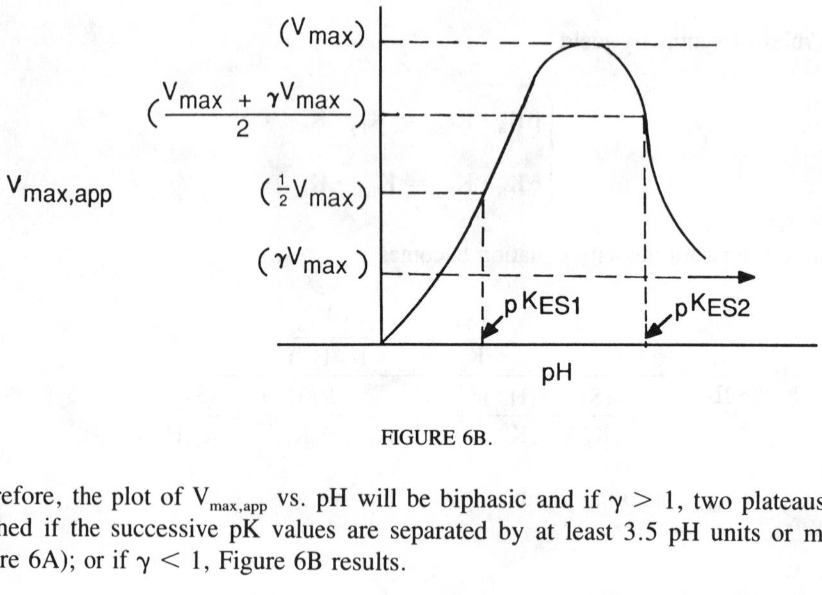

FIGURE 6B.

Therefore, the plot of $V_{max,app}$ vs. pH will be biphasic and if $\gamma > 1$, two plateaus will be reached if the successive pK values are separated by at least 3.5 pH units or more (see Figure 6A); or if $\gamma < 1$, Figure 6B results.

d. All Forms of E Bind S^0, and All Forms of ES Yield Product P^0

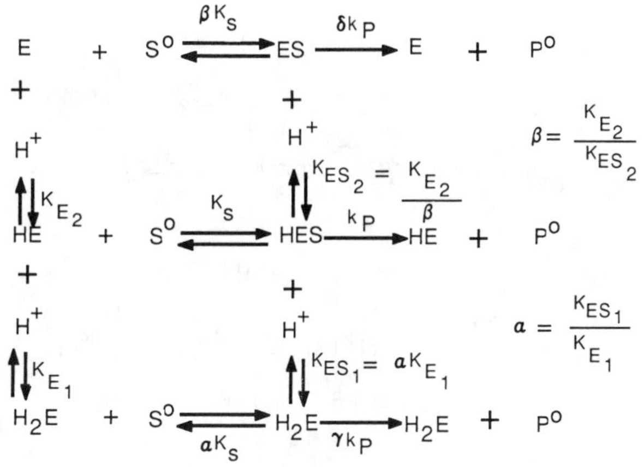

SCHEME III.

The initial forward velocity equation becomes

$$\frac{v}{E_t} = \frac{k_p \dfrac{(S)}{K_S} + \dfrac{\gamma k_p(S)(H^+)}{K_S K_{ES_1}} + \dfrac{\delta k_p(S)K_{ES_2}}{K_S(H^+)}}{1 + \dfrac{(H^+)}{K_{E_1}} + \dfrac{K_{E_2}}{(H^+)} + \dfrac{(S)}{K_S} + \dfrac{(S)(H^+)}{K_S K_{ES_1}} + \dfrac{(S)K_{ES_2}}{K_S(H^+)}\Big)} \tag{32}$$

Factoring, setting $k_p\ E_t = V_{max}$, and rearranging

$$\frac{v}{V_{max}} = \frac{(S)}{K_S\left[1 + \dfrac{(H^+)}{K_{E_1}} + \dfrac{K_{E_2}}{(H^+)}\right] \Big/ \left[1 + \dfrac{\gamma(H^+)}{K_{ES_1}} + \dfrac{\delta K_{ES_2}}{(H^+)}\right] + (S)\left[1 + \dfrac{(H^+)}{K_{ES_1}} + \dfrac{K_{ES_2}}{(H^+)}\right) \Big/ \left(1 + \dfrac{\gamma(H^+)}{K_{ES_1}} + \dfrac{\delta K_{ES_2}}{(H^+)}\right)} \tag{33}$$

e. System C (Scheme II) with the Addition of an ES ⇌ EP Step

The addition of an ES ⇌ EP step has already been derived for the H^+-independent case and under steady-state conditions, that is (Scheme II in Chapter 2),

(See Chapter 2, Equation 8 for the solution)

$$\frac{v}{E_t} = \frac{k_{+1}k_{+2}k_{+3}(S) - k_{-1}k_{-2}k_{-3}(P)}{(k_{-1}k_{-2} + k_{-1}k_{+3} + k_{+2}k_{+3}) + k_{+1}(S)(k_{+2} + k_{-2} + k_{+3}) + k_{-3}(P)(k_{-1} + k_{+2} + k_{-2})} \tag{2-8}$$

For simplicity, assume as given in Scheme IV that only HE binds S, but that once HES forms, it can also ionize to ES or add a proton to H_2ES. Assume that only HES will yield HEP, which can either ionize to EP or add a proton to H_2EP, or that HEP ⟶ HE + P, only.

This yields, therefore, for Scheme IV, the forward initial velocity (after setting P = O).

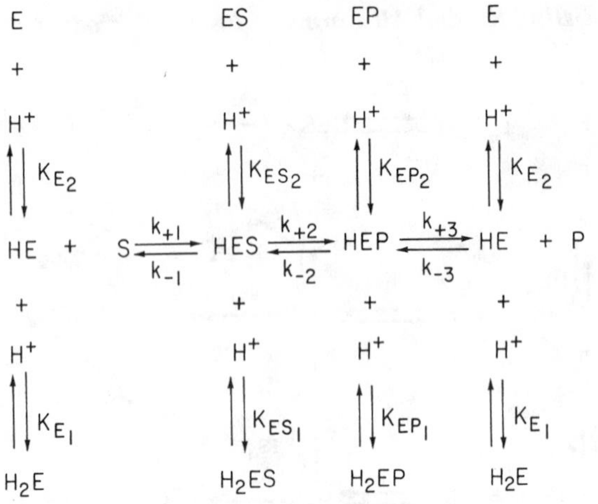

SCHEME IV.

With the case for ionization of EP and, under steady-state conditions,

$$\frac{v}{E_t} = \frac{k_{+1}k_{+2}k_{+3}(S)}{(k_{-1}k_{-2} + k_{-1}k_{+3} + k_{+2}k_{+3})\left(1 + \frac{(H^+)}{K_{E_1}} + \frac{K_{E_2}}{(H^+)}\right) + k_{+1}(S)\left[(k_{-2} + k_{+3})\left(1 + \frac{(H^+)}{K_{ES_1}} + \frac{K_{ES_2}}{(H^+)}\right) + k_{+2}\left(1 + \frac{(H^+)}{K_{EP_1}} + \frac{K_{EP_2}}{(H^+)}\right)\right]} \tag{34}$$

Therefore,

$$V_{max,app} = \frac{k_{+2}k_{+3}E_t}{(k_{-2} + k_{+3})\left[1 + \frac{(H^+)}{K_{ES_1}} + \frac{K_{ES_2}}{(H^+)}\right] + k_{+2}\left[1 + \frac{(H^+)}{K_{EP_1}} + \frac{K_{EP_2}}{(H^+)}\right]} \tag{35}$$

$$\frac{V_{max,app}}{K_{m,app}} = \frac{k_{+1}k_{+2}k_{+3}E_t}{(k_{-1}k_{-2} + k_{-1}k_{+3} + k_{+2}k_{+3})\left(1 + \frac{(H^+)}{K_{E_1}} + \frac{K_{E_2}}{(H^+)}\right)} \tag{36}$$

$$K_{m,app} = \frac{(k_{-1}k_{-2} + k_{-1}k_{+3} + k_{+2}k_{+3})\left(1 + \frac{(H^+)}{K_{E_1}} + \frac{K_{E_2}}{(H^+)}\right)}{\left[k_{+1}(k_{-2} + k_{+3})\left(1 + \frac{(H^+)}{K_{ES_1}} + \frac{K_{ES_2}}{(H^+)}\right) + k_{+2}\left(1 + \frac{(H^+)}{K_{EP_1}} + \frac{K_{EP_2}}{(H^+)}\right)\right]} \tag{37}$$

A derivation of a steady-state system where all E forms bind S and all ES forms yield their respective EP form, all of which yield product, for a unireactant system involving a two-protonic equilibria, i.e., a so-called diprotonic system, has actually been derived by Ottolenghi[72] and as might be expected it is a formidable equation with 117 terms in the denominator. Thus, to simplify the understanding of the measured pKs, i.e., the identities of these assigned ionizable groups, their heats of ionization, i.e., ΔH_{ioniz} may be estimated from the studies of the effects of temperature or from the use of site-specific reagents.

2. A Two-Substrate Reaction Illustrated by Brewers' Yeast Glucose 6-Phosphate Dehydrogenase

Although a two-substrate reaction might be considered too complex for analysis, in the case of the brewers' yeast glucose 6-phosphate dehydrogenase, Kuby and Roy[71] deduced a self-consistent mechanism which allowed for the random binding of the two fully ionized substrates to the enzyme via two major pathways, and product formation by both $E \cdot A^- \cdot B^-$ and $HE \cdot A^- \cdot B^-$ (but not from $H_2E \cdot A^- \cdot B^-$), when the protein was stabilized in solution as only the two-subunit catalyst. Thus, in the absence of superimposed association-dissociation reactions, the $V_{max,app}$ data obtained between pH 5 and 10 and 18 and 32°C in several buffers, led to the postulate that at least two sets of protonic equilibria may govern the catalysis (one near pH 5.7 at 25°C and another near pH 9.2). Furthermore, two pathways for product formation (i.e., two $V_{max's}$) appear to be required to explain the biphasic nature of the log $V_{max,app}$ vs. pH curves, with $V_{max}^{basic} > V_{max}^{acidic\,+\,neutral}$. Of the several buffers explored, either a uniform degree of interaction or a minimal degree of buffer species interaction could be assessed from the enthalpy changes associated with the derived values for ionization constants attributed to the protonic equilibria in the enzyme-substrates ternary complexes, for the case of Tris-acetate-EDTA buffers, at constant ionic strength. With the selection of this buffer at 0.1(Γ/2) and 25 and 32°C, the kinetic mechanism emerged as shown in Scheme V.

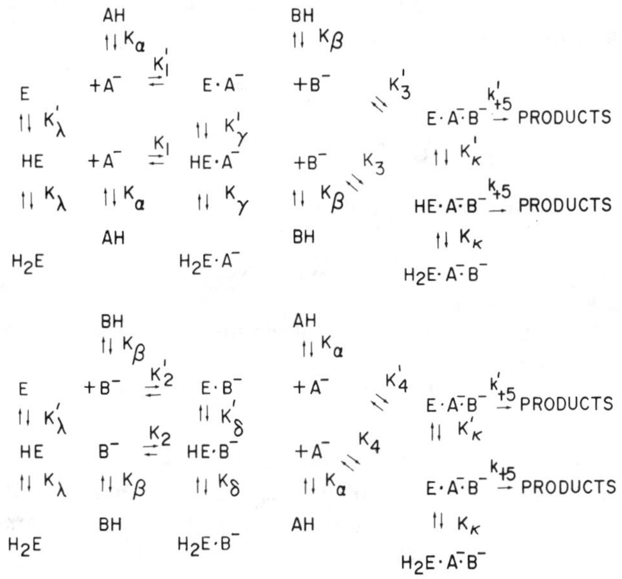

SCHEME V. Postulated kinetic mechanism for the two-subunit glucose 6-phosphate dehydrogenase species. The various ionic, i.e., protonated, states of the enzyme are denoted simply by H_2E, HE, and E.

A quasi-equilibrium is presumed, with rate-limiting steps (k_{+5} and k'_{+5}), at the interconversion of the ternary complexes.

Assuming the quasi-equilibrium for the forward reactions $v_{o(overall)} = v_{o(acidic\,+\,neutral)} + v_{o(basic)}$, it follows that

$$v_{o(acidic\,+\,neutral)} = \frac{(k_{+5}E_t)}{\left[1 + \frac{(H^+)}{K_\kappa} + \frac{K'_\kappa}{(H^+)}\right] + \frac{K_3}{(B^-)}\left[1 + \frac{(H^+)}{K_\gamma} + \frac{K'_\gamma}{(H^+)}\right] + \frac{\overline{K}_{A^-}}{(A^-)}\left[1 + \frac{(H^+)}{K_\delta} + \frac{K'_\delta}{(H^+)}\right] + \frac{K_1K_3}{(A^-)(B^-)}\left(1 + \frac{(H^+)}{K\lambda} + \frac{K'_\lambda}{(H^+)}\right)\Bigg]} \quad (38)$$

where

$$K_1K_3 = K_2K_4 \tag{39}$$

and

$$V_{o(basic)} = \frac{(k'_{+5}E_t)}{\left[1 + \frac{(H^+)}{K'_\kappa} + \frac{(H^+)^2}{K_\kappa K'_\kappa}\right] + \frac{K'_3}{(B^-)}\left[1 + \frac{(H^+)}{K'_\gamma} + \frac{(H^+)^2}{K_\gamma K'_\gamma}\right] + \frac{\overline{K}_{A^-}}{(A^-)}\left[1 + \frac{(H^+)}{K'_\delta} + \frac{(H^+)^2}{K_\delta K'_\delta}\right] + \frac{K'_1K'_3}{(A^-)(B^-)}\left[1 + \frac{(H^+)}{K'_\lambda} + \frac{(H^+)^2}{K_\lambda K'_\lambda}\right]} \tag{40}$$

and where $K'_1\ K'_3 = K'_2K'_4$; also from the rules of symmetry it follows that

$$K'_\lambda K_1 = K'_1K'_\gamma$$
$$K'_\lambda K_2 = K'_2K'_\delta$$
$$K'_\delta K'_3K_4 = K'_\gamma K'_4K_3 \tag{41}$$

As in a previous study, an independent binding mechanism may be assumed[58] but was checked at the various pH values to ensure it held throughout the pH range studied. Therefore, the following equalities are set:

$$\overline{K}_{A^-} = K_1 = K'_1 = K_4 = K'_4$$
$$\overline{K}_{B^-} = K_2 = K'_2 = K_3 = K'_3 \tag{42}$$

and it follows from the equalities of Equation 41 that

$$K'_Y = K'_\delta = K'_\lambda \tag{43}$$

which was defined as equal to K'_H; therefore, it was logical to assume also that

$$K_\lambda = K_\delta = K_\gamma = K_{H_2} \tag{44}$$

Thus, both first (K_{H_2}) and second ionizations (K'_H) of the enzyme's substrate (s) binding sites were assumed to be independent of the binding of either A^- or B^- in the binary complexes, $E \cdot A^-$ or $E \cdot B^-$, or in the free enzyme E, but not in the ternary complexes.

Therefore, Equations 38 and 40 become

$$V_{o(acidic + neutral)} = \frac{k_{+5}E_t}{\left[1 + \frac{(H^+)}{K'_\kappa} + \frac{K'_\kappa}{(H^+)}\right] + \frac{\overline{K}_{B^-}}{(B^-)}\left[1 + \frac{(H^+)}{K_{H_2}} + \frac{K'_H}{(H^+)}\right] + \frac{\overline{K}_{A^-}}{(A^-)}\left[1 + \frac{(H^+)}{K_{H_2}} + \frac{K'_H}{(H^+)}\right] + \frac{\overline{K}_{A^-}\overline{K}_{B^-}}{(A^-)(B^-)}\left[1 + \frac{(H^+)}{K_{H_2}} + \frac{K'_H}{(H^+)}\right]} \tag{45}$$

and

$$V_{o(basic)} = \frac{k'_{+5}E_t}{\left[1 + \frac{(H^+)}{K'_\kappa} + \frac{(H^+)^2}{K_\kappa K'_\kappa}\right] + \frac{\overline{K}_{B^-}}{(B^-)}\left[1 + \frac{(H^+)}{K'_H} + \frac{(H^+)^2}{K_{H_2}K'_H}\right] + \frac{\overline{K}_{A^-}}{(A^-)}\left[1 + \frac{(H^+)}{K'_H} + \frac{(H^+)^2}{K_{H_2}K'_H}\right] + \frac{\overline{K}_{A^-}\overline{K}_{B^-}}{(A^-)(B^-)}\left[1 + \frac{(H^+)}{K'_H} + \frac{(H^+)^2}{K_{H_2}K'_H}\right]} \tag{46}$$

This reduces the mechanism shown in Scheme V to that shown in Scheme VI where the number of parameters to be estimated are considerably reduced

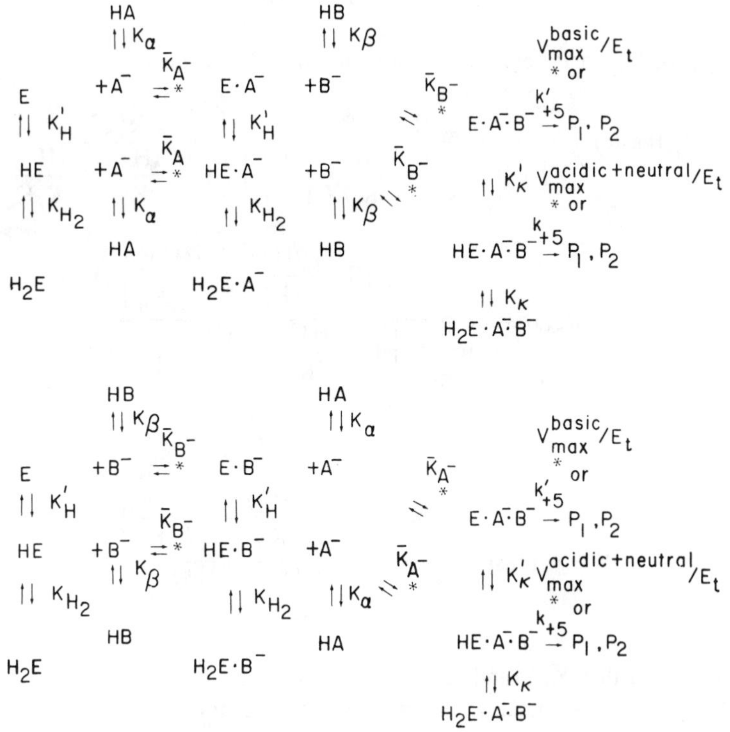

SCHEME VI.

The following pH-dependent parameters may be expressed in terms of their pH-independent constants and as a function of the (H^+) ion concentration:

$$V_{max,app}(\text{acidic} + \text{neutral}) = \frac{V^*_{max}(\text{acidic} + \text{neutral})}{\left(1 + \frac{(H^+)}{K_\kappa} + \frac{K'_\kappa}{(H^+)}\right)} = \frac{k_{+5}E_t}{\left(1 + \frac{(H^+)}{K_\kappa} + \frac{K'_\kappa}{(H^+)}\right)} \tag{47}$$

$$\overline{K}_{B^-,app}(\text{acidic} + \text{neutral}) = \frac{\overline{K}_{*B^-}\left[1 + \frac{(H^+)}{K_{H_2}} + \frac{K'_H}{(H^+)}\right]}{\left[1 + \frac{(H^+)}{K_\kappa} + \frac{K'_\kappa}{(H^+)}\right]} \tag{48}$$

and

$$\overline{K}_{A^-,app}(\text{acidic} + \text{neutral}) = \frac{\overline{K}_{*A^-}\left[1 + \frac{(H^+)}{K_{H_2}} + \frac{K'_H}{(H^+)}\right]}{\left[1 + \frac{(H^+)}{K_\kappa} + \frac{K'_\kappa}{(H^+)}\right]} \tag{49}$$

and after substitution into Equation 45,

$$v_{o(acidic + neutral)} = \frac{V_{max,app}(acidic + neutral)}{\left[1 + \frac{\overline{K}_{B^-,app}}{(B^-)} + \frac{\overline{K}_{A^-,app}}{(A^-)}\left(1 + \frac{\overline{K}_{*B^-}}{(B^-)}\right)\right]}$$

Similarly, for the basic region, the following may be defined:

$$V_{max,app}(basic) = \frac{V_{*max}(basic)}{\left[1 + \frac{(H^+)}{K'_\kappa} + \frac{(H^+)^2}{K_\kappa K'_\kappa}\right]} = \frac{k'_{+5}E_t}{\left[\left(1 + \frac{(H^+)}{K'_\kappa} + \frac{(H^+)^2}{K_\kappa K'_\kappa}\right]\right.} \quad (50)$$

$$\overline{K}_{B,app}(basic) = \frac{\overline{K}_{*B^-}\left[1 + \frac{(H^+)}{K'_H} + \frac{(H^+)^2}{K_{H_2}K'_H}\right]}{\left[1 + \frac{(H^+)}{K'_\kappa} + \frac{(H^+)^2}{K_\kappa K'_\kappa}\right]} \quad (51)$$

and

$$\overline{K}_{A^-,app}(basic) = \frac{\overline{K}_{*A^-}\left[1 + \frac{(H^+)}{K'_H} + \frac{(H^+)^2}{K_{H_2}K'_H}\right]}{\left[1 + \frac{(H^+)}{K'_\kappa} + \frac{(H^+)^2}{K_\kappa K'_\kappa}\right]} \quad (52)$$

It will be noted that $\overline{K}_{*_A}{}^-$ and $\overline{K}_{*_B}{}^-$ for the basic and acidic + neutral range are the same, but V_{*max} (acidic + neutral) and V_{*max} (basic) are not.

After substitution into Equation 46

$$v_{o(basic)} = \frac{V_{max,app}(basic)}{1 + \frac{\overline{K}_{B^-,app}(basic)}{(B^-)} + \frac{\overline{K}_{A^-,app}(basic)}{(A^-)}\left(1 + \frac{\overline{K}_{*B^-}}{(B^-)}\right)} \quad (53)$$

Finally, in terms of total substrate species A_t and B_t and recalling that only A^- and B^- are the active species,

$$(A^-) = \frac{A_t}{\left(1 + \frac{(H^+)}{K_\alpha}\right)} \quad \text{and} \quad (B^-) = \frac{B_t}{\left(1 + \frac{(H^+)}{K_\beta}\right)} \quad (54)$$

and since

$$v_{o(overall)} = v_{o(acidic + neutral)} + v_{o(basic)} \quad (55)$$

TABLE 1
pH-Dependent Kinetic Parameters and Ionization Constants Determined at 25 and 32°C in 0.1(Γ/2) Tris-Acetate-EDTA Buffers for Brewers' Yeast 6-Phosphate Dehydrogenase[71]

Derived kinetic parameters or ionization constants	Defined reaction	Values	
		At 25°C	At 32°C
pK_α^a	Titration (ionization) constant to yield $NADP^{3-}$	6.13	6.13
$pK_\beta{}^a$	2nd ionization (titration) constant to yield G 6-P^2	6.08	6.08
$\overline{K}_{*A}{}^{-b}$	Intrinsic dissociation constant of enzyme-$NADP^{3-}$ binary or ternary complexes	3.1_4 (±1.56) × 10^{-6} *M*	5.6_8 (±2.96) × 10^{-6} *M*
$\overline{K}_{*B}{}^{-b}$	Intrinsic dissociation constant of enzyme G 6-P^{2-} binary or ternary complexes	1.1_5 (±0.37) × 10^{-5} *M*	1.4_7 (±0.76) × 10^{-5} *M*
pK_{H_2}	1st ionization constant from enzyme binary complexes or free enzyme	5.23	5.43
pK'_H	2nd ionization constant from enzyme binary complexes or free enzyme	8.19	8.13
pK_κ	1st ionization constant from enzyme-substrates ternary complex	5.69	5.60
pK'_κ	2nd ionization constant from enzyme-substrates ternary complex	9.08	8.90
$V_{*max\ (acidic\ +\ neutral)}$	$k'_{+5}\ E_t$ or maximal velocity in acidic + neutral range	420 (±65) μmol/min/mg	525 (±177)
$V_{*max\ (basic)}$	$k'_{+5}\ E_t$ or maximal velocity in basic range	905 (±115) μmol/min/mg	1810 (±900)
k_{+5}	1st order velocity coefficient for rate-limiting step in acidic + neutral range	357 (±55) s^{-1}	446
k'_{+5}	1st order velocity coefficient for rate-limiting step in basic range	769 (±98) s^{-1}	1539 (±764)

[a] Titration constants to yield fully ionized species of $NADP^{3-}$ and $G6-P^{2-}$; values determined in 0.1 (Γ/2).
[b] pH-Independent intrinsic dissociation constants for enzyme-fully ionized substrate; in the case of NADP, in its fully ionized state, it has a net charge of 3^-.[9]

$$V_{o(overall)} = \frac{V_{max,app}(\text{acidic + neutral})}{\left\{1 + \frac{\overline{K}_{B^-,app}(\text{acidic})}{B_t}\left(1 + \frac{(H^+)}{K_\beta}\right) + \frac{\overline{K}_{A^-,app}(\text{acidic})}{A_t}\left(1 + \frac{(H^+)}{K_\alpha}\right)\left[1 + \frac{\overline{K}_{*B^-}}{B_t}\left(1 + \frac{(H^+)}{K_\beta}\right)\right]\right\}} + \frac{V_{max,app}(\text{basic})}{\left\{1 + \frac{\overline{K}_{B^-,app}(\text{basic})}{B_t}\left(1 + \frac{(H^+)}{K_\beta}\right) + \frac{\overline{K}_{A^-,app}(\text{basic})}{A_t}\left(1 + \frac{(H^+)}{K_\alpha}\right)\left[1 + \frac{\overline{K}_{*B^-}}{B_t}\left(1 + \frac{(H^+)}{K_\beta}\right)\right]\right\}} \quad (56)$$

To evaluate the various kinetic parameters, the graphical procedures already described in principle are utilized after making the necessary approximations from the equations for the pH-dependent parameters (Equations 47 to 51) and rearrangements to yield linear plots.[71]

In Table 1, the final evaluated data are summarized. As a check on the overall evaluation, theoretical plots were drawn with the aid of the following expressions:

$$V_{max,app}(\text{overall}) = V_{max,app}(\text{acidic} + \text{neutral}) + V_{max,app}(\text{basic}) \tag{57}$$

$$V_{max,app}(\text{overall}) = \frac{V_{*max}(\text{acidic} + \text{neutral})}{\left(1 + \frac{(H^+)}{K'_\kappa} + \frac{K'_\kappa}{(H^+)}\right)} + \frac{V_{*max}(\text{basic})}{\left(1 + \frac{(H^+)}{K'_\kappa} + \frac{(H^+)^2}{K_\kappa K'_\kappa}\right)} \tag{58}$$

$$\overline{K}_{B,app}(\text{overall}) = \overline{K}_{B,app}(\text{acidic} + \text{neutral}) + \overline{K}_{B,app}(\text{basic}) \tag{59}$$

$$\overline{K}_{A,app}(\text{overall}) = \overline{K}_{A,app}(\text{acidic} + \text{neutral}) + \overline{K}_{A,app}(\text{basic}) \tag{60}$$

$$\overline{K}_{B,app}(\text{overall}) = (\overline{K}_{*B^-})\left(1 + \frac{(H^+)}{K_\beta}\right)\left[\frac{\left[1 + \frac{(H^+)}{K_{H_2}} + \frac{K'_H}{(H^+)}\right]}{\left[1 + \frac{(H^+)}{K_\kappa} + \frac{K'_\kappa}{(H^+)}\right]} + \frac{\left[1 + \frac{(H^+)}{K'_H} + \frac{(H^+)^2}{K_{H_2}K'_H}\right]}{\left[1 + \frac{(H^+)}{K'_\kappa} + \frac{(H^+)^2}{K_\kappa K'_\kappa}\right]}\right] \tag{61}$$

$$\overline{K}_{A,app}(\text{overall}) = (\overline{K}_{*A^-})\left(1 + \frac{(H^+)}{K_\alpha}\right)\left[\frac{\left[1 + \frac{(H^+)}{K_{H_2}} + \frac{K'_H}{(H^+)}\right]}{\left[1 + \frac{(H^+)}{K_\kappa} + \frac{K'_\kappa}{(H^+)}\right]} + \frac{\left[1 + \frac{(H^+)}{K'_H} + \frac{(H^+)^2}{K_{H_2}K'_H}\right]}{\left[1 + \frac{(H^+)}{K'_\kappa} + \frac{(H^+)^2}{K_\kappa K'_\kappa}\right]}\right] \tag{62}$$

and compared in Figure 7 with the experimental data[71] where satisfactory agreement is observed. Thus, whereas two pH-independent maximal velocities, V_{*max} (acidic + neutral) and V_{*max} (basic), are required as a minimum assumption to account for the data, only one value is required for each of the pH-independent intrinsic dissociation constants for the ionized substrates, i.e., $\overline{K}_{*_A-}$ or $\overline{K}_{*_B-}$, throughout the entire pH range.

In an analogous study, Viola[846] recently studied the pH variation of kinetic parameters of glucose 6-phosphate dehydrogenase from *Leuconostoc mesenteroides*, an enzyme which differs from the brewers' yeast enzyme of Kuby and Roy[71] in that it may utilize either NAD$^+$ or NADP$^+$.[847]

II. EFFECT OF TEMPERATURE ON THE REACTION VELOCITY (SEE CHAPTER 1, PART VI)

Some of the most useful pieces of information to be derived from studies on the effect of temperature are the thermodynamic parameters which may be estimated from the intrinsic dissociation or ionization constants, i.e., from those values derived kinetically and estimated at two or more temperatures (e.g., from Table 1).

Thus, from the van't Hoff equation:

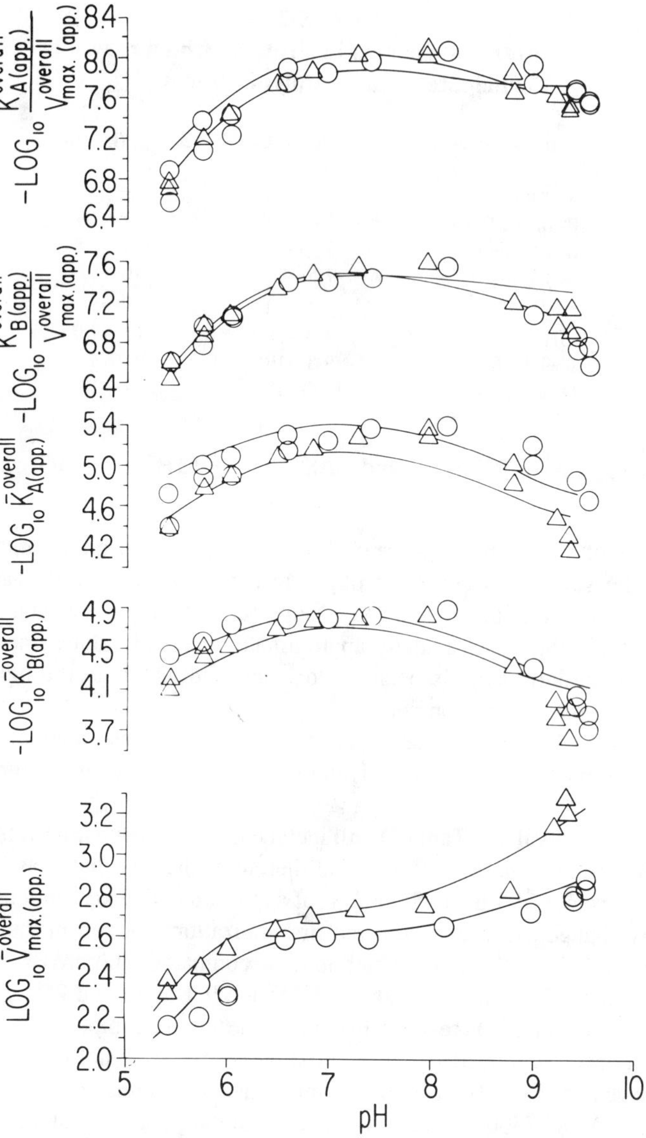

FIGURE 7. Effect of pH on the Michaelis constants and the maximal velocities at 25 and 32°C, in 0.1 (Γ/2) Tris-acetate-EDTA buffers for the two-chain enzyme species of glucose-6-phosphate dehydrogenase. In all cases, (○) data at 25°C and (△) data at 32°C.[71]

$$\frac{d\ln(K_{eq})}{dT} = \left(\frac{\Delta H^\circ_{ioniz}}{RT^2}\right) \tag{63}$$

which may be conveniently modified for measurements at two temperatures to give

$$2.303[pK_{a1} - pK_{a2}] = \left(\frac{\Delta H^\circ_{ioniz}}{R}\right)\left[\frac{(T_2 - T_1)}{T_1 \times T_2}\right] \tag{64}$$

from which the standard enthalpy of ionization may be estimated. Values for ΔG°_{ioniz} and ΔS°_{ioniz} (the standard free energy and entropy of ionization) may be obtained from

TABLE 2
Prototropic Groups in Proteins which may Participate in the Catalytic Process[73,74]

Ionizable group	pK_a 25°C	ΔH_{ioniz} (kcal/mol)
α-COO^-	3.0—3.2	±1.5
β- or γ-COO^-	3.0—5.0	±1.5
Imidazolium	5.5—7.0	6.9—7.5
α-NH_3^+	7.5—8.5	10—13
ϵ-NH_3^+	9.4—10.6	10—13
-SH	8.0—8.5	6.5—7.0
Phelolic hydroxyl	9.8—10.5	6
Guanidinium	11.6—12.6	12—13

$$\Delta G^\circ_{ioniz} = -RT\ln K_T \quad \text{and} \quad \Delta S^\circ_{ioniz} = \frac{1}{T}(\Delta H^\circ_{ioniz} - \Delta G^\circ_{ioniz}) \tag{65}$$

A comparison of the estimated value of ΔH°_{ioniz} and its intrinisic ionization constant with those listed[73,74] for several prototropic groups present in proteins, e.g., Table 2, provides information on the nature of those ionizable groups participating in the catalytic mechanism. The values listed in Table 2 are actually approximate ranges of values since they represent the measured (not the intrinsic) dissociation constants and which accordingly will be influenced by their electrostatic environments.

Kuby and Roy,[71] for the glucose 6-phosphate dehydrogenase reaction, calculated the thermodynamic parameters for the data given in Table 1 and these are summarized in Table 3.

In some cases these values (Table 3) will include the thermodynamic value for ionization of an interacting group or buffer species, and therefore, they are listed as "total" values.

From the pK_κ values of 5.6 and 5.7 a ΔH° of approximately 5 kcal/mol may be estimated for the enthalpy change associated with the first ionization reaction in the enzyme-ternary compound (i.e., to $HE{\cdot}A^-{\cdot}B^-$), and which may be compared with a ΔH° of approximately 11 kcal/mol estimated from the pK'_κ value (of 8.9 and 9.1 at 32 and 25°C, respectively) as ascribed to the second ionization reaction in the enzyme ternary compound (i.e., to $E{\cdot}A^-{\cdot}B^-$). A pK_{a_1} of 5.7 at 25°C, with a ΔH°_{ioniz} of approximately 5.1 kcal/mol may also be compared with a likely range of 5.5 to 7.0 for imidazolium groups in proteins with a range of associated ΔH° values of 6900 to 7500 cal/mol (Table 2). Similarly, a pK_{a_2} of 9.1 at 25°C with a ΔH°_{ioniz} of 10.8 kcal/mol may be compared with the estimates for an ϵ-ammonium group of a lysyl residue, that is, a pK_a range of 9.4 to 10.6, with a ΔH° range of 10 to 13 kcal/mol (Table 2).

Interestingly, pK_κ decreases at 25°C from 5.7 in the enzyme ternary compound to 5.2 in the case of pK_{H_2} for the enzyme binary compound or the free enzyme; similarly, pK'_κ decreases from 9.1 at 25°C to 8.2 in pK'_H. Moreover, the associated values for ΔH^{ioniz}_{total} become −12 and 4 kcal/mol in the case of pK_{H_2} and pK'_H, respectively.

If one assumes that the shifts in pK_as are due to interaction with an adjacent functional group in either the enzyme protein or in the enzyme-substrate binary compound and that this interaction is diminished on binding the second substrate, one can estimate the degree of interaction with this group from the enthalpy changes. Thus, in the basic range, a value may be calculated equal to 6.8 kcal/mol for the enthalpy of ionization at 25°C for this hypothetical interacting group in the enzyme protein. This value for ΔH°_{ioniz} may be compared with a value of ΔH° = 6.9 kcal/mol obtained by Benesch and Benesch[75] for the ionization of the -SH group in thioglycolic acid.

Moreover, Kuby et al.[58] had found that the rates of reaction with DTNB and a single

TABLE 3
Thermodynamic Parameters for the pH-Independent Ionization Constants and Intrinsic Dissociation Constants[71]

pH independent or intrinsic dissociation constant	T (Kelvin)	Value	ΔH^{ioniz}_{total} or ΔH°_{ioniz} (kcal/mol)	ΔG^{ioniz}_{total} or ΔG°_{ioniz} (25°C) (kcal/mol)	ΔS^{ioniz}_{total} (25°C) or ΔS°_{ioniz} (cal/deg/mol)
pK_{H_2}	305.2	5.43	−11.8	7.1	−63.6
	298.2	5.23			
pK_H'	305.2	8.13	4.0	11.2	−24.1
	298.2	8.19			
pK_K	305.2	5.60	5.1	7.8	−9.1
	298.2	5.69			
pK_K'	305.2	8.90	10.8	12.4	−5.4
	298.2	9.08			
$p\overline{K}_{*B^-}$	305.2	4.83	6.3	6.7	−1.4
	298.2	4.94			
$p\overline{K}_{*A^-}$	305.2	5.25	15.3	7.5	26.1
	298.2	5.50			

Note: In 0.1 ionic strength Tris-acetate-EDTA buffers of the glucose 6-phosphate dehydrogenase-catalyzed reaction. From Kuby, S. A. and Roy, R. N., *Biochemistry,* 15, 1975, 1976. With permission.)

thiol group per subunit, as revealed by the rate of inactivation of the enzyme, could be reduced in the presence of either substrate, $NADP^+$ or D-glucose 6- phosphate^{2-}, an observation which implicates a thiol group in the active conformation of the enzyme.

It is of interest to note that from the intrinsic dissociation constants of the ionized substrate and the enzyme, $\overline{K}_{*_B^-}$ and $\overline{K}_{*_A^-}$, ΔH° values of 6.3 and 15 kcal/mol are obtained with only a minimal value for ΔS° (25°C) in the case $\overline{K}_{*_B^-}$ (−1.4 ℓ.u.), but with a much larger and positive entropy value of 26 ℓ.u. associated with $p\overline{K}_{*_A}$. The ΔG° (25°C) in both cases remain fairly constant at 6.7 and 7.5 kcal/mol for $\overline{K}_{*_B^-}$ and $\overline{K}_{*_A^-}$, respectively. These data will be discussed further (see Chapter 5) in terms of their mechanistic implications.

B. THE MEASURED ARRHENIUS ENERGY OF ACTIVATION IS DETERMINED FROM DATA AT ONE GIVEN pH VALUE

From the $V_{max,app}$ data at one given pH value as a function of temperature, the measured Arrhenius energy of activation (ΔE_{act}) may be obtained from the slope of log $V_{max,app}$ vs. $\frac{1}{T}$ plots, according to

$$d(\log V_{max,app}) = \frac{-\Delta E_{act}}{2.303R} d\left(\frac{1}{T}\right) \tag{66}$$

derived from the Arrhenius equation,

$$\frac{d\ln k}{dT} = \frac{\Delta E_{act}}{RT^2} \tag{67}$$

After recalculation of the apparent first-order velocity constant (eg., $k_{+5,app}$ or $K'_{+5,\ app}$) from $\frac{V_{max,app}}{E_t}$ in units of sec^{-1} (i.e., moles of NADPH per mole of catalytic sites per second, in the case of the glucose 6-phosphate dehydrogenase with two catalytic sites per mole of apoenzyme), the thermodynamic parameters for activation according to Eyring's absolute rate reaction theory[76] may be calculated (see Chapter 1, Part 6). Thus, $\Delta H^{\ddagger}$ is obtained from

$$\Delta H^{\ddagger} = \Delta E_{act} - RT \tag{68}$$

$\Delta S^{\ddagger}$ is, in turn, obtained from

$$\Delta S^{\ddagger} = \left(\frac{\Delta H^{\ddagger}}{T}\right) + 2.303R\left[(\log k_{+5,app}) - \log\left(\frac{RT}{Nh}\right)\right] \tag{69}$$

which is derived from

$$k = \left(\frac{RT}{Nh}\right)_{\ell} - \Delta H^{\ddagger}/RT_{\ell}\Delta S^{\ddagger}/R \tag{70}$$

and finally, $\Delta G^{\ddagger}$ from

$$\Delta G^{\ddagger} = \Delta H^{\ddagger} - T\Delta S^{\ddagger} \tag{71}$$

In Table 4, the above data for the activation energy parameters are calculated for the glucose 6-phosphate dehydrogenase-catalyzed reaction from the data in Table 1 for the pH-independent maximal velocities.

Thus, the velocity data[71] required the incorporation into the mechanism of two pH-independent maximal velocities, one for the acidic and neutral region, V_{max} (acidic + neutral) (with a value of, e.g., approximately 525 at 32°C), and a second maximal velocity in the basic region, V_{*max} (basic) (with a value more than three times its counterpart in the acidic region, e.g., 1810 μmol min^{-1}mg^{-1} at 32°C). Derived values for k_{+5} and k'_{+5} are also listed in Table 4 together with their apparent ΔE_{act} and values for $\Delta H^{\ddagger}$, $\Delta G^{\ddagger}$, $\Delta S^{\ddagger}$ at 25°C. Interestingly, ΔE_{act} differs for the two sets of "turnover numbers", with only about 5.8 kcal/mol in the acidic and neutral region compared with almost 18 kcal/mol (or approximately 17 kcal/mol for $\Delta H^{\ddagger}$) in the basic region. $\Delta G^{\ddagger}$ (25°C) remains fairly constant in both sets, that is, approximately 14 kcal/mole; however, $\Delta S^{\ddagger}$ (25°C) is almost −30 ℓ.u. for the acidic region compared with +12.7 ℓ.u. for the basic region, with implications of larger entropic influences in the acidic and neutral region, but with a greater potential energy barrier in the basic region. If one accepts Pauling's estimates of hydrogen-bond energies,[77] that is, approximately 1 to 8 kcal/mol, then from the energetics of the system as described by $\Delta H^{\ddagger}$ (25°C) $\simeq$ 5 kcal/mol on the acidic and neutral side, it would permit at least one hydrogen bond to be broken in the transition state, and possibly three on the basic side. These data, as given in Tables 1, 3, and 4 will have a bearing on the postulated mechanism of action, proposed by Kuby and Roy[71] and to be described later, for dehydrogenation by the brewers' yeast D-glucose 6-phosphate dehydrogenase (refer to Figure 5 of Kuby and Roy;[71] see also Chapter 8, Figure 7).

C. ENZYME INACTIVATION KINETICS

Sometimes useful information on the nature of the structure of the enzyme, in solution, may be gleaned from a study of the kinetics of inactivation under defined conditions of pH and temperature.

Thus, acid-inactivation kinetics of the dimeric human isoenzymes (muscle type and brain type) of ATP-creatine transphosphorylase (creatine kinase)[288] were studied at pH values form 4 to 6.5 at five temperatures from 2 to 40°C.[286] Under each particular condition of pH and temperature, plots of the $\log_{10}$ of remaining activity (%) vs. time proved linear, up to at least 60% inactivation and often to 90% inactivation. From the statistically drawn slopes

TABLE 4
Effect of Temperature on the pH-Independent Maximal Velocities of the Glucose 6-Phosphate Dehydrogenase Catalyzed Reaction in 0.1(Γ/2) Tris-Acetate-EDTA Buffers[71]

pH-independent maximal velocity	T (Kelvin)	V_{*max} (μmol · min^{-1} mg^{-1})	k_{+5} or k_{+5}' (s^{-1})	ΔE_{act} (kcal/mol)	$\Delta H^{\ddagger}$ (25°C) (kcal/mol)	$\Delta G^{\ddagger}$ (25°C) (kcal/mol)	$\Delta S^{\ddagger}$ (25°C) (cal/deg/mol)
$V_{*max\ (acidic\ +\ neutral)}$	298.2	402	357	5.8	5.2	14.0	−29.5
	305.2	525	446				
$V_{*max\ (basic)}$	298.2	905	769	17.9	17.3	13.5	12.7
	305.2	1810	1539				

[Slope $= \frac{-k_{app}}{2.303}$] the apparent first order velocity constants were calculated.[286] Plots of log k'_{app} vs. pH also proved to be linear at the five temperatures studied, from pH 4.5 to 6.3, approximately, for the human brain isoenzyme and from pH 4 to 5.8, approximately, for the muscle isoenzyme. To permit an evaluation of the ΔE_{act} and thermodynamic values for activation according to Eyring's absolute rate reaction theory[76] and a comparison of these thermodynamic parameters for both isoenzymes, from plots of the $\log_{10} k'_{app}$ vs. pH for several fixed values of T, values of $\log_{10} k'_{app}$ (in s^{-1}) were extrapolated to a reference pH of 5.6 at all temperatures, even though the muscle enzyme might be essentially stable at this pH at 2°C (i.e., $k'_{app} < 10^{-4}$), or too slow for accurate measurements at 15°C, but measureable at 36 or 41°C. For the more acid-labile brain enzyme, measurements could be extended to this reference pH value and above, to approximately pH 6 to 6.3, at almost all temperatures explored. It is noteworthy that pH 5.6 is the isoelectric point extrapolated to zero ionic strength (pI_0) for the calf brain enzyme[287] and very likely is not far removed from the pI_0 of the human brain enzyme.[288]

From the calculated ΔE_{act} obtained from the various slopes [slope $= \frac{-\Delta E_{act}}{2.303R}$] of plots of $\log_{10} k'_{app}$ vs. $\frac{1}{T}$ according to

$$d(\log_{10} k'_{app}) = \frac{-\Delta E_{act}}{2.303R} d\left(\frac{1}{T}\right) \tag{72}$$

the thermodynamic parameters for activation according to Eyring's absolute rate reaction theory,[76] were calculated. Thus, $\Delta H^\ddagger$ was obtained from

$$\Delta H^\ddagger = \Delta E_{act} - RT \tag{73}$$

$\Delta S^\ddagger$ was in turn, obtained from

$$\Delta S^\ddagger = \frac{\Delta H^\ddagger}{T} + 2.303R\left[\log_{10} k'_{app} - \log \frac{(RT)}{Nh}\right] \tag{74}$$

and finally, $\Delta G^\ddagger$ from

$$\Delta G^\ddagger = \Delta H^\ddagger - T\Delta S^\ddagger \tag{75}$$

(where k'_{app}, the apparent first order velocity constant for inactivation, is expressed in s^{-1}, and $\Delta H^\ddagger$, $\Delta S^\ddagger$, and $\Delta G^\ddagger$, are the enthalpy, entropy, and free energy changes, respectively, of activation for entering into the transition state required for denaturation). Calculations were also made at temperatures above and below that temperature which appeared to correspond to fitted points of intersection in the Arrhenius plots; those temperatures were selected for calculation which were found to be applicable for both cases, that is, for both the brain and muscle type (i.e., at 299.2 and 314.2 K, or 26 and 41°C, respectively).

Finally, it will be shown that all plots of $\log_{10} k'_{app}$ vs. pH, between 2 and 41°C, appeared to be linear within the narrow range of pH values studied (approximately 4 to 6.3); therefore, from the slopes of these plots, empirically,

$$\frac{d(\log k'_{app})}{d(pH)} = -\eta_{H^+} \tag{76}$$

or

$$k'_{app} = k'(H^+)^{\eta_{H^+}} \quad (77)$$

where $\eta_H{}^+$ would represent the kinetic order with respect to H^+ (see also References 289 to 291). These $\eta_H{}^+$ values were determined for the five different temperatures measured, for both brain and muscle type, and are presented as inserts in Figure 8A which illustrate the plots obtained for the apparent first order velocity constants as a function of pH, at several fixed temperatures (from 2 to 41°C), for both brain type and muscle type creatine kinase. The pH range which could be conveniently and accurately explored (for a range of log k′, in min^{-1}, values equal to approximately 0 to −3) was to a large degree dependent upon the fixed temperature employed.

Thus, for the surprisingly acid-labile brain type, at 2°C, a pH range of 4.5 to 5.5 could be explored; whereas, the comparable range was very much reduced at 41°C, to only about pH 5.8 to 6.2, if measurable rate constants were to be accurately obtained which covered the same 1000-fold range in k'_{app} values. Under all conditions, it may be ascertained from Figure 8A that the inactivation-rate constants obtained for the brain enzyme were very much larger than for the muscle enzyme. For example, at pH 5.3 and 26°C, interpolation of the data for brain and muscle type from the plots described in Figure 8A yield values of $k'_{app} \simeq 1.12 \times 10^{-1}\ min^{-1}$ and $1.62 \times 10^{-3}\ min^{-1}$ for brain and muscle type, respectively, or rates of inactivation almost 69-times larger for the brain enzyme than for the muscle enzyme. Similarly, at 2°C, values of $k'_{app} \simeq 4.9 \times 10^{-3}\ min^{-1}$ and approximately $1.59 \times 10^{-4}\ min^{-1}$ were obtained at pH 5.3 for brain and muscle enzyme, respectively, or again, almost a 31-fold more rapid rate of inactivation is observed for the labile brain enzyme. Furthermore, it may be observed that as the temperature is increased, at a given pH value, the ratio of the k'_{app} values for brain to muscle type becomes increasingly larger; e.g., at 36°C, and again at pH 5.3, this ratio becomes almost 500:1 in favor of the noticeably unstable brain enzyme.

At each given temperature, plots of $\log_{10} K'_{app}$ (in min^{-1}) vs. pH appear to be linear (see Figure 8A). Apparently, these data are at least one pH unit removed from the pK′ of any protonated ionizable group (or groups) responsible for initiating the sequence of events leading to inactivation and denaturation. Therefore, it is not possible to deduce the "critical" pK′ value(s) (see References 289 and 292) but only the order of reaction with respect to H^+ ion ($\eta_H{}^+$), as was done for the case of the H^+ ion-promoted kinetics of release of 5-thio-2-nitrobenzoate from the mixed disulfide, TNB-protein [formed by reaction between 5,5′-dithio-bis (2-nitrobenzoate) and the two normally reactive sulfhydryl groups per mole of calf brain ATP-creatine transphosphorylase.[289]]

Empirically, the acid-inactivation data all conform to the expression

$$k'_{app} = k'(H^+)^{\eta_{H^+}} \quad (77)$$

where η_{H^+} is the kinetic order with respect to H^+(ion). The values of η_{H^+}(or the negative slopes of the plots of log K'_{app} vs. pH) are summarized as inserts in the respective plots (Figure 8A) for the human brain- and muscle-type isoenzymes. Whereas, an integral number of H^+ ions[293] seem to be involved in the inactivation process between 2 and 26°C, and fixed at approximately 2 H^+ ions (i.e., 2.1 to 2.0) for the muscle enzyme, only possibly at 2°C does a comparable situation exist for the brain enzyme which shows a progressive rise in η_{H^+} values from 2.3 to almost 6.1 at 41°C. Between 26 and 41°C for the muscle type, values of η_{H^+} slowly increase from the previous temperature-invariant value of 2 to 3.2. The rates of acid denaturation become almost explosive in nature for the brain type at 41°C, where a decrease in one pH unit would result in an increase in the value of log k'_{app} by 6.1-fold or a 1.26×10^6-fold increase in the value of k'_{app}; this condition is in contrast to the case of the muscle type, where there appears to be only a 1.58×10^3-fold increase in k'_{app} at 41°C for a one-pH unit decrease. Therefore, at 41°C, whatever overall processes are involved in

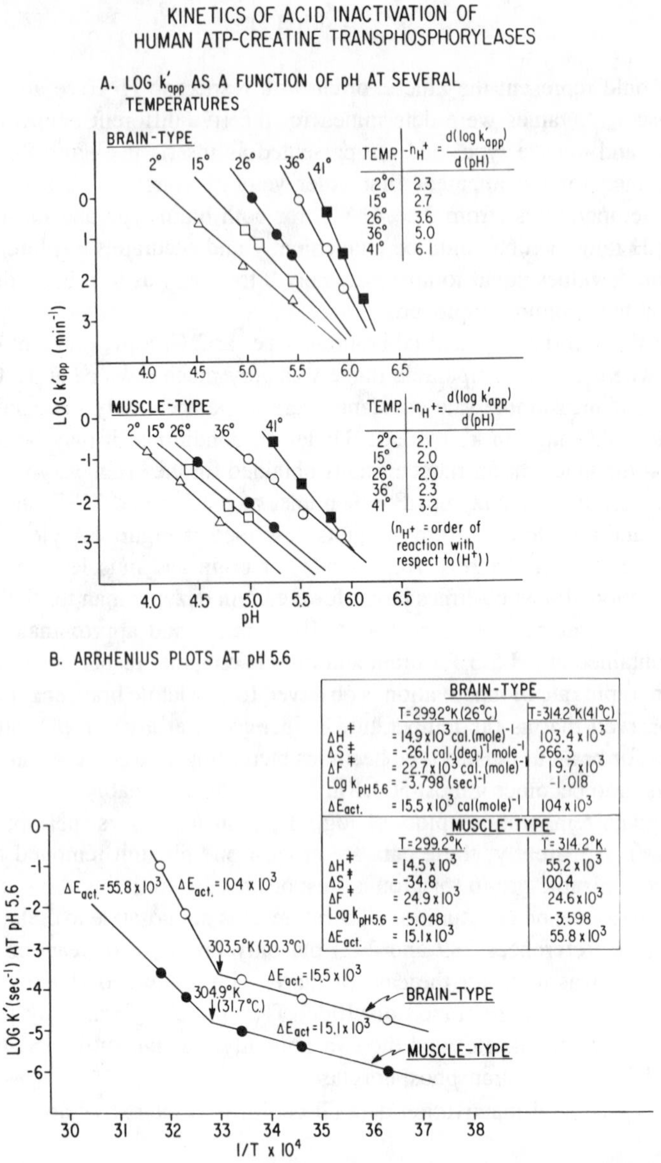

FIGURE 8. Kinetics of acid inactivation of the human ATP-creatine transphosphorylase isoenzymes. A. Plots of log k'_{app} (k'_{app} = the apparent first order velocity constants for inactivation in min^{-1}) vs. pH for the brain type (upper set of plots) and muscle type (lower set of plots). Inserts to both sets contain values of η_{H+}, corresponding to the several temperatures employed. B. Arrhenius plots of the above log k' values (in sec^{-1}) at a reference pH = 5.6 vs. (1)/(T) (in degrees absolute), for brain type (○--○) and muscle type (●--●). Arrows point to intersections of the fitted linear segments. Inserts for brain and muscle type contain respective calculated values at 299.2 and 314.2 K, for $\Delta H^{\ddagger}$, $\Delta S^{\ddagger}$, $\Delta G^{\ddagger}$, log k'_{app} (at pH 5.6), and the apparent Arrhenius energy of activation (ΔE_{act}).[286]

this complex pH-promoted inactivation, $\frac{-d(\log k'_{app})}{d(pH)}$ is some 800-fold larger for the brain enzyme than for the muscle enzyme, with the labile structure of the former obviously much more susceptible to environmental influences than the latter as noted also for the calf-brain vs. calf-muscle enzyme[49,289] by several other approaches. Nevertheless, it is of definite interest to note the close to integral numbers (2.0 to 2.1) of H^+ ions which may be involved, at least at lower temperatures, in the primary process of acid inactivation (especially for the muscle type). These two H^+ ions may be those protons which trigger the entire sequence of events leading to loss of the conformational integrity required for the enzymatic activity, especially in the case of the more rigid muscle-type structure.[49,289] Curiously, both the human muscle and brain types are dimeric structures,[288] as shown for the rabbit-muscle[294,295] and calf-brain enzymes,[287] and it may not be coincidental that only two protons may be required for initiating the events leading to the protonated transition state required for inactivation; where, in this case, inactivation would imply protonation of the one group per polypeptide chain leading to a loss in that geometrical structure necessary for enzymic catalytic activity. It should also be noticed that the human-muscle isoenzyme apparently differs to some degree from the reported acid inactivation results for the rabbit-muscle enzyme[291] or the porcine-muscle enzyme,[290] where η_{H+} values of 3.4 were given for both of these cases at 25°C, and which is a value approached here only at 41°C for the human muscle enzyme, but similar to the human-brain isoenzyme at 26°C (see Figure 8A). Further, Scopes[290] had found the value of η_{H+} to be independent of temperature from 10 to 50°C for the porcine-muscle enzyme. A comparison of the data of Birktoft and Ottesen [291] for the rabbit-muscle enzyme and those of Scopes for the porcine-muscle enzyme[290], with those given in Figure 8A[286] for the human isoenzymes of creatine kinase, reveal another interesting point, that is, that at some particular pH values, e.g., at pH 4.7 and 25°C, the value of k'_{app} for the human-muscle enzyme would be roughly one tenth to one fifth those reported for the porcine-muscle enzyme which in turn was reported to be one tenth of the k'_{app} value for the rabbit-muscle enzyme,[291] and finally, the values of k'_{app} for the human-brain enzyme would be at least ten times those obtained by Birktoft and Ottesen[291] for the rabbit-muscle enzyme. Thus, the order of acid stability, under one particular set of conditions, i.e., at pH 4.7 and 25°C, would be in inverse order of their k'_{app} values, and therefore, human-muscle enzyme > porcine-muscle enzyme > rabbit-muscle enzyme > human-brain enzyme.

Since the slope of the log k'_{app} vs. pH plot is approximately -2.0 for the human-muscle enzyme at 25°C, but -3.4 for the porcine and rabbit-muscle enzyme, there will, however, appear to be a transposition of this order as one approaches pH 5.6, where the porcine-muscle enzyme will then appear to be the most acid stable. Finally, it is of interest to note that Birktoft and Ottesen[291] demonstrated that an exact coincidence exists in the first-order rate constants for denaturation obtained from either a decrease in solubility, as described by Scopes,[290], or by a loss in enzymatic activity, the method employed for the data shown in Figure 8A[286] at 25°C; whether this coincidence applies at all the temperatures shown in Figure 8A has not, however, been demonstrated.

In Figure 8B, the results of the effect of temperature on the acid-promoted inactivation kinetics for both human muscle- and brain-type insoenzymes are graphically summarized in terms of Arrhenius plots for those data given in Figure 8A but extrapolated to a fixed reference pH value of 5.6. Thus, in Figure 8B, plots of log k'_{app} (in s^{-1}), extrapolated, if necessary, to pH 5.6 from the data given in Figure 8A (in min^{-1}), are now replotted vs. the reciprocal of the absolute temperature for both the human brain-type and muscle-type isoenzymes. Rather than a linear function, each of the plots may be "fitted" to two intersecting lines from which the individual slopes may be used to calculate the ΔE_{act}. The implication is that there may be two modes of inactivation corresponding to each linear segment of the Arrhenius plots and that the change in slopes may represent, therefore, some large conformational changes in the transition state to be associated with each change in slope of the Arrhenius plot. A similar observation was made by Scopes[290] in the case of the porcine-muscle enzyme. The temperatures at the intersection of the two fitted linear segments are indicated for both the brain type and muscle type in Figure 8B. Accordingly, the thermodynamic parameters are calculated for a given point below (26°C) and one above (41°C) the points of intersection and those temperature values were selected which would apply to either the muscle or brain type. It should not be meant to imply that if a sufficient number of points were obtained in the region of this intersection point, a smooth bend would not result.[296] The two temperatures at the intersection points are found to be 304.9 K (31.7°C) and 303.5 K (30.3°C) for the muscle and brain type, respectively, with apparently a slightly higher value for the relatively more acid- and thermally stable muscle type. The calculated Arrhenius ΔE_{act} is approximately the same value at 26°C (i.e., below the intersection point temperatures) for both isoenzymes, i.e., 15.1 and 15.5 kcal/mol for muscle and brain type, respectively. However, above the intersection point temperatures, the values for ΔE_{act} rise approximately 3.7- to 6.7-fold, respectively, i.e., to 55.8 kcal/mol and to 104 kcal/mol for muscle and brain type. The twofold difference in ΔE_{act} above the intersection point temperatures, between the two isoenzymes, may be correlated with the more flexible or looser structure deduced for the brain-type enzyme, particularly the calf brain-type, by the several techniques summarized in several reports.[49,287,289,297]

From the values for ΔE_{act} at 299.2 and at 314.2 K, the values for $\Delta H^{\ddagger}$, $\Delta S^{\ddagger}$, and $\Delta G^{\ddagger}$ were accordingly calculated for both isoenzymes; these thermodynamic parameters for activation are summarized in the insert to Figure 8B. It may be noted that the values for $\Delta G^{\ddagger}$ do not differ significantly at the two temperatures for the muscle type (24.9 and 24.6 kcal/mol) and are not too unusually different from the $\Delta G^{\ddagger}$ values of the brain-type (22.7 and 19.7 kcal/mol at 26 and 41°C, respectively). Since the enthalpy changes of activation are only about 0.6 kcal/mol smaller than the respective values for the apparent ΔE_{act}, the large increase in slopes in the temperature regions above the intersection point temperatures are actually reflected not only in the $\Delta H^{\ddagger}$ increases, but in the relatively enormous increases in the entropy changes. Thus, $\Delta S^{\ddagger}$ values rise precipitously, from -34.8 and -26.1 entropy units, at 299.2 K, for muscle and brain type, respectively, to 100.4 and 266.3 ℓ.u., respectively, at 314.2 K. The Arrhenius slopes corresponding to the lower temperature range are almost parallel for the two isoenzymes, but diverge in the upper temperature range, reflecting the 2.7-fold larger values of $\Delta S^{\ddagger}$ for the brain vs. muscle type. At this temperature of 41°C, the value of $-\eta_{+}$ has also risen to 6.1 for the brain type vs. 3.2 for the muscle type. Thus, not only H^+ ion-induced inactivation changes but also a severe disruption of structure appears to have taken place for the brain type under these conditions, and under both thermal and acidic entropic influences to result possibly in a final denatured random coil even in the transition state. The more rigid muscle-type structure appears to have more readily resisted these thermal and acidic entropic forces of disruption. The selection of a reference pH of 5.6 would appear to have much simplified the evaluation of the thermodynamic parameters of activation, especially since a pH of 5.6 would appear to be close to the pI_0 for the brain type, and thus any additional complicating electrostatic factors may have been largely eliminated in the case of the brain enzyme (but not completely eliminated for the muscle type which possesses a much more alkaline pI_0; a measured pI at 0.05 ($\Gamma/2$) = 9.0[288]). Nevertheless, a selection of other reference pH values would lead to other calculated values for the thermodynamic parameters for the transition state, and the absolute values for these parameters would not as yet appear to be attainable.[286]

Chapter 5

EFFECT OF METAL COFACTORS ON THE REACTION VELOCITY

I. UNIREACTANT CASES

Initially, we will discuss two cases applicable to unireactant cases: (1) where there is an exclusive or obligatory interaction of the free metal with the free enzyme and (2) where there is an exclusive interaction of the free metal with the free substrate and not the free enzyme.

A. CASE 1. EXCLUSIVE OR OBLIGATORY INTERACTION OF THE FREE METAL WITH THE FREE ENZYME

Several enzymes contain dissociable divalent cations as part of their active centers,[956] e.g., Mg^{2+} or Mn^{2+} in enolase. In addition, several enzymes require the presence of a monovalent cation, e.g., Na^+, K^+, or NH_4^+, either for the stabilization of a particular conformation responsible for maximal activity or as part of their active centers, e.g., β-D galactosidase from *Escherichia coli*.[78]

The following scheme would allow for activation in case (1)

$$E + M \underset{k_{-1}}{\overset{k_{+1}}{\rightleftharpoons}} E\text{-}M$$

$$S + E\text{-}M \underset{k_{-2}}{\overset{k_{+2}}{\rightleftharpoons}} S\text{-}E\text{-}M \underset{k_{-3}}{\overset{k_{+3}}{\rightleftharpoons}} P\text{-}E\text{-}M \underset{k_{-4}}{\overset{k_{+4}}{\rightleftharpoons}} E\text{-}M + P$$

SCHEME I.

and the initial-rate equations for forward or reverse reactions are (steady-state solutions)

$$v_o^f = \frac{V_{max}^f(S)(M)}{\overline{K}_M K_S + K_S(M) + (S)(M)}; \quad v_o^r = \frac{V_{max}^r(P)(M)}{\overline{K}_M K_P + K_P(M) + (P)(M)} \tag{1}$$

or

$$\frac{v_o^f}{V_{max}^f} = \frac{1}{1 + \frac{K_S}{(S)} + \frac{\overline{K}_M}{(M)}\frac{K_S}{(S)}}; \quad \text{and} \quad \frac{v_o^r}{V_{max}^r} = \frac{1}{1 + \frac{K_P}{(P)} + \frac{\overline{K}_M K_P}{(M)(P)}} \tag{2}$$

(where for the forward reaction, if $k_{-3} \rightarrow O$, $K_S = \frac{k_{+3} + k_{-2}}{k_{+2}}$; $\overline{K}_M = \frac{k_{-1}}{k_{+1}}$; similarly, for the reverse reaction, if $k_{+3} \rightarrow O$, $K_P = \frac{k_{+4} + k_{-3}}{k_{-4}}$.

B. CASE 2. PLUS A METAL-FREE (AND KINETICALLY ACTIVE) E-S COMPLEX

If Scheme I is extended to include in Scheme II the possibility of a *metal-free E-S complex* together with its *breakdown* to yield product by a slower reaction (k_p) than by the breakdown of M-E-S (βk_p), and if all reactions are assumed to be in quasi-equilibrium except for the breakdown of the two complexes (M-E-S and E-S) then

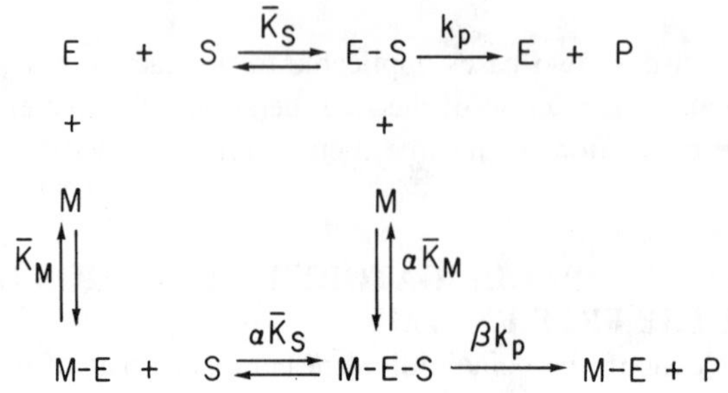

SCHEME II.

The initial velocity expression for the forward reaction becomes

$$v_o^f = k_p(\text{E-S}) + \beta k_p(\text{M-E-S}) \tag{3}$$

$$v_o^f = \frac{V_{max}\dfrac{(S)}{\overline{K}_S} + \dfrac{(\beta V_{max})(M)(S)}{\overline{K}_M(\alpha\overline{K}_S)}}{1 + \dfrac{(S)}{\overline{K}_S} + \dfrac{(M)}{\overline{K}_M} + \dfrac{(M)(S)}{\overline{K}_M(\alpha\overline{K}_S)}} \tag{4}$$

where $V_{max} = k_p E_t$ of the unactivated reaction and $\beta V_{max} = \beta k_p E_t$ of the activated reaction. If we factor out $(1 + \dfrac{\beta(M)}{\alpha\overline{K}_M}$ and divide through by this factor divided by $\overline{K}_S$, i.e.,

$$\frac{v_o^f}{V_{max}} = \frac{\dfrac{(S)}{\overline{K}_S}\left[1 + \dfrac{\beta(M)}{\alpha\overline{K}_M}\right]}{\left[\dfrac{(S)}{\overline{K}_S}\left(1 + \dfrac{(M)}{\overline{K}_M\alpha}\right) + \left(1 + \dfrac{(M)}{\overline{K}_M}\right)\right]} \tag{5}$$

$$\frac{v_o^f}{V_{max}} = \frac{(S)}{\left[\dfrac{(S)(1 + (M)/\overline{K}_M(\alpha))}{(1 + \beta(M)/\alpha\overline{K}_M)} + \overline{K}_S\dfrac{(1 + (M)/\overline{K}_M)}{(1 + \beta(M)/\alpha\overline{K}_M)}\right]} \tag{6}$$

or divide through by (S):

$$\frac{v_o^f}{V_{max}} = \frac{1}{\left[1 + \dfrac{(M)}{\alpha\overline{K}_M}\right]\Big/\left[1 + \dfrac{\beta(M)}{\alpha\overline{K}_M}\right] + \dfrac{\overline{K}_S}{(S)}\left[1 + \dfrac{(M)}{\overline{K}_M}\right]\Big/\left[1 + \dfrac{\beta(M)}{\alpha\overline{K}_M}\right]} \tag{7}$$

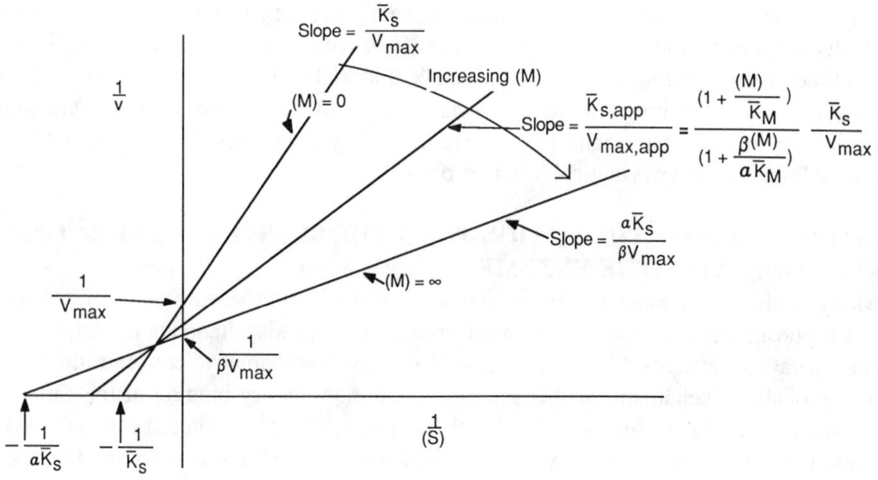

FIGURE 1.

$$\frac{1}{v_o^f} = \frac{1}{V_{max}} \frac{\left[1 + \frac{(M)}{\alpha\bar{K}_M}\right]}{\left[1 + \frac{\beta(M)}{\alpha\bar{K}_M}\right]} + \frac{\bar{K}_S}{V_{max}} \frac{1}{(S)} \frac{\left(1 + \frac{(M)}{\bar{K}_M}\right)}{\left(1 + \frac{\beta(M)}{\alpha\bar{K}_M}\right)} \tag{8}$$

An apparent $\bar{K}_{S,app}$ and $V_{max,app}$ can be defined at any (M) concentration.

$$\bar{K}_{S,app} = \bar{K}_S \frac{\left(1 + \frac{(M)}{\bar{K}_M}\right)}{\left(1 + \frac{(M)}{\alpha\bar{K}_M}\right)} \tag{9}$$

$$V_{max,app} = V_{max} \frac{\left(1 + \frac{\beta(M)}{\alpha\bar{K}_M}\right)}{\left(1 + \frac{(M)}{\alpha\bar{K}_M}\right)}; \quad \frac{\bar{K}_{S,app}}{V_{max,app}} = \frac{\bar{K}_S}{V_{max}} \frac{\left(1 + \frac{(M)}{\bar{K}_M}\right)}{\left(1 + \frac{\beta(M)}{\alpha\bar{K}_M}\right)} \tag{10}$$

Therefore,

$$\frac{1}{V_{max}} = \frac{1}{V_{max,app}} + \frac{\bar{K}_{S,app}}{V_{max,app}} \frac{1}{(S)} \tag{11}$$

When

$$(M) \to \infty; \quad \begin{cases} \bar{K}_{S,app} = \alpha\bar{K}_S \\ V_{max,app} = \beta V_{max} \end{cases} \tag{12}$$

A double-reciprocal plot for this case is given in Figure 1. An enzyme which appears to obey the mechanism, with $\beta k_p \simeq 0$, i.e., $\beta \simeq O$, is methylaspartase.[79,281]

The dissociation constant for the enzyme-metal complex can therefore be estimated kinetically from Equations 4 to 12 for Scheme II, or from Equations 1 and 2, if Scheme I holds. Direct ligand binding measurements can be undertaken to confirm these kinetic results by a variety of equilibrium binding methods which will be discussed later. One should, however, bear in mind that ligand binding studies will yield values of n_i and $\overline{K}_i$ for binding sites in addition to the enzymatically active ones.

B. INTERACTION OF THE CATION WITH THE SUBSTRATE PRIOR TO REACTION WITH THE ENZYME

Many of those unireactant cases which involve nucleotide substrates (e.g., nucleotide di- and triphosphates) usually require what appears to be a simultaneous participation of a divalent metal ion (usually Mg^{2+}, Ca^{2+}, or Mn^{2+}) in stoichiometric concentrations.

The probable mechanism for this activation is the obligatory binding of the cation with the substrate prior to its interaction with the enzyme.[80,87] Therefore, the metal-substrate complex, rather than the uncomplexed substrate, appears to be the true reactant. The reaction mechanism is given in Scheme III for this case.

$$S + M \underset{k_{-1}}{\overset{k_{+1}}{\rightleftarrows}} M\text{-}S + E \underset{k_{-2}}{\overset{k_{+2}}{\rightleftarrows}} E\text{-}MS \underset{k_{-3}}{\overset{k_{+3}}{\rightleftarrows}} E\text{-}MP \underset{k_{-4}}{\overset{k_{+4}}{\rightleftarrows}} E + M\text{-}P$$

$$M\text{-}P \underset{k_{-5}}{\overset{k_{+5}}{\rightleftarrows}} P + M$$

SCHEME III.

Interesting features of this mechanism are that the substrate (S) can exist in more than one kinetically important form (e.g., S and M-S) and that the total concentration of the substrate added cannot be taken as equal to the kinetically active species; rather, it is the concentration of the metal complex which must be calculated and inserted in the relevant rate equation.

Provided the equilibrium between the substrate and metal is established rapidly (i.e.,
the dissociation constant for M-S $\underset{k_{+1}}{\overset{k_{-1}}{\rightleftarrows}}$ M + S is K_0), then if a quasi-equilibrium is presumed to hold (although this is not a necessary requirement, since the steady-state derivation for Scheme III yields the identical expression as for Scheme IV).

$$S + M \overset{K_0}{\rightleftarrows} M\text{-}S$$

$$+$$

$$E$$

$$\Big\updownarrow K_{MS} \qquad\qquad M + P$$

$$\qquad\qquad\qquad \Big\updownarrow$$

$$E\text{-}MS \underset{k_{-p}}{\overset{k_{+p}}{\rightleftarrows}} E\text{-}MP \rightleftarrows E + M\text{-}P\text{;}$$

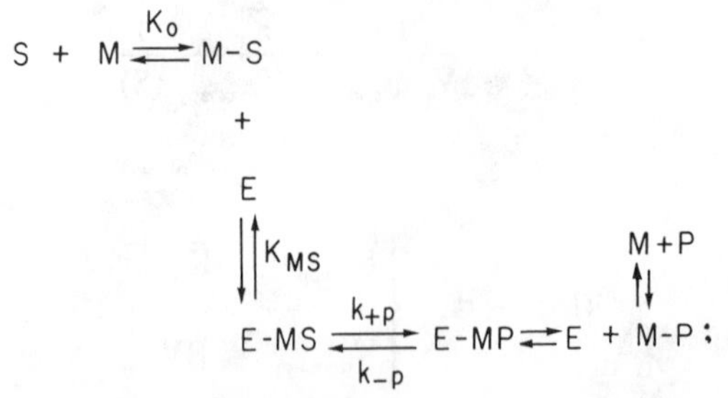

SCHEME IV.

For the initial forward reaction only,

$$\frac{v}{V_{max}} = \frac{(M\text{-}S)}{K_{M\text{-}S} + (M\text{-}S)} \tag{13}$$

or the usual Michaelis equation holds with the substrate replaced by the metal complex (M-S)

$$\frac{v}{V_{max}} = \frac{1}{\left(\frac{K_{M\text{-}S}}{(M\text{-}S)} + 1\right)} \tag{14}$$

which must be calculated from the total concentration of substrate, S_t, total concentration of metal, M_t, and the dissociation constant, K_0, for the M-S complex.

One can also express the concentration of the complex M-S in terms of the uncomplexed metal, M, and uncomplexed substrate, S, and K_0, however, the expression would be of less interest since M-S is the active substrate. Nevertheless, set $K_0 = \frac{(M)\ (S)}{(M - S)}$, and after substitution of (M-S) in Equation 14 it leads to

$$\frac{v}{V_{max}} = \frac{(M)(S)}{K_o K_{M\text{-}S} + (M)(S)} \tag{15}$$

which may be rearranged (Equation 16) to yield the expression for varied free (M) (Equation 17) or for varied free (S) (Equation 18).

$$\frac{v_o^f}{V_{max}} = \frac{1}{\frac{K_o K_{M\text{-}S}}{(M)(S)} + 1} \tag{16}$$

$$\frac{1}{v_o^f} = \frac{1}{V_{max}} + \frac{1}{V_{max}} \frac{K_{M\text{-}S}}{(S)} \left[\frac{K_o}{(M)}\right] \tag{17}$$

$$\frac{1}{v_o^f} = \frac{1}{V_{max}} + \frac{1}{V_{max}} \frac{K_{M\text{-}S}}{(M)} \left[\frac{K_o}{(S)}\right] \tag{18}$$

Thus, the velocity dependence of the reaction will be symmetrical with respect to free (M) and free (S); what is more significant, it will also be symmetrical with respect to total M (M_t) and total S (S_t), as we shall see shortly.

Thus, if the reciprocal plots are linear over the concentration range explored, then the $K_{S,app}$ for the substrate (when determined at a fixed $[M_t]$) will equal the $K_{M,app}$ for metal (when determined at a fixed $[S_t]$ equal to the $[M_t]$ concentration used in the first plot). Researchers[10,87] found in the case of creatine kinase that the apparent K_m for Mg_t in the presence of a certain concentration of ADP_t (adenosine diphosphate) was the same as that for ADP_t in the presence of the same concentration of Mg_t provided the double-reciprocal plots were linear in this range.

To prove this, first express the velocity equation in terms of total concentrations, define the following conservation expressions for

$$\begin{cases} M_t = M + M\text{-}S \\ S_t = S + M\text{-}S \end{cases} \tag{19}$$

and, after combining and substitution with $K_0 = \frac{(M)\,(S)}{(M-S)}$, the resulting quadratic equation is solved for (M) (S) (remembering that the concentration of (M-S) cannot be greater than (S_t) or (M_t), whichever is less). Substitution in Equation 18 leads to

$$\frac{v_o^f}{V_{max}} = \frac{1}{1 + \frac{2K_{M\text{-}S}}{[(M_t) + (S_t) + K_o] - [((M_t) + (S_t) + K_o)^2 - 4(M_t)(S_t)]^{1/2}}} \tag{20}$$

The symmetry of the velocity expression with respect to total concentrations is evident from Equation 20; thus, interchanging values of (M_t) and (S_t) will yield identical values of (v_0/V_{max}). If the concentration of (S_t) or $(M_t) >> K_0$, then (M-S) → (S_t) (in the presence of excess M) or (M-S) → (M_t) (in the presence of excess $[S_t]$). Then the system can be analyzed from Equation 13 (or 14) in reciprocal form to obtain $K_{M\text{-}S}$.

In the case of $MgATP^{2-}$, where $K_0 \simeq 0.01$ m*M*, when (ATP_t) and Mg_t are large compared to K_0, the velocity will appear to be almost independent of (Mg_t) at relatively low values of (S_t) and will appear to plateau at individual values of $V_{max,app}$, starting in the vicinity of $Mg_t \simeq (ATP)_t$, so long as the reactive (M-S) complex equals approximately the concentration of the limiting ligand. When all of the $(ATP)_t$ has been converted to $MgATP^{2-}$ for each fixed value of Mg_t, increasing $(ATP)_t$ will then have no effect on the initial velocity and a $V_{max,app}$ will be attained; or, a $V_{max,app}$ will then appear to be a function of $(ATP)_t$ at each fixed concentration of Mg_t:[81]

$$V_{max,app} = \frac{(ATP)_t}{K_{MgATP^{2-}} + (ATP)_t} \tag{21}$$

Because of the fortuitously low dissociation constant for $MgATP^{2-}$, varying (M_t) and $(ATP)_t$ simultaneously almost converts all of the $(ATP)_t$ into $MgATP^{2-}$ and usually leads accordingly to linear double-reciprocal plots in concentration ranges where $(ATP)_t$ and $(Mg_t) > K_0$; also, it accounts for the fact that many of the kinase reactions appear to plateau or reach a maximum in the vicinity of $(Mg_t) = (ATP)_t$.[80,87] A better practice, as experience has shown, is to fix by calculation the free Mg^{2+} at, e.g., 1m*M* [82-84,831] and then systematically explore the effect of the initial velocity on the $MgATP^{2-}$ concentration (by calculation). This reduces the (ATP^{4-}) concentration to a negligible concentration so that it cannot interfere, for example, by inhibition.[49]

II. BIREACTANT CASES

Creatine kinase has proved to be an excellent model for the study of the effect of metal cofactors. It had been proposed very early that the active substrates were actually $MgATP^{2-}$ and $Cr^{\pm}$ for the forward reaction and $MgADP^{-}$ and $Cr{\sim}P^{2-}$ for the reverse reaction.[80] Confirmation of this proposal came from equilibrium binding studies[59] where it was shown that Mg^{2+} was bound only slightly; however, the binding was enhanced in the presence of $(ATP)_t$ thus pointing to $MgATP^{2-}$ as the bound species and similarly for $MgADP^{-}$.

Under the several conditions studied, the rabbit muscle enzyme appears to follow a quasi-equilibrium mechanism; however, under the specific conditions of Kuby et al.,[87] it apparently followed a random mechanism with independent binding of the substrates. Under the special conditions of Morrison and James[83] (see also References 84 and 831), apparently the mechanism has broadened so that the condition of independent binding could not be invoked. Interestingly, the calf brain enzyme at low ionic strengths and pH 8.8 appeared also to follow the mechanism of Morrison and James (see Reference 49), whereas at pH 7.4 it reduced to a mechanism which approached that of the calf or rabbit muscle enzymes. More recently, Schimerlik and Cleland[66] have claimed that at pH 7.0 the rabbit muscle enzyme reverts to an ordered rapid-equilibrium mechanism in contrast to those studies by Morrison and Cleland[84] at pH 8.0, where the rabbit muscle enzyme followed a random rapid-equilibrium mechanism (refer to References 832 and 835). Thus, the enzyme is capable of subtle changes with respect to its environment, especially pH, to allow for some minor changes in its mechanism. The major overall quasi-equilibrium mechanism is apparently an invariant condition of the creatine kinase system. We will discuss first the simpler description of the random quasi-equilibrium mechanism with independent substrate binding. This will then follow with the more complex case, that is, that of the random quasi-equilibrium mechanism with a rate-limiting step at the interconversion of the ternary complexes without independent binding of the substrates. In this last case, mention can also be made of the adenylate kinase mechanism, which is very similar[50] and which will be discussed in detail later (Section III.B; see also Reference 964). Finally, we will discuss the extraordinarily complex case, that is, that of the evaluation of the thermodynamic equilibrium constant for the creatine kinase-catalyzed reaction where almost 17 ionization and complexation constants, 6 conservation expressions for total species, and almost 31 individual substrate species and magnesium-associated species had to be considered simultaneously (Section IV).

A. CASE I: RANDOM QUASI-EQUILIBRIUM WITH INDEPENDENT BINDING OF SUBSTRATES

Two general mechanisms will be considered: (1) without enzyme-magnesium (EM) complexes and product inhibition and (2) with enzyme-magnesium complexes.

1. Without Enzyme-Magnesium Complexes and Product Inhibition

(1) Assume that a "quasi-equilibrium" mechanism where all possible enzyme complexes (except EM and associated reactions) exist in significant concentration compared to the total enzyme concentrations.

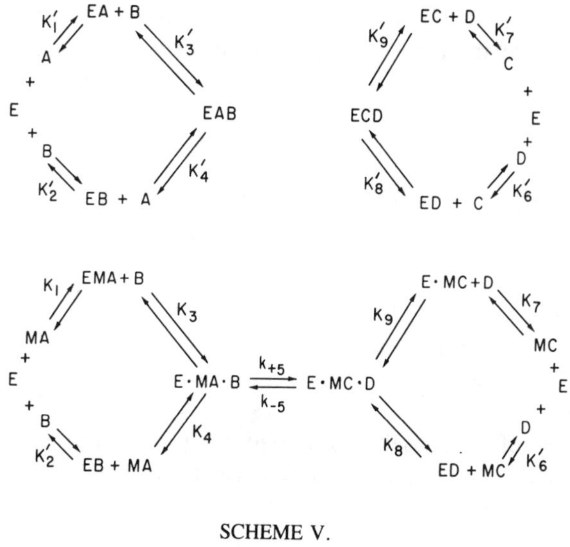

SCHEME V.

In addition, consider the following equilibria:

$$M + A \underset{}{\overset{K_o^f}{\rightleftarrows}} MA$$

$$M + C \underset{}{\overset{K_o^r}{\rightleftarrows}} MC \qquad (22)$$

(where K_o^f and K_o^r are written as dissociation constants).
Consider the following conservation expressions:

$$A_t = \Sigma(A + MA + \ldots)$$

$$C_t = \Sigma(C + MC + \ldots)$$

$$M_t = \Sigma(M + MA + \ldots + MC + \ldots) \qquad (23)$$

(2) Assume that all equilibria, except the equation for $E \cdot MA \cdot B \underset{k_{-5}}{\overset{k_{+5}}{\rightleftarrows}} E \cdot MC \cdot D$ are adjusted rapidly, (3) that the affinity of the enzyme for (A) or (MA) is not influenced by (B), but that (A) influences (MA), i.e., (A) and (MA) compete for binding independent of (B), and (4) that (C) and (MC) compete for binding, but the (D) does not influence the binding of (C) or (MC), where the symbols for the creatine kinase-catalyzed reaction are set as follows:

$$ATP^{4-} + Cr^{\pm} \rightleftarrows ADP^{3-} + Cr \sim P^{2-}$$

$$A + B \rightleftarrows C + D \qquad (24)$$

and

$$MgATP^{2-} + Cr^{\pm} \rightleftarrows MgADP^{-} + Cr \sim P^{2-}$$

$$MA + B \rightleftarrows MC + D \qquad (25)$$

where the following relationships exist between the constants (by assumptions 3 and 4), where all equilibrium constants, including K_O, are dissociation constants, and the unprimed constants indicate reactions with the metal complexes of (A) or (C), i.e., (MA) or (MC), and the primes indicate reactions with the uncomplexed species (except where there are redundancies).

$$K_1 = K_4 = \overline{K}_{MA} \qquad K_6' = K_9' = K_9 = \overline{K}_D$$

$$K_1' = K_4' = \overline{K}_A \qquad K_7' = K_8' = \overline{K}_C$$

$$K_2' = K_3' = K_3 = \overline{K}_B \qquad K_8 = K_7 = \overline{K}_{MC} \qquad (26)$$

Assume also (5) that k_{+5} ($E \cdot MA \cdot B$) is rate limiting and that the same complex, $E \cdot MA \cdot B$, is formed regardless of the pathway, i.e., $MA \cdot E \cdot B$ is the same as $E \cdot MA \cdot B$ and (6) that for the forward reaction, measurements of initial velocities virtually reduce k_{-5} ($E \cdot MC \cdot D$) to zero.

After application of the conservation equation for total enzyme concentration and substitution of the equations describing the equilibria and equalities, the velocity for the forward reaction may be written as

$$v = \frac{V_{max}}{1 + \frac{\overline{K}_B}{(B)} + \frac{\overline{K}_{MA}}{(MA)}\left(1 + \frac{(A)}{\overline{K}_A}\right) + \frac{\overline{K}_{MA}\overline{K}_B}{(MA)(B)}\left(1 + \frac{(A)}{\overline{K}_A}\right)} \tag{27}$$

which is the equation for simple competitive inhibition[51] (see also Chapter 1, Equations 39 and 41), where (A) is the competitive inhibitor of the substrate (MA). By setting

$$V'_{max} = \frac{V_{max}}{1 + \frac{\overline{K}_B}{(B)}} \tag{28}$$

then

$$\frac{V'_{max}}{v} = 1 + \frac{\overline{K}_{MA}}{(MA)}\left(1 + \frac{(A)}{\overline{K}_A}\right) \tag{29}$$

if (B) is held constant. If (A) can be reduced to zero at high $Mg_t = M_t$, e.g., by Equation 22 for K_0^f then

$$\frac{V'_{max}}{v} = 1 + \frac{\overline{K}_{MA}}{(MA)} \tag{30}$$

and $\overline{K}_{MA}$ can be evaluated by $\frac{1}{v}$ vs. $\frac{1}{(MA)}$ plots at fixed (B).

$$\frac{(MA)}{(A)}\left(\frac{V'_{max}}{v} - 1\right) = \frac{\overline{K}_{MA}}{\overline{K}_A} + \frac{\overline{K}_{MA}}{(A)} \tag{31}$$

and a plot of the left-hand side of Equation 31 vs. $\frac{1}{(A)}$ yields a slope $= \overline{K}_{MA}$ and an ordinate intercept to $\overline{K}_{MA}/\overline{K}_A$.

If (A) and (MA) are held constant and (B) is varied, Equation 27 may be rearranged to yield

$$\frac{1}{v} = \frac{1}{V''_{max}}\frac{\overline{K}_B}{(B)} + \frac{1}{V''_{max}} \tag{32}$$

where

$$\frac{1}{V''_{max}} = \frac{1}{V_{max}}\left[\frac{\overline{K}_{MA}}{(MA)}\left(1 + \frac{(A)}{\overline{K}_A}\right) + 1\right] \tag{33}$$

A plot of $\frac{1}{v}$ vs. $\frac{1}{(B)}$ will then yield a value of $\overline{K}_B$ (from the slope and the ordinate intercept) for any fixed value of (A) and (MA), i.e., when A_t and M_t are also fixed.

a. Without Enzyme-Magnesium Complexes, with Product Inhibition, and Formation of "Dead-End" Complexes

Assume the same mechanism as under case 1 except that product inhibition by (C) and (MC) exists for the forward course of the reaction where (C) and (MC) compete for the catalytic binding sites of (A) and (MA) but not for the site of (B), i.e., a case of a "dead-end complex" which forms $E \cdot B \cdot MC$ and also $E \cdot B \cdot C$. The following equations then have to be considered, in addition to those already presented under case 1, which also include equations described as K_7 and K_7' under Scheme V.

$$
\begin{array}{ccccc}
E + B & & & & E + MC \\
\updownarrow K_2' & & & & \updownarrow K_7 \\
EB + MC & \overset{K_\alpha}{\rightleftharpoons} & E \cdot B \cdot MC & \overset{K_\beta}{\rightleftharpoons} & E \cdot MC + B \\
& \overset{K_\alpha'}{\rightleftharpoons} & E \cdot B \cdot C & \overset{K_\beta'}{\rightleftharpoons} & \\
EB + C & & & & EC + B \\
\updownarrow K_2' & & & & \updownarrow K_7' \\
E + B & & & & E + C
\end{array}
$$

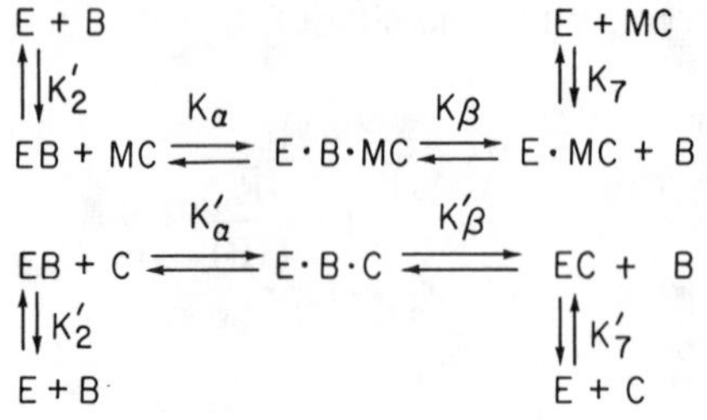

SCHEME VI.

The equality equations are now extended to

$$K_7' = K_8' = K_\alpha' = \overline{K}_C$$

$$K_7 = K_8 = K_\alpha = \overline{K}_{MC}$$

$$K_2' = K_3 = K_3' = K_\beta' = K_\beta = \overline{K}_B \tag{34}$$

Then, for the forward reaction:

$$
v = \cfrac{V_{max}}{1 + \cfrac{\overline{K}_B}{(B)} + \cfrac{\overline{K}_{MA}}{(MA)}\left(1 + \cfrac{(A)}{\overline{K}_A} + \cfrac{(C)}{\overline{K}_C} + \cfrac{(MC)}{\overline{K}_{MC}}\right)}
$$
$$
+ \cfrac{\overline{K}_{MA}}{(MA)}\cfrac{\overline{K}_B}{(B)}\left(1 + \cfrac{(A)}{\overline{K}_A} + \cfrac{(C)}{\overline{K}_C} + \cfrac{(MC)}{\overline{K}_{MC}}\right) \tag{35}
$$

If (C) and (MC) approach zero, then Equation 35 becomes Equation 27 without product inhibition.

If (B) is the variable substrate, then in reciprocal form:

$$
\frac{V_{max}}{v} = \frac{\overline{K}_B}{(B)}\left[1 + \frac{\overline{K}_{MA}}{(MA)}\left(1 + \frac{(A)}{\overline{K}_A} + \frac{(C)}{\overline{K}_C} + \frac{(MC)}{\overline{K}_{MC}}\right)\right]
$$
$$
+ \frac{\overline{K}_{MA}}{(MA)}\left(1 + \frac{(A)}{\overline{K}_A} + \frac{(C)}{\overline{K}_C} + \frac{(MC)}{\overline{K}_{MC}}\right) + 1 \tag{36}
$$

A plot of $\dfrac{1}{v}$ vs. $\dfrac{1}{(B)}$ [with fixed concentrations of (A), (MA), (C), and (MC)] will result in both slope and ordinate intercept varying as a function of the term in the brackets of Equation 36, and the inhibition of (B) would appear to be noncompetitive with respect to (C) and (MC) (see Chapter 1, Equations 47 and 48). The value of $\overline{K}_B$ could be obtained from the ratio of slope to ordinate intercept, and $\overline{K}_B$ would be independent of (A), (MA), (C), or (MC), provided all were held fixed by fixing A_t, M_t, and C_t or if M_t can be fixed large enough to render $A + C \rightarrow$ zero.

$$\frac{V_{max}}{v} = \frac{\overline{K}_B}{(B)}\left[1 + \frac{\overline{K}_{MA}}{(MA)}\left(1 + \frac{(MC)}{\overline{K}_{MC}}\right)\right] + \frac{\overline{K}_{MA}}{(MA)}\left(1 + \frac{(MC)}{\overline{K}_{MC}}\right) + 1 \quad (37)$$

If, on the other hand, (MA) is the variable substrate, i.e., if (A) and (MA) are varied and (B) is held constant, Equation 35 rearranged in reciprocal form would now be

$$\frac{1}{v} - \frac{V_{max}}{\left[1 + \frac{\overline{K}_B}{(B)}\right]} = 1 + \frac{\overline{K}_{MA}}{(MA)}\left[1 + \frac{(A)}{\overline{K}_A} + \frac{(C)}{\overline{K}_C} + \frac{(MC)}{\overline{K}_{MC}}\right] \quad (38)$$

Thus, with (B) constant, (C) and (MC) will appear to be competitive inhibitors of the variable substrate (MA), since only the slope will vary, but not the intercept $\frac{\left[1 + \frac{\overline{K}_B}{(B)}\right]}{V_{max}}$.

If M_t can be fixed so large that (A) and (C) → 0 by virtue of their formation constants, then Equation 38 reduces to

$$\frac{1}{v}\frac{V_{max}}{\left[1 + \frac{\overline{K}_B}{(B)}\right]} = 1 + \frac{\overline{K}_{MA}}{(MA)}\left[1 + \frac{(MC)}{\overline{K}_{MC}}\right] \quad (39)$$

and (MC) is seen to be a simple competitive inhibitor of (MA) in contrast to the noncompetitive type of inhibition to be expected by (MC) with (B) as the variable substrate (Equations 36 and 37).

b. Without Enzyme-Magnesium Complexes, with Product Inhibition, but "Dead-End" Complexes are not Formed

Assume the same conditions as under case a, but that the "dead-end" complex E · MC · B (or E · C · B) does *not* form, i.e., K_α, K'_α, K_β, and K'_β are not operating, but only K_7 and K'_7 are proceeding. In this case it is assumed that (B) cannot combine with EC or E · MC and (C) or (MC) cannot combine with E · B. For the former reaction, the equalities expressed by Equation 26 (except, of course, for $\overline{K}_D$) have to be considered.
Then

$$v = \frac{V_{max}}{1 + \frac{\overline{K}_B}{(B)} + \frac{\overline{K}_{MA}}{(MA)}\left(1 + \frac{(A)}{\overline{K}_A}\right) + \frac{\overline{K}_{MA}\overline{K}_B}{(MA)(B)}\left(1 + \frac{(A)}{\overline{K}_A} + \frac{(C)}{\overline{K}_C} + \frac{(MC)}{\overline{K}_{MC}}\right)} \quad (40)$$

Thus, competitive inhibition will result in the case of reciprocal plots of $\frac{1}{v}$ vs. $\frac{1}{(S)}$ for both variable substrate creatine or the variable substrate (MA) with the ordinate intercept being constant for a set of arbitrary conditions, and with the slope varying and dependent on (MC) and (C).

Again, holding M_t so large as to reduce (A) = (C) = 0, leaves

$$v = \frac{V_{max}}{1 + \frac{\overline{K}_B}{(B)} + \frac{\overline{K}_{MA}}{(MA)} + \frac{\overline{K}_{MA}\overline{K}_B}{(MA)(B)}\left(1 + \frac{(MC)}{\overline{K}_{MC}}\right)} \quad (41)$$

which reveals (MC) as a competitive product inhibitor of either variable (B) or variable (MA), with its paired substrate held fixed.

2. With Enzyme-Magnesium Complexes and Associated Reactions are Significant Compared to Total Enzyme Concentration

Assume that the EM complex and associated reactions are present and significant compared to the total enzyme concentration. The following equations have to be considered in addition to those given in Scheme V.

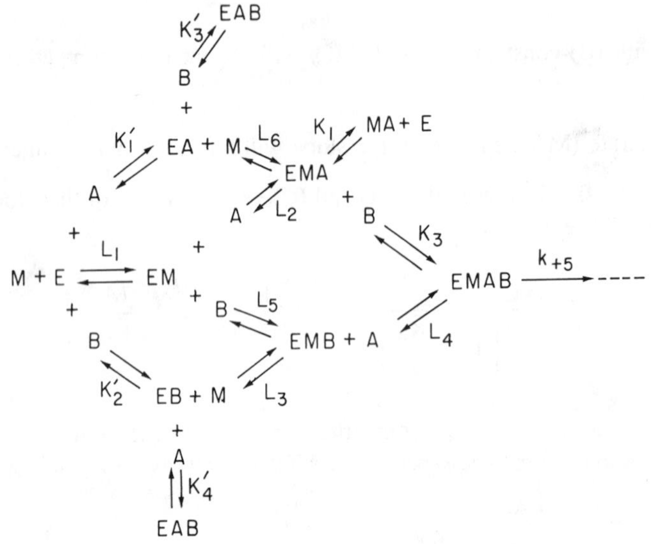

SCHEME VII.

Assume, furthermore, (1) that (B) does not influence the binding of (A) or (MA) and that (A) and (MA) do not influence the binding of (B), (2) that (A) and (MA) influence the binding of each other, (3) that (M) influences the binding of (A) in a sense that (MA) can form (but for the sake of generality it is not considered that $L_1 = L_6$ and K_1' may not equal (L_2), (4) that (M) does not influence the binding of (B), and (5) that the same complex EMA is formed regardless of the pathway.

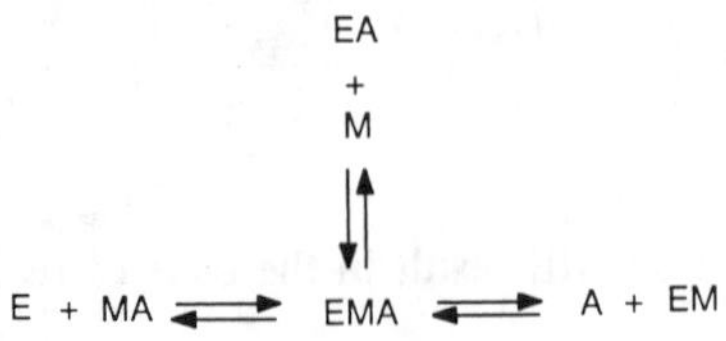

SCHEME VIII.

The dangers of this last assumption are evident, and Dixon and Webb[10] have clearly pointed out the implications involved.

The equality equations are extended to Equations 42 and 43 and include also Equation 26.

$$L_1 = L_3 = \overline{K}_M \tag{42}$$

$$K'_2 = K'_3 = K_3 = L_5 = \overline{K}_B \tag{43}$$

The velocity equation may then be written as follows:

$$v = \frac{V_{max}}{1 + \frac{\overline{K}_B}{(B)} + \frac{\overline{K}_{MA}}{(MA)}\left(1 + \frac{(A)}{\overline{K}_A}\right) + \frac{\overline{K}_{MA}\overline{K}_B}{(MA)(B)}\left(1 + \frac{(A)}{\overline{K}_A}\right) + \frac{\overline{K}_{MA}(M)}{(MA)\overline{K}_M}\left(1 + \frac{\overline{K}_B}{(B)}\right)} \tag{44}$$

If Equation 44 is now compared with Equation 27, it is readily seen that they differ only by the additional term in the denominator:

$$\frac{\overline{K}_{MA}(M)}{(MA)\overline{K}_M}\left(1 + \frac{\overline{K}_B}{(B)}\right)$$

If $\frac{(M)}{\overline{K}_M} \rightarrow 0$, i.e., if $\overline{K}_M$ is very large compared to M, or if $M \rightarrow 0$, then Equation 44 reduces to Equation 27 where the EM terms were neglected.

If, at high (M_t), $(MA) \rightarrow (A_t)$ and $(A) \rightarrow 0$, then, for this special case,

$$\frac{1}{v}\frac{V_{max}}{\left(1 + \frac{\overline{K}_B}{(B)}\right)} \simeq 1 + \frac{\overline{K}_{MA}}{(MA)}\left(1 + \frac{(M)}{\overline{K}_M}\right) \tag{45}$$

This approximation [at constant (B)] might also permit evaluation of $\overline{K}_M$, even if $\overline{K}_M$ were very large numerically, provided that values for $\overline{K}_{MA}$ and $\overline{K}_B$ could be assigned.

Equation 44 may be written in reciprocal form as: Equation 46,

$$\frac{1}{v}\frac{V_{max}}{\left(1 + \frac{\overline{K}_B}{(B)}\right)} = 1 + \frac{\overline{K}_{MA}}{(MA)}\left(1 + \frac{(A)}{\overline{K}_A} + \frac{(M)}{\overline{K}_M}\right) \tag{46}$$

which will reduce to the comparable Equation 29 if

$$\frac{(M)}{\overline{K}_M}\frac{\overline{K}_{MA}}{(MA)} \rightarrow 0$$

Substitution of $\frac{(M)\,(A)}{K_o}$ for (MA) (Equation 22) in Equation 44 leads to Equation 47 in terms of (M) and (A) only.

$$v = \frac{V_{max}}{\left(1 + \frac{\overline{K}_B}{(B)}\right)\left(1 + \frac{\overline{K}_{MA}}{(A)}\frac{K_o}{\overline{K}_M}\right) + \frac{1}{(M)}\left[\frac{\overline{K}_{MA}\overline{K}_o}{(A)}\left(1 + \frac{(A)}{\overline{K}_A}\right) + \frac{\overline{K}_{MA}K_o}{(A)}\frac{\overline{K}_B}{(B)}\left(1 + \frac{(A)}{\overline{K}_A}\right)\right]} \tag{47}$$

In reciprocal form, after rearrangement, one obtains

$$\frac{1}{v}\frac{V_{max}}{\left(1+\frac{\bar{K}_B}{(B)}\right)} = 1 + \frac{\bar{K}_{MA}K_o}{(A)\bar{K}_M} + \frac{1}{(M)}\left(\frac{\bar{K}_{MA}K_o}{(A)} + \frac{\bar{K}_{MA}K_o}{\bar{K}_A}\right) \tag{48}$$

If (A) is kept constant while (M) is varied, i.e., by varying A_t since

$$A_t \simeq A + MA, \quad \text{or} \quad A \simeq \left(\frac{K_oA_t}{M + K_o}\right) \tag{49}$$

this, of course, results in a variation of (MA) (for simplicity, only A_t has been set equal to (MA) + (A), and all other complexes have been neglected for these treatments. These treatments, however, can easily be extended to other metal complexes. Thus, to increase (M), A_t will have to be increased to maintain (A) constant while (MA) is also increased. Thus, a plot of $\frac{1}{v}$ vs. $\frac{1}{(M)}$ at constant (A) and (B) does not yield a direct value for K_M, rather for K_M apparent, which is equal to Equation 50.

$$\bar{K}_{M_{apparent}} = \frac{\left(\frac{\bar{K}_{MA}K_o}{(A)} + \frac{\bar{K}_{MA}K_o}{\bar{K}_A}\right)}{1 + \frac{\bar{K}_{MA}}{(A)}\frac{K_o}{\bar{K}_M}} \tag{50}$$

and $\bar{K}_M$ can only be evaluated with the knowledge of $\bar{K}_{MA}$, $\bar{K}_A$, and K_0. In the special case where $\frac{\bar{K}_M(A)}{K_o} << K_{MA}$

$$\bar{K}_{M_{app}} \simeq \bar{K}_M\left(1 + \frac{(A)}{\bar{K}_A}\right) \tag{51}$$

The numerical values calculated for M and A (where A_t had been varied to maintain (A) constant) and consequently the final calculated values for both the slope and the ordinate intercept hinge on the numerical value for K_0; therefore, $\bar{K}_M$ will be dependent on a value for K_0. Moreover, should K_0 be arbitrarily selected, the "meaning" of the reciprocal plot can be lost and the computed values for the derived constants can be in serious error. Only in the very special case where

$$\frac{(A)}{\bar{K}_A} \ll 1, \quad \text{and} \quad \frac{\bar{K}_M(A)}{K_o} \ll \bar{K}_{MA}, \quad \text{will} \quad \bar{K}_{M_{app}} \simeq \bar{K}_M \tag{52}$$

If (A) is held at very high concentration, then at constant (B) the slope approximates Equation 53.

$$\text{Slope} \simeq \frac{\bar{K}_{MA}K_o}{\bar{K}_A}\left(1 + \frac{\bar{K}_B}{(B)}\right)\frac{1}{V_{max}} \tag{53}$$

and

$$\text{ordinate intercept} \simeq \left(1 + \frac{\overline{K}_B}{(B)}\right)\frac{1}{V_{max}} \tag{54}$$

so that

$$\overline{K}_{M_{app}} \simeq \frac{\overline{K}_{MA}K_o}{\overline{K}_A} \tag{55}$$

which does not yield a value for K_M.

At this point, it is pertinent to return to the case where EM was considered negligible. Substitution of $(MA) = \dfrac{(M)\,(A)}{K_o}$ into Equation 27 leads to Equation 56.

$$\frac{1}{v}\frac{V_{max}}{\left(1 + \dfrac{\overline{K}_B}{(B)}\right)} = 1 + \frac{1}{M}\left(\frac{\overline{K}_{MA}K_o}{(A)} + \frac{\overline{K}_{MA}K_o}{\overline{K}_A}\right) \tag{56}$$

and differs from the case where EM was not negligible (Equation 48) by lacking the term $\dfrac{\overline{K}_{MA}\,K_o}{(A)\,\overline{K}_M}$ in the ordinate intercept. With (A) and (B) held constant (as before, by varying A_t), then only the slope of a plot of $\dfrac{1}{v}$ vs. $\dfrac{1}{(M)}$ will vary with constant (A), and the intercept will not vary. $\overline{K}_{M_{app}}$ will then be equal to the ratio

$$\overline{K}_{M_{app}} = \frac{\text{Slope}}{\text{ordinate intercept}} = \frac{\overline{K}_{MA}K_o}{\overline{K}_A} + \frac{\overline{K}_{MA}K_o}{(A)} \tag{57}$$

Obviously, in this case, the numerical value for K_M itself has no meaning (from the original mechanism). Moreover, if (A) is held very high, $\overline{K}_{M_{app}} \simeq \dfrac{\overline{K}_{MA}\,K_o}{\overline{K}_A}$, which is the same composite term (at high (A), Equation 55) as in the case where EM was considered.

Finally, it is necessary to consider the analogous plots $\dfrac{1}{v}$ vs. $\dfrac{1}{(A)}$, where (M) is held constant by varying M_t (and simultaneously, of course, (MA)). If (EM) is included, then Equation 44, after substitution of (MA) from Equation 22 [(M) (A)/K_0], and rearrangement in reciprocal form yields Equation 48 and after rearrangement, Equation 58 is obtained for fixed (M) and (B) and varied (A).

$$\frac{1}{v}\frac{V_{max}}{\left(1 + \dfrac{\overline{K}_B}{(B)}\right)} = \left(1 + \frac{1}{(M)}\frac{\overline{K}_{MA}K_o}{\overline{K}_A}\right) + \frac{1}{(A)}\left(\frac{\overline{K}_{MA}K_o}{\overline{K}_M} + \frac{\overline{K}_{MA}K_o}{(M)}\right) \tag{58}$$

For the ratio of slope to ordinate intercept, defined as $\overline{K}_{A_{app}}$, at constant (B):

$$\overline{K}_{A_{app}} = \left(\frac{\overline{K}_{MA}K_o}{\overline{K}_M} + \frac{\overline{K}_{MA}K_o}{(M)}\right)\Big/\left(1 + \frac{1}{(M)}\frac{\overline{K}_{MA}K_o}{\overline{K}_A}\right) \tag{59}$$

Both the slope and the ordinate intercept are functions of (M). Only in the very limiting case, where (M) is held very high, will

$$\overline{K}_{A_{app}} \simeq \overline{K}_{MA}K_o/\overline{K}_M \tag{60}$$

and will not include the $\overline{K}_A$ term, but yields nevertheless a real and positive value.

Similarly, for the case where EM was neglected, Equation 56 may be written as

$$\frac{V_{max}}{v}\frac{1}{\left(1+\dfrac{\overline{K}_B}{(B)}\right)} = \frac{1}{(M)}\frac{\overline{K}_{MA}K_o}{\overline{K}_A} + \frac{1}{(A)}\frac{\overline{K}_{MA}K_o}{(M)} \tag{61}$$

A plot of $\frac{1}{v}$ vs. $\frac{1}{(A)}$ at constant (M) and (B) yields

$$\overline{K}_{A_{app}} = \frac{\overline{K}_{MA}K_o}{(M) + \dfrac{\overline{K}_{MA}K_o}{\overline{K}_A}} \tag{62}$$

If (M) is held high so that when

$$(M) \gg \frac{\overline{K}_{MA}K_o}{\overline{K}_A}, \quad \text{then} \quad \overline{K}_{A_{app}} \simeq \frac{\overline{K}_{MA}K_o}{(M)} \tag{63}$$

or in the case where (M) $\rightarrow \infty$, $\overline{K}_{A_{app}} \rightarrow 0$, i.e., $K_{A_{app}}$ vanishes.

Conversely, at low concentrations of (M), where (M) $<< \frac{\overline{K}_{MA}\ K_O}{\overline{K}_A}$, $\overline{K}_{A_{app}} \rightarrow \overline{K}_A$, and similarly for the case where (EM) is considered (Equation 59), when (M) is small, i.e., M $\ll \overline{K}_M$ and (M) $\frac{\overline{K}_{MA}\ K_O}{\overline{K}_A}$, $\overline{K}_{A_{app}} \rightarrow \overline{K}_A$.

The same comments made above with respect to K_0 apply here equally well; however, by assuming K_0 to be accurately known, a criterion for the contribution of (EM) to the velocity is available: $\overline{K}_{A_{app}}$ will vanish at extremely high concentrations of (M) in the case where (EM) is negligible, but will yield a real and positive value where (EM) is not negligible (i.e., approximating $\frac{\overline{K}_{MA}\ K_O}{\overline{K}_M}$, Equation 60).

Rearranging the $\overline{K}_{A_{app}}$ equation for the case with negligible (EM) (Equation 62) leads to

$$\overline{K}_{A_{app}}\left(1 + \frac{1}{(M)}\frac{\overline{K}_{MA}K_o}{\overline{K}_A}\right) = \frac{1}{(M)}(\overline{K}_{MA}K_o) \tag{64}$$

and a plot of the left-hand side of this equation vs. $(\frac{1}{(M)})$ yields an ordinate intercept at zero and a slope of $\overline{K}_{MA}\ K_O$.

On the other hand, rearrangement of the $K_{A_{app}}$ equation for the case where (EM) was not negligible (Equation 59) results in Equation 65.

$$\overline{K}_{A_{app}}\left(1 + \frac{1}{(M)}\frac{\overline{K}_{MA}K_o}{\overline{K}_A}\right) = \frac{1}{(M)}\overline{K}_{MA}K_o + \frac{\overline{K}_{MA}K_o}{\overline{K}_M} \tag{65}$$

An analogous plot to the one above leads to a real and positive ordinate intercept, $\frac{\overline{K}_{MA} K_O}{\overline{K}_M}$, where the slope, $\overline{K}_{MA} K_0$, is identical in both cases considered. Thus, an estimate of $\overline{K}_M$ [and a criterion for the contribution of (EM)] can be made in principle, provided $\overline{K}_A$ is known (see left-hand side of equation), which, of course, is dependent ultimately on a knowledge of K_0. The ordinate intercepts of the $\frac{1}{v}$ vs. $\frac{1}{(A)}$ plots for the two mechanisms considered here prove to be identical, i.e., with or without the (EM) complex:

$$\frac{1}{V_{max(A)_{app}}} = \frac{1}{V'_{max}} + \frac{1}{V'_{max}} \frac{1}{(M)} \frac{\overline{K}_{MA} K_o}{\overline{K}_A} \tag{66}$$

refer to Equation 58 with (EM), and Equation 61 without (EM)) where

$$V'_{max} = \frac{V_{max}}{\left(1 + \frac{\overline{K}_B}{(B)}\right)} \tag{67}$$

Similar equations may be readily set up for the $\frac{1}{v}$ vs. $\frac{1}{(M)}$ plots in terms of $K_{M_{app}}$ and $V_{max(M)_{app}}$ for the two general cases considered.

Dixon and Webb,[10] in an analysis of similar problems, have derived kinetic equations for six hypothetical mechanisms. The above treatment may be taken as complementary to theirs and is based largely on Kuby and Noltmann.[48]

B. CASE II. RANDOM QUASI-EQUILIBRIUM WITHOUT INDEPENDENT BINDING OF SUBSTRATES TWO GENERAL MECHANISMS WILL BE CONSIDERED HERE, CASE 1 WITHOUT "DEAD-END" COMPLEXES AND CASE 2 WITH "DEAD-END" COMPLEXES

1. Without "Dead-End" Complexes

Consider the mechanism of Scheme IX (*vide infra*) without "dead-end" complexes where (MA) = $MgATP^{2-}$, (B) = creatine$^{\pm}$, (MC) = $MgADP^{-}$, (D) = $Cr{\sim}P^{2-}$, and the rate-limiting step is assumed to be at the interconversion of the ternary complexes ($E \cdot MA \cdot B \rightleftarrows E \cdot MC \cdot D$) and where free Mg^{2+} = M does not participate significantly in the mechanism.

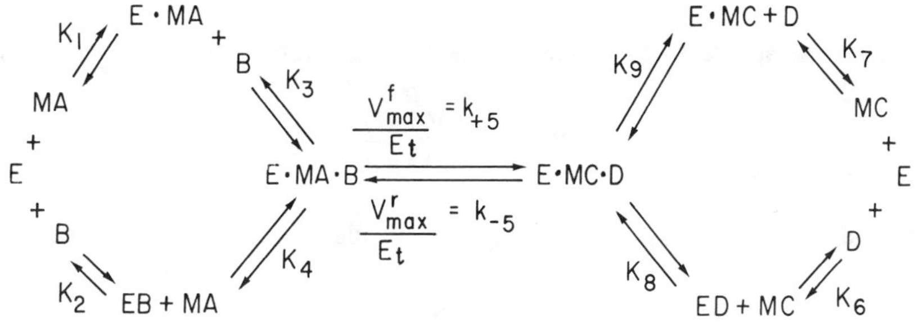

SCHEME IX.

Calculations of $MgATP^{2-}$ as a variable or fixed substrate in the forward direction are made, in principle, by employing a set of conservation equations and assigned values for the complexation and dissociation constants.[48]

Thus, a given value of Mg_t and ATP_t (and Na_t, if present) will determine the concentration of $MgATP^{2-}$. If free Mg^{2+} is to be fixed also, then ATP^{4-} may be rendered negligible in the forward direction but not necessarily ADP^{3-} in the reverse direction.

Thus, the set of approximate conservation equations which were adequate for calculations of $MgATP^{2-}$ in the forward reaction under the substrate range conditions (at pH 7.4 to 8.8) explored by Jacobs and Kuby[49] for the calf brain ATP-creatine transphosphorylase are

$$\begin{aligned} ATP_t &\simeq MgATP^{2-} + ATP^{4-} + HATP^{3-} + MgHATP^{-} + NaATP^{3-} \\ Mg_t &\simeq Mg^{2+} + MgATP^{2-} + MgHATP^{-} \\ Na_t &\simeq Na^{+} + NaATP^{3-} \end{aligned} \tag{68}$$

For the reverse direction, for calculation of $MgADP^-$ and $Cr{\sim}P^{2-}$ (creatine phosphate^{2-}), the set of conservation equations could be adequately approximated by

$$\begin{aligned} ADP_t &\simeq MgADP^{-} + HADP^{2-} + ADP^{3-} + MgHADP + NaADP^{2-} \\ Cr \sim P_t &\simeq Cr \sim P^{2-} + HCr \sim P^{-} + MgCr \sim P + NaCr \sim P^{-} \\ Mg_t &\simeq Mg^{2+} + MgADP^{1-} + MgHADP + MgCr \sim P \\ Na_t &\simeq Na^{+} + NaADP^{2-} + NaCr \sim P^{-} \end{aligned} \tag{69}$$

Subscript t implies total concentration and appropriate valences are assigned to each of the ionizable or complex species.

For the mechanism, given in Scheme IX without "dead-end" complexes:

$$v_o^f = \frac{V_{max}^f\left[(MA)(B) - \dfrac{(MC)(D)}{K'_{equil}}\right]}{\left[K_1K_3 + K_3(MA) + K_4(B) + (MA)(B) + \dfrac{K_1K_3(MC)(D)}{K_7K_9} + \dfrac{K_1K_3(MC)}{K_7} + \dfrac{K_1K_3K_8(D)}{K_7K_9}\right]} \tag{70}$$

and

$$K_1K_3 = K_2K_4; \quad K_6K_8 = K_7K_9 \tag{71}$$

where K'_{equil} is the apparent equilibrium constant at a given pH value,

$$K'_{equil} = \frac{(MC)(D)}{(MA)(B)} \tag{72}$$

and to be distinguished from the defined thermodynamic overall equilibrium constant of the system, i.e.,

$$K_{eq} = \frac{(MC)(D)}{(MA)(B)}(H^+) \tag{73}$$

K_{equil} had been assigned a value of 2.81×10^{-10} by Kuby and Noltmann[48] with use of the same values selected for the intrinsic dissociation and metal complexation constants and more than 100 sets of equilibrium data. This will be briefly discussed below (see Table 12).

It must be stressed as indicated by Kuby and Noltmann[48] that the absolute values of the derived kinetic parameters will hinge to some degree on the values selected for intrinsic dissociation and metal complexation constants. However, since the same complexation and dissociation values were employed in both the thermodynamic and kinetic calculations, as a measure of self-consistency, the value of the equilibrium constant obtained from any of the four Haldane relations may be compared with the calculated thermodynamic value of 2.81×10^{-10}. For example, consider the pH 8.8 data of Jacobs and Kuby[49] on the calf brain ATP-creatine transphosphorylase:

$$K_{equil} = \frac{V^f_{max}K_7K_9(H^+)}{V^r_{max}K_1K_3} = 3.16 \times 10^{-10}$$

$$K_{equil} = \frac{V^f_{max}K_6K_8(H^+)}{V^r_{max}K_2K_4} = 2.31 \times 10^{-10}$$

$$K_{equil} = \frac{V^f_{max}K_7K_9(H^+)}{V^r_{max}K_2K_4} = 2.78 \times 10^{-10}$$

$$K_{equil} = \frac{V^f_{max}K_6K_8(H^+)}{V^r_{max}K_1K_3} = 2.61 \times 10^{-10} \qquad (74)$$

or an average value of $K^{equil}_{average} = 2.72\ (\pm 0.35) \times 10^{-10}$, surprisingly close to the assigned average value of 2.81×10^{-10}.

For the limiting cases where the concentrations of products may be set to zero initially, the forward and reverse initial velocities are, respectively,

$$v^f_o = V^f_{max}\left[1 + \frac{K_4}{(MA)} + \frac{K_3}{(B)} + \frac{K_1K_3}{(MA)(B)}\right]^{-1} \qquad (75)$$

$$K_1 \cdot K_3 = K_2 \cdot K_4 \qquad (76)$$

$$v^r_o = V^r_{max}\left[1 + \frac{K_8}{(MC)} + \frac{K_9}{(D)} + \frac{K_7K_9}{(MC)(D)}\right]^{-1} \qquad (77)$$

$$K_7 \cdot K_9 = K_6 \cdot K_8 \qquad (78)$$

For variable $MgATP^{2-}$ (MA) and fixed creatine (B), For example, a primary plot of $\frac{1}{v_0}$ vs. $\frac{1}{(MA)}$ would yield

$$\frac{1}{v^f_o} = \frac{1}{V^f_{max}}\left(1 + \frac{K_3}{(B)}\right) + \frac{K_4}{(MA)}\frac{1}{V^f_{max}}\left(1 + \frac{K_2}{(B)}\right) \qquad (79)$$

and values for K_1 to K_4 and V^f_{max} may be estimated from secondary plots of slopes and ordinate intercepts, and the relation Equation 76. Thus,

$$\underset{(1/v_O \text{ vs. } 1/(MA))}{\text{Slope}} = \frac{K_4}{V^f_{max}} + \frac{K_2K_4}{V^f_{max}}\frac{1}{(B)} \qquad (80)$$

$$\text{Y-intercept} = \frac{1}{V^f_{max}} + \frac{K_3}{V^f_{max}}\frac{1}{(B)} \qquad (81)$$

Also, in the case of variable (B) and fixed (MA), again all four constants (and V^f_{max}) may be estimated for the forward reaction. Similarly, from Equation 77 in reciprocal form and appropriate secondary plots and relation Equation 78, all four constants for the reverse direction and V^r_{max} may be obtained.

If (D) = 0 and (MC) the product were added back to the forward reaction, i.e.,

$$v^f_o = V^f_{max}\left[\frac{K_1K_3}{(MA)(B)} + \frac{K_3}{(B)} + \frac{K_4}{(MA)} + \frac{K_1K_3(MC)}{K_7(MA)(B)} + 1\right]^{-1} \quad (82)$$

(MC) the product (i.e., $MgADP^{1-}$) would appear to be a competitive product inhibitor of either (MA) or (B) (i.e., of $MgATP^{2-}$ or $Cr^{\pm}$).

2. With "Dead-End" Complexes

Consider the mechanism in Scheme IX but add the reaction Equations 83 and 84 involving the "dead-end" complexes E · B · MC or E · MA · D.

$$(E \cdot B) + (MC) \overset{K_\alpha}{\rightleftharpoons} E \cdot B \cdot MC \overset{K_\beta}{\rightleftharpoons} E \cdot MC + (B) \quad (83)$$

$$E \cdot MA + D \overset{K_\gamma}{\rightleftharpoons} E \cdot MA \cdot D \overset{K_\delta}{\rightleftharpoons} E \cdot D + (MA) \quad (84)$$

and the equalities (Equation 71) are extended to include

$$K_\alpha K_2 = K_\beta K_7$$

$$K_\gamma K_1 = K_\delta K_8 \quad (85)$$

which leads to the overall velocity expression given by

$$v^f_o = \frac{V^f_{max}\left[(MA)(B) - \frac{(MC)(D)}{K'_{eq}}\right]}{K_1K_3 + K_3(MA) + K_4(B) + (MA)(B) + \frac{K_1K_3(MC)(D)}{K_7K_9} + \frac{K_1K_3(MC)}{K_7} + \frac{K_1K_3K_9(D)}{K_7K_9} + \frac{K_1K_3(MC)(B)}{K_2K_\alpha} + \frac{K_3(D)(MA)}{K_\gamma}} \quad (86)$$

If (MC) = (D) → 0, Equation 86 reduces also to the limiting forward velocity expression, i.e., Equation 75. If (MA) = (B) → 0, Equation 86 also reduces to the limiting reverse velocity expression, Equation 77, for the cases where there are no dead-end complexes. Thus, initial velocity data alone will not allow either mechanism to be distinguished.

On the other hand, for the case of product (MC) inhibition with the "dead-end" complexes and (D) = 0, Equation 86 reduces to Equation 87 for the initial forward velocity,

$$\frac{V^f_{max}}{v^f_o} = 1 + \frac{K_3}{(B)} + \frac{K_3}{(B)}\left[\frac{K_1}{(MA)} + \frac{K_1(MC)}{K_7(A)}\right] + \frac{K_4}{(MA)}\left[1 + \frac{(MC)}{K_\alpha}\right] \quad (87)$$

which after rearrangement, will show that (MC) would be *competitive* with respect to variable substrate (MA), but *noncompetitive* with respect to variable substrate (B).

Interestingly, the uncomplexed species may compete for binding with the metal complexed substrate species.

Thus, consider that an ADP^{3-} (C) species competes for binding with $MgADP^{-}$ (MC) and not with $Cr{\sim}P^{2-}$ (D), and that an abortive and inactive complex E · C · D forms in a random manner as for the basic mechanism (leading to another value for $K_{Cr{\sim}P^{2-}}$ for the abortive ternary complex with ADP^{3-}, that is, K_9') i.e., for the reverse direction,

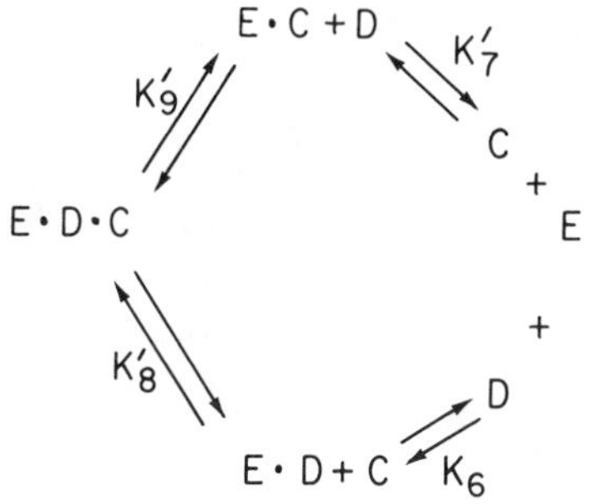

SCHEME X.

and which equilibria (designated by primed values) should be added to the basic mechanism (for K_1 to K_9) given above in Scheme IX. Note that K_6 (or $\overline{K}_{Cr{\sim}P^{2-}}$) remains the same. Thus,

$$K_7'K_9' = K_6K_8' \tag{88}$$

in addition to

$$K_7K_9 = K_6K_8 \tag{89}$$

and the limiting velocity expression for the reverse reaction,

$$v_o^r = V_{max}^r \left[1 + \frac{K_8}{(MC)} + \frac{K_9}{(D)} + \frac{K_7K_9}{(MC)(D)}\right]^{-1} \tag{90}$$

is amended to

$$v_o^r = V_{max}^r \left[1 + \frac{K_9}{(D)} + \frac{K_8}{(MC)}\left(1 + \frac{(C)}{K_8'}\right) + \frac{K_7K_9}{(MC)(D)}\left(1 + \frac{(C)}{K_7'}\right)\right]^{-1} \tag{91}$$

To distinguish intrinsic dissociation constants of a particular substrate from the binary or ternary complex, the symbolism $\overline{K}_S$ and K_S, respectively, has been found convenient.

Thus, the limiting velocity expressions for the forward (Equation 75) and reverse reaction (Equation 77) may be written in this terminology:

$$\frac{v_o^f}{V_{max}^f} = \frac{1}{\left[1 + \frac{K_{MA}}{(MA)} + \frac{K_B}{(B)} + \frac{\overline{K}_{MA}}{(MA)}\frac{K_B}{(B)}\right]} \tag{92}$$

$$\frac{v_o^r}{V_{max}^r} = \frac{1}{\left[1 + \frac{K_{MC}}{(MC)} + \frac{K_D}{(D)} + \frac{\overline{K}_{MC}}{(MC)}\frac{K_D}{(D)}\right]} \tag{93}$$

TABLE 1
Values for Derived Kinetic Parameters at pH 8.8 (30°C) for ATP-Creatine Transphosphorylase from Calf Brain[49]

Derived kinetic parameters	Defined equilibrium	Defined intrinsic dissociation constant	Value (*M*)
K_1	$E \cdot MA \rightleftarrows E + MA$	$\overline{K}_{MgATP2-}$ [a]	$0.93\ (\pm 0.14) \times 10^{-3}$
K_2	$E \cdot B \rightleftarrows E + B$	$\overline{K}_{creatine}$	$2.9\ (\pm 1.1) \times 10^{-2}$
K_3	$E \cdot MA \cdot B \rightleftarrows E \cdot MA + B$	$K_{creatine}$	$3.7\ (\pm 0.4) \times 10^{-3}$
K_4	$E \cdot MA \cdot B \rightleftarrows E \cdot B + MA$	$K_{MgATP2-}$ [b]	$1.35\ (\pm 0.45) \times 10^{-4}$
K_6	$E \cdot D \rightleftarrows E + D$	$\overline{K}_{Cr\sim P2-}$	$2.0\ (\pm 0.6) \times 10^{-2}$
K_7	$E \cdot MC \rightleftarrows E + MC$	$\overline{K}_{MgADP1-}$	$1.2\ (\pm 0.1) \times 10^{-4}$
K_8	$E \cdot MC \cdot D \rightleftarrows E \cdot D + MC$	$K_{MgADP1-}$	$1.0\ (\pm 0.3) \times 10^{-5}$
K_9	$E \cdot MC \cdot D \rightleftarrows E \cdot MC + D$	$K_{Cr\sim P2-}$	$2.0\ (\pm 0.1) \times 10^{-3}$
$\frac{V^f_{max}}{E_t}$	$E \cdot MA \cdot B \rightleftarrows E \cdot MC \cdot D$		$2.5\ (\pm 0.1) \times 10^2$ μmol · min^{-1} mg^{-1}
$\frac{V^r_{max}}{E_t}$	$E \cdot MC \cdot D \rightleftarrows E \cdot MA \cdot B$		$8.8\ (\pm 0.2) \times 10^1$ μmol · min^{-1} mg^{-1}

[a] $\overline{K}_S$ denotes an intrinsic dissociation constant of the particular substrate from the binary complex.
[b] K_S denotes a dissociation constant of the particular substrate from the ternary complex.

In Table 1, some data on the calf brain ATP-creatine transphosphorylase taken from Jacobs and Kuby[49] are listed. Note that in general $K_S << \overline{K}_S$, implying some cooperative type of binding, whereby the binding of the second substrate enhances greatly the binding of the first substrate.

III. STEADY-STATE KINETIC ANALYSIS OF KINASE-TYPE OF CATALYZED REACTIONS

A. ILLUSTRATED BY THE ISOENZYMES OF CALF AND HUMAN ATP-CREATINE TRANSPHOSPHORYLASE (CREATINE KINASE)

Thus, accepting the analyses of the data given in Table 1[49] for the calf brain creatine kinase, at pH 8.8 and 30°, it is of interest that $\overline{K}_{s_1}$ (for $ES_1 \rightleftarrows E + S_1$) values are consistently much larger (7- to 12-fold larger) than the corresponding K_{s_1} (for $ES_1S_2 \rightleftarrows ES_2 + S_1$) values for $MgATP^{2-}$, for creatine$^{\pm}$, or in the reverse direction, for $MgADP^-$ and for $Cr \sim P^{2-}$ (e.g., compare K_7 and K_8 for $MgADP^-$, K_1 and K_4 for $MgATP^{2-}$, K_2 and K_3 for creatine$^{\pm}$, K_6 and K_9 for $Cr \sim P^{2-}$). A plausible explanation might involve a substrate pair-induced conformational change occurring within the ternary complexes ($E \cdot MA \cdot B$ or $E \cdot MC \cdot D$) to permit tighter binding of the magnesium nucleotide complex or its nonnucleotide substrate to the enzyme, than in the case where only a single member of each substrate pair is associated with the enzyme. Morrison and James[83] have invoked this hypothesis for the rabbit muscle ATP-creatine transphosphorylase. It is of interest that the relative differences between K_{s_1} and $\overline{K}_{s_1}$ values (i.e., the intrinsic dissociation constant for a particular substrate from the ternary vs. binary complex, Table 1) are comparatively large for each substrate, compared to the rabbit muscle enzyme where the comparable situation was considered negligible or relatively small in former analyses on the rabbit muscle creatine kinase enzyme,[48,298] and as will be shown below for the calf muscle creatine kinase.

As in the case of the rabbit muscle creatine kinase,[48,59] and shown by equilibrium-binding measurements, it appears that the uncomplexed species ADP^{3-} can compete for binding with $MgADP^-$; however, only very tentative values may be assigned for the calf brain creatine kinase at pH 8.8, with relatively large uncertainties (better values can be provided at pH 7.4 where the inhibition by relatively high values of ADP_0 is more pronounced — see below). Only the order of magnitude is considered significant for the present, that is, $K_7' \cong 0.6 \times 10^{-3}$ *M*; $K_8' \cong 1.5 \times 10^{-4}$ *M*; $K_9' \simeq 5 \times 10^{-3}$ *M* (refer to the assigned equilibria given in Scheme X). If these values prove approximately correct, it would appear that at pH 8.8 the binding coefficients of ADP^{3-} to the free enzyme or in the ternary complex are approximately one order of magnitude less than for the case of $MgADP^-$ ($K_7 = 1.2 \times 10^{-4}$; $K_8 \simeq 1.0 \times 10^{-5}$) with the value for $K'_{Cr\sim P}$ for the abortive complex (K_9') increasing only slightly and probably not significantly over the reactive ternary complex ($K_9 = 2 \times 10^{-3}$ *M*).

It is of interest, at this point, to present a comparison of some kinetic data obtained[299] for the other creatine kinase isoenzymes (see Table 2) i.e., for the calf muscle isoenzyme and the calf hybrid-isoenzyme of ATP-creatine transphosphorylase.[286] If it may be assumed that the kinetic analyses, as given above for the calf brain isoenzyme, holds also for the hybrid and muscle isoenzymes, then the kinetic data may be evaluated in a similar manner and summarized in Table 2 for the three calf isoenzymes, obtained at the same pH of 8.8 (at 30°) and under similar conditions. Firstly, a comparison of the muscle type with the brain type reveals the following significant differences, or similarities, in the derived kinetic parameters. (1) For the forward reaction constants, the muscle type possesses much larger dissociation constants for the dissociation of the substrates form the ternary complexes, e.g., larger values for K_3 ($K_{creatine}$) and K_4 ($K_{MgATP^{2-}}$); however, similar values for K_1 and K_2 (for dissociation from the binary complexes) are observable for both isoenzymes and surprisingly they possess similar V^f_{max} values. (2) The differences between $\overline{K}_{s_1}$ and K_{s_1} values for the forward reaction catalyzed by the muscle enzyme are either insignificant (refer to $\overline{K}_{MgATP^{2-}}$ and $K_{MgATP^{2-}}$) or are decreased by only 2.5-fold from $\overline{K}_{creatine}$ to $K_{creatine}$ (K_2 vs. K_3), in contrast to the very much larger differences, that is approximately one order of magnitude, observed for the brain type. (3) For the reverse direction, $K_{Cr\sim P^{2-}}$ is almost 12 times larger for the muscle type than the brain type and K_{MgADP^-} is almost 10 times larger, whereas the equivalent values for $\overline{K}_{s_1}$ are again not too dissimilar for the two isoenzymes, which again reduce the differences between K_{s_1} and $\overline{K}_{s_1}$ to about 2-fold or less. The V^r_{max} for the reverse direction for the brain-type is surprisingly low compared to its V^f_{max} (and apparently a characteristic of all brain-type enzymes studied thus far; see Table 3 below for the human isoenzymes of creatine kinase) and within a factor of two of the V^r_{max} for the muscle type. Parenthetically, it may be remarked at this point that both isoenzymes, muscle type and brain type, appear to be nicely adapted to their respective environments, provided correlations may be drawn between rates determined by both K_{s_1} and $\overline{K}_{s_1}$ and the physiological substrate concentrations under resting or certain steady-state conditions (see Reference 285 and references cited therein). Curiously, the hybrid type (which contains one muscle type and one brain type of polypeptide chain) does not lie precisely intermediate in its kinetic behavior between the muscle type and brain type, but in some respects resembles more the brain type. Compare the approximately six- to sevenfold decrease in values of K_4 ($K_{MgADP^{2-}}$) over K_1 ($\overline{K}_{MgADP^{2-}}$) and K_3 ($K_{creatine}$) over K_2 ($\overline{K}_{creatine}$) in the forward direction; however, there is only a three- to fourfold decrease in values in the reverse direction of K_8 (K_{MgADP^-}) over K_7 ($\overline{K}_{MgADP^-}$) and K_9 ($K_{Cr\sim P^{2-}}$) over K_6($\overline{K}_{Cr\sim P^{2-}}$). Values for K_3 ($K_{creatine}$), K_4 ($K_{MgADP^{2-}}$), and K_8 (K_{MgADP^-}) for the hybrid type are almost identical, or very similar, to those values for the brain type. These similarities between brain and hybrid type of isoenzymes possibly illustrate and effect of interactions between catalytically active but nonidentical polypeptide chains.

In Table 3,[300] values for the derived kinetic parameters at pH 8.8 (30°C) are described for the three normal human isoenzymes of creatine kinase[286] and their dystrophic counterparts, i.e., the five isoenzymes isolated from Duchenne dystrophic human tissues,[301] for seven enzyme species, namely, for the normal and dystrophic muscle types from the muscle tissue,

TABLE 2
Values for the Derived Kinetic Parameters at pH 8.8 (30°C) for the ATP-Creatine Transphosphorylase (Creatine Kinase) Isoenzymes of the Calf[299]

Derived kinetic parameters	Defined equilibrium	Defined intrinsic dissociation constant	Brain type	Hybrid type (M)	Muscle type (M)
K_1	$E \cdot MA \rightleftarrows E + MA$	$(\overline{K}_{MgATP^{2-}})$[a]	0.93 (±0.14) × 10^{-3} M	0.77 × 10^{-3}	0.97 (±0.18) × 10^{-3}
K_2	$E \cdot B \rightleftarrows E + B$	$(\overline{K}_{creatine})$	2.9 (±1.1) × 10^{-2} M	2.9 × 10^{-2}	5.3 (±3.8) × 10^{-2}
K_3	$E \cdot MA \cdot B \rightleftarrows E \cdot MA + B$	$(K_{creatine})$[b]	3.7 (±0.4) × 10^{-3} M	4.3 × 10^{-3}	2.1 (±0.4) × 10^{-2}
K_4	$E \cdot MA \cdot B \rightleftarrows E \cdot B + MA$	$(K_{MgATP^{2-}})$	1.3 (±0.45) × 10^{-4} M	1.2 × 10^{-4}	7.8 (±5.6) × 10^{-4}
K_6	$E \cdot D \rightleftarrows E + D$	$(\overline{K}_{Cr\sim P^{2-}})$	2.0 (±0.6) × 10^{-2} M	3.7 × 10^{-2}	4.5 (±0.42) × 10^{-2}
K_7	$E \cdot MC \rightleftarrows E + MC$	$(\overline{K}_{MgADP^{-}})$	1.2 (±0.1) × 10^{-4} M	0.98 × 10^{-4}	1.7 (±0.22) × 10^{-4}
K_8	$E \cdot MC \cdot D \rightleftarrows E \cdot D + MC$	$(K_{MgADP^{-}})$	1.0 (±0.3) × 10^{-5} M	3.2 × 10^{-5}	9.4 (±0.44) × 10^{-5}
K_9	$E \cdot MC \cdot D \rightleftarrows E \cdot MC + D$	$(K_{Cr\sim P^{2-}})$	2.0 (±0.1) × 10^{-3} M	1.1 × 10^{-2}	2.3 (±0.63) × 10^{-2}
$\frac{V^f_{max}}{E_t} = k_{+5}$			2.5 (±0.1) × 10^2 µmol min^{-1} mg^{-1}	1.1 × 10^2 µmol min^{-1} mg^{-1}	2.2 (±0.025) × 10^2 µmol min^{-1} mg^{-1}
$\frac{V^r_{max}}{E_t} = k_{+5}$			8.8 (±0.1) × 10^1 µmol min^{-1} mg^{-1}	1.2 × 10^2 µmol min^{-1} mg^{-1}	1.3 (±0.42) × 10^2 µmol min^{-1} mg^{-1}

[a] $\overline{K}_s$ denotes an intrinsic dissociation constant of the particular substrate from the binary complex.
[b] K_s denotes an intrinsic dissociation constant of the particular substrate from the ternary complex.

TABLE 3
Values for the Derived Kinetic Parameters at pH 8.8 (30°C) for ATP-Creatine Transphosphorylases from Normal and Dystrophic Human Tissues[300]

	Muscle type (mol/l)		Hybrid type (mol/l)		Brain type (mol/l)		
	Normal muscle	Dystrophic muscle	Normal preparation	Dystrophic muscle	Normal brain	Dystrophic brain	Dystrophic muscle
K_1 ($\overline{K}_{MgATP^{2-}}$)[a]	6.3 (±0.5) × 10^{-4}	7.0 (±0.9) × 10^{-4}	5.3 (±0.7) × 10^{-4}	8.9 (±2.9) × 10^{-4}	6.6 (±0.4) × 10^{-4}	4.7 (±0.02) × 10^{-4}	3.7 (±0.2) × 10^{-4}
K_2 ($\overline{K}_{creatine\pm}$)	4.9 (±1.4) × 10^{-2}	3.4 (±0.9) × 10^{-2}	5.9 (±4.3) × 10^{-2}	4.6 (±3.4) × 10^{-2}	1.7 (±0.2) × 10^{-2}	1.3 (±0.4) × 10^{-2}	4.8 (±0.3) × 10^{-3}
K_3 ($K_{creatine\pm}$)[b]	1.9 (±0.2) × 10^{-2}	1.4 (±0.003) × 10^{-2}	6.6 (±0.6) × 10^{-3}	4.3 (±2.0) × 10^{-3}	2.5 (±0.1) × 10^{-3}	3.2 (±0.03) × 10^{-3}	1.9 (±0.1) × 10^{-3}
K_4 ($K_{MgATP^{2-}}$)	2.6 (±0.8) × 10^{-4}	2.8 (±0.06) × 10^{-4}	1.3 (±0.9) × 10^{-4}	2.1 (±1.7) × 10^{-4}	0.99 (±0.11) × 10^{-4}	1.3 (±0.3) × 10^{-4}	1.5 (±0.1) × 10^{-4}
K_6 ($\overline{K}_{Cr\sim P^{2-}}$)	7.2 (±1.0) × 10^{-2}	4.6 (±0.4) × 10^{-2}	3.5 (±1.3) × 10^{-2}	4.1 (±1.2) × 10^{-2}	1.1 (±0.2) × 10^{-2}	1.4 (±0.6) × 10^{-2}	0.56 (±0.11) × 10^{-2}
K_7 ($\overline{K}_{MgADP^{-}}$)	1.8 (±0.3) × 10^{-4}	1.3 (±0.03) × 10^{-4}	1.4 (±0.003) × 10^{-4}	1.3 (±0.1) × 10^{-4}	1.4 (±0.01) × 10^{-4}	1.1 (±0.1) × 10^{-4}	1.4 (±0.03) × 10^{-4}
K_8 ($K_{MgADP^{-}}$)	5.9 (±2.3) × 10^{-5}	6.8 (±0.6) × 10^{-5}	1.4 (±0.6) × 10^{-5}	1.1 (±0.3) × 10^{-5}	2.2 (±0.6) × 10^{-5}	1.7 (±0.7) × 10^{-5}	5.1 (±1.1) × 10^{-5}
K_9 ($K_{Cr\sim P^{2-}}$)	3.2 (±2.0) × 10^{-2}	2.3 (±0.06) × 10^{-2}	3.0 (±0.2) × 10^{-3}	3.1 (±0.3) × 10^{-3}	1.9 (±0.003) × 10^{-3}	1.7 (±0.3) × 10^{-3}	2.0 (±0.1) × 10^{-3}
V^f_{max}/E_t	202 (±20) μmol min^{-1} mg^{-1}	169 (±13) μmol min^{-1} mg^{-1}	159 (±5) μmol min^{-1} mg^{-1}	117 (±32) μmol min^{-1} mg^{-1}	168 (±2) μmol min^{-1} mg^{-1}	145 (±4) μmol min^{-1} mg^{-1}	147 (±5) μmol min^{-1} mg^{-1}
V^F_{max}/E_t	147 (±38) μmol min^{-1} mg^{-1}	124 (±2) μmol min^{-1} mg^{-1}	87 (±6) μmol min^{-1} mg^{-1}	79 (±2) μmol min^{-1} mg^{-1}	47 (±1) μmol min^{-1} mg^{-1}	42 (±2) μmol min^{-1} mg^{-1}	55 (±2) μmol min^{-1} mg^{-1}

[a] $\overline{K}_s$ denotes an intrinsic dissociation constant of the particular substrate from the binary complex.
[b] K_s denotes a dissociation constant of the particular substrate from the ternary complex.

for the normal prepared hybrid type and the dystrophic hybrid type from the muscle tissue, for the normal brain type, dystrophic brain type from the brain tissue, and for the dystrophic brain type from the muscle tissue. For each of the derived kinetic parameters (K_1 to K_9, V^f_{max}, and V^r_{max}) estimates of their standard deviations for 95% confidence limits are provided. In the case of the muscle types only the value of K_6 (i.e., $\overline{K}_{Cr\sim P^{2-}}$) appears significantly lower for the dystrophic muscle type compared to its normal counterpart. For the hybrid types, V^f_{max} appears significantly lower for the dystrophic isoenzyme (from the muscle) than its normal prepared counterpart; for all other kinetic parameters there are no other real significant differences. In the case of the brain types, (1) V^f_{max} is significantly higher for the normal than the other two dystrophic brain types (from the brain and from the muscle), (2) K_1 ($\overline{K}_{MgADP^{2-}}$) is significantly lower for both the dystrophic brain type from the muscle and from the brain compared to the normal brain type, (3) K_2 ($\overline{K}_{creatine^{\pm}}$), K_8 (K_{MgADP^-}), in the case of the brain type from dystrophic muscle, differ significantly from the normal brain type, and K_6 ($\overline{K}_{Cr\sim P^2}$), K_4 ($K_{MgADP^{2-}}$), and K_3 ($K_{creatine^{\pm}}$) differ slightly. However, in general, there are no real major differences discernible by this kinetic analysis between the normal and the dystrophic counterpart, and only between the normal and dystrophic isoenzymes of the brain type are significant differences to be found, although minor in nature. In this connection, it is of interest to note that relatively few differences, if any, were detected by physical-chemical means[288] between the normal and dystrophic muscle types; however, a few significant differences did manifest themselves in the case of the brain type.

Interestingly, for the normal human brain type creatine kinase, at pH 8.8, the intrinsic dissociation constants for dissociation of the substrate from the ternary complex (termed $K_{substrate}$) are far smaller than the dissociation constant of the same substrate from its binary complex (termed $\overline{K}_{substrate}$), a phenomenon also found in the case of the calf brain ATP-creatine transphosphorylase[49,299] (see Tables 1 and 2). Thus, compare the appropriate values for creatine$^{\pm}$ (K_3 vs. K_2), or for $MgADP^-$ (K_8 vs. K_7), or for $Cr \sim P^{2-}$ (K_8 vs. K_7), all of which show decreases by approximately six- to sevenfold. Much smaller differences between K_s and $\overline{K}_s$ are manifested by the normal human muscle-type enzyme (with only $2^1/_2$- to 3-fold changes in intrinsic constants), e.g., compare the data in Table 2 for $Cr \sim P^{2-}$ (K_9 vs. K_6); a similar situation was apparent for the calf muscle creatine kinase (see Table 2). It would appear, in general, that the addition of the second substrate to the enzyme binary complex greatly increases the binding constant for the brain-type enzyme; however, it is only slightly increased for the muscle-type enzyme under the same conditions. Curiously, the normal human hybrid-type isoenzyme of creatine kinase does not lie precisely intermediate in its kinetic behavior (compare also the calf hybrid-type creatine kinase, Table 2 above), but in some respects resembles more the brain type, e.g., compare in Table 3 the 12- to 13-fold decrease in values of K_9 ($K_{Cr\sim P^{2-}}$) over K_6 ($\overline{K}_{Cr\sim P^{2-}}$), which again might possibly illustrate the effect of interactions between catalytically active but nonidentical polypeptide chains.

To return to a discussion of the kinetics of the calf isoenzymes of creatine kinase, in addition to the results presented above for pH 8.8 (see Tables 1 and 2), kinetic measurements were also conducted at pH 7.4 and 30°C. These results are shown in Table 4[49] for the calf muscle-type and brain-type isoenzymes. For the calf brain-type creatine kinase, at pH 7.4, there are some quantitative differences in the intrinsic constants as well as a relatively large increase (almost one order of magnitude) in V^r_{max} as compared to pH 8.8 (see Tables 2 and 4). Thus, kinetically operable ionization constant or constants or pH-related conformational changes (or both) in the ternary complex, may have to be entertained in the final analyses of its kinetic mechanism (see below for a further discussion of this point). The gathering of these data at pH 7.4[49,299] as well as at pH 8.8 had been prompted by two reasons. (1) The brain enzyme displayed a surprisingly low value of V^r_{max} at pH 8.8 compared to V^f_{max} (Table 1) and which might therefore be the result of an ionizable function kinetically operable in the ternary complex (i.e., $H^+ \cdot E \cdot MC \cdot D$). (Parenthetically, one may note again that this

TABLE 4
Values for the Derived Kinetic Parameters at pH 7.4 (30°C) for ATP-Creatine Transphosphorylase (Creatine Kinase) from Calf Brain and Calf Muscle[49,299]

Derived kinetic parameter	Defined intrinsic dissociation constant	Calf brain type (*M*)	Calf muscle type (*M*)
K_1	$\bar{K}_{MgATP^{2-}}$ [a]	1.8 (±0.3) × 10^{-4}	2.8_4 (±0.24) × 10^{-4}
K_2	$\bar{K}_{creatine}$	1.3 (±0.5) × 10^{-2}	3.5 (±0.35) × 10^{-2}
K_3	$K_{creatine}$ [b]	3.0 (±0.4) × 10^{-3}	1.6_5 (±0.05) × 10^{-2}
K_4	$\bar{K}_{MgATP^{2-}}$	4.7 (±1.5) × 10^{-5}	1.3 (±0.06) × 10^{-4}
K_6	$\bar{K}_{Cr\sim P^{2-}}$	6.7 (±1.4) × 10^{-4}	4.6 (±0.05) × 10^{-3}
K_7	$\bar{K}_{MgADP^{-}}$	3.9 (±0.3) × 10^{-5}	4.1 (±0.32) × 10^{-5}
K_8	$K_{MgADP^{-}}$	3.0 (±0.3) × 10^{-5}	1.7 (±0.37) × 10^{-5}
K_9	$K_{Cr\sim P^{2-}}$	5.2 (±1.2) × 10^{-4}	1.8 (±0.03) × 10^{-3}
$V^f_{max}/E_t = k_{+5}$		1.5 (±0.2) × 10^2 μmol min^{-1} mg^{-1}	1.5 (±0.01) × 10^2 μmol min^{-1} mg^{-1}
$V^r_{max}/E_t = k_{-5}$		6.4 (±0.5) × 10^2 μmol min^{-1} mg^{-1}	3.3 (±0.075) × 10^2 μmol min^{-1} mg^{-1}

[a] $\bar{K}_s$ denotes an intrinsic dissociation constant of the particular substrate from the binary complex.

[b] K_s denotes an intrinsic dissociation constant of the particular substrate from the ternary complex.

feature of a relatively low value of V^r_{max} at pH 8.8 seems to be characteristic of the brain-type isoenzyme in contrast to the muscle-type creatine kinase isoenzyme — see Table 2 for the calf isoenzymes and Table 3 for the human isoenzymes of creatine kinase.) (2)The general lability of the brain-type enzyme[286,287,289,297] might be reflected in pH-dependent conformational changes which in turn might lead to secondary changes in the kinetic parameters.

Since that time, a number of amino acid residues have been implicated in the substrate(s) binding sites and catalysis.[856] Thus, a *reactive* sulfhydryl group had been implicated in the catalytic mechanism as a result of complete enzyme inactivation by iodoacetamide,[309,310] iodoacetic acid,[289,311] 2,4-dinitrofluorobenzene[311] or 5,5′-dithiobis (2-nitrobenzoic acid).[289] Smith and Kenyon,[312] however, have inferred from their studies with the smaller sulfhydryl reagent, methyl methane thiosulfonate, that the sulfhydryl group is a nonessential group. An arginine residue was placed at the nucleotide binding site by chemical modification studies with diacetyl[313] and by nuclear Overhauser studies of the adenine protons of ADP.[314,315] The quenching of the protein fluorescence by nucleotides[315] also implicated one or more aromatic residues (e.g., tryptophan) at the nucleotide binding site. At the creatine binding site, evidence for a carboxylate residue was obtained from a study of the effect of pH on the kinetic parameters of creatine and phosphocreatine[316] and by affinity labeling with epoxycreatine.[317] The sequence around a reactive lysine residue was obtained by use of a bifunctional reagent, 1,5-difluoro 2,4-dinitrobenzene, to link to the ''essential thiol'', and was shown to be asp-ser-pro-val-*lys*-leu-leu-phe-leu.[318] A lysyl residue was also suggested to be at the reaction center and to interact with the transferred phosphoryl group on the basis of nuclear Overhauser measurements of what the authors called a ''transition-state analogue complex'', that is, enzyme MgADP·formate·creatine.[319] Chemical modification studies with diethylpyrocarbonate had implicated one or more histidines in the catalysis.[321,337] Cook et al.,[316] from studies of pH on the initial rate, also suggested that a histidine residue functioned at the reaction center as a general acid-base catalyst to protonate phosphocreatine and to deprotonate creatine during the phosphoryl transfer reactions. This suggestion is mindful of the original hypothesis of Rabin and Watts[322] who had invoked an imidazole group at the active site, hydrogen bonded to an S atom of an adjacent cysteinyl residue. Perhaps a reaction with the molecularly small sulfhydryl reagents of Kenyon[312] would not completely prevent interaction of the reactive -SH sulfur of Watts[309,310] to any adjacent imidazole group, whereas, the bulky alkylating agents would.[289,299-311] Mildvan et al.[323] reported recently on an NMR study of the 16 histidine residues per subunit of rabbit muscle creatine kinase as a function of pH and substrate concentration, and in response to the racemic, stable, paramagnetic metal-nucleotide complex, β,γ-bidentate Cr^{3+} ATP (the chromium ATP complex of Cleland[324]; from these observations, Mildvan et al.[323] implicated three histidine residues at or near the active site. With the primary sequence of the rabbit muscle creatine kinase finally obtained[848] the way is now open for the final identification of those residues involved in substrate binding and in the catalytic mechanism of creatine kinase.[856,981]

In Table 4, where the derived kinetic parameters at pH 7.4 (30°C) for the calf muscle-type and brain-type creatine kinase are presented, we note the quantitative differences in the intrinsic constants at pH 7.4 compared to pH 8.8 (Tables 1 and 2, as well as the relatively huge increase of almost one order of magnitude) in V^r_{max} for the reverse direction, as compared to the values for pH 8.8. Thus, although it now appears likely *(vide supra)* that there exist kinetically operable ionization constants,[309,323] pH-related conformational changes in the ternary complexes also require consideration, and steady-state kinetic data alone (e.g., see Tables 2 and 4) do not provide an unambiguous assignment of either of these two factors. However, the data of Okabe et al.[289] on the chemical reactivity of the -SH group of the calf brain vs. the calf muscle creatine kinase as a function of pH do provide additional support for pH-dependent conformational changes in the calf brain creatine kinase. Thus, the mixed disulfide (TNB-protein) (TNB, 5-thio-2-nitrobenzoate) formed by reaction between DTNB (5,5′-dithiobis 2-nitrobenzoic acid) and the two normally reactive sulfhydryl groups per mole (or one per subunit), could be prepared from both calf isoenzymes at 0°C. Although the TNB-muscle protein appeared to be stable between pH 5.5 and pH 8.5, the TNB group could be released in a pseudo-first order fashion from the brain isoenzyme derivative, under mildly acidic conditions, and in the absence of any external reductant (see Figure 2). It should be noted that in contrast to the calf muscle isoenzyme, with a total of eight sulfhydryl groups per mole, or four per subunit, there are ten sulfhydryl groups per mole, or five per subunit, in the brain isoenzyme. At 30°C ionic strengths above 0.2, the kinetics of TNB release seemed to be approximately first order (0.85 measured) in H^+ over a narrow pH range of 5.5 to 6.5, as shown in Figure 2. This H^+-initiated process presumably is the result of a pH-dependent unfolding which leads to a thiol-disulfide reaction between the mixed disulfide and a normally unreactive or inaccessible sulfhydryl group within the brain isoenzyme protein. In Figure 2 the intersection of the straight-line portions of the graph would lead to a pK′ of about 6.5 at 30°C for this conformationally operable ionizable group, provided electrostatic interactions did not alter the pK′ values involved. One candidate for this group would be an imidazolium ion[74] and it would be tempting to consider the possibility of a hydrogen bond between this ''sheltered'' type of sulfhydryl group and a suitable base, the latter being the site for the proton attack. In summary, the simplest explanation of the data in Figure 2 would have a rate-limiting H^+-promoted or -induced conformational change which would bring about a geometric transfer of a normally unreactive sulfhydryl group, or groups, to the vicinity of the disulfide so as to permit their reaction with the final liberation of the TNB mercaptide ion. This relatively labile enzyme structure in the brain enzyme is in contrast to the relatively stable muscle enzyme, over the same pH-range and temperature, and it does not show a pH-dependent release of the TNB group from its mixed disulfide (see Figure 2). To return to the data at pH 7.4,[300] which are summarized in Table 4 in terms of calculated values for the derived kinetic parameters for the calf brain and muscle isoen-

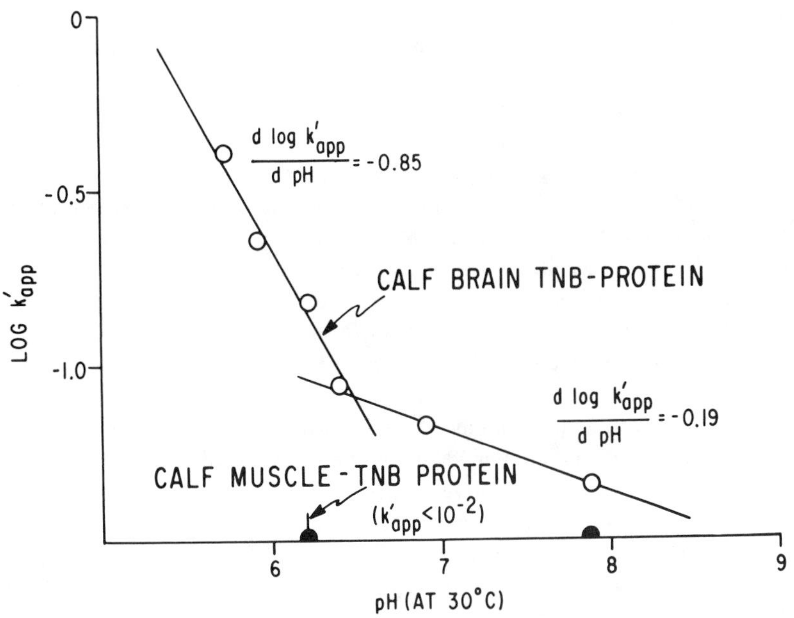

FIGURE 2. pH dependence of the kinetics of release, at 30°, of 2-nitro-5-thiobenzoate (TNB) from the mixed disulfide, TNB-protein. ○---○, $\log_{10} k'_{app}$, the logarithm of the apparent first order rate constants for TNB release plotted as function of pH (measured at 30°, after the reaction) in reaction mixtures containing initially calf brain TNB-protein$_0$, 2.99×10^{-6} *M*, and several 0.05 *M* imidazole-0.2 *M* KCl buffers, at various pH values (either imidazole-acetic acid, pH 5.7 to 6.2, or imidazole-HCl from pH 5.9 to 7.9). ●---●, two points obtained with the calf muscle TNB-protein, with k'_{app} values $<10^{-2}$ min^{-1}, at ph 6.2 and pH 7.9, i.e., displaying a relatively stable behavior within this pH range.[289]

zymes, it may be noted that whereas V^f_{max} for the calf brain isoenzyme is within 60% of the value at pH 8.8 (see Table 1) V^r_{max} has increased at pH 7.4 over 7-fold to a value of 640 μmol (min)$^{-1}$ (mg)$^{-1}$ or to 27,250 mol (min)$^{-1}$ (mol catalytic sites)$^{-1}$ (assuming that the two polypeptide chains in the dimer of molecular weight 85,200[287] contain two catalytic sites). It may be remarked, parenthetically, that this velocity would be more than sufficient to account for its physiological role in the central nervous system, that is, that of a rapid resynthesis of ATP_0 from ADP_0 and Cr ~ P_0,[325,326] and that the kinetic parameters as given in Table 4 make the brain enzyme, in contrast to the muscle enzyme, fairly well adapted to the brain milieu and attendant steady-state substrate concentrations. This is especially striking in the lower values for $K_{\quad 2-}$ and $K_{creatine}$ in the case of the brain enzyme, and note the surprisingly large change (decrease) of almost 30-fold in the values of $\overline{K}_{Cr \sim P}$ (K_6) at pH 7.4 (Table 4) vs. pH. 8.8 (Table 1). It is of further interest to note here that the relatively large differences in the values of K_{s_1} vs. $\overline{K}_{s_1}$ at pH 8.8 (ranging from 7- to 12-fold increases; see Table 1 for the brain isoenzyme and all four substrates) have decreased at pH 7.4 (see Table 4) to much smaller differences which range only from an insignificant 1.3-fold change in values for the reverse reaction constants to about 4-fold change in the difference between the forward reaction constants. If the above interpretation at pH 8.8 has merit, i.e., involving substrate pair-induced conformational changes occurring within the ternary complexes (so as to enhance the binding of each substrate in the ternary complex compared to the binary complex), then the argument would likely be extended to another pH-dependent conformational change of the brain isoenzyme itself, so as to permit a tightening of the protein structure at pH 7.4 vs. pH 8.8 and near the substrate-binding sites, i.e., to a more rigid structure at pH 7.4 which would better resist environmental influences and which would retain a structure similar to the final kinetically active ternary complex. Thus, the pH-dependent conformational change would lead, at pH 7.4, to smaller or nonexistent cooperative substrate effects by having the substrate-induced conformational changes overshadowed by this pH-dependent conformational change (see Table 4 and compare Table 2); this facet of the brain enzyme kinetics at pH 7.4 (that is, K_{s_1} values approaching $\overline{K}_{s_1}$) therefore approaches muscle enzyme kinetics.

One final argument to support this point of view may be seen in the ADP^{3-} inhibition constants. As indicated above, the inhibition by ADP_0 is much more pronounced at pH 7.4 compared to pH 8.8. If analyses by curve fitting and successive approximations are conducted in the same manner as above, for K'_7, K'_8, and K'_9, highly tentative values of 0.3×10^{-4}, 0.8×10^{-4}, and 1.6×10^{-3} *M*, respectively, are obtained. In contrast to the case for pH 8.8, at pH 7.4 the values of $K_{ADP^{3-}}$ (K'_8) for the abortive ternary complex and K_{MgADP^-} (3×10^{-5}) for the active ternary complex do not differ by at least the same order of magnitude to be found at pH 8.8; moreover, the dissociation of $Cr \sim P^{2-}$ from the abortive $E \cdot ADP^{3-}$ $Cr \sim P^{2-}$ ternary complex, i.e., K'_9, and K_9 ($K_{Cr \sim P^{2-}} = 5.2 \times 10^{-4}$) also differ only by a factor of 3-fold, as well as the binding of ADP^{3-} and $MgADP^{3-}$ to the enzyme ($K_{ADP}{}^{3-}$ and K_{MgADP^-}) which approach similar values within a factor of 3-fold (K'_7 vs. K_7) at pH 7.4, provided these very highly tentative values derived by curve fitting may be taken as significant. In any case, the enhanced inhibition by ADP_0 is explainable qualitatively by the decreased values for K'_7, K'_8, and K'_9, at pH 7.4 vs. pH 8.8, relative to the values of K_6 to K_9. Such enhancement in binding perhaps could be the result of the pH-dependent conformational change in the protein so as to permit tighter binding at pH 7.4 vs. pH 8.8 of the uncomplexed ADP^{3-} (vs. $MgADP^-$) to the enzyme and also within the abortive ternary complex containing $Cr \sim P^{2-}$. As far as the calf muscle-type enzyme is concerned, a comparison of the kinetics at pH 7.4 (Table 4) vs. pH 8.8 (Table 2) reveals that (1) the values of K_{s_1} vs. $\overline{K}_{s_1}$, for all of the four substrates, are well within 2- to 2.5-fold of each other, or even smaller changes at pH 7.4 than at 8.8, (2) V^f_{max} has decreased only by 32%, but V^r_{max} has increased 2.6-fold in going from pH 8.8 to 7.4, (3) K_1 and K_4 ($\overline{K}_{MgADP^{2-}}$ and $K_{MgADP^{2-}}$) have decreased 3.4- to 5.8-fold, respectively, at pH 7.4 vs. 8.8; however, K_2 and K_3 (for $\overline{K}_{creatine}$ and $K_{creatine}$) show essentially no significant changes between pH 8.8 and 7.4, and (4) for the reverse direction, K_7 and K_8 (for $\overline{K}_{MgADp^-}$ and K_{MgADP^-}) show a 4.1- to 5.5-fold decrease, respectively; however, K_6 and K_9 (for $\overline{K}_{Cr \sim P^{2-}}$ and $K_{Cr \sim P^{2-}}$) show a most dramatic change of a 10- to 12-fold decrease in going from pH 8.8 to pH 7.4. A one-order of magnitude decrease in the intrinsic dissociation constants for creatine phosphate would indicate the titration of an ionizable group in the calf muscle isoenzyme for substrate binding with a pK′ at least within one pH unit of 7.4.

In summary, then, it is highly probable that both kinetically operable ionization constants within the pH region of 7 to 9 and superimposed pH-related conformational changes in the ternary complex may have to be considered in the final analysis of the kinetic mechanism of the calf brain isoenzyme (kinetic data are obtained at only pH values of 8.8 and 7.4, and, for such a complex system, are insufficient to make a final assignment of these ionization constants). Moreover, cooperative substrate-binding effects at pH 8.8 are likely the result of paired substrate-induced conformational changes in the ternary complex containing the magnesium nucleotide and guanidine base substrate. However, much smaller effects are to be observed at pH 7.4, which is very likely the result of a general tightening of the tertiary structure of this unusual protein which appears easily altered in its properties by environmental influences. That the native two-chain brain enzyme[287] contains a "looser" or less restrained geometric structure than the calf muscle isoenzyme seems evident from (1) the kinetics of -SH titrations under denaturing conditions,[287,289,297] (2) the possibility of the relatively large conformational changes discussed above which may occur during kinetic formation of the ternary complexes and relatively smaller ones to be found in the calf muscle enzyme (e.g., see Table 2), and (3) the general lability of the enzyme, e.g., under treatments at acidic and alkaline pH values,[287,297] especially under electrophoretic conditions.[287] This brain enzyme may then contain within it the direct means of studying conformationally dependent processes (and hinted at from these kinetic analyses), and perhaps interactions between catalytic sites, presumably one per subunit for the noncovalently linked two polypeptide chain structure,[287] whereas such studies seem difficult indeed in the case of the more "tightly knit" and stable rabbit muscle enzyme structure.[294] Also, in the case of the more refractory rabbit muscle enzyme, efforts to observe directly, by several physical means, conformational changes accompanying substrate-binding or substrate-dependent conformational changes, have not been too successful or revealed only relatively small or minor changes, e.g., in sedimentation coefficients, hydrogen-deuterium exchange, intrinsic viscosity, optical rotatory dispersion,[327-329] or in the lack of spectrophotometric changes accompanying binding of creatine.[330] Such postulated conformational changes for the rabbit muscle enzyme, accompanying substrate binding and suggested from the steady-state kinetics,[83,331,332] have been deduced only indirectly, e.g., from alterations in reactivity of protein derivatization reagents when in the presence of substrates or equilibrium mixtures of substrates,[327,329,333-335] from susceptibility to tryptic digestions,[327,339] from antienzyme inhibition studies[336] or from solvent proton nuclear magnetic resonance (NMR) relaxation rates in the presence of paramagnetic ions such as Mn^{+2}.[334,341] It is of interest in this regard that in a kinetic study on the rabbit muscle enzyme[338] by temperature-jump means (to be discussed in Chapter 10 and References 849 to 851) with a pH indicator to monitor rapid pH changes in the approach to equilibrium after a temperature perturbation, Hammes and Hurst[338] had interpreted their results in terms of conformational changes following the binding of either ADP^{3-} or ATP^{4-} or their metal complexes to the enzyme; whereas, no conformational change could be assigned to the binding of creatine or creatine phosphate. Any conformational changes (or isomerizations[338]) in the ternary complex or complexes could not be precisely defined[338] because of the complexity of the relaxation process involved when in the presence of metal nucleotides. Whether there are at least two types of isomerizations or conformational changes, one affecting the reactivity of certain reactive groups (e.g., -SH groups[289]) and the other affecting the pK values of certain ionizable residues possibly involved in the cooperative substrate effects, is equally tenable as discussed by Hammes and Hurst.[338] It is of interest to note in passing that at pH 7.6 and 11°C, Hammes and Hurst[338] could detect no difference between rates of isomerization of the ADP^{3-} · enzyme complex and the metal nucleotide-enzyme complex (as revealed by pH changes) and is thus not in disagreement with the deductions reached here for the calf brain enzyme at pH 7.4. However, it should be remarked that such rapid isomerizations would, in the steady-state, not affect the steady-state analysis based here on a rate-limiting step at the interconversion of the ternary complexes, and would thus not permit any unique assignments of isomerization other than those at the rate-limiting steps.

A kinetic comparison has been made of the brain, the muscle, and hybrid isoenzymes of calf and man (Tables 2 and 3); it is of interest to note that these isoenzymes display subtle differences in their gross physical and chemical properties which seem to be reflected in their kinetics, and their kinetic parameters, for the most part, may be correlated with their steady-state intracellular environments. Also, further insights may be gained into the problem of interactions between like and unlike catalytic sites, minifestly reflected in the hybrid enzyme whose kinetic parameters do not lie precisely intermediate between muscle and brain type, and some additional and general understanding permitted of the overall mechanism of the ATP-creatine transphosphorylases.

It is of interest that Rao and Cohn[342] have recently reported that by using ^{31}P NMR measurements, they attempted to measure the equilibrium constant for the step involving the interconversion of the ternary complexes in the rabbit muscle creatine kinase-catalyzed reaction. In the nomenclature employed here, this would be described by

$$K_5 = \frac{k^{+5}}{k^{-5}} = \frac{V^f_{max}/E_t}{V^r_{max}/E_t}$$

Under these conditions (pH 7.8, 4°C, ~10% D_20, approximately 4-m*M* enzyme (sites) concentration, 0.15 *M* Hepes (K^+) buffer, and initially 3.9 m*M* ATP_t, 3.8 m*M* creatine, and 5.1 mM $Mg(Ac)_2$) the ratio of (E · MgADP · Cr~P)/(E · MgATP ~ Cr) ≃ 1 ± 0.3.

From Table 2, at pH 8.8 (30°C), $\frac{k_{+5}}{k_{-5}}$ for calf muscle type, calf hybrid type, and calf brain type, equals 1.7, 0.92, and 2.8, respectively.

From Table 4, at pH 7.4 (30°C), $\frac{k_{+5}}{k_{-5}}$ = 0.45 and 0.23 for the calf muscle type and brain type, respectively. Thus, at pH 8.8, the ratios of $\frac{k_{+5}}{k_{-5}}$ are within a factor of 2 to 3 of Rao and Cohn's[342] estimate, and within a factor 2 to 4 at pH 7.4 for all three isoenzymes; however, at 30°C vs. the 4°C, used for Rao and Cohn's determinations; a temperature difference which would certainly affect the equilibrium.

In an interesting application of ^{31}P NMR techniques applied to the perfused rat heart at 37°C, the kinetic constants were obtained for the phosphate exchange and relaxation rates of the transferred phosphate in both substrates.[870] Their results indicated that the creatine kinase reaction in the rat perfused heart is at equilibrium and that its rate is not limited by the diffusion of its substrates between different locations of the several isoenzymes in the heart. Therefore, they could find no indication of compartmentation of the substrates of the creatine kinase reaction.

B. ILLUSTRATED BY THE ADENYLATE KINASES FROM RABBIT AND CALF MUSCLE AND FROM CALF LIVER

At this point, it is pertinent to turn to a discussion of an analogous enzyme (another kinase), that is, ATP-AMP transphosphorylase, or adenylate kinase, which catalyzes the reversible reaction $ATP_t + AMP_t \overset{Mg^{2+}}{\rightleftarrows} 2ADP_t$. It is very likely, as will be described below, that the enzyme actually catalyzes the following reaction:[50]

$$MgATP^{2-} + AMP^{2-} \rightleftarrows MgADP^{-} + ADP^{3-} \qquad (94)$$

where the active substrates are the magnesium complexes of ATP and ADP, and where uncomplexed ADP and AMP may act as the other pair of substrates. Thus, adenylate kinase appears to catalyze a reaction between two forms of ADP, that is, between ADP^{3-} and $MgADP^-$ in the reverse reaction.[50]

The kinetic properties of the adenylate kinases, from several sources, have been the subject of a number of investigations (see Reference 343 for a summary), with the first definitive attempt by Noda on the rabbit muscle myokinase.[344] However, because in the early investigations metal complexation by the nucleotide substrates were not quantitatively considered, conflicting interpretations of the steady-state data were often encountered[345,346] By a careful study of the isotope exchange rates at equilibrium (a technique to be described in Chapter 6), Rhoads and Lowenstein[347] were able to exclude an ordered mechanism as well as a mechanism involving enzyme covalent intermediates for the rabbit muscle myokinase. Moreover, they added support for the hypothesis[346] that there were two binding sites for the myokinase — one for the magnesium complexes of the nucleotides and one for the uncomplexed nucleotides.[348] Similarly, Su and Russell concluded that although the yeast adenylate kinase[349] followed a random mechanism, a rate-limiting step existed at the interconversion of the ternary complexes, whereas Rhoads and Lowenstein[347] felt that phosphoryl group transfer was not at the site of the rate-limiting reaction. This conclusion of Rhoads and Lowenstein[347] seemed to be confirmed by ^{31}P NMR studies on the porcine muscle adenylate kinase.[350] To add to the conflict in mechanistic interpretations, Markland and Wadkins[351] had tentatively interpreted their isotope exchange studies on the bovine mitochondrial adenylate kinase as consistent with an ordered mechanism. More recently, in a systematic study on two genetic variants of the human erythrocyte adenylate kinase, with a careful examination of the effects of the metal complexes of the substrates on the reaction velocity, i.e., of $MgATP^{2-}$ and $MgADP^{-}$, Brownson and Spencer[352] concluded that a rapid equilibrium random mechanism was followed by these enzymes with a rate-limiting step at the interconversion of the ternary complexes, but for a case without independent binding of the substrates.

Since both isoenzymes of adenylate kinase from the calf were recently isolated in crystalline form, that is, the muscle type and liver type,[29] an opportunity presented itself to compare them and the rabbit myokinase kinetically.[50] Later, (in Chapter 8) the structure-equilibrium substrate binding relationships for the muscle type (rabbit and calf) will be presented[348] which will complement and support the conclusions reached kinetically.[50,964]

Hamada and Kuby,[50] in their treatment of the data, made calculations of $MgATP^{2-}$ and AMP^{2-} for the forward reaction ($MgATP^{2-} + AMP^{2-} \longrightarrow$) by employing the set of approximate conservation equations.

$$ATP_0 \simeq MgATP^{2-} + ATP^{4-} + HATP^{3-} + MgHATP^{-} + KATP^{3-} + NaATP^{3-}$$

$$AMP_0 \simeq MgAMP + AMP^{2-} + HAMP^{-} + KAMP^{-} + NaAMP^{-}$$

$$Mg_0 \simeq Mg^{2-} + MgATP^{2-} + MgHATP^{-} + MgAMP$$

$$Na_0 \simeq Na^{+} + NaATP^{3-} + NaAMP^{-}$$

$$K_0 \simeq K^{+} + KATP^{3-} + KAMP^{-} \qquad (95)$$

Subscript 0 implies total concentrations and appropriate valences are assigned to each of the ionizable or complex species.

For the reverse direction ($MgADP^{-} + ADP^{3-} \rightarrow$), for calculation of $MgADP^{-}$, the set of conservation equations could be adequately approximated by

$$ADP_0 \simeq MgADP^{-} + ADP^{3-} + HADP^{2-} + MgHADP + NaADP^{2-}$$

$$Mg_0 \simeq Mg^{2+} + MgADP^{-} + MgHADP$$

$$Na_0 \simeq Na^{+} + NaADP^{2-} \qquad (96)$$

The values assigned for the chelation and dissociation constants are given in Table 5. The method of calculations was similar to that described in detail for the rabbit muscle ATP-creatine transphosphorylase.[48] Calculations employed a fixed concentration of free Mg^{2+} of 1 m*M* for the kinetic studies on the forward reaction, where either $MgATP^{2-}$ or AMP^{2-} were variable substrates with each substrate varied at fixed concentrations of the complementary substrate. For the reverse direction, where it proved impossible to maintain a fixed concentration of uncomplexed Mg^{2+}, either $MgADP^{-}$ or ADP^{3-} was the variable substrate, with its paired substrate fixed.

As will be seen, under these conditions of a reasonably fixed ionic strength at 25°C and pH 7.4, the steady-state kinetics conformed to first-degree velocity expressions[8] for two substrate reactions. For the quasi-equilibrium random mechanism similar to the one proposed for the calf brain ATP-creatine transphosphorylase[49] and where independent binding was not invoked, the following intrinsic constants may be given as shown in Scheme XI

TABLE 5
Stability Constants of the Various Ionic and Complex Substrate Species[50]

Species	Formation (stability) constant selected (mol/l)	Ref.
$MgATP^{2-}$	6×10^4	48
$HATP^{3-}$	$10^{6.9}$	48
$MgHATP^-$	5.5×10^2	48
$NaATP^{3-}$	12	48
$KATP^{3-}$	11	353
MgAMP	49	372
$HAMP^-$	$10^{5.5}$	353
$NaAMP^-$	3	353
$KAMP^-$	2	353
$MgADP^-$	2.5×10^3	48
$HADP^{2-}$	$10^{6.7}$	48
MgHADP	35	48
$NaADP^{2-}$	6	48

Note: Selected for 25°C and $(\Gamma/2) = 0.16$ to 0.18.

POSTULATED KINETIC MECHANISM OF ATP-AMP TRANSPHOSPHORYLASE (MYOKINASE) FROM RABBIT MUSCLE

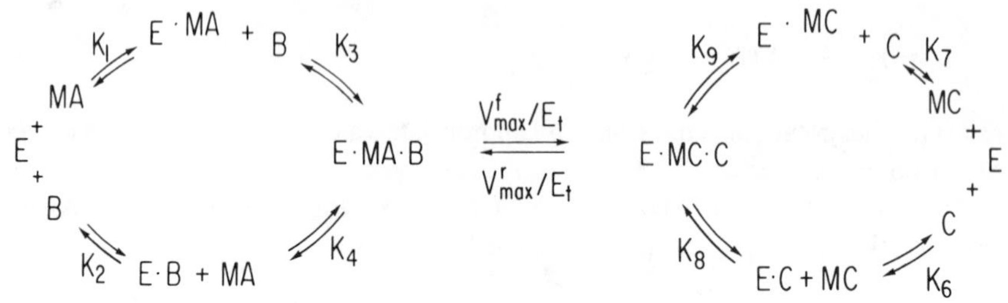

SCHEME XI.

where MA = $MgATP^{2-}$, B = AMP^{2-}, MC = $MgADP^-$, and C = ADP^{3-}. For the limiting cases, where the concentrations of products may be set to zero initially (see text), then for the forward and reverse initial velocities, respectively:

$$v_o^f = V_{max}^f\left[1 + \frac{K_4}{(MA)} + \frac{K_3}{(B)} + \frac{K_1K_3}{(MA)(B)}\right]^{-1} \tag{97}$$

$$K_1K_3 = K_2K_4 \tag{98}$$

$$v_o^r = V_{max}^r\left[1 + \frac{K_8}{(MC)} + \frac{K_9}{(C)} + \frac{K_7K_9}{(MC)(C)}\right]^{-1} \tag{99}$$

$$K_7K_9 = K_6K_8 \tag{100}$$

where v_0^f and v_0^r denote initial velocities for forward and reverse reactions.

For variable $MgATP^{2-}$ (MA) and fixed AMP^{2-} (B), for example, a primary plot of $1/v_0^f$ vs. $1/(MA)$ would yield

$$\frac{1}{v_o^f} = \frac{1}{V_{max}^f}\left(1 + \frac{K_3}{(B)}\right) + \frac{K_4}{(MA)}\frac{1}{V_{max}^f}\left(1 + \frac{K_2}{(B)}\right) \tag{101}$$

and from the appropriate secondary plots of slopes and ordinate-intercepts and relation 98 values K_1 to K_4 may be estimated for the forward reaction. Thus,

$$\underset{(1/v_o \text{ vs. } 1/(MA))}{\text{Slope}} = \frac{K_4}{V_{max}^f} + \frac{K_2K_4}{V_{max}^f}\frac{1}{(B)};$$

$$\text{Y-intercept} = \frac{1}{V_{max}^f} + \frac{K_3}{V_{max}^f}\frac{1}{(B)} \tag{102}$$

Also, for variable (B) and fixed (MA), again all four constants (and V_{max}^f) may be estimated for the forward reaction. Within the experimental errors and calculation uncertainties, the values for the kinetic parameters proved to be the same and average values are presented together with their ranges of uncertainty. Similarly, appropriate secondary plots and relation (Equation 100) yield values for all four constants for the reverse direction.[44,53]

As a measure of self-consistency, estimation of a defined overall equilibrium constant of the system, i.e.,

$$K_{eq} = \frac{(MC)(C)}{(MA)(B)} \tag{103}$$

may be obtained from any of four Haldane relations (see text) and compared with a calculated thermodynamic value from estimated concentrations of total reactants and products at equilibrium.

However, it must be stressed, as it was before in the case of the calculated thermodynamic equilibrium constant for the ATP-creatine transphosphorylase catalyzed reaction,[48] that the absolute values of the kinetic parameters hinge to some degree on the values selected for intrinsic dissociation and metal complexation constants; moreover, since all values for kinetic parameters in principle represent derived values, the numerical estimation contain within themselves some inherent uncertainties. The probable errors listed below in the tables will also contain within them an estimate of this uncertainty.

1. Kinetic Analysis

Earlier studies on the rabbit muscle enzyme[347] by isotope exchange studies at equilibrium clearly ruled out an ordered mechanism and suggested a random mechanism with the formation of possible "dead-end" complexes containing ADP^{3-} in the forward reaction and AMP^{2-} in the reverse reaction with a rate-limiting step somewhere at the dissociation of the binary complexes. However, rigorous product inhibition studies were carried out only by Brownson and Spencer[352] on the human erythrocyte enzymes and were clearly consistent with a rapid equilibrium random mechanism with a rate-limiting step at the interconversion of the ternary complexes. One notes that the isotope exchange data had been carried out at low ionic strengths of approximately 0.02 at pH 8;[343] however, measurements of initial velocities were either at ionic strengths of approximately 0.06 (reverse reaction) or at approximately 0.14 (forward reaction)[347] or at approximately 0.17 (reverse reaction).[352] In this study at fixed ionic strengths of 0.16 to 0.18 and 25°C, the data presented for both calf isoenzymes and for the rabbit muscle enzyme would tend to support the proposed mechanism of Reference 352. Furthermore, binary enzyme-substrate complexes for forward or reverse reactions have been demonstrated and measured for all three enzymes (rabbit muscle, calf muscle, and calf liver adenylate kinases).[348] Therefore, the possibility of any ordered mechanism could again be considered unlikely if it presumed that binary enzyme-substrate complexes become insignificant in the steady-state. Consequently, the kinetic analysis which evolved was that of the quasi-equilibrium random mechanism as presented in Scheme XI. Additional confidence in this mechanism for the case of the calf isoenzymes and for the rabbit muscle enzyme may be gained in the future when, it is hoped, additional product inhibition may be obtained; these data alone will not exclude the possibility of a random mechanism in which product release is rate limiting.[347,350]

However, regardless as to any assumptions made of "dead-end" complexes in the reaction mechanism, it may now be shown from overall velocity expressions that the methods of graphical analysis employed above will yield valid measurements of the basic kinetic parameters.

Independent of whether "dead-end" complexes are formed, the overall velocity expression for either case indicates that the limiting expressions when both products approach zero (i.e., by initial velocity studies), for either the forward or reverse reaction, reduce to Equations 97 and 99; however, the product inhibition patterns will be altered. Thus, consider the mechanism as given above (Scheme XI) and without "dead-end" complexes, then

$$v_o^f = \frac{V_{max}^f\left[(MA)(B) - \frac{(MC)(C)}{K_{eq}}\right]}{K_1K_3 + K_3(MA) + K_4(B) + (MA)(B) + \frac{K_1K_3(MC)(C)}{K_7K_9} + \frac{K_1K_3(MC)}{K_7} + \frac{K_1K_3K_8(C)}{K_7K_9}} \tag{104}$$

and $K_1K_3 = K_2K_4$ and $K_6K_8 = K_7K_9$ where K_{eq} is given by Equation 103. In the limits of $(MC) = (C) \rightarrow 0$, Equation 104 reduces to Equation 97 for v_0^f. On the other hand, if either or both dead-end complexes, $E \cdot B \cdot MC$ or $E \cdot MA \cdot C$, could form, then in addition to the reactions involving K_1 to K_9 (inclusive) described above, one may add

$$E \cdot B + MC \overset{K_\alpha}{\rightleftarrows} E \cdot B \cdot MC \overset{K_\beta}{\rightleftarrows} E \cdot MC + B;$$

$$E \cdot MA + C \overset{K_\gamma}{\rightleftarrows} E \cdot MA \cdot C \overset{K_\delta}{\rightleftarrows} E \cdot C + MA$$

SCHEME XII

and the equalities are extended to include

$$\begin{bmatrix} K_\alpha K_2 = K_\beta K_7 \\ K_\gamma K_1 = K_\delta K_6 \end{bmatrix} \tag{105}$$

which leads to the overall velocity expression shown in Equation 106.

$$v_o^f = \frac{V_{max}^f\left[(MA)(B) - \frac{(MC)(C)}{K_{eq}}\right]}{K_1K_3 + K_3(MA) + K_4(B) + (MA)(B) + \frac{K_1K_3(MC)(C)}{K_7K_9} + \frac{K_1K_3(MC)}{K_7} + \frac{K_1K_3K_8(C)}{K_7K_9} + \frac{K_1K_3(MC)(B)}{K_2K_\alpha} + \frac{K_3(C)(MA)}{K_\gamma}} \tag{106}$$

Equation 106 also reduces, in the case where (MC) = (C) $\rightarrow$ 0, to the limiting forward velocity expression given above (Equation 97).

Thus, the limiting velocity expressions for the mechanism presented above under Scheme XI will permit an evaluation of K_1 to K_9 values by suitable measurement of v_0^f and v_0^r as a function of either the metal-nucleotide substrate concentration at various fixed values of the uncomplexed nucleotide substrate, or the converse, provided, of course, that other species of the several possible ionizable and metal complexes of the substrates in solution do not seriously interfere by inhibition or activation. To facilitate this initial set of kinetic analyses[50] a fixed concentration of free Mg^{2+} = 1 m*M* was employed for the forward reaction ($MgATP^{2-}$ + AMP^{2-} $\rightarrow$) at pH 7.4 and 25°C. This will help reduce the concentration of ATP^{4-} to a very small value over most of the ranges of $MgATP^{2-}$ employed. However, at the very highest concentrations of ATP_0 and Mg_0 studied, one species, $MgHATP^-$, may become kinetically significant in value. For the reverse reaction ($MgADP^{1-}$ + ADP^{3-} $\rightarrow$), fixing the free Mg^{2+} concentration was not feasible, and so it may vary over a considerable range in values.

2. Kinetics of the Rabbit Muscle ATP-AMP Transphosphorylase

Systematic measurement at pH 7.4, approximately 0.18 ($\Gamma/2$), and 25°C of the initial velocities of the forward reaction as a function of one substrate concentration (with its paired substrate held fixed, and Mg^{2+}_{free} fixed at 1 m*M*) yield, except at the highest concentrations, essentially linear double-reciprocal plots and therefore approximately first-degree (8) velocity expressions. The results (in terms of double reciprocal plots of $1/v_0^f$ vs. 1/substrate) for the forward reactions are given in Figures 3A and B for variable $MgATP^{2-}$ at several fixed values of AMP^{2-} and for AMP^{2-} at several fixed values of $MgATP^{2-}$.[50]

At the highest concentration of $MgATP^{2-}$ studied, a type of acceleration in initial velocity is evident, possibly as a result of $MgHATP^-$ assuming a kinetically significant concentration. However, at the highest AMP^{2-} concentration explored, a very slight inhibition just becomes evident, in fact, much less noticeable than that observed by Rhoads and Lowenstein[347] under their conditions. The common intersection for Figure 3A, by extrapolation from the linear portions, extends into the third quadrant and is similar to that for Figure 3B. The inserts to Figures 3A and B contain the secondary plots derived from the slopes and ordinate-intercepts of the primary plots.

Similarly, Figures 3C and D contain the data obtained for the reverse direction at pH 7.4, approximately 0.16 ($\Gamma/2$), and 25°C, in terms of plots for the variable substrate $MgADP^-$ and ADP^{3-}, respectively. In this case, the primary plots of Figures 3C and D are essentially a set of linear curves which intersect in the second quadrant and in the case of variable $MgADP^-$, varied over a 20-fold range, the common intersection is almost at the ordinate axis. For the reverse direction, there does not appear to be any significant inhibition by excess substrate (either ADP^{3-} or $MgADP^-$) over the range of concentrations studied. Inserts to Figure 3C and D again provide the secondary plots derived from the slopes and ordinate intercepts of the primary plots. Results of the analyses of Figures 3A to D are given in Table 6 (first column) in terms of the calculated values for the derived kinetic parameters at pH 7.4, 25°C, and ($\Gamma/2$) = 0.16 to 0.18. Accepting the analyses for the moment, it is of interest that for the forward direction, $\overline{K}_{s,1}$ (for $ES_1 \rightleftarrows E + S_1$) values are two- to threefold smaller than the corresponding $K_{s,1}$ (for $ES_1S_2 \rightleftarrows ES_2 + S_1$) values for $MgATP^{2-}$ and for AMP^{2-}, respectively. Whereas, for the reverse direction, $\overline{K}_{s,1}$ values are approximately three times the corresponding $K_{s,1}$ value for $MgADP^-$ or for ADP^{3-}. If a head-to-tail alignment of the myokinase molecule must take place prior to catalysis, [348] such a large substrate pair-induced conformational change occurring within the ternary complexes (E · MA · B or E · MC · C) could allow an increase in binding of each susbstrate of the reverse direction to the enzyme within its associated ternary complex, but could also permit inhibition of the

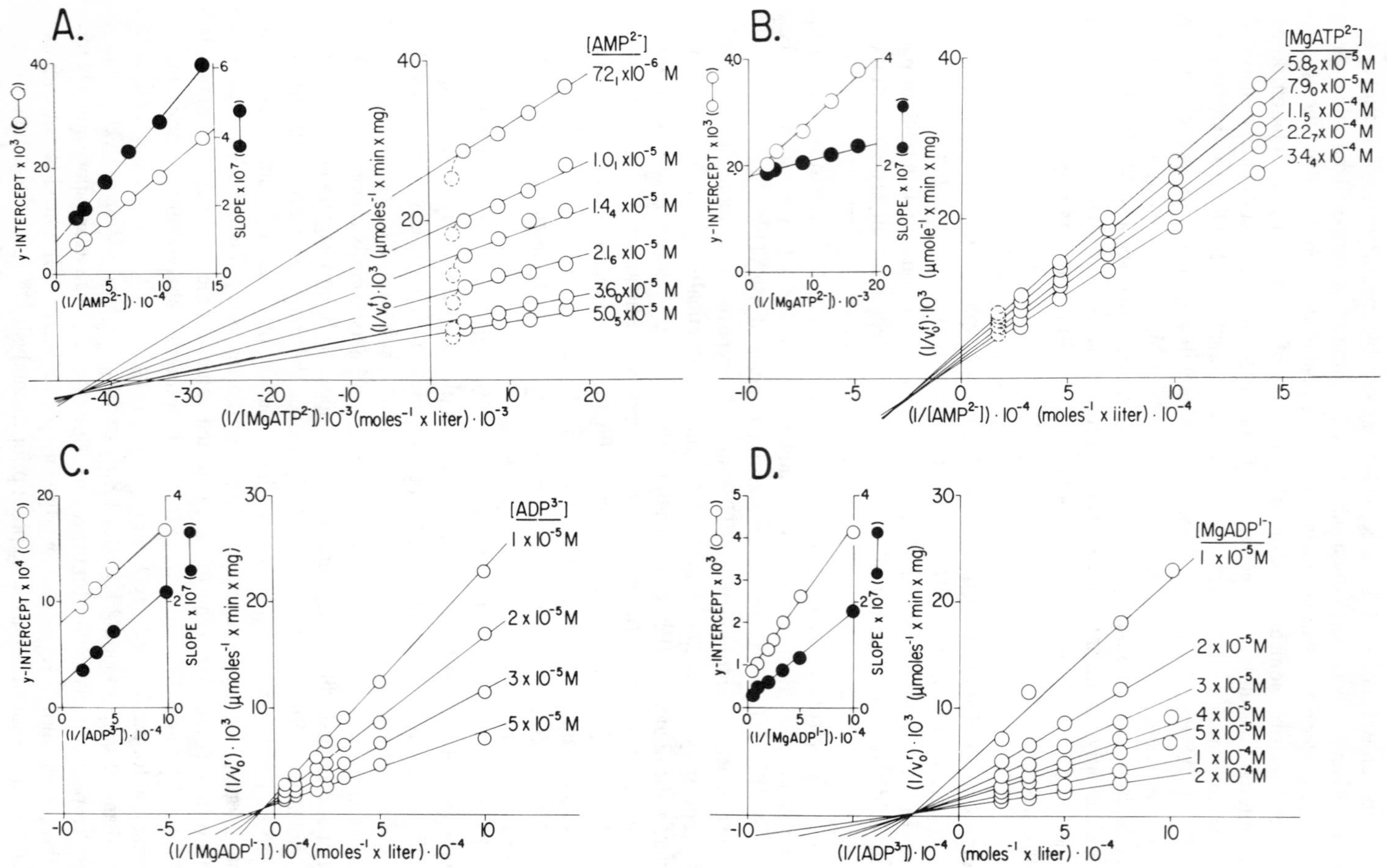

FIGURE 3. Steady-state kinetics at pH 7.4, 0.16—0.18 (Γ/2), of the forward reaction, $MgATP^{2-} + AMP^{2-} \rightarrow MgADP^{-} + ADP^{3-}$, and of the reverse reaction, $MgADP^{-} + ADP^{3-} \rightarrow MgATP^{2-}$, catalyzed by ATP-AMP transphosphorylase from rabbit muscle. (Taken from Reference 50.)

TABLE 6

Comparison of Values for Derived Kinetic Parameters at PH 7.4, 25°C, and 0.16 (Γ/2) for the Forward and Reverse Reactions, Catalyzed by ATP-AMP Transphosphorylase from Rabbit Muscle[50]

$K_1 = (\overline{K}_{MgATP^{2-}})^a$ 2.0 (± 0.4) × 10^{-5} *M*	$(\overline{K}_{MgdATP^{2-}})$ 2.1_5 (±0.9_8) × 10^{-4} *M*	$(\overline{K}_{MgATP^{2-}})$ 1.6 (±0.3) × 10^{-4} *M*	$(\overline{K}_{MgdATP^{2-}})$ 2.2 (±1.5) × 10^{-4} *M*	$(\overline{K}_{Mg\epsilon ATP^{2-}})$ 9.8 (±0.8) × 10^{-5} *M*
$K_2 = (\overline{K}_{AMP^{2-}})$ 3.5 (±1.1) × 10^{-5} *M*	$(\overline{K}_{dAMP^{2-}})$ 1.4_4 (±0.04_4) × 10^{-4} *M*	$(\overline{K}_{dAMP^{2-}})$ 6.5 (±2.5) × 10^{-5} *M*	$(\overline{K}_{AMP^{2-}})$ 6.2 (±4.8) × 10^{-5} *M*	$(\overline{K}_{AMP^{2-}})$ 8.0 (±0.8) × 10^{-5} *M*
				$K_i(\overline{K}_{\epsilon AMP^{2-}})$ 2.29 (±0.10) × 10^{-4} *M*
$K_3 = (K_{AMP^{2-}})^b$ 9.4 (±1.0) × 10^{-5} *M*	$(K_{dAMP^{2-}})$ 3.6_3 (±1.5_9) × 10^{-5} *M*	$(K_{dAMP^{2-}})$ 5.3 (±0.7) × 10^{-5} *M*	$(K_{AMP^{2-}})$ 3.0 (±1.3) × 10^{-5} *M*	$(K_{AMP^{2-}})$ 8.1 (±1.6) × 10^{-5} *M*
$K_4 = (K_{MgATP^{2-}})^b$ 5.4 (±1.1) × 10^{-5} *M*	$(K_{MgdATP^{2-}})$ 4.8_7 (±0.0_4) × 10^{-5} *M*	$(K_{MgATP^{2-}})$ 1.4 (±0.4) × 10^{-4} *M*	$(K_{MgdATP^{2-}})$ 1.1 (±0.5) × 10^{-4} *M*	$(K_{Mg\epsilon ATP^{2-}})$ 9.9 (±1.7) × 10^{-5} *M*
$K_6 = (\overline{K}_{ADP^{3-}})$ 4.0 (±0.9) × 10^{-5} *M*	$(\overline{K}_{dADP^{3-}})$ 1.3_3 (±0.9_3) × 10^{-4} *M*			
$K_7 = (\overline{K}_{MgADP^{-}})$ 1.3_4 (±0.52) × 10^{-4} *M*	$(\overline{K}_{MgdADP^{-}})$ 3.0_7 (±0.4_4) × 10^{-5} *M*			
	K_i = 2.2_7 (±0.02) × 10^{-5} *M*			
$K_8 = (K_{MgADP^{-}})$ 5.1 (±0.9) × 10^{-5} *M*	$(K_{MgdADP^{-}})$ 0.99 (±0.59) × 10^{-5} *M*			
$K_9 = (K_{ADP^{3-}})$ 1.6 (±0.7) × 10^{-5} *M*	$(K_{dADP^{3-}})$ 3.4_8 (±0.9_0) × 10^{-5} *M*			
V^f_{max}/E_t 5.5 (±0.5) × 10^2 μmol min^{-1} mg^{-1}; 1.1_8 × 10^4 mol · min^{-1} (mol protein)$^{-1}$; 1.9_6 × 10^2 s^{-1}	0.17 (±0.1) × 10^2 μmol min^{-1} mg^{-1}	0.63 (±0.06) × 10^2 μmol min^{-1} mg^{-1}	4.3 (±0.1) × 10^2 μmol min^{-1} mg^{-1}	3.7 (±0.1) × 10^2 μmol min^{-1} mg^{-1}
V^r_{max}/E_t 1.3 (±0.1) × 10^3 μmol min^{-1} mg^{-1}; 2.7_8 × 10^4 mol · min^{-1} (mol protein)$^{-1}$; 4.6_4 × 10^2 s^{-1}	0.18 (±0.04) × 10^2 μmol min^{-1} mg^{-1}			

[a] $\overline{K}_s$ denotes an intrinsic dissociation constant of the particular substrate from the binary complex.
[b] K_s denotes a dissociation constant of the particular substrate from the ternary complex.

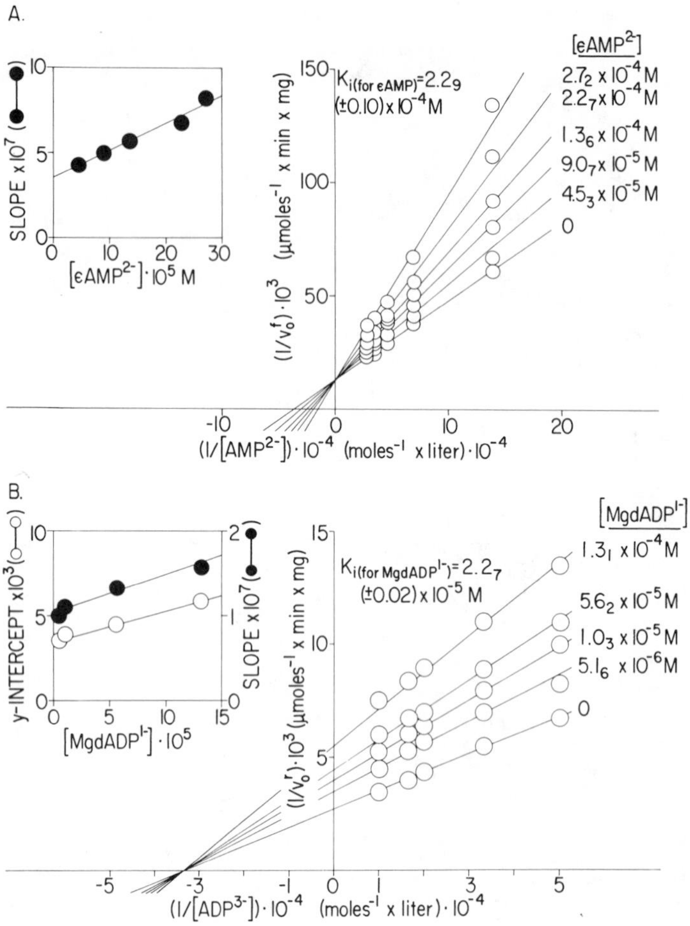

FIGURE 4. Inhibition of rabbit muscle myokinase by ϵAMP^{2-} in the forward direction and by $MgADP^{-}$ in the reverse direction. (Taken from Reference 50.)

binding of each substrate of the forward direction to the enzyme within its associated ternary complex.

The rabbit muscle enzyme will also catalyze a reaction between the following pairs of substrates for the forward reaction: $MgdATP^{2-}$ + $dAMP^{2-}$, $Mg\epsilon ATP^{2-}$ + AMP^{2-}, $MgdATP^{2-}$ + AMP^{2-}, and $MgATP^{2-}$ + $dAMP^{2-}$.[50] However, only the reverse direction between $MgdADP^{-}$ + $dADP^{3-}$ could be conveniently studied; catalysis did not appear to take place at any reasonable rate with ϵADP_0 (plus Mg_0) in the reverse direction, nor with ϵAMP_0 as a substrate (plus ATP_0 and Mg_0) in the forward direction.[50] However, ϵAMP^{2-} did act as a competitive inhibitor for AMP^{2-} (Figure 4) and its K_i (approximately 10^{-4} *M*) could be estimated. Linear double-reciprocal plots were observed for the above substrate pairs, except at the highest concentrations of variable $MgATP^{2-}$ with fixed $dAMP^{2-}$, where acceleration in rates were observed. At millimolar concentrations of AMP_0 or $dAMP_0$, only approximately 60% of the total nucleoside monophosphate is uncomplexed (i.e., $dAMP^{2-}$ or AMP^{2-}) and almost 27% of the total nucleoside monophosphate exists as $NaAMP^{1-}$ (or $NadAMP^{1-}$) in the presence of 0.15 *M* NaCl, but only approximately 3% is present as MgAMP (or MgdAMP) when Mg^{2+}_{free} has been fixed at 1 m*M* and $MgATP^{2}$ = 1 m*M* (by the addition or Mg_0 = 2.05 m*M*, ATP_0 = 1.06 m*M*, and AMP_0 (or $dAMP_0$) = 1.665 m*M*). In this case, only $NaATP^{3-}$ (equal to 0.030 m*M*) and ATP^{4-} (equal to 0.017 m*M*) rise to a few percent of the total nucleoside triphosphate (with $MgHATP^{-}$ only at approximately 0.003 m*M*). Therefore, the sodium salts of $NadAMP^-$ and/or $NaATP^{2-}$ appear to play some kinetic role (if only as activators) when at high concentrations of total nucleotide substrate (either $MgATP^{2-}$ or $dAMP^{2-}$).

In Table 6, the derived kinetic parameters for all the data are summarized in columns 2 to 5, and compared with the data already presented for the rabbit muscle myokinase in column 1 (i.e., for the catalyzed reactions $MgATP^{2-} + AMP^{2-} \rightleftharpoons MgADP^- + ADP^{3-}$).

Only in column 2 is a complete set of data available for the catalyzed reaction between the 2-deoxy analogues of the substrates (i.e., between $MgdATP^{2-} + dAMP^{2-} \rightleftharpoons MgdADP^- + dADP^{3-}$) which may be compared with the reactions between the ribonucleotide substrates (column 1). The V_{max}s for the deoxynucleotide substrates are surprisingly small compared to the V_{max}s for the ribonucleotide substrates ($V^f_{max} \simeq 17$ vs. 550 and $V^r_{max} \simeq 18$ vs. 1300), but that the intrinsic dissociation constants from the ternary complexes are of the same order of magnitude. In the case of the data given in column 2, for the deoxy nucleotide substrates, $\overline{K}_{s,1}$ is always larger (three- to fivefold) than its corresponding value for $K_{s,1}$ for each given substrate in the forward or reverse direction. One may infer that the substrate(s) induced-conformational change resulted in a much tighter binding of the second substrate within the ternary-complex, retarding its dissociation and allowing only a relatively slow interconversion of the ternary complexes. In Figure 4B, $MgdADP^-$ is demonstrated to be noncompetitive with respect to ADP^{-3} with a K_i calculated to be approximately 2×10^{-5} *M* and similar to the value for $\overline{K}_{MgdADP^-}$ found directly, that is, 3×10^{-5} *M*. Thus, support is provided for the hypothesis of two separate sites, one for ADP^{3-} and one for $MgADP^-$ (or $MgdADP^-$).

For either variable $Mg\epsilon ATP^{2-}$ or AMP^{2-}, the point of intersection of the set of linear curves in the double-reciprocal plot was almost on the negative abscissa[50] and as given in Table 6 (last column) there is no significant difference between $\overline{K}_{AMP^{2-}}$ and $K_{AMP^{2-}}$ (approximately 8×10^{-5} *M*, column 5) and between $\overline{K}_{Mg\epsilon ATP^{2-}}$ and $K_{Mg\epsilon ATP^{2-}}$. Also, $K_{AMP^{2-}}$ was found to be identical within experimental error (i.e., approximately 9×10^{-5} *M*, column 1) with the case when $MgATP^{2-}$ was the paired substrate. Therefore, with $Mg\epsilon ATP^{2-}$ and AMP^{2-} as a substrate pair, little if any substrate-induced conformational change appears to have taken place within the ternary compound and the reaction proceeds fairly rapidly ($V^f_{max} \simeq 370$). Moreover, the values for $K_{AMP^{2-}}$, or dissociation of AMP^{2-} from the ternary complex with either $Mg\epsilon ATP^{2-}$ or with $MgATP^{2-}$, are identical and therefore in support of the hypothesis that a rate-limiting step exists at the interconversion of the ternary complexes. In Figure 4A, ϵAMP^{2-}, although apparently not a substrate, binds fairly tightly to the enzyme and in a competitive fashion with respect to AMP^{2-} (with $K_i \simeq 10^{-4}$ *M*), pointing to an individual binding site for uncomplexed AMP^{2-} and illustrating the high degree of substrate specificity shown by the myokinase, i.e., for the 6-NH_2 group of the adenine moiety.

In column 4 (Table 6), for the substrate pair $MgdATP^{2-}/AMP^{2-}$, again $\overline{K}_{s,1}$ is larger than $K_{s,1}$, but only by twofold, and within experimental error $\overline{K}_{AMP^{2-}}$ (6.2 (±4.8) $\times 10^{-5}$ *M*) is identical to that obtained with the substrate pair $MgATP^{2+}/AMP^{2-}$ (3.5 (±1.1) $\times 10^{-5}$ *M*). Moreover, V^f_{max} (430) approaches that obtained for its natural substrate pair (550). Finally, in the case of the substrate pair $MgATP^{2-}/dAMP^{2-}$ (column 3, Table 1 (P), the V^f_{max} decreases dramatically (to only 63), pointing to the rigid requirements for a "good" substrate at the AMP^{2-} site. In this last case, there is only a very slight, and probably not significant, difference between $K_{s,1}$ and $\overline{K}_{s,1}$ and the value for $K_{MgATP^{2-}}$ is almost the same as for the natural pair of substrates. In conclusion, the evidence is consistent with a mechanism involving a random quasi-equilibrium mechanism with a rate-limiting step largely at the interconversion of the ternary complexes; for the case of its natural substrate pair ($MgATP^{2-}/AMP^{2-}$ or $MgADP^-/ADP^{3-}$) a relatively large substrate(s)-induced conformational change may take place within the ternary complexes. Furthermore, and in support of the conclusion of Rhoads and Lowenstein,[347] there appear to be separate sites for the binding of the mag-

TABLE 7
Values for Derived Kinetic Parameters at pH 7.4, 25°C, and 0.16 (Γ/2) for ATP-AMP Transphosphorylase from Calf Muscle[50]

Defined kinetic parameters	Defined equilibrium	Defined intrinsic dissociation constant	Value (M)[50]
K_1	$E \cdot MA \rightleftarrows E + MA$	$(\bar{K}_{MgATP^{2-}})^a$	$5.1\ (\pm 0.3) \times 10^{-5}$
K_2	$E \cdot B \rightleftarrows E + B$	$(\bar{K}_{AMP^{2-}})$	$5.3\ (\pm 0.1) \times 10^{-5}$
K_3	$E \cdot MA \cdot B \rightleftarrows E \cdot MA + B$	$(K_{AMP^{2-}})^b$	$1.1\ (\pm 0.6) \times 10^{-4}$
K_4	$E \cdot MA \cdot B \rightleftarrows E \cdot B + MA$	$(K_{MgATP^{2-}})$	$5.9\ (\pm 2.9) \times 10^{-5}$
K_6	$E \cdot C \rightleftarrows E + C$	$(\bar{K}_{ADP^{3-}})$	$4.8\ (\pm 1.0) \times 10^{-5}$
K_7	$E \cdot MC \rightleftarrows E + MC$	$(\bar{K}_{ADP^{-}})$	$1.9\ (\pm 0.2) \times 10^{-5}$
K_8	$E \cdot MC \cdot C \rightleftarrows E \cdot C + MC$	$(K_{MgADP^{-}})$	$2.1\ (\pm 0.5) \times 10^{-5}$
K_9	$E \cdot MC \cdot C \rightleftarrows E \cdot MC + C$	$(K_{ADP^{3-}})$	$5.1\ (\pm 0.7) \times 10^{-5}$
V^f_{max}/E_t	$E \cdot MA \cdot B \rightleftarrows E \cdot MC \cdot C$		$6.3_6\ (\pm 0.30) \times 10^2$ μmol min^{-1} mg^{-1}; $1.3_6 \times 10^4$ mol. min^{-1} (mol protein)$^{-1}$ $2.2_7 \times 10^2\ s^{-1}$
V^r_{max}/E_t	$E \cdot MC \cdot C \rightleftarrows E \cdot MA \cdot B$		$7.5_0\ (\pm 0.83) \times 10^2$ μmol min^{-1} mg^{-1}; $1.6_1 \times 10^4$ mol. min^{-1} (mol protein)$^{-1}$ $2.6_8 \times 10^2\ s^{-1}$

a $\bar{K}_s$ denotes an intrinsic dissociation constant of the particular substrate from the binary complex.
b K_s denotes a dissociation constant of the particular substrate from the ternary complex.

nesium complexes of the nucleotides, i.e., $MgATP^{2-}$ and $MgADP^{-}$, and for the binding of the uncomplexed substrates, AMP^{2-} and ADP^{3-}. There is, moreover, a rigid requirement for the 6-NH_2 group of AMP^{2-} which allows it to be a "good" or "poor" substrate.

3. Kinetics of the Calf Muscle and Calf Liver ATP-AMP Transphosphorylase Isoenzymes

The data in terms of double-reciprocal plots for the forward reaction (with $MgATP^{2-}$/AMP^{2-} substrate pair) and for the reverse reaction ($MgADP^{-}$/ADP^{3-} substrate pair) for the calf muscle and for the calf liver adenylate kinases, and their derived kinetic parameters at pH 7.4 (25°C) and 0.16 to 0.18 (Γ/2) are summarized in Table 7 for the calf muscle enzyme and in Table 8 for the calf liver isoenzyme.

Interestingly, V^r_{max} decreases to only approximately 208 (Table 8) for the calf liver enzyme compared to 750 for the calf muscle enzyme (Table 7) and 1300 for the rabbit muscle enzyme. However, V^f_{max} for the forward reaction catalyzed by the calf liver enzyme is actually larger (1080 μmol $(min)^{-1}$ mg^{-1}) than for either muscle-type enzyme (V^f_{max} for calf muscle = 636; V^f_{max} for rabbit muscle enzyme = 550) thus favoring, within the mitochondrion, a reaction between $MgATP^{2-}$ and AMP^{2-} to yield the acceptor for oxidative phosphorylation, that is, $MgADP^{-}$/ADP^{3-}.

Whereas, there is at most a twofold difference in $\bar{K}_{AMP^{2-}}$ vs. $K_{AMP^{2-}}$ for the calf muscle enzyme (with $\bar{K}_{AMP^{2-}} < K_{AMP^{2-}}$) and the other kinetic parameters showing little, if any, significant changes between $\bar{K}_{s,1}$ and $K_{s,1}$ (Table 7); in the liver enzyme (Table 8) in the forward reaction, $\bar{K}_{s,1}$, is almost one third that of $K_{s,1}$ with the second substrate inhibiting

TABLE 8
Values for Derived Kinetic Parameters at pH 7.4, 25°C, and 0.16 (Γ/2) for ATP-AMP Transphosphorylase from Calf Liver[50]

Defined kinetic parameters	Defined equilibrium	Defined intrinsic dissociation constant	Value (M)[50]
K_1	$E \cdot MA \rightleftarrows E + MA$	$(\overline{K}_{MgATP^{2-}})$[a]	$5.4\ (\pm 0.5) \times 10^{-5}$
K_2	$E \cdot B \rightleftarrows E + B$	$(\overline{K}_{AMP^{2-}})$	$5.4\ (\pm 0.9) \times 10^{-5}$
K_3	$E \cdot MA \cdot B \rightleftarrows E \cdot MA + B$	$(K_{AMP^{2-}})$[b]	$1.7\ (\pm 0.1) \times 10^{-4}$
K_4	$E \cdot MA \cdot B \rightleftarrows E \cdot B + MA$	$(K_{MgATP^{2-}})$	$1.7\ (\pm 0.1) \times 10^{-4}$
K_6	$E \cdot C \rightleftarrows E + C$	$(\overline{K}_{ADP^{3-}})$	$2.3\ (\pm 1.4) \times 10^{-5}$
K_7	$E \cdot MC \rightleftarrows E + MC$	$(\overline{K}_{ADP^{-}})$	$1.9\ (\pm 0.7) \times 10^{-5}$
K_8	$E \cdot MC \cdot C \rightleftarrows E \cdot C + MC$	$(K_{MgADP^{-}})$	$1.6\ (\pm 0.9) \times 10^{-5}$
K_9	$E \cdot MC \cdot C \rightleftarrows E \cdot MC + C$	$(K_{ADP^{2-}})$	$1.8\ (\pm 0.5) \times 10^{-5}$
V^f_{max}/E_t	$E \cdot MA \cdot B \rightleftarrows E \cdot MC \cdot C$		$1.0_8\ (\pm 0.03) \times 10^3$ μmol min^{-1} mg^{-1}; $2.7_6 \times 10^4$ mol. min^{-1} (mol protein)$^{-1}$ $4.6_1 \times 10^2\ s^{-1}$
V^r_{max}/E_t	$E \cdot MC \cdot C \rightleftarrows E \cdot MA \cdot B$		$2.0_8\ (\pm 0.04) \times 10^2$ μmol min^{-1} mg^{-1}; $5.3_2 \times 10^2$ mol. min^{-1} (mol protein)$^{-1}$ $8.8_8 \times 10^2\ s^{-1}$

[a] $\overline{K}_s$ denotes an intrinsic dissociation constant of the particular substrate from the binary complex.
[b] K_s denotes a dissociation constant of the particular substrate from the ternary complex.

the binding of the first, in the ternary complex. In the reverse reaction, however, $\overline{K}_{s,1} \simeq K_{s,1}$, within experimental error.

The absolute values for either $\overline{K}_{s,1}$ or $K_{s,1}$ in the forward reaction are within a factor of two, almost the same for either isoenzyme, and in the reverse direction they are also similar within a factor of two or three, with, e.g., $K_{ADP^{3-}}$ for the liver enzyme being almost one third that of the muscle enzyme.

Interestingly, phosphoenolpyruvate (PEP) acts as an inhibitor of the calf liver enzyme, but apparently not of the calf muscle enzyme, nor of the rabbit muscle enzyme. PEP was shown to be a competitive inhibitor of $MgATP^{2-}$ (with a $K_i \simeq 1.2$ mM), but a noncompetitive inhibitor with respect to AMP^{2-}, $MgADP^{-}$, and ADP^{3-} (with a $K_i \simeq 1.2$ mM), Thus, at the 4-mM concentrations of PEP usually employed in the coupled-enzyme assay systems for the forward direction,[50] the liver enzyme would be severely inhibited. For this reason the concentration of PEP was decreased to 0.25 mM where the inhibition would be insignificant. Whether the inhibition by PEP could play any physiological role in the liver mitochondrion is an interesting point of speculation.

Although very large attendant errors are involved in the above analyses of derived kinetic parameters (Tables 6 to 8), which hinge on the values for the magnesium complexation and dissociation constants selected in Table 5 (compare a discussion on the analogous case for the rabbit muscle ATP-creatine transphosphorylase[48]), nevertheless, if identical values were utilized in an evaluation of the thermodynamic equilibrium value for the reaction in Equation 94 (see below, Table 9), a self-consistent mechanism would require that the estimate of K_{eq} from kinetic analyses at least yields a similar value. From the overall velocity Equations 104 or 106 and the equalities Equations 98 and 100 as given above, the four Haldane

TABLE 9
Kinetically Derived Values for the Thermodynamic Equilibrium Constant at 25°C, and 0.16 to 0.18 (Γ/2)[50]

Haldane expression	Calf muscle adenylate kinase	Calf liver adenylate kinase	Rabbit muscle adenylate kinase
$K_{eq} = \frac{V^f_{max}}{V^r_{max}} \frac{K_7K_9}{K_1K_3} =$	2.03×10^{-1}	1.98×10^{-1}	4.88×10^{-1}
$K_{eq} = \frac{V^f_{max}}{V^r_{max}} \frac{K_6K_8}{K_2K_4} =$	2.74×10^{-1}	2.07×10^{-1}	4.48×10^{-1}
$K_{eq} = \frac{V^f_{max}}{V^r_{max}} \frac{K_7K_9}{K_2K_4} =$	2.66×10^{-1}	1.96×10^{-1}	4.78×10^{-1}
$K_{eq} = \frac{V^f_{max}}{V^r_{max}} \frac{K_6K_8}{K_1K_3} =$	1.80×10^{-1}	2.09×10^{-1}	4.57×10^{-1}
Average $K_{eq} =$	2.31×10^{-1}	2.03×10^{-1}	4.68×10^{-1}

Overall average $K_{eq} = 3.01\ (\pm 1.46)^a \times 10^{-1}$ where $K_{eq} = \frac{(MgADP^-)\ (ADP^{3-})}{(MgATP^{2-})\ (AMP^{2-})}$

[a] Standard deviation $= \sigma = [\sum (X - \overline{X})^2/(n - 1)]^{1/2}$.

expressions follow from the definition of this assigned equilibrium constant, Equation 103. Thus, at 25°C and 0.16 (Γ/2), Table 9 provides the estimates of K_{eq} derived kinetically (from the four Haldane relationships shown) for each of the three adenylate kinases studied, together with their average values. Taking the overall average of the three sets of determinations yields an average $K_{eq} = 3.01\ (\pm\ 1.46) \times 10^{-1}$ (for $\pm 1\sigma$), which may be compared with that given below in Table 10, that is, $2.74\ (\pm 0.62) \times 10^{-1}$ (for $\pm\ 1\sigma$) for the direct estimate of the equilibrium constant. As will be seen, in the same Table 10, illustrations are also given of the type of calculations necessary for this estimation where only the pertinent species are given. Incidentally, one notes that under these same conditions less than two thirds of the AMP_0 is in the form of AMP^{2-} (with the remainder largely as $NaAMP^-$ and MgAMP, not shown in Table 10).

Although the agreement in calculated equilibrium constants seems almost fortuitously excellent, the range of estimations of the kinetically derived values between the three separate sets, however, is large, with 1σ encompassing almost one half the average value (Table 9). Nevertheless, within each set there appears to be close agreement and excellent precision, and one may conclude that there is at least a self-consistent analysis for each separate set.

With a careful and systematic control over the effects of metal complexation at 25°C and under conditions where the ionic strength did not vary by too large a degree, a mechanism for the rabbit muscle myokinase similar to that proposed for the human erythrocyte adenylate kinases[352] has emerged (see Scheme XI). Thus, the data presented in Figures 3 and 4 for the rabbit muscle enzyme and for the calf muscle and calf liver adenylate kinases[50] are consistent with a quasi-equilibrium random mechanism. In agreement with Rhoads and Lowenstein,[347] mechanisms involving covalent enzyme intermediates are unlikely in view of the sets of intersecting double-reciprocal plots. The fact that either $K_{AMP^{2-}}$ or $\overline{K}_{AMP^{2-}}$ may be estimated from three substrate pairs (i.e., with $MgATP^{2-}$, $MgdATP^{2-}$, and $Mg\epsilon ATP^{2-}$) and yield similar values ($K_{AMP^{2-}}$ = (3 to 9) $\times 10^{-5}$ *M* and $\overline{K}_{AMP^{2-}}$ = (3.5 to 8) $\times 10^{-5}$ *M*); (see Table 6) supports the hypothesis that there is a rate-limiting step which exists largely at the interconversion of the ternary complexes and is therefore similar to that of the human erythrocyte enzymes[352] at equivalent ionic strengths.

The competitive inhibition of AMP^{2-} by ϵAMP^{2-} (and fixed $MgATP^{2-}$) and the non-competitive inhibition of ADP^{3-} by $MgdADP^-$ (and fixed $MgADP^-$) support the contention

TABLE 10
Concentrations at Equilibrium (Moles/Liter) and Thermodynamic Equilibrium Constant at 25°C and 0.16 (Γ/2)[50]

Species	Calf muscle adenylate kinase	Calf liver adenylate kinase	Rabbit muscle adenylate kinase
Mg_0	5.288×10^{-3}	5.288×10^{-3}	5.288×10^{-3}
ATP_0	2.309×10^{-4}	1.982×10^{-4}	2.551×10^{-4}
ADP_0	3.536×10^{-4}	7.380×10^{-4}	3.503×10^{-4}
AMP_0	2.309×10^{-4}	1.982×10^{-4}	2.551×10^{-4}
Mg^{2+}	4.74×10^{-3}	4.72×10^{-3}	4.70×10^{-3}
ATP^{4-}	8.001×10^{-7}	6.896×10^{-7}	8.913×10^{-7}
$MgATP^{2-}$	2.275×10^{-4}	1.953×10^{-4}	2.514×10^{-4}
ADP^{3-}	2.529×10^{-5}	2.452×10^{-5}	2.523×10^{-5}
$MgADP^{-}$	2.997×10^{-4}	2.894×10^{-4}	2.965×10^{-4}
AMP^{2-}	1.249×10^{-4}	1.073×10^{-4}	1.382×10^{-4}
K_{eq}	2.667×10^{-1}	3.386×10^{-1}	2.155×10^{-1}

Overall average $K_{eq} = 2.74\ (\pm 0.62)^{a} \times 10^{-1}$ where $K_{eq} = \dfrac{(MgADP^{-})\ (ADP^{3-})}{(MgATP^{2-})\ (AMP^{2-})}$

Note: Initial reaction mixture (0.500 ml): 10 m*M* Tris (Cl^-), pH 7.4 at 25°C, 150 m*M* NaCl, 8 m*M* DTE, 0.1 m*M* EDTA, 1 mg/ml albumin, 1 m*M* $MgCl_2$, 4.288 m*M* $MgAc_2$, 0.7 to 0.9 m*M* ADP_0; 1.5 μg, 1.04 μg, or 4.8 μg calf liver, calf muscle, or rabbit muscle adenylate kinase, respectively. A control reaction mixture without enzyme was conducted simultaneously. At equilibrium (1/4 to 1/2 h), 0.25 ml added to 1.00 ml ice-cold 25% (w/v) perchloric acid (final concentration = 20%) and allowed to stand 60 min at 0°C. After centrifugation and neutralization with 0.47 ml of 3 *N* KOH, followed by clarification, aliquots of the supernatant liquid were analyzed enzymatically for ADP_0 with use of the above reaction mixture but containing (1.0 ml vol): 50 m*M* KCl, 1 m*M* PEP, 10 m*M* $MgCl_2$, 0.15 m*M* NADH, 0.2 units of lactate dehydrogenase and pyruvate kinase. The supernatant liquids were devoid of myokinase activity and 1 mg/ml albumin facilitated the coprecipitation of the myokinase, since at lower concentrations of perchloric acid or in the absence of albumin, myokinase activity was detected in the neutralized supernatant.

[a] Standard deviation $= \sigma = [\sum (X - \overline{X})^2/(n - 1)]^{1/2}$.

of Rhoads and Lowenstein[347] that there is a separate site for binding the metal-complexed nucleotide substrates (i.e., $MgATP^{2-}$ and $MgADP^{-}$) and one for the uncomplexed nucleotide substrates (APD^{3-} and AMP^{2-}). Earlier, direct equilibrium binding measurements of ADP^{3-} at 3°C (pH 7.9 and 0.16 (Γ/2)) to the rabbit muscle enzyme had yielded a value of 7×10^{-5} *M* for the intrinsic dissociation constant (see Chapter 8, Table 5) and is in reasonable agreement with the kinetic estimate of $\overline{K}_{ADP^{3-}}$ found here at 25°C (pH 7.4 and 0.16 (Γ/2)), i.e., 4×10^{-5} *M* (Table 6) again support the mechanism presented here, that of a quasi-equilibrium random mechanism, but without the requirement of independently bound substrates.

Finally, the self-consistency of the kinetic analyses for all three adenylate kinases is reflected in the precision of the calculated Haldane constants for each set and in the agreement between the overall average (from all three sets) of the kinetically calculated equilibrium constant and the direct estimation of the thermodynamic equilibrium constant (Equations 94 and 103).

Parenthetically, there appears to be a wide divergence of estimates for the apparent equilibrium constant for the myokinase catalyzed reaction.[343] The value presented here, 2.7 $\times 10^{-1}$ at 25°C and at 0.16 to 0.18 (Γ/2) for $\frac{(MgADP^-)(ADP^{3-})}{(MgATP^{2-})(AMP^{2-})}$ may represent one of the few attempts at the systematic estimation of such a thermodynamic value. Earlier, Rhoads and Lowenstein had approximated the above equilibrium constant from the radioactivities at isotopic exchange equilibrium also to be approximately 2.7 $\times 10^{-1}$ at 25°C (but at 0.02 (Γ/2), a value which is in exact agreement with that given here. Recently, Rao et al.[350] have estimated the overall equilibrium constant to be 3.8 $\times 10^{-1}$ at 4°C, pH 7.0, 10% D_20, and an ionic strength in excess of 0.07.

In Reference 29, an interesting series of observations on the interaction of the multisubstrate inhibitor, P^1, P^5-di(adenosine-5′)pentaphosphate (or AP_5A) was presented. Whereas inhibition of the rabbit and calf muscle myokinase (cytoplasmic enzymes) was very significant at approximately 10^{-8} *M* and similar to that reported earlier,[355] inhibition of the liver-type adenylate kinase (a mitochondrial enzyme) required almost two orders of magnitude higher concentration, approximately 10^{-6} *M*, for similar extents of inhibition. Moreover, for the forward reactions, AP_5A acted as a competitive inhibitor with respect to either substrate for all three enzymes, but for the reverse direction AP_5A appeared to be noncompetitive with respect to either substrate. An explanation for its action as a noncompetitive inhibitor to ADP^{3-}, for example, or its competitive nature with respect to $MgATP^{2-}$, was sought in the structure of the magnesium complex of $Mg_2(AP_5A)^-$ which may or may not contain within it an analogous structure to the particular substrate being varied. These observations also tend to give credence to the idea that there is a separate site for binding each of the two types of substrates: (1) ADP^{3-}/AMP^{2-} or (2) $MgADP^-/MgATP^{2-}$. PEP also seemed to be unique in that it inhibited the liver enzyme, but not the muscle enzymes. Other than the above two observations, there were only a few major differences that could be detected kinetically between the calf isoenzymes (Tables 7 and 8). One case is the huge differences in V_{max}, with V^r_{max} being much smaller for the liver isoenzyme, an observation which might be of physiological importance if the mitochondrial adenylate kinase plays any regulatory role in oxidative phosphorylation.

From substrate binding studies on fragments of the rabbit muscle myokinase it has been deduced that residues 1 to 44 of the rabbit muscle myokinase may contain the binding sites for $MgATP^{2-}$ and $MgADP^-$ and the terminal sequence of residues 171 to 194 may contain the binding sites for AMP^{2-} and ADP^{3-}. Therefore, for interaction between the two sites to occur prior to catalysis and phosphoryl group transfer, a head-to-tail interaction is mandatory. The enzyme may or may not exist in solution with such a preferred conformation, however, it is consistent with the X-ray-deduced crystal structure of the porcine muscle adenylate kinase,[356-358] and the remarkable space-filling model of Steitz and co-workers[359] as well as the model derived from NMR studies by Fry et al.[861,862,965] See also References 964 and 982. The substrates, in turn, might induce such a large conformational change or further induce smaller changes within the ternary complexes. The fact that AP_5A is such a poor inhibitor for the liver enzyme, relatively speaking, compared to the muscle enzyme, might imply that either the atomic distances, in which the two binding sites must approach one another, differ in the liver enzyme compared to the muscle enzyme or that possibly the liver enzyme may exist in a less-flexible structure to allow the necessary binding and required juxtaposition of the two binding sites. The fact that $V^r_{max} << V^f_{max}$ in the liver enzyme seems to indicate that the enzyme is already partially fixed in some preferred conformation so as to facilitate phosphoryl group transfer to AMP^{2-}; this may be its important role in the mitochondrion.

A review of the kinetic properties of the adenylate kinases will be presented in Volume II, (Chapter 17, Reference 964).

C. SOME COMMENTS ON OTHER ATP TRANSPHOSPHORYLASES OR KINASES

Pyruvate kinase catalyzes the physiological phosphorylation of MgADP by PEP and, in the other direction, the phosphorylation of pyruvate by MgATP: PEP + MgADP $\leftrightarrows$ MgATP + pyruvate, where the charges of the various ionic species have been omitted. NMR distance measurements suggest that there exists a molecular contact between the carbonyl oxygen of pyruvate and the γ-phosphate of ATP.[885] The stereochemical studies of Blättler and Knowles[886] with chiral [γ^{16}O, γ-^{17}O γ-^{18}O]-ATP unambiguously demonstrated that an inversion of configuration around the phosphorus occurred during the reaction. Thus, the phosphoryl-transfer reaction step appears to occur by a direct, in-line displacement.

The enzyme also catalyzes the phosphorylation of a number of other divergent substrates, e.g., fluoride (in the presence of bicarbonate[873]), hydroxylamine (as its CO_2 adduct, N-hydroxycarbamate[433,874,875]), and glycolate[876] as well as the decarboxylation of oxalacetate[877,878] and the detritiation of pyruvate.[879] In addition, and unlike many other kinases, pyruvate kinase requires an enzyme-bound divalent metal ion for catalysis, in addition to the metal ion required to complex the nucleotide substrate[880] as well as its monovalent, e.g., K^+, requirements.[887,888] Gupta et al.[880] also showed that the substitution of inert Cr ATP (chromium ATP) would activate the detritiation of pyruvate only in the presence of added divalent cation.

Further evidence for the dual divalent metal ion requirement of pyruvate kinase has come from NMR chemical shift titrations of enzyme-bound ATP with Mg_t, which indicated that at least two equivalents of Mg_t are required for saturation,[881] and from the synergistic activation of pyruvate kinase by mixed divalent metal ions.[882]

Dougherty and Cleland[883] recently reported the pH profiles for several alternate reactions catalyzed by pyruvate kinase (i.e., with the alternate substrates oxalacetate, glycolate, hydroxylamine, and fluoride). For the decarboxylation of oxacetate, for example, equilibrium-ordered kinetics was followed with Mg^{2+} adding before oxalacetate. The pH profiles showed pKs of only two groups, one of which they interpreted as a ligand for the enzyme-bound Mg^{2+} and the other as water coordinate to this Mg^{2+}. They also reported [884] the pH profiles for the natural substrates PEP and pyruvate, as well as the slow alternate substrate phosphoenol-α-ketobutyrate and a number of competitive inhibitors. pH profiles for those substrates in which an enolate-keto conversion may be involved showed a third pK which they interpreted as corresponding to an acid-base catalyst for the enolization process and which they believed to be a lysyl residue.[884] They proposed a chemical mechanism for the natural substrates which involved a phosphoryl transfer between MgADP and a Mg^{2+}-bound enolate, with metal coordination of the enolate serving to make it a good leaving group.[884] They concluded that there may be three types of kinase reactions in which the leaving group (1) is transferred to a metal ion, as with pyruvate kinase, (2) is transferred to a proton, as with hexokinase, or (3) has a sufficiently low pK so that catalysis is not needed, as in the case of the adenylate or acetate kinases.[844]

Protein covalent modification is a major control mechanism of biological systems with many enzymes either activated or inactivated[940-945] e.g., by phosphorylation through specific kinases or dephosphorylation through phosphatases. However, the maintenance of a steady-state level of the particular modified protein requires a continuous expenditure of energy, e.g., by ATP utilization in the steady state. Schacter et al.[946,947] considered the energy utilization in a reversible phosphorylation system with the use of a synthetic nonapeptide substrate as a model for covalent modification. Recently, Goldbeter and Koshland[948] evaluated, from a general standpoint, the ATP turnover (or energy consumption) for the particular cases: (1) when an effector controls the modifying enzyme, (2) when an effector controls the demodifying enzyme, and (3) when an effector activates one of these modifying enzymes and inhibits the other.

IV. CALCULATIONS OF THE EQUILIBRIUM CONCENTRATIONS OF THE VARIOUS IONIC AND COMPLEX SPECIES OF THE SUBSTRATES AND THERMODYNAMIC FUNCTIONS ASSOCIATED WITH THE CATALYZED EQUILIBRIA OF KINASE TYPE OF REACTIONS

As indicated in this section of metal cofactors, it sometimes proves essential to consider metal complexation of the substrate in order to evaluate not only the kinetics of the reaction, but even its thermodynamic equilibrium constant. The following treatment as illustrative of this point is taken from Kuby and Noltmann[48] based on the original set of measured equilibria of Noda et al.[87] on the rabbit muscle ATP-creatine transphosphorylase-catalyzed reaction which were measured at several pH values, and Mg_t values at a number of initial concentrations of either ATP_t or Cr_t, or of ADP_t and $Cr \sim P_t$. These data illustrated pointedly the dependence of the calculated equilibrium coefficients (in terms of total species) on the total magnesium concentration and on the pH.

The conservation equations employed for the recalculation of the data of Noda et al.[87] for the various substrates for the pH range studied are (where "B" here represents the buffer species[87])

$$\begin{aligned} ATP_t &\simeq ATP^{4-} + HATP^{3-} + MgATP^{2-} + MgHATP^{-} + Mg_2ATP \\ &\quad + 2Mg(ATP)_2^{6-} + NaATP^{3-} \\ ADP_t &\simeq ADP^{3-} + HADP^{2-} + MgADP^{-} + MgADP + NaADP^{2-} \\ Cr \sim P_t &\simeq Cr \sim P^{2-} + HCr \sim P^{-} + MgCr \sim P + NaCr \sim P^{-} \\ Cr_t &\simeq Cr^{\pm} \\ Mg_t &\simeq Mg^{2+} + MgATP^{2-} + MgHATP^{-} + 2Mg_2ATP + Mg(ATP)_2^{6-} \\ &\quad + MgADP^{-} + MgCr \sim P + MgB^{+} + MgOH^{+} \\ Na_t &\simeq Na^{+} + NaATP^{3-} + NaADP^{2-} + NaCr \sim P^{-} \end{aligned} \tag{107}$$

Those species have been neglected (including possible magnesium complexes of Cr) which are not present in significant concentrations between pH 6 and 10, where the estimates of their stability constants would suggest that they be ignored for the present, or where no data are available to assess the order of magnitude of the constants.

After substitution from the defined acid dissociation and stability constants (see below), the equations may be written as

$$\begin{aligned} ATP_t &\simeq ATP^{4-}\left[1 + \frac{(H^+)}{K_7} + \frac{K_2}{K_7}(Mg^{2+})(H^+) + K_1(Mg^{2+}) + K_1K_3(Mg^{2+})^2 \right. \\ &\quad \left. + 2K_1K_4(Mg^{2+})(ATP^{4-}) + K_6(Na^+)\right] \\ ADP_t &\simeq ADP^{3-}\left[1 + \frac{(H^+)}{L_7} + L_1(Mg^{2+}) + \frac{L_2}{L_7}(Mg^{2+})(H^+) + L_6(Na^+)\right] \\ Cr \sim P_t &\simeq Cr \sim P^{2-}\left[1 + \frac{(H^+)}{Q_7} + Q_1(Mg^{2+}) + Q_6(Na^+)\right] \end{aligned}$$

$$Cr_t \simeq Cr^{\pm}$$

$$B_t \simeq B^-\left[1 + \frac{(H^+)}{U_2} + U_1(Mg^{2+})\right]$$

$$Mg_t \simeq Mg^{2+}\left[1 + K_1(ATP^{4-}) + \frac{K_2}{K_7}(H^+)(ATP^{4-}) + 2K_1K_3(Mg^{2+})(ATP^{4-}) \right.$$

$$+ K_1K_4(ATP^{4-})^{2+} + L_1(ADP^{3-}) + \frac{L_2}{L_7}(H^+)(ADP^{3-})$$

$$\left. + Q_1(Cr \sim P^{2-}) + U_1(B^-) + \frac{K_mK_w}{(H^+)}\right]$$

$$Na_t \simeq Na^+[1 + K_6(ATP^{4-}) + L_6(ADP^{3-}) + Q_6(Cr \sim P^{2-})] \tag{108}$$

where the various equilibria are defined as

$$K_1 = \frac{(MgATP^{2-})}{(Mg^{2+})(ATP^{4-})}; \quad K_2 = \frac{MgHATP^-}{(Mg^{2+})(HATP^{3-})};$$

$$K_3 = \frac{Mg_2ATP}{(Mg^{2+})(MgATP^{2-})}; \quad K_4 = \frac{(Mg(ATP_2^{6-})}{(MgATP^{2-})(ATP^{4-})};$$

$$K_6 = \frac{(NaATP^{3-})}{(Na^+)(ATP^{4-})}; \quad K_7 = \frac{(ATP^{4-})(H^+)}{(HATP^{3-})};$$

$$L_1 = \frac{MgADP^-}{(Mg^{2+})(ADP^{3-})}; \quad L_2 = \frac{MgHADP}{(Mg^{2+})(HADP^{2-})};$$

$$L_6 = \frac{(NaADP^{2-})}{(Na^+)(ADP^{3-})}; \quad L_7 = \frac{(ADP^{3-})(H^+)}{(HADP^{2-})};$$

$$Q_1 = \frac{(MgCr \sim P)}{(Mg^{2+})(Cr \sim P^{2-})}; \quad Q_6 = \frac{(NaCr \sim P^-)}{(Na^+)(Cr \sim P^{2-})};$$

$$Q_7 = \frac{(Cr \sim P^{2-})(H^+)}{(HCr \sim P^-)};$$

$$U_1 = \frac{(MgB^+)}{(Mg^{2+})(B^-)}; \quad U_2 = \frac{(B^-)(H^+)}{(B^{\pm})};$$

$$K_m = \frac{(MgOH^+)}{(OH^-)(Mg^{2+})}; \quad K_w = (H^+)(OH^-) \tag{109}$$

In this case, not only the equilibrium concentrations of the various ionic and metal complex species were desired, but also a calculation of a measure of the thermodynamic equilibrium state from a set of almost 100 pieces of data at equilibrium.

At a given temperature and pH, a large number of defined mass-action constants can be expressed in terms of various combinations of these species (e.g., uncomplexed, Na complexes, and Mg chelates).

Three constants, K_a, K_b, and K_c were defined to demonstrate their numerical magnitudes.

$$K'_a = \frac{(ADP^{3-})(Cr \sim P^{2-})}{(ATP^{4-})(Cr^{\pm})} \tag{110}$$

$$K_a = K'_a(H^+) \tag{111}$$

$$K'_b = \frac{(MgADP^{-})(Cr \sim P^{2-})}{(MgATP^{2-})(Cr^{\pm})} \tag{112}$$

$$K_b = K'_b(H^+) \tag{113}$$

$$K'_c = \frac{(MgADP^{-})(Cr \sim P^{2-})(Mg^{2+})}{(Mg_2ATP)(Cr^{\pm})} \tag{114}$$

$$K_c = K'_c(H^+) \tag{115}$$

Reaction "a" is expressed by

$$ATP^{4-} + Cr^{\pm} \rightleftarrows ADP^{3-} + Cr \sim P^{2-} + H^+ \tag{116}$$

reaction "b" by

$$MgATP^{2-} + Cr^{\pm} \rightleftarrows MgADP^{-} + Cr \sim P^{2-} + H^+ \tag{117}$$

and reaction "c" by

$$Mg_2ATP + Cr^{\pm} \rightleftarrows MgADP^{-} + Cr \sim P^{2-} + Mg^{2+} + H^+ \tag{118}$$

where K_a, K_b, and K_c are the pH-independent equilibrium constants.

K_a and K_b are the constants of intrinsic interest in this discussion, whereas K_c is more of academic interest because of the previous mention of the possibility of such an equilibrium.[87] (However, see References 85 and 91.)

The procedure for selecting the numerical values of the stability and acid dissociation constants consisted in first deciding on the most reliable range of the values reported or which could be estimated.[48] Many of the stability constants were then systematically varied to determine what influence they would have on the final calculated equilibrium constants so as to yield numerical values most consistent with the 100 sets of equilibrium data available with Mg present.[87] The statistical analysis of variance was used as a guide for determining the best fit (see Addendum in Reference 48). The two dominating constants, K and L, were, in addition, varied in pairs. The final selected values are listed in Table 11 and, in most cases, rounded off to one significant figure. Monk's values for the buffers[86] have never been seriously contested (in contrast with the published metal stability values of the nucleotides, which vary enormously[353] and are continuously being remeasured or reevaluated). Monk's values were used directly after rounding off to one significant figure and correcting a typographical error for the pK value of glycine = 2.08.

Table 12 summarizes the average values of the calculated equilibrium constants of the ATP-creatine transphosphorylase-catalyzed reaction for the several conditions of pH and

TABLE 11
Selected Values of the Stability Constants and Acid Dissociation Constants for the Nucleotide and Nonnucleotide Species Involved in the ATP-Creatine Transphosphorylase-Catalyzed Equilibria[48]

		Formation constant (M^{-1}) for complexation or dissociation constant (M) for ionization		
Species	**Constant**	**Value selected for 30°C**	**Value selected for 20°C**	**Value selected for 38°C**
$MgATP^{2-}$	K_1	7×10^4	5×10^4	9×10^4
$MgHATP^-$	K_2	6×10^2	5×10^2	8×10^2
Mg_2ATP	K_3	2×10^1	2×10^1	2×10^1
$Mg(ATP)_2^{6-}$	K_4	3×10^2	2×10^2	4×10^2
$NaATP^{3-}$	K_6	1.4×10^1	1.0×10^1	1.7×10^1
$HATP^{3-}$	K_7	$10^{-6.9}$	$10^{-6.9}$	$10^{-6.9}$
$MgADP^-$	L_1	3×10^3	2×10^3	4×10^3
MgHADP	L_2	4×10^1	3×10^1	6×10^1
$NaADP^{2-}$	L_6	7	5	9
$HADP^{2-}$	L_7	$10^{-6.7}$	$10^{-6.7}$	$10^{-6.7}$
MgCr ~ P	Q_1	2.4×10^1	1.7×10^1	3.0×10^1
$NaCr \sim P^-$	Q_6	2	1	2
$HCr \sim P^-$	Q_7	$10^{-4.5}$	$10^{-4.5}$	$10^{-4.5}$
$MgOH^+$	K_m	4×10^2	4×10^2	4×10^2
	K_w	$10^{-13.8}$	$10^{-14.2}$	$10^{-13.6}$
Mg-glycine	U_1	1×10^2	1×10^2	1×10^2
	U_2	$10^{-9.65}$	$10^{-9.91}$	$10^{-9.45}$
Mg-glycyl glycine	U_1	1×10^1		
	U_2	$10^{-8.05}$		

temperature measured together with the standard deviations for the equilibrium constants obtained at the individual pH values and temperatures.

The statistical analyses revealed that, after all the trends had been eliminated from the estimated values, an error of approximately $\pm 6\%$ could be assigned to these equilibrium values. Probably the most accurate sets of data are those obtained at pH 8.0.[48,310,831] Thus, the relatively large standard deviation of the pH 9.8 data justified their omission from the final averages at 30°C which are listed together with an overall standard deviation from the mean of approximately $\pm 14\%$. In view of the complicated nature of the system, the final analysis may be considered quite satisfactory. (See also References 852 and 853).

By utilizing the average equilibrium constants at 20, 30, and 38°C obtained at pH values of 9.0, 8.9, and 8.8, respectively, the thermodynamic functions were calculated as shown in Table 13. Because of the relatively large σ values of the 20 and 38°C data, $\Delta H°$ values calculated from the average equilibrium constants can be uncertain to the extent of almost ± 5 kcal within these stated deviations. Since $\Delta S°$ is calculated from $\Delta H°$ and $\Delta G°$, it may also be uncertain to the extent of approximately ± 10 entropy units. On the other hand, $\Delta G°$ values would not vary by more than ± 0.1 kcal/mol within the same stated range of uncertainties.

The free energy of hydrolysis of $Cr \sim P^{2-}$ may be calculated next from the $\Delta G°$ values listed in Table 13 and the data of Burton and associates[88] or of Robbins and Boyer.[89]

The reaction

$$MgATP^{2-} + H_2O \rightleftarrows MgADP^- + HPO_4^{2-} + H^+ \quad (124)$$

at pH 7 was estimated by Burton to be -7.0 kcal/mol at 37°C or by Robbins and Boyer to

TABLE 12
Equilibrium Constants of ATP-Creatine Transphosphorylase-Catalyzed Reaction[48]

pH	Temp (°C)	No. of determinations	$K_a' (H^+) = K_a$	$K_b' (H^+) = K_b$	$K_c' (H^+) = K_c$	α, SD from the mean for K_a, K_b, K_c (%)
7.4	30°	12	$7.2\ (\pm 0.88) \times 10^{-9}$	$3.08\ (\pm 0.38) \times 10^{-10}$	$1.54\ (\pm 0.19) \times 10^{-11}$	12
8.0	30°	12	$7.04\ (\pm 0.43) \times 10^{-9}$	$3.02\ (\pm 0.18) \times 10^{-10}$	$1.51\ (\pm 0.09) \times 10^{-11}$	6
8.9	30°	16	$5.71\ (\pm 0.62) \times 10^{-9}$	$2.45\ (\pm 0.27) \times 10^{-10}$	$1.22\ (\pm 0.13) \times 10^{-11}$	11
Average	30°	40	$6.56\ (\pm 0.95) \times 10^{-9}$	$2.81\ (\pm 0.41) \times 10^{-10}$	$1.40\ (\pm 0.20) \times 10^{-11}$	14
9.0	20°	16	$6.30\ (\pm 1.26) \times 10^{-9}$	$2.52\ (\pm 0.52) \times 10^{-10}$	$1.26\ (\pm 0.26) \times 10^{-11}$	20
8.8	38°	16	$5.47\ (\pm 0.85) \times 10^{-9}$	$2.43\ (\pm 0.38) \times 10^{-10}$	$1.22\ (\pm 0.19) \times 10^{-11}$	16
9.8	30°	16	$2.98\ (\pm 0.73) \times 10^{-9}$	$1.28\ (\pm 0.31) \times 10^{-11}$	$0.644\ (\pm 0.157) \times 10^{-11}$	24

Note: σ (standard deviation from the mean) = $\left[\frac{\sum (x-\bar{x})^2}{(n-1)}\right]^{1/2}$ (119)

TABLE 13
Thermodynamic Data for the ATP-Creatine Transphosphorylase-Associated Reactions

	$\Delta G°$ (kcal/mol)				
	20°	30°	38°	$\Delta H°$ (kcal/mol)	$\Delta S°$ (cal/degree/mol)
Reaction "a" (uncomplexed nucleotides)	11.0	11.4	11.8	−1.4	−33
Reaction "b" (magnesium nucleotides)	12.9	13.3	13.7	−0.5	−46

Note: $\Delta H°$ from the integrated van't Hoff equation: $\Delta G°$ is calculated from $\Delta G° = -RT \ln K$,

$$\frac{d\ln K}{dT} = \frac{\Delta H°}{RT^2} \tag{121}$$

$$\ln\frac{K_2}{K_1} = \frac{-\Delta H°}{R}\left(\frac{1}{T_2} - \frac{1}{T_1}\right) \tag{122}$$

and

$$-\Delta S° = \frac{(\Delta G° - \Delta H°)}{T} \tag{123}$$

be −7.8 kcal/mol at 30°C. For reaction "b" (Equation 117), the average value for $K_b'(H^+)$ was 2.81×10^{-10} for 30°C, or calculated with a $-\Delta H°$ of 0.5 kcal/mol to be 2.76×10^{-10} for 37°C. The K_b' values at pH 7.0 are thus 2.81×10^{-3} and 2.76×10^{-3}, respectively; $\Delta G°$ at pH 7.0 may be calculated to be +3.5 and +3.6 kcal/mol at 30 and 37°C, respectively. Subtracting the hydrolysis equation (Equation 124) for $MgATP^{2-}$ from reaction "b" (Equation 117) at pH 7.0, and using the $\Delta G°$ value of Burton, one obtains a $\Delta G°$ of 10.6 kcal/mol for the reaction.

$$HPO_4^{2-} + Cr^{\pm} \rightleftarrows Cr \sim P^{-2} + H_2O \tag{125}$$

or −10.6 kcal/mol for the free energy of hydrolysis of $Cr \sim P^{2-}$ at 37°C and −11.3 kcal/mol at pH 7.0 and 30°C with the use of Robbins and Boyer's value.

Similarly, the data of Burton at 37°C for the hydrolysis of ATP^{4-} at pH 7.0 and pH 7.5, −8.6 and −9.3 kcal/mol, respectively, may be used to compute the $\Delta G°$ of $Cr \sim P^{2-}$ hydrolysis from the values for reaction "a" (Equation 116) at 37°C, i.e., +1.7 and +1.0 kcal/mol at pH 7.0 and 7.5 respectively. $\Delta G°$ for $Cr \sim P^{2-}$ hydrolysis at 37°C is then calculated to be −10.3 kcal/mol for both pH 7.0 and 7.5; this value appears to be the best estimate for 37°C.

Finally, the data support the idea that a proton is released by the equilibrium reactions "a" or "b" (Equations 116 and 117). Thus, whereas the values for K_a' and K_b' vary considerably over the pH range studied (several hundredfold), both K_a and K_b show only a small trend with pH, which may not be significant in view of the experimental error and the large number of assumptions and approximations made in the estimations.

It is of interest also to compute from the data available the average number of H^+ equivalents liberated per 1 mol of each of the reactants at a given pH and temperature, defined by the symbol ν_{H^+}. Such theoretical calculations have already been made by Alberty et al.[90] and by Smith[374] for the hydrolysis of ATP^{4-}, and analogous calculations can be made from their generalized reaction:

$$A^a + B^b \rightarrow C^c + D^d + [(a + b) - (c + d)](H^+) \tag{126}$$

where a, b, c, and d are the average ionic charges (not necessarily integers) of the reactants and products...." and the term [(a + b) − (c + d)] represents the number of H^+ ions produced by the reaction.[90] As an example, see reaction "b" (Equation 117).

$$MgATP^{2-} + Cr^{\pm} \rightleftarrows MgADP^{-} + Cr \sim P^{2-} + V_{H^+}(H^+) \qquad (127)$$

at concentrations of total species of ATP_t, Cr_t, ADP_t, and $Cr \sim P_t$, and at a particular pH and temperature (pH maintained by neutralization of the acid equivalents liberated by either a buffer or by the addition of a strong base during the course of the reaction to equilibrium as in a pH-stat), the equation may be written as

$$\nu_{H^+} = \begin{pmatrix}\text{Net charge} \\ \text{of MgADP}^-\end{pmatrix} \frac{(MgADP^-)}{(ADP_t)} + \begin{pmatrix}\text{Net charge} \\ \text{of Cr} \sim P^{2-}\end{pmatrix} \frac{(Cr \sim P^{2-})}{(Cr \sim P_t)}$$

$$- \begin{pmatrix}\text{Net charge} \\ \text{of MgATP}^{2-}\end{pmatrix} \frac{(MgATP^{2-})}{(ATP_t)} - \begin{pmatrix}\text{Net charge} \\ \text{of Cr}^{\pm}\end{pmatrix} \frac{(Cr^{\pm})}{(Cr_t)} \qquad (128)$$

or

$$\nu_{H^+} = (1)f_{MgADP^-} + (2)f_{CR\sim P^{2-}} - (2)f_{MgATP^{2-}} \qquad (129)$$

where f denotes the fraction of these species in their particular complex or ionic form, and ν_{H+} is the number of acid equivalents required to maintain the electrical equality of the system per mole or reactants.

With the pKs as listed in Table 11, theoretically, at the pH values studies, the value of ν_{H+} should approach unity at a pH of approximately 8, as demonstrated by Kuby and Noltmann[48] from the calculated data of the various ionic and complex species at equilibrium.

Direct determinations of ν_{H+} have been forthcoming from pH-stat experiments. Thus, with the stoichiometric ratio, ν_{H+} defined as equal to

$$\frac{(H^+)_{formed}}{(ATP_t)_{disappeared}} = \nu_{H^+} \qquad (130)$$

at pH 8.8 and 30°C, ν_{H+} was determined under pH-stat conditions for the forward reaction in the presence of Mg_t to be 1.00 ± 0.02;[92] for the reverse reaction in the presence of Ca^{2+} and K^+ at 25°C, a value of −0.98 ± 0.02 was found for ν_{H+} by manual titration[93] and substantiates the above conclusion. As the pH decreases below pH 8, ν_{H+} should depart from 1.0, and, for example, Jacobs and Kuby[49] found $\nu_{H+} = 0.95 \pm 0.02$ at pH 7.4 and 30°C.

Recently, Garfinkel and Garfinkel[809] provided an algorithm for the calculation of free Mg^{2+} and of total ATP concentrations, which may be carried out with a hand calculator. Their calculations agreed with those of Gupta and Moore,[810] that is, that the level of free Mg^{2+} in frog skeletal muscle is approximately 0.6 m*M*. Refer also to References 809 and 811 for more recent values of stability constants of MgATP. Also, note the very recent estimates of the stability constants of Mg^{2+} and Cd^{2+} complexes of the adenine nucleotides and thionucleotides (as well as the rate constants for formation and dissociation of MgATP and MgADP) by Pecoraro et al.,[816] which they determined at 30°C and $\mu = 0.1M$ (by the addition of KNO_3) with the use of ^{31}P NMR.

APPENDIX

I. DERIVATION OF EQUATIONS 20 AND 24 FROM CHAPTER 1 CORRESPONDING TO THE MECHANISM OF SCHEME II FOR INHIBITION BY HIGH SUBSTRATE CONCENTRATION[41]

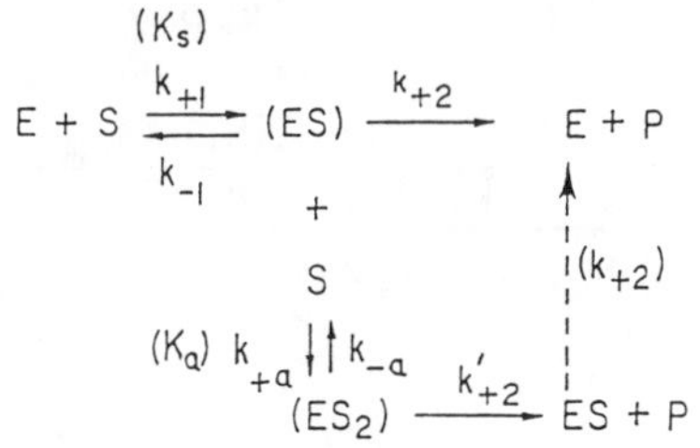

STRUCTURE I.

$$E_t = E + ES + ES_2 \tag{b}$$

$$0 = \frac{d(ES)}{dt} = k_{+1}(E)(S) - [k_{-1} + k_{+2} + k_{+a}(S)](ES) + k_{-a}(ES_2) \tag{c}$$

$$0 = \frac{d(ES_2)}{dt} = k_{+a}(ES)(S) - (k_{-a} + k'_{+2})(ES_2) \tag{d}$$

$$\therefore \quad (E) = \frac{k_{-1} + k_{+2} + k_{+a}(S))(ES) - k_{-a}(ES_2)}{k_{+1}(S)} \tag{e}$$

$$(ES_2) = \frac{k_{+a}(ES)(S)}{k_{-a} + k'_{+2}} \tag{f}$$

$$\therefore \quad (E) = \frac{(k_{-1} + k_{+2} + k_{+a}(S))(ES) - \left(\dfrac{k_{-a}k_{+a}(ES)(S)}{k_{-a} + k'_{+2}}\right)}{k_{+1}(S)} \tag{g}$$

$$(E) = \frac{(k_{-1} + k_{+2} + k_{+a}(S))(ES)}{k_{+1}(S)} - \frac{k_{-a}k_{+a}(ES)}{k_{+1}(k_{-a} + k'_{+2})} \tag{h}$$

insertion into Equation (b):

$$\therefore \quad E_t = \left[\frac{k_{-1} + k_{+2} + k_{+a}(S)}{k_{+1}(S)}\right](ES) - \frac{k_{-a}k_{+a}(ES)}{k_{-1}(k_{-a} + k'_{+2})}$$
$$+ ES + \frac{k_{+a}(ES)(S)}{k_{-a} + k'_{+2}} \tag{i}$$

$$\therefore \quad E_t = (ES)\left[\frac{k_{-1} + k_{+2} + k_{+a}(S)}{k_{+1}(S)} - \frac{k_{-a}k_{+a}}{k_{+1}(k_{-a} + k'_{+2})} + 1 + \frac{k_{+a}(S)}{k_{-a} + k'_{+2}}\right] \qquad (j)$$

Case (a): if $k_{+2}' = 0$, $v_0 = k_{+2}$ (ES) and, therefore,

$$v_o = \frac{k_{+2}E_t}{\left[\frac{k_{-1} + k_{+2} + k_{+a}(S)}{k_{+1}(S)} - \frac{k_{-a}k_{+a}}{k_{+1}k_{-a}} + 1 + \frac{k_{+a}(S)}{k_{-a}}\right]} \qquad (k)$$

$$v_o = \frac{k_{+2}E_t(S)}{\left[\left(\frac{k_{-1} + k_{+2} + k_{+a}(S)}{k_{+1}}\right) - \frac{k_{+a}(S)}{k_{+1}} + (S) + \frac{k_{+a}(S)^2}{k_{-a}}\right]} \qquad (l)$$

$$v_o = \frac{k_{+2}E_t(S)}{\left(\frac{k_{-1} + k_{+2}}{k_{+1}}\right) + (S) + \left(\frac{k_{+a}}{k_{-a}}\right)(S)^2} \qquad (m)$$

$$v_o = \frac{k_{+2}E_t(S)}{\left[\frac{k_{-1}}{k_{+1}} + \left(\frac{k_{+2}}{k_{+1}}\right) + (S) + \frac{k_{+a}}{k_{-a}}(S)^2\right]} \qquad (n)$$

$$v_o = \frac{k_{+2}E_t(S)}{\left(\frac{k_{-1}}{k_{+1}} + \frac{k_{+2}}{k_{+1}}\right) + (S) + \frac{k_{+a}}{k_{-a}}(S)^2} \qquad (o)$$

$$v_o = \frac{k_{+2}E_t(S)}{\left(\frac{k_{-1} + k_{+2}}{k_{+1}}\right) + (S) + \frac{k_{+a}}{k_{-a}}(S)^2} \qquad (p)$$

$$K_m = \frac{k_{-1} + k_{+2}}{k_{+1}} \quad \text{and} \quad K_a = \frac{k_{+a}}{k_{-a}}, \quad v_o = \frac{k_{+2}E(S)}{(K_m + (S) + K_a(S)^2)} \qquad (q)$$

and

$$k_{+2}E_t = V_{max}$$

$$v_o = \frac{V_{max}(S)}{K_m + (S) + K_a(S)^2} \qquad (r)$$

Thus, at high (S), the $(S)^2$ term dominates in the denominator, and the rate becomes zero at high concentrations of substrate.

A maximum rate is obtained at a substrate concentration, when

$$\frac{dv}{d(S)} = 0 \quad \text{or} \quad (S)_{max} = \left(\frac{K_m}{K_a}\right)^2$$

Proof:

$$0 = \frac{d(v_o)}{d(S)} = \frac{[K_m + (S) + K_a(S)^2]V_{max} - V_{max}(S)[1 + 2K_a(S)]}{[K_m + (S) + K_a(S)^2]^2}$$

$$= \frac{V_{max}K_m + V_{max}(S) + V_{max}K_a(S)^2 - V_{max}(S) - 2V_{max}K_a(S)^2}{[K_m + (S) + K_a(S)^2]^2}$$

$$0 = \frac{V_{max}K_m - V_{max}K_a(S)^2}{[K_m + (S) + K_a(S)^2]^2}$$

$$\therefore \; V_{max}K_m - V_{max}K_a(S)^2 = 0; \quad \text{and} \quad (S)_{max} = (K_m/K_a)^{1/2}$$

Case (b): if $k'_{+2} \neq 0$,

$$v_{overall} = k_{+2}(ES) + k'_{+2}(ES_2) \tag{s}$$

$\therefore$ from Equation (f),

$$= k_{+2}(ES) + k'_{+2}\left(\frac{k_{+a}(ES)(S)}{k_{-a} + k'_{+2}}\right) \tag{t}$$

$$v = \left\{k_{+2} + \frac{k'_{+2}(k_{+a})(S)}{k_{-a} + k'_{+2}}\right\}(ES) \tag{u}$$

Substitute Equation (j) for (ES) into Equation (u)

$$v = \frac{\left\{k_{+2} + \dfrac{k'_{+2}(k_{+a})(S)}{(k_{-a} + k'_{+2})}\right\}E_t}{\left\{\dfrac{k_{-1} + k_{+2} + k_{+a}(S)}{k_{+1}(S)} - \dfrac{k_{-a}k_{+a}}{k_{+1}(k_{-a} + k'_{+2})} + 1 + \left(\dfrac{k_{+a}(S)}{k_{-a} + k'_{+2}}\right)\right\}} \tag{v}$$

$$v = \frac{\left[k_{+2} + \dfrac{k'_{+2}(k_{+2})(S)}{k_{-a} + k'_{+2}}\right]E_t(S)}{\left[\dfrac{k_{-1} + k_{+2} + k_{+a}(S)}{k_{+1}}\right] - \dfrac{k_{-a}k_{+a}(S)}{k_{+1}(k_{-a} + k'_{+2})} + (S) + \left(\dfrac{k_{+a}(S)^2}{k_{-a} + k'_{+2}}\right)} \tag{w}$$

$$v = \frac{\left\{k_{+2}\dfrac{\dfrac{k'_{+2}(S)}{(k_{-a} + k'_{+2})}}{k_{+2}}\right\}E_t(S)}{\left(\dfrac{k_{-1} + k_{+2}}{k_{+1}}\right) + \dfrac{k_{+a}(S)}{k_{+1}} - \dfrac{k_{-a}}{k_{+1}}\dfrac{(S)}{[(k_{-a} + k'_{+2})/k_{+a}]} + (S) + \dfrac{(S)^2}{[(k_{-a} + k'_{+2})/k_{+a}]}} \tag{x}$$

$$K'_m = \frac{k_{-a} + k'_{+2}}{k_{+a}} \quad \text{and} \quad K_m = \frac{k_{-1} + k_{+2}}{k_{+1}}$$

$$v = \frac{\left\{k_{+2} + \dfrac{k'_{+2}(S)}{k'_m}\right\}E_t(S)}{K_m + \left(\dfrac{k_{+a}}{k_{+1}} - \dfrac{k_{-a}}{k_{+1}K'_m} + 1\right)(S) + \dfrac{(S)^2}{k'_m}} \tag{y}$$

At sufficiently high (S), $v \rightarrow v_{lim}$, given by

$$v_{lim} = \frac{\frac{k'_{+2}(S)}{K'_m} E_t(S)}{\frac{(S)^2}{K'_m}} = \frac{\frac{k'_{+2}(S)^2 E_t}{K'_m}}{\frac{(S)^2}{K'_m}} = k'_{+2}E_t = V_{max,2}$$

II. DERIVATION OF EQUATION 79 FROM CHAPTER 1 FOR INHIBITION OF ENZYME (E) BY ANTIENZYME (I) WITH A STOICHIOMETRY OF 2E:1 I[29]

$$E_2I \overset{K_i}{\rightleftarrows} 2E + I \tag{a}$$

$$(1 - \alpha)\ (2\alpha) \tag{b}$$

$$K'_d = \frac{(E)^2(I)}{(E_2I)} \tag{c}$$

α = fraction dissociated; $(1 - \alpha)$ = fraction undissociated.

$$i = \left(1 - \frac{v_i}{v_o}\right) = \text{fraction inhibited} \tag{d}$$

where v_0 = initial velocity in absence of I and v_i = initial velocity in presence of I. Conservation equations:

$$(I_t) = (I) + (E_2 \cdot I) \tag{e}$$

$$(E_t) = (E) + 2(E_2 \cdot I) \tag{f}$$

$$E_t \propto (1 - \alpha) + 2\alpha = 1 + \alpha \tag{g}$$

$$\frac{E}{E_t} = \frac{2\alpha}{(1 + \alpha)}\ ; \quad \frac{E_2I}{E_t} = \frac{(1 - \alpha)}{(1 + \alpha)} \tag{h}$$

Let

$$\text{Let} \quad (1 - \alpha) = i \tag{i}$$

that is, the fraction inhibited is proportional to (E_2I) or to the fraction undissociated. Therefore,

$$\alpha = (1 - i) \tag{j}$$

and

$$2\alpha = 2(1 - i) \tag{k}$$

which is proportional to the free E or to the fraction dissociated, and therefore

$$E_t \propto i + 2(1 - i) = 2 - i \quad \text{or} \quad (1 + \alpha) = 2 - i \tag{l}$$

Therefore,

$$\frac{E_2I}{E_t} = \frac{(1 - \alpha)}{(1 + \alpha)} = \frac{i}{(2 - i)} \tag{m}$$

and

$$\frac{E}{E_t} = \frac{2\alpha}{(1 + \alpha)} = \frac{2(1 - i)}{(2 - i)} \tag{n}$$

Therefore,

$$(E_2I) = \left(\frac{i}{2 - i}\right)(E_t) \tag{o}$$

and

$$(E) = \frac{(2 - 2i)}{(2 - i)}(E_t) \tag{p}$$

$$K'_d = \frac{(E)^2(I)}{(E_2I)} = \frac{\left(\frac{2 - 2i}{2 - i}\right)^2 E_t^2}{\left(\frac{i}{2 - i}\right)E_t}(I) = (E_t)\frac{(2 - 2i)^2}{(2 - i)}\frac{(I)}{(i)} \tag{q}$$

and

$$(I) = \frac{K'_d(2 - i)(i)}{E_t(2 - 2i)^2} \tag{r}$$

$$(I_t) = (I) + (E_2I) = \frac{K'_d(2 - i)(i)}{E_t(2 - 2i)^2} + \left(\frac{i}{2 - i}\right)E_t \tag{s}$$

Therefore,

$$(I_t)\frac{(2 - i)}{(i)} = \frac{(2 - i)^2}{(2 - 2i)^2}\frac{K'_d}{(E_t)} + (E_t) \tag{t}$$

Section II
Enzyme Kinetics and Substrate Binding

The action of enzymes seems to be dependent on two factors: one geometric, the other energetic. The geometric factor is determined by the spatial relationships of the substrate and the corresponding enzyme . . . The question of the energetics of enzyme-catalyzed reactions is [however] one about which we know the least. David Rittenberg, On the bigness of enzymes, in *Essays in Biochemistry*, Graff, S., ed., John Wiley & Sons, New York, 1956, 232—233.

Chapter 6

ISOTOPE EXCHANGE STUDIES AND KINETIC ISOTOPE EFFECTS

I. ISOTOPIC EXCHANGE STUDIES

The technique of isotopic exchange, conducted at chemical equilibrium (although chemical equilibrium is not a requirement for useful application of the technique), is a powerful tool for evaluation of enzymatic mechanisms and to confirm the mechanism which may emerge from initial-rate and product inhibition studies. Boyer[94] was one of the first to introduce the technique to enzyme kinetics. In most cases, unfortunately, the velocity equations for isotope exchange are far too complicated to permit the determination of the usual kinetic constants; however, of course, the technique can be utilized to determine the overall equilibrium constant of the system. Also, even traces of contaminating enzymes with side reactions may obscure the expected results so that purity of the enzymatic catalyst sometimes becomes a determining factor as to whether the technique will yield useful information. Thus, although in its present development the technique is not a quantitative one, it may be used qualitatively. By observing the effects of varying reactant concentrations on the exchange rates, the kinetic mechanism can be deduced or confirmed. The two basic questions to be asked in the case of multireactant kinetic mechanisms are (1) does the enzyme catalyze and exchange between a given substrate-product pair in the absence of all other reactants and (2) is substrate inhibition observed when a particular substrate-product pair is varied, and if it is, is this inhibition total or partial?

We will illustrate the approach for one case, that is, an ordered two-substrate, two-product reaction.

In order to apply the results of isotope exchange experiments, two important assumptions are normally made, even if only implied. (1) A reaction involving radioactively labeled substrates follows the same mechanism as the normal reaction and with the same rate constants. This assumes that many isotope effects are negligible, which is ordinary a safe assumption, provided tritium is not used as a radioactive atom and especially that the tritium atom is not directly involved in the reaction or in the binding of the substrate to the enzyme (Section II). (2) The concentrations of all radioactive species are so small as to have a negligible effect on the concentrations of unlabeled species; otherwise, their concentrations should be included in the calculated concentrations.

The isotope exchange can be more readily understood in relation to examples, and we will consider the transfer, at chemical equilibrium, of a radioactive atom (represented by an asterisk) from (1) A* to P* and (2) A* to Q* in the compulsory-order ternary complex mechanism of an ordered BiBi system. (See Chapter 2, Section I.D.7 and Equations 142 and 147.)

A. A*-P EXCHANGE IN AN ORDERED BiBi SYSTEM AT CHEMICAL EQUILIBRIUM

The basic King-Altman figure is as follows (Scheme I).

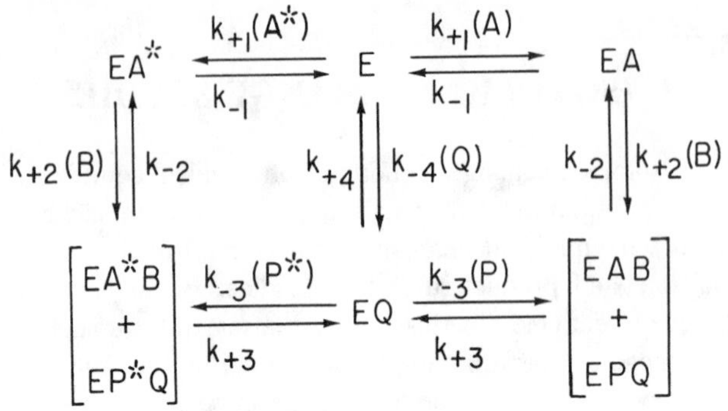

Scheme I

In Scheme I, the A*-P exchange requires the binding of A* to E, which, of course, can occur only if there is a significant concentration of free E. Therefore, this exchange reaction will be inhibited by a high concentration of either unlabeled (A) or (Q), as they will compete with labeled A* for E. The effects of unlabeled (B) and (P) are more subtle. Thus, the exchange reaction of A*-P* include the step of B binding to EA*, and therefore, a finite concentration of B is required. However, if both (B) and (P) are present at very high concentrations, the enzyme will exist largely as (EAB + EPQ) and so there will, in essence, be no E to which A* can bind. Therefore, at low values of (B) (with (B/P) ratio fixed so as to maintain the chemical equilibrium), the velocity of A*-P exchange, $v^*_{A^*\text{-}P}$, will appear to rise, but as (B) (together with (P)) is further increased, the velocity will reach a maximum and then fall to zero.

One can solve for the rate equation using a King-Altman procedure[95] a modified King-Altman procedure[82], or algebraically by solving the appropriate set of simultaneous differential steady-state equations.[97]

Since we are interested, for this example, only in the approach, we will consider only the initial velocity for exchange A* into P, at chemical equilibrium, and the algebraic procedure is simpler.

Thus, the rates of change of the concentrations of the labeled intermediates can be written in the usual way, and then set to zero according to the usual steady-state assumption.

$$\frac{d(EA^*)}{dt} = k_{+1}(E)(A^*) - (k_{-1} + k_{+2}(B))(EA^*) + k_{-2}(EA^*B) = 0 \quad (1)$$

$$\frac{d(EA^*B)}{dt} = k_{+2}(B)(EA^*) - (k_{-2} + k_{+3})(EA^*B) + k_{-3}(P^*)(EQ) = 0 \quad (2)$$

Since at t = 0, P* = 0, and solving for EA*B from these two equations,

$$(EA^*B) = \frac{k_{+1}k_{+2}(E)(A^*)(B)}{k_{-1}(k_{-2} + k_{+3}) + k_{+2}k_{+3}(B)} \quad (3)$$

The initial rate of exchange

$$v^*_{A^*\text{-}P} = k_{+3}(EA^*B) \quad (4)$$

Therefore,

$$v^*_{A^*\text{-}P} = \frac{k_{+1}k_{+2}k_{+3}(E)(A^*)(B)}{k_{-1}(k_{-2} + k_{+3}) + k_{+2}k_{+3}(B)} \tag{5}$$

For this expression to be used, (E) must be known; however, this would be available only if all the kinetic constants of the unlabeled reaction have been determined. For this reason, it is simpler, and is the usual case, to study the isotope exchange under chemical equilibrium conditions where the unlabeled reactants are maintained at their equilibrium concentrations. At equilibrium, and utilizing the conservation equation for E_t in terms of E,

$$E_t = E + EA + EAB + EQ = E + \frac{E(A)}{K_1} + \frac{E(A)(B)}{K_1K_2} + \frac{E(Q)}{K_4} \tag{6}$$

and

$$\frac{E}{E_t} = \frac{1}{\left[1 + \dfrac{(A)}{\left(\dfrac{k_{-1}}{k_{+1}}\right)} + \dfrac{(A)(B)}{\left(\dfrac{k_{-1}}{k_{+1}}\right)\left(\dfrac{k_{-2}}{k_{+2}}\right)} + \dfrac{(Q)}{\left(\dfrac{k_{+4}}{k_{-4}}\right)}\right]} \tag{7}$$

where

$$K_1 = \frac{k_{-1}}{k_{+1}}, \quad K_2 = \frac{k_{-2}}{k_{+2}} \quad \text{and} \quad K_4 = \frac{k_{+4}}{k_{-4}} \tag{8}$$

Substitute E/E_t from Equation 7 into Equation 5 to yield Equation 9. Therefore,

$$v^*_{A^*\text{-}P} = \frac{k_{+1}k_{+2}k_{+3}E_t(A^*)(B)}{[k_{-1}(k_{-2} + k_{+3}) + k_{+2}k_{+3}(B)]\left[1 + \dfrac{k_{+1}(A)}{k_{+1}} + \dfrac{k_{+1}k_{+2}(A)(B)}{k_{-1}k_{-2}} + \dfrac{k_{-4}(Q)}{k_{+4}}\right]} \tag{9}$$

To place in terms of the kinetic parameters, multiply numerator and denominator by k_{+4},

$$v^*_{A^*\text{-}P} = \frac{k_{+1}k_{+2}k_{+3}k_{+4}E_t(A^*)(B)}{k_{+4}[k_{-1}(k_{-2} + k_{+3}) + k_{+2}k_{+3}(B)]\left[1 + \dfrac{(A)}{K_{ia}} + \dfrac{(A)(B)}{K_{ia}K_{\underline{\underline{B}}}} + \dfrac{(Q)}{K_{iq}}\right]} \tag{10}$$

where, we define as before,

$$\frac{\text{num}_1}{\text{coef}_{AB}} = V^f_{max} = \frac{k_{+1}k_{+2}k_{+3}k_{+4}E_t}{k_{+1}k_{+2}(k_{+3} + k_{+4})} = \frac{k_{+3}k_{+4}}{k_{+3} + k_{+4}}E_t = \frac{E_t}{\dfrac{1}{k_{+3}} + \dfrac{1}{k_{+4}}} \tag{10a}$$

$$\frac{k_{-1}}{k_{+1}} = K_{ia}, \quad \frac{k_{-2}}{k_{+2}} = K_{\underline{\underline{B}}} \quad \text{and} \quad \frac{k_{+4}}{k_{-4}} = K_{iq} \tag{11}$$

$$\therefore \quad k_{+1}k_{+2}k_{+3}k_{+4}E_t = V^f_{max}(k_{+1}k_{+2})(k_{+3} + k_{+4}) \tag{12}$$

$$\therefore \quad v^*_{A^*\text{-}P} = \frac{V_{max}k_{+1}k_{+2}(k_{+3} + k_{+4})(A^*)(B)}{[k_{+4}k_{-1}(k_{-2} + k_{+3}) + k_{+4}k_{+2}k_{+3}(B)]\left[1 + \frac{(A)}{K_{ia}} + \frac{(A)(B)}{K_{ia}K_{\underline{B}}} + \frac{(Q)}{K_{iq}}\right]} \tag{13}$$

$$v^*_{A^*\text{-}P} = \frac{V^f_{max}(A^*)(B)}{\left[\frac{(k_{+1}/k_{+1})k_{+4}k_{-1}(k_{-2} + k_{+3}) + k_{+2}k_{+3}k_{+4}(B)}{k_{+1}k_{+2}(k_{+3} + k_{+4})}\right]\left[1 + \frac{(A)}{K_{ia}} + \frac{(A)(B)}{K_{ia}K_{\underline{B}}} + \frac{(Q)}{K_{iq}}\right]} \tag{14}$$

(Where we multiplied by $\frac{k_{+1}}{k_{+1}}$ the first term in the denominator.) Since

$$K_{m_B} = \frac{coef_A}{coef_{AB}} \qquad coef_{AB} = k_{+1}k_{+2}(k_{+3} + k_{+4})$$

$$coef_A = k_{+1}k_{+4}(k_{-2} + k_{+3})$$

$$K_{m_A} = \frac{coef_B}{coef_{AB}} \qquad coef_B = k_{+2}k_{+3}k_{+4}$$

$$K_{ia} = \frac{k_{-1}}{k_{+1}} \tag{15}$$

therefore,

$$v^*_{A^*\text{-}P} = \frac{V^f_{max}(A^*)(B)}{[K_{ia}K_{m_B} + K_{m_A}(B)]\left[1 + \frac{(A)}{K_{ia}} + \frac{(A)(B)}{K_{ia}K_{\underline{B}}} + \frac{(Q)}{K_{iq}}\right]} \tag{16}$$

To summarize, define

$$\frac{k_{-1}}{k_{+1}} = K_{ia}, \quad \frac{k_{-2}}{k_{+2}} = K_{\underline{B}} \quad \text{and} \quad \frac{k_{+4}}{k_{-4}} = K_{iq}$$

$$\frac{num}{coef_{AB}} = V^f_{max} = \frac{k_{+1}k_{+2}k_{+3}k_{+4}E_t}{k_{+1}k_{+2}(k_{+3} + k_{+4})}$$

$$K_{m_A} = \frac{coef_A}{coef_{AB}} = \frac{k_{+2}k_{+3}k_{+4}}{k_{+1}k_{+2}(k_{+3} + k_{+4})}$$

$$K_{m_B} = \frac{coef_B}{coef_{AB}} = \frac{k_{+1}k_{+4}(k_{-2} + k_{+3})}{k_{+1}k_{+2}(k_{+3} + k_{+4})} \tag{17}$$

After substitution of definitions and multiplying through,

$$v^*_{A^*\text{-}P} = \frac{V^f_{max}(A^*)(B)}{\left[K_{ia}K_{m_B} + K_{m_B}(A) + \frac{K_{m_A}(A)(B)}{K_{\underline{B}}} + \frac{K_{ia}K_{m_B}(Q)}{K_{iq}} + K_{m_A}(B) + \frac{K_{m_A}(B)(A)}{K_{ia}} + \frac{K_{m_A}(A)(B)^2}{K_{ia}K_{\underline{B}}} + \frac{K_{m_A}(A)(Q)}{K_{iq}}\right]} \tag{18}$$

We note that P does not have to be included in Equation 18 for $v^*_{A^*\text{-}P}$, since, if equilibrium is to be maintained, only three out of the four reactant concentrations can be chosen at will. Any one of the reactants, (A), (B) or (Q) can be replaced by (P), since the following identity holds.

$$K_{eq} = \frac{k_{+1}k_{+2}k_{+3}k_{+4}}{k_{-1}k_{-2}k_{-3}k_{-4}} = \frac{(P)(Q)}{(A)(B)} \tag{19}$$

If (B) and (P) are varied in a constant ratio (i.e., $\frac{B}{P} = x$, in order to maintain equilibrium) at fixed values of (A) and (Q), the effect on the exchange rate can be estimated by noting that the denominator of Equation 18 is quadratic in (B) (i.e., a $(B)^2$ term appears), but the numerator is directly proportional to (B).

This equation is of the same form as an equation for substrate inhibition (i.e., for a single-substrate reaction where a second-substrate molecule might bind to the ES complex to yield SES, and at low substrate concentration the S^2 term is not significant, but at high concentration of S, S^2 becomes significant and the velocity rises to a maximum and then $\rightarrow 0$, instead of V_{max}).

$$v_f = \frac{V_{max}}{\left(\frac{K_m}{(S)} + 1 + \frac{(S)}{K_{is}}\right)} = \frac{V_{max}(S)}{\left(K_m + (S) + \frac{(S)^2}{K_{is}}\right)} \tag{20}$$

for

$$\begin{array}{ccccc} E + S & \underset{k_{-1}}{\overset{k_{+1}}{\rightleftarrows}} & (ES + EP) & \underset{k_{-2}}{\overset{k_{+2}}{\rightleftarrows}} & E + P \\ & & + & & \\ & & S & & \\ & & \downarrow\uparrow K_{is} & & \\ & & SES & & \end{array}$$

SCHEME II

(See also Chapter 7, Equations 44 and 45, and Chapter 1, Section IV.B.) Thus, the exchange velocity for A*-P, as (B) and (P) are increased at a constant ratio from zero to saturation, rises to maximum and then decreases to zero.

Varying (A) and (Q) where (Q) = X(A) does not introduce any $(A)^2$ terms into the denominator and, consequently, no substrate inhibition is observed.

Since no (P) terms appear in Equation 18 (see Equation 9), varying (A) and (P) (at constant ratio) does not lead to substrate inhibition. This is verified by substituting

$$\frac{(P)(Q)}{K_{ip}K_{\underline{\underline{P}}}} = \frac{(A)(B)}{K_{ia}K_{\underline{\underline{B}}}} \quad \text{and} \quad \frac{(A)(B)K_{\underline{\underline{P}}}}{K_{ia}K_{\underline{\underline{B}}}(P)} = \frac{(Q)}{K_{iq}} \tag{21}$$

where $K_{\underline{\underline{P}}} = \frac{k_{+3}}{k_{-3}}$ in the second bracket term in the denominator of Equation 16 and therefore providing an exchange velocity equation containing denominator (P) terms.

However, there are no (A) (P) terms to become $(A)^2$ when (P) = X(A). If we rearrange Equation 18 to

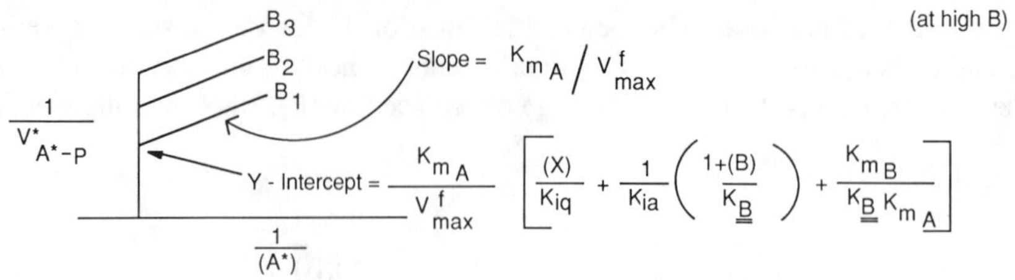

FIGURE 1.

$$\frac{v^*_{A^*\text{-}P}}{V^f_{max}} = \frac{1}{\frac{K_{m_A}}{(A^*)}\left(1 + \frac{(Q)}{K_{iq}}\right)\left(1 + \frac{K_{ia}K_{m_B}}{(A^*)}\right) + \frac{(A)}{(A^*)}\left(\frac{K_{m_B}}{K_B} + \frac{K_{m_B}}{(B)} + \frac{K_{m_A}}{K_{ia}} + \frac{K_{m_A}(B)}{K_{ia}K_B}\right)} \tag{22}$$

where (A) = unlabeled concentration of A at equilibrium and (A*) = labeled concentration of A.

Thus, where (A) and (Q) are varied $\left(\frac{A}{Q} = \text{constant}\right)$, at fixed concentrations of B and P, and $\frac{1}{v^*_{A^*\text{-}P}}$ plotted vs. $\frac{1}{(A^*)}$, eventually as (B) becomes high (B/P fixed), a type of uncompetitive inhibition sets in at relatively low values of A such that (A/A*) → 1, whereas, when (A) and (P) $\left(\frac{A}{P} = \text{constant}\right)$ are varied at fixed B and Q $\left(\frac{B}{Q} \text{ fixed}\right)$ such that when B becomes high (B/Q fixed), the a mixed type of noncompetitive inhibition results when (A/A*) → 1.

Thus, consider the two cases discussed for the A*-P exchange reaction.

Case (1): (A) and (Q) varied [with $\left(\frac{A}{Q}\right)$ = constant and (Q) = x(A)] at fixed concentrations of (B) and (P) [$\left(\frac{B}{P}\right)$ = constant].

At high ($\underline{\underline{B}}$), the following approximation may be derived:

$$\frac{1}{v^*_{A^*\text{-}P}} \cong \frac{K_{m_A}}{V^f_{max}(A^*)}\left(1 + \frac{(Q)}{K_{iq}}\right) + \frac{1}{V^f_{max}}\frac{(A)}{(A^*)K_{\underline{\underline{B}}}}\left\{\frac{K_{m_A}(B)}{K_{ia}} + K_{m_B} + \frac{K_{m_A}}{K_{ia}}\right\} \tag{23}$$

If, $\left(\frac{A}{A^*}\right) \to 1$ (i.e., at low A),

$$\frac{1}{v^*_{A^*\text{-}P}} \cong \frac{K_{m_A}}{V^f_{max}(A^*)} + \frac{K_{m_A}}{V^f_{max}}\left[\frac{X}{K_{iq}} + \frac{1}{K_{ia}}\left(1 + \frac{(B)}{K_{\underline{\underline{B}}}}\right) + \frac{K_{m_B}}{K_{\underline{\underline{B}}}K_{m_A}}\right] \tag{24}$$

and a plot of $\frac{1}{v^*_{A^*\text{-}P}}$ vs. $\left(\frac{1}{A^*}\right)$ is given in Figure 1.

Case 2: (A) and (P) varied [$\left(\frac{A}{P}\right)$ = constant, (P) = x(A)] at fixed (B) and (Q) $\left(\frac{B}{Q}\right)$ = constant].

At high (B), the approximate Equation 25 may be derived.

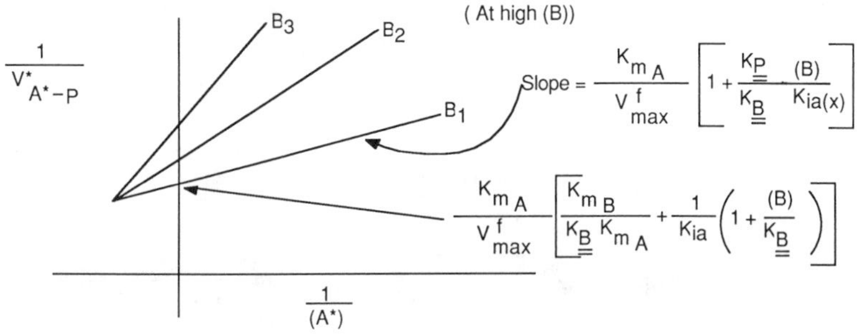

FIGURE 2.

$$\frac{1}{v^*_{A^*\text{-}P}} \cong \frac{1}{(A^*)}\frac{K_{m_A}}{V^f_{max}}\left[1 + \frac{K_{\underline{P}}}{K_{\underline{\underline{B}}}}\frac{(B)}{K_{ia}(X)}\right] + \frac{(A)}{(A^*)}\frac{K_{m_A}}{V^f_{max}}\left[\frac{K_{m_B}}{K_{\underline{\underline{B}}}K_{m_A}} + \frac{1}{K_{ia}}\left(1 + \frac{(B)}{K_{\underline{\underline{B}}}}\right)\right] \quad (25)$$

If $\left(\frac{A}{A^*}\right) \to 1$, i.e., at low (A), the plot given in Figure 2 results.

The equations for the other exchange reactions for this ordered BiBi system, at chemical equilibrium can be derived in a similar manner.

Thus, the velocity equation for an A*-Q exchange at chemical equilibrium is

$$v^*_{A^*\text{-}Q} = \frac{V^f_{max}(A^*)(B)}{\left(K_{ia}K_{m_B} + K_{m_A}(B) + \frac{K_{ia}K_{m_B}K_{m_Q}(P)}{K_{m_P}K_{iq}}\right)\left(1 + \frac{(A)}{K_{ia}} + \frac{(A)(B)}{K_{ia}K_{\underline{\underline{B}}}} + \frac{(Q)}{K_{iq}}\right)} \quad (26)$$

If the denominator is multiplied through, it would contain $(B)^2$ terms, so that when (B) is varied along with P (and $\left(\frac{B}{P}\right)$ = constant) or with Q (and $\left(\frac{B}{Q}\right)$ = constant), substrate inhibition would eventually be observed, i.e., when $v^*_{A^*\text{-}Q}$ is followed as a function of (B) complete inhibition by the substrate of the exchange velocity eventually will result, as described above for the $v_{A^*\text{-}P}$ exchange rate with (A/P) fixed and (A) varied.

Also, the denominator of Equation 26 contain (A) (P) and (A) (B) (P) terms. Thus, when (A) and (P) are varied with (P) = X(A), $(A)^2$ terms result in the denominator with only an (A) term in the numerator. Therefore, substrate inhibition is observed. When (A) and (Q) are varied at (Q) = X(A), no $(A)^2$ terms appear in the denominator, and there is no substrate inhibition.

Finally, consider the B*-P exchange in this same ordered two-substrate reaction, at chemical equilibrium:

$$\frac{v^*_{B^*\text{-}P}}{V^f_{max}} = \frac{(A)(B^*)}{\left[K_{ia}K_{m_B}\left(1 + \frac{(A)}{K_{ia}} + \frac{(A)(B)}{K_{ia}K_{\underline{\underline{B}}}} + \frac{(Q)}{K_{iq}}\right)\right]} \quad (27)$$

where the lack of inhibition by any reactant is observed.

Thus, consider the case where (A) and (Q) are varied [(Q) = x(A) and $\frac{A}{Q}$ = constant] with (B) and (P) fixed $\frac{(B)}{(P)}$ = constant.

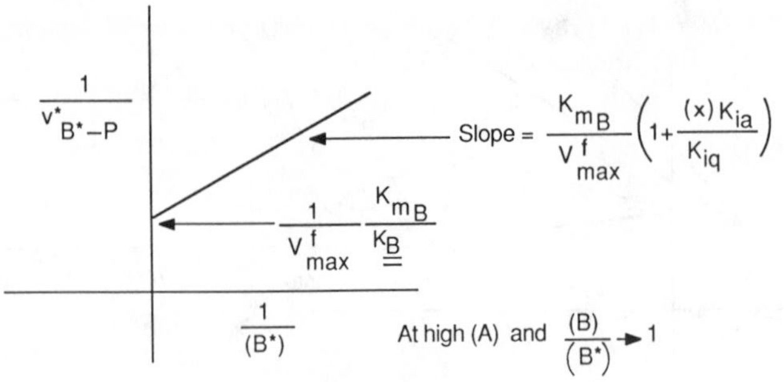

FIGURE 3.

$$\frac{V^f_{max}}{v^*_{B^*\text{-}P}} = \frac{K_{ia}K_{mB}}{(A)(B^*)} + \frac{K_{mB}}{(B^*)} + \frac{(B)}{(B^*)}\frac{K_{mB}}{K_{\underline{\underline{B}}}} + \frac{K_{ia}K_{mB}(x)}{K_{iq}(B^*)} \tag{28}$$

$$\frac{V^f_{max}}{v^*_{B^*\text{-}P}} = \frac{K_{mB}}{(B^*)}\left(1 + \frac{(x)K_{ia}}{K_{iq}} + \frac{K_{ia}}{(A)}\right) + \frac{(B)}{(B^*)}\frac{K_{mB}}{K_{\underline{\underline{B}}}} \tag{29}$$

At high (A), and if $\frac{(B)}{(B^*)} \rightarrow 1$ (or at low (B)),

$$\frac{V^f_{max}}{v^*_{B^*\text{-}P}} = \frac{K_{mB}}{(B^*)}\left(1 + \frac{(x)K_{ia}}{K_{iq}}\right) + \frac{K_{mB}}{K_{\underline{\underline{B}}}} \tag{30}$$

Thus, obviously there is no inhibition by (A) (or Q), and the following plot of $\frac{1}{v^*_{B^*\text{-}P}}$ vs. $\frac{1}{(B^*)}$ results (Figure 3).

Qualitatively, we could have arrived at the same conclusion that in the compulsory-ordered ternary complex mechanism, exchange, from B* to P* or to Q*, is not inhibited by (A), because saturating concentrations of (A) do not remove (EA), but in fact, increases the (EA) concentration. Similar results apply to the reverse reaction, where exchange from Q* is inhibited by excess (P), but exchange from P* is not inhibited by excess (Q). Thus, the substrate (B) and its product (P) are identified by this procedure, where B* exchange is not inhibited by excess A, and P* exchange is not inhibited by excess (Q).

The random-order mechanism is distinguished from the compulsory-ordered mechanism by the fact that no exchange can be completely inhibited by the paired substrate. For example, if (B) is present in excess, (A*) cannot bind to E since E is saturated with B to form EB; however, EB can still bind A* to give EA*B, which eventually yields the products P* or Q*

Since measurements of radioactivity can be made extremely sensitive, the technique of isotope exchange is capable of detecting very minor alternative pathways. However, one should exercise caution since these isotope exchange experiments demand very highly purified enzymes, in contrast to conventional nonisotopic kinetic experiments, if valid conclusions are to be reached. As an example, consider that in a dehydrogenase reaction with ordinary kinetic measurements, contaminating enzymes are unlikely to affect the overall kinetics since it is not likely that there will be two dehydrogenases which catalyze the same overall reaction. However, exchange studies between NAD^+ (or $NADP^+$) and NADH (or NADPH) is another matter, since there are several known enzymes which can catalyze this exchange, and to deduce, for example, from these exchange studies, whether the mechanism is random or ordered will require unusually pure enzymes, free of contaminating exchange catalysts for the pyridine nucleotides.

One final point, isotopic exchange permits a very useful simplification in the case of the substituted enzyme mechanism (i.e., ping-pong), in that one can study one half of the reaction only, i.e., each half-reaction may be studied independently. Thus, consider an A*-P exchange in the "Ping-Pong BiBi" system, whose basic King-Altman figure is

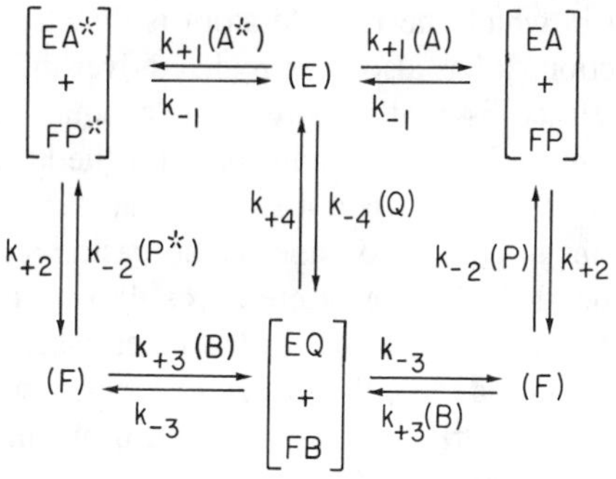

SCHEME III.

At chemical equilibrium,

$$v^*_{A^*\text{-}P} = \frac{V^f_{max}(A^*)}{K_{m_A} + \dfrac{K_{m_A}(A)}{K_{ia}} + \dfrac{K_{m_A}K_{ip}(A)}{K_{ia}(P)} + \dfrac{K_{m_A}K_{ip}(A)(B)}{K_{ia}K_{ip}(P)}} \tag{31}$$

B is obviously inhibitory when varied along with Q, but not when it is varied along with P at a constant ratio. (Note in the last term, a $\left(\frac{B}{P}\right)$ ratio exists, and therefore it will not inhibit the $v^*_{A^*\text{-}P}$ exchange rate, for $\frac{(B)}{(P)}$ = constant if varied simultaneously, and set (B) = x(P)).

There is no (B) term in the numerator, so that B is not required for the A*-P exchange. If (B) = 0, i.e., for the first partial reaction, the Equation 31 is reduced to

$$v^*_{A^*\text{-}P} = \frac{V^f_{max}\dfrac{K_{ia}}{K_{m_A}}(A^*)(P)}{K_{ia}(P) + K_{ip}(A) + (A)(P)} = \frac{V^f_{max^*}(A^*)(P)}{K_{ia}(P) + K_{ip}(A) + (A)(P)} \tag{32}$$

where $V^f_{max^*} = V^f_{max}\dfrac{K_{ia}}{K_{m_A}}$. Thus, the A*-P exchange takes place at a $v^*_{A^*\text{-}P}$, in the absence of (B), by Equation 32. This, of course, provides the important qualitative distinction between substituted enzyme and ternary-complex mechanisms, since in the so-called sequential ternary-complex mechanism no exchange can occur unless the system is complete. This comparatively easy and convenient method of distinguishing the ping-pong from the sequential mechanism was used and discussed very early in the development of the isotopic technique, in general, by Doudoroff et al.[98] and by Koshland.[99]

The possibility of studying only parts of the mechanism is a valuable asset in the ping-pong type of mechanism, especially in those complex cases of three or more substrates (e.g., see Reference 100).

II. KINETIC ISOTOPE EFFECTS

An isotopic substitution, e.g., replacement of a hydrogen atom by deuterium, can lead to kinetic effects which can, in turn, provide information concerning the nature of the process which is occurring. For example, in processes in which an H or D atom is being transferred, is is often observed that the H atom is transferred six to ten times more rapidly than the D atom. It is often thought that a kinetic isotope effect of such magnitude provides good evidence for a mechanism in which the H or D atom is directly involved in bond making or breaking during the reaction. It has also been realized over the past several years that the theory of kinetic isotope effects is vastly more complex than previously considered and current ideas inject great care in drawing conclusions, for the kinetic isotope effect may be far smaller than indicated above, even when a H or D transfer is involved in the process.

This section will attempt to give an outline of the basic theory of isotope effects (for further information on general theory, see References 486 to 489) and indicate how the theory may be applicable to enzyme systems. In the latter case, Northrop's method[490] will be described. As a means of introducing the theory of kinetic isotope effects, it is perhaps easier and simpler to consider at first the less involved problem of the isotope effects on chemical equilibria, briefly introduced in Chapter 1.

A. ISOTOPE EFFECTS ON CHEMICAL EQUILIBRIA

Given the most simple equilibrium occurring in the gas phase.

$$H_2 \rightleftharpoons 2H \tag{33}$$

From statistical mechanics, the equilibrium constant is given by Equation 105 in Chapter 1.

$$K_{eq} = \frac{(Q^o)^{\nu}_{products}}{(Q^o)^{\nu}_{reactants}} \ell^{-\Delta E^o_0/RT} \tag{1-105}$$

by the ratio of the partition functions for products and reactants, multiplied by a Boltzmann term which involves the difference between the zero-point levels of the reactant and product molecules, and for Equation 33, this may be given by

$$K_{H_2} = \frac{(Q)^2_H}{(Q)_{H_2}} \ell^{-\Delta E^o_0/RT} \tag{34}$$

where the Qs are the partition functions and, as shown in Figure 4, $\Delta E^o_0 = 103.2$ kcal/mol. The partition function for a hydrogen atom relates only to translational freedom and involves the mass of the atom, m_H.

$$Q_H = \frac{(2\pi m_H \underline{\underline{k}}\, T)^{3/2}}{h^3} \tag{35}$$

where $\underline{\underline{k}}$ is the Boltzmann constant, and h is the Planck constant (see Equation 96 in Chapter 1 for the translational partition function (Q_{tr}) for the activated complex).

The total partition function of the H_2 molecule also has a translational factor, which is now multiplied by a rotational factor (involving the moment of inertia, I, of the molecule), and by a vibrational factor, given by q_ν, thus,

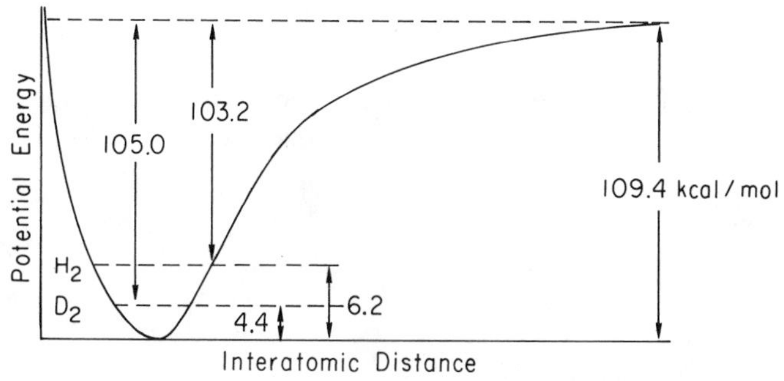

FIGURE 4. Schematic potential energy curves for H_2 and D_2, which depict the zero-point levels. All energy values are given in kcal mol^{-1}. From Laidler, K. J. and Bunting, P. S., in *The Chemical Kinetics of Enzyme Action,* 2nd ed., Clarendon Press, Oxford, 1973, 44. With permission.

$$Q_{H_2}\frac{(2\pi m_{H_2}\underline{\underline{k}}\,T)^{3/2}}{h^3}\cdot\frac{(8\pi^2 I\,\underline{\underline{k}}\,T)}{2h^2}\,q_\nu \tag{36}$$

The I for a homonuclear diatomic molecule is related to the masses as follows:

$$I = \frac{m_H d_{H_2}^2}{2} \tag{37}$$

where d_{H_2} is the distance between the atoms.

For the present discussion of kinetic isotope effects, we will be concerned only with the mass dependency of the pre-exponential part of the expression for K_{H_2}, which after substitution of Q_H and Q_{H_2}, now may be written as

$$K_{H_2} = \frac{\dfrac{(2\pi m_H\,\underline{\underline{k}}\,T)^3}{h^6}\,\ell^{-103,200/RT}}{\dfrac{(2\pi_{H_2}\underline{\underline{k}}\,T)^{3/2}}{h^3}\,\dfrac{(8\pi^2 I\,\underline{\underline{k}}\,T)_\nu}{2h^2}} \tag{38}$$

after substitution of Equation 37 for I, and noting that the vibrational factor q_ν is practically independent of mass at ordinary temperatures, it may readily be seen that

$$K_{H_2} \propto \frac{m_H^3 \ell^{-103,200/RT}}{m_{H_2}^{3/2} m_H} \tag{39}$$

or, since

$$m_{H_2} = 2m_H \tag{40}$$

$$\therefore\quad K_{H_2} \propto m_H^{1/2}\ell^{-103,200/RT} \tag{41}$$

Now apply the same treatment to the dissociation of the D_2 molecule, i.e.,

$$D_2 \rightleftarrows 2D \tag{42}$$

which leads consequently to the analogous result

$$K_{D_2} \propto m_D^{1/2}\ell^{-105,000/RT} \qquad (43)$$

The exponential term now differs, since the zero-point energy level for the D_2 molecule lies below that for the H_2 molecule (Figure 4). If one takes the ratio of the equilibrium constants

$$\frac{K_{H_2}}{K_{D_2}} = \left(\frac{m_H}{m_D}\right)^{1/2} \ell^{1,800/RT} \qquad (44)$$

at 300 K, this ratio is computed to be

$$\frac{K_{H_2}}{K_{D_2}} = \left(\frac{1}{2}\right)^{1/2} 10^{1.315} = 0.71 \times 20.7 = 14.7 \qquad (45)$$

This isotope effect is thus largely the result of the difference in zero-point energies of H_2 and D_2 (with H_2 nearer the activated state) leading to the factor of 20.7 in the equilibrium constant at 300 K.

This isotope effect will be even smaller if only one atom in an equilibrium is replaced, e.g.,

$$\left.\begin{array}{rl} \text{compare} & R\text{–}H \rightleftarrows R + H \\ \text{with} & R\text{–}D \rightleftarrows R + D \end{array}\right\} \qquad (46)$$

where R might be an organic radical. If R is considerably heavier than H or D, the isotope effect on the pre-exponential factor becomes negligible, leaving the major effect as due to the difference in zero-point energies. A typical zero-point energy difference for the C–H and C–D bonds is 1.15 kcal mol^{-1}, and therefore,

$$\frac{K_{RH}}{K_{RD}} \propto \ell^{1,150/RT} \cong 6.9 \quad \text{at} \quad 300\text{ K} \qquad (47)$$

With the above discussion in mind, an important generality is evident, i.e., substitution with a heavier atom will favor the formation of a stronger bond.[482] The stronger the bond, the greater the vibrational frequency, ν, so that there is wider separation between zero-point levels for the two isotopes, with these levels being $^1/_2\ h\nu$ higher than the classical ground states.

A relevant problem is that of acid dissociations in H_2O vs. D_2O. It is often observed that the rates of acid-catalyzed reactions are two to three times faster in D_2O than in H_2O, an observation which may be explained by the following mechanism:

$$A + H^+ \underset{}{\overset{\text{fast}}{\rightleftarrows}} AH^+$$

$$AH^+ \xrightarrow{\text{slow}} \text{products} \qquad (48)$$

Assume that replacement of the H by D may shift the first equilibrium more to the right, with little effect on the second reaction. This assumption implies that acid dissociation constants are smaller in D_2O than in H_2O; i.e., for the following acid dissociations:

$$\mathrm{HA} \rightleftarrows \mathrm{A}^- + \mathrm{H}^+, \qquad K_H = \frac{(A^-)(H^+)}{(HA)}$$

$$\mathrm{DA} \rightleftarrows \mathrm{A}^- + \mathrm{D}^+, \qquad K_D = \frac{(A^-)(D^+)}{(AD)} \tag{49}$$

K_H is greater than K_D. It is evident that if an acid HA is dissolved in D_2O, the rapid exchange results in the acid existing largely as DA.

Experimentally, acids are stronger in H_2O than in D_2O, with the

$$\Delta pK = pK_D - pK_H \cong 0.6 \tag{50}$$

in many cases.[491,492]

B. ISOTOPE EFFECTS ON RATES OF CHEMICAL REACTIONS

1. Primary Isotope Effects

A number of detailed treatments of this problem have been presented (see References 486, 487, 489, 493, and 494. Consider, at first, the pair of simple reactions

$$\begin{cases} \mathrm{H + H\text{–}H \rightarrow H\text{–}H + H} \\ \mathrm{H + D\text{–}H \rightarrow H\text{–}D + H} \end{cases} \tag{50a}$$

This comparison has the advantage that this is one of the very pairs of reactions where a rigorous treatment of the isotope effect is possible (if quantum-mechanical tunneling is ignored). The potential-energy surface is the same for both reactions, with only the zero-point levels differing. If one applies the absolute rate reaction theory or activated-complex theory (see Chapter 1, Section 6, Equation 107), the rate constant for the H + H_2 reaction is given by

$$k_{H+H_2} = \frac{(\underline{\underline{k}}\,T)}{(h)} \frac{Q^{\ddagger}}{Q_H Q_{H_2}} \ell^{-\Delta E_0^{\ddagger}/RT} \tag{51}$$

where the Qs are the partition functions and $\Delta E_0^{\ddagger}$ is the difference between the zero-point levels for initial and activated states. A similar expression for the H + D–H reaction is as follows:

$$k_{H+DH} = \frac{(\underline{\underline{k}}\,T)}{(h)} \frac{Q^{\ddagger\prime}}{Q_H Q_{HD}} \ell^{-\Delta E_{0'}^{\ddagger}/RT} \tag{52}$$

Although the values for $\Delta E_0^{\ddagger}$ and $\Delta E_{0'}^{\ddagger}$ are uncertain (approximately 8 to 10 kcal/mol), their difference is precisely known, as shown in Figure 5, i.e., 6.2 to 5.4 = 0.8 kcal/mol.

The rate constant ratio is therefore

$$\frac{k_{H+H_2}}{k_{H+DH}} = \frac{Q^{\ddagger\prime} Q_{HD} \ell^{800/RT}}{Q^{\ddagger\prime} QH_2} \tag{53}$$

and the dependencies of the partition functions on the masses are

$$Q_{H_2} \propto \left(\frac{m_{H_2}^{3/2}}{m_H}\right), \quad Q_{HD} \propto m_{HD}^{3/2}\left(\frac{m_H m_D}{m_H + m_D}\right), \quad Q^{\ddagger} \propto m_{H_3}^{3/2} \quad \text{and} \quad Q_{\prime}^{\ddagger} \propto m_{HDH}^{3/2} \tag{54}$$

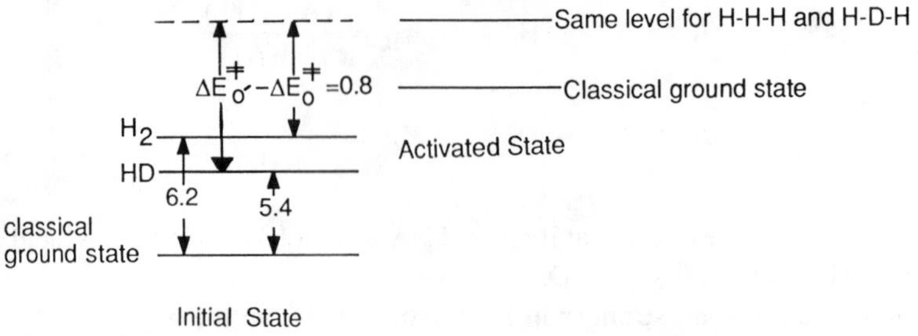

FIGURE 5. Zero-point levels in the initial and activated states for H-H-H and H-D-H. From Laidler, K. J. and Bunting, P. S., in *The Chemical Kinetics of Enzyme Action*, 2nd ed., Clarendon Press, Oxford, 1973, 44. With permission.

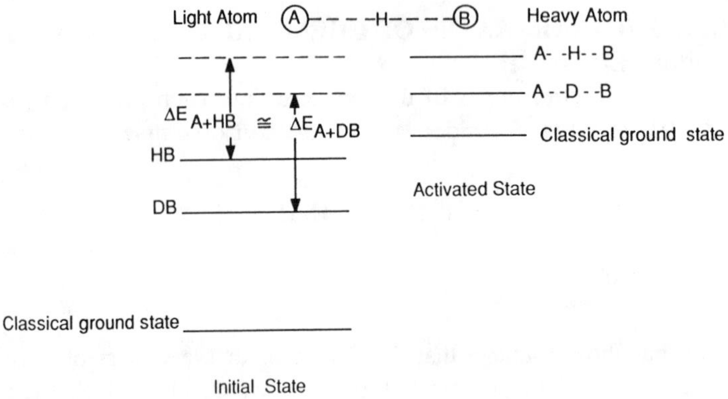

FIGURE 6. Reaction between A + HB where A is a light atom and B is a heavy one. The kinetic isotope effect is reduced because of the asymmetry of the activated state. From Laidler, K. J. and Bunting, P. S., in *The Chemical Kinetics of Enzyme Action*, 2nd ed., Clarendon Press, Oxford, 1973, 44. With permission.

and the ratios are

$$\frac{Q_{HD}}{Q_{H_2}} = \left(\frac{3}{2}\right)^{3/2} \cdot \frac{2}{3} = \left(\frac{3}{2}\right)^{1/2}; \quad \text{and} \quad \frac{Q^{\ddagger}}{Q^{\ddagger\prime}} = \left(\frac{3}{4}\right)^{3/2} \tag{55}$$

Thus,

$$\frac{k_{H+H_2}}{k_{H+DH}} = \left(\frac{3}{2}\right)^{1/2}\left(\frac{3}{4}\right)^{3/2} e^{800/RT} = 0.8 \times 3.35 = 2.68 \quad \text{at} \quad 300\ \text{K} \tag{56}$$

The isotope effect, therefore, is the result mainly from the difference in zero-point energy levels of the reactants since the pre-exponential factor exerts only a minor effect. This case was particularly simple in that the zero-point levels were the same in the two activated states, i.e., for H–H–H and H–D–H. This came about as a result of the symmetry of the activated state, where the central H or D atom lies at the center of gravity of the activated complex and consequently does not influence the frequencies of the symmetric vibrations. Asymmetry in the activated state can arise in two ways: (1) the force constants of the two bonds may be different, or (2) the atom being transferred may be attached to groups of differing masses. Consider, e.g., the activated complex as given in Figure 6, i.e., Ⓐ---H---Ⓑ. In this case the central atom is attached to a light atom and to a heavy atom; however, the force constants are the same. By solving the dynamical equations, it may be shown that the frequency of

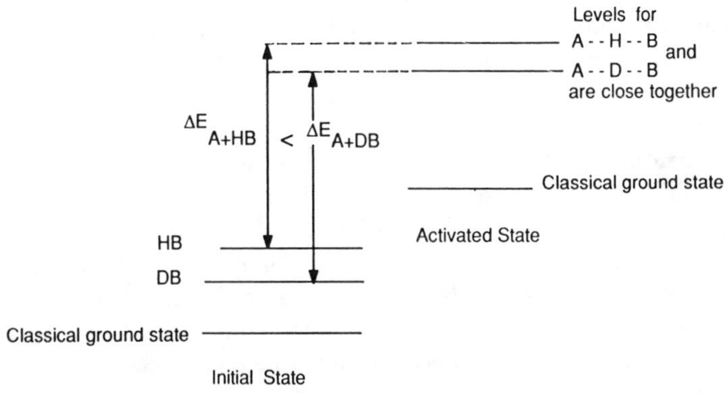

FIGURE 7. Reaction between A + HB, in which both A and B are much heavier than H. A significant isotope effect results. From Laidler, K. J. and Bunting, P. S., in *The Chemical Kinetics of Enzyme Action,* 2nd ed., Clarendon Press, Oxford, 1973, 44. With permission.

the stretching vibration is reduced if D replaces the H atom. The effect of the asymmetry is to decrease the difference between $\Delta E_{A\,+\,HB}$ and $\Delta E_{A\,+\,DB}$, and consequently to reduce the kinetic isotope effect. A similar result is obtained if asymmetry results from unequal force constants for the A–H and H–B bonds. However, if the two atoms, A and B, are both much heavier than the central atom, H or D, a different situation results (see Figure 7).

After solving the dynamical equations, the result obtained is that the frequency of the symmetric vibration is practically independent of the mass of the central atom. Regardless of the relative magnitude of the force constants involved, the situation is the same. This case is of interest in the discussion of organic reactions and enzymic-catalyzed reactions, since the atoms A and B can be carbon, oxygen, or nitrogen atoms, all of which are much heavier than H and D atoms. In these systems, the zero-point energy levels in the activated state may be close together for H and D atoms, leading to the result that the dominant factor is due largely to the zero-point energies of the of the reactants. Thus, the situation may be similar to that schematically represented in Figure 7, and where there will be a significant kinetic isotope effect.

In conclusion, therefore, the rate constant for a reaction with a H-atom transfer in an enzyme system should be greater than that for a reaction with a D-atom transfer. The difference between the zero-point energy levels for an organic reaction R–H compared to R–D is approximately 1.15 kcal/mol, a value larger than the difference between H–H and H–D, a value which leads to the following approximate estimate of the isotope effect (at 300 K).

$$\frac{k_H}{k_D} \cong \ell^{1,150/RT} = 6.9 \quad \text{(at 300 K)} \qquad (57)$$

Such ratios are often found experimentally. If the ratios are much smaller in value, then it is possible that the reaction mechanism under study may not involve a H- or D-atom transfer.

The above discussion ignored the possibility of quantum-mechanical tunneling through the potential-energy barrier;[487] rather, is assumed that, from classical mechanics, the system passes over the top of the barrier. However, quantum-mechanical theory does allow for the situation where a system with too little energy to surmount the barrier may nevertheless pass from the initial to final state by tunneling or leaking through the barrier. Tunneling becomes important for particles of small mass, and when the barrier is low and narrow. From theoretical treatments, is seems that tunneling may, for certain barriers, be quite important for the transfer of H atoms or of H^+ or H^- ions; it is probably not important for D, D^+, and D^-. Tunneling is likely of negligible importance for heavier atoms. It follows from the possibility that tunneling may occur with H but not with D that the kinetic isotope ratio may be much larger than the value of approximately 7 given above (Equation 57), and values over 20 have been observed. Even for a reaction as simple as $H + H_2$, the quantitative theory of tunneling is extremely difficult, since the rate of tunneling depends strongly on the exact shape of the potential energy barrier, which is still not reliably known for any system. Information about tunneling therefore rests largely on experimental evidence rather than from a sound theoretical basis. Nor is it easy to obtain reliable unambiguous evidence for tunneling, since, although theory predicts that certain effects will be obtained when tunneling occurs, these same effects may sometimes be due to other causes. The four major effects predicted if tunneling does occur are as follows.

1. Arrhenius plots will display a deviation from linearity (see Figure 27 in Chapter 1 for a case of nonlinear Arrhenius plots).

   Tunneling is fairly insensitive to temperature and becomes relatively more important as the temperature is decreased. Nonlinear Arrhenius plots have been observed for a number of reactions, including $H + H_2 \rightarrow H_2 + H$ and $D + H_2 \rightarrow DH + H$, as well as a number of reactions in solution. One of the most thoroughly investigated examples is the base-catalyzed bromination, in heavy water, of the cyclic ketone 2-carbethoxycyclopentanone; this reaction is likely controlled by a proton transfer from the substrate to the base catalyst. For this reaction, catalyzed by F^- ions and other bases, Bell and co-workers[495,496] found a significant deviation from linearity in the Arrhenius plot which could not be explained by other means than tunneling.

   They interpreted their data in terms of the equations for the penetration of a parabolic barrier whose width at its base was calculated as 1.17 Å, and whose height was about 20% larger than the observed activation energy. Similar deviations from linear Arrhenius behavior and a similar explanation of tunneling were found by Caldin and co-workers[320,497,759] for reactions of the trinitrobenzyl ion with acetic acid and hydrofluoric acid, and for the reaction of 4-nitrobenzyl cyanide with ethoxide ions. It is important to note in this context that this type of deviation from Arrhenius behavior will also be observed if two reactions are occurring simultaneously. Care must be exercised to eliminate the possibility of parallel reactions, as well as of consecutive reactions before invoking the explanation of tunneling.
2. This second predicted effect is related to the first in that if tunneling is presumed to occur, it is assumed that the frequency factor (A) in the Arrhenius equation (see Chapter 1, Equation 91) will also be abnormally small. Thus, in the work on the bromination of 2-carbethoxycyclopentanone, A_D/A_H was found to be 24 at low temperatures.
3. A third predicted effect in the case of tunneling is the observation of unusually large isotope effects, with the lighter isotope reacting much more rapidly than expected. Often a value of $k_H/k_D \gg 10$ is taken as evidence for tunneling.
4. Extension of experiments to include tritium (T) as well as D and H provides the fourth indication of tunneling. As we have seen, the theory of kinetic isotope effects where tunneling was neglected would predict that the effects are largely due to differences in zero-point levels, which depends on the masses of the atoms involved. In the case without tunneling, it can be shown that k_H, k_D, and k_T are related as follows (see Equations 76 and 79).

$$\log\left(\frac{k_H}{k_T}\right) = 1.44 \log\left(\frac{k_H}{k_D}\right) \quad (58)$$

   However, if tunneling for H occurs, the k_H value will be unusually large, with the result that the right-hand side factor will be less than 1.44. Thus, a low value for the right-hand side factor is considered to be evidence for tunneling.

In a recent study of the primary and secondary protium-to-tritium (H/T) and deuterium-to-tritium (D/T) kinetic isotope effects for the catalytic oxidation of benzyl alcohol to benzaldehyde by yeast alcohol dehydrogenase, it was concluded that a significant contribution from hydrogen tunneling takes place in the hydrogen transfer step, i.e., in the rate-limiting hydride transfer step.[1001] Note that these authors felt that a significant deviation from $k_H/k_T = (k_D/k_T)^{3.26}$, rather than Equation 58, should be used as the criterion for hydrogen tunneling.[1001]

When hydrogen-deuterium isotope effects are studied in solution, careful attention must be given to the rapid exchange reactions with the solvent of hydrogen atoms attached to oxygen, nitrogen, and certain other atoms, because of the rapid nature of ionization reactions which may occur, e.g., $\text{–O–H} \overset{\text{fast}}{\rightleftharpoons} -O^- + H^+$. Even a change of solvent alone, i.e., D_2O for H_2O, may have a significant effect on the rate in a manner which is not easy to predict. Thus, the viscosity of D_2O at 25°C is 23% greater than H_2O; the latent heat of vaporization is higher[498], 1.50 compared to 1.44 kcal/mol at the freezing points, and the dielectric constant is slightly lower,[499] 77.9 at 25°C, compared to to 78.3 for H_2O. Evidence has been presented that liquid D_2O possesses more structure than liquid H_2O. Caution should be exercised in interpreting kinetic isotope effects involving exchangeable hydrogen atoms, when the solvent is, of necessity, changed. However, this particular solvent effect cannot be large and Bender et al.[500] have suggested that a k_H/k_D ratio of only 1.0 to 1.5 results from this cause. These difficulties, of course, do not arise if the hydrogen atom is attached to a carbon atom, in which case it may not be exchangeable; however, such hydrogen atoms are not often involved in enzymatic reactions. Schowen[504] has dealt in detail with solvent effects on enzymic reactions and the significance of fractionation factors.*

One may ask whether the magnitude of an H–D kinetic isotope effect will allow one to decide whether the reaction mechanism involves a proton (H^+), hydride ion (H^-), or a hydrogen atom (H). Because of the complex nature of the situation, Jencks[494] in fact, has taken the view that it is difficult to draw valid conclusions based solely on kinetic isotope effects which would allow one to distinguish between these three possibilities. However,

* Consider a compound with exchangeable hydrogenic sites which is dissolved in a mixture of heavy and light water. At equilibrium, the compound will contain in each site a mixture of deuterium and protium. The ratio of deuterium to protium in each particular site will only be equal to the ratio in the bulk solvent, if the binding in that particular site equals the binding in an average water site. Protium will accumulate at the solute site if it binds its hydrogen more weakly, whereas, if the binding is tighter, deuterium will be preferred. The so-called "Isotopic fractionation Factor" is a quantitative measure of the deuterium preference (and accordingly, of the tightness of binding) at any hydrogenic site. In Equation 59, the isotopic fractionation factor, ϕ_i is defined:

$$\phi_i = \frac{[D_i]}{[H_i]} \Big/ [\eta/(1 - \eta)] \qquad (59)$$

where η is the atom fraction of deuterium in a mixed isotopic solvent. ϕ_i is the deuterium-to-protium ration at the ith hydrogenic site, relative to the ratio for water. Therefore, ϕ_i is the equilibrium constant for the exchange reaction of Equation 60, where L is either H or D.

$$H_i + LOD \rightleftharpoons D_i + LOH \qquad (60)$$

For the significance of ϕ_i and a list of fractionation factors which are of use in mechanistic interpretations, see Reference 504.

TABLE 1
Typical Kinetic Isotope Ratios at 25°C

Isotope forms	Ratios
H, D	7[a]
H, T	17[a]
^{12}C, ^{13}C	1.04
^{12}C, ^{14}C	1.08
^{14}N, ^{16}N	1.04
^{16}O, ^{18}O	1.04

[a] If tunneling occurs, significantly larger ratios are to be expected. O'Leary[503] has recently discussed a number of studies of enzyme reaction mechanisms by means of heavy-atom isotope effects.

From Laidler, K. J. and Bunting, P. S., in *The Chemical Kinetics of Enzyme Action,* 2nd ed., Clarendon Press, Oxford, 1973, 44. With permission.

on the other hand, Swain et al.,[501] after consideration of the types of activated states to be expected to arise for H^+, H^-, and H transfers, suggested that there are some significant differences to be expected. They concluded that when a H^- is transferred, the bonds in the activated state may be shorter and stronger than for H^+ transfer; therefore, for a H^- transfer they predicted a smaller k_H/k_D ratio, and possibly an inverse isotope effect (i.e., $k_H/k_D < 1$). They felt that the effect of an H transfer should lie in between that for H^- and H^+ transfer reactions. In addition, an isotope effect with H^- should be less sensitive to structural changes than those with H^+ (with H again lying in between) because the activated state is less polarized. Belleau and Moran[502] attempted to apply these criteria to enzyme systems. However, Jenck's crititicism seems justified in most cases, and alternative evidence in addition to kinetic isotope studies should be applied in order to make a reliable distinction between H^+, H^-, and H transfer mechanisms.

The discussion until now has been restricted to hydrogen isotope effects and we will now briefly postpone a further discussion of these effects (as well as a discussion Northrop's approach[490]) in order to mention a few investigations (including enzyme systems) concerned with isotope effects with heavier atoms, e.g., carbon (C). The most abundant isotope of C is ^{12}C, and studies have dealt with ^{13}C and ^{14}C. The effects to be predicted are much smaller for the heavier elements since the ratio of their masses is smaller. In Table 1, a some typical values for kinetic isotope ratios are listed for several elements. However, it should be stressed that, for reasons discussed above, one may also expect to find some wide variations from these values.

2. Secondary Isotope Effects

All of the above discussion has dealt with so-called primary kinetic isotope effects, in which the isotopic substitution involves an atom directly attached to a bond which is broken or formed during the reaction. In addition to the primary effects, it has also been found that secondary effects may arise when there is substitution to an atom which is not directly involved in the bond which is being formed or broken — such effects are known as secondary isotope effects. Many of the secondary hydrogen kinetic isotope effects in enzyme-catalyzed reactions arise as a result of the following generalized reaction:

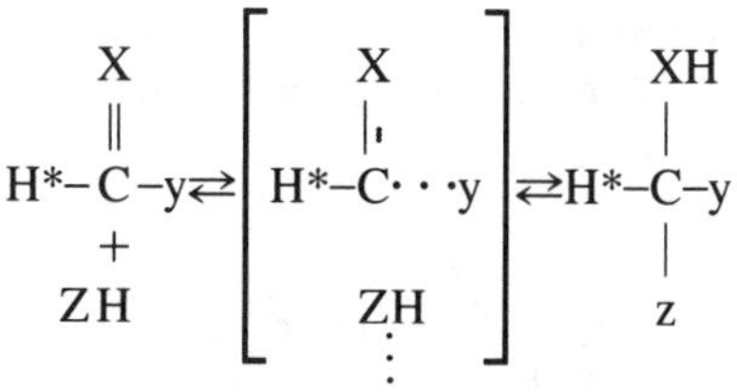

$$
\begin{matrix}
\text{X} \\
\| \\
\text{H*-C-y} \\
+ \\
\text{ZH}
\end{matrix}
\rightleftharpoons
\left[
\begin{matrix}
\text{X} \\
| \\
\text{H*-C} \cdots \text{y} \\
\\
\text{ZH}
\end{matrix}
\right]
\rightleftharpoons
\begin{matrix}
\text{XH} \\
| \\
\text{H*-C-y} \\
| \\
\text{z}
\end{matrix}
\tag{61}
$$

where, in this reaction, trigonally hybridized carbon is converted (through a transition state) to its tetragonal form. The measured values of k_H/k_D are called the α-deuterium isotope effects. Whereas, the primary isotope effects for hydrogen normally lie in the range $k_H/k_D = 2$ to 10, the secondary isotope effects are far smaller in magnitude and usually have limiting values of 1.02 to 1.40. Substitution at even more remote positions than shown in Equation 61 may also lead to rate differences, but probably at the low end of the limiting values given above for k_H/k_D.

Kirsch[505] has described a semiempirical formulation of the origin of secondary isotope effects as originally provided by Streitwieser et al.[506] and Melander.[507] Essentially, secondary isotope effects are explained in the same way as the primary effects, i.e., in terms of changes in vibrational frequencies and hence zero-point levels. Thus, the secondary effects found in S_N substitution reactions can be explained in terms of the frequency changes resulting from a change of hybridization where essentially the process may be represented as

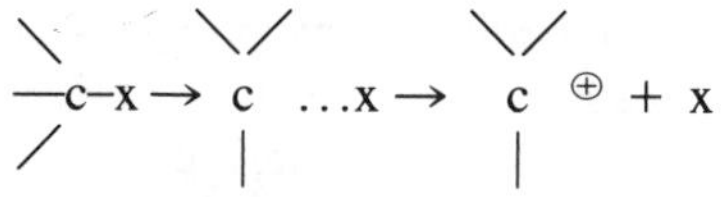

$$
\begin{matrix}
\diagdown \\
-\text{c-x} \\
\diagup
\end{matrix}
\rightarrow
\begin{matrix}
\diagdown\diagup \\
\text{c} \ldots \text{x} \\
|
\end{matrix}
\rightarrow
\begin{matrix}
\diagdown\diagup \\
\text{c} \quad \oplus \\
|
\end{matrix}
+ \text{x}
\tag{62}
$$

There is initially sp^3 hybridization, and finally there is sp^2 hybridization. In the activated state, the hybridization lies between sp^3 and sp^2. A decrease in the bending frequencies occurs when the change $sp^3 \rightarrow sp^2$ takes place, and accordingly there is a decrease with the formation of the activated complex or transition state. The D and H levels for the activated state are therefore closer together than in the initial state and the energy diagram resembles that given in Figure 7 (but for different reasons) with a resulting positive kinetic effect. One may predict a value for k_H/k_D of 1.3 to 1.4. In several cases it has been found preferable to interpret secondary kinetic isotope effects in terms of inductive effects, steric effects, or hyperconjugation.

Inductive effects — Consider the pair of ionizations

$$
HCOOH \rightleftharpoons HCOO^- + H^+, \quad K_H
$$

$$
DCOOH \rightleftharpoons DCOO^- + H^+, \quad K_D
\tag{63}
$$

experimentally, $K_H \cong 1.12 \, K_D$ and $\Delta pK \cong 0.035$. One explanation would have D as slightly more electron donating than H; this explanation may be invoked to explain differences in rate.

Steric effects — On an average, the C–H bond is slightly larger than the C–D bond as a result of the anharmonicity of the vibration and the result of the fact that the amplitude of the vibration of the C–H bond is larger than that of the C–D bond. Thus, the $-CH_3$ group, for example, displays less steric hindrance than the $-CD_3$ group.

Hyperconjugation — In a solvolysis reaction of the type

$$
\begin{matrix}
\text{H} \quad \text{H} \\
| \quad | \\
-\text{C-C-X} \\
| \quad |
\end{matrix}
\rightarrow
\begin{matrix}
\text{H} \quad \text{H} \\
| \quad | \\
-\text{C-C}^\oplus \\
| \quad |
\end{matrix}
+ X^-
$$

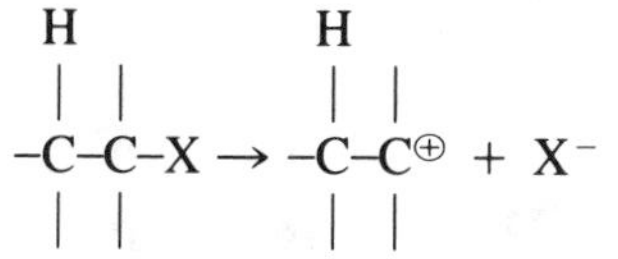

replacement of the H by D results in a decrease in rate, a fact explained in terms of hyperconjugation in the initial state, i.e., resonance between the structures

$$
\begin{array}{ccc}
\text{H} & & \\
| \ | & & \text{H}^+ \ | \\
-\text{C}-\text{C}-\text{X} \rightleftharpoons - & \text{C} = \text{C} & + \ \text{X}^- \\
| \ | & & | \\
\end{array}
$$

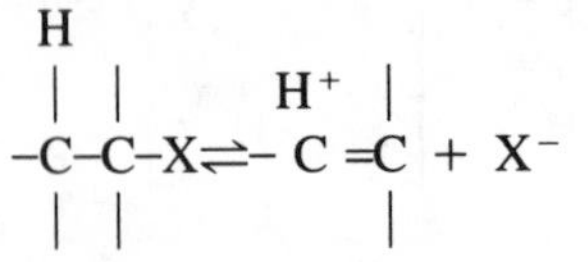

which assists the reaction. Resonance is less important for D than for H because of the loss of zero-point energy in hyperconjugation. From the above discussions, it is clear that the theoretical interpretation of kinetic isotope effects is a difficult problem especially in view of a lack of knowledge of potential-energy surfaces. Thus, one should be cautious in drawing conclusions from the magnitudes alone of the effects. A careful kinetic study is mandatory in order to separate the individual rate constants; composite rate constants will lead to a great deal of confusion. Any pre-equilibria, such as ionization, will also have a considerable effect on the results and on the interpretations.

C. THE NORTHROP METHOD FOR DETERMINATION OF THE ABSOLUTE MAGNITUDE OF HYDROGEN ISOTOPE EFFECTS IN ENZYME-CATALYZED REACTIONS

Based on a fixed relationship between deuterium and tritium isotope effects, Northrop[490] devised a method for calculating the absolute magnitude of intrinsic isotope effects.[508] By an intrinsic isotope effect, it is meant that the effect which originates from a single isotopically sensitive step of the catalysis, exclusive of interference from isotopically insensitive steps. The treatment will, in the beginning, deal with enzymatic reactions involving only a direct transfer of deuterium and tritium between substrates, and it is assumed that only primary hydrogen isotope effects are present which affect only a single step of catalysis. The nomenclature of Northrop will be strictly adhered to, as well as his reaction scheme numbering, in order to avoid any deviation in his treatment which might lead to confusion in applying his method.

1. A Simple Enzyme Kinetic Mechanism with a Single Substrate-Single Product and Single ES Intermediate

Consider the simplest enzyme kinetic mechanism (an irreversible single substrate-single product reaction) as given in Scheme IV with a single ES intermediate.

$$
\text{E} \ \underset{k_2}{\overset{k_1 S}{\rightleftharpoons}} \ \text{ES} \ \overset{k_3}{\rightarrow} \ \text{EP} \ \overset{k_5}{\rightarrow} \ \text{E} + \text{P}
$$

SCHEME IV

The pertinent steady-state kinetic expressions, for this discussion, from Scheme IV are

$$
V = \frac{k_3 k_5 E^t}{k_3 + k_5} \tag{65}
$$

$$
\frac{V}{K} = \frac{k_1 k_3 E^t}{k_2 + k_3} \tag{66}
$$

$$
K = \frac{k_5(k_2 + k_3)}{k_1(k_3 + k_5)} \tag{67}
$$

$$v = \frac{k_1k_3k_5[S]E^t}{k_5(k_2 + k_3) + k_1(k_3 + k_5)[S]} \tag{68}$$

where V represents the maximal velocity, K the Michaelis constant, and v the initial velocity at one initial concentration of substrate, S. One notes that the expressions for V and $\frac{V}{K}$ contain the fewest number of rate constants and accordingly would be the kinetic parameters of interest in measurements of isotope effects on enzyme-catalyzed reactions. Dividing Equation 65 by an analogous one representing a reaction with a deuterium-containing substrate, Equation 69 results.

$$\frac{V_H}{V_D} = \frac{(k_{3H}/k_{3D}) + (k_{3H}/k_5)}{1 + (k_{3H}/k_5)} \tag{69}$$

(Note that $k_{5H} = k_{5D} = k_5$ since the isotope affects only a single step of catalysis, i.e., k_3). The size of the apparent isotope effect on the maximal velocity (V_H/V_D) relative to the size of the intrinsic isotope effect (k_{3H}/k_{3D}) is dependent on the ratio of the rate constant for catalysis to the rate constant for product release (k_{3H}/k_5). Similarly from Equation 66, the following equation is obtained for the apparent isotope effect on V/K:

$$\frac{(V/K)_H}{(V/K)_D} = \frac{(k_{3H}/k_{3D}) + (k_{3H}/k_2)}{1 + (k_{3H}/k_2)} \tag{70}$$

where $k_{2H} = k_{2D} = k_2$ and $k_{1H} = k_{1D} = k_1$). The size of the apparent isotope effect on (V/K) relative to the intrinsic isotope effect is now dependent on the ratio of the rate constant for catalysis to that for substrate release (k_{3H}/k_2).

a. Nomenclature for Apparent Isotope Effects

In referring to the case of the "deuterium isotope effect on V_{max}", it also is meant to imply both the hydrogen rate and its ratio to a deuterium rate. In the following symbolism, therefore, only the isotope and kinetic parameter of interest are identified; the unstated rate constant and ratio of constants are implied. Thus, Equations 69 and 70 are rewritten in the following general form.

$$D_V = \frac{D_k + R}{1 + R} \tag{71}$$

$$D_{(V/K)} = \frac{D_K + C}{1 + C} \tag{72}$$

The intrinsic isotope effect on the catalysis (D_k) is written without the subscript denoting the particular reaction step involved because it is implied by the primary isotope effect. R represents the "ratio of catalysis", i.e., the ratio of the rate constants for the catalytic step to that for other forward steps contributing to the maximal velocity. C, or "commitment to catalysis", represents the tendency of the enzyme-substrate complex to proceed through catalysis as opposed to its tendency to dissociate into free enzyme and substrate. R and C have also been referred to as "partitioning factors".

b. Calculation of Intrinsic Isotope Effects

Subtraction of 1 from both sides of Equation 72 yields

$$D_{(V/K)} - 1 = \left[\frac{D_K + C}{1 + C}\right] - 1$$

$$D_{(V/K)} - 1 = \frac{D_K - 1}{1 + C} \tag{73}$$

Division of Equation 73 by a comparable one for the tritium isotope effect leads to

$$\frac{D_{(V/K)} - 1}{T_{(V/K)} - 1} = \frac{\dfrac{D_k - 1}{1 + C}}{\dfrac{T_k - 1}{1 + C}} \tag{74}$$

$$\frac{D_{(V/K)} - 1}{T_{(V/K)} - 1} = \frac{D_k - 1}{T_k - 1} \tag{75}$$

Thus, only the apparent isotope effects are expressed as a function of the intrinsic isotope effects. Swain et al.[509] calculated a fixed relationship between the intrinsic isotope effects of deuterium and tritium. Substitution of the Swain relationship

$$(T_k = D_k^{1.442}) \tag{76}$$

into Equation 75 yields finally

$$\frac{D_{(V/K)} - 1}{T_{(V/K)} - 1} = \frac{D_k - 1}{D_k^{1.442} - 1} \tag{77}$$

(See also Equation 58.)

The above equation provides a relationship between the apparent deuterium and tritium isotope effects on the measured ratio of kinetic parameters, (V/K), and the single intrinsic isotope effect, which accordingly may now be calculated. By means of tabulated values (see Appendix D of Reference 490) for experimental values of

$$\frac{(V/K)_H/(V/K)_D - 1}{(V/K)_H/(V/K)_T - 1} = \frac{k_H/k_D - 1}{(k_H/k_D)^{1.442} - 1} \tag{78}$$

where

$$(k_H/k_T) = (k_H/k_D)^{1.442} \tag{79}$$

values for k_H/k_D and k_H/k_T are obtained for the bond-breaking step. Clearly, Equation 77 hinges on the validity of the Swain relationship. The theoretical derivation by Swain et al.[509] had employed a single quantum mechanical model, i.e., only stretching vibrations were considered and these were assumed to be in harmonic oscillation; they also assumed that the isotopic differences arose only from differences in zero-point vibrational energies, with tunneling effects excluded. This derivation led to the following expression where the related magnitude of the deuterium vs. tritium isotope effects was expressed as an exponential function of the isotopic masses.

$$\frac{k_H}{k_T} = \left(\frac{k_H}{k_D}\right)^{\left[\frac{1/m_1^{1/2} - 1/m_3^{1/2}}{1/m_1^{1/2} - 1/m_2^{1/2}}\right]} \tag{80}$$

With use of the following values for the atomic masses, $m_1 = 1.0081$, $m_2 = 2.0147$, and $m_3 = 3.0170$, Equation 79 occurs. Swain et al.[509] felt that their relation would hold as long as the temperature remained low enough to fix X–H bonds in their lowest vibrational state (e.g., 0 to 100°C). Others have considered effects of transition-state vibrational energy differences as tunneling on the Swain relationship. Bigeleisen,[510] with a more complete quantum-mechanical model, which included tunneling, suggested a range of values of 1.33 to 1.56 for the exponent in Equation 80.

Later Stern and associates[511,512] considered 180 model-reaction systems as well as incorporating secondary isotope effects and even extending the temperature range to 20 to 1000 K, and concluded that Bigeleisen's limits to the mass exponent would probably not be exceeded. Only the possibility of steric isotope effects are likely to seriously affect the Swain relationship.[512,513]

Thermodynamic isotope effects of steric origin, such as effects on substrate binding to the enzyme are probably negligible. However, an isotope effect is possible on those conformational changes which may lead to an enzyme-substrate transition state, especially when one considers a tight fit between the substrate and an active site "pocket" or "fold" of the enzyme. As the catalysis becomes more rate limiting, however, the steric kinetic isotope effect is reduced, since a slower rate of catalysis allows conformational changes to approach equilibrium.

Therefore, steric kinetic isotope effects are likely to present a problem only when apparent isotope effects are small. Thus far, there does not appear to be experimental evidence of a significant deviation from the Swain relationship (See Reference 1001).

2. An Enzyme Kinetic Mechanism with Reversible Steps and Multiple Intermediate Complexes

Consider an extension of a single-substrate enzyme kinetic mechanism given in Scheme IV to the more realistic and generalized model here in Scheme V, which incorporates reversible steps and three additional intermediate enzyme substrate complexes,

$$E_1 \underset{k_2}{\overset{k_1B}{\rightleftarrows}} E_2 \underset{k_4}{\overset{k_3}{\rightleftarrows}} E_3 \underset{k_6^*}{\overset{k_5^*}{\rightleftarrows}} E_4 \underset{k_8}{\overset{k_7}{\rightleftarrows}} E_5 \underset{k_{10}P}{\overset{k_9}{\rightleftarrows}} E_6 \xrightarrow{k_{11}} E_1$$

SCHEME V

where step 1 represents substrate B binding, steps 2 and 4 represent conformational changes of the enzyme and/or additional catalytic steps, step 3 is the isotopically sensitive step (denoted by asterisks) with k_5 (or k_6) as the catalytic step, step 5 is the release of the first product, and step 6 represents the sum of all those steps required for the generation of free enzyme, E_1. As before, the apparent (V/K) isotope effect with respect to substrate B maybe expressed in the following form:

$$D_{\left(\frac{V}{K_B}\right)} = \frac{D_{k_5} + C_f + C_r}{1 + C_f + C_r} \tag{81}$$

$$C_f = \frac{k_5}{k_4}\left(1 + \frac{k_3}{k_2}\right) \tag{82}$$

$$C_r = \frac{k_6}{k_7}\left(1 + \frac{k_8}{k_9}\right) \tag{83}$$

The C term has two parts: C_f, the forward commitment to catalysis, and C_r, the reverse commitment to catalysis. If, however, k_7, rather then k_5, represents catalysis, the expression for the forward commitment to catalysis becomes

$$C_f = \frac{k_7}{k_6}\left[1 + \frac{k_5}{k_4}\left(1 + \frac{k_3}{k_2}\right)\right] \tag{84}$$

A comparison of Equation 84 with Equation 82 and with Equation 72 for Scheme IV reveals a pattern which allows by simple inspection of a proposed mechanism[514] to write kinetic expressions for (V/K) isotope effects. Note that because both C_f and C_r are present in Equation 81, in the absence of an equilibrium isotope effect, it is necessary that

$$D_{(V/K_B)} = D_{(V/K_P)} \tag{85}$$

which provides a useful check on the internal consistency of the data. The apparent isotope effect on V is of the form

$$D_V = \left(\frac{D_{k_5} + R_f + C_r}{1 + R_f + C_r}\right) \tag{86}$$

where R_f is the forward ratio of catalysis, i.e.,

$$R_f = \left(\frac{k_5}{k_3} + \frac{k_5}{k_7}\left(1 + \frac{k_8}{k_9}\right) + \frac{k_5}{k_9} + \frac{k_5}{k_{11}}\right)\left(\frac{k_3}{k_3 + k_4}\right) \tag{87}$$

By substracting 1 from each side of Equation 81, the commitment factors are eliminated from the numerator of the $D_{(V/K)}$ expression:

$$D_{(V/K)} - 1 = \left(\frac{D_{k_5} + C_f + C_r}{1 + C_f + C_r}\right) - 1 \tag{88}$$

$$D_{(V/K)} - 1 = \left(\frac{D_{k_5} - 1}{1 + C_f + C_r}\right) \tag{89}$$

and division by the tritium expression removes the remaining commitment factors.

$$\frac{D_{(V/K)} - 1}{T_{(V/K)} - 1} = \frac{\left(\dfrac{D_{k_5} - 1}{1 + C_f + C_r}\right)}{\left(\dfrac{T_{k_5} - 1}{1 + C_f + C_r}\right)} \tag{90}$$

$$\frac{D_{(V/K)} - 1}{T_{(V/K)} - 1} = \frac{D_{k_5} - 1}{T_{k_5} - 1} \tag{91}$$

$$\frac{D_{(V/K)} - 1}{T_{(V/K)} - 1} = \frac{D_{k_5} - 1}{D_{k_5}^{1.442} - 1} \tag{92}$$

Hence, despite the increased complexity of the kinetic equations for Scheme V, the technique of Northrop is still very much applicable and the addition of extra reversible steps to the enzymatic mechanism does not influence the final calculation of the intrinsic isotope effects.

3. Isotope Effects on the Equilibrium Constant

Isotope effects on catalysis may not be equal in both directions of a reaction, and if they are unequal, there will be an isotope effect on the equilibrium constant. Define

$$D_{K_{eq}} = \frac{D_{k_f}}{D_{k_r}} \tag{93}$$

where $D_{K_{eq}}$ is the deuterium isotope effect on the equilibrium constant, and D_{k_f} and D_{k_r} are the forward and reverse intrinsic isotope effects. The generalized equations for the apparent isotope effects are, for the forward direction of the reaction,

$$D_{V_f} = \frac{D_{k_f} + R_f + C_r \cdot D_{K_{eq}}}{1 + R_f + C_r} \tag{94}$$

$$D_{(V/K)_f} = \frac{D_{k_f} + C_f + C_r \cdot D_{K_{eq}}}{1 + C_f + C_r} \tag{95}$$

$$\frac{D_{(V/K)_f} - 1}{T_{(V/K)_f} - 1} = \frac{D_{k_f} - 1 + C_r(D_{K_{eq}} - 1)}{D_{k_f}^{1.44} - 1 + C_r(D_{K_{eq}}^{1.44} - 1)} \tag{96}$$

and for the reverse direction,

$$D_{V_r} = \frac{D_{k_r} + R_r + C_f/D_{K_{eq}}}{1 + R_r + C_f} \tag{97}$$

$$D_{(V/K)_r} = \frac{D_{k_r} + C_f/D_{K_{eq}} + C_r}{1 + C_f + C_r} \tag{98}$$

$$\frac{D_{(V/K)_r} - 1}{T_{(V/K)_r} - 1} = \frac{D_{k_r} - 1 + C_f\left(\frac{1}{D_{K_{eq}}} - 1\right)}{D_{k_r}^{1.44} - 1 + C_f\left(\frac{1}{D_{K_{eq}}^{1.44}} - 1\right)} \tag{99}$$

A comparison of Equation 95 with Equation 98 reveals that

$$\frac{D_{(V/K)_f}}{T_{(V/K)_r}} = D_{K_{eq}} \tag{100}$$

an equation which provides a useful check on the self-consistency of the results. Substitution of the above relation for $D_{K_{eq}}$ into Equation 99 shows that

$$\frac{D_{(V/K)_f}D_{K_{eq}} - 1}{T_{(V/K)_f}D_{K_{eq}}^{1.44} - 1} = \frac{D_{(V/K)_r} - 1}{T_{(V/K)_r} - 1} \tag{101}$$

These equations show that the influence of equilibrium isotope effects are dependent on the values for the commitment factors (C_f and C_r). Since these are unknown, any errors introduced into the calculated intrinsic isotope effects are also unknown. Nevertheless, as shown by Northrop[490], a set of limits of error can be obtained by comparison of the results of the forward and reverse intrinsic isotope effects.

4. Apparent Kinetic Isotope Effects in Bisubstrate Reactions

Ping-pong mechanism — Given the mechanism shown in Scheme VI:

$$E + A \underset{k_2}{\overset{k_1}{\rightleftarrows}} EA \underset{k_4}{\overset{k_3}{\rightleftarrows}} FB \overset{k_5}{\rightarrow} F + P$$

$$F + B \underset{k_8}{\overset{k_7}{\rightleftarrows}} FB \underset{k_{10}}{\overset{k_9}{\rightleftarrows}} EQ \overset{k_{11}}{\rightarrow} E + Q$$

SCHEME VI

The V/K isotope effect with respect to substrate A is given in the following equation:

$$D_{(V/K_a)} = \frac{D_{k_3} + C_f + C_r}{1 + C_f + C_r} \qquad (102)$$

where $C_f = k_3/k_2$ and $C_r = k_4/k_5$. $D_{(V/K_a)}$ is therefore independent of the second half-reaction and independent of the substrate B concentration.

Ordered mechanism — Consider the mechanism of Scheme VII.

$$\begin{array}{ccccc} & A & B & P & Q \\ & \downarrow & \downarrow & \uparrow & \uparrow \\ \hline E & EA & (x) & EQ & E \end{array}$$

SCHEME VII

This mechanism and all other sequential reactions yield isotopic data which are dependent on both subtrates. Deuterium-label experiments are straight forward. In tritium-label experiments, however, it is more complex and it makes a difference as to which substrate, A or B, is labeled. If B is labeled with T, $T_{(V/K_B)}$ is obtained directly. If A is labeled, then the apparent isotope effect varies as a function of B concentration, and it must be extrapolated to zero concentration of B. Tritium discrimination as a function of substrate B concentration obeys the following relationship, where v is the initial velocity and K_{ia} is the inhibition constant for substrate A;

$$T_v = T_{(V)/K_B} \left\{ \frac{1 + \dfrac{T_{K_a}[B]}{T_{K_{ia}} \cdot T_{K_b}}}{1 + \dfrac{K_a(B)}{K_{ia} \cdot K_b}} \right\} \qquad (103)$$

after subtraction of 1 from both sides of the equation, followed by a rearrangement with the assumption that there is no isotope effect on T_{Kia} (see above, Scheme VII and note that K_{ia} is the dissociation constant for A), one obtains

$$T_v - 1 = \frac{T_{(V/K_b)} - 1}{1 + \dfrac{K_a(B)}{K_{ia} \cdot K_b}} \qquad (104)$$

After taking the reciprocals, it assumes the following form;

$$\frac{1}{T_v - 1} = \frac{1}{T_{(V/K_b)} - 1} + \frac{1}{T_{(V/K_b)} - 1} \cdot \frac{K_a(B)}{K_{ia} \cdot K_b} \tag{105}$$

and a plot of $\frac{1}{T_v - 1}$ vs. (B) should be linear, yielding the (V/K) isotope effect from the intercept.[775]

Random mechanism — Similar to ordered mechanism, the concentration of the cosubstrate influences the size of the apparent (V/K) isotope effect; however, whereas, saturation by one substrate abolishes the (V/K) isotope effect of the other in the ordered mechanism, such is not the case in the random mechanism. Limiting values for deuterium isotope effects on V/K may be obtained from double-reciprocal plots, with one substrate varied and the other fixed. Tritium discrimination, however, is again more complex and in contrast to the ordered mechanism, plots of $\frac{1}{T_v - 1}$ vs. substrate concentration are not linear and extrapolation may be difficult. In the limiting case, however, of a rapid equilibrium random mechanism, the following holds.

$$D_V = D_{(V/K_a)} = D_{(V/K_b)} \tag{106}$$

5. Some Applications of the Northrop Approach or Modifications Thereof

A few applications may be noted here. In an earlier study of the malic enzyme by Schimerlik et al.,[515] they had observed an equilibrium isotope effect of only 1.2; thus, only limiting values could be set for the intrinsic isotope effects, which were 5 and 8 for the oxidation of malate, and 4 and 6.5 for the reverse reaction. These large ranges were the result of the small apparent isotope effect on (V/K) of 1.5 for the deuterium, and 2.0 for tritium. In a recent study of the same enzyme, but with an additional study of the ^{13}C isotope effect on (V/K) with a deuterated and unlabeled substrate, Hermes et al.[581] found with malic enzyme (and with TPN as cosubstrate), ^{13}C isotope effects of 1.031 and 1.025 with malate-2-d, plus a deuterium isotope effect of 1.47. In the deuterium-sensitive step the hydride transfer comes first and reverse hydride transfer is 6 to 12 times faster than decarboxylation (the ^{13}C sensitive step). With 3-acetylpyridine-TPN, however, Hermes et al.[581] found a deuterium isotope effect of 2.18 and a ^{13}C isotope effect of 1.0037 which they felt showed that if the intrinsic isotope effects were similar to those with TPN, the reverse hydride transfer was slowed relative to decarboxylation by a factor of 25, possibly as a result of the more positive redox potential of 3-acetylpyridine-TPN.

Conversely, with the same use of these multiple isotope effects, Hermes et al.[581] found in the case of glucose 6-phosphate dehydrogenase from *Leuconostoc mesenteroides,* that both ^{13}C and deuterium isotope effects resulting from a substitution at C-1 of glucose 6-phosphate are on the same step in the reaction mechanism. In addition, by also measuring the α-secondary deuterium isotope effect at C-4 of TPN, and the effect of this substitution on the ^{13}C isotope effect, they were able to simultaneously solve the resulting five equations for the intrinsic primary ^{13}C, primary deuterium, and α-secondary deuterium isotope effects, as well as the forward and reverse commitments.[581]

Rose[517] has discussed several studies dealing with hydrogen isotope effects in proton transfer to and from carbon. In particular, he discussed studies dealing with pyruvate kinase, aconitase, and aldolase. Recently, Rose and Iyenger[518] also demonstrated the efficacy of employing rapid-mixing and rapid-stop methods (see Chapter 10) in the study of H/D exchange rates of triose phosphate isomerase (see Reference 984 for a review of this enzyme's mechanism). They suggested that the H/D exchange on the enzyme, which occurs more slowly than catalysis, is required for the isotope effect to be seen. They proposed that a conformational change of the Enz-D-glyceraldehyde-3-phosphate complex prior to the enolization is the step which requires a proton transfer. They felt that their work[518] demonstrated the value of rapid measurements of the rate of D_2O effect to distinguish between enzyme-bound protons and the protons of the medium as the cause of the effect. They noted in the important studies of Bender[519,520] on the deacylation of chymotrypsin and papain that the isotope effect in D_2O was lost within 10 sec of dilution into H_2O; they had therefore concluded that the effect was the result of the D_2O medium rather than from a deuterium-substituted enzyme. Rose and Iyenger however, point out that 10 sec is a long time period for all except the hydrogen-bonded backbone hydrogens. With rapid mixing devices they felt that this time interval could be shortened by at least a factor of 10^3 and therefore placing it in the range of many group ionizations.[518] (See also Reference 823.)

The concept of a rate-limiting step for a reaction which proceeds through a sequence of intermediates has been one of the fundamental ideas of classical physical-organic chemistry. However, recently, Northrop[606] questioned the generality of the concept of a rate-limiting step in enzymic reactions. He appeared to demonstrate, in certain simulated isotopic experiments, that alterations of the step identified as rate limiting by current definitions do not consistently affect V_{max} in the expected manner. For example, in a few specific cases (but not all cases) even if the rate constant for the isotopic step was smaller than that for any other step, no isotopic effect on the overall rate would be observed. This is contrary to what might be expected from classical treatments of simple (nonenzymic) systems. Ray,[608] on the other hand, interpreted Northrop's conclusion as simply due to the fact that current definitions, when applied to rate-limiting steps, would appear to lead to inconsistent conclusions.

Ray,[608] therefore, provided a definition for the rate-limiting steps that eliminated any inconsistencies and still retained the classical concept of a rate-limiting step in a reaction which proceeded by way of a sequence of intermediates. Thus, for any enzymic steady-state process which involves a linear reaction sequence of reactions from free reactants to free products, the rate-limiting step is taken as the most sensitive step, or the step which, if perturbed in isotopic studies, results in the largest change in the overall velocity, v. This definition gives rise to a sensitivity function (S.F.) which can be formulated for each forward step in a reaction and used to specify the extent to which each step is rate limiting. In both a V_{max}/E_t system (i.e., an enzymic reaction at saturating substrate (S))and a V_{max}/K_m system (i.e., an enzymic reaction conducted at a specified substrate concentration well below its K_m), the most sensitive step may then be identified by the relative magnitude of the S.F. for the various forward steps.

Recently, in an excellent study by Vanoni and Matthews,[817] the kinetic isotopic effects on the oxidation of reduced nicotinamide adenine dinucleotide phosphate by the porcine liver flavoenzyme, methylenetetrahydrofolate reductase, were described in some detail. This enzyme catalyzes the irreversible NADPH-linked reduction of CH_2-H_4 folate (5,10-methylenetetrahydrofolate) to CH_3-H_3 folate (5 methyltetrahydrofolate). This reaction commits H_4 folate (tetrahydrofolate)-bound one-carbon units to the pathways of *S*-adenosylmethionine-dependent biological methylations. In mammalian cells the enzymatic activity is regulated by the allosteric effector, *S*-adenosylmethionine.[818,819] Previous work by Matthews et al.[820,821] had shown that the enzyme catalyzes the transfer of reducing equivalents from NADPH to suitable electron acceptors, e.g., CH_2-H_4 folate and menadione, by ping-pong BiBi mechanisms, and the reduction of the enzyme by NADPH is irreversible, whereas the reoxidation of the reduced flavoprotein by CH_2-H_4 folate is readily reversible. In the presence of a saturating concentration of menadione, the reductive half-reaction appears to be rate limiting in the steady-state turnover,[821] i.e., it appears to be the most sensitive half-reaction in the more accurate terminology introduced by Ray.[608]

Vanoni and Matthews[817] found that the methylenetetrahydrofolate reductase stereospecifically removes the *pro*-S hydrogen from the 4-position of NADPH. During the oxidation of [4(S)-^{3}H] NADPH, a kinetic isotope effect on V/K_{NADPH} of 10.8 ± 0.4 was observed.

A comparison of the rates of oxidation of [4(S)-^{2}H] NADPH and [4(S)-^{1}H] NADPH yielded a kinetic isotope effect on V of 4.78 ± 0.15 and on V/K_{NADPH} of 4.54 ± 0.59. From a comparison of the oxidation of [4(R)-^{2}H] NADPH and [4 R-^{1}H] NADPH, they obtained the secondary kinetic isotope effect on V, or 1.04 ± 0.01. When the NADPH menadione oxidoreductase reaction is carried out in tritiated water, no incorporation of the solvent tritium into residual NADPH was observed. They concluded from these observations that the oxidation of NADPH is largely or entirely rate limiting in the reductive half-reaction at saturating menadione concentrations.[817] In the presence of saturating NADPH, the flavin reduction proceeds with a rate constant of 160 sec^{-1}, a value which is 29-fold slower than estimates of the lower limit for the diffusion-limited rate constant characterizing NADPH binding to the enzyme under physiological conditions.

Alberty and Knowles[822] had proposed a series of criteria for the catalytic efficiency of an enzyme. They suggested that diffusion of physiological concentrations of the less thermodynamically stable substrate to the active center of the enzyme should be rate limiting in catalysis by a perfectly efficient enzyme, i.e., it should be the most difficult step in the terminology of Ray.[608] They suggested, in addition, that evolution would act to maximize catalytic efficiency in enzymes, provided that they were not those enzymes involved in metabolic control; however, they also considered that the imperatives of catalytic efficiency might be sacrificed to the imperatives of metabolic control and that those enzymes involved in control might not show optimal catalytic efficiency.

Methylenetetrahydrofolate reductase is an allosterically regulated enzyme acting at a branch point of metabolism, and Vanoni and Matthews[817] have now attempted to assess the catalytic efficiency (of an irreversible segment of the overall reaction) of an allosterically regulated enzyme involved in metabolic control. It now appears that in the evolution of methylenetetrahydrofolate reductase, some catalytic power appears to have been sacrificed in the interest of metabolic control since the reduction of the enzyme-bound flavin by physiological concentrations of NADPH contributes to rate limitation in turnover with the physiological electron acceptor methylenetetrahydrofolate. However, the evolutionary pressure on the reductive half-reaction is minimized because this half-reaction is characterized by a pseudo first-order rate constant (160 sec^{-1}) that is approximately three times greater than the rate constant for reoxidation of the enzyme by saturating concentrations of methylenetetrahydrofolate (50 sec^{-1}).[817]

As noted, there has recently been described a method (by estimating multiple isotope effects) for determination of the intrinsic isotope effects on the bond-breaking step of an enzyme-catalyzed reaction by measuring the ^{13}C isotope effect with the substrate that was either unlabeled or deuterated in positions giving a primary or secondary isotope effect. Together with the primary and α-secondary deuterium isotope effects and the various equilibrium isotope effects on the reaction, these values allowed a simultaneous solution of five equations for the intrinsic ^{13}C and primary and α-secondary deuterium isotope effects and the forward and reverse commitments for the bond-breaking step. The method was illustrated with the data obtained for the oxidation of glucose 6-phosphate by TPN, catalyzed by *L. mesenteroides* glucose 6-phosphate dehydrogenase in H_2O,[581] and very recently in D_2O by Hermes and Cleland.[825] Although the intrinsic ^{13}C isotope effect on the hydride-transfer step did not appreciably change in H_2O vs. D_2O,[825] a large decrease in the primary deuterium isotope effect was found, and also an apparent decrease in the α-secondary effect. Hermes and Cleland[825] interpreted their data interestingly in terms of a transition state which allowed coupled hydrogen motions for the three hydrogens in the transition state and tunneling of the hydrogen motions, whereas, that of the carbon did not show tunneling.

Chapter 7

NON-MICHAELIS-MENTEN KINETICS AND ALLOSTERIC KINETICS

I. SOME GENERAL REMARKS

Although many enzyme-catalyzed reactions show Michaelis-Menten kinetics as exhibited by a hyperbolic relationship between the initial rate and the initial substrate concentration, there are, however, some unusual types of behavior as exemplified by so-called sigmoidal kinetics.[169,482] The simplest equation for representing a sigmoidal type of kinetics, involving a single substrate is

$$v = \frac{\alpha(A)^2}{\beta + \gamma(A)^2} \tag{1}$$

where α, β and γ are constants. Thus, at low substrate concentration, $v = \left(\frac{\alpha}{\beta}\right)(A)^2$ and the slope, or $\frac{dv}{d(A)} = 2\left(\frac{\alpha}{\beta}\right)(A)$, increases with an increase in (A); whereas, at higher values of (A), the $\gamma(A)^2$ term becomes important, and the curve finally levels off at a value of v $= \frac{\alpha}{\gamma}$, leading in essence to a sigmoidal relation between v and A (Figure 1). In reciprocal form,

$$\frac{1}{v} = \frac{\beta + \gamma(A)^2}{\alpha(A)^2} = \left(\frac{\beta}{\alpha}\right)\frac{1}{(A)^2} + \left(\frac{\gamma}{\alpha}\right) \tag{2}$$

and a straight line is obtained if $\frac{1}{v}$ is plotted against $(1/A^2)$. (see Figure 2). Ferdinand[101] has used a more general and more useful equation which can be fitted to most of the cases of sigmoidal kinetics.

$$v = \frac{\alpha(A) + \beta(A)^2}{\gamma + \delta(A) + \epsilon(A)^2} \tag{3}$$

and if $\alpha = \delta = 0$, Equation 3 reduces in form to that given by Equation 1. Also, the necessary condition for sigmoidal kinetics arises when

$$\beta\gamma > \alpha\delta \tag{4}$$

However, a more general equation for representing sigmoidal kinetics is

$$v = \frac{\alpha + \beta(A)^n}{\gamma + \delta(A)^n} \tag{5}$$

with n equal to 2 or greater, and α and γ being functions of (A) involving powers not higher than $n - 1$. At high substrate concentrations the limiting rate becomes

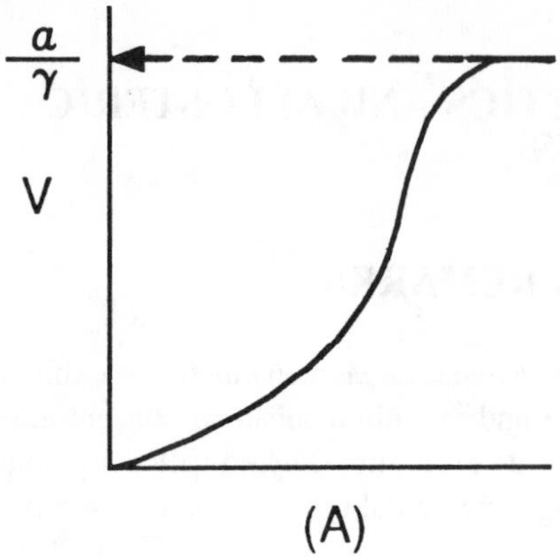

FIGURE 1.

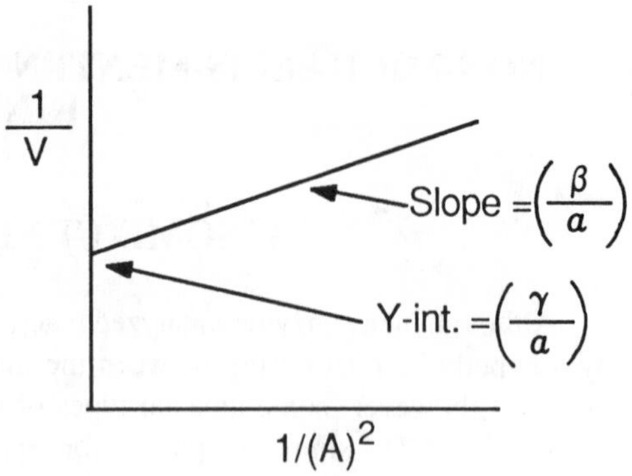

FIGURE 2.

$$\underset{(A \to \infty)}{V_{max}} = \beta/\delta \tag{6}$$

and

$$(V_{max} - v) = \frac{\beta}{\delta} - \frac{\alpha + \beta(A)^n}{\gamma + \delta(A)^n} \tag{7}$$

$$\therefore \quad (V_{max} - v) = \frac{(\beta\gamma - \alpha\delta)}{\delta(\gamma + \delta(A)^n)} \tag{8}$$

and

$$\left(\frac{v}{V_{max} - v}\right) = \frac{\delta(\alpha + \beta(A)^n)}{(\beta\gamma - \alpha\delta)} \tag{9}$$

At high values of (A) a plot of log $\left(\frac{v}{V_{max} - v}\right)$ vs. log (A) will have a slope of n; such a plot is commonly known as a ''Hill plot'' since it was first employed by A. V. Hill in his studies on the binding of O_2 to hemoglobin.[102]

II. MECHANISMS WHICH MAY LEAD TO NON-MICHAELIS-MENTEN BEHAVIOR, BUT DO NOT YIELD SIGMOIDAL KINETICS

A. MICHAELIS-MENTEN KINETICS INVOLVING A SERIES OF INTERMEDIATES WHICH YIELD HYPERBOLIC REACTION VELOCITY EXPRESSIONS

These intermediates result from the scheme

$$E + A \rightleftarrows EA \rightleftarrows EA' \rightleftarrows EA'' \ldots \rightleftarrows EP \rightleftarrows E + P$$

SCHEME I

leads to an equation of the form

$$v_f = \frac{V^f_{max}(A)}{K_{m_A} + (A)} \tag{9a}$$

which is of a hyperbolic nature. (See Figure 2 in Chapter 1.)

We have already considered the case of two intermediates before. (See Chapter 2, Scheme I and Equation 42).

B. CATALYSIS BY TWO ENZYMES WHICH MAY LEAD TO NONHYPERBOLIC KINETICS BUT NOT SIGMOIDAL KINETICS

Catalysis by two enzymes, or by an enzyme with two or more catalytic sites per molecule, may lead to nonhyperbolic kinetics, but which are not sigmoidal. Thus,

$$v = \frac{V^f_{max}(A)}{K_{m_A} + (A)} + \frac{V'_{max}(A)}{K'_{m_A} + (A)} \tag{10}$$

is applicable to either case and which rearranges to

$$v = \frac{(K'_{m_A}V_{max} + K_{m_A}V'_{max})(A) + (V_{max} + V'_{max})(A)^2}{K_{m_A}K'_{m_A} + (K_{m_A} + K'_{m_A})(A) + (A)^2} \tag{11}$$

This Equation 11 is of the same form as the generalized Equation 3, which requires that for sigmoidal kinetics to arise that the condition in Equation 4 hold, or that

$$K_{m_A}K'_{m_A}(V_{max} + V'_{max}) > (K'_{m_A}V_{max} + K_{m_A}V'_{max})(K_{m_A} + K'_{m_A}) \tag{12}$$

Obviously, this condition given by Equation 12 cannot hold since the right-hand side contains the terms of the left-hand side, plus two additional terms; thus, sigmoidal kinetics will not arise from this mechanism, which will, however, show deviation from hyperbolic kinetics.

C. IF THE SUBSTRATE CONTAINS AN IMPURITY THAT FORMS AN INACTIVE COMPLEX WITH IT, SIGMOIDAL KINETICS MAY NOT RESULT

The substrate, A, contains an impurity, L, that forms an inactive complex with it (e.g., a heavy-metal nucleotide complex).

$$L + A \rightleftarrows LA \tag{13}$$

$$K_{L_A} = \frac{(L)(A)}{LA} \tag{14}$$

L_{total} is related to A_{total} by

$$L_{total} = a\ A_{total} \tag{15}$$

i.e., and assume that $A_t >> L_t$, so that

$$a = \frac{L_{total}}{A_{total}} \tag{16}$$

$$A_t \simeq A \tag{17}$$

$$L_t = L + LA \tag{18}$$

Therefore

$$K_{L_A} = \frac{(L_t - LA)(A_t)}{(LA)} \tag{19}$$

$$(LA)K_{L_A} = A_tL_t - (LA)A_t \tag{20}$$

$$(LA)(K_{L_A} + A_t) = A_tL_t \tag{21}$$

$$(LA) = \frac{A_tL_t}{K_{L_A} + A_t} = \frac{A_t(aA_t)}{K_{L_A} + A_t} = \frac{aA_t^2}{K_{L_A} + A_t} \tag{22}$$

Since

$$A_t = A + LA \tag{23}$$

even though LA << A,

$$A = A_t - LA = A_t - \frac{aA_t^2}{K_{L_A} + A_t} = \frac{A_t(K_{L_A} + A_t) - A_t^2}{K_{L_A} + A_t} \tag{24}$$

$$(A) = \frac{A_tK_{L_A} + A_t^2 - aA_t^2}{K_{L_A} + A_t} = \frac{A_tK_{L_A} + A_t^2(1 - a)}{K_{L_A} + A_t} \tag{25}$$

If the rate of the enzyme-catalyzed reaction is given by

$$v = \frac{V_{max}(A)}{K_{m_A} + (A)} \tag{9a}$$

substitute (A) from Equation 25.
Therefore,

$$v = \frac{V_{max}[A_tK_{L_A} + A_t^2(1 - a)]/[K_{L_A} + A_t]}{K_{m_A} + \dfrac{[A_tK_{L_A} + A_t^2(1 - a)]}{(K_{L_A} + A_t)}} \tag{26}$$

$$= \frac{V_{max}[A_tK_{L_A} + A_t^2(1 - a)]}{K_{m_A}(K_{L_A} + A_t) + A_tK_{L_A} + A_t^2(1 - a)} \tag{27}$$

Therefore,

$$v_f = \frac{V_{max}[A_tK_{L_A} + A_t^2(1 - a)]}{K_{m_A}K_{L_A} + A_t(K_{m_A} + K_{L_A}) + A_t^2(1 - a)}$$

$$= \frac{V_{max}K_{L_A}A_t + V_{max}(1 - a)A_t^2}{K_{m_A}K_{L_A} + (K_{m_A} + K_{L_A})A_t + (1 - a)A_t^2} \tag{28}$$

which is of the same form as Equation 3.

For the condition $\beta\gamma > \alpha\delta$ to hold, it is required that

$$[V_{max}(1 - a)][K_{m_A}K_{L_A}] > (V_{max}K_{L_A})[K_{m_A} + K_{L_A}] \tag{29}$$

but since this is clearly not possible, sigmoidal kinetics cannot result from this mechanism.

D. IF THE SUBSTRATE CONTAINS AN IMPURITY WHICH COMBINES WITH THE ENZYME, FORMING INACTIVE EI, SIGMOIDAL KINETICS MAY NOT OCCUR

The substrate S contains an impurity, I, which combines with the enzyme forming an inactive EI complex.

The general equation for the forward velocity containing an inhibitor, I, is

$$\frac{1}{v^f} = \frac{1}{V_{max}}\left(1 + \frac{(I)}{\alpha K_i}\right) + \frac{K_{ms}}{V_{max}}\frac{1}{(S_o)}\left(1 + \frac{(I)}{K_i}\right) \tag{1-77}$$

or

$$\frac{V^f_{max}}{v^f} = \left(1 + \frac{(I)}{\alpha K_i}\right) + K_{ms}\frac{1}{S_o}\left(1 + \frac{(I)}{K_i}\right) \tag{30}$$

$$\frac{v^f}{V^f_{max}} = \frac{1}{\left(1 + \frac{(I)}{\alpha K_i}\right) + \frac{K_{ms}}{(S_o)}\left(1 + \frac{(I)}{K_i}\right)} \tag{31}$$

$$v^f = \frac{V^f_{max}(S_o)}{S_o\left(1 + \frac{(I)}{\alpha K_i}\right) + K_{ms}\left(1 + \frac{(I)}{K_i}\right)} \tag{32}$$

Since I is an impurity in S, assume Equation 33 holds, i.e.,

$$a = \frac{I_t}{S_t} \quad (\text{or} \quad I_t = aS_t) \tag{33}$$

and

$$I_t = I + EI \simeq I \tag{34}$$

since EI is very small.

Therefore,

$$v^f = \frac{V^f_{max}(S_t)}{S_t\left(1 + \frac{aS_t}{\alpha K_i}\right) + K_{ms}\left(1 + \frac{aS_t}{K_i}\right)} \tag{35}$$

$$= \frac{V^f_{max}(S_t)}{S_t\left(\frac{\alpha K_i + aS_t}{\alpha K_i}\right) + \alpha K_{ms}\left(\frac{K_i + aS_t}{\alpha K_i}\right)} \tag{36}$$

$$= \frac{\alpha K_i V^f_{max}(S_t)}{S_t(\alpha K_i + aS_t) + \alpha K_{ms}(K_i + aS_t)} \tag{37}$$

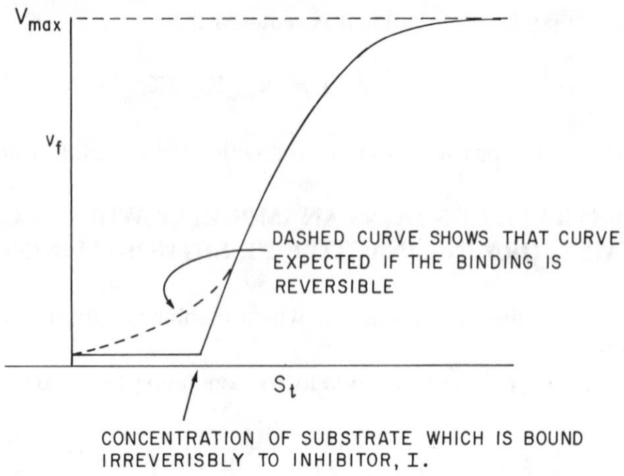

FIGURE 3.

$$= \frac{\alpha K_i V^f_{max}(S_t)}{\alpha K_i S_t + a S_t^2 + \alpha K_{ms} K_i + \alpha a S_t} \tag{38}$$

$$v^f = \frac{\alpha K_i V^f_{max}(S_t)}{\alpha K_i K_{ms} + (\alpha K_i + \alpha a) S_t + a A_t^2} \tag{39}$$

Since this Equation 39 is not of the same form as Equation 3 (β would be zero) and therefore, the sigmoidal condition, $\beta\gamma > \alpha\delta$, cannot be fulfilled.

E. THE ENZYME CONTAINS AN IMPURITY WHICH FORMS AN INACTIVE EI

The enzyme contains an impurity, I, which inhibits by combining with the enzyme. I is now a constant at constant E_t, and I_t is no longer proportional to S_t (Equation 33). Therefore, Equation 32 applies here, that is,

$$v^f = \frac{V^f_{max}(S_t)}{S_t\left(1 + \frac{(I)}{\alpha K_i}\right) + K_{ms}\left(1 + \frac{(I)}{K_i}\right)} \tag{32}$$

which is obviously not of the correct sigmoidal form, but rather yields hyperbolic kinetics.

III. MECHANISMS WHICH MAY LEAD TO A NON-MICHAELIS-MENTEN BEHAVIOR AND WHICH ARE SIGMOIDAL IN NATURE

A. THE ENZYME CONTAINS AN IMPURITY WHICH MAY COMBINE WITH THE SUBSTRATE (THIS CASE WAS CONSIDERED BY J. WESTLEY.[103])

In the extreme situation, the inhibitor is contained as an impurity in the enzyme preparation which will combine irreversibly with the substrate, rendering it inactive, and the effective substrate concentration is therefore less than S_t by a fixed amount. If there are no other complications, the velocity vs. S_t curve is expected to be hyperbolic, but is displaced to the right by an amount equal to the concentration of the substrate bound to the inhibitor. If the binding is not irreversible, a rounding off of the curve is to be expected as shown in Figure 3 corresponding, therefore, to a situation involving sigmoidal kinetics.

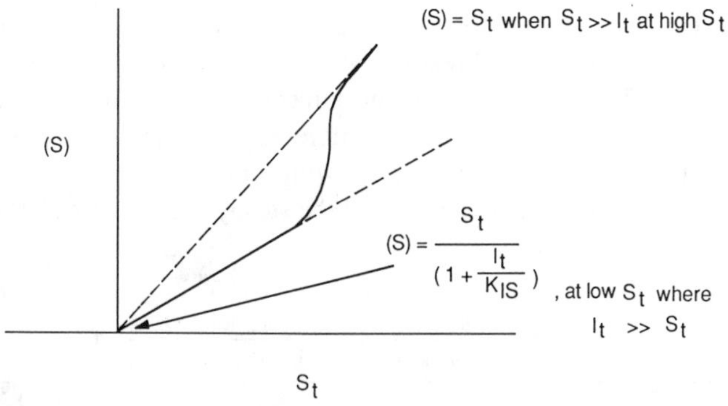

FIGURE 4.

The steady-state treatment of the above case involves a solution to quadratic equations which are unwieldy in form. Therefore, consider the following simple treatment: consider that the total substrate can rise so high that $S_t >> I_t$, and that $S_t \simeq S$. But at a relatively low concentration of S_t, where now $I_t >> S_t$, and $I_t \simeq I$,

$$S_t = (S) + IS \tag{40}$$

remembering that

$$K_{IS} = \frac{(I)(S)}{(IS)} \tag{41}$$

Therefore

$$S_t = (S) + \frac{(I)(S)}{K_{IS}} = (S)\left(1 + \frac{(I)}{K_{IS}}\right) = (S)\left(1 + \frac{I_t}{K_{IS}}\right) \tag{42}$$

Therefore, if we plot now the concentration of S, the free substrate, vs. the concentration of total substrate S_t, (Figure 4),

Since at low S_t the initial slope is $\left(\dfrac{1}{1 + \dfrac{I_t}{K_{IS}}}\right)$ of the (S) vs. S_t plot, and since the rate at low S_t is proportional to (S), sigmoidal kinetics will appear to result.

B. TWO FORMS OF THE SAME ENZYME

Thus, sigmoidal behavior results if the enzyme can exist in two forms (conformers, or e.g., an acylated enzyme) with two different activities. Rabin[104] has suggested the following mechanism of this type (see also References 105 to 107):

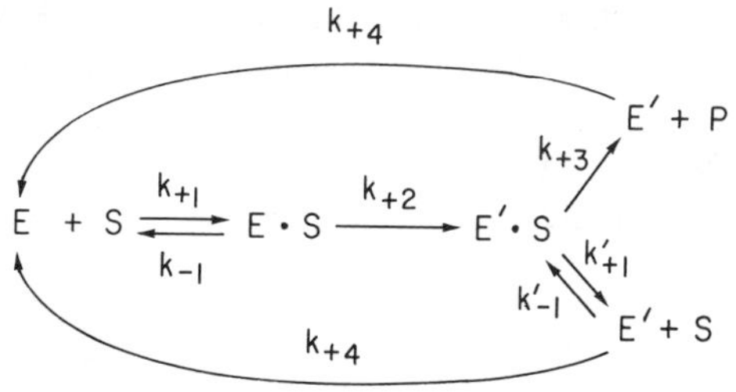

SCHEME II.

The enzyme exists, therefore, in form E and E′; E cannot give rise directly to the product P, but must first form E′S which then breaks down to E′ and product, with E′ interconverting to E. With an increase in the initial substrate concentration, there is an unusually large increase in rate, since S drives E into E′, the more active of the two enzyme species. At low substrate concentrations, the rate thus varies with S raised to a higher power than unity, so that sigmoidal kinetics results. Applications of the steady-state approach leads to Equation 43

$$v = k_{+1}k_{+3}\frac{\left\{\frac{k_{+2}}{k_{+3} + k'_{+1}}(S) + \frac{k_{+2}k'_{-1}(S)^2}{k_{+4}(k_{+3} + k'_{+1})}\right\}E_t}{\left[k_{-1} + k_{+2} + k_{+1}\left\{1 + \frac{k_{+2}}{k_{+4}} + \frac{k_{+2}}{k_{+3} + k'_{+1}}\right\}(S) + \frac{k_{+1}k_{+2}k'_{-1}(S)^2}{k_{+4}(k_{+3} + k'_{+1})}\right]} \tag{43}$$

which is now of the same form as Equation 3 and provided that the rate constants satisfy the condition of Equation 4, sigmoidal kinetics will result.

C. CONSIDER THE CASE WHERE TWO MOLECULES OF SUBSTRATE MAY BIND TO ONE MOLECULE OF ENZYME (cf. EQUATION 20 IN CHAPTER 6)

Sigmoidal kinetics can result if the enzyme can form the complex ES_2 which gives rise to product formation more rapidly than ES. Another special case is when ES itself is inactive, i.e.,

$$E + S \underset{k_{-1}}{\overset{k_{+1}}{\rightleftarrows}} ES \xrightarrow{k_{+2}(S)} ES_2 \xrightarrow{k_{+3}} E + S + P$$

SCHEME III

the steady-state solution of which is

$$v^f = \frac{k_{+1}k_{+2}k_{+3}E_t(S)^2}{k_{-1}k_{+3} + (k_{+1} + k_{+2})k_{+3}(S) + k_{+1}k_{+2}(S)^2} \tag{44}$$

This equation is of the same form as Equation 3 but with $\alpha = 0$, and therefore the condition of Equation 4 is necessarily satisfied.

Worcel et al.[108] found that this mechanism applies to the enzyme, NAD oxidase. (See also References 109 and 110 for a discussion of a more general mechanism, involving the attachment of a molecule of substrate.)

One may note that this mechanism implies site interaction for if the enzyme were to bind a molecule of substrate at each of two or more identical sites, with no interaction between the sites, normal hyperbolic behavior is to be expected.

D. THE CASE OF THE SUBSTRATE ALSO ACTING AS A MODIFIER

Frieden[111] has suggested the following mechanism in which the substrate can form an active ES complex and also an inactive SE complex by binding at another site, with the possibility of binding at both sites to yield SES.

Thus, consider the quasi-equilibrium mechanism:

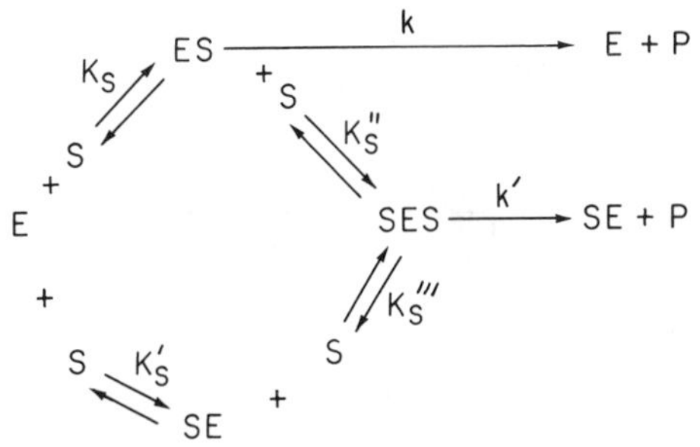

SCHEME IV.

which leads to the rate expression for the forward velocity.

$$v_f = \frac{E_t\left\{k(S) + \dfrac{k'(S)^2}{K_S''}\right\}}{K_S + \left\{1 + \left(\dfrac{K_S}{K_S'}\right)\right\}(S) + \dfrac{(S)^2}{K_S''}} \tag{45}$$

Since this is of the same form as Equation 3, sigmoidal kinetics are possible provided the condition of Equaiton 4 is fulfilled, that is,

$$\left(\frac{K_S}{K_S''}\right)k' > \left\{1 + \frac{K_S}{K_S'}\right\}k \tag{46}$$

which requires that the reaction of SES is of sufficient rapidity compared with that of ES.

E. TWO-SUBSTRATE RANDOM STEADY-STATE SYSTEMS

If two substrates, (A) and (B), undergo a steady-state reaction by a random ternary complex mechanism (Chapter 2, Scheme 5); the general form of the forward reaction velocity is given by Equation 60 in Chapter 2, that is,

$$\frac{v_o^f}{E_t} = \frac{K_1(A)(B) + K_2(A)^2(B) + K_3(A)(B)^2}{K_7 + K_8(A) + K_9(B) + K_{10}(A)(B) + K_{11}(A)^2} \tag{2-60}$$

$$+ K_{12}(B)^2 + K_{13}(A)^2B + K_{14}(A)(B)^2$$

With (B) held fixed and (A) varied, Equation 65 in Chapter 2 results,

$$\frac{v_f}{E_t} = \frac{i(A)^2 + j(A)}{k + \ell(A)^2 + m(A)} \tag{2-60}$$

which is of the same form as Equation 3 and therefore, may result in sigmoidal kinetics. (Atkinson and Walton[112] have dealt with this mechanism experimentally).

F. TWO-SUBSTRATE MECHANISMS WITH ALTERNATIVE PATHWAYS

Two-substrate systems will also lead to sigmoidal kinetics if they proceed via two or more alternative pathways. As an example, consider that a reaction might occur by both a ping-pong mechanism and a ternary complex mechanism, e.g., as shown in Scheme V.

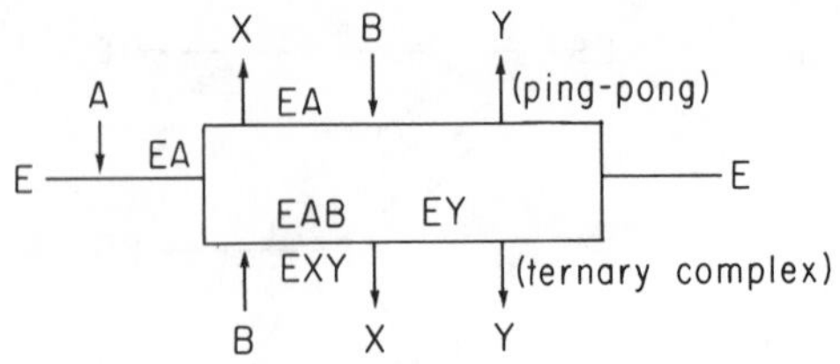

SCHEME V.

Several researchers have considered several combinations of different mechanisms which may lead to sigmoidal behavior.[113-115]

G. THE CASE OF INTERACTING SUBUNITS AND ALLOSTERIC KINETICS

There have been several explanations for sigmoidal kinetics based on the assumption that those enzymes with subunit structures may show interaction between the component subunits. One of the first such theories was proposed by J. Monod et al.[116] in their classical paper which also proposed the concept of allosteric kinetics. In this term, they referred to the attachment of a substrate or a modifier at a site other than the catalytic site, with resulting modification of the enzyme. One should bear in mind that the concept of ''allostery'' in itself does not provide the explanation for sigmoidal kinetics, for there are enzymes with subunit structures which do not display the allosteric effect. Rather, it is the existence of interacting subunits which, under certain conditins, may give rise to sigmoidal kinetics.

Monod's et al.[116] treatment follows, based on the following postulates:

1. Enzyme molecules which display allosteric kinetics are oligomers, consisting of a number of identical subunits known as protomers.
2. All of the subunits ina given molecule are presumed to be of the same conformation. This is considered to be an extreme case of subunit interaction, for if one subunit alters its conformation, all the others in the same molecule presumably must do so as well.
3. If each subunit in a given molecule may exist in two conformations, there are presumably two forms, therefore, for the enzyme molecule, E^r and E^t, where in each of the two forms, all of the subunits are in either one or the other conformation.

Sigmoidal kinetics will follow from these postulates, for, if the equilibrium constant, L (referred to as the ''allosteric constant'' by Monod et al.[116]) is defined as

$$L = \frac{[E^t]}{[E^r]} \tag{47}$$

where $[E^t]$ and $[E^r]$ denote the concentrations of the two conformers of free enzyme (without bound substrate). Each subunit is assumed to be capable of binding one substrate molecule, the dissociation constants for binding being identical, apart from statistical factors. The various equilibria for binding the ligand (A) are represented in Scheme VI in which one conformer of subunits is depicted as a circle and the other as a square; the statistical factors are also given in the scheme. As an illustration, consider the expression for $\frac{(E^rA)(A)}{(E^rA_2)}$ These are three ways for A to bind to E^rA, but two ways for E^rA_2 to lose an A, which leads to the statistical factor 2/3.

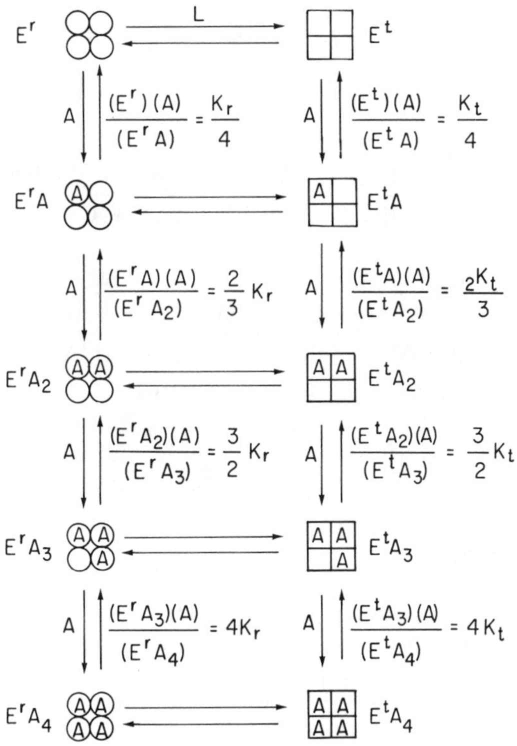

SCHEME VI. A case of a 4-subunit interacting model, according to Monod et al.[106]

The fraction of sites which have substrate molecules bound to them, or the so-called "saturation function", $\bar{Y}_A$ is given by

$$\bar{Y}_A = \frac{\left\{ \begin{array}{l} (E^rA) + 2[E^rA_2] + 3[E^rA_3] + 4[E^rA_4] \\ + (E^tA) + 2(E^tA_2) + 3(E^tA_3) + 4(E^tA_4) \end{array} \right\}}{\begin{array}{l} 4\{(E^r) + (E^rA) + (E^rA_2) + [E^rA_3] + [E^rA_4] \\ + (E^t) + (E^tA) + (E^tA_2) + (E^tA_3) + (E^tA_4)\} \end{array}} \tag{48}$$

$$\bar{Y}_A = \frac{\left(\begin{array}{l} (E^rA)\left\{1 + \dfrac{3(A)}{K_r} + \dfrac{3(A)^2}{K_r^2} + \dfrac{(A)^3}{K_r^3}\right\} \\ + (E^tA)\left\{1 + \dfrac{3(A)}{K_t} + \dfrac{3(A)^2}{K_t^2} + \dfrac{(A)^3}{K_t^3}\right\} \end{array} \right)}{\begin{array}{l} 4\left[(E^r)\left\{1 + \dfrac{4(A)}{K_r} + \dfrac{6(A)^2}{K_r^2} + \dfrac{4(A)^3}{K_r^3} + \dfrac{(A)^4}{K_r^4}\right\}\right. \\ \left. + (E^t)\left\{1 + \dfrac{4(A)}{K_t} + \dfrac{6(A)^2}{K_t^2} + \dfrac{4(A)^3}{K_t^3} + \dfrac{(A)^4}{K_t^4}\right\}\right] \end{array}} \tag{49}$$

Define

$$\alpha = \frac{(A)}{K_r} \quad \text{and} \quad C = \frac{K_r}{K_t} \quad \left(\therefore\ C = \frac{1(A)}{\alpha K_t} \quad \text{or} \quad \frac{(A)}{K_t} = C\alpha\right) \tag{50}$$

Therefore,

$$\bar{Y}_A = \frac{(E^rA)(1 + \alpha)^3 + (E^tA)(1 + C\alpha)^3}{4[(E^r)(1 + \alpha)^4 + (E^t)(1 + C\alpha)^4]} \tag{51}$$

$$C = \frac{K_r}{K_t} = \frac{\frac{(E^r)(A)}{(E^rA)}}{\frac{(E^t)(A)}{(E^tA)}} = \frac{E^r}{(E^rA)}\frac{(E^tA)}{E^t} = \frac{\frac{(E^r)}{(E^t}}{1/L}\frac{(E^tA)}{(E^rA)} = \frac{1}{(L)}\frac{(E^tA)}{(E^rA)} \tag{52}$$

$$\therefore \quad \frac{(E^tA)}{(E^rA)} = CL = L\frac{K_r}{K_t} = \frac{(E^t)}{(E^r)}\frac{K_r}{K_t} \tag{53}$$

Substitute into Equation 51,

$$(E^tA) = (E^rA)CL \tag{54}$$

$$L = \frac{(E^t)}{(E^r)} \quad \text{and} \quad (E^t) = L(E^r) \tag{55}$$

$$\therefore \quad \bar{Y}_A = \frac{(E^rA)(1 + \alpha)^3 + (E^tA)(1 + C\alpha)^3}{4\{(E^r)(1 + \alpha)^4 + (E^t)(1 + C\alpha)^4\}} \tag{56}$$

$$\bar{Y}_A = \frac{(E^rA)(1 + \alpha)^3 + (E^rA)CL(1 + C\alpha)^3}{4\{(E^r)(1 + \alpha)^4 + L(E^r)(1 + C\alpha)^4\}} \tag{57}$$

$$= \frac{(E^rA)\{(1 + \alpha)^3 + CL(1 + C\alpha)^3\}}{4(E^r)\{(1 + \alpha)^4 + L(1 + C\alpha)^4\}} \tag{58}$$

Since

$$\frac{K_r}{4} = \frac{(E^r)(A)}{(E^rA)} \quad \text{and} \quad \frac{(E^rA)}{4(E^r)} = \frac{(A)}{K_r} = \alpha \tag{59}$$

Therefore,

$$\bar{Y}_A = \frac{\alpha(1 + \alpha)^3 + \alpha CL(1 + C\alpha)^3}{(1 + \alpha)^4 + L(1 + C\alpha)^4} \tag{60}$$

The general solution for n subunits is therefore:

$$\bar{Y}_A = \frac{\alpha(1 + \alpha)^{n-1} + \alpha CL(1 + C\alpha)^{n-1}}{(1 + \alpha)^n + L(1 + C\alpha)^n} \tag{61}$$

Under certain conditions, the plot of $\bar{Y}_A$ vs. α (note that α is proportional to (A); α =

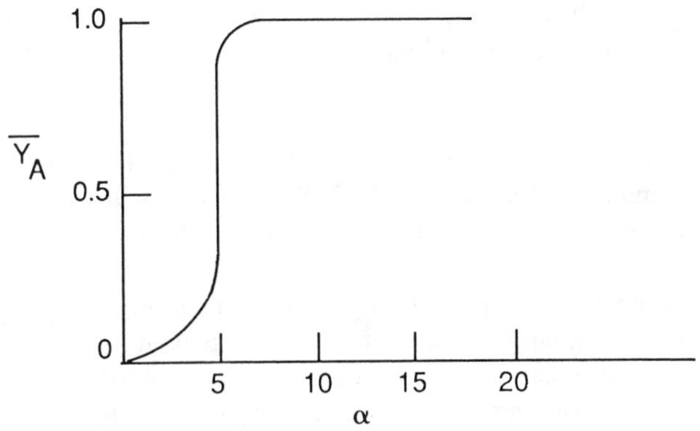

FIGURE 5.

$\frac{(A)}{K_r}$ is sigmoidal; e.g., when c is small and L is large. Thus if $K_t >> K_r$ and L is large (or $E^t >> E^r$); or, since $C = \frac{K_r}{K_t}$ and if $C \rightarrow 0$, $K_t >> K_r$ (e.g., n = 4; L = 10^3, and C = 0.10 as in Figure 5).

Cooperativity, i.e., activation, or curvature of the lower part of the curve, is the greatest when (1) $(E^t) >> (E^r)$, i.e., L is large; and $K_t >> K_r$, i.e., C is small or (2) $(E^r) >> (E^t)$, i.e., L is small, and $K_r >> K_t$, i.e., C is large.

When, however, C = 1, or when $L \rightarrow 0$,

$$\overline{Y}_A = \left(\frac{\alpha}{1 + \alpha}\right) = \frac{(A)}{K_r + (A)} \tag{62}$$

hyperbolic behavior results. Also, when

$$L \rightarrow \infty, \quad \overline{Y}_A = \frac{C\alpha}{1 + C\alpha} \tag{63}$$

again hyperbolic behavior results.

Frieden[117] has noted that the above treatment involves the tacit assumption that the chemical reaction is so slow that it does not disturb the equilibria represented in Scheme VI. Wong et al.[118,119] have attempted a steady-state treatment of the problem. The equation of Monod et al. has been successfully applied to several systems, notably hemoglobin[116] and aspartate transcarbamylase.[120,121]

For some systems, however, this treatment has not proved satisfactory and an alternative model has been proposed for the case of interacting subunits.

Moreover, since the above system is derived on an equilibrium basis, the assumption that "steady-state" binding holds as well is often not proved and leads to the final assumption which is difficult to prove, that is

$$\overline{Y}_A = \frac{v_o}{V_{max} \cdot n} \tag{64}$$

where v is the steady-state velocity and V_{max} is the asymptotic velocity at saturation for n equivalent sites.[117] If, however, the assumption holds, then

$$v = \frac{V_{max}[\alpha(1 + \alpha)^{n-1} + \alpha CL(1 + C\alpha)^{n-1}]}{(1 + \alpha)^n + L(1 + C\alpha)^n} \tag{65}$$

However, later (Chapter 9, Equation 5) we will show how a true steady-state solution may be arrived at, and without treating the interconversion step of the enzyme as a rapid equilibrium step.

H. THE "INDUCED-FIT" MODEL FOR INTERACTING SUBUNITS

Implicit in the model of Monod et al.[116] is the hypothesis that all of the subunits in a given protein molecule exist in the same conformation at the same point in time and that a change in conformation of one subunit requires a change in all the others. As indicated above, this is an extreme case of interaction. Koshland and co-workers[122-125] have developed mechanisms of the same kind, but based upon a more general hypothesis in which hybrid conformational states can exist, i.e., one subunit can alter its conformation without producing a change in the others. However, interaction between subunits in a given molecule exists since a change in the conformation of one subunit changes the relative stability of conformations of neighboring subunits. The affinity of the two conformations for the substrate differs and the addition of the substrate, therefore, disturbs the equilibrium between the conformational forms, giving rise consequently to sigmoidal kinetics under suitable conditions.

Koshland and co-workers proposed four basic interacting subunit mechanisms, which are given in Scheme VII.

The Four Basic Interacting-subunit Mechanisms of Koshland *et al.* [122 - 125]

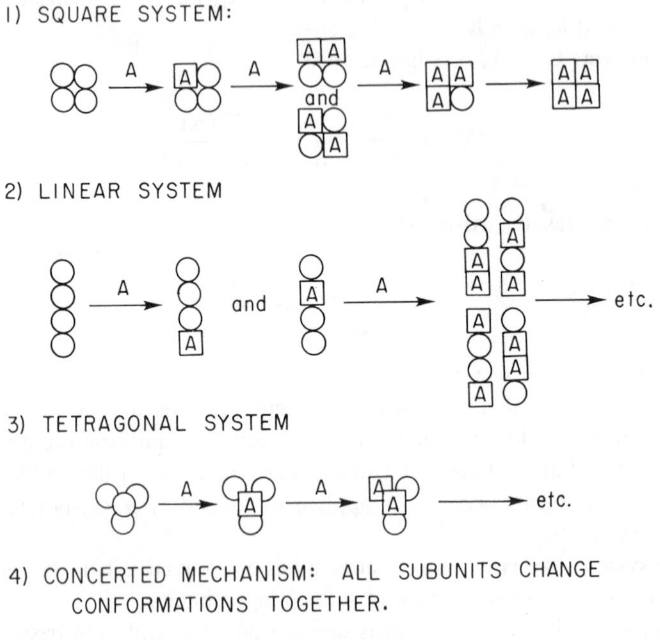

SCHEME VII.

It is assumed that when a substrate molecule becomes attached to a subunit in one conformation, represented by a circle, it induces its conversion into the other conformation, represented by a square. When there is a conformational change in one subunit, it favors (but does not compel) a corresponding change in a neighboring subunit, but *not* in a non-neighboring subunit. In the square system, for example, it is assumed that there are no diagonal interactions, each subunit interacting with only two others. In the tetragonal system on the other hand, each subunit interacts with each of the other three. Finally, the "concerted mechanism" (Number 4 of Scheme VII) is equivalent to that of Monod et al.

In Scheme VIII, a more general model for subunit interaction is presented[132,724,803] and the so-called "concerted-symmetry" and "sequential-interaction" models briefly mentioned above may be considered extreme cases of this more general model (Scheme VIIIa). However, even this more general model is a simplified representation, since states with more than one bound ligand molecule can assume different geometrical configurations. Thus, in Scheme VIIIb we note that the $T_3\,R(S)_2$ complex can exist in two nonequivalent forms.

The concerted symmetry model, as represented by the first and fourth columns of Scheme VIIIa can be further extended if ligand-induced association-dissociation of multi-subunit enzymes are included as a type of possible allosteric interaction.

For example, assume (1) that association-dissociation is a concerted process (i.e., there are no intermediate oligomers with less than n monomers) and (2) that the monomer and oligomer each have different affinities for a given ligand or that they have different intrinsic catalytic activities. The monomer might represent the T-state and the oligomer the R-state. This system possesses the additional characteristic in that its kinetic properties would vary with the enzyme dilution. At low enzyme concentrations, dissociation might be favored, whereas, at high enzyme concentrations the equilibrium might be shifted in favor of the oligomer as, for example, in the following tetrameric association-dissociation system.

$$4E \rightleftarrows E_4 \tag{66}$$

where $K_{eq} = \dfrac{[\text{tetramer}]}{[\text{monomer}]^4}$. Frieden[725,726] has derived the saturation equation for a dimer that undergoes a rapid and reversible association-dissociation reaction (see also Reference 727) i.e.,

$$\overline{Y}_A = \frac{\alpha(1 + \alpha)^{n-1} + 2K_{eq}[E]c\alpha(1 + c\alpha)^{2n-1}}{(1 + \alpha)^n + 2K_{eq}[E](1 + c\alpha)^{2n}} \tag{67}$$

where

$$\alpha = \frac{(A)}{K_E} \tag{68}$$

for the monomer, C is the ratio of monomer to dimer substrate dissociation constants, i.e.,

$$\frac{K_E}{K_{E_2}} = C \tag{69}$$

[E] is the concentration of free monomer, and K_{eq} is the equilibrium constant for the dimerization reaction, i.e.,

$$K_{eq} = \frac{E_2}{(E)^2} \tag{70}$$

The concentration of the free monomer, [E] is described by Equation 71

$$[E] = \frac{-(1 + \alpha)^n + [(1 + \alpha)^{2n} + 4[E]_o K_{eq}(1 + c\alpha)^{2n}]^{1/2}}{2K_{eq}(1 + c\alpha)^{2n}} \tag{71}$$

where $[E]_0$ is the total enzyme concentration in terms of the monomer.

If the intrinsic catalytic activities in both the monomer and dimer are identical, then

$$\overline{Y}_A = \frac{v}{n \cdot V_{max}} \tag{72}$$

(see also Chapter 8, Section III and Equations 8-32 and 8-33); otherwise, if k_1 and k_2 are the rate constants for breakdown of the enzyme substrate complex of monomer and dimer, respectively, then

$$\frac{v_o}{nV_{max}} = \frac{k_1\alpha(1 + \alpha)^{n-1} + 2k_2K_{eq}(E)C\alpha(1 + c\alpha)^{2n-1}}{(1 + \alpha)^n + K_{eq}(E)(1 + c\alpha)^{2n}} \tag{73}$$

For higher degrees of polymerization, the equations become more complex, and beyond a tetramer there does not appear to be a unique solution for [E]. The case becomes more complicated if the number of binding sites does not linearly increase with the number of monomer molecules, i.e., some sites might become inaccessible as a result of subunit polymerization. This situation becomes even more complex if a number of polymeric species coexist in equilibrium (e.g., E, E_2, E_3, and E_4).

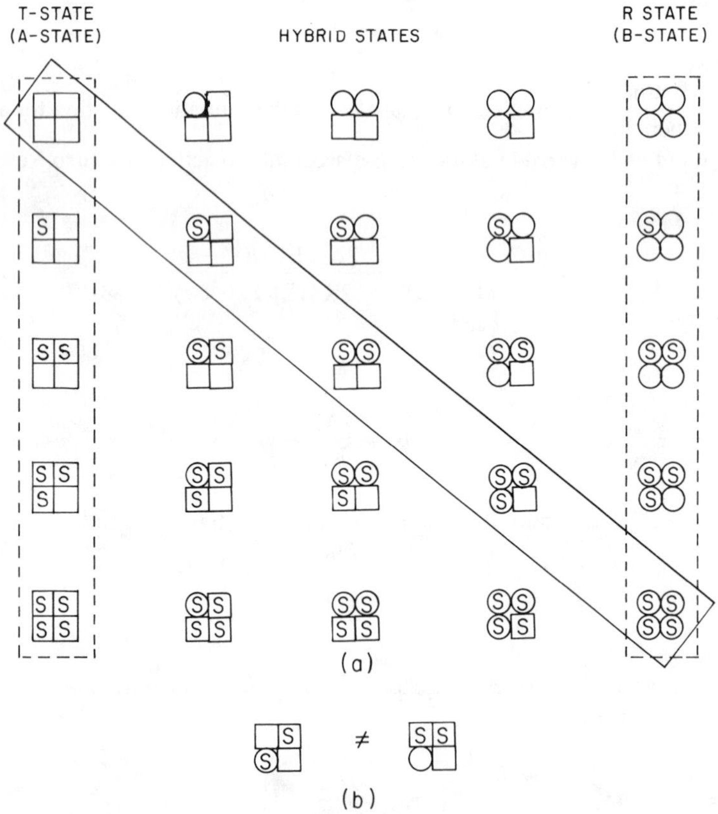

SCHEME VIII. (a) General model for a four-site allosteric enzyme involving hybrid oligomers. The first and fourth columns represent the concerted symmetry model. The diagonal represents the "sequential interaction" model. As shown, there are 25 different types of enzyme forms. (b) If the potential nonequivalent complexes are included (e.g., two different T_3 RS_2 forms), then there are 44 possible enzyme forms. (Redrawn from Hammes, G. G. and Wu, C.-W., *Science*, 172, 1205, 1971. With permission.)

I. SOME OBSERVATIONS ON ALLOSTERIC KINETICS

There are some enzymes which appear to be influenced by substances which are not substrate analogs and which appear not to be attached to the active site of the enzyme, i.e., not to the substrate binding site. These substances are referred to as "modifiers" or "effectors", and may act as either activators or inhibitors. Since they appear to be attached to sites different from the substrate binding site, this phenomenon is referred to as an "allosteric" situation (from the Greek "allos", or other, and "steros", or solid).

One may observe that with this definition of allosteric kinetics, the original or basic model of Monod, et al. (Scheme VI) as such, is not actually an allosteric model since there is no postulate for sites other than those to which the substrate molecules are attached. When the treatment, however, is extended to include additional sites for attachment of the modifier with its resulting effect on the catalytic activity, the mechanism does become an allosteric kinetic model.

Sigmoidal kinetics per se do not provide evidence that an allosteric situation exists, as shown above. The most direct evidence for allosteric kinetics is that the enzyme may be made insensitive to the modifier (i.e., effector) without loss in catalytic activity, thereby, providing evidence that the modifier binding site is different from the substrate binding site or catalytic site. The modifier should act by inducing a conformational change which in turn alters the activity at the catalytic site.

Some of the best studied cases where ideas on allosteric kinetics have been attempted are the aspartate transcarbamylase and glyceraldehyde 3-phosphate dehydrogenase.

A good review of ligand interactions at equilibrium and of allosterism may be found in the recent text by Cantor and Schimmel,[770] and an excellent summary and a formal approach to the study of cooperative binding to macromolecules (the case where binding of one ligand influences the binding strength of the macromolecule towards a subsequent ligand or ligands) may be found in the review by Perlmutter-Hayman.[938] Madsen will discuss in detail two enzymes which display allosteric and non-Michaelis-Menten behavior in Volume II of this work,[983] i.e., glycogen phosphorylase and glycogen synthetase.

Determination of the three-dimensional structure of glycogen phosphorylases a and b by X-ray crystallography has permitted the identification and characterization of several discrete ligand binding sites.[664-667] These sites have been probed by a variety of physical-chemical and kinetic techniques, and some understanding of the specificity and interactions between these sites is slowly being obtained. Thus, the activator site in the N-terminal domain is near serine-14 phosphate and has a preferred specificity for adenosine 5′-phosphate (AMP) but will also accept inosine 5′-phosphate (IMP) as an activator.[668,669] Crystallographic studies on phosphorylase b have demonstrated that the allosteric inhibitors, α-D-glucose 6-phosphate and ATP bind at this site, which accounts for their competition with AMP, and stabilize the inactive "T" conformation of this enzyme[670,671] At the interface between the N-terminal and C-terminal domains, which are 30 Å apart, the catalytic site is situated and contains the coenzyme, pyridoxal phosphate, and binds the substrate, α-D-glucose 1-phosphate, or orthophosphate. Two types of inhibitors for this site are recognized: glucose or its analogues which compete with glucose-1-P but stabilize the inactive "T" conformation[672] and analogues of glucose-1-P which stabilize the "R" conformation.[673] The latter include UDPG and glucose cyclic 1,2-phosphate, both of which have been demonstrated to bind to the catalytic site by crystallographic means.[673,674]

A third site whose existence was recognized by calorimetric and kinetic means[675,676] was shown by X-ray crystallography to be located 10 Å from the active site.[666,678] A detailed characterization suggests that compounds which bind at this site will be inhibitors.[679] A wide variety of compounds may bind to this site but they may have in common a condensed ring system which intercalates (stacks) between the aromatic side chains of Phe-285 and Tyr-612. The major source of binding energy is hydrophobic in nature (which includes the stacking energy as a hydrophobic contribution) and there is a correlation with the loss of solvent-accessible surface of the conjugated-ring system of the bound ligand. Sprang et al.[679] have critically discussed the thermodynamics and structural parameters of this binding site.

Recently, Madsen et al.,[680] by kinetic analyses supplemented by X-ray crystallographic binding studies, reported on the interactions between the above three sites in glycogen phosphorylase b by ligands specific for each site on the phosphorylase b monomer and consequently acquired a quantitative description of the so-called heterotropic site-site interactions on each monomer. They chose conditions to minimize the homotropic interactions between sites on the two monomers of the molecule (see References 681 and 983). The fourth major site on each monomer, that for glycogen storage, was not studied in this report, could be ignored by conducting their kinetic studies at saturating glycogen concentrations. Their analyses permitted an assignment of each ligand to a primary binding site, as well as a determination of its dissociation constant and interaction with ligands which bind to the other sites.

In the case of another allosteric system, Wada et al.[986] recently studied the allosteric kinetics of cAMP hydrolysis by a purified calf liver cGMP-stimulated cyclic nucleotide phosphodiesterase in the presence and absence of several competitive inhibitors of the methylxanthine type. Their data were analyzed according to a rapid-equilibrium two-site competitive model for allosteric enzymes, with random binding of the substrates, effectors, or inhibitors to the ligand-free state.[169]

For references and reviews regarding aspartate transcarbamylase, see References 120 and 126 to 132. For reviews on glyceraldehyde 3-phosphate dehydrogenase, i.e., on the yeast enzyme, see References 133 to 138, and on the rabbit-muscle enzyme, see References 139 to 142.

Chapter 8

EQUILIBRIUM LIGAND BINDING — MULTIPLE EQUILIBRIA

As indicated in the previous chapters (and see Appendix 1 (Part B)), measurements of equilibrium substrate binding often provide a means of confirmation or selection of plausible kinetic mechanisms. In principle, it is also the means of understanding of cooperative substrate effects (sigmoidal substrate binding or sigmoidal kinetics (Chapter 7). Consequently, a discussion of equilibrium ligand binding measurements rightfully belongs in the scope of this text.

Any means whereby quantitative measurements of enzyme-substrate interactions or complex formation have been employed in the study of equilibrium substrate binding will be included.

A quantitative theory of ''multiple equilibria'' has been proposed[143,145] and is discussed below.

Consider the protein, P, which combines with n molecules of the substrate, A, to form the following complexes, PA...PA_n as represented by the following equilibria:

$$\begin{array}{rcl} P + A & \rightleftarrows & PA \\ PA + A & \rightleftarrows & PA_2 \\ \vdots \quad \vdots & \rightleftarrows & \vdots \\ PA_{i-1} + A & \rightleftarrows & PA_i \\ \vdots \quad \vdots & \rightleftarrows & \vdots \\ PA_{n-1} + A & \rightleftarrows & PA_n \end{array}$$

SCHEME I

These equilibria may be represented by their equilibrium constants.

$$\frac{(PA)}{(P)(A)} = K_1$$

$$\frac{(PA_2)}{(PA)(A)} = K_2$$

$$\frac{(PA_i)}{(PA_{i-1})(A)} = K_i \tag{1}$$

$$\frac{(PA_n)}{(PA_{n-1})(A)} = K_n$$

From these equations, the following relations occur.

$$\begin{aligned} (PA) &= K_1(P)(A) \\ (PA_2) &= K_2(PA)(A) = K_1K_2(P)(A)^2 \\ (PA_i) &= K_i(PA_{i-1})(A) = (K_1K_2\ldots K_i)(P)(A)^i \\ (PA_n) &= K_n(PA_{n-1})(A) = (K_1K_2\ldots K_i\ldots K_n)(P)(A)^n \end{aligned} \tag{2}$$

I. THE RELATIONSHIP BETWEEN THE EQUILIBRIUM CONSTANTS FOR LIGAND BINDING IN THE ABSENCE OF INTERACTIONS

The relationship between the equilibrium constants in the *absence* of interactions may be deduced first. Thus, if each site is uninfluenced by its neighbor and if each has the same intrinsic affinity for A, then the equilibrium constants K_1 ... K_n are not independent, but bear definite relationships to each other which may be deduced by the rules of combinations and permutations.

Consider the complex PA; obviously, the n possible forms of PA are dependent upon the particular site on P to which A is attached. These complexes may be distinguished by use of the notation $_1PA$, $_2PA$, ... $_nPA$, where the subscript at the left indicates the particular site on P to which A is bound.

If the intrinsic affinity of each site for A is identical, then the equilibrium constant, K, for the association

$$P + A \overset{K}{\rightleftarrows} {_1PA} \text{ is the same as for the reaction}$$

$$P + A \overset{K}{\rightleftarrows} {_2PA} \tag{3}$$

Therefore,

$$K = \frac{(_1PA)}{(P)(A)} = \frac{(_2PA)}{(P)(A)} = \ldots = \frac{(_nPA)}{(P)(A)} \tag{4}$$

Since

$$(PA) = (_1PA) + (_2PA) + \ldots + (_nPA) \tag{5}$$

Therefore,

$$K_1 = \frac{(PA)}{(P)(A)} = \frac{(_1PA) + (_2PA) + \ldots + (_nPA)}{(P)(A)} \tag{6}$$

and

$$K_1 = K + K + \ldots + K = nK \tag{7}$$

Thus, the relation is shown between K_1 and the intrinsic constant K. A similar relation may be obtained between K_2 and K.

Now, if the PA_2 complex is examined, there are now $\frac{n(n-1)}{2}$ possible forms, dependent on the particular combination of two sites on P to which the two As are attached. The quantity $\frac{n(n-1)}{2}$ is derived from the relationship for the number of possible combinations of n sites taken m at a time.

$$_nC_m = \frac{n!}{m!(n - m)!} \tag{8}$$

Thus, for $m = 2$,

$$ {}_nC_2 = \frac{n!}{2!(n-2)!} = \frac{n(n-1)(n-2)(n-3)\ldots}{2(n-2)(n-3)\ldots} = \frac{n(n-1)}{2} \tag{9} $$

The different complexes of type PA_2 may be denoted by the symbols ${}_{1,2}PA_2$, ${}_{1,3}PA_2$, … ${}_{j,\ell}PA_2$, where the subscripts at the left indicate the sites on P at which the two As are attached. It follows from Equation 4

$$ ({}_1PA) = ({}_2PA) = \ldots = ({}_nPA) \tag{10} $$

since the values of K, (P), and (A), respectively, are identical.
An analogous set of equilibrium expressions can be written for the complexes involving ${}_{j,\ell}PA_2$, from which it is a simple matter to prove that

$$ ({}_{1,2}PA_2) = ({}_{1,3}PA_2) = \ldots = ({}_{j,\ell}PA_2) \tag{11} $$

if it is assumed as before that the intrinsic affinity constant, K, is the same for any single site taking up one A. Therefore, it follows that

$$ K_2 = \frac{(PA_2)}{(PA)(A)} = \frac{[({}_{1,2}PA_2) + \ldots + ({}_{j,\ell}PA_2) + \ldots]}{[({}_1PA) + \ldots ({}_iPA) + \ldots](A)} \tag{12} $$

$$ = \frac{\frac{n(n-1)}{2}({}_{1,2}PA_2)}{n({}_1PA)(A)} \tag{13} $$

$$ = \left(\frac{n-1}{2}\right)K \tag{14} $$

This type of treatment may be generalized by extension to the constant K_i. Start with

$$ K_i = \frac{(PA_i)}{(PA_{i-1})(A)} $$

(from Equation 1) and to obtain a relation between K_i and the intrinsic constant, K, first determine the number of possible forms of the complex PA_i. The problem is now essentially how many combinations of n binding sites are possible if i of them at a time is taken. Application of Equation 8 leads to

$$ {}_nC_i = \frac{n!}{i!(n-i)!} \tag{15} $$

or there are $\frac{n!}{i!(n-i)!}$ different forms of the complex PA_i. Each one of these forms, ${}_\lambda PA_i$, is present in a concentration equal to any other form of PA_i.

Similar operations lead us to the conclusion that there are

$$ \frac{n!}{(i-1)![n-(i-1)]!} \tag{16} $$

different forms of the complex PA_{i-1}.
Therefore,

TABLE 1
Statistical Factors in the Equlibrium Constants for a Complex with Ten Sites[145]

$K_1 = 10\ K$	$K_6 = 5/6\ K$
$K_2 = 9/2\ K$	$K_7 = 4/7\ K$
$K_3 = 8/3\ K$	$K_8 = 3/8\ K$
$K_4 = 7/4\ K$	$K_9 = 2/9\ K$
$K_5 = 6/5\ K$	$K_{10} = 1/10\ K$

$$K_i = \left\{ \frac{\dfrac{n!}{i!(n-i)!}}{\dfrac{n!}{(i-1)!(n-i+1)!}} \right\} \frac{\lambda PA_i}{(\delta PA_{i-1})(A)} \tag{17}$$

$$K_i = \frac{(i-1)!(n-i+1)!}{i!(n-i)!} K \tag{18}$$

$$= \frac{[(i-1)(i-2)\ldots][(n-i+1)(n-1)(n-i-1)\ldots]}{[i(i-1)(i-2)\ldots][(n-i)(n-i-1)\ldots]} K \tag{19}$$

$$K_i = \left(\frac{n-i+1}{i}\right) K \tag{20}$$

This is a general relationship between the constant K_i for the formation of the ith complex, PA_i, and the intrinsic affinity constant, K.

In Table 1 the relationships between K_i and K for a ten-site model have been calculated. These factors can be derived also on a statistical basis (i.e., $(n - i + 1)/i$ of Equation 20) and are frequently referred to as the statistical factors.

II. SOME GENERAL EQUATIONS FOR LIGAND BINDING FOR THE CASE OF A SINGLE SET OF SITES WITH THE SAME INTRINSIC BINDING CONSTANT

A convenient measure of the extent of combination of A with the protein is the quantity $\bar{r}_A$, where

$$\bar{r}_A = \frac{\text{moles of bound A}}{\text{moles of total protein}} = \frac{A_b}{P_t} \tag{21}$$

or the average number of moles of A bound (A_b) per mole of total protein (P_t), and which may be expressed as follows:

$$\bar{r}_A = \frac{(PA) + 2(PA)_2 + \ldots + i(PA_i) + \ldots + n(PA_n)}{(P) + (PA) + (PA_2) + \ldots + (PA_i) + \ldots + PA_n} \tag{22}$$

However, from Equation 2, it is necessary that

$$\bar{r}_A = \frac{K_1(P)(A) + 2K_1K_2(P)(A)^2 + \ldots\ i(K_1K_2\ldots K_i)(P)(A)^i + \ldots}{(P) + K_1(P)(A) + \ldots + (K_1K_2\ldots K_i)(P)(A)^i + \ldots} \tag{23}$$

or

$$\bar{r}_A = \frac{K_1(A) + 2K_1K_2(A)^2 + \ldots + i(K_1K_2 \ldots K_i)(A)^i + \ldots + n(K_1K_2 \ldots K_n)(A)^n}{1 + K_1(A) + K_1K_2(A)^2 + \ldots + (K_1K_2 \ldots K_i)(A)^i + \ldots + (K_1K_2 \ldots K_n)(A)^n} \tag{24}$$

III. SOME REDUCED EQUATIONS FOR A SINGLE SET OF BINDING SITES WITH NO INTERACTIONS

If each site has the same intrinsic affinity for A and if each is uninfluenced by its neighbors, then the individual constants, K_1, K_2, ... K_n of Equation 24 may be replaced by the appropriate special case of Equation 20 to give

$$\bar{r}_A = \frac{\left(nK(A) + \frac{2n(n-1)}{2!}K^2(A)^2 + \ldots + i\frac{n(n-1)\ldots(n-i+1)}{i!}K^i(A)^i + \ldots + n\frac{n!}{n!}K^n(A)^n\right)}{\left(1 + nK(A) + \frac{n(n-1)}{2!}K^2(A)^2 + \ldots + \frac{n(n-1)\ldots(n-i+1)}{i!}K^i(A)^i + \ldots + \frac{n!K^n(A)^n}{n!}\right)} \tag{25}$$

From the binomial theorem, the denominator of Equation 25 is the expansion of

$$[1 + K(A)]^n \tag{26}$$

In addition, if (A) is factored out in the numerator of Equation 25, the derivative of the denominator is obtained with respect to (A), as can be verified by term-by-term differentiation. Therefore,

$$\text{numerator} = (A)\frac{\partial}{\partial(A)}(\text{denominator}) = (A)\frac{\partial}{\partial(A)}[1 + K(A)]^n \tag{27}$$

$$= (A)nK[1 + K(A)]^{n-1} \tag{28}$$

Consideration of the relations between Equations 26 and 28 leads to

$$\bar{r}_A = \frac{(A)nK[1 + K(A)]^{n-1}}{[1 + K(A)]^n} \tag{29}$$

or

$$\bar{r}_A = \frac{nK(A)}{1 + K(A)} \quad \text{for n equivalent and *} \tag{30}$$

for n equivalent and noninteracting sites for ligand or substrate A on a protein enzyme, neglecting electrostatic effects.*

* Note that whereas $\bar{r}_A = \frac{A_b}{P_t}$ (Equation 21), $\overline{Y}_A$ (Equation 48, Chapter 7), the "saturation function" described in Chapter 7, Section III.G, is acutally

$$\overline{Y}_A = \frac{A_b}{n \cdot P_t} \tag{32}$$

and therefore,

$$\overline{Y}_A = \frac{\bar{r}_A}{n} \tag{33}$$

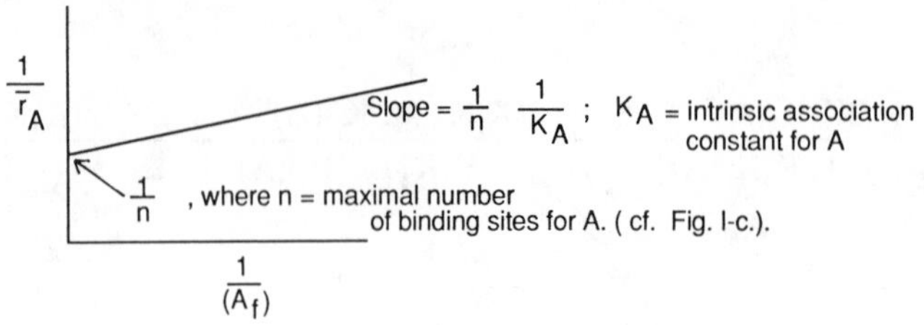

FIGURE 1.

Thus, for the case of n *independent* binding sites of a single type capable of binding species A, the equation for the extent of binding is simply n times that for a single site, with the same intrinsic association constant, K.

Taking the reciprocal of Equation 30,

$$\frac{1}{\bar{r}_A} = \frac{1}{nK}\frac{1}{(A)} + \frac{1}{n} \tag{31}$$

where a simple plot of $\frac{1}{\bar{r}_A}$ vs. $\frac{1}{(A)}$, where $(A) = (A_f)$, or the unbound concentration of (A), yields $\frac{1}{n}$ from the Y-intercept and $\frac{1}{nK}$ from the slope; K is obtained from $\frac{\text{Y-intercept}}{\text{slope}}$.

Another useful form is obtained from Equation 30.

$$\bar{r}_A = \frac{nK(A)}{1 + K(A)}$$

$$\bar{r}_A + K(A)\bar{r}_A = nK(A)$$

$$\frac{\bar{r}_A}{(A)} + K\bar{r}_A = nK$$

$$\frac{\bar{r}_A}{(A_f)} = nK - K\bar{r}_A, \tag{34}$$

which leads to a plot of

commonly referred to as a Scatchard plot[144] where the slope = −K, the Y-intercept = nK, (y-intercept/slope) = −n, and X-intercept = n (see Figure 2).

IV. COMPETITIVE BINDING OF TWO LIGANDS, A AND B, AT THE SAME SITES OF THE PROTEIN.

Equation 30 was the result of n independent binding sites of a single type capable of binding species A (neglecting electrostatic effects),[145] i.e.,

$$\bar{r}_A = \frac{A_b}{P_t} = \frac{\sum_{i=1}^{n} iPA_i}{\sum_{i=0}^{n} PA_i} = \frac{nK_A(A_f)}{1 + K_A(A_f)} \tag{35}$$

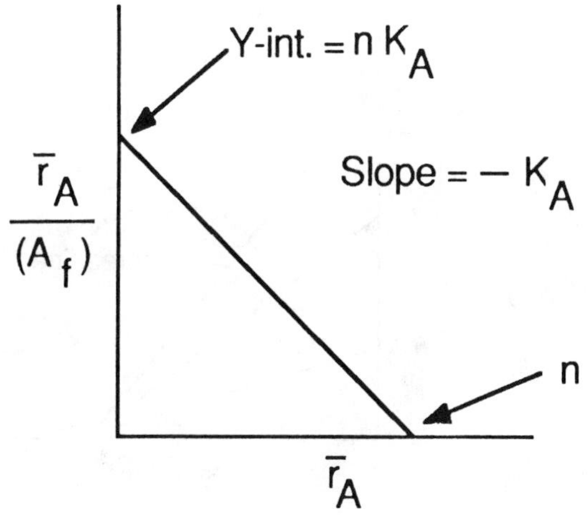

FIGURE 2.

If two species, A and B, may be bound competitively at the same sites of the protein,[146]

$$\bar{r}_A = \frac{\sum_{i=i}^{n} \sum_{j=0}^{n-i} i(PA_iB_j)}{\sum_{i=0}^{n} \sum_{j=0}^{n-i} (PA_iB_j)}, \; i + j \not> n \tag{36}$$

and

$$\bar{r}_A = \frac{n(A_f)K_A}{1 + K_A(A_f) + K_B(B_f)} \tag{37}$$

$$\bar{r}_A[1 + K_A(A_f) + K_B(B_f)] = n(A_f)K_A \tag{38}$$

$$\frac{\bar{r}_A}{(A_f)}[1 + K_B(B_f)] = nK_A - \bar{r}_AK_A \tag{39}$$

$$\frac{\bar{r}_A}{A_f} = \frac{nK_A}{[1 + K_B(B_f)]} - \frac{\bar{r}_AK_A}{[1 + K_B(B_f)]} \tag{40}$$

$$\frac{\bar{r}_A}{A_f} = nK_{A,app} - \bar{r}_AK_{A,app} \tag{41}$$

where

$$K_{A,app} = \frac{K_A}{[1 + K_B(B_f)]} \tag{42}$$

(See Figure 3 and, for example, Kuby et al.[59] who had found in the case of the rabbit muscle creatine kinase that ADP^{3-} or ATP^{4-} were competitive binders with respect to $MgADP^-$ or $MgATP^{2-}$, respectively).

Similarly, for competitive binding of A by several species,[147]

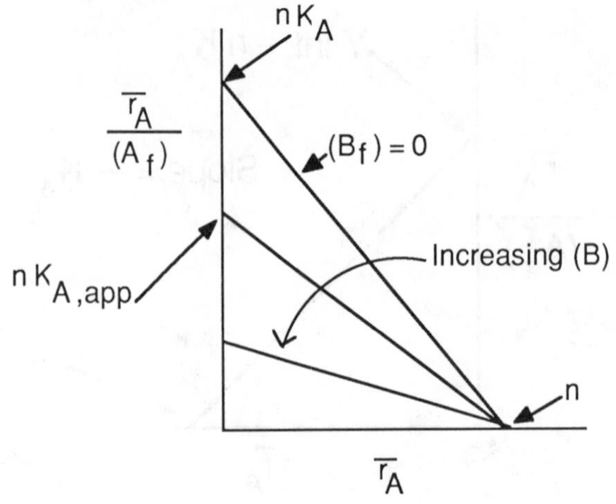

FIGURE 3.

$$\bar{r}_A = \frac{n(A_f)K_A}{1 + K_A(A_f) + K_B(B_f) + K_C(C_f) + \ldots} \tag{43}$$

V. CASE OF MULTIPLE CLASSES OF BINDING SITES

When there are two or more independent classes of sites, $\frac{1}{\bar{r}_A}$ vs. $\frac{1}{(A_f)}$ or $\frac{\bar{r}}{(A_f)}$ vs. $(\bar{r}_A)$ plots will no longer be linear. Nevertheless, it is still possible to determine limiting slopes and intercepts. The relation of these graphical parameters to site-binding constants, however, is not what seems to have been assumed on intuitive grounds; therefore, Klotz and Hunston[148] derived the general expression for the graphical parameters. The special case of a two-site system, so often encountered experimentally, is dealt with in detail.

A. THE CASE OF m CLASSES OF INDEPENDENT SITES FOR BINDING A

Thus, for the case of m classes of independent sites for binding A and each class, i, having n_i sites with an intrinsic binding constant K_i, Equation 35 may be generalized to

$$\bar{r}_A = \sum_{i=1}^{m} \frac{n_i K_i (A_f)}{1 + K_i(A_f)} \tag{44}$$

The total number of sites,

$$n_t = \sum_{i=1}^{m} n_i \tag{45}$$

Define certain average binding constants by

$$\langle K \rangle_\gamma = \frac{\sum_{i=1}^{m} n_i K_i^{\gamma}}{\sum_{i=1}^{m} n_i K_i^{\gamma-1}} \tag{46}$$

where γ takes on values of 0, 1, or 2. From Equation 44, the following may be written:

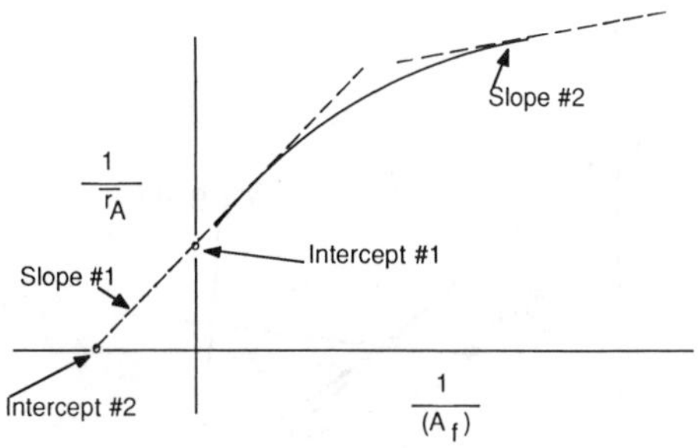

FIGURE 4.

$$\frac{1}{\bar{r}_A} = \left[\sum_{i=1}^{m} \frac{n_iK_iA_f}{1 + K_iA_f}\right]^{-1} = \left[\sum_{i=1}^{m} \frac{n_iK_i}{\left(\frac{1}{(A_f)} + K_i\right)}\right]^{-1} \tag{47}$$

from which the following results for the two intercepts and two limiting slopes given in Figure 4 above may be derived.

Y-intercept or intercept #1:

$$\underset{1/(A_f)\to 0}{\text{LIM}} \left(\frac{1}{\bar{r}_A}\right) \left[\sum_{i=1}^{m} \frac{n_iK_i}{K_i}\right]^{-1} = \frac{1}{\sum_{i=1}^{m} n_i} = \frac{1}{n_t} \tag{48}$$

Slope #1 (or initial slope):

$$\frac{d\left(\frac{1}{\bar{r}_A}\right)}{d\left(\frac{1}{A_f}\right)} = (-)\left[\sum_{i=1}^{m} \frac{n_iK_i}{\left(\frac{1}{(A)} + K_i\right)}\right]^{-2} (-1)\left[\sum_{i=1}^{m} \frac{n_iK_i}{\left(\frac{1}{A_f} + K_i\right)^2}\right] \tag{49}$$

$$= \frac{\sum_{i=1}^{m} \frac{n_iK_i}{(1/A_f + K_i)^2}}{\left[\sum_{i=1}^{m} \frac{nK_i}{(1/A_f + K_i)}\right]^2} \tag{50}$$

$$\underset{1/A_f\to 0}{\text{LIM}} \left[\frac{d\left(\frac{1}{\bar{r}_A}\right)}{d\left(\frac{1}{A_f}\right)}\right] = \frac{\sum_{i=1}^{m} \left(\frac{n_i}{K_i}\right)}{\left[\sum_{i=1}^{m} n_i\right]^2} = \frac{1}{n_t\langle K\rangle_{\gamma=0}} \tag{51}$$

Intercept #2 (X-intercept):

$$\text{Slope } \#1 = -\frac{\text{Intercept } 1}{\text{Intercept } 2} \tag{52}$$

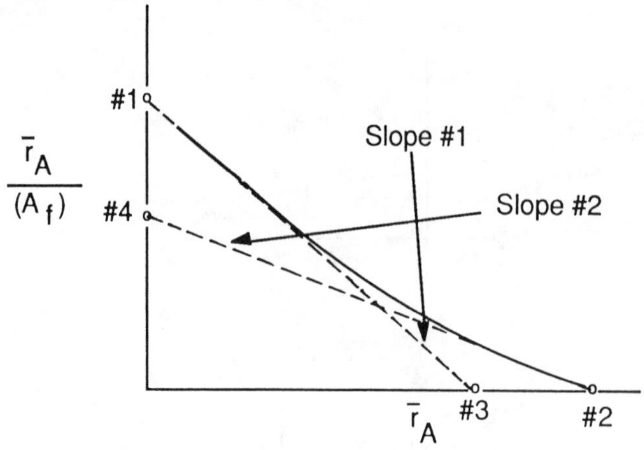

FIGURE 5 A

$$\text{Intercept \#2} = -\frac{\frac{1}{n_t}}{\frac{1}{n_t\langle K\rangle_{\gamma=0}}} = -\langle K\rangle_{\gamma=0} \qquad (53)$$

Slope #2 (limiting slope at large $\frac{1}{A_f}$ values)

$$\frac{d\left(\frac{1}{\bar{r}_A}\right)}{d\left(\frac{1}{A_f}\right)} = \frac{\left[\sum_{i=1}^{m}\frac{n_iK_i}{(1/A_f + K_i)^2}\right]\frac{1}{(A_f^2)}}{\left[\sum_{i=1}^{m}\frac{nK_i}{(1/A_f + K_i)}\right]^2\frac{1}{(A_f^2)}} = \frac{\sum_{i=1}^{m}\frac{n_iK_i}{(1 + K_iA_f)^2}}{\left[\sum_{i=1}^{m}\frac{n_iK_i}{1 + K_iA_f}\right]^2} \qquad (54)$$

Since $\frac{1}{A_f} \rightarrow \infty$ as $(A_f) \rightarrow 0$,

$$\lim_{A_f\to 0}\left[\frac{d\left(\frac{1}{\bar{r}_A}\right)}{d\left(\frac{1}{A_f}\right)}\right] = \frac{\sum_{i=1}^{m} n_iK_i}{\left[\sum_{i=1}^{m} n_iK_i\right]^2} = \frac{1}{\sum_{i=1}^{m} n_iK_i} = \frac{1}{n_t\langle K\rangle_{\gamma=1}} \qquad (55)$$

For a graph with $\frac{\bar{r}}{A_f}$ vs. $\bar{r}_A$ as variables, see Figures 5A and B. From Equation 44 it follows that

$$\frac{\bar{r}_A}{(A_f)} = \sum_{i=1}^{m}\frac{n_iK_i}{1 + K_iA_f} \qquad (56)$$

which leads to the following relations from the graphical intercepts and slopes given in Figures 5A and B.

For intercept #1 (Y-intercept), As $\bar{r}_A \rightarrow 0$, $A_f \rightarrow 0$. Therefore,

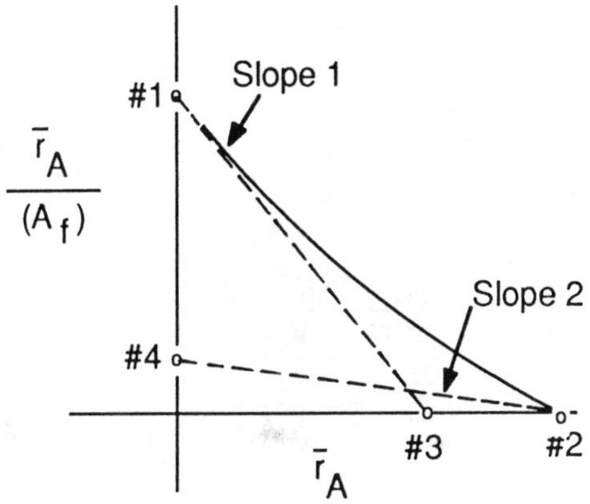

FIGURE 5B

$$\lim_{(A_f)\to 0}\left(\frac{\bar{r}_A}{A_f}\right) = \sum_{i=1}^{m} n_iK_i = n_t \frac{\sum_{i=1}^{m} n_iK_i}{\sum_{i=1}^{m} n_i} = n_t \langle K\rangle_{\gamma=1} \quad (57)$$

For intercept #2 (X-intercept), $\left(\frac{\bar{r}}{A_f}\right) \to 0$ as $(A_f) \to \infty$

$$\bar{r}_A = \sum_{i=1}^{m} \frac{(n_iK_iA_f)}{(1 + K_iA_f)} \frac{\left(\frac{1}{A_f}\right)}{\left(\frac{1}{A_f}\right)} = \sum_{i=1}^{m} \frac{n_iK_i}{\left(\frac{1}{A_f} + K_i\right)} \quad (58)$$

Therefore,

$$\lim_{A_f\to\infty} \bar{r}_A = \sum_{i=1}^{m} \frac{n_iK_i}{K_i} = \sum_{i=1}^{m} n_i = n_t \quad (59)$$

For Slope #1,

$$\frac{d\left(\frac{\bar{r}_A}{A_f}\right)}{d\bar{r}_A} = \frac{\frac{d(\bar{r}_A/A_f)}{d(A_f)}}{\frac{d\bar{r}_A}{d(A_f)}} \quad (60)$$

$$\frac{d\left(\frac{\bar{r}_A}{A_f}\right)}{dA_f} = \sum_{i=1}^{m} (-1) \frac{n_iK_i}{(1 + K_iA_f)^2K_i} = -\sum_{i=1}^{m} \frac{n_iK_i^2}{(1 + K_iA_f)^2} \quad (61)$$

and

$$\frac{d\bar{r}_A}{dA_f} = \sum_{i=1}^{m} \left[\frac{n_iK_i}{1 + K_iA} + (-1)\frac{n_iK_iA_fK_i}{(1 + K_iA_f)^2}\right. \quad (62)$$

$$= \sum_{i=1}^{m} \left[\frac{n_iK_i(1 + K_iA_f)}{(1 + K_iA_f)(1 + K_iA_f)} - \frac{n_iK_i^2A_f}{(1 + K_iA_f)^2} \right] \tag{63}$$

$$= \sum_{i=1}^{m} \frac{n_iK_i}{(1 + K_iA)^2} \tag{64}$$

$$\therefore \frac{d\left(\frac{\bar{r}_A}{A_f}\right)}{d\bar{r}_A} = - \frac{\sum_{i=1}^{m} \frac{n_iK_i^2}{(1 + K_iA_f)^2}}{\sum_{i=1}^{m} \frac{n_iK_i}{(1 + K_iA_f)^2}} \tag{65}$$

$$\underset{A_f \to 0}{\mathrm{LIM}} \left[\frac{d\left(\frac{\bar{r}_A}{A_f}\right)}{d\bar{r}_A} \right] = - \frac{\sum_{i=1}^{m} n_iK_i^2}{\sum_{i=1}^{m} n_iK_i} = -\langle K \rangle_{\gamma=2} \tag{66}$$

For intercept #3,

$$\text{Slope } \#1 = - \frac{\text{Intercept } \#1}{\text{Intercept } \#3} \tag{67}$$

$$\text{Intercept } \#3 = \frac{-n_t\langle K \rangle_{\gamma=1}}{-\langle K \rangle_{\gamma=2}} = \frac{n_t\langle K \rangle_{\gamma=1}}{\langle K \rangle_{\gamma=2}} \tag{68}$$

For Slope #2,

$$\frac{d\left(\frac{\bar{r}_A}{A_f}\right)}{d\bar{r}_A} = - \frac{\sum_{i=1}^{m} \frac{n_iK_i^2}{(1 + K_iA_f)^2} \frac{1}{A_f^{-2}}}{\sum_{i=1}^{m} \frac{n_iK_i}{(1 + K_iA)^2} \frac{1}{A_f^{-2}}} = - \frac{\sum_{i=1}^{m} \frac{n_iK_i^2}{(1/A_f + K_i)^2}}{\sum_{i=1}^{m} \frac{n_iK_i}{(1/A_f + K_i)^2}} \tag{69}$$

$$\underset{A \to \infty}{\mathrm{LIM}} \left[\frac{d\left(\frac{\bar{r}_A}{A}\right)}{d\bar{r}_A} \right] = - \frac{\sum_{i=1}^{m} n_i}{\sum_{i=1}^{m} \frac{n_i}{K_i}} = -\langle K \rangle_{\gamma=0} \tag{70}$$

For intercept #4,

$$\text{Slope } \#2 = - \frac{\text{Intercept } \#4}{\text{Intercept } \#2} \tag{71}$$

$$\therefore \text{Intercept } \#4 = -(-\langle K \rangle_{\gamma=0})n_t = n_t\langle K \rangle_{\gamma=0} \tag{72}$$

One may note that these intercepts and slopes for multiple classes are not always the quantities one may have expected intuitively. Thus, in Figures 5A and B, intercept #3 is *not* the number of sites in the first of m classes, but rather

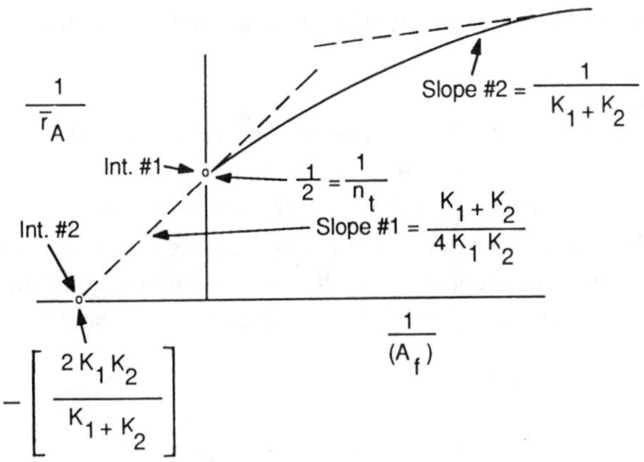

A

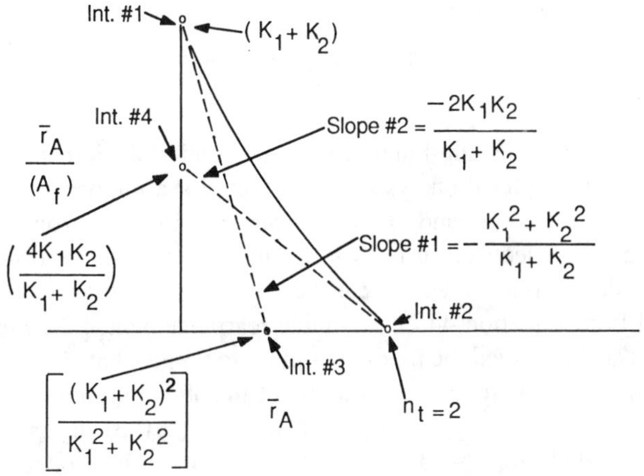

B

FIGURE 6.

$$\frac{n_t\langle K\rangle_{\gamma=1}}{\langle K\rangle_{\gamma=2}} \tag{68}$$

Similarly, intercept #4 is not the site-binding constant for the mth class of sites, but rather

$$n_t\langle K\rangle_{\gamma=0} \tag{72}$$

B. THE CASE FOR TWO INDEPENDENT SITES

The following is derived for two independent sites: with $n_1 = 1$ with K_1; and $n_2 = 1$ with K_2 (see Figures 6A and B).

In Figure 6B we see that although intercept #2 is equal to the total number of binding sites (two in this case), intercept $\#1 = K_1 + K_2$ and is not equal to the binding constant for the first site.

Similarly, intercept $\#3 = \dfrac{(K_1 + K_2)^2}{(K_1^2 + K_2^2)}$ and is not equal to the number of sites in class 1 (one in this case); on the contrary, intercept #3 will always be greater than unity.

Furthermore, intercept #4 is not equal to K_2 (the binding constant for the second site), but rather $\frac{4K_1K_2}{K_1 + K_2}$.

If however, K_1 and K_2 differ by a factor of 100, then some of the intercepts approach the "intuitively" expected values. Thus, in Figure 6B if $K_1 = 100\ K_2$, intercept #1 = $1.01\ K_1$, intercept #2 = 2, intercept #3 = 1.02, and intercept #4 = $0.04\ K_1$.

The relationship of the macroscopic- (i.e., classical $\tilde{K}$) or stoichiometric-binding constants are related to the microscopic- (i.e., K) or site-binding constants; for this two-site system, the relationships are analogous to those derived for a bifuncitonal proton-dissociating molecule,[149,150] that is,

$$\tilde{K}_1 = K_1 = K_2 \text{ and } \tilde{K}_2 = \frac{K_1K_2}{K_1 + K_2} \tag{73}$$

For a multisite system, general relations between stoichiometric- and site-binding constants have been given by Fletcher et al.[151] Referring to Figure 6B, we see that intercept #1 = $K_1 + K_2$, and does equal $\tilde{K}_1$, however, intercept #4 = $4 \times \frac{K_1K_2}{K_1+K_2}$ and is actually $4 \times \tilde{K}_2$.

In conclusion, the various site-binding constants and stoichiometric-binding constants can be deduced from a graphical analysis of a two-site (or a multisite) system; however, in general, the graphical intercepts and slopes contain contributions from both sites, and the individual values must be deduced, usually by an iterative procedure to separate the values. Final selection of the individual values can be assigned by fitting to the overall binding curve constructed from Equation 44 or a suitable rearrangement, i.e., Equation 46 or 56.

Thus, Kuby et al.[58] analyzed the binding of glucose 6-phosphate by the 2-subunit species of glucose 6-phosphate dehydrogenase, in terms of the above graphical procedure of Klotz and Hunston, leading to $n_1 + n_2 = 4\ (\pm 0.8)$, $n_1 = 2\ (\pm 0.3)$, $n_2 = 2\ (\pm 0.5)$, $K'_{B,1} = 3.4_5\ (\pm 1.8) \times 10^4\ M^{-1}$, $K'_{B,2} = 3.9\ (\pm 3.0) \times 10^3\ M^{-1}$. It is of interest that the $K'_{B,1}$ agreed approximately with the value for K'_{assoc}, estimated by protection offered by glucose 6-phosphate against the rate of inactivation by DTNB; thus $K'_{B,1}$ for set one, with the two major equivalent binding sites, apparently encompasses the substrate (glucose 6-phosphate) binding sites, with one site per subunit of 51,000 Da.

These data, plus the equilibrium ligand binding data on NADP (see Table 2) which confirmed the kinetic mechanism (see Chapter 2, Scheme XIX and Equation 120) by the agreement in the estimated intrinsic dissociation constants for the substrates estimated both kinetically (Chapter 4, Table 1) and thermodynamically (Table 2). The kinetic studies discussed earlier, especially the effects of pH and temperature (see Chapter 4, Tables 1, 3, and 4) led to the final postulated kinetic mechanism given below (Figure 7).

The data presented in Table 2 also illustrated several methods commonly employed for the measurement of equilibrium substrate binding, e.g., difference spectroscopy, equilibrium dialysis, gel filtration, and substrate protection against rate of inactivation.

It is common in molecular receptor binding studies, radioimmunoassays, and in facilitated diffusion experiments (e.g., ligand binding to membranous preparations of erythrocyte glucose transporter[655] to make use of Scatchard plots[623] in the form of either $(L)_B/(L)_F$ vs. $(L)_B$ (see Figure 8 A and B) where $(L)_B = \frac{(A_b)}{(P_t)_A} = \bar{r}_A$ (see Equations 21 and 30) expressed e.g., in moles of ligand A bound per unit (dry weight) concentration of membrane protein and $(L)_F$ = free molar concentration of ligand A; or $\frac{(A_b)}{(A_f)}$ vs. (A_b), where (A_b) and $A_f)$ are

TABLE 2
Equilibrium Ligand Binding of the Substrates to Glucose 6-Phosphate Dehydrogenase[58]

Substrate species	Method	Conditions	K'_{assoc}	K'_{dissoc}	n (maximal number moles bound per mole monomer protein
$NADP^+$	Difference spectroscopy	1×10^{-2} *M* EDTA, 5×10^{-3} *M* Tris (Cl^-), 0.15 *M* KCl, pH 7.5, 29°C	$3.8 (\pm 0.8) \times 10^5$	$2.65 (\pm 0.54) \times 10^{-6}$ (refer to $3.1_4 (\pm 1.56) \times 10^{-6}$ *M*[a] at 25°C, Chapter 4, Table 1)	(2)
$NADP^+$	Equilibrium dialysis	1×10^{-2} *M* EDTA, 5×10^{-3} *M* Tris (Cl^-), 0.15 *M* KCl, pH 7.8, 3°C	$8.7_6 (\pm 3.82) \times 10^5$	$1.14 (\pm .54) \times 10^{-6}$	2.18 (±0.21)
$NADP^+$	Gel filitration (Sephadex® G-25)	Same as equilibrium dialysis	$8.0_5 (\pm 0.21) \times 10^5$	$1.24 (\pm 0.21) \times 10^{-6}$	$2.0_4 (\pm 0.04)$
$NADP^+$	Protecton against [c]Nbs_2 inactivation kinetics	1×10^{-3} *M* EDTA, 5×10^{-2} *M* Tris (Ac^-), pH 8.0, 29°C	$7.0 (\pm 5) \times 10^5$	$1.4_3 (\pm 1.0) \times 10^{-6}$	(2)[b]
Glucose 6-phosphate	Equilibrium dialysis	5×10^{-3} *M* EDTA, 5×10^{-3} *M* Tris (Cl^-), 0.15 *M* KCl, pH 7.8, 3°C	$K_1 = 3.4_5 (\pm 1.8) \times 10^4$ $K_2 = 3.9 (\pm 3.0) \times 10^3$	$K_1 = 2.9 (\pm 1.5) \times 10^{-5}$ $K_2 = 2.6 (\pm 2.0) \times 10^{-4}$	$n_1 + n_2 = 4 (\pm 0.8)$ $n_1 = 2 (\pm 0.3)$ $n_2 = 2 (\pm 0.5)$
Glucose 6-phosphate	Protection against [c]Nbs_2 inactivation kinetics	1×10^{-3} *M* EDTA, 5×10^{-2} *M* Tris (Ac^-), pH 8.0, 29°C	$2 (\pm 1.5) \times 10^4$	$5 (\pm 3.8) \times 10^{-5}$ (refer to $1.1_5 (\pm 0.37) \times 10^{-5}$[a] at 25°C, Chapter 3, Table 4)	(2)[c]

Note: [c]Nbs_2, 5,5′-dithiobis (2-nitrobenzoic acid).

[a] pH-independent intrinsic dissociation constant estimated at 25°C and 0.1(Γ/2) Tris-acetate-EDTA for the two-subunit species of glucose 6-phosphate dehydrogenase.
[b] Value of *n* assigned from equilibrium dialysis or gel-filtration studies.
[c] Value of *n* assigned from equilibrium dialysis studies.

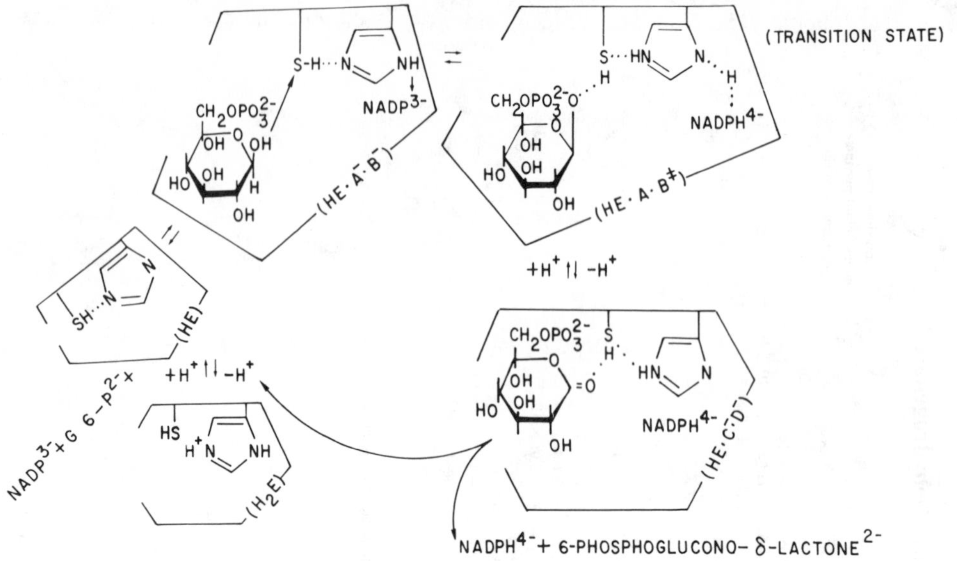

FIGURE 7. Postulated kinetic mechanism of dehydrogenation by D-glucose-6-phosphate dehydrogenase in the acidic and neutral pH range.[58]

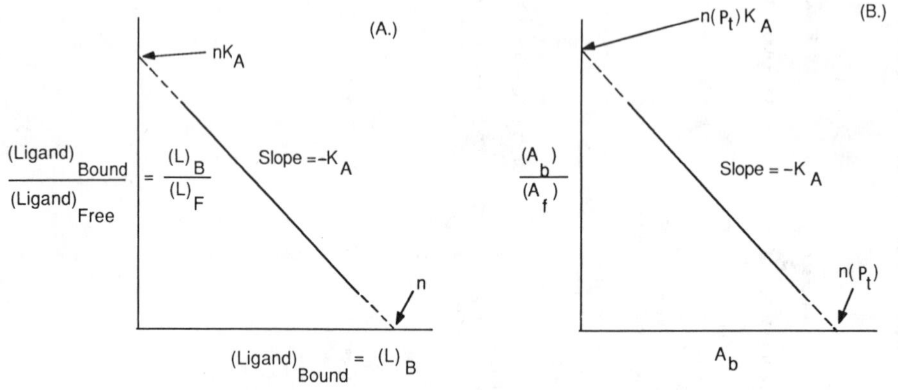

FIGURE 8. Scatchard plots[623] as might be employed in ligand binding to membraneous receptors or transporters (see also Reference 655).

the moles bound of ligand A and the molar free concentrations of A, respectively, and derived from

$$\frac{(L)_B}{(L)_F} = nK_A - K_A(L)_B \tag{74}$$

and

$$\frac{(A_b)}{(A_f)} = n(P_t)K_A - K_A(A_b) \tag{75}$$

Klotz[620] took experimentalists to task who rely on such Scatchard plots (Figure 8) to estimate receptor capacity in binding studies (e.g., see Reference 661). Theoretically, the

X-intercept should yield total binding receptor capacity (either as n or n × (P_t), see Figure 8).

Experimentally, with membranous preparations in particular, however, extrapolation may be difficult, because true saturation may never be approached or achieved. Klotz[620] noted that there is a geometric distortion inherent in the ordinate scale (i.e., in the $\frac{(L)_B}{(L)_F}$ axis) which may cause data to appear to approach the abscissa axis quite closely by eye, whereas, the binding might be very short of saturation. Thus, subjective error leads to an underestimation of the value of n × P_t (or n), i.e., in total receptor capacity. Klotz, therefore, suggested a graph of $(L)_B$ vs. $\log_{10}$ $(L)_F$, where the binding curve should rise sigmoidally towards an upper plateau which should represent the value of n. Klotz maintained that unless the data pass through an inflection point and approach a plateau, no information about total receptor capacity can be derived. Feldman[621] statistically fitted the same data criticized by Klotz[620,661] to the plots described in Figure 8 employing several mathematical binding models, e.g., the independent-site model of ligand binding,[623,624,662,663], with the use of a weighted least-squares regression analysis, and did derive quantitative measures of the uncertainty in receptor capacity. "The uncertainty was within such bounds that an approximate picture of the nature and number of sites was recoverable."[621]

VI. EQUILIBRIUM SUBSTRATE BINDING AS ILLUSTRATED BY THE ATP-TRANSPHOSPHORYLASE ENZYMES (E.G., CREATINE KINASE AND ADENYLATE KINASE)

As we turn now to a discussion of the kinases, let us consider those enzyme-substrate interactions of rabbit muscle ATP-creatine (Cr) transphosphorylase (creatine kinase) and rabbit muscle myokinase (adenylate kinase) as measured by the technique of equilibrium dialysis and gradient sedimentation, respectively.[59]

Rabbit muscle creatine kinase is a two-subunit dimeric protein[294] with a subunit size of about 42,000 Da whose primary stucture was established from the sequence of its cDNA clones,[848] whereas the rabbit-muscle myokinase is a monomeric protein[362] whose covalent structure has recently been elucidated.[363,854]

Molecular weights of 8.1×10^4 for ATP-Cr transphosphorylase[364,365,855] and 2.13×10^4 for myokinase[366] were used for calculation of the average number of moles bound per mole of total protein.[59]

The seleciton of pH 7.9, a temperature of 3°C, and an ionic strength of at least 0.16, in the study by Kuby et al.,[59] was a compromise for the study and comparison of both the proteins and was guided by the following considerations.

Theoretical titration curves constructed from the amino acid analyses of the two proteins[365,366] indicated at pH 7.9 net charges of approximately −5/mol for ATP-Cr transphosphorylase and +4 for myokinase. Although these values cannot be taken as absolute, since they are subject to electrostatic influences, they nevertheless provide a good index of the possible magnitude of the Donnan ionic asymmetric distributions. Calculation of the theoretical osmotic pressure as a function of the salt concentration, protein concentration, and pH[367] favored pH 7.9 as a compromise pH to study both proteins and revealed that at a salt concentration of at least 0.15 *M*, errors due to the theoretical Donnan osmotic pressure at this pH and in the presence of 50 mg ATP-Cr transphosphorylase protein per milliliter would not be greater than 3%. In contrast, at pH 9 the errors could be considerably larger for ATP-Cr transphosphorylase.

For reasons of stability of both the substrates and the proteins, a temperature close to 0°C and a slightly alkaline pH was employed. Preliminary binding measurements had indicated that the binding coefficients in general had pK values less than 5 so that relatively high protein concentrations had to be employed. Final concentrations studied ranged from 1 to 7% for ATP-Cr transphosphorylase and up to 3% for myokinase. To adjust the ionic strength and to reduce the Donnan corrections, as discussed above, the neutral 1:1 electrolyte KCl was selected.

As discussed in (Chapter 5, Section IV,) each of the adenine nucleotides studied[59] is presumed to exist in solution, expecially in the presence of Mg, in the form of a large number of ionic and complex species,[48,353] each of which could be considered to be in equilibrium with the protein. The concentrations of many of these species, however, are very small under the conditions selected for study so that their contribution to the binding picture could be neglected. The most important species of ATP at pH 7.9 are ATP^{4-} and $MgATP^{2-}$ and to a lesser degree $KATP^{3-}$ and $HATP^{3-}$, and, similarly, the analogous species for ADP and AMP. To what degree the results could be affected by neglecting all the other species is a possible source of error, which will have to be justified by the final analysis and the overall experimental error. For that study, only the dominant species were considered.[59] The following binding equations from References 145, 147, 150, and 368) are applicable to the study.

Case 1 — For *n* equivalent and noninteracting sites on a protein, neglecting electrostatic effects, i.e., *n* independent sites of a single type capable of binding species A,

$$\bar{r}'_A = \frac{A_b}{P_0} = \frac{\sum_{i=1}^{n} iPA_i}{\sum_{i=0}^{n} PA_i} = \frac{nK_A(A_f)}{1 + K_A(A_f)} \tag{76}$$

where, $\bar{r}'A$ is the average number of moles of A bound (A_b) per mole of total protein (P_0), *n* is the maximum number of moles of species A bound per mole of total protein, A_f is the concentration of unbound A, and K_A is the intrinsic association constant.

Case 2 — If two species, A and B, may be bound competitively at the same sites of the protein.

$$\bar{r}'_A = \frac{\sum_{i=1}^{n}\sum_{j=0}^{n-i} i(PA_iB_j)}{\sum_{i=1}^{n}\sum_{j=0}^{n-i} (PA_iB_j)}, \quad i + j \ngtr n \tag{77}$$

and

$$\bar{r}'_A = \frac{n(A_f)K_A}{1 + K_A(A_f) + K_B(B_f)} \tag{78}$$

and similarly for competition by several species.

$$\bar{r}'_A = \frac{n(A_f)K_A}{1 + K_A(A_f) + K_B(B_f) + K_C(C_f) + \ldots} \tag{79}$$

If ATP^{4-} and $MgATP^{2-}$ are considered the only species of ATP which are significantly bound to the protein, and if their binding is competitive at the same *n* sites, then, following case 2

$$\bar{r}'_{ATP^{4-}} = \frac{nK_{ATP^{4-}}(ATP^{4-})_f}{1 + K_{ATP^{4-}}(ATP^{4-})_f + K_{MgATP^{2-}}(MgATP^{2-})_f} \tag{80}$$

and

$$\bar{r}'_{MgATP^{2-}} = \frac{nK_{MgATP^{2-}}(MgATP^{2-})_f}{1 + K_{ATP^{4-}}(ATP^{4-})_f + K_{MgATP^{2-}}(MgATP^{2-})_f} \tag{81}$$

if

$$\bar{r}'_{A+B} = \bar{r}'_{ATP^{4-}} + \bar{r}'_{MgATP^{2-}} = \bar{r}'_{measured} \tag{82}$$

then

$$\bar{r}'_{A+B} = \frac{n[K_{ATP^{4-}}(ATP^{4-})_f + K_{MgATP^{2-}}(MgATP^{2-})_f]}{1 + K_{ATP^{4-}}(ATP^{4-})_f + K_{MgATP^{2-}}(MgATP^{2-})_f} \tag{83}$$

or

$$\frac{\bar{r}'_{A+B}}{K_A(ATP^{4-})_f + K_B(MgATP^{2-})_f} = n - \bar{r}'_{A+B} \tag{84}$$

where $K_A = K_{ATP^{4-}}$ and $K_B = K_{MgATP^{2-}}$. A plot of the left-hand side of Equation 84 vs. $\bar{r}'_{A+B}$ should be linear for the postulates and for the assumptions made with both x- and y-intercepts yielding an identical value of n and a theoretical slope of 1.0.

The equation may be rearranged to read

$$\frac{\bar{r}'_{A+B}}{(ATP^{4-} + MgATP^{2-})_f} = n\left(K_A \frac{(ATP^{4-})}{(ATP^{4-} + MgATP^{2-})_f} + K_B \frac{(MgATP^{2-})}{(ATP^{4-} + MgATP^{2-})_f}\right)$$

$$- \bar{r}'_{A+B}\left(K_A \frac{(ATP^{4-})_f}{ATP^{4-} + MgATP^{2-})_f} + K_B \frac{(MgATP^{2-})_f}{(ATP^{4-} + MgATP^{2-})_f}\right) \tag{85}$$

which is of the form

$$\frac{\bar{r}'_{A+B}}{(ATP^4 + MgATP^2)_f} = n\phi(K_{A,B}) - \bar{r}'_{A+B}\phi(K_{A,B}) \tag{86}$$

and which is similar to the Scatchard rearrangement of the equation given for case 1, that is

$$\frac{\bar{r}'_A}{A_f} = nK_A - \bar{r}'_A K_A \tag{87}$$

For a series of determinations, if the mole fraction of the unbound species (which is *not* equal to the initial concentration before the start of the run),

$$\left(\frac{ATP^{4-}}{ATP^{4-} + MgATP^{2-}}\right)_f \text{ or } \left(\frac{MgATP^{2-}}{ATP^{4-} + MgATP^{2-}}\right)_f$$

could be held constant, then $\phi(K_{A,B})$ will be held constant and a plot of $\frac{\bar{r}'_{A+B}}{(ATP^{4-} + MgATP^{2-})}$ vs. $\bar{r}'_{A+B}$ will yield $\phi(K_{A,B})$ as the slope, $n\,\phi(K_{A,B})$ as the ordinate intercept, and n as the abscissa intercept.

If K_A (for ATP^{4-}) can be evaluated from experiments without Mg, i.e. where $MgATP^{2-}$ is 0, then K_B (for $MgATP^{2-}$) may be evaluated from the $\phi(K_{A,B})$ values. In the special case where

$$K_A = K_B = K, \tag{88}$$

$\phi(K_{A,B})$ equals K, and regardless of the mole fractions

$$\frac{(ATP^{4-})_f}{A''_f} \text{ or } \frac{(MATP^{2-})_f}{A''_f} \tag{89}$$

(where $A_f'' = ATP^{4-} + MgATP^{2-}$), $\phi(K_{A,B})$ will be invariant and identical to K_A. A value of K_B could be arrived at very simply for this special case.

In the case where $K_B >> K_A$, as the mole fraction of $\frac{(MgATP^2)_f}{A_f''}$ increases, $n\phi(K_{A,B})$ increases. For the converse case, where $K_b << K_A$, $\phi(K_{A,B)}$ decreases with an increase in $\frac{(MgATP^{2-})_f}{A_f''}$.

In principle, $(ATP^{4-})_f$ or $(MgATP^{2-})_f$ can be evaluated from measurements of total unbound ATP $[(ATP)_{0,f}]$ and the total unbound Mg ($Mg_{0,f}$) when Mg has been added to the system. To calculate the species, the following approximate conservation equations were employed for pH 7.9 and 3°C in the presence of KCl.[59]
For ATP, in the absence of Mg,

$$(ATP_{0,f} \simeq (ATP^{4-})_f + (HATP^{3-})_f + (KATP^{3-})_f \tag{90}$$

or

$$(ATP)_{0,f} \simeq (ATP^{4-})_f\left(1 + \frac{(H^+)}{K_7} + K_5(K^+)\right) \tag{91}$$

(Definitions of the acid dissociation and complex stability constants are given in Table 3; see also Chapter 5, Section IV and Table 11.)

For ATP, with Mg added,

$$(ATP)_{0,f} \simeq (ATP^{4-})_f + (MgATP^{2-})_f + (HATP^{3-})_f + (KATP^{3-})_f \tag{92}$$

$$(Mg)_{0,f} \simeq (Mg^{2+})_f + (MgATP^{2-})_f \tag{93}$$

where, e.g., such magnesium complexes as $MgHATP^-$, $Mg(ATP)_2^{6-}$, and Mg_2ATP have been neglected at the pH and the range of concentrations of Mg_0 and ATP_0 employed.

After rearrangement in terms of (ATP^4), the equation (with Mg present) may be written as

$$[(ATP^{4-})_f]^2 + \frac{(ATP^{4-})_f}{K_1\left(1 + \frac{(H^+)}{K_7} + K_5(K^+)_f\right)} \times \left(1 + \frac{(H^+)}{K_7} + K_5(K^+) + K_1[(Mg)_{0,f} - (ATP)_{0,f}]\right) - \frac{(ATP)_{0,f}}{K_1\left(1 + \frac{(H^+)}{K_7} + K_5(K^+)\right)} \simeq 0 \tag{94}$$

TABLE 3
Stability Constants of the Various Ionic and Complex Substrate Species Selected For 3°C and $\mu = 0.2$[59]

Species		Formation (stability) constant selected (mol/l)[1]
$MgATP^{2-}$	(K_1)	2×10^4 [a]
$KATP^{3-}$	(K_5)	4
$HATP^{3-}$	$\frac{1}{K_7}$	$10^{6.9}$
$MgADP^-$		1×10^3 [b]
$KADP^{2-}$		2
$HADP^{2-}$		$10^{6.7}$
MgCr-P		9
$KCr\text{-}P^-$		0.3
$HCr\text{-}P^-$		$10^{4.5}$
MgAMP		20
$KAMP^-$		0.7
$HAMP^-$		$10^{6.5}$

Note: The values for the stability constants selected here are based upon the data of References 369-374 and see Chapter 5, Tables 5 and 11).

[a] Estimated from a value of 7×10^4, selected for 30°C and $\mu = 0.1$[48] with a ΔH of 5 kcal/mol (to yield therefore $3 \times 10^4\ M^{-1}$ at 3°C for K_1) and corrected to a μ of 0.2 by applying an ionic strength contribution of approximately 30%.

[b] Estimated from a value of 3×10^3 [48] selected for 30°C and $\mu = 0.1$ with a ΔH of 6 kcal/mol[369] of about 20%[369] and rounding off to one significant figure.

and

$$(Mg^{2+})_f = \frac{(Mg)_{0,f}}{1 + K_1(ATP^{4-})_f} \tag{95}$$

and

$$(MgATP^{2-})_f = K_1(Mg^{2+})_f(ATP^{4-})_f \tag{96}$$

Similar equations can be set up for ADP, AMP, and Cr-P, considering only the equivalent species, as for ATP.

The numerical values for the stability constants of the various ionic and complex species of the substrates involved have been the subject of much investigation, and a wide range of values have been reported (e.g., see References 353, 360, and 361, and Chapter 5, Section IV). The reasons for selecting the particular metal stability constants employed here (Table 3) have been reviewed elsewhere[48] for 30°C. Except for the acid dissociation constants, they have been converted to 3°C and 0.2 ionic strength with the aid of ΔH of 5 kcal/mol for the ATP species,[369] 6 kcal/mol for the ADP species,[369] and 6 kcal/mol for AMP and Cr-P species, assuming the values that Smith[374,372] found for $MgHPO_4$ and $KHPO_4^-$. A value of 300 cal or less for ΔH[375] was assumed to apply for all acid dissociation constants.

TABLE 4
Apparent Intrinsic Dissociation Constants (K'_D) and Maximal Numbers of Moles Bound per Mole of Protein (n) for Various Substrate Species of ATP-Cr Transphosphorylase at 0.01 *M* Tris/0.15 *M* KCl, pH 7.9, and 3°C[59]

	Equilibrium dialysis		Gradient sedimentation	
Species	K_D' (mol/l)	n	K_D' (mol/l)	n
$MgATP^{2-}$	1×10^{-4}	1.9	3×10^{-4}	1.8
ATP^{4-}	3×10^{-4}		5×10^{-4}	
$MgADP^{-}$	6×10^{-5}	1.9	7×10^{-5}	1.8
ADP^{3-}	1×10^{-4}		1×10^{-4}	
$Cr\text{-}P^{2-}$			pK = 2 to 3	⩾2 (?)
Cr	pK ⩽ 2 (?)	(?)		
Mg^{2+}	pK ≃ 2	⩾4 (?)	pK ≃ 2	⩾4 (?)

Note: K_D' values are the reciprocals of the intrinsic association constants for equilibrium dialysis and gradient sedimentation. The values are significant to one figure only or, where the pK is listed, to only one order of magnitude.

TABLE 5
Apparent Intrinsic Dissociation Constants for Binding of the Various Substrate Species to ATP-AMP Transphosphorylase (Myokinase) Measured by the Ultracentrifuge Technique at 0.01 *M* Tris/0.15 *M* KCl, pH 7.9, 3°C[59]

Species	Intrinsic dissociation constant[a] (mol/l)	Maximal number of moles bound per mole of enzyme (n)
ATP^{4-}	1×10^{-4}	1.8 (as average)
$MgATP^{2-}$	1×10^{-4}	
ADP^{3-}	7×10^{-5}	1.6
AMP^{2-}	6×10^{-4}	2.3
Mg^{2+}	pK ⩽ 2 (?)	⩾ 3 (?)

[a] Reciprocals of the intrinsic association constants. The values are significant to one figure only or, where pK is listed, to an order of magnitude only. For Mg^{2+}, only the minimal value can be assigned.

A summary of all the derived constants for equilibrium binding to rabbit-muscle ATP-Cr transphosphorylase is presented in Table 4 obtained by the two methods, equilibrium dialysis and gradient sedimentation.[59]

Table 5 summarizes the derived constants for rabbit muscle myokinase obtained by the sedimentation method. These calculated intrinsic constants are expressed as dissociation constants (the reciprocal of association constants) in order to facilitate a comparison with kinetically derived constants discussed elsewhere (Chapter 5, Section II, III.A, and III.B) for the same enzymes (e.g., See References 48 to 50, 83, 84, 299, 300, 310, 343, 346, 831, 832, 835, 856, 964, and 981).

The order of binding affinity of the substrate species, expressed in terms of the intrinsic association constants, may thus be written for the two enzymes as follows:

For ATP-Cr transphosphorylase,

$$MgADP^{-} \geq \begin{Bmatrix} MgATP^{2-} \\ ADP^{3-} \end{Bmatrix} > ATP^{4-} >$$

$$Cr\text{-}P^{2-} > \begin{Bmatrix} Mg^{2+} \\ Cr^{\pm} \end{Bmatrix}$$

For myokinase,

$$ADP^{3-} \geq \begin{Bmatrix} MgATP^{2-} \\ ATP^{4-} \end{Bmatrix} > AMP^{2-} > Mg^{2+}$$

SCHEME II.

The value for $MgADP^{-}$ in the case of myokinase is not known.

It is of interest to note that the affinity of both these enzymes for Mg^{2+} (pK $\simeq$ 2) at pH 7.9 and 3°C is quite small compared to that of the Mg nucleotide complexes for ATP or ADP (pK $\simeq$ 4).

The agreement between the two methods applied to ATP-Cr transphosphorylase (Table 4) may be taken as quite satisfactory if one considers the large experimental errors inherent in those measurements as well as in the simplifying assumptions and approximations made for calculation of the derived constants.

It is considered that the simple competitive binding mechanism as outlined here in terms of the dominant nucleotide species is sufficient to explain the experimental data.

In spite of the experimental precautions exercised, the results show a degree of scatter which will not permit evaluation of second-order effects either in terms of electrostatic contribution of charged ions or in terms of activity coefficients. Although other species than those considered here may play a role in the competitive binding picture, the experimental error did not allow the introduction of additional binding constants for the subordinate species not containing Mg, e.g., the potassium species. Only simple concentration corrections for the presence of small amounts of these species (which are rendered even smaller in the presence of Mg) were therefore made. An attempt was made to determine what error would arise from the use of metal complex stability constants other than those listed in Table 3. For example, a twofold increase in the stability values selected, in general, would affect the intrinsic binding coefficients for $MgATP^{2-}$ by less than 20% and slightly more for $MgADP^{-}$ under these experimental conditions, and would thus only alter the second figure. In view of these uncertainties, the data in Tables 4 and 5 are presented only with a significance of one figure. The possible contributions of H^{+}, K^{+}, Cl^{-}, and $Tris^{+}$ to the binding picture remain to be established.

The binding data are of great usefulness in a comparison with kinetically derived constants. The data presented here are sufficiently accurate to provide restraints, on the one hand, and confirmation, on the other hand, for several plausible kinetic mechanisms which may be proposed. For example, any kinetic mechanism which postulates that an enzyme-Mg intermediate has an intrinsic dissociation constant smaller or equal to that of the magnesium complexes of the nucleotides[376] must be considered highly unlikely. The kinetically derived constants presented by Kuby and Noltmann[48] for the rabbit muscle creatine kinase are in qualitative agreement with the binding constants presented here in Table 4. Similarly, Noda's[346] values for the rabbit muscle myokinase reaction, recalculated from the kinetic data of Callaghan and Weber[345] and Noda[344] approximate the binding data (Table 5) for this enzyme (see also Reference 50 and Chapter 5, Table 6).

Finally, the binding data indicate that a value of $n \simeq 2$ may be taken as significant for the nucleotides and their Mg complexes for *both* ATP-Cr transphosphorylase and myokinase. Within the experimental limitations, it appears that for ATP-Cr transphosphorylase the binding sites are equivalent with little interaction, whereas for myokinase this point is in doubt. Thus, in Table 5, the data for ADP^{3-} obtained by Kuby et al.[59] yielded an extrapolated value of $n_{ADP^{3-}}$ of only approximately 1.6, but within the calculated experimental error;[59] it would have also been possible to fit the data to an equation for two distinct sets of binding sites (with $n = 1$ for each set; see Reference 150). However, because of the limited number of experimental points in the critical range at relatively high values for $\bar{r}'$ approaching the x-intercept, a more complicated nonlinear plot to distinguish the two constants did not appear justified at that time. This point will be further discussed below. Also, under the conditions of measurements (at 3°C pH 7.9, and (γ/2) and attendant errors[59] for the rabbit muscle myokinase, an extrapolated value of $n = 1.8$, as an average, was estimated for both ATP^{4-} and $MgATP^{2-}$.

By nuclear magnetic relaxation and electron paramagnetic resonance techniques, Price et al.[377] had deduced that there is only one binding site for MnATP or ATP per mole of porcine muscle adenylate kinase. On the other hand, by ^{31}P-NMR, Rao et al.[350] deduced that both ATP and ADP can bind at *both* nucleotide binding sites of porcine myokinase, and that one site (the so-called AMP site) binds only to the uncomplexed nucleotides and the other site binds to ADP and ATP with or without the presence of Mg.

A further investigation of the substrate (or substrate analogue) properties of the myokinase seemed worthwhile with the recent availability of suitable analogues of the adenine nucleotides[378] and with the detailed kinetics study made by Hamada and Kuby[50] on their substrate or inhibitor properties of these analogues (see Chapter 5, Table 6). Furthermore, during the course of determination of the amino acid sequences of the rabbit and calf muscle myokinases,[363,854] certain relatively large peptide fragments became available for ligand-binding studies by Hamada et al.[348] (one of which includes the His-36 in the porcine enzyme implicated by McDonald et al.[379] as interacting with ATP), a clearer picture of the two distinct substrate binding sites (but one catalytic site) for myokinase emerged. Before entering into discussion of these studies[348] which supports the contention (see above) that the reaction catalyzed by myokinase is actually

$$MgATP^{2-} + AMP^{2-} \rightleftarrows MgADP^{-} + ADP^{3-} \tag{97}$$

it is important to note the X-ray crystallographic studies of Schulz et al.[356] on the porcine myokinase, whose sequence had been elucidated by Heil et al.[380] Unfortunately, all attempts by Schulz et al. to bind substrates or substrate analogues to their crystals (later defined as crystal form A[357] apparently had failed, and therefore, X-ray crystallographic evidence for the substrate binding sites were not convincing. Largely, by comparison with the results of Ap_5A [p′,p⁵-di(adenosine-5′)-pentaphosphate] bound to crystals A, Pai et al.[358] indirectly assigned the sites for binding of AMP and ATP to crystal form B by X-ray diffraction analysis. However, no bound AMP could be detected at their assigned "AMP site", although $MnATP^2$ could be made to bind to it; no ATP could be made to bind to their assigned "ATP-binding site".[358] In fact, Rao et al.,[350] largely by ^{31}P-NMR studies, seemed to exclude $MgATP^{2-}$ from the putative "AMP-binding site" of the porcine myokinase, casting doubt therefore on the identity of the AMP-binding site occupied by $MnATP^2$, which had been assigned by X-ray diffraction analysis.[358] It is of interest in this regard to note that Ap_5A (as we have seen above) was found to act as a competitive inhibitor with respect to either substrate of the forward reaction (i.e., $MgATP^{2-} + AMP^{2-} \rightarrow$) catalyzed by both calf and rabbit muscle myokinase, but acted as a noncompetitive inhibitor (mixed type) of either substrate of the reverse reaction (i.e., $MgADP^{-} + ADP^{3} \rightarrow$),[50] a finding which might tend to make any unambiguous assignment of Ap_5A to the myokinase substrate binding sites difficult.

Both a fluorescence-quenching technique and a UV-difference spectral method were used to study the binding of the fluorescent 1,N⁶-etheno analogues of the adenine nucleotides

(ϵATP, ϵADP, and ϵAMP)[378] to crystalline rabbit and calf muscle ATP-AMP transphosphorylase in the presence and absence of Mg^{2+}, at 0.16(Γ/2), 25°C, and pH 7.4.[348] In addition, Hamada et al.[348] studied the binding of the ϵ-analogues of the adenine nucleotides to two S-[^{14}C] carboxymethylated peptide fragments of the rabbit muscle myokinase (residues 1 to 44 = MT-I, residues 172 to 194 = MT-XII).

The *treatment of the data* for the fluorescence-quenching measurements[348] is as follows. Based on the Stern-Volmer formulation,[381] one may define

$$\frac{\Delta F_{cor}}{\Delta F_{max}} = \frac{L_b}{nP_t} = \frac{\bar{r}'_A}{n} \text{ and } L_{total} = L_b = L_t = A_f \tag{97A}$$

where

$$\Delta F_{cor} = F_{measured\ in\ control} - F_{measured\ in\ sample} = Quenching_{cor} \tag{98}$$

for $F_{control} > F_{sample}$) and where F and F_{max} are the measured fluorescence and maximum fluorescence intensity. ΔF_{cor} is the fluorescence quenching corrected for dilution and ΔF_{max} is the quenching value extrapolated to a maximally bound ligand concentration (L = ligand concentration, A = concentration of species A, P = protein or peptide concentration, and subscripts b, f, and t refer to bound, free, and total concentration, respectively).

Since

$$\frac{\bar{r}'_A}{A_f} = nK'_A - \bar{r}'_A K'_A \tag{99}$$

for ligand A at equivalent and independent binding sites[144,146] where K'_A is the association constant for ligand A, n is the number of moles of A maximally bound per mole protein and $\bar{r}'_A$ is the average number moles of A bound per mole of protein with the assumption

$$\left(\frac{\Delta F_{cor}}{\Delta F_{max}}\right) = \text{fraction of sites occupied} = \frac{L_b}{nP_t} = \frac{\bar{r}'_A}{n} \tag{100}$$

$$\frac{\dfrac{\Delta F_{cor}}{\Delta A_{max}}}{A_f} = K'_A - \frac{\Delta F_{cor}}{\Delta F_{max}} K'_A \tag{101}$$

or

$$\frac{\Delta F_{cor}}{A_f} = \Delta F_{max} K'_A - \Delta F_{cor} K'_A \tag{102}$$

A plot of $\Delta F_{cor}/A_f$ vs. ΔF_{cor} (with A_f estimated with a preliminary assignment of n) yields as y-intercept = $\Delta F_{max} \cdot K'_A$ and x-intercept = ΔF_{max} and initial estimates of the values for K'_A and ΔF_{max} are obtained. At high values of titrant (i.e., as $A_t \rightarrow A_f$ (or $A_t > P_t$)), then $(\Delta F/A_f) \rightarrow (\Delta F/A_t) \simeq \Delta F_{max} \cdot K'_A - \Delta F \cdot K'_A$ or

$$\frac{1}{\Delta F} = \frac{1}{A_t} \frac{1}{\Delta F_{max} K'_A} + \frac{1}{\Delta F_{max}} \tag{103}$$

and a plot of $1/\Delta F$ vs. $1/A_t$ yields the first preliminary estimate of $1/\Delta F_{max}$ as the y-intercept.

Substitution of $n\ \Delta F_{cor}/\Delta F_{max} = \bar{r}'_A$ into Equation 99 and a plot of $\bar{r}'_A/A_f$ vs. $\bar{r}'_A$, fitted to Equation 99, yields as y-intercept $= nK'_A$, slope $= -K'_A$ and x-intercept $= n$, now with second and better approximations of n and K'_A. In principle, therefore, an iterative procedure is used to evaluate the binding parameters, and the final Scatchard plot, fitted statistically to Equation 99 with the aid of a Hewlett-Packard® 9820A programmable calculator, programmed to yield estimates of the confidence limits for the binding parameters. The program yields the standard deviations for 95% confidence limits, calculated by the statistical analysis of variance of regression[382] after application of the *t* test at a 95% confidence level.

In the case of ADP^{3-}, the simplest model derived from the experimental data fitted the case for *two sets* of nonequivalent but independent binding sites, with quenching at each site assumed to be similar, where[144,147]

$$\bar{r}'_{total} = \frac{L_b}{P_t} = \sum_{i=1}^{2} \frac{n_i K_i(A_f)}{1 + K_i(A_f)} \tag{104}$$

and where $n_1 + n_2$ was found to be equal to approximately 2 and $K_1 > K_2$. The analysis of this two-site system was facilitated by the estimation of the limiting slopes and intercepts and the use of the correct relationships of these graphical parameters to the site-binding constants as described by Klotz and Hunston[148] (see above):

$$\bar{r}'_{total} = \frac{n_1 K_i(A_f)}{1 + K_i(A_f)} + \frac{nK_2(A_f)}{1 + K_2(A_f)}$$

$$\frac{\bar{r}'_{total}}{A_f} = n_1K_1 + n_2K_2 + (A_f)K_1K_2(n_1 + n_2) - \bar{r}'_{total}\,[K_1 + K_2 + A_fK_1K_2] \tag{105}$$

Since significant concentrations of ethenoanalogue of the adenine nucleotide will be present in the solutions near the equivalence point of the titrations, a fluorescence correction for the self-absorption of the incident light by the $1,N^6$-etheno analogue of the adenine nucleotide proved necessary.

From Reference 381, the so-called inner-filter effect for self-absorption of the emitted fluorescence light may be expressed as

$$\left\{\frac{F_{cor}}{F_{obs}}\right\} = \frac{2.303(A_{\lambda excit})(d_2 - d_1)}{10^{-A_{\lambda excit}d_1} - 10^{A_{\lambda excit}d_2}} \tag{106}$$

and with depth d_1 set instrumentally small, and d_2, the liquid depth, set at 1 cm, for 3.5-cm thin window cells with a right-angled viewing path, the equation for self-absorption reduces to

$$F_{cor} = F_{obs}\frac{2.303A_{\lambda excit}}{1 - 10^{-A_{\lambda excit}}} \tag{107}$$

In practice, the optical density at 324 nm* was measured in the same cells in a Cary-

* For the etheno analogues of the adenine nucleotides, the fluorescence excitation wavelength maximum is at approximately 306 (±5) nm (uncorrected with the instrument employed) and it is a relatively broad spectrum in this range. However, since optimal conditions for fluorescence titrations proved to be at approximately 0.1 m*M* protein or peptide, a significant amount of absorption of the exciting light would take place at 306 nm by the protein or peptide; therefore, a 324-nm excitation wavelength was selected for the fluorescence titrations where light absorption by the portein or peptide was very small, while still retaining the 416 (±5)-nm emission maximum (uncorrected), with only a slight loss in quantum efficiency.

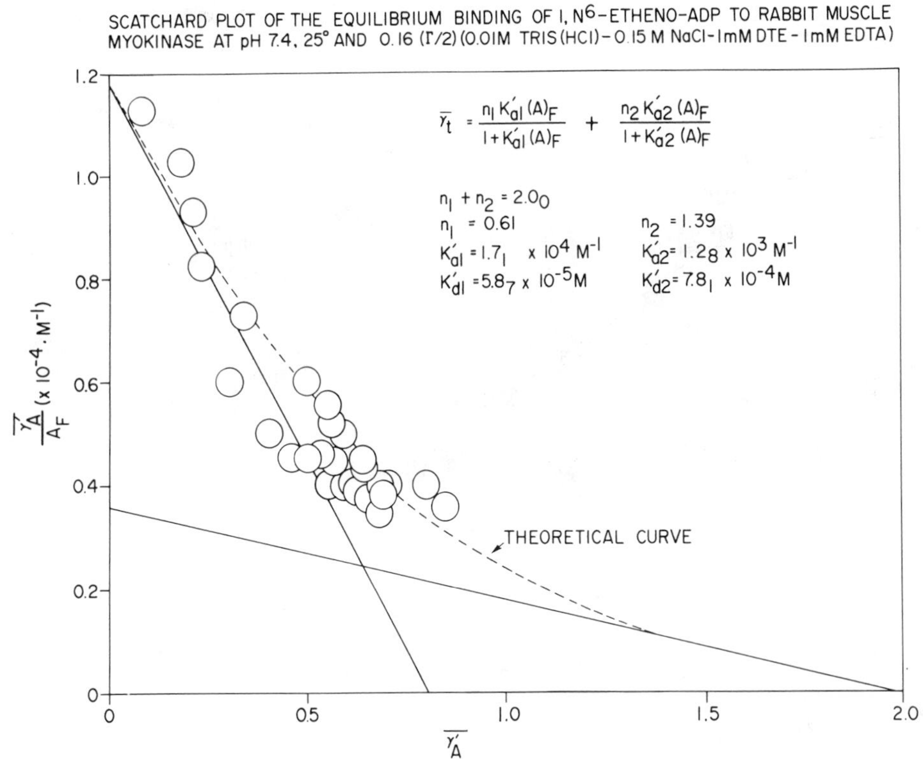

A

FIGURE 9. Equilibrium binding of 1, N^6-ethano analogues of adenine nucleotides to rabbit muscle and calf muscle myokinase as measured by flourescence quenching.

14® spectrophotometer at concentrations sufficiently low to avoid a fluorescence error. A table was then constructed for each point in the titration corresponding to each concentration of etheno analogue and the factor calculated to correct the observed fluorescence (F_{obs}) to yield the value of F_{cor}, now corrected for the inner-filter contribution.

For the *ultraviolet-difference spectra titrations*, the difference spectra were measured in two 1.0-cm tandem double-compartment cylindrical cells with use of a thermostated Cary-14® spectrophotometer, as described by Herskovits and Laskowski.[383,384] The difference spectra, of, for example, enzyme ϵATP^{4-} complex minus enzyme, could be used as a measure of the binary enzyme-ligand complex formation. In particular, the values of $\Delta A_{275\text{-}400\ nm}$ were found to be a direct measure of binding. After correction of the ΔA for dilution, and extrapolation of $1/\Delta A$ vs. $1/A_t$, $1/\Delta A_{max}$ could be estimated as a first approximation; estimations of L_b, A_f, and $\bar{r}'_A$ followed in the manner described above for the fluorescence titrations, with the assumption that $\frac{\Delta A}{\Delta A_{max}}$ = fraction of sites occupied

$$= \frac{L_b}{P_t} = \frac{\bar{r}'_A}{n};\ L_{total} - L_b = L_f = A_f \qquad (108)$$

In Figure 9 are shown some typical Scatchard plots obtained from fluorescence-quenching titrations for the binding by rabbit muscle myokinase of 1,N^6-etheno ADP (ϵADP^{3-}) (left-hand side), of $Mg\epsilon ADP^-$ and of ϵAMP^{2-} (middle plots), and for binding by calf muscle myokinase of $Mg\epsilon ATP^{2-}$ and of ϵATP^{4-}.

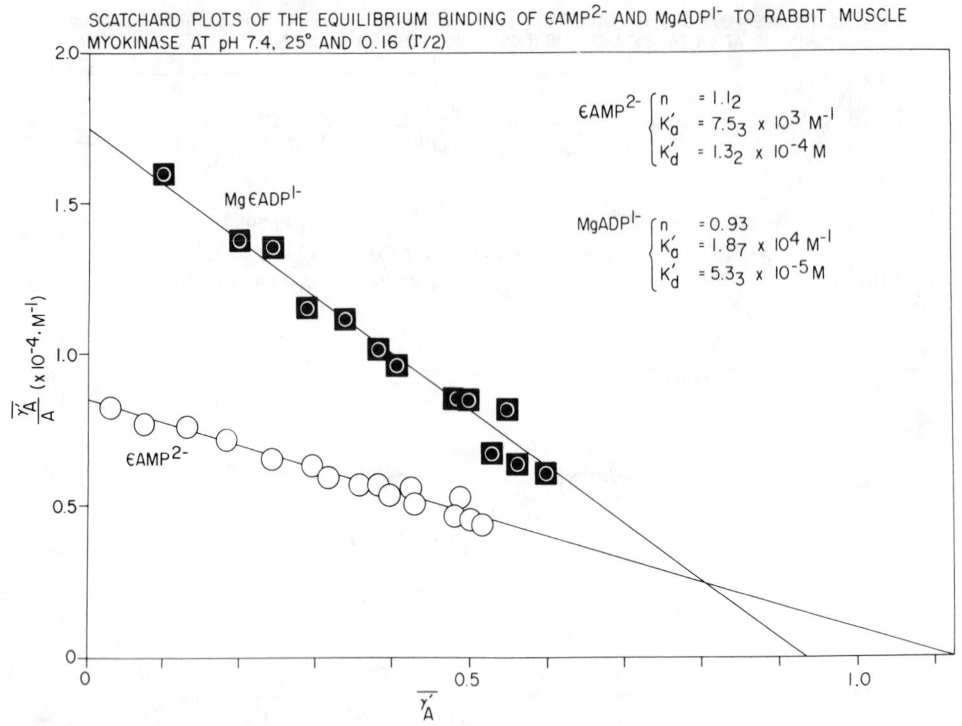
SCATCHARD PLOTS OF THE EQUILIBRIUM BINDING OF εAMP2- AND MgADP1- TO RABBIT MUSCLE MYOKINASE AT pH 7.4, 25° AND 0.16 (Γ/2)
εAMP2- n = 1.12, K'a = 7.53 x 10^3 M^-1, K'd = 1.32 x 10^-4 M
MgADP1- n = 0.93, K'a = 1.87 x 10^4 M^-1, K'd = 5.33 x 10^-5 M
MgεADP1-
εAMP2-
γ'A/A (x 10^-4. M^-1)
γ'A

FIGURE 9B

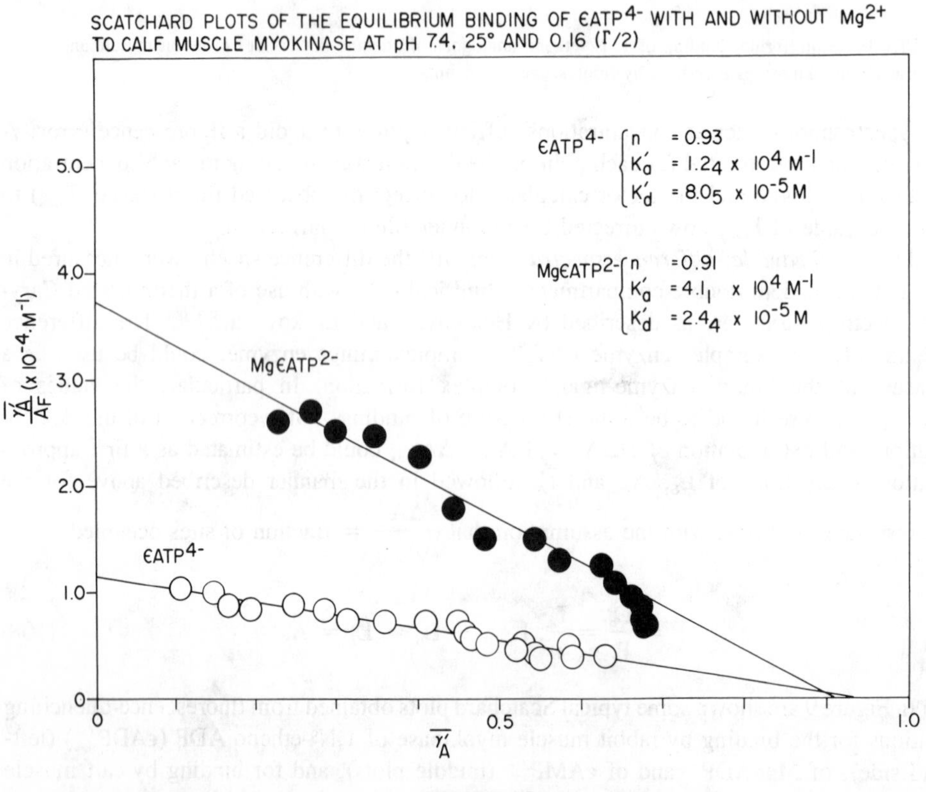
SCATCHARD PLOTS OF THE EQUILIBRIUM BINDING OF εATP4- WITH AND WITHOUT Mg2+ TO CALF MUSCLE MYOKINASE AT pH 7.4, 25° AND 0.16 (Γ/2)
εATP4- n = 0.93, K'a = 1.24 x 10^4 M^-1, K'd = 8.05 x 10^-5 M
MgεATP2- n = 0.91, K'a = 4.11 x 10^4 M^-1, K'd = 2.44 x 10^-5 M
MgεATP2-
εATP4-
γ'A/AF (x 10^-4. M^-1)
γ'A

FIGURE 9C

These data are summarized in Tables 6 and 7 for the rabbit and calf muscle myokinase, respectively, together with their standard deviations, estimated for a 95% confidence level.

For the ϵADP^{3-} data, application of the *t* test[382] at a 90 or 95% confidence level revealed that the quality of the statistical fit for an assumed single binding site case significantly improved if the case of two classes of sites were assumed. Therefore, the data for ϵADP^{3-} were fitted to the case for $n \simeq 2$. It will be noted, qualitatively, that a value of $n \simeq 2.0$ for uncomplexed ϵADP^{3-} is in agreement with the earlier binding data of Kuby et al. on ADP^{3-}.[59] Moreover, it appears that both sites are *not* identical with a 13-fold difference in the estimated association constants ($K'_{a,1} \simeq 1.7_1\ (\pm 0.36) \times 10^4\ M^{-1}$, and $K'_{a,2} \simeq 1.3\ (\pm 0.30) \times 10^3\ M^{-1}$ for the rabbit muscle myokinase, for example) and similar to the calf muscle enzyme. On the other hand, $n \simeq 1$ for $Mg\epsilon ADP^-$ with $K'_a = 1.8_7\ (\pm 0.47) \times 10^4$ and is almost identical with $K'_{a,1}$ (the tight binding site) for ϵADP^{3-} and the rabbit muscle myokinase, for example, and also similar to the calf muscle enzyme myokinase. Further, $n \simeq 1$ for ϵAMP^{2-} **, and $n \simeq 1$ for ϵATP^{4-} or for $Mg\epsilon ATP^{2-}$, in the case of both rabbit and calf muscle myokinase and in agreement with the data on $MnATP^-$ obtained by Price et al.[377] for the porcine enzyme. The binding of $Mg\epsilon AMP$ to either enzyme proved too slight to measure.

In Figure 10, the interesting observation is presented that an S-carboxymethylated peptide fragment containing only the *first 44 residues* of the rabbit muscle enzyme, that is MT-I, may bind either ϵATP^{4-} or $Mg\epsilon ATP^{2-}$, with values of $n \cong 1$ or bind $\epsilon\ ADP^{3-}$ or $Mg\epsilon ADP^-$ (see References 348 and 964) again with values of $n \simeq 1$ and with values comparable to the intrinsic dissociation constants for the native molecule, or that the binding of ϵAMP^{2-} or of ϵADP^{3-} may take place to *another* S-carboxymethylated peptide fragment containing only the *last 23 residues* (MT-XII), again with values of $n \simeq 1$, and with values of K'_d similar to that of the native molecule.[348,964]

The binding data for MT-I and MT-XII were summarized in Reference 348. MT-I does *not* bind ϵAMP^{2-} significantly, nor does MT-XII bind significantly to $Mg\epsilon AMP$, $Mg\epsilon ADP^-$, ATP^{4-}. However, MT-I binds stoichiometrically to $Mg\epsilon ATP^{2-}$, $Mg\epsilon ADP^-$, ϵADP^{3-}, or ϵATP^{4-} and MT-XII binds stoichiometrically to ϵAMP^{2-} or to uncomplexed ϵADP^{3-}. Other peptide fragments in the rabbit muscle enzyme molecule, that is, fragments MT-IV or MT-VI, did not bind significantly to any of the etheno analogues, nor did insulin or serum albumin. Thus, the fragments MT-I and MT-XII appear to be remarkably specific.

In Reference 348, the binding data were summarized for the binding of etheno analogues to an *equimolar* mixture of the two peptide fragments, MT-I plus MT-XII (see also Reference 964). This *mixture* of peptides largely duplicated the binding pattern of the entire native molecule, both qualitatively and, except for ϵATP^{4-} or $Mg\epsilon ATP^{2-}$, even quantitatively. In the case of the ϵATP analogues, the values for the intrinsic association constants for binding to the synthetic equimolar mixture of peptides were approximately one tenth those obtained with the intact native myokinase molecule.

A. EQUILIBRIUM BINDING OF 1,N^6-ETHENO ANALOGUES OF ADENINE NUCLEOTIDES AS MEASURED BY UV-DIFFERENCE SPECTROSCOPY

All except $Mg\epsilon AMP$ (not shown here) generate difference spectra (of the enzyme etheno analogue complex minus the enxyme)[348] and the values of $\Delta A_{(275\text{-}400)}$ appear to be a direct measure of ligand binding or complex formation.

Scatchard plots of these UV-difference spectra data are presented in Figure 11 which are then summarized in Table 8. A comparison of the data obtained by the UV-difference spectroscopic technique (Table 8) with data of Table 6, obtained by fluorescence quenching, shows that (except for uncomplexed ϵATP^{4-} where there appears to be an unexplained

** Where the charges are depicted for the etheno analogs, e.g., $Mg\epsilon ATP^{2-}$, $Mg\epsilon ADP^-$, ϵADP^{3-}, ϵAMP^{2-}, the net charge refers only to the phosphate groups, since the ring may bear a positive charge.

TABLE 6
Equilibrium Binding of 1,N^6-Etheno Analogs of Adenine Nucleotides to Rabbit Muscle Myokinase

	εAMP	εADP $n_1 + n_2 = 2.00\ (\pm 0.05)$		εATP
Without Mg^{2+}	$n = 1.12\ (\pm 0.80)$ $K'_a = 7.5_2\ (\pm 0.77) \times 10^3\ M^{-1}$ $K'_d = 1.3_3\ (\pm 0.14) \times 10^{-4}\ M$	$n_1 = 0.61\ (\pm 0.13)$ $K'_{a_1} = 1.7_1\ (\pm 0.36) \times 10^4\ M^{-1}$ $K'_{d_1} = 5.8_7\ (\pm 1.24) \times 10^{-5}\ M$	$n_2 = 1.3\ (\pm 0.30)$ $K'_{a_2} = 1.3\ (\pm 0.4) \times 10^3\ M^{-1}$ $K'_{d_2} = 7.8\ (\pm 2.4) \times 10^{-4}\ M^{-1}$	$n = 0.81\ (\pm 0.09)$ $K'_a = 3.3_6\ (\pm 0.42) \times 10^4\ M^{-1}$ $K'_d = 2.9_8\ (\pm 0.37) \times 10^{-5}\ M$
With Mg^{2+}	$K'_a \leq 6 \times 10^2\ M^{-1}$		$n = 0.93\ (\pm 0.40)$ $K'_a = 1.87\ (\pm 0.47) \times 10^4\ M^{-1}$ $K'_d = 5.3_3\ (\pm 1.33) \times 10^{-5}\ M$	$n = 0.97\ (\pm 0.04)$ $K'_a = 3.8_0\ (\pm 0.20) \times 10^4\ M^{-1}$ $K'_d = 2.6_3\ (\pm 0.14) \times 10^{-5}\ M$

Note: n = maximal number of moles bound per mole of protein, K'_a = intrinsic association constant, and K'_d = intrinsic dissociation constant.

TABLE 7
Equilibrium Binding of 1,N^6-Etheno Analogs of Adenine Nucleotides to Calf Muscle Myokinase

	εAMP	εADP $n_1 + n_2 = 2.00\ (\pm 0.10)$		εATP
Without Mg^{2+}	$n = 0.91\ (\pm 0.11)$ $K'_a = 1.4_8\ (\pm 0.18) \times 10^4\ M^{-1}$ $K'_d = 6.7_4\ (\pm 0.82) \times 10^{-5}\ M$	$n_1 = 0.42\ (\pm 0.13)$ $K'_{a_1} = 1.2_1\ (\pm 0.37) \times 10^4\ M^{-1}$ $K'_{d_1} = 8.2\ (\pm 2.5) \times 10^{-5}\ M$	$n^1 = 1.5_8\ (\pm 0.29)$ $K'_{a_2} = 1.2\ (\pm 0.2) \times 10^3\ M^{-1}$ $K'_{d_2} = 8.3\ (\pm 1.4) \times 10^{-4}\ M^{-1}$	$n = 0.93\ (\pm 0.08)$ $K'_a = 1.2_4\ (\pm 0.15) \times 10^4\ M$ $K'_d = 8.0_6\ (\pm 0.98) \times 10^{-5}\ M$
With Mg^{2+}	$K'_a \leq 3 \times 10^2\ M^{-1}$		$n = 1.0_5\ (\pm 0.30)$ $K'_a = 4.4_9\ (\pm 2.41) \times 10^3\ M^{-1}$ $K'_d = 2.2_3\ (\pm 1.20) \times 10^{-4}\ M$	$n = 0.91\ (\pm 0.17)$ $K'_a = 4.1_1\ (\pm 0.54) \times 10^4\ M^{-1}$ $K'_d = 2.4_3\ (\pm 0.32) \times 10^{-5}\ M$

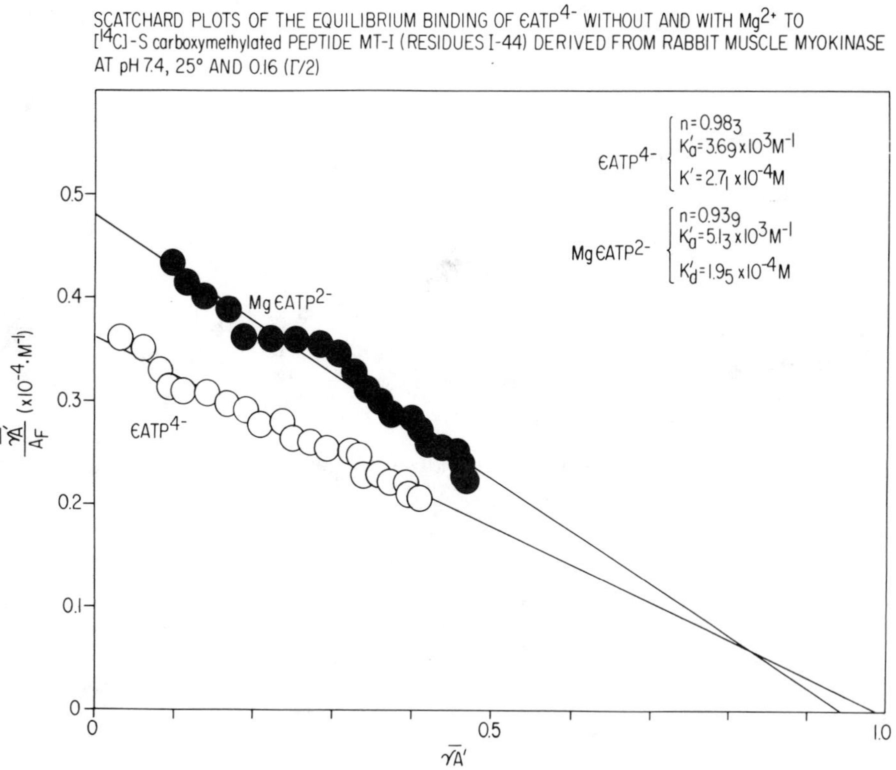

FIGURE 10. Equilibrium binding of 1,N[6]-etheno analogues of adenine nucleotides to ^{14}C -*S*-carboxymethylated peptides MT-I (residues 1 to 44).[348]

eightfold difference) there is good agreement in the estimations of the binding parameters for AMP^{2-}, ϵADP^{3-}, $Mg\epsilon ADP^{-}$, and $Mg\epsilon ATP^{2-}$. Therefore, these data lend additional confidence to the data gathered by fluorescence quenching, both qualitatively and quantitatively.

The etheno analogues of the adenine nucleotides[378] have proved to be extremely valuable tools for the study of the substrate interactions with the adenylate kinase enzyme protein. Thus, the use of these analogues permitted the application of two convenient and precise physical techniques, that is, fluorescence-quenching and UV-difference spectrophotometric methods, in the study of their equilibrium binding to the ATP-AMP transphosphorylase. The agreement by these two methods in the estimation of the binding parameters to the rabbit muscle adenylate kinase has provided a measure of confidence in their reliability and accuracy (see Reference 964).

A detailed examination was made of the steady-state kinetics of the rabbit muscle adenylate kinase[50] (see also Chapter 5, Table 6). Interestingly, although the enzyme catalyzed the forward reaction (at a relatively rapid rate) with the substrate pair $Mg\epsilon ATP^{2-} + AMP^{2-}$ ($V/E_t = 370$ μmol min^{-1} mg^{-1} vs. 550 for the physiological substrate pair, $MgATP^{2-} + AMP^{2-}$), the combination of etheno analogues $Mg\epsilon ATP^{2-} + \epsilon AMP^{2-}$ proved largely unreactive, in agreement with Secrist et al.[378], and emphasized the unusual requirement of the enzyme for an intact 6-NH_2 group in the AMP^{2-} molecule, but not in $MgATP^{2-}$. Kinetic estimation (at 25°C, pH 7.4, and $(\Gamma/2) = 0.16$) of the intrinsic dissociation constants from the enzyme substrate binary complexes[50] led to the values of $\overline{K}_{Mg\epsilon ATP^{2}} = 9.8\ (\pm 0.8) \times 10^{-5}\ M$ and K_i ($\overline{K}_{\epsilon AMP^{2-}}$) $= 2.2_9\ (\pm 0.10) \times 10^{-4}\ M$. By UV-difference spectroscopy (Table 8) values of K_d' for $Mg\epsilon ATP^{2-} = 3.9_7\ (\pm 1.84) \times 10^{-5}\ M$ ($n_{Mg\epsilon ATP^{2}} = 0.96$) and for $\epsilon AMP^{2-} = 1.6_2\ (\pm 0.34) \times 10^{-4}\ M$ ($n_{\epsilon AMP^{2-}} = 0.92$) were obtained; or by fluorescence

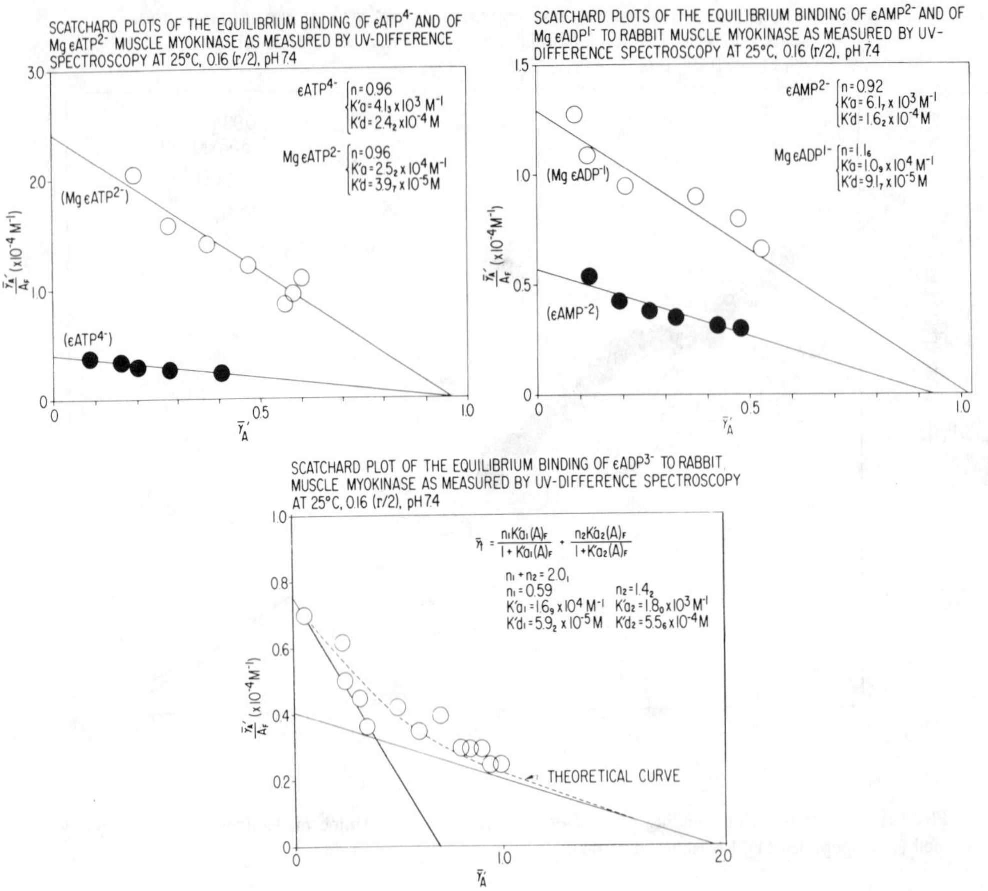

FIGURE 11. Scatchard plots of the equilibrium binding of etheno analogues of adenine nucleotides of rabbit muscle myokinase as measured by UV-difference spectroscopy at 25°C, 0.16 (Γ/2), and pH 7.4. Upper left-hand plot: Scatchard plots of the equilibrium binding ATP^{4-} and of $Mg\epsilon ATP^{2-}$ to rabbit muscle myokinase. Plots fitted to Equation 99. *Lower plot:* Scatchard plot of the equilibrium binding of ADP^{3-} to rabbit muscle myokinase. The dashed line is the theoretical curve fitted to Equation 104 for two nonequivalent but independent sites. The two solid lines are drawn according to Equation 105 for the limiting slopes and intercepts of Klotz and Hunston.[148]

quenching (Table 6) $K'_{Mg\epsilon ATP^{2-}} = 2.6_3\ (\pm 0.14) \times 10^{-5}\ M$ ($n = 0.97$) and $K'_{\epsilon AMP^{2-}} = 1.3_3\ (\pm 0.14) \times 10^{-4}\ M$ ($n = 1.1_2$). Both sets are in fair agreement with the kinetically derived values, lending some credence to the kinetic evaluation, based on a random quasi-equilibrium type of mechanism with a rate-limiting step largely at the interconversion of the ternary complexes.[50] The earlier estimate[59] of $K'_{ADP^3} \simeq 7 \times 10^{-5}\ M$ (see Table 5) for an overall value of $n = 1.6$ at 3°C, pH 7.9, and $(\Gamma/2) \simeq 0.2$ proved to be in surprising agreement with the $\overline{K}_{ADP^{3-}} \simeq 4.0\ (\pm 0.9) \times 10^{-5}\ M$, estimated kinetically at 25°C.[50] The present binding data on ϵADP^{3-} fit very well to a two-site, independent, but nonequivalent, model, i.e., $n_1 + n_2 = 2.0$ for both rabbit and calf muscle adenylate kinases (Tables 6 and 7). In the presence of an excess of Mg^{2+}, however, the binding data for $Mg\epsilon ADP^{-}$ reduce to a single-site model, with $n \simeq 0.93$ and 1.05, for rabbit and calf muscle myokinase, respectively (Table 6 and 7). Most interesting is the fact that in the presence of excess Mg^{2+}, the binding of ϵAMP^{2-} becomes too slight to measure accurately (Tables 6, 7, and 8), whereas, in the absence of Mg^{2+}, the binding of ϵAMP^{2-} is stoichiometric, with values of $n \simeq 1$ (Tables 6, 7, and 8).

It would appear from these ligand binding data given above, that the kinetic evidence for a two-substrate binding-site model for myokinase[50,343,346,347] i.e., one site for the Mg

TABLE 8
Equilibrium Binding of 1,N⁶-Etheno Analogs of Adenine Nucleotides to Rabbit Muscle Myokinase by UV-Difference Spectroscopy

	εAMP	εADP $n_1 + n_2 = 2.00\ (\pm 0.15)$		εATP
Without Mg^{2+}	$n = 0.92\ (\pm 0.11)$ $K'_a = 6.1_7\ (\pm 1.29) \times 10^3\ M^{-1}$ $K'_d = 1.6_2\ (\pm 0.34) \times 10^{-4}\ M$	$n_1 = 0.59\ (\pm 0.07)$ $K'_{a_1} = 1.6_9\ (\pm 0.20) \times 10^4\ M^{-1}$ $K'_{d_1} = 5.9_2\ (\pm 0.69) \times 10^{-5}\ M$	$n_2 = 1.4_2\ (\pm 0.15)$ $K'_{a_2} = 1.8\ (\pm 0.2) \times 10^3\ M^{-1}$ $K'_{d_2} = 5.6\ (\pm 0.7) \times 10^{-4}\ M^{-1}$	$n = 0.96\ (\pm 0.11)$ $K'_a = 4.1_3\ (\pm 0.44) \times 10^3\ M^{-1}$ $K'_d = 2.4_2\ (\pm 0.26) \times 10^{-4}\ M$
With Mg^{2+}	$K'_a \leq 2 \times 10^2\ M^{-1}$		$n = 1.1_6\ (\pm 0.11)$ $K'_a = 1.0_9\ (\pm 0.16) \times 10^4\ M^{-1}$ $K'_d = 9.1_7\ (\pm 1.35) \times 10^{-5}\ M$	$n = 0.96\ (\pm 0.30)$ $K'_a = 2.5_2\ (\pm 01.17) \times 10^4\ M^{-1}$ $K'_d = 3.9_7\ (\pm 1.84) \times 10^{-5}\ M$

complexes of the adenine nucleotide and one site for the uncomplexed nucleotides rests on more solid ground. But the strongest support for this hypothesis of Rhoads and Lowenstein[347] comes from the observation that two tryptic fragments derived from the S-[^{14}C]carboxymethylated, maleylated rabbit muscle myokinase protein showed remarkable and specific binding to certain 1,N^6-etheno analogues of the analogues of the adenine nucleotides.[348,964]

Thus, the binding properties of these unusual peptide fragments (MT-I and MT-XII), derived from the head and the tail of the molecule, respectively, now have confirmed the hypothesis of Rhoads and Lowenstein[347] deduced kinetically, that two separate and distinct binding sites exist (1) for the Mg complexes of the nucleotide substrates, $MgATP^{2-}$ and $MgADP^{-}$, and (2) for the uncomplexed nucleotide substrates, AMP^{2-} and ADP^{3-} (see References 343 and 346) but with *one* overall catalytic site. Therefore, the catalyzed reaction for myokinase may actually be written as Equation 97.[50]

For interaction between the two sites to occur prior to catalysis and phosphoryl-group transfer, a head-to-tail interaction is mandatory. The enzyme may or may not exist in solution with such a preferred conformation, but it is consistent with the X-ray-deduced crystal structure of the porcine muscle adenylate kinase;[356] such a conformational change has been implied by NMR studies as a result of binding of Ap_5A to human muscle adenylate kinase.[857]

In fact, the substrates in turn might induce such a large conformational change, or further induce smaller changes within the ternary complexes, as was deduced kinetically[50] from the fact that estimated values for $\overline{K}_{s,1}$ and $K_{s,1}$ (intrinsic dissociation constants of the substrates from the binary and ternary complexes) differed. The fact that Ap_5A is such a poor inhibitor for the calf liver adenylate kinase,[29] relatively speaking, compared to the calf muscle or rabbit muscle adenylate kinase, might imply that either the atomic distances, in which the two binding sites must approach one another, differ in the liver enzyme compared to the muscle enzyme or that possibly the liver enzyme may exist in a less flexible structure to allow the necessary binding and required juxtaposition of the two binding sites. The fact that $V^r << V^f$ in the liver enzyme[29] seems to indicate that the enzyme is already partially fixed in some preferred conformation so as to facilitate a phosphoryl group transfer to AMP^{2-} (and, as suggested in Reference 29, this may be its important role in the mitochondrion).

The chemical requirements for binding at the two binding sites of myokinase are now amenable to a systematic elucidation, by the fortunate set of circumstances that they are distinct sites, occupying entirely different loci and sequences of amino acids within the molecules, i.e., near the head and near the tail of the protein molecule (see Reference 964).

Steitz and co-workers,[385] in a remarkable computer-drawn space-filled model of the porcine adenylate kinase (calculated from the data of Schulz[356] and Pai et al.[358] for crystal form B, clearly portrayed the "bilobal" character of this phosphoryl-group transfer enzyme with the deep cleft formed between the lobes. In their model, presumably $MgATP^{2-}$ and AMP^{2-} would bind at opposite ends of the cleft, followed by a conformational change in the muscle type which would bring the two substrate binding sites together and permit transphosphorylation to occur.

Fry et al.,[862] in a computer graphics representation of rabbit muscle adenylate kinase, showed the location of enzyme bound MgATP (Plate 1)* which was located by a combination of NMR[861] and X-ray diffraction studies.[356,358]

Three segments of the enzyme exhibiting sequence homology to other ATP- and GTP-binding proteins (see also Figure 13 and References 860 to 862) are shown in pink, and the "ATP" molecule is shown in red. In Figure 12, this is also depicted by an ORTEP (computer graphics program) with the representation showing MgATP and the three homologous segments of adenylate kinase.

* Plate 1 appears after page 336.

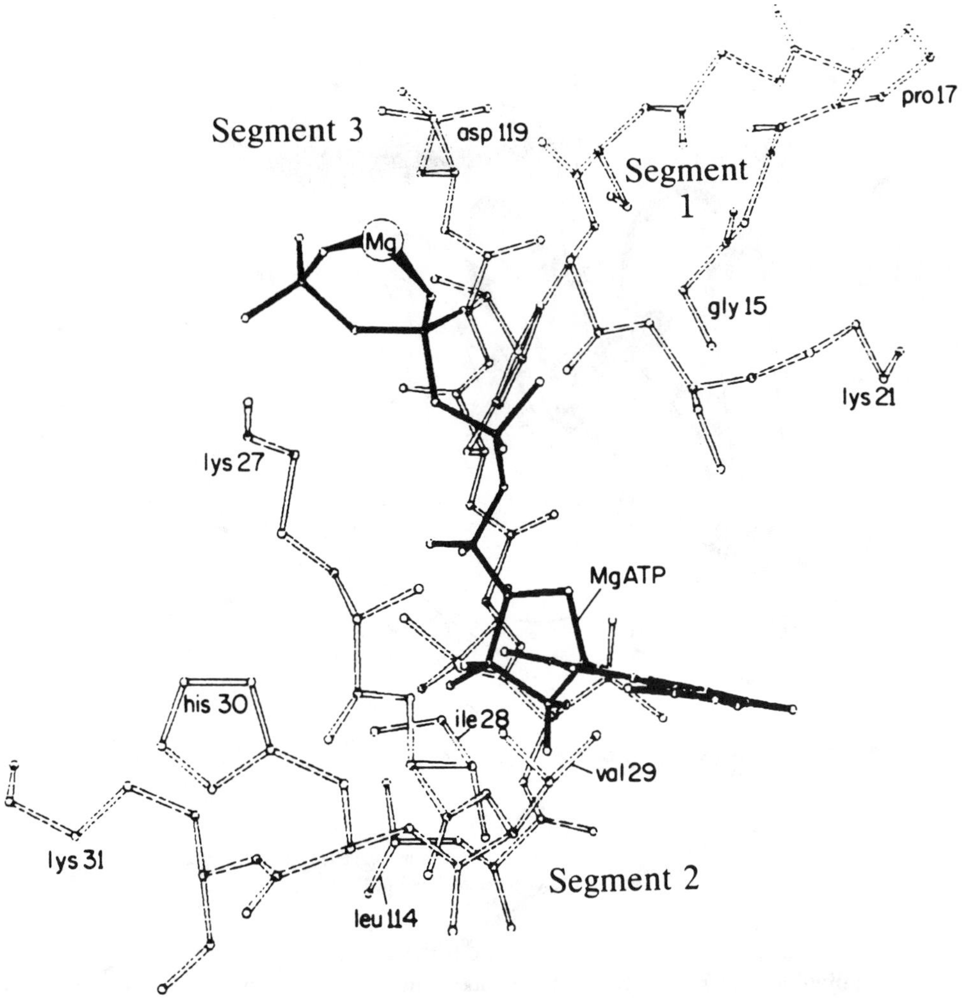

FIGURE 12. ORTEP (computer graphics program) representation showing MgATP and the three homologous segments of adenylate kinase (see Figure 13).[862]

Before leaving a discussion of equilibrium substrate binding as illustrated by the ATP-transphosphorylases enzymes, it is relevant to point out the recent interesting fluorescence study of the tryptophan residues of rabbit muscle creatine kinase by Messmer and Kägi.[889] Spectroscopic studies of rabbit skeletal muscle creatine kinase and its complexes with the adenine nucleotides had earlier suggested the possibility a tryptophan residue at or near the nucleotide binding sites (e.g., see References 890 and 891). This suggestion was supported by the nuclear Overhauser effect ^{1}H NMR studies of Vašák et al.[892] which indicated that there were "through-space" interactions between the protons of the adenine ring of bound ADP and one or more aromatic side chains of the protein. Additional evidence for a tryptophan residue in the environment of the active site was recently obtained[889] by a fluorescence-quenching study with the use of acrylamide and iodide as external quenchers (see References 893 and 911). Thus, whereas the addition of iodide reduces the tryptophan fluorescence of unliganded creatine kinase to approximately 75% of the unquenched control, no such effect is observed on addition of this quencher to the creatine kinase ·ADP or creatine kinase ·ATP complexes.[889] Similarly, the relative effectiveness of quenching of the creatine kinase · nucleotide complexes by acrylamide is only 60% of that measured in the unliganded enzyme.

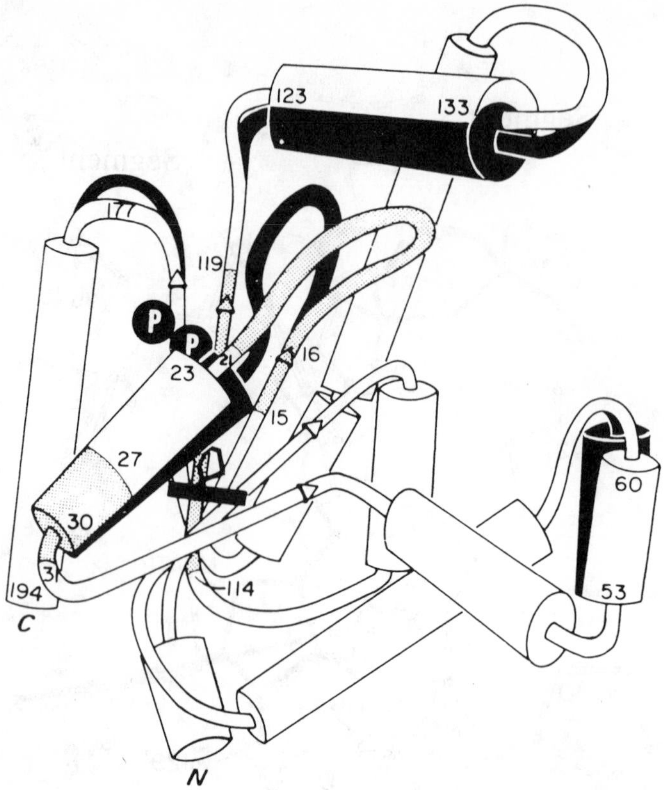

FIGURE 13. Metal-ATP binding site of rabbit muscle adenylate kinase (taken from Reference 862; see also References 860 and 864).

Both these data and the spectral characteristics of the quenched fluorescence suggested that nucleotide binding perturbs a tryptophan residue which is close to the active site and which is partially exposed to the solvent. The differential effectiveness of external quenchers on unliganded and liganded creatine kinase allowed Messmer and Kägi[889] to determine the ligand binding equilibria by a ''fluorescence-quenchability'' titration. The values obtained for complexes of creatine kinase with ATP and with ADP were in good agreement with those obtained by Kuby et al.[59] (See Table 4), especially by the gradient sedimentation technique.

VII. STRUCTURAL RELATIONSHIPS OF THE NUCLEOTIDE-MAGNESIUM COMPLEXES TO THE TRANSPHOSPHORYLASE ENZYMES

It is appropriate at this point to present a brief discussion of the structures proposed for the ATP molecule and its metal complexes, which might in turn display very specific effects on enzymic mechanisms. This point has been briefly alluded to above, for example as described in Reference 348. An older but still applicable discussion was presented by Kuby and Noltmann[48] and is briefly summarized here.

Thus, as early as 1954, Melchior[371] had proposed that particular configurations of the ATP molecule and its metal complexes might specifically affect the enzymatic mechanism; these configurations in solution and their distribution would depend on the net charge (and therefore the pH) and the metal or proton bound to the ATP. Atom models were used to

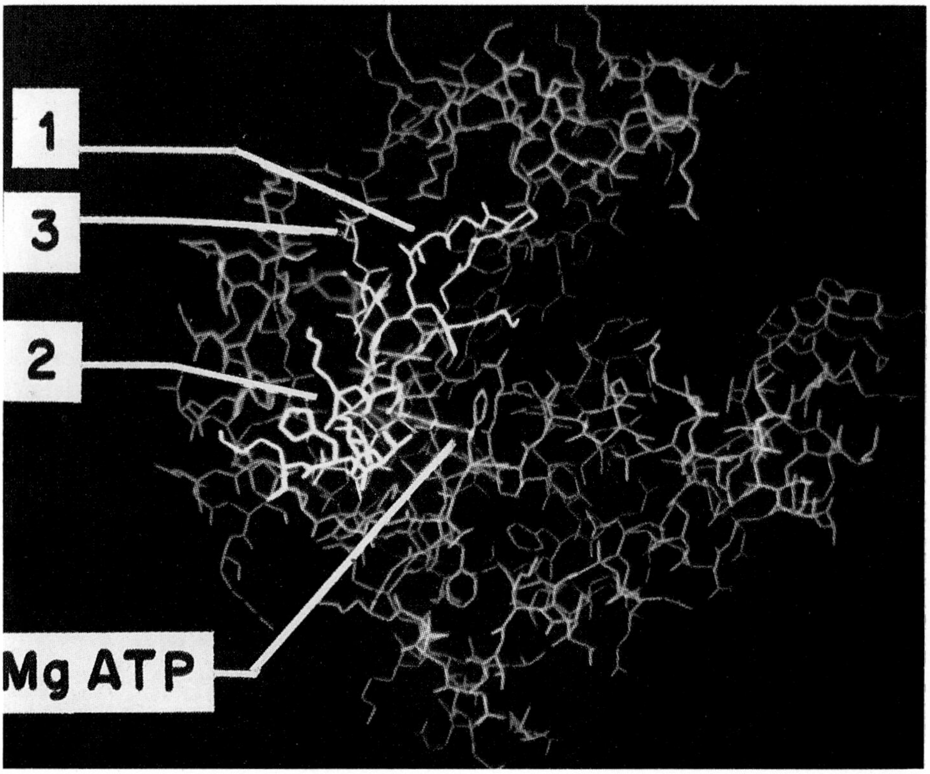

PLATE 1. A computer graphics representation of rabbit muscle adenylate kinase showing the location of bound metal ATP. The three segments of the enzyme exhibiting sequence homology to several ATP- and GTP-binding proteins.[861,862] are shown in pink, and the MgATP molecule is shown in red. The X-ray coordinates of conformation A of porcine adenylate kinase[356] were used with a substitution of a histidine residue for glutamine at position 30. The metal ATP was fit into the enzyme structure using a set of distances obtained by NMR.[861] Taken from Reference 862.

show the possibilities of folded vs. extended structures of ATP^{4-} and $HATP^{3-}$ and their metal complexes,[371] and it was demonstrated that if complex formation took place only via the polyphosphate chain of ATP^{4-}, the Mg ion is almost completely surrounded.

This model of Melchior for $MgATP^{2-}$, if applicable today, would imply, for example, that a direct attachment of the complexed Mg to the ATP-creatine transphosphorylase (see Table 4) or to myokinase (see Table 5) would be unlikely. It is of interest that Kuby and Noltmann[48] had pointed out early that "...the conformation(s) of the Mg chelates of ATP or ADP in solution is (are) not proved yet, since evidence for either folded[371,386-389] or extended[391] structure has been presented, nor has the fundamental nature of the chelate structure(s) itself been definitely established.[353,390,392-394] However, it was recently concluded[395] from a study of NMR and proton spectra that magnesium is bonded to only the β- and γ-phosphates of ATP, and similarly Mg is bonded to both phosphate groups of ADP. These investigations[395] provide one feasible means of eventual characterization of the metal chelates of ATP and ADP in solution; however, the influence of the enzyme itself, by attachment, on the distribution of the substrate configurations is, at present, a problem not easily amenable to direct experimentation (but see Reference 396)." These comments by Kuby and Noltmann made over two decades ago are still very pertinent today. For in the past two decades this problem has been extensively investigated (especially in the case of metals other than Mg) by the ^{31}P-relaxation method, which allows calculation of the distances between metal ions and phosphorus nuclei on the basis of the effect of paramagnetic metal ions on ^{31}P-relaxation times.[260,262,397,398,965,982] Recently, Cleland and co-workers prepared various isomers of "substitution-inert"[399] nucleotide complexes of Co^{3+} and Cr^{3+} and established their structure by X-ray, NMR, and circular dichroism methods.[400-403] These complexes of known structure were then tested for their activity with some enzymes in order to probe the chelation pattern of metal-nucleotide complexes in certain enzyme-catalyzed reactions.[404] Jaffe and Cohn[405] have further suggested, recently, that a metal-dependent stereospecificity reversal (e.g., in the case of the diastereoisomers of ATPβs[adenosine 5′-(2-thiotriphosphate)] is an indication for the involvement of that particular phosphoryl group in metal chelation during the catalysis.

Unfortunately, the most important and ubiquitous metal ion in enzyme-catalyzed phosphoryl transfer reactions is still Mg^{2+}, a metal ion which is neither paramagnetic nor substitution inert; even today, there is no direct or unambiguous method to observe the binding of diamagnetic metal ions with nucleotides, even in nonenzymatic systems. However, very recently, Huang and Tsai[406] proposed a new approach, by the use of ^{17}O NMR, to study such problems, and, in particular, whether or not Mg^{2+} ion interacts with the α-phosphate group of ATP. They[406] reviewed the available methods employed and their inadequacies in defining the chelation pattern of the Mg^{2+} complex with ATP. Thus, although other methods, such as IR,[407] have been employed, the most popular method currently in use is ^{31}P-NMR. On the basis of ^{31}P chemical shift changes (as mentioned above), Cohn and Hughes[395] had first reported that MgATP is a β,γ-bidentate (at pH 8.0). However, on the basis of essentially the same ^{31}P-NMR data, Kuntz and Swift[408] concluded that MgATP is an α,β,γ-tridentate, whereas Tran-Dinh et al.[409] concluded that MgATP is a β-monodentate. Gupta and Mildvan[410] argued that since the chemical shift and coupling constant of the P_α signal of ATP behave similarly to those of the P_α signal of ADP, and since MgADP is believed to be an α,β-bidentate, Mg^{2+} should also interact with the α-phosphate of ATP. Finally, to illustrate the present controversy, recently Ramirez and Marecek[411] suggested that MgATP is actually a mixture of α,β,-β,γ-; and α,γ-bidentates; whereas Bishop et al.[412] feel that MgATP is predominantly an α,β,γ-tridentate. It was Jaffe and Cohn[413] who actually criticized the ^{31}P chemical shift method. In fact, they felt[413] that there is no compelling reason for expecting the magnitude of the chemical shift change to be related to the site of Mg^{2+} binding, since the magnitude of the chemical shift change has been shown to be unrelated to the site of binding of a proton.[414,415] On the basis of the finding by Tsai et al.[416] that Mg^{2+} causes the ^{17}O-NMR signal of $[\gamma\text{-}^{17}O_3]$ATP to broaden, Huang and Tsai[406] employed the ^{17}O-NMR method to investigate the binding of Mg^{2+} with ATP and ADP. They concluded from their results that Mg^{2+} interacts with both α- and β-phosphates of ADP, and with all three (α-,β-, and γ-)phosphates of ATP; but the extent of α-coordination in MgATP may be smaller than the β- and γ-coordination. They felt that their results established the "macroscopic" structure of MgADP and MgATP, but the "microscopic" structures remained to be determined.

Recently, in a significant report, Sabat et al[898] determined the crystal structure of the α,β,γ-tridentate manganese complex of ATP cocrystallized with 2,2′-dipyridylamine (DPA). The MgATP complex had long eluded crystallization attempts, until Cini et al.[899,900] reported the preparation and crystallizations of the complexes of ATP with Mg^{2+}, Ca^{2+}, Sr^{2+}, Mn^{2+}, Co^{2+}, Cu^{2+}, and Zn^{2+}, in the presence of 2, 2′- dipyridylamine. A preliminary note appeared on the structures of these complexes of ATP with Mg^{2+}, Ca^{2+}, Mn^{2+}, and Co^{2+},[900] and the refined structures of MgATP and CaATP have recently been published.[901] However, apparently the MnATP crystals were of substantially better quality than the crystals of any of the other isomorphous metal complexes and thus permitted Sabat et al[898] the opportunity to make a very accurate determination of the crystal structure of the Mn complex of ATP which cocrystallized with 2,2′-DPA. Thus, they found that the 1:1:1 complex of Mn^{2+}, ATP, and 2,2′-DPA crystallizes as $Mn(HATP)_2 \cdot Mn(H_2O)_6 \cdot (HDPA)_2 \cdot 12H_2O$. The structure is composed of two ATP molecules which share a common Mn atom, and the metal exhibits α,β,γ coordination to the triphosphate chains of two dyad-related ATP molecules, which results in a hexacoordinated Mn^{2+} ion surrounded by six phosphate groups. The metal to oxygen distances they measured were 2.205, 2.156, and 2.144 Å for the α-,β-, and γ-phosphate groups, respectively. No metal-base interactions were observed. A second hexaaqua-coordinated Mn^{2+} ion was also located on a dyad axis. The hydrated Mn ions appeared to sandwich the phosphate-coordinated Mn ions in the crystal with a metal-metal distance of 5.322 Å. The ATP molecule appeared to be protonated on the N(1) site of the adenine base and to exhibit the anticonformation ($X = 66.0°$); the ribofuranose ring was in the ${}^{2}_{3}T$ conformation with pseudorotation parameters $P = 179$ and $\tau_m = 34.1°$.

Although it is a most elegant illustration of the application of X-ray crystallography, it is difficult to extrapolate this particular complex of MnATP and DPA to the environment of many ATP-transphosphorylases; however, it may have a more direct bearing in the case of those kinases like pyruvate kinase where a second metal ion may be involved, besides the metal-ATP substrate, in substrate binding and enzyme catalysis.[880,881,902,903]

Recently, Jarori et al.[867] made attempts to study the structure of metal nucleotide complexes bound to rabbit muscle creatine kinase by making ^{31}P NMR measurements, using Mn(II) and Co(II), and in agreement with Leyh et al[868] who had made use of superhyperfine coupling between Mn(II) and ^{17}O substituted on the phosphate groups, they were inclined to a case for direct coordination of Mn(II) and Co(II) with all three phosphates of ATP on the enzyme (see also Reference 985).

It is clear from this discussion that even after 2 decades, the basic problem posed by Kuby and Noltmann,[48] that is, the nature and kind of structures of the Mg chelates of the nucleotides, in solution and after binding to the enzyme, is still an exciting and formidable problem for the enzymologist interested in the mode of action of the physiologically important enzymes, termed ATP-transphosphorylases.

It is pertinent therefore to mention some recent studies conducted on rabbit muscle adenylate kinase and on an ATP-binding peptide synthesized by the Merrifield approach.[865,866]

A synthetic peptide $I_{1\text{-}45}$ residues 1 to 45 of rabbit muscle adenylate kinase (see Reference 855 for amino acid [a.a.] sequence), was investigated by NMR techniques to map out the topography of the $MgATP^{2-}$ binding site with the use of paramagnetic probes (e.g., Cr^{3+} ATP) and to deduce the conformation of the bound $MgATP^{2-}$.[858-862]

Thus, proton NMR was used to study the interactions of β,γ-bidentate Cr^{3+} ATP and $MgATP^{2-}$ with rabbit muscle adenylate kinase, which has 194 amino acid residues, and with the synthetic peptide $I_{1\text{-}45}$. The peptide is globular and binds Cr^{3+} ATP competitively with $MgATP^{2-}$ with a dissociation constant, $K_D(Cr^{3+}$ ATP) = 35 μM, comparable to that of the complete enzyme $K_{I(Cr5^{3+}ATP)}$ = 12 μM. Time-dependent nuclear Overhauser effects (NOEs) were used to measure interproton distances on enzyme- and peptide-bound MgATP. The correlation time was measured directly for peptide-bound MgATP by studying the frequency dependence of the NOEs at 250 MHz and 500 MHz. The H2′ to H1′ distance so obtained (3.07 Å) was within the range established by X-ray and model-building studies of nucleotides (2.9 ± 0.2 Å). Interproton distances yielded conformations of enzyme- and peptide-bound MgATP with indistinguishable anti glycosyl torsional angles (X = 63 ± 12°) and 3′-endo/01′-endo ribose puckers (δ = 96 ± 12°). Enzyme- and peptide-bound MgATP molecules exhibited different C4′-C5′ torsional angles (γ) of 170 and 50°, respectively. Ten intermolecular NOEs from protons of the enzyme and four such NOEs from protons of the peptide to protons of bound MgATP were detected, indicating proximity of the adenine ribose moiety to the same residues on both the enzyme and the peptide. Paramagnetic effects of β,γ-bidentate Cr^{3+} ATP on the longitudinal relaxation rates protons of the peptide provided a set of distances to the side chains of five residues, which allowed the location of the bound Cr^{3+} atom to be uniquely defined. Distances from enzyme-bound Cr^{3+} ATP to the side chains of three residues of the protein agreed with those measured for the peptide. The mutual consistency of interproton and Cr^{3+} to proton distances obtained in metal-ATP complexes of both the enzyme and the peptide suggested that the conformation of the peptide was very similar to that of residues 1 to 45 of the enzyme. Assuming this to be the case and using molecular models and a computer graphics system, MgATP could be fit into the X-ray structure of porcine muscle adenylate kinase[356-358] in a unique manner such that all of the distances determined by NMR were accommodated (see Figure 13). The adenine ribose moiety is bound in a hydrophobic pocket consisting of residues ile 28, val 29, his 36, leu 37, and leu 91, while the Mg^{2+} triphosphate portion binds near lys 21 and gln 24. In this complex, the γ-phosphoryl group of MgATP is directed toward residues 172 to 194 which, in the form of a peptide, has previously been shown to bind ε-AMP[348] and later by NMR techniques to bind AMP in an unusual conformation.[965] The MgATP binding site determined by NMR is also consistent with the results of chemical modification[863] and kinetic studies[50] and with structural and sequence homologies among many nucleotide-binding proteins.[858-862] Three segments of the rabbit muscle adentlate kinase sequence which are homologous to several ATP- and GTP-binding proteins, were at or near this site (stippled). Segment 1 (a.a. 15 to 21: G-G-P-G-S-G-K) shows homology to F_1ATPase, myosin, rec A protein, transducin, and *ras* p21; segment 2 (a.a. 27 to 31: K-I-V-H-K) shows homology to cAMP- and cGMP-dependent protein kinases and *src* tyr kinase; and segment 3 (a.a. 114 to 119: L-L-L-Y-V-D) shows homology to F_1ATPase and PFK. In rabbit muscle adenylate kinase lys 21 is near Pα, lys 27 is near Pβ and Pγ, segment 3 is a hydrophobic shield, and the glycine-rich loop (a.a. 15 to 21) may control access to the site[858-862] (see Figure 12).

Chapter 9

SOME COMPLEX KINETIC MECHANISMS AND TREATMENT OF ENZYME KINETIC DATA

I. SOME COMPLEX KINETIC MECHANISMS

A. SIMPLIFICATION PROCEDURES APPLICABLE TO COMPLEX KING-ALTMAN PATTERNS

As the number of enzyme species increases, for example, in complex kinetic mechanisms, then the number of King-Altman interconversion patterns increases accordingly. Consequently, attempts have been made to reduce the tedium of the method in complex situations, and several variations of the basic King-Altman procedure have been suggested which lead to some simplification in the use of the method. We will discuss those suggested by Volkenstein and Goldstein.[157] (See also References 156, 158, 159, and 758; also, several computer programs have been written for the derivation of kinetic expressions for enzyme systems.[160-163;812])

Recently, graphic theory has been applied to the derivation of complex rate expressions for product formation in steady-state enzyme-catalyzed systems, and for the transient concentrations of the several enzyme-containing species in non-steady-state systems.[1007]

1. Addition of Multiple Lines Connecting Two Corners

The easiest modification of the King-Altman method is restricted to cases having multiple lines connecting two points in one or both directions. The lines for a given direction may simply be added.

a. Steady-State Equation for Monod, Wyman, and Changeux Model

Thus, if we consider the model of Monod, Wyman, and Changeux[116] (Chapter 7, Scheme VI) for a unireactant case, but operating entirely under steady-state conditions, including the interconversion of the enzyme conformers. Thus, consider the case of a single-substrate single-site enzyme existing in two forms, R and T.

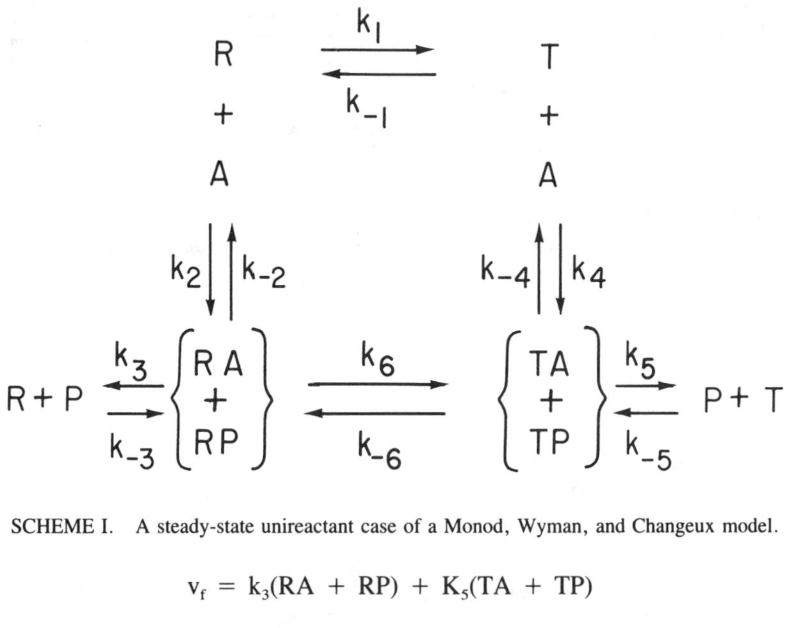

SCHEME I. A steady-state unireactant case of a Monod, Wyman, and Changeux model.

$$v_f = k_3(RA + RP) + K_5(TA + TP) \qquad (1)$$

(Note that the central complexes containing R and T enzyme species have been combined.)

The basic King-Altman figure may be shown as Scheme II.

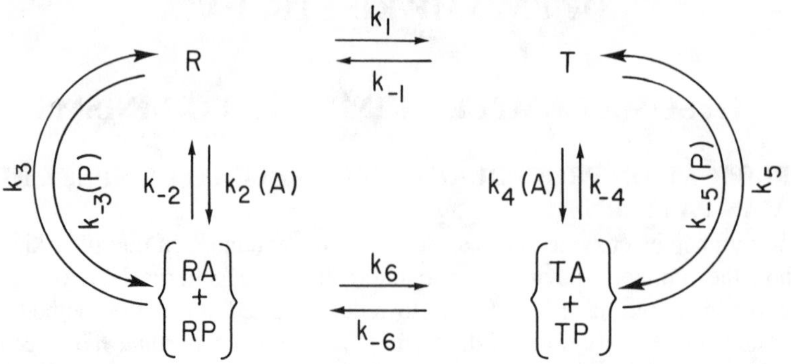

SCHEME II. King-Altman figure for Scheme I.

The basic figure contains four corners (n = 4) and six lines (m = 6). Equation 47 in Chapter 2 predicts 20 patterns with three lines.

$$_{m}C_{n-1} = \frac{(m)!}{(n - 1)!\,(m - n + 1)!} = \frac{(6)!}{(4 - 1)!\,(6 - 4 + 1)!} = 20 \qquad (2)$$

However, the 20 patterns include all the *three-lined* patterns containing the two different *two-line loops* shown below.

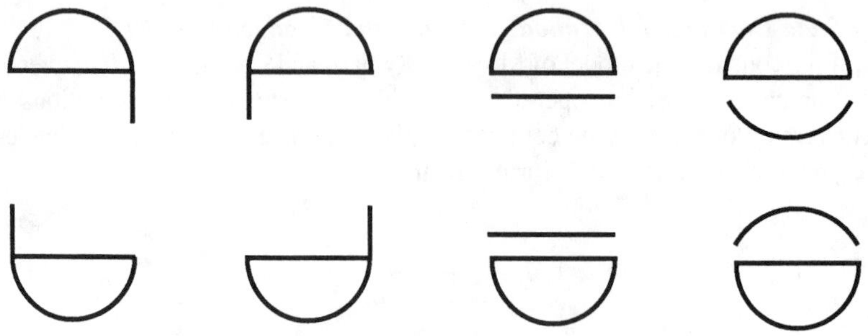

SCHEME III.

(Note the view of Scheme II after rotating it by 90°.

Thus, by Equation 48 in Chapter 2 or by inspection, we would eliminate four three-lined patterns by each kind of loop (i.e., we would reject eight patterns). The twelve remaining and valid interconversion patterns are as follows.

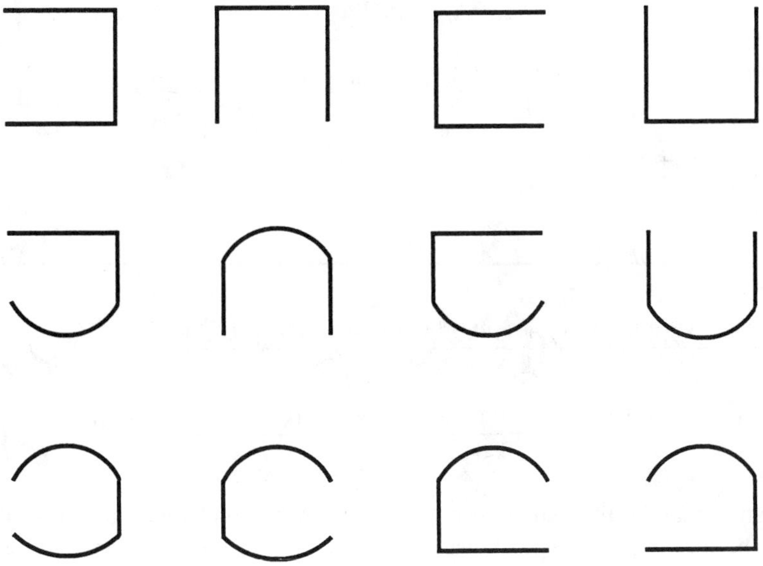

SCHEME IV.

(Again, rotate Scheme II by 90°.)

However, instead of having to deal with twelve patterns, the basic figure can be simplified by adding the lines connecting R with RA and T with TA. In general, therefore, when multiple lines connect two enzyme species (corners) in one or both directions, the lines can be added. Thus, the basic figure is then simplified to Scheme V.

$$R \quad \xrightleftharpoons[k_{-1}]{k_1} \quad T$$

$$\left(k_{-2}+k_3\right) \Big\uparrow\Big\downarrow \left(k_2(A)+k_{-3}(P)\right) \quad \left(k_4(A)+k_{-5}(P)\right) \Big\uparrow\Big\downarrow \left(k_{-4}+k_5\right)$$

$$\left\{\begin{matrix}RA\\+\\RP\end{matrix}\right\} \quad \xrightleftharpoons[k_{-6}]{k_6} \quad \left\{\begin{matrix}TA\\+\\TP\end{matrix}\right\}$$

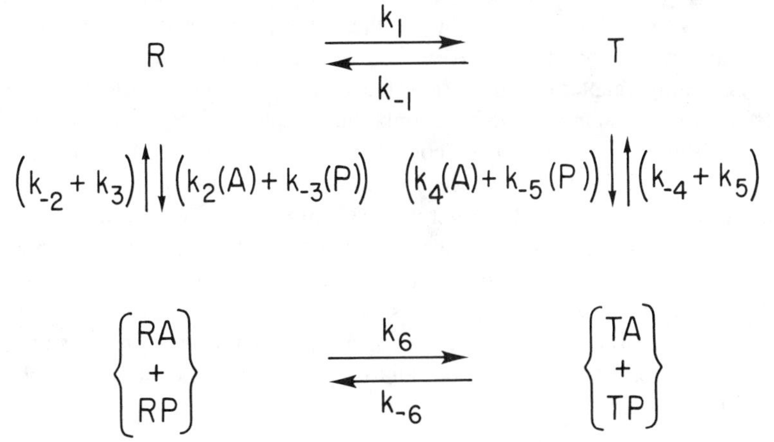

SCHEME V.

The resulting equation, Equation 5, in the absence of P, i.e., setting $P = 0$, was given by Plowman[164] where

$$v_f = k_3(RA + RP) + k_5(TA + TP) \tag{3}$$

$$v_f = k_3(RA) + k_5(TA) \tag{4}$$

$$\frac{v_f}{E_t} = \frac{\left[\frac{k_{-1}k_3[k_{-4} + k_5 + (k_5/k_3)k_6 + k_{-6}]}{k_4(k_6 + k_{-6})} + \frac{k_1k_5[k_{-2} + k_{-3} + k_6 + (k_3/k_5)k_{-6}]}{k_2(k_6 + k_{-6})}\right](A) + \left[\frac{k_5k_6 + k_3k_{-6}}{k_6 + k_{-6}}\right](A)^2}{\left[\frac{(k_1 + k_{-1})[(k_{-2} + k_3 + k_6)(k_{-4} + k_5) + (k_{-2} + k_3)k_{-6}]}{k_2k_4(k_6 + k_{-6})}\right] + \left[\frac{(k_{-1}k_2 + k_1k_4)}{k_2k_4} + \frac{(k_1 + k_{-6})(k_{-2} + k_3)}{k_2(k_6 + k_{-6})} + \frac{(k_{-1} + k_6)(k_{-4} + k_5)}{k_4(k_6 + k_{-6})}\right](A) + (A)^2} \tag{5}$$

This equation may be reduced further to Equation 6 by combining rate constants, and it proves to be *identical in form* to Equation 65 in Chapter 2) derived earlier for a Random Bi-Uni steady-state system, that is,

$$\frac{v_f}{E_t} = \frac{i'[A]^2 + j'[A]}{k' + l'[A]^2 + m'[A]} \tag{6}$$

except for the definitions of the constants; also, it is of the same form as Equation 3 in Chapter 7 and therefore may result in sigmoidal kinetics.

It is of the form of a 2/1 function in the Cleland description with, however, $n = 1$ (i.e., one enzyme with one active site on one protomer, but with the capability of existing in two different active states). It is also evident that the Equation 6 would be of the same form, if only one of the central complexes, RA or TA, were catalytically active.

In the preceding chapter on *allosteric* effects and here, we have discussed only homotropic effects by unireactant systems, i.e., only the complexities resulting from the substrate interacting with the enzyme have been considered. Heterotropic effects (i.e., those effects involving nonidentical ligands) would lead to another dimension in our discussion and will be postponed to Volume II (see also Reference 983) of this treatise (on enzyme mechanisms) where they may be discussed in context with their mechanistic implications. It may be noted here that one class of the heterotropic effectors would be positive or negative modifiers which, although they do *not* participate as substrates or products in the reaction, they act by shifting the equilibrium between the two or more enzyme forms in ways to change the apparent rate equation. Another class of heterotropic effectors might act also as substrates or products in the reaction.

2. Reduction of Matrices to a Point

Another simplification of certain complex patterns in the King-Altman method involves the reduction of portions of a complex pattern to a point represented by the appropriate matrix solution. To illustrate the technique we will consider Scheme 6.

a. A Case of an Alternative Substrate, I, which can Substitute for Substrate A in a Bisubstrate Steady-State Mechanism such as a "Ping-Pong BiBi" Mechanism

Fromm[165-167,758,812] has shown that the use of an alternative substrate can provide information about the prevailing kinetic pathway. In this technique, a two-substrate enzyme is provided with two different first (or second) substrates in addition to the second substrate, so that we have two simultaneous reactions occurring; either one or both alternative products may be measured.

Consider the following reaction Scheme VI where I is an alternative substrate of enzyme E which can substitute for substrate A in the mechanism; P, the first product released, is common to both A and I, however the second product is different. The Cleland figure is

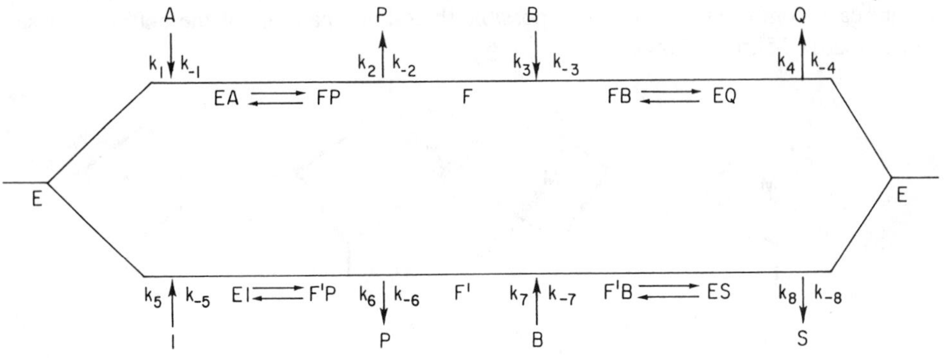

SCHEME VI.

There are seven different enzyme species when both I and A substrates (i.e., the first substrate) are present. The basic King-Altman figure is accordingly Scheme VII with n = 7 and m = 8; note the upper four-sided figure is denoted as M and its lower four-sided figure as N.

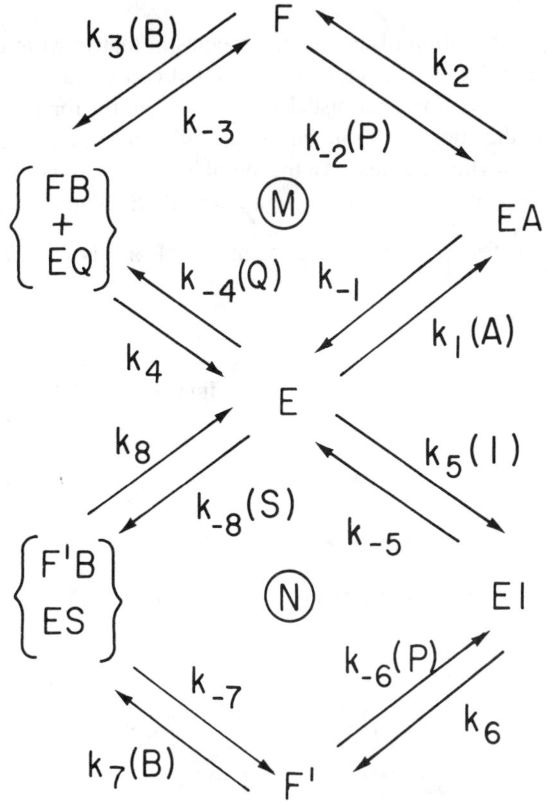

SCHEME VII. King-Altman figure for Scheme VI.

By Equation 47 in Chapter 2 there are 16 possible 6-lined valid interconversion patterns. A simplification is possible by reduction of portions of the pattern to a point represented by the appropriate matrix solution. Thus, the basic figure (Scheme VII) could be considered as two separate matrices (M and N) joined through one common point. In this way only 2 sets of 4 simple patterns would be required instead of the 16 complex patterns from the usual procedure. The four possible three-lined patterns for the top four-sided figure (matrix M) are each combined with the four possible three-lined patterns of the bottom four-sided figure (matrix M), as follows:

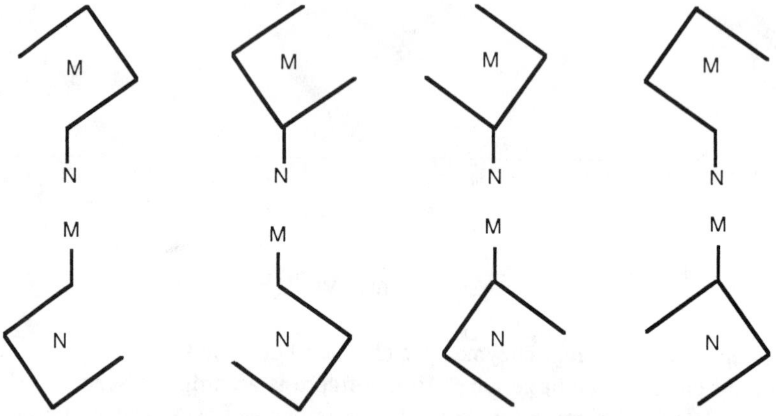

SCHEME VIII.

The fractional concentration of an enzyme species in matrix M is given by N times the usual M-matrix solution for that species; the fractional concentration of an enzyme species in matrix N is given by M times the usual N-matrix solution for that species. As before, the matrix solution is the sum of all rate constants and concentration factors read along the lines leading to the enzyme species. At the point common to both matrices, where the enzyme species (E) lies, the fractional concentration of (E) is given by M × N, where M = the matrix solution for $\frac{(E)}{E_t}$ determined in matrix M and N = the matrix solution for $\frac{(E)}{E_t}$ determined in matrix N.

Thus, for $\frac{(E)}{E_t}$, the four three-lined patterns ending on E in "M" are

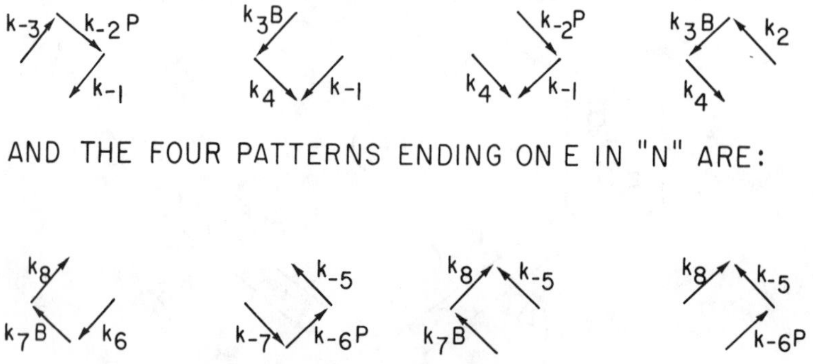

SCHEME IX.

This leads to Equation 7.

$$\frac{(E)}{E_t} = M \times N = \frac{\begin{gathered}(k_{-3}k_{-2}(P)k_{-1} + k_3(B)k_4k_{-1} + k_4k_{-2}(P)k_{-1} + k_2k_3(B)k_4)\\ \times\ (k_6k_7(B)k_8 + k_{-7}k_{-6}(P)k_{-5} + k_{-5}k_7(B)k_8 + k_8k_{-6}(P)k_{-5})\end{gathered}}{\text{denominator}} \tag{7}$$

Similarly, the distribution equations for the remaining six enzyme species are

$$\frac{(EA)}{E_t} = \frac{N[k_{-3}k_{-2}(P)k_1(A) + k_3(B)k_4k_1(A) + k_{-2}(P)k_4k_1(A) + k_{-4}(Q)k_{-3}k_{-2}(P)]}{\text{denominator}}$$

$$\frac{(F)}{E_t} = \frac{N[k_{-3}k_1(A)k_2 + k_{-1}k_{-4}(Q)k_{-3} + k_4k_1(A)k_2 + k_2k_{-4}(Q)k_{-3}]}{\text{denominator}}$$

$$\frac{(FB + EQ)}{E_t} = \frac{N[k_1(A)k_2k_3(B) + k_3(B)k_{-1}k_{-4}(Q) + k_{-2}(P)k_{-1}k_{-4}(Q) + k_{-4}(Q)k_2k_3(B)]}{\text{denominator}}$$

$$\frac{(EI)}{E_t} = \frac{M[k_{-8}(S)k_{-7}k_{-6}(P) + k_5(I)k_{-7}k_{-6}(P) + k_7(B)k_8k_5(I) + k_{-6}(P)k_8k_5(I)]}{\text{denominator}}$$

$$\frac{(F')}{E_t} = \frac{M[k_6k_{-8}(S)k_{-7} + k_{-7}k_5(I)k_6 + k_{-5}k_{-8}(S)k_{-7} + k_8k_5(I)k_6]}{\text{denominator}}$$

$$\frac{(F'B + ES)}{E_t} = \frac{M[k_{-8}(S)k_6k_7(B) + k_5(I)k_6k_7(B) + k_7(B)k_{-5}k_{-8}(S) + k_{-6}(P)k_{-5}k_{-8}(S)]}{\text{denominator}} \tag{8}$$

If we set

$$v = \frac{d(Q)}{dt} = k_4(FB + EQ) \tag{9}$$

i.e., if the velocity of the reaction is taken as the rate of Q production (i.e., the rate of appearance of the unique product of A) and in the absence of products, then

$$\frac{v}{E_t} = \frac{k_4(FB + EQ)}{(E) + (EA) + (F) + (FB + EQ) + (EI) + (F') + (F'B + ES)} \tag{10}$$

Substituting the King-Altman solution and in Cleland's coefficient form,

$$v_f = \frac{\text{Num}(A)(B)^2}{\text{Coef}_{AB}(A)(B) + \text{Coef}_{BB}(B)^2 + \text{Coef}_{ABB}(A)(B)^2 + \text{Coef}_{IB}(I)(B) + \text{Coef}_{IBB}(I)(B)^2} \tag{11}$$

and, after (B) is canceled in numerator and denominator, and redefining the coefficients,

$$v_f = \frac{\text{Num}(A)(B)}{\text{Coef}_A(A) + \text{Coef}_B(B) + \text{Coef}_{AB}(A)(B) + \text{Coef}_I(I) + \text{Coef}_{IB}(I)(B)} \tag{12}$$

Finally, dividing both numerator and denominator by Coef_{AB} and defining terms in the usual manner with

$$K_{m_I} = \frac{\text{Coef}_B}{\text{Coef}_{IB}} \tag{13}$$

which is the K_m for I as a substrate and

$$K'_{m_B} = \frac{Coef_I}{Coef_{IB}} \qquad (13a)$$

which is the K_m for B when I is the substrate.

$$\frac{v^f}{V^f_{max}} = \frac{1}{\frac{K_{m_A}}{(A)}\left(1 + \frac{K'_{m_B}(I)}{K_{m_I}(B)} + \frac{(I)}{K_{m_I}}\right) + 1 + \frac{K_{m_B}}{(B)}} \qquad (14)$$

with (A) as the varied first substrate and

$$\frac{v_f}{V^f_{max}} = \frac{1}{\frac{K_{m_B}}{(B)}\left(1 + \frac{K'_{m_B}K_{m_A}(I)}{K_{m_B}K_{m_I}(A)}\right) + 1 + \frac{K_{m_A}}{(A)} + \frac{K_{m_A}(I)}{K_{m_I}(A)}} \qquad (15)$$

with (B) as the varied second substrate.

3. Steady-State Systems with Rapid Equilibrium Segments

A reasonable assumption to make, when one or more portions of a multistep reaction sequence is relatively faster than the overall reaction, is that the reactions within those portions or segments approach equilibrium as the overall reaction reaches a steady state. Cha[168] demonstrated that the derivation of the complete velocity equation by the King-Altman method can be simplified if one considers all the enzyme species within the rapid equilibrium segment as a single entity (i.e., as a single corner in a King-Altman figure). The distribution of species within the rapid equilibrium segment is computed by the usual rapid equilibrium calculations.

a. An Ordered BiBi System where the Binding and Dissociation of the First Substrate (A) and the Last Product (Q) are Exceedingly Fast Compared to the Other Steps

The King-Altman figure may be represented as Scheme X with the area enclosed by dashed lines representing the rapid equilibrium segment, which is designated by X, where (X) = (E) + (EA) + (EQ).

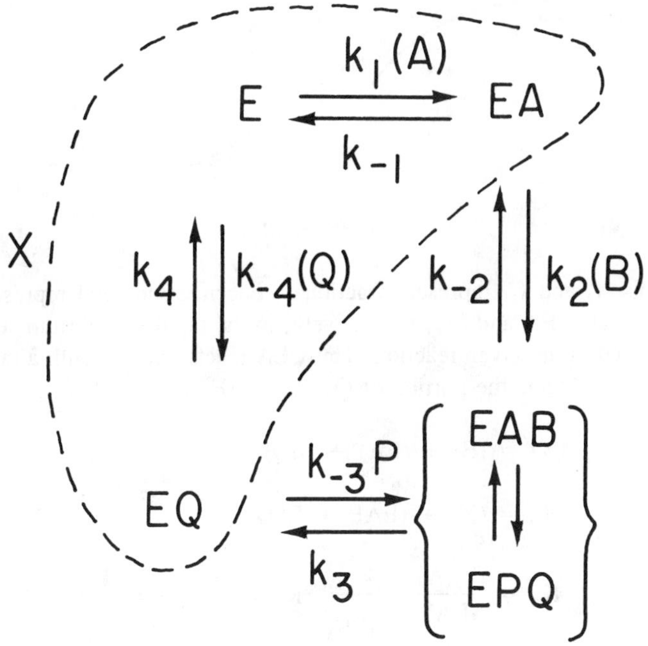

SCHEME X.

The basic figure in Scheme X may now be reduced to two lines connecting X and the central complexes, EAB + EPQ.

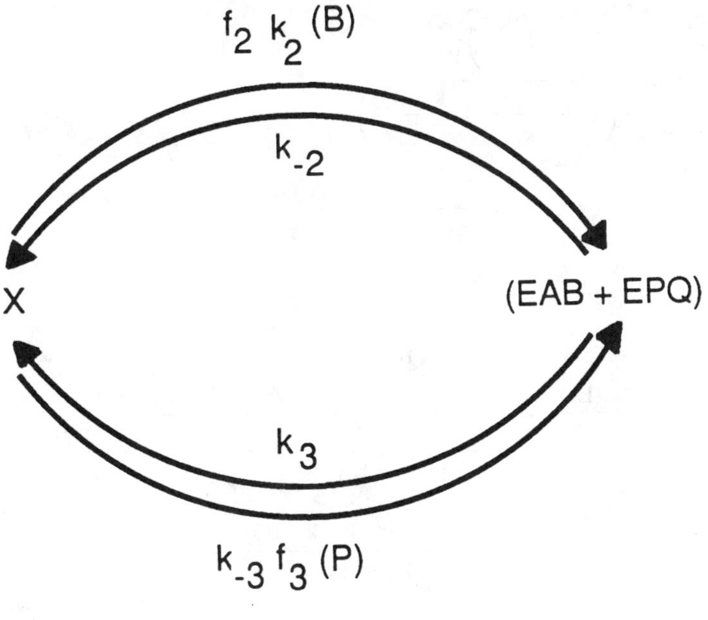

SCHEME XI.

Now, the interconversion patterns used in the King-Altman treatment are

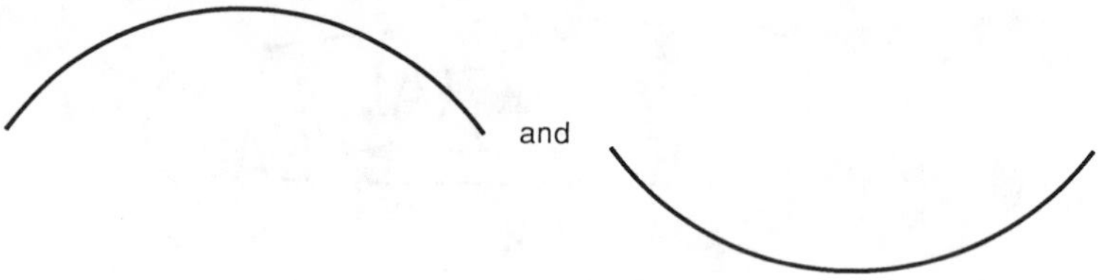

SCHEME XII.

The symbols, f_2 and f_3, represent fractional concentrations and represent the relative proportion of X that is EA and EQ, respectively, in the rapid equilibrium segment, X, that is actually involved in the given reaction. Thus, EA reacts with B (with a rate constant, k_2) to yield EAB: f_2 represents the portion of (X) that is (EA).

$$(X) = (E) + (EA) + (EQ) \tag{16}$$

$$E_t = (X) + (EAB + EPQ) \tag{17}$$

$$K_A = \frac{(E)(A)}{(EA)} = \frac{k_{-1}}{k_1}; \quad K_Q = \frac{(E)(Q)}{(EQ)} = \frac{k_4}{k_{-4}} \tag{18}$$

$$f_2 = \frac{(EA)}{(X)} = \frac{(E)\frac{(A)}{K_A}}{(E)\left(1 + \frac{(A)}{K_A} + \frac{(Q)}{K_Q}\right)} = \frac{\frac{(A)}{K_A}}{\left(1 + \frac{(A)}{K_A} + \frac{(Q)}{K_Q}\right)}$$

$$= \frac{\left(\frac{k_1}{k_{-1}}\right)\frac{(A)}{K_A}}{\left(1 + \frac{k_1(A)}{k_{-1}} + \frac{k_{-4}(Q)}{k_4}\right)} \tag{19}$$

$$\therefore \quad f_2 = \frac{k_1k_4(A)}{k_{-1}k_4 + k_1k_4(A) + k_{-1}k_{-4}(Q)} \tag{20}$$

Similarly,

$$f_3 = \frac{(EQ)}{(X)} = \frac{\frac{(Q)}{K_Q}}{1 + \frac{(A)}{K_A} + \frac{(Q)}{K_Q}} = \frac{\left(\frac{k_{-4}(Q)}{k_4}\right)}{1 + \frac{k_1(A)}{k_{-1}} + \frac{k_{-4}(Q)}{k_4}} \tag{21}$$

$$\therefore \quad f_3 = \frac{k_{-1}k_{-4}(Q)}{k_{-1}k_4 + k_1k_4 + k_{-1}k_{-4}(Q)} \tag{22}$$

The King-Altman treatment therefore yields

Ending on X:

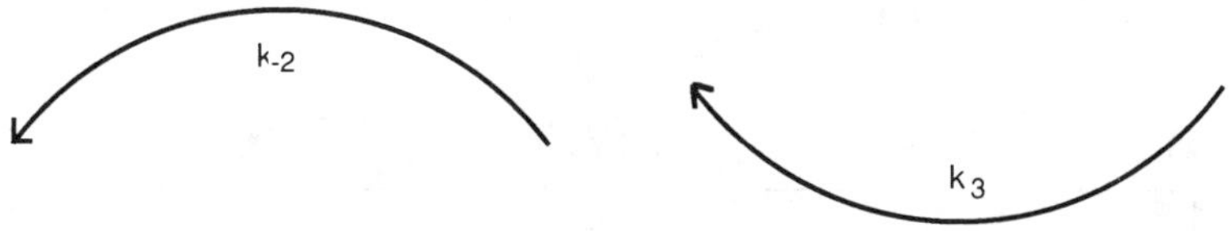

SCHEME XIII.

$$\frac{X}{E_t} = \frac{k_{-2} + k_3}{\text{denominator}} \tag{23}$$

Ending on (EAB + EPQ) :

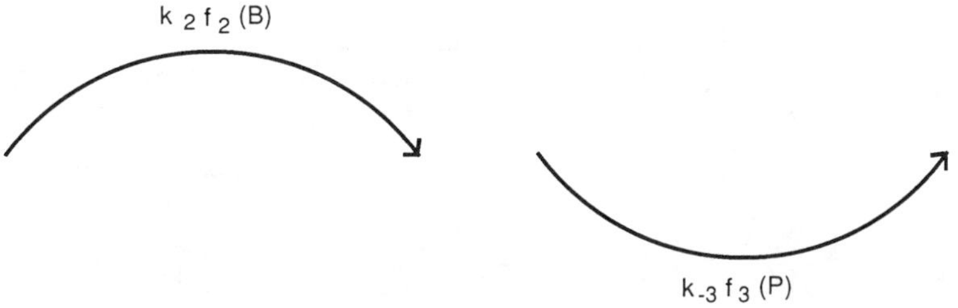

SCHEME XIV.

$$\frac{(EAB + EPQ)}{E_t} = \frac{k_2f_2(B) + k_{-3}f_3(P)}{\text{denominator}} \tag{24}$$

where the denominator is the sum of both numerators, and where k_2f_2 and $k_{-3}f_3$ represent effective rate constants. The velocity equation is obtained after substitution of the above relationships into the steady-state equations

$$v = k_2(EA)(B) - k_{-2}(EAB + EPQ) \tag{25}$$

and remembering that $EA = f_2(X)$;

$$\therefore \quad v = k_2f_2(X)(B) - k_{-2}(EAB + EPQ) \tag{26}$$

$$\frac{v}{E_t} = \frac{k_2k_3f_2(B) - k_{-2}k_{-3}f_3(P)}{k_{-2} + k_3 + k_2f_2(B) + k_{-3}f_3(P)} \tag{27}$$

After substitution of f_2 and f_3 (Equations 20 and 22), and simplifying with use of Cleland's coefficient approach,

$$v = \frac{V_fV_r\left\{(A)(B) - \frac{(P)(Q)}{K_{eq}}\right\}}{\left[V_rK_{ia}K_{m_B} + V_rK_{m_B}(A) + V_r(A)(B) + \frac{V_fK_{m_P}(Q)}{K_{eq}} + \frac{V_f(P)(Q)}{K_{eq}}\right]} \tag{28}$$

The velocity equation (Equation 28) has the same form as the total rapid equilibrium ordered BiBi system (see Equation 29) except that the parameters associated with B and P are K_ms instead of dissociation constants (e.g., see Chapter 2, Scheme XXVIII and Equations 162 and 163). Thus, the velocity equation for a total rapid equilibrium ordered bisubstrate system is Equation 29.

$$v = \frac{V^f_{max}V^r_{max}\left\{(A)(B) - \frac{(P)(Q)}{K_{eq}}\right\}}{\left[V^r_{max}\bar{K}_AK_B + V^r_{max}K_B(A) + V^r_{max}(A)(B) + \frac{V^f_{max}K_P(Q)}{K_{eq}} + \frac{V^f_{max}(P)(Q)}{K_{eq}}\right]} \tag{29}$$

B. HYBRID AND MIXED MECHANISMS

1. Hybrid "Ping-Pong — Random" BiBi System

Segel[169] has considered the following mechanism (Scheme XV) in which substrates A and B add randomly to form the ternary complex, EAB; however, the binary complex, EA, can still undergo a partial ping-pong type of reaction.

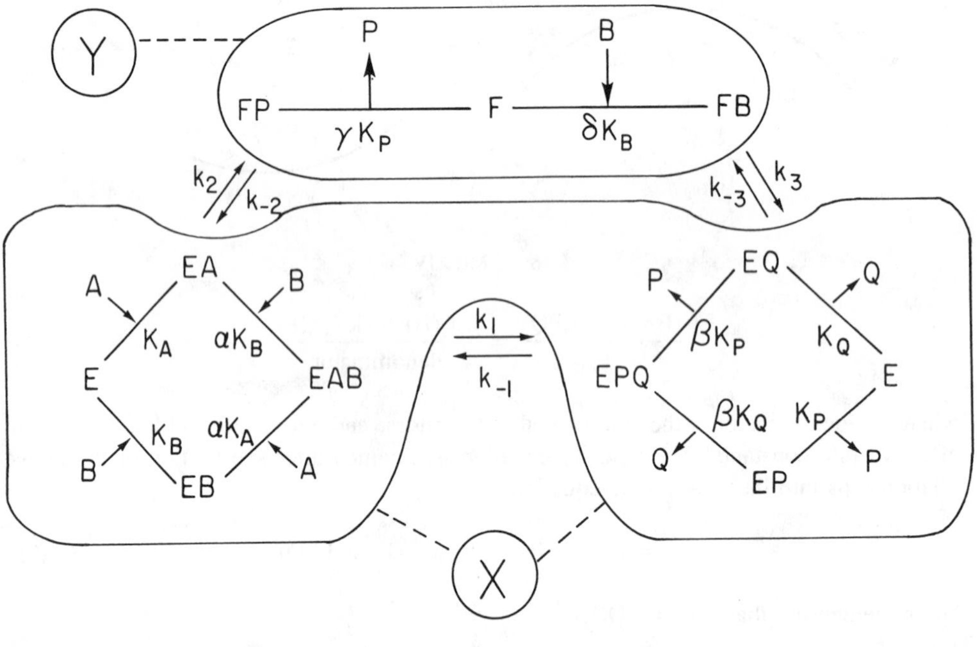

SCHEME XV. A hybrid ping-pong random Bi Bi system.

If the substrates were to add in a true random fashion, the velocity equation would contain $(A)^2$ and $(B)^3$ terms since there are two A-adding and three B-adding steps. However, by making the following assumptions, the equation will be considerably simpler: (1) assume that the substrates (ligands) add in a rapid equilibrium fashion and (2) that there are rate-limiting steps at the catalytic steps

$$EA \rightleftarrows FP, \quad FB \rightleftarrows EQ, \quad \text{and} \quad EAB \rightleftarrows EPQ$$

Let X represent the random BiBi segments and Y represent the ordered Uni Uni segment of the ping-pong sequence, then the basic King-Altman figure may be represented in Scheme XVI.

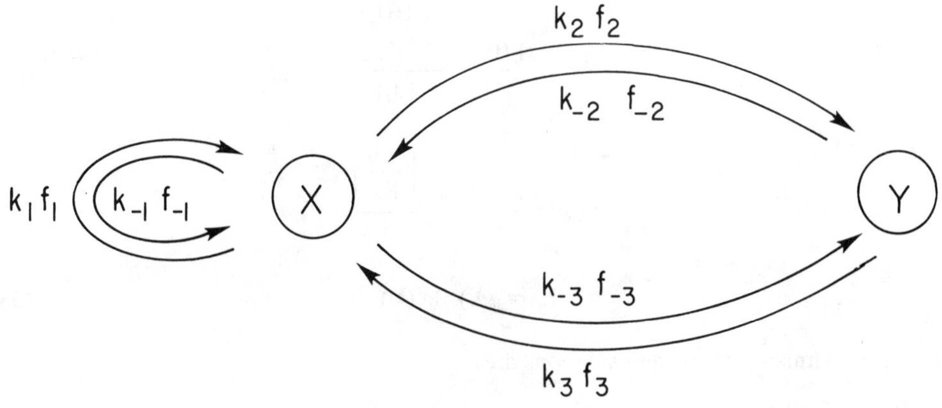

SCHEME XVI. King-Altman figure for Scheme XV.

The total net velocity is given by Equation 30,

$$v = [k_1(EAB) - k_{-1}(EPQ)] + [k_2(EA) - k_{-2}(FP)] \tag{30}$$

where the terms in the second set of brackets represent that portion of the net velocity through the ping-pong segment.

Thus, the equation can be written as Equation 31,

$$v = k_1f_1(X) - k_{-1}f_{-1}(X) + k_2f_2(X) - k_{-2}f_{-2}(Y) \tag{31}$$

where,

$$f_1 = \frac{(EAB)}{(X)} = \frac{\left[\dfrac{(A)(B)}{\alpha K_A K_B}\right]}{\left[1 + \dfrac{(A)}{K_A} + \dfrac{(B)}{K_B} + \dfrac{(A)(B)}{\alpha K_A K_B} + \dfrac{(P)}{K_P} + \dfrac{(Q)}{K_Q} + \dfrac{(P)(Q)}{\beta K_P K_Q}\right]} = \frac{\left[\dfrac{(A)(B)}{\alpha K_A K_B}\right]}{(d_1)} \tag{32}$$

$$f_{-1} = \frac{(EPQ)}{(X)} = \frac{\left[\dfrac{(P)(Q)}{\beta K_P K_Q}\right]}{(d_1)} \tag{33}$$

$$f_2 = \frac{(EA)}{X} = \frac{\left[\dfrac{(A)}{K_A}\right]}{(d_1)} \tag{34}$$

$$f_{-2} = \frac{(FP)}{Y} = \frac{\left[\dfrac{(P)}{\gamma K_P}\right]}{\left[1 + \left[\dfrac{(B)}{\delta K_B}\right] + \left[\dfrac{(P)}{\gamma K_P}\right]\right]} = \frac{\left[\dfrac{(P)}{\gamma K_P}\right]}{(d_2)} \tag{35}$$

$$f_3 = \frac{(FB)}{Y} = \frac{\left[\frac{(B)}{\delta K_B}\right]}{(d_1)} \tag{36}$$

$$f_{-3} = \frac{(EQ)}{X} = \frac{\left[\frac{(Q)}{K_Q}\right]}{(d_1)} \tag{37}$$

$$E_t = (X) + (Y) \tag{38}$$

The King-Altman patterns and solutions are

$$\frac{(X)}{E_t} = \frac{[k_{-2}f_{-2} + k_3f_3]}{[k_{-2}f_{-2} + k_3f_3 + k_2f_2 + k_{-3}f_{-3}} \tag{39}$$

$$\frac{(Y)}{E_t} = \frac{[k_2f_2 + k_{-3}f_{-3}]}{[k_{-2}f_{-2} + k_3f_3 + k_2f_2 + k_{-3}f_{-3}]} \tag{40}$$

(Note that the internal loop in segment X of Scheme XVI, i.e., the $k_1f_1/k_{-1}f_{-1}$ loop is not included in the King-Altman solutions).

From Equations 31, 39, and 40 (with solutions for f_1, f_{-1}, f_2, and f_{-2} given in Equations 32 to 35), the final velocity equation for the overall forward reaction in the absence of products can be derived as in Equation 41,

$$v_f = \frac{\left(\frac{(B)}{\delta K_B}\right)\left[\frac{V^f_{max_1}(A)(B)}{\alpha K_A K_B} + \frac{V^f_{max_2}(A)}{K_A}\right]}{\left(\frac{(B)}{\delta K_B}\right)\left[1 + \frac{(A)}{K_A} + \frac{(B)}{K_B} + \frac{(A)(B)}{\alpha K_A K_B}\right] + \left(\frac{k_2(A)}{k_3 K_A}\right) \cdot \left[1 + \left(\frac{(B)}{\delta K_B}\right)\right]} \tag{41}$$

after defining

$$V^f_{max_1} = k_1 E_t \tag{42}$$

equals maximal forward velocity through EAB and

$$V^f_{max_2} = k_3 E_t \tag{42a}$$

equals maximal forward velocity through FB or

$$v_f = \frac{\left[\frac{V^f_{max_1}(A)(B)}{\alpha K_A K_B} + \frac{V^f_{max_2}(A)}{K_A}\right]}{\left[1 + \frac{(A)}{K_A} + \frac{(B)}{K_B} + \frac{(A)(B)}{\alpha K_A K_B}\right] + \frac{[k_2(A)/k_3 k_A]}{(B)/\delta K_B}\left[1 + \frac{(B)}{\delta K_B}\right]} \tag{43}$$

$$v_f = \frac{\left[\frac{V^f_{max_1}(A)(B)}{\alpha K_A K_B} + \frac{V^f_{max_2}(A)}{K_A}\right]}{\left[1 + \frac{(A)}{K_A} + \frac{(B)}{K_B} + \frac{(A)(B)}{\alpha K_A K_B}\right] + \frac{\delta k_2 K_B(A)}{k_3 K_A(B)}\left[1 + \frac{(B)}{\delta K_B}\right]} \tag{44}$$

If (A) were the variable substrate, normal hyperbolic plots and linear double-reciprocal plots would result under assumed rapid equilibrium conditions. If the flux through F were insignificant, the velocity equation Equation 44 would reduce to a rapid equilibrium random system. Conversely, if the flux through EAB were insignificant, then the velocity equation would reduce to a rapid equilibrium ping-pong BiBi system.

2. An Ordered Three-Substrate System with a Rapid Equilibrium Random Sequence in Ligands A and B and in R and Q

Consider Scheme XVII below.

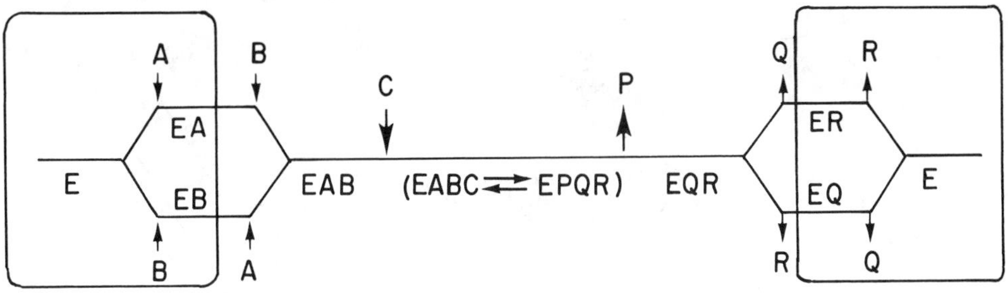

SCHEME XVII.

Assume that the dissociations of EA and EB are extremely rapid compared to the rates of the forward steps; therefore, the first ligands to add in either direction (A and B) are in a state of quasi-equilibrium with their respective complexes and the free enzyme, E. Assume, also, that R and Q, the last two products to leave are in a state of rapid equilibrium.

In order to simplify the derivation, consider only the forward reaction.

In Scheme XVIII, the basic King-Altman figure is given, with E, EA, and EB represented by a single corner, X.

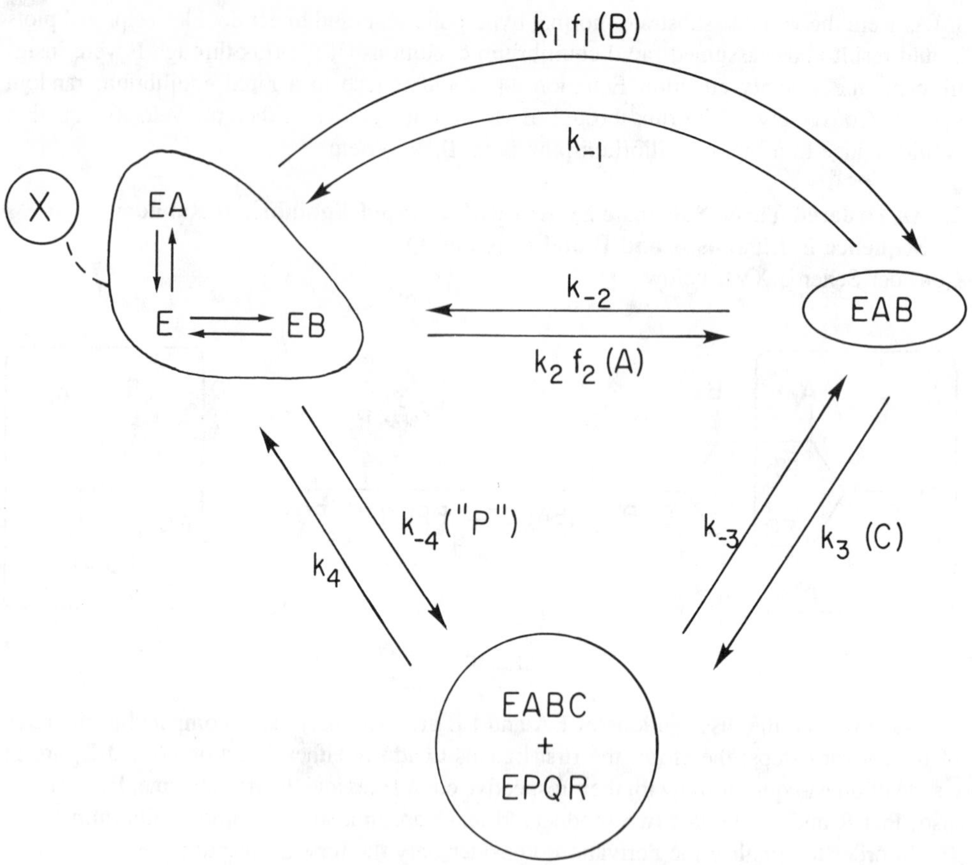

SCHEME XVIII.

In Scheme XVIII, EPQR dissociates apparently to E in a single step, k_4; however, the k_4 step is actually a composite of several product-release steps leading eventually from EPQR to E. Consequently, the arrow in the reverse direction labeled k_{-4}(''P'') is a composite of several product-addition steps leading back from E to EPQR. Thus, since the product-release and -addition steps have been omitted, the basic figure could describe any ordered ter reactant system (ter ter, ter Bi, etc.).

If the two lines connecting EAB with X are now combined, then Scheme XIX results.

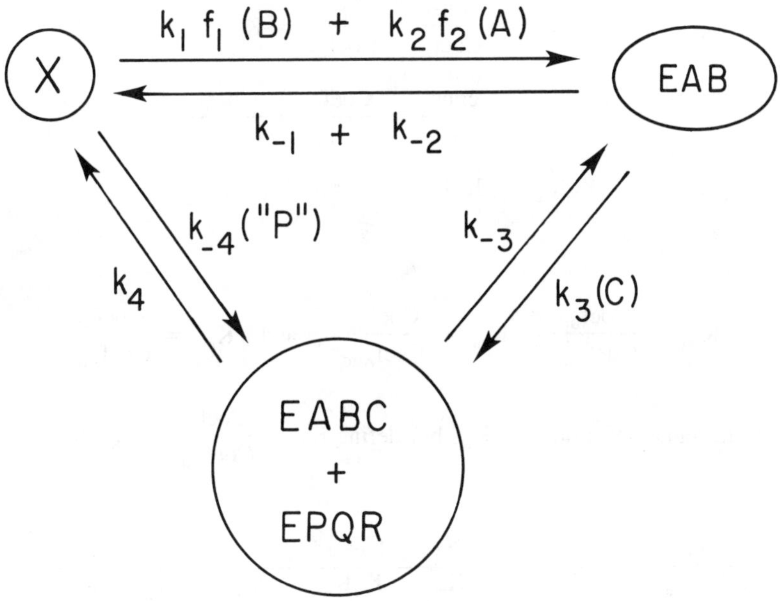

SCHEME XIX.

where

$$f_1 = \frac{(EA)}{(X)} = \frac{\left(\frac{(A)}{K_A}\right)}{\left[1 + \frac{(A)}{K_A} + \frac{(B)}{K_B}\right]} \tag{45}$$

$$f_2 = \frac{EB}{(X)} = \frac{\left(\frac{(B)}{K_B}\right)}{\left[1 + \frac{(A)}{K_A} + \frac{(B)}{K_B}\right]} \tag{46}$$

and

$$v^f = k_3(C)(EAB) - k_{-3}(EABC + EPQR) \tag{47}$$

After substitution and rearranging,

$$v^f = \frac{(A)(B)(C)V^f_{max}}{K_{ia}K_{m_B}K_{ic} + K_{m_B}K_{ic}(A) + K_{m_A}K_{ic}(B) + K_{ia}K_{m_B}(C) + K_{m_B}(A)(C) + K_{m_A}(B)(C) + K_{m_C}(A)(B) + (A)(B)(C)} \tag{48}$$

where

$$K_{ia} = \frac{const}{Coef_A} = \frac{Coef_C}{Coef_{AC}} = K_A$$

$$K_{ib} = \frac{\text{const}}{\text{Coef}_B} = \frac{\text{Coef}_C}{\text{Coef}_{BC}} = K_B$$

$$K_{ic} = \frac{\text{const}}{\text{Coef}_C} = \frac{\text{Coef}_B}{\text{Coef}_{BC}} = \frac{\text{Coef}_A}{\text{Coef}_{AC}}$$

$$K_{ia}K_{m_B} = K_{ib}K_{m_A} \tag{49}$$

and

$$K_{m_A} = \frac{\text{Coef}_{BC}}{\text{Coef}_{ABC}}, \quad K_{m_B} = \frac{\text{Coef}_{AC}}{\text{Coef}_{ABC}}, \quad \text{and} \quad K_{m_C} = \frac{\text{Coef}_{AB}}{\text{Coef}_{ABC}} \tag{50}$$

If we retain the same definition for K_{ia}, but define $K_{ib} = \dfrac{\text{Coef}_A}{\text{Coef}_{AB}} \neq K_B$, then Equation 48 becomes

$$\frac{v_o^f}{V_{max}^f} = \frac{1}{\left[\frac{K_{ia}K_{ib}K_{mC}}{(A)(B)(C)} + \frac{K_{ib}K_{mC}}{(B)(C)} + \frac{K_{ia}K_{mC}}{(A)(C)} + \frac{K_{ia}K_{mB}}{(A)(B)} + \frac{K_{mB}}{(B)} + \frac{K_{mA}}{(A)} + \frac{K_{mC}}{(C)} + 1\right]} \tag{51}$$

A comparison of the above equation with Equation 200 in Chapter 2, a steady-state ordered three-substrate reaction, reveals an additional term in the denominator of Equation 51, that is, $\dfrac{K_{ia}\ K_{mC}}{(A)\quad (C)}$.

Under certain restrictive conditions and ranges of substrate concentrations, the lower pathway of Scheme XVII will appear to drop out, leaving a simple ordered sequence of reactions, with A adding first, followed by B, and then C, to yield the products of reaction.

Thus, consider A as the variable substrate, B as a constant fixed substrate, and C as the varying, but fixed, substrate.

Rearranging Equation 51 to fit these conditions,

$$\frac{V_{max}^f}{v_o^f} = \left[1 + \frac{K_{mB}}{(B)} + \frac{K_{mC}}{(C)}\left(1 + \frac{K_{ib}}{(B)}\right)\right] + \frac{K_{mA}}{(A)}\left[1 + \frac{K_{ia}K_{mB}}{K_{mA}(B)} + \frac{K_{ia}K_{ib}K_{mC}}{K_{mA}(B)(C)} + \frac{K_{ia}K_{mC}}{K_{mA}(C)}\right] \tag{52}$$

or

$$\frac{V_{max}^f}{v_o^f} = \left[1 + \frac{K_{mB}}{(B)} + \frac{K_{mC}}{(C)}\left(1 + \frac{K_{ib}}{(B)}\right)\right] + \frac{K_{mA}}{(A)}\left[1 + \frac{K_{ia}K_{mB}}{K_{mA}(B)}\left(1 + \frac{K_{ib}K_{mC}}{K_{mB}(C)}\right) + \frac{K_{ia}K_{mC}}{K_{mA}}\right] \tag{53}$$

If $(B) \gg K_{ib}$, $(B) \gg K_{mB}$, and $(B) \gg (K_{ia}\ K_{mB}/K_{mA}$, as in Chapter 2, Section I.E for NADPH-Cytochrome *c*(Cyt *c*) reductase, and (B) = FAD with a $K_{ib} \cong 10^{-8}$,[828] then FAD could be fixed readily at approximately 10^{-5} to 10^{-6} *M*.[829]

Thus, Equation 53 would reduce to

$$\frac{V^f_{max}}{v^f_o} \rightarrow \left[1 + \frac{K_{m_C}}{(C)}\right] + \frac{K_{m_A}}{(A)}\left[1 + \frac{K_{ia}K_{m_C}}{K_{m_A}(C)}\right] \quad (54)$$

but a plot of $\frac{1}{v_0}$ vs. $\frac{1}{(A)}$ (e.g., where A might be TPNH) at varied but fixed concentrations of C (e.g., Cyt *c*) and at 10^{-5} *M* FAD (i.e., B) would yield a family of converging straight lines. However, if the fixed value of (C) were held sufficiently high so that (C) = $\frac{K_{ia}K_{m_C}}{K_{m_A}}$, then Equation 54 would reduce further to

$$\frac{V^f_{max}}{v^f_o} \rightarrow \left[1 + \frac{K_{m_C}}{(C)}\right] + \frac{K_{m_A}}{(A)} \quad (55)$$

where now, plots of $\frac{1}{v}$ vs. $\frac{1}{(A)}$ would yield parallel plots and appear to be ping-pong,[829] identical in form to Equation 185, Chapter 2 with C replacing B (however, notice that the chemistry of the two mechanisms differ). It is conceivable that if a wide-enough concentration range of (C) would be explored, the family of plots would appear to go from a converging system to a parallel system (see Reference 844).

Similarly, if (C) were the variable substrate, under similar conditions of (B) and K_{ib}, then

$$\frac{V^f_{max}}{v^f_o} \rightarrow \left[1 + \frac{K_{m_A}}{(A)}\right] + \frac{K_{m_C}}{(C)}\left(1 + \frac{K_{ia}}{(A)}\right) \quad (56)$$

Now, if (A) becomes larger than K_{ia}, to reduce $\frac{K_{ia}}{(A)}$ to a small value compared to 1.0, then the system would appear to be parallel in nature; whereas, at small values in (A), a converging family of curves would result. These conditions appear to be met by NADH (NADPH) Cyt *c* reductase, an FMN-containing enzyme.[682,871,872] Also, see Reference 988 where a review of this enzyme is given in Volume II of this work.

II. TREATMENT OF ENZYME KINETIC DATA

Some references on statistics which may prove useful for this section are References 170 to 179.

A. GRAPHICAL ANALYSES FOR OBTAINING PROVISIONAL ESTIMATES OF THE KINETIC PARAMETERS

Wilkinson[32] discussed from a statistical point of view the three most commonly employed graphical methods for estimation of the kinetic parameters, K_m and V_{max}, based on various linear forms of the Michaelis-Menten relation,

$$v_0 = \frac{V_{max}}{\frac{K_m}{(S_0)} + 1} \quad (57)$$

The first, and most commonly used, the double-reciprocal form of Lineweaver and Burk[31] is based on Equation 14 from Chapter 1,

$$\frac{1}{v_0} = \frac{K_m}{V_{max}(S_0)} + \frac{1}{V_{max}} \tag{1-14}$$

(see Chapter 1, Figure 3 for analyses of this graphical plot), and yields from a plot of

$$\frac{1}{v_0} \text{ vs. } \frac{1}{(S_0)}, \quad \text{Slope} = \frac{K_m}{V_{max}},$$

$$\text{Y-intercept} = \frac{1}{V_{max}}, \quad \text{and} \quad \text{X-intercept} = \frac{1}{-K_m} \tag{58}$$

The second, suggested by Hanes[180] and employed by Hultin[39], is obtained by multiplying Equation 14 from Chapter 1 through by (S_0).

$$\frac{(S_0)}{v_0} = \frac{K_m}{V_{max}} + \frac{(S_0)}{V_{max}} \tag{59}$$

Here, a plot of $\frac{(S_0)}{v_0}$ vs. (S_0) is analogous to a Scatchard[144] plot of equilibrium substrate binding data (Chapter 8, Equation 30) fitted to $\frac{\bar{r}_A}{(A)} = nK_A - K_A\bar{r}_A$. Thus, an $\frac{(S_0)}{v_0}$ vs. (S_0) plot yields: Slope $= \frac{1}{V_{max}}$, Y-intercept $= \frac{K_m}{V_{max}}$, and x-intercept $= -K_m$. Scatchard[144] and Hultin[39] were of the opinion that this type of plot was superior to the double-reciprocal type of plot since it permitted a more even distribution of the data. Moreover, Wilkinson[32] pointed out that it appeared to be statistically superior and he believed that more accurate subjective estimates were likely to be obtained by the Hanes-Hultin type of plot. If v_0 were considered to be reasonably homogeneous in variance, $\frac{1}{v_0}$ would then appear to exhibit a much greater variation in accuracy over the practical ranges of substrate concentration than would $[(S_0)/v_0]$. Wilkinson showed[32] that a relative weight variation of about 80-fold would have to be considered in a $\frac{1}{v_0}$ vs. $\frac{1}{(S_0)}$ plot, compared to a relative weight variation of only 2-fold for the $\frac{(S_0)}{v_0}$ vs. (S_0) plot, to provide a refined least-square plot so as to obtain statistically accurate measures of the parameters.

Thus, over a relatively narrow range of substrate concentrations corresponding to one-third to three times the value of K_m, a linear least-square plot of $\frac{(S_0)}{v_0}$ vs. (S_0) will often suffice to yield reasonable measures of uncertainties in the K_ms and V_{max}s. This procedure was employed by Kuby et al.[58] in the case of the bisubstrate reaction catalyzed by glucose 6-phosphate dehydrogenase.

A third graphical method is common, usually called the Hofstee plot.[37,38] In this plot, v_0 is plotted vs. $\frac{v_0}{(S_0)}$ and corresponds to the rearrangement given in Equation 60.

$$v_0 = V_{max} - K_m \frac{v_0}{(S_0)} \tag{60}$$

and yields as slope = $-K_m$, Y-intercept = V_{max}, and X-intercept = $\frac{V_{max}}{K_m}$.

Dowd and Riggs[181] made a study of the effects of errors in measurements of the initial velocity on the estimates of K_m and V_{max}. They concluded, as did Wilkinson,[32] that the double-reciprocal plot was the least-reliable plot; however, the choice between the other two methods (the Hanes-Hultin and the Hofstee plot) depended, to some extent, on the nature of the error in v_0. The Hofstee plot seemed to be most sensitive to deviations from linearity. They pointed out[181] that the inferiority of the double-reciprocal plot could be ameliorated, to a large extent, by the proper weighting of the experimental values. However, of course, none of these plots can provide the statistical accuracy that can be obtained by the use of statistical methods to fit the data directly to the rectangular hyperbola described in the Michaelis-Menten equation (Equation 57). Such statistical methods have been described by several workers,[32,33,182] Cleland has provided a Fortran® program,[34] and Leatherbarrow has written a Basic program.[1006]

Before continuing the discussion on the statistical evaluation (and the use of computers) for the evaluation of kinetic parameters it should be emphasized that initially all the kinetic data should be plotted in the simplest fashion and one of the above three plots made before a more sophisticated analysis is attempted. Thus, if the data appear to be greatly scattered, a statistical analysis will not yield any advantage or superior accuracy. Similarly, if the results do not appear to obey the Michaelis-Menten relation, misleading results will be obtained if attempts are made to fit the data statistically to that equation. Moreover, if some region, e.g., a double-reciprocal plot, is obviously nonlinear (possibly because of "substrate inhibition" at high concentrations of substrate), then this region should not be employed in any attempt to fit it to the simple Michaelis-Menten equation. Thus, the reader is cautioned in the use of computers to evaluate results that he has not plotted at least "by eye", for they can lend what might be a spurious validity to the interpretation. It is equally obvious however that the judicious use of computer statistical treatments can lend great precision to the analysis of data, in particular if it is necessary to process a large amount of data (see, e.g., Reference 48 addendum) or to distinguish between intersection points (i.e., whether or not a family of curves intersect above or on one of the axes). The ability of a computer-aided analysis to make distinctions will also hinge on the experimental accuracy; if it is necessary to distinguish, for example, between types of inhibition (e.g., a mixed inhibition or uncompetitive), it may prove necessary experimentally to repeat the sets of analyses so that an evaluation of the standard errors may be used as an aid in distinguishing these types of inhibition.

One other point before continuing, during the study of the glucose 6-phosphate dehydrogenase kinetics,[58] the data which fit the Michaelis-Menten formation were subjected to a statistical analysis [in terms of $(v_0^f)^{-1}$ vs. $(S_1)^{-1}$ at several fixed values of (S_2)]. A *weighted* least-square plot was used (Kuby, 1974, unpublished data) to evaluate the slopes, X- and Y-intercepts, and their standard errors, where the weights were determined experimentally from repetitive analyses (i.e., of v_0 at a given value of S_1 and fixed S_2). A comparison of the statistically evaluated parameters with those obtained from the several alternative plots described above revealed that, in agreement with Hultin's conclusion,[39] the original plot of Hanes,[180] [i.e., in terms of $\frac{(S_1)}{V_0^f}$ vs. (S_1) at fixed values of (S_2)] would actually permit application of a relatively simple linear least-square analysis with essentially the same evaluation for the parameters (and their standard errors) as with the more complex weighted least-square plot.

Before entering into a discussion of statistical analyses of enzymatic initial rate vs. substrate data, it behooves us to mention a novel method of evaluating the data which literally attempts to bypass any cumbersome statistical methodology. Eisenthal and Cornish-Bowden[771]

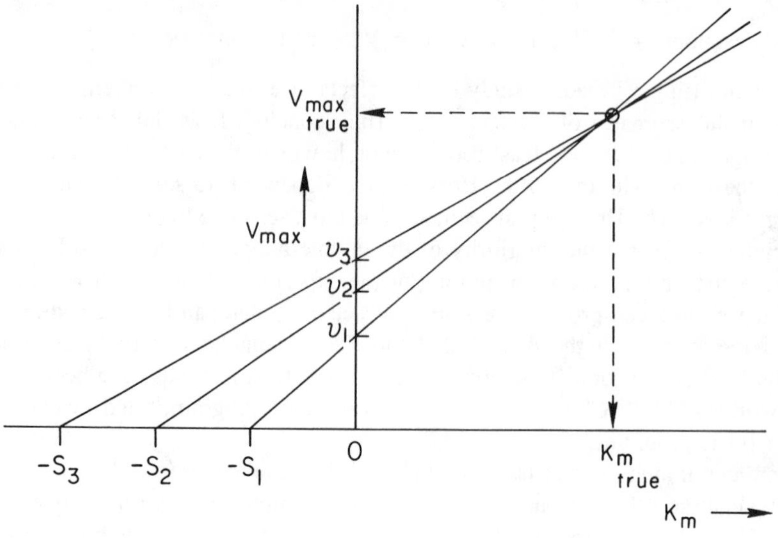

FIGURE 1. Idealized "direct linear plot" of V_{max} vs. K_m according to Eisenthal and Cornish Bowden[771] for a single-substrate reaction where v_1, v_2, v_3, etc. are the observed velocities plotted on the V_{max} axis for the corresponding initial substrate concentrations, S_1, S_2, S_3, etc. plotted on the $-K_m$ axis. The interaction point yields the $K_{m_{true}}$ and $V_{max_{true}}$ values, as shown.

introduced their simple graphical treatment of enzymatic steady-state kinetic data which they described as the "direct linear plot" (see also Reference 772). They rearranged the Michaelis-Menten equation for a single-substrate reaction to the form given in Equation 60a,

$$V_{max} = v_0 + \frac{v_0}{(S_0)} K_m \tag{60a}$$

and then graphically plotted the data as shown in Figure 1. A quotation from Eisenthal and Cornish-Bowden[771] describes their comparatively simple procedure: "Set up axes, K_m, and V [V_{max}], corresponding to the familiar x and y axes, respectively. For each observation (S,v) [the measured substrate concentration and the measured velocity] mark off the points $K_m = -S$ on the K_m axis [we recall that K_m has units of concentration], V = v on the V axis, and draw a line through the two points, extending it into the first quadrant. When this is done for all observations, the lines intersect at a common point whose coordinates (K_m,V) provide the value of K_m and V that satisfy the Michaelis-Menten equation for every observation."

This direct linear plot also provides a very convenient and useful feature, that is, a measure of the *experimental error*. For, in most experiments, the intersection of the lines in the first quadrant in Figure 1 usually does not occur precisely at a single point, bur rather over a small range. The graphically evaluated $K_{m_{true}}$ and $V_{max_{true}}$ values are actually their *median values* along the K_m axis and along the V_{max} axis, respectively; the ranges in values along each of the two axes provide a measure of the experimental error in K_m and in V_{max}, as shown in Figure 2.

Cornish-Bowden and Endrenyi[934] pointed out that the above median method (see also Reference 935) could not readily be generalized to equations of more than two parameters, therefore, another method was described recently[934,936] for fitting equations to enzyme kinetic data, which they entitled the "robust regression" method. This method of fitting the data neither required prior knowledge of weights nor an error distribution, but the corresponding

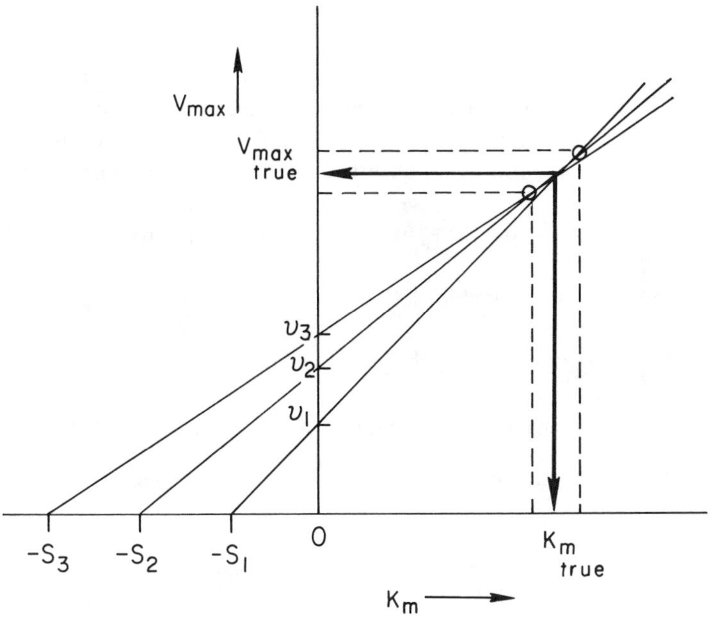

FIGURE 2. Analysis of the experimental error in $K_{m_{true}}$ and $V_{max_{true}}$ by means of the direct linear plot.[771] The errors in K_m and V_{max} are estimated by first taking their respective median values of the ranges over which the lines intersect. The extremes in either direction, as indicated by the dashed lines, yield their respective experimental errors.

computer program[937] is, however, restricted to equations of not more than four parameters and three variables, which are linear when written in reciprocal form.[934,936,937]

B. STATISTICAL ANALYSES

The main objects of any statistical analysis of enzyme kinetics data are (1) to obtain accurate estimates of the kinetic parameters of the selected mathematical model and (2) to obtain a measure of the precision of the results. It would be essential to obtain some knowledge of the variance of K_m and V_{max} before these parameters would be compared to other K_ms and V_{max}s obtained either under different experimental conditions or calculated for different selected mathematical models. Thus, Plowman and Cleland[183] fitted their experimental data (initial velocity studies) on ATP-citrate lyase to the velocity equations for several likely three-substrate reaction mechanisms and compared the estimated values for the kinetic parameters and their standard errors of the fitted constants together with the calculated values of the variance (or the average square of the difference of any experimental velocity from the fitted curve). These calculations were conducted by the Fortran® computer program of Cleland[34] and facilitated a choice in selection of the rapid-equilibrium mechanism, that is, of substrate A adding in an ordered fashion followed by substrates B and C in a random mechanism.

1. The Least-Square Method

The method is commonly used to fix the "best" numerical values of the constants in a formula and proceeds from the assumption that the most probable values of the constants are those for which the sum of the squares of the differences between the observed and the calculated results are minimized, provided the errors are random.

Assume that the general formula to be fitted is

$$y = a + bx \tag{61}$$

and which would correspond to one of the three linear forms of the Michaelis-Menten relation (i.e., Equation 14 from Chapter 1 and Equations 59 and 60 from this chapter discussed above. The problem is to compute the most probable values of a and b (and later, their standard errors). Ideally, the following observation equations would be obtained.

$$a + bx_1 - y_1 = 0, \quad a + bx_2 - y_2 = 0, \ldots a + bx_n - y_n = 0 \tag{62}$$

Since, in practice, this is unattainable, let

$$a + bx_1 - y_1 = \delta_1, \quad a + bx_2 - y_2 = \delta_2, \ldots a + bx_n - y_n = \delta_n \tag{63}$$

where $\delta_i \neq 0$. Therefore, determine the numerical values of the constants, a and b, so as to make $\sum_{i=1}^{n} (\delta_i^2)$ a minimum, i.e.,

$$\sum_{i=1}^{n} (\delta_i^2) = \delta_1^2 + \delta_2^2 + \ldots + \delta_n^2 \tag{64}$$

is to be minimized, or, choose such values of a and b that will make the summation, $\sum_{n=1}^{n=n} (a + bx_n - y_n)^2$, as small as possible. To do this, we apply the usual conditions for minimizing a function of several variables and equate to zero the partial derivatives of $\sum_{i=1}^{n} (\delta_i^2)$ with respect to a and b.

Thus,

$$\frac{\partial}{\partial a} \Sigma(a + bx - y)^2 = 0, \quad \therefore \quad \Sigma(a + bx - y) = 0,$$

$$\frac{\partial}{\partial b} \Sigma(a + bx - y)^2 = 0, \quad \therefore \quad \Sigma x(a + bx - y) = 0 \tag{65}$$

If there are n observation equations, there are na's and

$$\Sigma(a) = na \tag{66}$$

$$\therefore \quad \begin{cases} na + b\Sigma(x) - \Sigma(y) = 0 \\ a\Sigma(x) + b\Sigma(x^2) - \Sigma(xy) = 0 \end{cases} \tag{67}$$

Equations 67 shows two simultaneous linear equations whose solution for a and b present no difficulty where

$$a = \frac{\Sigma(x) \cdot \Sigma(xy) - \Sigma(x^2) \cdot \Sigma(y)}{[\Sigma(x)]^2 - n\Sigma(x^2)} = \bar{y} - b\bar{x}$$

$$b = \frac{\Sigma(x)\Sigma(y) - n\Sigma(xy)}{[\Sigma(x)]^2 - n\Sigma(x^2)} \tag{68}$$

where $\bar{y}$ and $\bar{x}$ are the means of y and x, respectively, and where s^2 = residual mean square, which may be taken equal to σ^2 (the variance, note σ = standard deviation).

$$s^2 = \frac{[n\Sigma x^2 - (\Sigma x)^2][n\Sigma y^2 - (\Sigma y)^2] - [n\Sigma xy - (\Sigma x)(\Sigma y)]^2}{(n - 2)n[n\Sigma x^2 - (\Sigma x)^2]} \tag{69}$$

Suppose, however, that instead of the general formula given by Equation 61 we have

$$y = a + bx + cx^2 \tag{70}$$

where a, b, and c are the constants to be determined. Analogous to Equation 68, the solutions for b and c (omitting a) are

$$b = \frac{\Sigma(x^4) \cdot \Sigma(xy) - \Sigma(x^3) \cdot \Sigma(x^2y)}{\Sigma(x^2) \cdot \Sigma x^4 - [\Sigma(x^3)]^2}$$

$$c = \frac{\Sigma(x^2) \cdot \Sigma(x^2y) - \Sigma(x^3) \cdot \Sigma(xy)}{\Sigma(x^2) \cdot (x^4) - [\Sigma(x^3)]^2} \tag{71}$$

where a is a constant to be determined when x = 0 and a = y_0.

Consider *Gauss*' method for solving a set of linear observation equations; in this case let x, y, and z represent the unknowns to be evaluated and let $a_1, a_2, \ldots, b_1, b_2, \ldots, c_1, c_2, \ldots, R_1, R_2, \ldots$ represent actual numbers whose values have been determined by the series of observation set forth in the following observation equations.

$$\left.\begin{aligned} a_1x + b_1y + c_1z &= R_1 \\ a_2x + b_2y + c_2z &= R_2 \\ a_3x + b_3y + c_3z &= R_3 \\ a_4x + b_4y + c_4z &= R_4 \end{aligned}\right\} \tag{72}$$

The problem is to find the best possible values of x, y, and z, which will satisfy the four given observation equations. We have selected four equations and three unknowns for simplicity, but any number may be included in the calculation. At this point assume that the observation equations have the same degree of accuracy, otherwise "weights" must be assigned to correct each of the equations into a set having the same degree of accuracy. (Weighted least square plots and weights, in general, will be dealt with below.)

To convert the observation equations into a set of "normal" equations solvable by ordinary algebraic processes: multiply the first equations by a_1, the second by a_2, third by a_3, and the fourth by a_4. Add the four results. Treat the four equations in the same way with b_1, b_2, b_3, and b_4, and with c_1, c_2, c_3, and c_4. Now write for brevity

$$\begin{aligned} [aa]_1 &= a_1^2 + a_2^2 + a_3^2 + a_4^2 \\ [ab]_1 &= a_1b_1 + a_2b_2 + a_3b_3 + a_4b_4 \\ [aR]_1 &= a_1R_1 + a_2R_2 + a_3R_3 + a_4R_4 \\ [bb]_1 &= b_1^2 + b_2^2 + b_3^2 + b_4^2 \\ [ac]_1 &= a_1c_1 + a_2c_2 + a_3c_3 + a_4c_4 \\ [aR]_1 &= b_1R_1 + b_2R_2 + b_3R_3 + b_4R_4 \end{aligned} \tag{73}$$

and likewise for $[cc]_1$, $[bc]_1$, and $[cR]_1$. The resulting equations are

$$\begin{cases} [aa]_1x + [ab]_1y + [ac]_1z = [aR]_1 \\ [ab]_1x + [bb]_1y + [bc]_1z = [bR]_1 \\ [ac]_1x + [bc]_1y + [cc]_1z = [cR]_1 \end{cases} \tag{74}$$

These three equations are called *normal* equations (first set) in x, y, and z.

The values of x, y, and z from this set of simultaneous equations may be solved by determinants or by substitution whose solutions are

$$x = \frac{-[ab]_{1y}}{[aa]_1} - \frac{[ac]_{1z}}{[aa]_1} + \frac{[aR]_1}{[aa]_1}$$

$$y = \frac{-[bc]_{2z}}{[bb]_2} + \frac{[bR]_2}{[bb]_2}$$

$$z = \frac{[cR]_3}{[cc]_3} \tag{75}$$

Collectively, the three parts of Equation 75 are called a set of elimination equations, where $[bb]_2$, $[bc]_2$, . . . $[cc]_3$, $[bR]_2$, $[cR]_3$, . . . are called the auxiliaries and are defined by

$$[bb]_2 = [bb]_1 - \frac{[ab]_1[ab]_1}{[aa]_1}$$

$$[bc]_2 = [bc]_1 - \frac{[ac]_1[ab]_1}{[aa]_1}$$

$$[cc]_2 = [cc]_1 - \frac{[ac]_1[ac]_1}{[aa]_1}$$

$$[bR]_2 = [bR]_1 - \frac{[ab]_1[aR]_1}{[aa]_1}$$

$$[cR]_2 = [cR]_1 - \frac{[ac]_1[aR]_1}{[aa]_1}$$

$$[cc]_3 = [cc]_2 - \frac{[bc]_2[bc]_2}{[bb]_2}$$

$$[cR]_3 = [cR]_2 - \frac{[bc]_2[bR]_2}{[bb]_2} \tag{76}$$

Note that the last equation of Equation 75 (above), i.e., $z = \frac{[cR]_3}{[cc]_3}$ gives the value of z directly; the second of Equation 75 provides a value of y when z is known, and the first equation gives the value of x when values of y and z are known.

2. Weighted Least-Square Plots and Variances

Now turn to the case for *unequal weights* in a least-square plot. Thus, the values for y = f(x) may not be homogeneous in variance. For example, σ^2 (the *variance of* y, or the *average value* of the *squared deviations from* μ_y, the *mean of* y) may vary systematically with μ_y, or that the observations are of intrinsically differing accuracy as a result of experimental procedure or technique.

In this case the usual method is to fit the function so as to make the weighted sum of the squares (the deviations, $\sum_{n=1}^{n=n} w(a + bx_n = y_n)^2$ a minimum, where the relative weights, w, are inversely proportional to the variances of the y values. Thus, the weight-regression procedure is equivalent to fitting an unweighted regression to an enlarged set of points where each point of the original set is repeated the appropriate w times.

Consider the two cases below where the true regression is linear and also dependent on either one or two determining variables.

Case (1):

$$\mu_y = \alpha + \beta_x \tag{77}$$

Case (2):

$$\mu_y = \alpha + \beta_1 x_1 + \beta_2 x_2 \tag{78}$$

Case (1): As before the fitted regression is of the form

$$y = a + bx \tag{79}$$

where now the weighted means for x ($\bar{x}$) or y ($\bar{y}$) are

$$\bar{x} = \frac{\Sigma wx}{\Sigma w}; \qquad \bar{y} = \frac{\Sigma wy}{\Sigma w} \tag{80}$$

and the estimated regression coefficient, b (the slope), and constant term, a (the intercept), are

$$b = \frac{\Sigma w(x - \bar{x})(y - \bar{y})}{\Sigma w(x - \bar{x})^2} = \frac{\left[\Sigma wxy - \frac{(\Sigma wx)(\Sigma wy)}{\Sigma w}\right]}{\left[\Sigma wx^2 - \frac{(\Sigma wx)^2}{\Sigma w}\right]}$$

$$a = \bar{y} - b\bar{x} \tag{81}$$

If the variances of the observations in y are

$$\frac{\sigma y^2}{w} \tag{82}$$

(whose square root is the standard deviation or standard error) then the variance factor σ_{y^2} is estimated by the so-called residual mean square, $s_y{}^2$, where

$$s_y^2 = \frac{\sum_{n=1}^{n=n} w(a + bx_n - y_n)^2}{n - 2} \quad \text{or} \quad s_y^2 = \frac{\Sigma wy^2 - \bar{y}\Sigma wy - b\Sigma w(x - \bar{x})(y - \bar{y})}{n - 2} \tag{83}$$

and the divisor of the sum of the squares of the deviations (or residual sum of squares) is the number of degrees of freedom of the estimated variance, which, in turn, equals the number of observations minus two.

The variances of a and b (in $y = a + bx$) are as follows

$$V(a) = \text{variance of a} = \sigma^2\left[\frac{1}{\Sigma w} + \frac{\bar{x}^2}{\Sigma w(x - \bar{x})^2}\right] \tag{84}$$

$$V(b) = \text{variance of b} = \frac{\sigma^2}{\Sigma w(x - \bar{x})^2}$$

Note that b and $\bar{y}$ are considered statistically independent and the estimated variances are obtained by substituting s^2 for σ^2 above (i.e., from Equation 80).

Case (2): The fitted regression equation now is of the form

$$y = a + b_1x_1 + b_2x_2 \tag{85}$$

and the weighted means, $\bar{y}_1$, $\bar{x}_1$, and $\bar{x}_2$ are as follows:

$$\bar{y} = \frac{\Sigma wy}{\Sigma w}, \quad \bar{x}_1 = \frac{\Sigma wx_1}{\Sigma w}, \quad \text{and} \quad \bar{x}_2 = \frac{\Sigma wx_2}{\Sigma w} \tag{86}$$

There are now two equations for the regression coefficients b_1 and b_2, that is,

$$\begin{cases} a_{11}b_1 + a_{12}b_2 = p_1 \\ a_{12}b_1 + a_{22}b_2 = p_2 \end{cases} \tag{87}$$

where

$$a_{11} = \Sigma w(x_1 - \bar{x}_1)^2, \quad a_{12} = \Sigma w(x_1 - \bar{x}_1)(x_2 - \bar{x}_2), \quad a_{22} = \Sigma w(x_2 - \bar{x}_2)^2,$$

$$p_1 = \Sigma w(x_1 - \bar{x}_1)y, \quad \text{and} \quad p_2 = \Sigma w(x_2 - \bar{x}_2)y \tag{88}$$

and the solutions of the two equations are

$$\begin{cases} b_1 = c_{11}p_1 + c_{12}p_2 \\ b_2 = c_{12}p_1 + c_{22}p_2 \end{cases} \tag{89}$$

where

$$c_{11} = \frac{a_{22}}{\Delta}, \; c_{22} = \frac{a_{11}}{\Delta}, \; c_{12} = \frac{a_{12}}{\Delta}, \text{ and } \Delta = a_{11}\, a_{22} - a_{12}^2$$

The solution of the constant term is simply

$$a = \bar{y} - b_1\bar{x}_1 - b_2\bar{x}_2 \tag{90}$$

As in case (1), the variance factor σ_y^2 is estimated by the residual mean square s_y^2, where

$$s_y^2 = \frac{\Sigma w(a + b_1x_1 + b_2x_2 - y_n)^2}{n - 3} = \frac{\Sigma wy^2 - \bar{y}\Sigma wy - b_1p_1 - b_2p_2}{n - 3} \tag{91}$$

A summary of the variances and covariances (the average value of the product of the corresponding deviations for all possible pairs of values, y_1, y_2) is as follows:

$$\text{Variance of a} = V(a) = \sigma_y^2\left(\frac{1}{\Sigma w} + c_{11}\bar{x}_1^2 + c_{22}\bar{x}_2^2 + c_{12}\bar{x}_1\bar{x}_2\right)$$

$$\text{Variance of } b_1 = V(b_1) = c_{11}\sigma_y^2$$

$$\text{Variance of } b_2 = V(b_2) = c_{22}\sigma_y^2$$

$$\text{Covariance }(b_1, b_2) = c_{12}\sigma^2,\ \text{Covariance }(b_1, \bar{y}) = 0,\ \text{and Covariance }(b_2, \bar{y}) = 0 \quad (92)$$

If the relative weights are not known precisely, then estimated values for the weights can be provided (as described below) in the initial statistical analysis. With improved estimates of the weights (through repetitive analyses), more accurate estimates of the statistical parameters are then forthcoming in a refined (and second) analysis.

In the use of *nonlinear* regression functions, if provisional estimates of parameters are provided, then linear regression methods may be employed for fitting the expressions. Thus, if a function f(x,c) is nonlinear in the parameter c (which is to be estimated), but if a provisional value c_0 is at hand, then the following linear approximation may be employed (derived from a Taylor expansion, with only the first two terms employed[32]).

$$f(x, c) \simeq f(x, c_0) + (c - c_0)[f_c'(x, c_0)] \quad (93)$$

where f_c' is the first derivative of f with respect to c, evaluated at c_0. A linear-regression analysis, in terms of provisional values of f and f_c' corresponding to the values of x, yields $(c - c_0)$ or the linear regression coefficient. If necessary, the statistical estimate for c may be further refined by repeating the process with the adjusted estimate of c_0 now provided as the new provisional value.

Wilkinson[32] showed that

$$V[f(y)] \simeq [f'(\mu_y)]^2 V(y) \quad (94)$$

where $f(\mu_y)$ is f(y) at μ_y and $f'(\mu_y)$ is the first derivative of f(y) at μ_y. (Recall that the true mean, or the expected value of a random variate y, is the average value of y in the population and is denoted by μ_y. The true mean is to be distinguished from the sample mean, $\bar{y}$, of a series of observations.) This equation is useful for approximating the variance in a function of y (i.e., V[f(y)]) from the variance in y, i.e., σ_y^2 [or V(y)], and the square of the first derivative of the function of y when $y = \mu_y$ (i.e., $[f'(\mu_y)]^2$); thus,

$$V(y) \simeq \frac{V[f(y)]}{[f'(\mu_y)]^2} = \frac{\sigma_y^2}{[f'(\mu_y)]^2} \quad (95)$$

For example, consider Equation 14 from Chapter 1,

$$\frac{1}{v_0} = \frac{K_m}{V_{max}(S_0)}\frac{1}{(S_0)} + \frac{1}{V_{max}} = f\left(\frac{1}{v_0}\right)$$

$$\therefore\ V(y) = \frac{\sigma^2}{\mu_v^4} \quad (96)$$

or, take Equation 59 where now

$$V\left(\frac{(S_0)}{(v_0)}\right) = \frac{\sigma^2(S_0)^2}{\mu_v^4} \tag{97}$$

Since, in the case of these linear regressions, the variances of the observations, y, are σ^2/w, therefore, the correct *relative weights* are $\mu_v{}^4$ and $\frac{\mu_v{}^4}{(S_0)^2}$, for these two cases, respectively.[32]

Also, since the true values for the rates μ_v are not known in advance, a provisional fit of the linear equation in use can be obtained if one uses the observed rate, v_0, for μ_v. This approximation provides the relative weights to be used in Equation 14 from Chapter 1 for $\frac{1}{v_0}$ as $(v_0)^4$, and the relative weight for $\frac{(S_0)}{(v_0)}$ in Equation 59 as $\frac{(v_0)^4}{(S_0)^2}$. Moreover, the weighted means (see Equation 80) are for Equation 14 (Chapter 1):

$$\overline{\left(\frac{1}{v_0}\right)} = \frac{\Sigma wy}{\Sigma w} = \frac{\Sigma\left[(v_0)^4\left(\frac{1}{v_0}\right)\right]}{\Sigma(v_0)^4} = \frac{\Sigma(v_0)^3}{\Sigma(v_0)^4} \tag{98}$$

$$\overline{\left(\frac{1}{S_0}\right)} = \frac{\Sigma wx}{\Sigma w} = \frac{\Sigma\left[(v_0)^4\left(\frac{1}{S_0}\right)\right]}{\Sigma(v_0)^4} = \frac{\Sigma[(v_0)^4/(S_0)]}{\Sigma(v_0)^4} \tag{99}$$

and for Equation 59:

$$\overline{\left(\frac{(S_0)}{(v_0)}\right)} = \frac{\Sigma\left(\frac{(v_0)^4}{(S_0)^2}\left(\frac{(S_0)}{v_0}\right)\right)}{\Sigma\left(\frac{(v_0)^4}{(S_0)^2}\right)} = \frac{\Sigma\left(\frac{(v_0)^3}{(S_0)}\right)}{\Sigma\left(\frac{(v_0)^4}{(S_0)^2}\right)} \tag{100}$$

$$\overline{(S_0)} = \frac{\Sigma\left(\frac{(v_0)^4(S_0)}{(S_0)^2}\right)}{\Sigma\left(\frac{(v_0)^4}{(S_0)^2}\right)} = \frac{\Sigma\left(\frac{(v_0)^4}{(S_0)}\right)}{\Sigma\left(\frac{(v_0)^4}{(S_0)^2}\right)} \tag{101}$$

Thus, the weighted fit, for example, of the straight line, $y = a + bx = \left(\frac{1}{v_0}\right) = a + b\left(\frac{1}{S_0}\right)$, leads to the following formulae for estimates of K_m and V_{max}.

$$K_m = \frac{\left[\Sigma(v_0)^4 \cdot \Sigma\left(\frac{(v_0)^3}{(S_0)}\right) - \Sigma(v_0)^3 \cdot \Sigma\frac{(v_0)^4}{(S_0)}\right]}{\left[\Sigma(v_0)^3 \cdot \Sigma\left(\frac{(v_0)^4}{(S_0)^2}\right) - \Sigma\left(\frac{(v_0)^3}{(S_0)}\right) \cdot \Sigma\frac{(v_0)^4}{(S_0)}\right]} \tag{102}$$

and

$$V_{max} = \frac{\left[\Sigma(v_0)^4 \cdot \Sigma\left(\frac{(v_0)^4}{(S_0)^2}\right) - \left(\Sigma\left[\frac{(v_0)^4}{(S_0)}\right]\right)^2\right]}{\left[\Sigma(v_0)^3 \cdot \Sigma\left(\frac{(v_0)^4}{(S_0)^2}\right) - \Sigma\left(\frac{(v_0)^3}{(S_0)}\right) \cdot \Sigma\left(\frac{(v_0)^4}{(S_0)}\right)\right]} \tag{103}$$

Alternatively, a weighted least-square fit may be obtained also from plots of $(\frac{S_0}{v_0})$ vs. (S_0) and leads to the same formulas for V_{max} and K_m, as above, since the weighted analyses are equivalent.

After obtaining a provisional estimate of K_m and V_{max} by one of the weighted least-square plots described above, refined and usually final estimates of these kinetic parameters, and their standard errors, may be obtained as Wilkinson showed[32] by fitting the data then directly to the Michaelis-Menten equation (Equation 57) in its hyperbolic form, with the use of these first provisional estimates of the kinetic parameters, denoted K_m and V_{max}, and now with an *unweighted analysis*.

Thus, with the use of Equation 93, the Michaelis-Menten equation may be expressed approximately in the linear form,

$$v_0 \simeq \frac{V_{max}}{V^0_{max}} \left\{ \frac{(S_0)V^0_{max}}{(S_0) + K^0_m} - (K_m - K^0_m) \frac{(S_0)V^0_{max}}{((S_0) + K^0_m)^2} \right\} \tag{104}$$

or

$$v_0 \simeq b_1 f(S_0) + b_2 f'_{K_m}(S_0) \tag{105}$$

where $b_1 = \frac{V_{max}}{V^0_{max}}$ and $b_2 = b_1 (K_m = K^0_m)$.

$f(S_0)$ is the provisional fit of the Michaelis-Menten equation or

$$f(S_0) = \frac{(S_0)V^0_{max}}{(S_0) + K^0_m} \tag{106}$$

with the provisional values provided for K_m and V_{max} (i.e., K^0_m and V^0_{max}), and $f'_{Km}(S_0)$ is the first derivative of the provisional fit with respect to

$$K_m(-(S_0)V^0_{max}/[(S_0) + K^0_m]^2) \tag{107}$$

Regression analysis of Equation 105 follows that described above for case (2), where a = 0, and leads to values for the regression coefficients, b_1 and b_2, and further refined estimates for K_m and V_{max} (refer to Equation 105), i.e.,

$$K_m = K^0_m + \frac{b_2}{b_1} \tag{108}$$

$$V_{max} = b_1 V^0_{max} \tag{109}$$

With repetition of the process, i.e., with these second values for K_m and V_{max} now set to K^0_m and V^0_{max}, as above, b_1 and b_2 approach unity and zero, respectively. This iterative least-square fit process leads to accurate estimates of K_m and V_{max} and their variances or standard errors.

Thus, $V_{max} = b_1 V^0_{max}$ and $K_m = K^0_m + \frac{b_2}{b_1}$ where

$$b_1 = \frac{[\Sigma f'_{K_m}(S_0)^2 \cdot \Sigma((v_0) \cdot f(S_0)) - \Sigma(f(S_0) \cdot f'_{K_m}(S_0)) \cdot \Sigma(v \cdot f'_{K_m}(S_0)]}{[\Sigma f(S_0)^2 \cdot \Sigma f'_{K_m}(S_0)^2 - (\Sigma f(S_0) \cdot f'_{K_m}(S_0))^2]} \tag{110}$$

$$b_2 = \frac{\Sigma f(S_0)^2 \cdot \Sigma v f'_{K_m}(S_0) - \Sigma(f(S_0) \cdot f'_{K_m}(S_0)) \cdot \Sigma((v_0) \cdot f(S_0))}{[\Sigma f(S_0)^2 \cdot \Sigma f'_{K_m}(S_0)^2 - (\Sigma f(S_0) \cdot f'_{K_m}(S_0))^2]} \tag{111}$$

$$s_v^2 = \frac{(\Sigma(v_0)^2 - b_1\Sigma((v_0) \cdot f(S_0)) - b_2 \cdot \Sigma((v_0) \cdot f'_{K_m}(S_0)))}{(n - 2)} \tag{112}$$

[$s^{2\,1/2} = s$ = standard deviation = σ_v.] The standard errors for K_m and V_{max} are, respectively,

$$K_m = \sigma_{K_m} = \frac{s}{b_1}\left(\frac{\Sigma f(S_0)^2}{\Sigma f(S_0)^2 \cdot \Sigma f'_{K_m}(S_0)^2 - (\Sigma f(S_0) \cdot f'_{K_m}(S_0))^2}\right)^{1/2}$$

$$V_{max} = V^0_{max} s\left\{\frac{\Sigma f'_{K_m}(S_0))^2}{\Sigma f(S_0)^2 \cdot \Sigma f'_{K_m}(S_0)^2 - (\Sigma f(S_0) \cdot f'_{K_m}(S_0))^2}\right\}^{1/2} \tag{113}$$

3. Some Comments on Statistical Analysis of Multisubstrate Reactions

For bisubstrate reactions, if possible, statistical evaluation of the apparent kinetic parameters may be made with either plots of $(v_0)^{-1}$ vs. $(S_1)^{-1}$ at several fixed values of (S_2), or from Hultin-type plots [$\frac{(S_1)}{(v_0)}$ vs. (S_1) at several fixed values of (S_2)], employing a weighted least-square analysis (especially if repetitive analyses are available to provide the weights), or unweighted if the first provisional estimates are to be made (e.g., by a Hultin-type plot) to test the kinetic model selected. These apparent kinetic parameters may then be used in the secondary replots to yield the kinetic parameters in the reaction mechanism as given, for example, in Chapters 2 and 3. Secondary replots may be conducted by the least-square methodology (with similar weights, or 1/variance as in the primary plots) to yield the initial provisional estimates for the kinetic parameters, which may then be inserted into the kinetic expression for the initial velocity and fitted by the least-square method, to yield the second estimates of the kinetic parameters. The iterative procedure is continued until no better estimates are obtained. This procedure will hold provided the primary plots and secondary replots are linear; otherwise, nonlinear areas are to be neglected or else the data are fitted to a suitable polynomial. Thus, after obtaining provisional estimates for $\overline{K}_A$, $\overline{K}_B$, and V_{max} by Hultin-plots and/or secondary replots for glucose 6-phosphate dehydrogenase,[58] these initial estimates were fitted to

$$v^f_o = V^f_{max}\left[\frac{\overline{K}_A\overline{K}_B}{(A)(B)} + \frac{\overline{K}_B}{(B)} + \frac{\overline{K}_A}{(A)} + 1\right]^{-1}$$

for each condition of temperature and pH to yield refined estimates of the kinetic parameters for each condition.[71]

The method (without secondary replots) may be illustrated as follows and is similar to that described by Cleland.[34] Thus, by the least-square method the following expression is to be minimized:

$$\Sigma w_i\left[\frac{V^f_{max}}{\left[\frac{\overline{K}_A\overline{K}_B}{(A)(B)} + \frac{\overline{K}_B}{(B)} + \frac{\overline{K}_A}{(A)} + 1\right]} - v_i\right]^2 \tag{114}$$

or written in double-reciprocal fashion to provide a linear relation between $\frac{1}{v}$ and $\frac{1}{(A)}$ or $\frac{1}{(B)}$, then the expression to be minimized is

$$\Sigma w_i\left[\left(\frac{1}{V_{max}}\left[1 + \frac{\overline{K}_B}{(B)}\right] + \frac{\overline{K}_A}{V_{max}}\frac{1}{(A)}\left[1 + \frac{\overline{K}_B}{(B)}\right]\right) - \frac{1}{v_i}\right]^2 \tag{115}$$

or

$$\Sigma w_i\left[\left(\frac{\overline{K}_A\overline{K}_B}{V_{max}(A)(B)} + \frac{\overline{K}_B}{V_{max}(B)} + \frac{\overline{K}_A}{V_{max}(A)} + \frac{1}{V_{max}}\right) - \frac{1}{v_i}\right]^2 = \Sigma w_i X^2 \tag{115a}$$

The above expression may be placed in the generalized form

$$\Sigma w_i X^2 = \Sigma w_i[ax_{1i} + bx_{2i} + cx_{3i} + dx_{4i} - x_{5i}]^2 \tag{116}$$

The partial derivative is taken of $(w_i X^2)$ with respect to a, b, c, and d and set to zero, thus yielding a set of n linear equations in n unknown constants.
Before continuing, note that

$$\overline{K}_A = \frac{c}{d} = \frac{a}{b}, \quad \overline{K}_B = \frac{b}{d} = \frac{a}{c}, \quad \text{and} \quad V_{max} = \frac{1}{d} \tag{117}$$

where $a = \frac{\overline{K}_A\overline{K}_B}{V_{max}}$, $b = \frac{\overline{K}_B}{V_{max}}$, $c = \frac{\overline{K}_A}{V_{max}}$ and $d = \frac{1}{V_{max}}$ and

$$X = ax_{1i} + bx_{2i} + cx_{3i} + dx_{4i} - x_{5i}) \tag{118}$$

Thus,

$$\frac{\partial(w_i X^2)}{\partial(a, b, c, d)} = 2w_i X \frac{\partial X}{\partial(a, b, c, d)} \tag{119}$$

and leads to the following results.

$$\frac{\partial(w_i X^2)}{\partial(a)} = 2w_i X\left(\frac{\partial X}{\partial a}\right) = 2w_i X(x_{1i}) = 0$$

$$= 2w_i[a\Sigma x_{1i}^2 + b\Sigma x_{1i}x_{2i} + c\Sigma x_{1i}x_{3i} + d\Sigma x_{1i}x_{4i} - \Sigma x_{1i}x_{5i}] \tag{120}$$

$$\frac{\partial(w_i X^2)}{\partial(b)} = 2w_i X\left(\frac{\partial X}{\partial b}\right) = 2w_i X(x_{2i}) = 0$$

$$= 2w_i[a\Sigma x_{1i}x_{2i} + b\Sigma x_{2i}^2 + c\Sigma x_{2i}x_{3i} + d\Sigma x_{2i}x_{4i} - \Sigma x_{2i}x_{5i}] \tag{121}$$

$$\frac{\partial(w_iX^2)}{\partial(c)} = 2w_iX\left(\frac{\partial X}{\partial c}\right) = 2w_iX(x_{3i}) = 0$$

$$= 2w_i[a\Sigma x_{1i}x_{3i} + b\Sigma x_{2i}x_{3i} + c\Sigma x_{3i}^2 + d\Sigma x_{3i}x_{4i} - \Sigma x_{3i}x_{5i}] \quad (122)$$

$$\frac{\partial(w_iX^2)}{\partial(d)} = 2w_iX\left(\frac{\partial X}{\partial d}\right) = 2w_iX(x_{4i}) = 0$$

$$= 2w_i[a\Sigma x_{1i}x_{4i} + b\Sigma x_{2i}x_{4i} + c\Sigma x_{3i}x_{4i} + d\Sigma x_{4i}^2 - \Sigma x_{4i}x_{5i}] \quad (123)$$

If determinants are used for the solution, then the appropriate positions in the matrix are conveniently represented by $D_{kj} = w_i x_{ki} x_{ji}$. The weighting factor to be used in the rate expression given in double-reciprocal form, if not known for this initial estimation of the kinetic parameters, may be assumed to be $w_i = \dfrac{v_i^4}{\text{variance of } v_i}$, if one assumes as Wilkinson did[32] that the variance is a constant for this set of initial rate measurements with the use of a fixed concentration of enzyme throughout. The substrate concentrations are assumed to be accurately known or at least relatively accurate. It is of course better to measure the variance experimentally at each substrate concentration and to make use of these empirical values as the weighting factors.

4. A Few Comments on Computer Fitting of pH and Initial Rate Data

As indicated above, the pH dependence of an enzyme-catalyzed reaction has become a popular approach to identifying ionizable active-site residues which play a role in substrate binding and in catalysis.[775-777]

It has, therefore, also become common to determine, by computer fitting, both the number of ionizing groups and their apparent pKa values. The pH data are fitted, with a suitable computer program, to the rate equation which appropriately describes the type of ionization involved.

Thus, in their very detailed examination of the ''solvent perturbation technique'' (see Chapter 4, Section I.A). Grace and Dunaway-Mariano[788] fitted the data they gathered on yeast hexokinase to the following equations by means of the Fortran® programs of Cleland.[34,776]

The initial velocity data were computer fitted to Equation 124,

$$v_0 = \frac{V_{max}(S)}{K_s + (S)} \quad (124)$$

the pH data to Equation 125,

$$\log Y = \frac{\log C}{1 + \dfrac{(H^+)}{K_a}} \quad (125)$$

and their inhibition data (e.g., with lyxose) to Equation 126,

$$v_0 = \frac{V_{max}(S)}{K_s\left(1 + \dfrac{(I)}{K_i}\right) + (S)} \quad (126)$$

where v_0 is the initial velocity, V_{max} the maximum velocity, K_s the Michaelis constant for the substrate, (I) is inhibitor concentration, (S) is the substrate concentration, (H^+) is the proton concentration, C is the pH-independent value of the parameter, Y is the observed v_0, $\frac{(V_{max})}{K_s}$, or K_i, and K_a is the acid dissociation constant for the groups which must be deprotonated for maximum activity.

In a very thorough study by Stone and Morrison (see also Reference 980) on the effect of pH on dihydrofolate reductase[795] they computer fitted their data with the use of programs written in BASIC® to perform weighted robust regression.[814,936] One advantage of these programs which is absent in the Fortran® programs[34,776] is that they included a procedure for iterative weighting that minimizes the effect of outlying points on the least-squares solution.[815] Thus, the data they obtained at each pH value by varying the concentration of 7,8-dihydrofolate (A) were fitted to Equation 127

$$V_0 = \frac{V_{max}[A]}{K_A + [A]} \tag{127}$$

to yield values for the maximum velocity . (V_{max}), the Michaelis constant (K_A), as well as for the ratio V_{max}/K_A.

In experiments where the concentration of an inhibitor (I) was varied, the data were fitted to Equation 128.

$$V_0 = \frac{V_{app}}{1 + (I)/K_{i_{app}}} \tag{128}$$

Analysis according to Equation 128, at each pH value, yielded an apparent inhibition constant $(K_{i_{app}})$ from which a value for K_i at the pH could be calculated by using the relationship given in Equation 129, together with the known substrate concentration (A) and the previously determined K_A.

$$K_i = \frac{K_{i_{app}}}{1 + [A]/K_A} \tag{129}$$

Values obtained for V_{max}, $\frac{V_{max}}{K_A}$, and $\frac{1}{K_i}$ were then weighted according to the inverse of their variances and fitted to Equation 130 or to Equation 131,

$$y = \frac{C}{1 + K_b/(H^+)} \tag{130}$$

$$y = \frac{C}{1 + (H^+)/K_a + K_b/(H^+)} \tag{131}$$

where y represents the value of V_{max}, or $\frac{V_{max}}{K_A}$, or $\frac{1}{K_i}$ at a particular pH value, C represents the pH-independent value of the parameter, and K_a and K_b are the acid dissociation constants.

The enthalpy of ionization (ΔH_{ion}) was determined by weighting pKa values obtained at different temperatures according to the inverse of their variances and fitting to Equation 132 gives

$$pK = \Delta H_{ion}/(2.303\ RT) \tag{132}$$

Isotopic effects (D_V and $D_{(V/K)}$ were determined by fitting data obtained with the unlabeled or deuterated pyridine nucleotide to Equation 127, and then calculating the appropriate ratio.

Two other applications of computer analyses may be cited here. (1) An interesting statistical analysis of competitive inhibition by a D-isomer inhibitor of an L-isomer substrate in a racemic mixture was conducted by Mares-Guia et al.[826] who made use of available computer programs[32,827] in order to demonstrate the competitive nature of the inhibition and also to statistically evaluate the K_i value. (2) As indicated above (Chapter 2, Section I.E.3), Garfinkel et al.[957] have recently described a computer program in BASIC® for microcomputers (e.g., the IBM PC), which ''fits'' enzyme kinetic data to enzyme kinetic models by a nonlinear regression analysis. Leatherbarrow also has a program (termed Enzfitter[1006]) commercially available for IBM PCs or compatibles.

Chapter 10

KINETICS OF THE TRANSIENT PHASE OR PRE-STEADY-STATE PHASE OF ENZYMIC REACTIONS

I. INTRODUCTION

Only in the simplest cases is it possible to determine the values of the individual rate constants from a study of the kinetics of an enzyme reaction with the use of initial rate measurements as we discussed in Chapter 1. All of our preceding discussions in Chapters 1 through 9 on enzyme kinetics have been based largely on the steady-state treatment or quasi- (rapid) equilibrium approach. In this chapter we will discuss the kinetic behavior before the actual steady state is reached or established. This early region of the reaction which has been termed the transient phase or pre-steady-state region is usually of a very short period in time, but techniques have been developed to study this period and they will be discussed under "Rapid or Fast Reaction Flow Techniques". Another approach is to perturb a system already at equilibrium and to follow the course of the system to its new equilibrium position, a technique to be discussed under "Relaxation Kinetics". A third approach, that is, that of the "Stirred Reactor"[6] will not be presented in detail, although it is a technique which, in principle, permits the accumulation in a true steady state of transient intermediates; unfortunately, it has rarely been applied to a study of enzyme kinetics.

In all of our previous discussions of initial rate studies, it was postulated in both the steady-state or quasi- (rapid) equilibrium approximations that during the linear (initial) rate phase of the reactions, the concentrations of the enzyme-substrate intermediates remain constant. However, as shown in Figure 1,[417] for a hypothetical single substrate-single product system before the steady state (e.g., with respect to [ES] − [ES]*) is established, there will be a brief period during which the concentrations of these complexes are changing. Measurements during this pre-steady-state or transient period will permit a number of the rate constants for the individual steps in the reaction to be determined.

Figure 1 is a computer representation of changes in intermediate and product concentrations during transient phases of the system:

$$E + S \underset{k_{-1}}{\overset{k_{+1}}{\rightleftarrows}} ES \underset{k_{-2}}{\overset{k_{+2}}{\rightleftarrows}} E^*S \underset{k_{-3}}{\overset{k_{+3}}{\rightleftarrows}} E^*P \underset{k_{-4}}{\overset{k_{+4}}{\rightleftarrows}} EP \underset{k_{-5}}{\overset{k_{+5}}{\rightleftarrows}} E + P$$

II. ANALYSIS OF RAPID REACTION RATE DATA

Before continuing, it is important to discuss a number of integrated reaction courses with time to facilitate the analyses of the reaction curves.

For the analysis of some "stopped-flow" measurements (the technique of stopped-flow will be discussed below), the rates of reaction are usually conducted at constant temperature and volume in a "closed system". For our present purposes, we may define a closed system as one in which any change in the quantity of the reactants and solvent is due solely to the chemical reaction. Let us, at first, consider only *simple reactions*, where a simple reaction is defined as one where the rate is determined exclusively by the rate of a single reaction step. In a simple reaction, whose stoichiometric equation is

$$aA + bB + cC + \ \ldots \rightleftarrows \ell L + mM + \ \ldots \tag{1}$$

it can be represented by a differential equation of the following form:

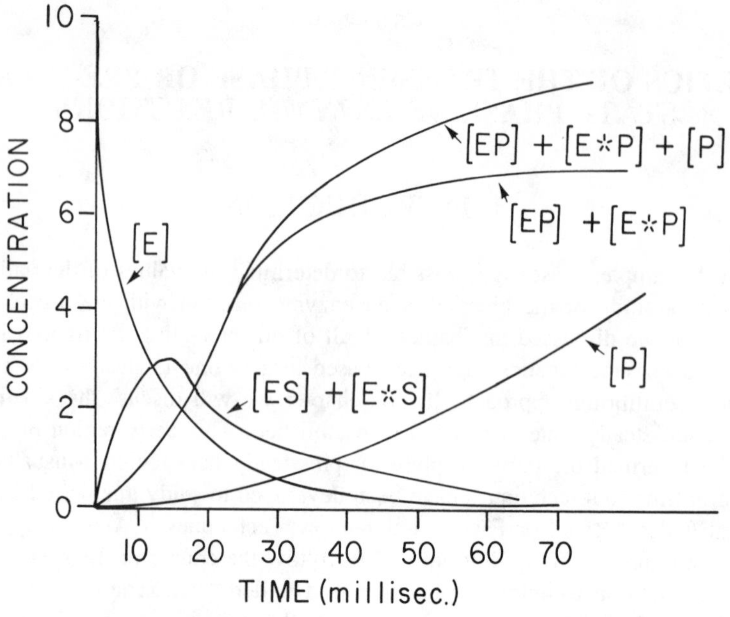

FIGURE 1. Computer representation of changes in intermediate and product concentrations during transient phases of the system:

$$E + S \underset{k_{-1}}{\overset{k_{+1}}{\rightleftarrows}} ES \underset{k_{-2}}{\overset{k_{+2}}{\rightleftarrows}} E^*S \underset{k_{-3}}{\overset{k_{+3}}{\rightleftarrows}} E^*P \underset{k_{-4}}{\overset{k_{+4}}{\rightleftarrows}} EP \underset{k_{-5}}{\overset{k_{+5}}{\rightleftarrows}} E + P.$$

Total enzyme (catalytic site) concentration = 10^{-7} *M*. Initial substrate concentration = 10^{-5} *M*. Constants: k_{-1}, k_{-2}, k_{-3}, k_{-4}, k_{-5}, all smaller than 10 s^{-1}. $k_{+1} = 10^{-7}\ M^{-1}\ s^{-1}$; $k_{+2} = 10^{-3}\ s^{-1}$; $k_{+4} = 10^{-2}\ s^{-1}$; $k_{+5} = 10\ s^{-1}$. (From Gutfreund, H., *Annu. Rev. Biochem.*, 40, 315, 1971. With permission.)

$$\frac{dx}{dt} = k(A_o - x)^j\left(B_o - \frac{b}{a}x\right)^k\left(C_o - \frac{c}{a}x\right)^\ell \ldots \tag{2}$$

In this equation, the exponents (j, k, etc.) are the simple positive integers 0, 1, 2, or 3, the coefficients (b/a, etc.) are determined by the stoichiometric equation, x is the decrease in the concentration of A due to the reaction, and the quantities A_0, etc., are the initial concentrations of the several reactants. It follows from these definitions

$$\frac{dx}{dt} = \frac{-d[A]}{dt} = \frac{-a}{b}\frac{d[B]}{dt} = \frac{a}{m}\frac{d[M]}{dt}, \quad \text{etc.} \tag{3}$$

The sum of the exponents $(j + k + \ell)$ is called the "order of the reaction". Similarly, j is the order of the reaction with respect to the substance A, k is the order with respect to B, etc.

In Table 1, some useful cases of integrated rate equations corresponding to Equation 2 are listed, together with their differential equations, half-lives, and dimensions of the rate constant.

Also, in Table 2, the kinetic laws for some complex reaction systems are given. For Mechanism 3 of Table 2, i.e., $A \xrightarrow{k_{+1}} B \xrightarrow{k_{+2}} C$, the solution is plotted, as shown in Figure 2, of the concentrations A, B, and C as a function of time, for a typical case where k_{+1} =

TABLE 1

Some Special Cases of Integrated Rate Expressions which Correspond to $\frac{dx}{dt} = k(A_0 - x)^j\left(B_0 - \frac{bx}{a}\right)^k\left(C_0 - \frac{cx}{a}\right)^\ell$

(j + k + l) Order	(j k ℓ) Suborder	(a b c) Stolchlometric coefficients	Differential equation	Definite integral	Half-life ($t_{1/2}$)	Units of k
1	100	1 — —	$\frac{dx}{dt} = k(A_0 - x)$	$\ln\left[\frac{A_0}{(A_0 - x)}\right] = kt$	$\frac{1}{k}\ln 2$	s^{-1}
2	200	1 — —	$\frac{dx}{dt} = k(A_0 - x)$	$\frac{x}{A_0(A_0 - x)} = kt$	$\frac{1}{kA_0}$	$s^{-1}M^{-1}$
2	110	1 1 —	$\frac{dx}{dt} = k(A_0 - x)(B_0 - x)$	$\frac{1}{(B_0 - A_0)}\ln\left[\frac{A_0(B_0 - x)}{B_0(A_0 - x)}\right] = kt$	$\frac{1}{k(B_0 - A_0)}\ln\left(\frac{2B_0 - A_0}{B_0}\right)$	$s^{-1}M^{-1}$
2	110	1 2 —	$\frac{dx}{dt} = k(A_0 - x)(B_0 - 2x)$	$\frac{1}{(B_0 - 2A_0)}\ln\left[\frac{A_0(B_0 - 2x)}{B_0(A_0 - x)}\right] = kt$	$\frac{1}{k(B_0 - 2A_0)}\ln\left[\frac{2(B_0 - A_0)}{B_0}\right]$	$s^{-1}M^{-1}$
3	300	1 — —	$\frac{dx}{dt} = k(A_0 - x)^3$	$\frac{2A_0x - x^2}{(A_0^2(A_0 - x)^2} = 2kt$	$\frac{3}{2kA_0^2}$	$s^{-1}M^{-2}$
3	111	1 1 1	$\frac{dx}{dt} = k(A_0 - x)$ $(B_0 - x)(C_0 - x)$	$\frac{(B_0 - C_0)}{\text{Denom.}^a}\ln\left(\frac{A_0}{A_0 - x}\right) + \frac{(C_0 - A_0)}{\text{Denom.}}\ln\left(\frac{B_0}{B_0 - x}\right) + \frac{(A_0 - B_0)}{\text{Denom.}}\ln\left(\frac{C_0}{C_0 - x}\right) = kt$	$\left(\frac{B_0 - C_0}{k\ \text{Denom.}}\right)\ln 2 + \left(\frac{C_0 - A_0}{k\ \text{Denom.}}\right)\ln\left(\frac{B_0}{B_0 - \frac{A_0}{2}}\right) + \left(\frac{A_0 - B_0}{k\ \text{Denom.}}\right)\ln\left(\frac{C_0}{C_0 - \frac{A_0}{2}}\right)$	$s^{-1}M^{-2}$
>1	j00	1 — —	$\frac{dx}{dt} = k(A_0 - x)^j$	$\frac{1}{(A_0 - x)^{j-1}} - \frac{1}{A_0^{j-1}} = (j - 1)kt$	$\frac{2^{j-1} - 1}{kA_0^{j-1}}$	$s^{-1}M^{1-j}$
0	000	1 — —	$\frac{dx}{dt} = k$	$x = kt$	$\frac{A_0}{2k}$	$s^{-1}M$
1/2	1/2 00	1 — —	$\frac{dx}{dt} = k(A_0 - x)^{1/2}$	$A_0^{1/2}(A_0 - x) = \frac{kt}{2}$	$\frac{3A_0^{1/2}}{2k}$	$s^{-1}M^{1/2}$

TABLE 1 (continued)

Some Special Cases of Integrated Rate Expressions which Correspond to $\frac{dx}{dt} = k(A_0 - x)^j\left(B_0 - \frac{bx}{a}\right)^k\left(C_0 - \frac{cx}{a}\right)^\ell$

Note: The subscript "0" indicates the concentration to be that at $t = 0$.

[a] Denom. $= A_0B_0(A_0 - B_0) + A_0C_0(C_0 - A_0) + B_0C_0(B_0 - C_0)$.

From Livingston, R., *Technique of Organic Chemistry, Investigation of Rates and Mechanisms of Reactions,* Vol. 8, Freiss, S. L. and Weissberger, A., Eds., Interscience, New York, 1953, chap. 4, 169. With permission.

TABLE 2
Kinetic Laws (Integrated Expressions) for Some Complex Reaction Systems

Mechanism		Integrated expressions
1. Parallel first-order reactions:	$A \xrightarrow{k_{+1}} U$ $A \xrightarrow{k_{+2}} V$ $A \xrightarrow{k_{+3}} W$	$\ln \frac{A_0}{A} = k^+$ or $A = A_0 e^{-kt}$ $U = U_0 + \frac{k_1 A_0}{k}(1 - e^{-kt})$ $V = V_0 + \frac{k_2 A_0}{k}(1 - e^{-kt})$ $W = W_0 + \frac{k_3 A_0}{k}(1 - e^{-kt})$
2. Parallel first-order reactions producing a common product:	$A \xrightarrow{k_{+1}} C + \ldots$ $B \xrightarrow{k_{+2}} C + \ldots$	$A = A_0 e^{-k_1 t}$ $B = B_0 e^{-k_2 t}$ $C = A_0 + B_0 - A_0 e^{-k_1 t} - B_0 e^{-k_2 t}$
3. Series first-order reactions:	$A \xrightarrow{k_{+1}} B \xrightarrow{k_{+2}} C$	$A = A_0 e^{-k_1 t}$ $B = \frac{A_0 k_1}{(k_2 - k_1)}(e^{-k_1 t} - e^{-k_2 t})$ if the initial concentrations of B_0 and C_0 are zero; therefore $A_0 = A + B + C$, $\therefore C = A_0 \left[1 + \frac{1}{k_{+1} - k_{+2}}(k_{-2} e^{-k_1 t} - k_{+1} e^{-k_{+2} t})\right]$
4. First-order reversible reaction:	$A \underset{k_{-1}}{\overset{k_{+1}}{\rightleftarrows}} B$	If only A is present at $t = 0$, $\ln \frac{k_{+1} A_0}{(k_{+1} + k_{-1})A - k_{-1} A_0} = (k_{+1} + k_{-1})t$ Or, if equilibrium concentrations (subscript "eq") are introduced, $k_{+1} A_{eq} = k_{-1} B_{eq} = k_{-1}(A_0 - A_{eq})$ or $A_{eq} = \left(\frac{k_{-1}}{k_{+1} + k_{-1}}\right) A_0$ and $\ln \left(\frac{A_0 - A_{eq}}{A - A_{eq}}\right) = (k_{+1} + k_{-1})t$
5. Second- and first-order reversible reactions:	$A + B \underset{k_{-1}}{\overset{k_{+1}}{\rightleftarrows}} C$	$k_{+1} t = \frac{C_{eq}}{(A_0^2 - C_{eq}^2)} \ln \left[\frac{C_{eq}(A_0^2 - C_{eq} C)}{A_0^2 (C_{eq} - C)}\right]$
6. Second-order reversible reaction:	$A + B \underset{k_{-1}}{\overset{k_{+1}}{\rightleftarrows}} C + D$	$k_{+1} t = \frac{C_{eq}}{2A_0(A_0 - C_{eq})} \ln \left[\frac{A_0 C_{eq} + C(A_0 - 2C_{eq})}{A_0(C_{eq} - C)}\right]$

From Livingston, R., *Technique of Organic Chemistry, Investigation or Rates and Mechanisms of Reactions*, Vol. 8, Freiss, S. L. and Weissberger, A., Eds., Interscience, New York, 1953, chap. 4, 169.

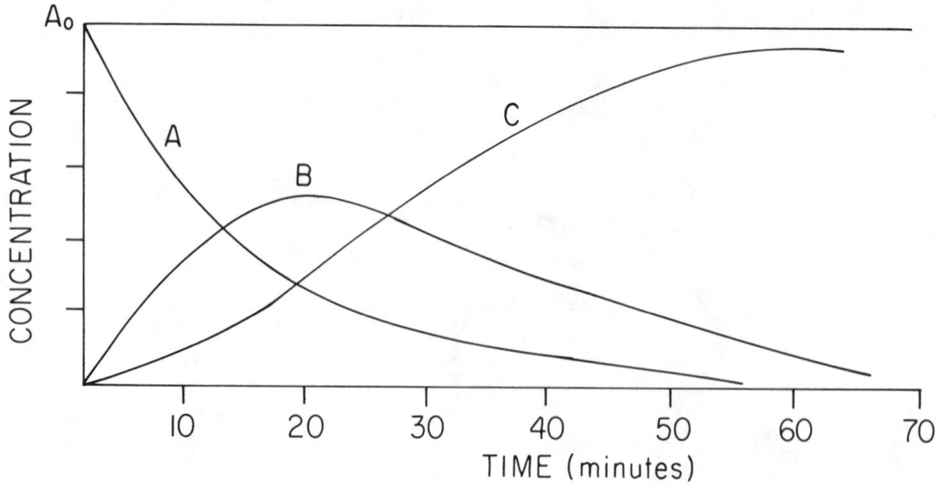

FIGURE 2. Theoretical concentration-time curves for A, B, and C in a series first-order reaction, $A \xrightarrow{k+1} B \xrightarrow{k+2} C$, where $k_{+1} = 0.1\ min^{-1}$ and $k_{+2} = 0.05\ min^{-1}$. (From Frost, A. H. and Pearson, R. G., *Kinetics and Mechanism,* John Wiley & Sons, New York, 1953, 154. With permission.)

$0.1\ min^{-1}$ and $k_{+2} = 0.05\ min^{-1}$. This is a case which shows a great deal of similarity to Figure 1 (Gutfreund[418] and Halford[419] have also treated this simple case of two consecutive steps).

It is of interest to note that the concentration of the intermediate B in Figure 2 goes through a maximum. By using the integral solution of B (see Table 2 and setting $\frac{dB}{dt} = 0$, it found that

$$(k_{+1}t)_{max} = \frac{1}{\left(\frac{k_{+2}}{k_{+1}} - 1\right)} \ln\left(\frac{k_{+2}}{k_{+1}}\right) \tag{4}$$

and also, the value of B at the maximum is

$$\frac{B_{max}}{A_0} = \left(\frac{k_{+2}}{k_{+1}}\right)^{\left(\frac{k_{+2}}{k_{+1}}\right)/\left(1 - \frac{k_{+2}}{k_{+1}}\right)} \tag{5}$$

which shows that the maximum becomes less pronounced as the ratio of the rate constants, $\left(\frac{k_{+2}}{k_{+1}}\right)$ gets larger.

In Mechanism 4 of Table 2, for a first-order reversible reaction, $A \underset{k_{+1}}{\overset{k_{+2}}{\rightleftarrows}} B$, the integrated solution with equilibrium concentrations explicitly introduced, shows that the approach to equilibrium is a first-order process with the effective rate constant equal to the sum of the constants for the forward and reverse directions.

III. KINETICS OF THE TRANSIENT PHASE

Several theoretical treatments of transient-phase enzyme kinetics have been given. The simplest enzyme reaction,

$$E + S \underset{k_{-1}}{\overset{k_{+1}}{\rightleftarrows}} ES \overset{k_{+2}}{\rightleftarrows} E + P \tag{6}$$

was considered by Gutfreund[418] and Roughton.[420] Thus, the rate of formation of the enzyme-substrate complex is given by

$$\begin{aligned} \frac{d(ES)}{dt} &= k_{+1}(E)(S) - (k_{-1} + k_{+2})(ES) \\ &= k_{+1}(S)[E_0 - (ES)] - (k_{-1} + k_{+2})(ES) \\ &= k_{+1}(S)E_0 - (k_{+1}(S) + k_{-1} + k_{+2})(ES) \end{aligned} \tag{7}$$

The rate of formation of the reaction product, P, is

$$\frac{d(P)}{dt} = k_{+2}(ES) \tag{8}$$

where

$$\frac{d^2(P)}{dt^2} = k_{+2} \frac{d}{dt}(ES) \tag{9}$$

Eliminate (ES) by substituting Equations 8 and 9 with Equation 7.

$$\frac{d^2P}{dt} + (k_{+1}(S) + k_{-1} + k_{+2}) \frac{dP}{dt} - k_{+2}k_{+1}(S)E_0 = 0 \tag{10}$$

If a high ratio of $(S)/E_0$ were to be employed, relatively little substrate would be consumed throughout the transient state, consequently, S would remain close to its zero-time value (i.e., S_0). With this provision, Equation 10 can be treated as a second-order differential equation with constant coefficients, leading to the solution obtained by Gutfreund and Roughton,[418,420] i.e.,

$$P = P_0 + \left[\frac{k_{+2}k_{+1}S_0E_0t}{k_{+1}S_0 + k_{-1} + k_{+2}}\right] + \frac{k_{+2}k_{+1}S_0E_0t}{(k_{+1}S_0 + k_{-1} + k_{+2})^2}[e^{-(k_{+1}S_0+k_{-1}+k_{+2})t} - 1] \tag{11}$$

where P is the initial concentration of product (included because the product measured by Gutfreund was H^+). If no reaction product were preadded to the system, $P_0 = 0$ in Equation 11. If the exponential term were written in the form of a power series of

$$e^x = 1 + x + \frac{x}{2!} + \frac{x}{3!} + \ldots \frac{x^n}{n!} + \ldots \tag{12}$$

when t is small, as early in the transient state, then one may neglect all terms beyond the third term for small values of t, and substituting only its first three terms into Equation 11, we obtain the simplified Equation 13 (or 14) valid only for the early part of the acceleration phase.

$$P - P_0 = \frac{k_{+1}k_{+2}S_0E_0t^2}{2} \tag{13}$$

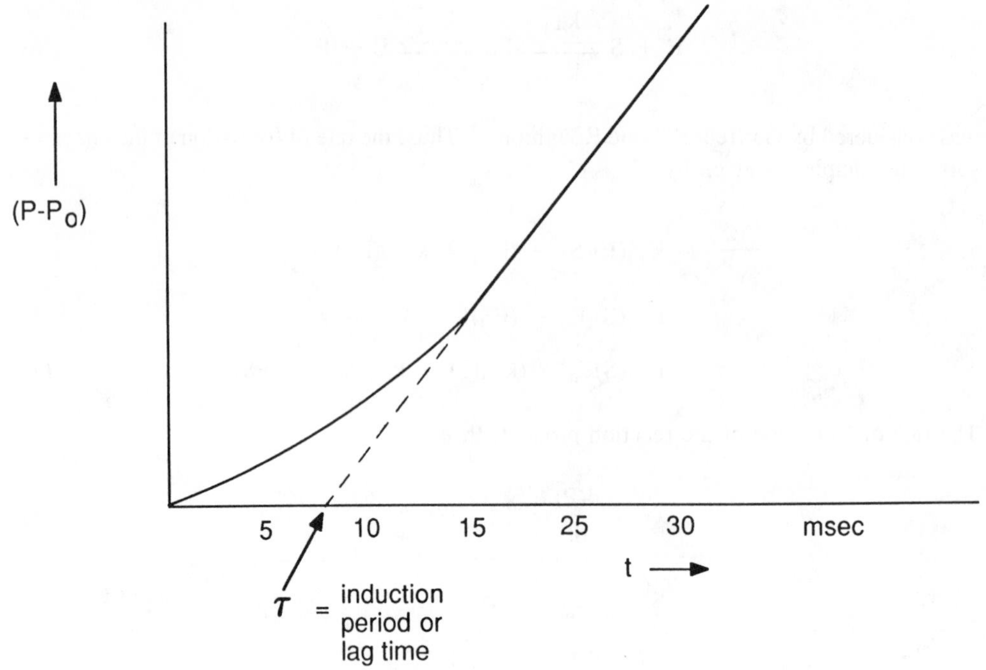

FIGURE 3. Pre-steady-state kinetics (formation of product during transient phase and a method of determining k_{+1}. (From Gutfreund, H., *Disc. Far. Soc.*, 10, 167, 1955. With permission.)

or if $P_0 = 0$,

$$P = \frac{k_{+1}k_{+2}E_0S_0t^2}{2} \tag{14}$$

Therefore, the reaction product accumulates exponentially with t early in the transient state, and by plotting P vs. t^2 (or $P - P_0$ vs. t^2) from Equations 13 or 14, a linear plot results whose slope

$$\frac{k_{+1}k_{+2}E_0S_0}{2} \tag{15}$$

and with S_0 known, $k_{+1}k_{+2}E_0$ may be obtained. From Chapter 1, Equations 6 to 10, if

$$V_{max} = k_{+2}E_0 \tag{16}$$

measured from steady-state kinetics and employing the Michaelis-Menten equation (Chapter 1, Equation 10), then k_{+1} may be calculated, while by measuring $K_m = (k_{-1} + k_{+2})/k_{+1}$ (Chapter 1, Equation 9), k_{-1} may be deduced; with a knowledge of the enzyme concentration, k_{+2} and all three rate constants are obtained. However, Equation 11 may be used in another way.[422] If $P - P_0$ is plotted vs. t as in Figure 3, a curve is obtained in accordance with Equation 11 where the curvature is due to the exponential term and at large values of t (where the exponential term may be neglected and the system is in the steady-state phase), then the curve follows the dotted straight line. Under these conditions, Equation 11 is reduced to Equation 17.

$$P - P_0 = \frac{k_{+2}k_{+1}S_0E_0t}{k_{+1}S_0 + k_1 + k_{+2}} - \frac{k_{+2}k_{+1}S_0E_0}{(k_{+1}S_0 + k_{-1} + k_{+2})^2}$$

$$= \frac{k_{+2}S_0E_0t}{S_0 + \left(\frac{k_{-1} + k_{+2}}{k_{+1}}\right)} - \frac{k_{+2}S_0E_0}{k_{+1}\left[S_0 + \left(\frac{k_{-1} + k_{+2}}{k_{+1}}\right)\right]^2}$$

$$\frac{k_{+2}S_0E_0t}{S_0 + K_m} - \frac{k_{+2}S_0E_0}{k_{+1}[S_0 + K_m]^2} \tag{17}$$

If $S_0 \gg K_m$, that is at saturation concentrations of substrate, Equation 17 may be approximated by

$$P - P_0 \simeq k_{+2}E_0t - \frac{k_{+2}E_0}{k_{+1}S_0} \tag{18}$$

and the x-intercept, or τ (when $P - P_0 = 0$) is

$$\tau = \frac{1}{k_{+1}S_0} \tag{19}$$

which provides a direct determination of k_{+1}.

It will be readily appreciated that the validity of Gutfreund's equation rests upon the assumption that K_m is represented by the Michaelis-Menten equation as derived by Briggs and Haldane for a steady state in (ES) (see Chapter 1), and that these equations accordingly apply to systems with a single intermediate enzyme-substrate complex ES, in which k_{+1} is the rate constant for its formation and which then breaks down directly to yield products. The above equations for the pre-steady state do not apply to systems in which there are two or more intermediate complexes where K_m, as we have seen is actually a more complex expression of rate constants than simply $\left(\frac{k_{-1}+k_{+2}}{k_{+1}}\right)$, and the rate of formation of product may be limited by one or more of the intermediate steps; in this case $1/\tau$ will no longer be a measure of k_{+1} but will likely be a function of several other rate constants as well as shown by Laidler and Bunting[423] for a double-intermediate mechanism. However, Gutfreund[422,424] found that in the case of rapid measurements in the pre-steady state of the hydrolysis of synthetic amino acid esters by a few proteolytic enzymes, his equations yielded approximate and valid measures of the rate constants. Thus, he found for the hydrolysis of benzoyl-L-arginine ethylester by trypsin, $k_{+1} \geqslant 4 \times 10^6$ and $k_{+2} \simeq 15\ s^{-1}$, since $K_m = 10^{-5}M$,[425] $k_{-1} \geqslant 25\ s^{-1}$. In those cases, Gutfreund[422,424,426,427] found $k_{+2} \ll k_{-1}$, so that $K_m = \frac{k_{-1}+k_{+2}}{k_{+1}}$ approached $\frac{k_{-1}}{k_{+1}}$, or in essence equivalent to the assumption which Michaelis and Menten had made in their original derivation (see Chapter 1).

The transient-state equations for more complex mechanisms than the single (ES) intermediate case discussed above have been treated in a general fashion by Darvey[428] and by Hijazi and Laidler[429] and summarized by Hammes and Schimmel[431] and by Laidler and Bunting.[423] Darvey has shown that τ (the induction period or lag time, see Figure 3) is independent of the enzyme concentration and may be expressed by a ratio of two polynomials in substrate concentration, S,

$$\tau = \frac{m_zS^z + m_{z-1}S^{z-1} + \ldots + m_1S + m_0}{d_zS^z + d_{z-1}S^{z-1} + \ldots + d_1S + d_0} \tag{20}$$

where the d-coefficients are all of the same sign, but the sign of the m-coefficients may vary. Therefore, τ may be either negative or positive, as had been observed by Ouellet and Stewart.[430] Hijazi and Laidler[429] have showed that the equations for several one-substrate, two-substrate, and inhibition mechanisms all obey the general form of

$$P = vt + \sum_{i=1}^{n} (B_i)e^{-\lambda_i t} - \sum_{i=1}^{n} \beta_i \qquad (21)$$

where v is the steady-state velocity, n is the number of exponential terms and which is equal to the number of enzyme species in the mechanism other than the free enzyme, and the sum of the exponents, $\lambda_1 + \lambda_2 + \lambda_3 + \ldots + \lambda_n$, is equal to the sum of all the reaction arrows in the mechanism (i.e., all of the monomolecular rate constants and all of the bimolecular rate constants each multiplied by its associated nonenzymatic ligand). Equation 21 was applied to a pre-steady-state kinetics study of chymotrypsin, alkaline phosphatase, and myosin, and provides a general procedure for estimating the bimolecular rate constants. Thus, whenever a ligand, S, reacts with only one enzyme species in the mechanism in a bimolecular k_s-step, i.e., $S + E_s \xrightarrow{ks} E_s \cdot S$, the rate constant, k_s, may be obtained from the slope of the linear plot of $(\lambda_1 + \lambda_2 + \lambda_3 + \ldots + \lambda_n)$ vs. S.

IV. RAPID FLOW METHODS

The conventional method for following the course of a chemical reaction is the so-called "static" method; the separate reactants are introduced into the thermostated reaction vessel, cuvet, pH-stat chamber, etc., mixed either manually or by a stirring device, and the concentration changes in that vessel are then monitored either by periodically sampling and analyzing the reaction mixture or by use of a continuous monitoring of the progress of the reaction via some change in one of the physical properties (e.g., spectrophotometric, fluorometric, etc.) of one of the reactants or products.

Difficulties will arise, however, if the reaction proceeds very rapidly, since the time required for mixing the reactants may be comparable to the half-life of the reaction, and the speed at which mixing may be accomplished sets an upper limit to the rate of the reaction steps that may be isolated and measured in this manner — by the static method — which is usually much too slow for the direct measurement of most enzymatic individual reaction steps. Measurements of rapid reactions became possible only after Hartridge and Roughton[354,432] in 1923 introduced the first of the continuous-flow techniques for studying fast reactions.

In this method, the enzyme and substrate solutions are forced into a specially designed "mixing chamber" at high speed (see Figure 4),[432] and the reaction mixture flows from the chamber into an observation tube. The time required by the reaction mixture to reach any point along the tube depends on the distance from the mixing chamber and on the combined flow velocity. The progress of the reaction over an interval of milliseconds can be monitored either by observation (usually by spectrophotometry) at different points along the tube at a single selected flow velocity, or by observation at a single point on the observation tube at different flow velocities.

There are a number of variations of this technique of continuous flow which will be discussed below.[437]

In general, there are in current use two types of rapid flow techniques: the continuous-flow method (or some variant of it, e.g., the chemical-stop procedure described below) and the "stopped-flow" techniques developed largely by Chance[434,441,468] and by Gibson.[435] Chance actually combined a variation of the two procedures in his "accelerated and stopped-flow" method[434,436,468] and successfully applied spectrophotometric methods of great sensitivity and stability in his monitoring devices.

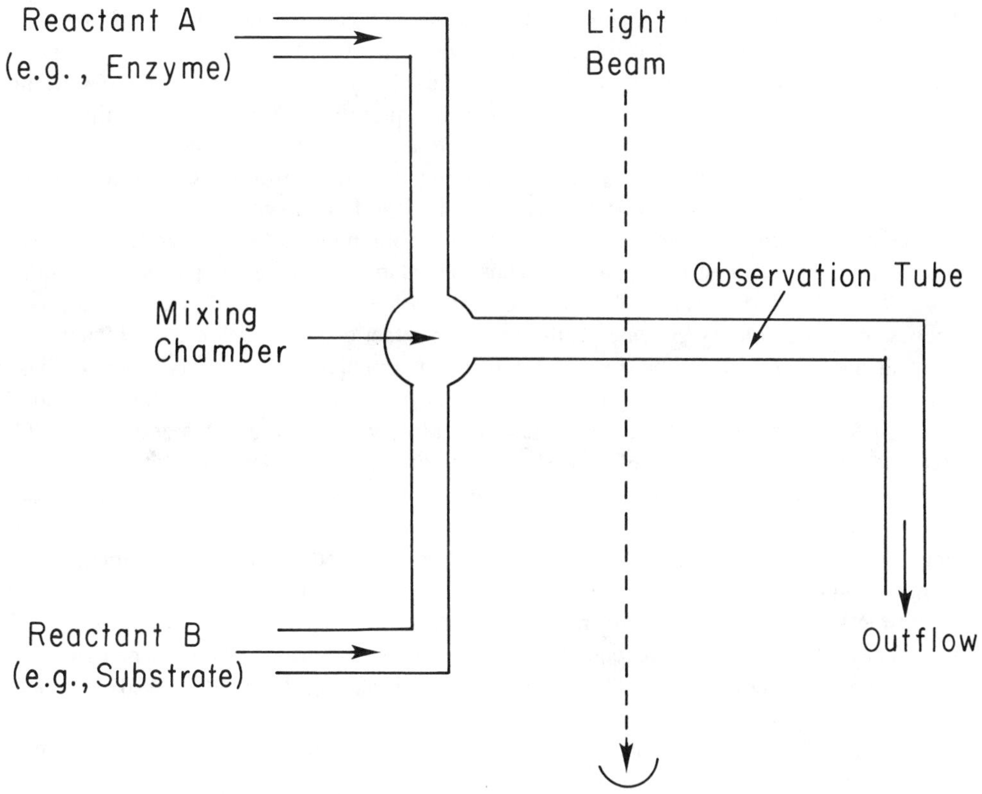

FIGURE 4. Schematic drawing of a continuous-flow apparatus.

In the stopped-flow technique, e.g., of the Gibson type, the solutions are rapidly injected, via syringe drives, into a mixing chamber, passed into an observation tube and then into a ''stopping-syringe'' which then brings the solution flow to a rapid stop; following the point of ''stopped-flow'', the progress of the reaction is monitored at a single observation point with a rapid recording of the optical (spectrophotometric or fluorometric) properties of the system made feasible by the use of a cathode-ray oscilloscope. Strictly speaking, the stopped-flow technique is not really a rapid-flow method, since the kinetic study is made of the static system with the fast flow utilized simply to bring about rapid mixing usually through the use of so-called ''turbulent-jet'' mixers. In general, the continuous-flow method, because it calls for a relatively prolonged flow of the reaction mixture (except in the case of the chemical-stop procedure, or in Chance's microadaptations of the accelerated-flow method) is more demanding than the stopped-flow technique with respect to enzymes and substrates. However, in both cases, advances in the engineering design of the flow apparatus have permitted an increasingly faster flow with more efficient and rapid mixing without cavitation problems and the resolution time now approaches 10^{-4} s.[438]

As an application of the continuous-flow technique, let us discuss a particular design of the apparatus which was utilized in a kinetic study of the oxidation by molecular oxygen of the cytochrome chain of intact yeast cells, by Ludwig et al.[439,440] For the study of this oxygen reaction, Chance[441,442] had derived and tested experimentally a relationship for the optimal design of a rapid-flow apparatus of the continuous-flow type employing spectrophotometric observations. These derivations clearly showed that the performance of the apparatus increased as the amount of material expended in a single observation was increased, and he suggested the design of a new type of flow apparatus in which this could be accomplished by a continuous recirculation of the reactants.[441] This regenerative-flow apparatus, either of a continuous or intermittent nature, was constructed and applied[442,443] to these studies. A reservoir for providing an adequate interval for the dissipation of the perturbing effect and the reestablishment of the initial state was provided. This is an especially useful device for the study of labile enzyme-substrate intermediates, provided the product, or one of the products, was innocuous, e.g., H_2O as in the case of cytochrome c oxidase, and the enzyme-substrate compounds were to decompose back to free enzyme at a convenient point past the observation site. The regenerated enzyme solution could then be recirculated back to the mixing chamber of the apparatus. Thus, an intermittent regenerative flow apparatus was selected by Ludwig et al.[439,440] as the best choice for these studies on the reaction kinetics of the respiratory enzymes with oxygen in intact cell suspensions.[439,440] Parenthetically, it may be remarked that in the search for intermediates of the cytochrome oxidase reaction, Chance had designed another rapid-flow device, a so-called pulsed-flow apparatus, capable of measurements of extremely fast reactions, with mixing times as short as 300 μsec and with the expenditure of modest volumes of material.[444] This apparatus, however, lacked the flexibility of design of the intermittent regenerative-flow apparatus, to be described, which allowed a systematic variation of over 200-fold in the initial oxygen concentration. Moreover, the sensitive spectrophotometer employed and the time-invariant nature of the continuous-flow design of the regenerative-flow apparatus permitted a scan of both the visible spectrum and the Soret region at a single oxygen concentration as low as 100 n*M*. Today, the intermittent regenerative-flow apparatus described herein remains one of the few rapid-flow devices capable of systematically measuring, both in the steady-state and in the pre-steady-state regions, the direct effect of a wide range in oxygen concentrations (in solution) on the intact cellular cytochrome components. The ''intermittent regenerative-flow'' apparatus is shown in schematic form in Figure 5.

The apparatus consists of basically two reservoirs (2 l in volume) joined to the specially designed mixing and observation chamber (Figure 5B)[442,443,445] inserted in place of the cuvette housing of the split-beam spectrophotometer.[446,447] The enzyme solution (or an anaerobic cell suspension rendered O_2 free by the metabolism of a suitable substrate) was propelled at a constant rate from the source reservoir by means of a nitrogen gas pressure drive (Figure 5A) through the mixing chamber and mixed with the substrate (e.g., O_2 dissolved in buffer) being injected at a constant rate into the main stream of the flowing enzyme, by means of a motor-driven syringe. Mixing takes place past the optical window, B (Figure 5B), at a fixed distance before the second optical window, A. Thus, one light beam traverses free enzyme (or reduced cytochrome), whereas, the second beam traverses the enzyme-substrate complex (or, e.g., steady-state oxidized cytochrome), with the split-beam spectrophotometer recording directly (in optical density units) the difference spectrum as a function of wavelength. The outflow is collected in the second reservoir bottle where the enzyme-substrate complex is allowed to decompose (and in the present experiments, the cytochromes are restored to the reduced state when the injected oxygen, i.e., the substrate for cytochrome oxidase, is eventually consumed, as the substrate reductant, e.g., ethanol, is enzymatically oxidized). The enzyme is then returned to the source reservoir for the next run. The flow velocity of the main stream was calibrated in terms of initial pressure, as measured on a manometer, and periodically checked by direct observations on the $\Delta h/\Delta t$ (i.e., change in height of liquid in the source reservoir with time, in turn calibrated in terms of Δ(volume)/Δt). The flow velocities of the main stream could be varied from approximately 30 cm^3 to 155 cm^3/s, with the lower limit set by the design of the turbulent mixer; laminar flow in the jets presumably diminishes the mixing efficiency.[442] A spacer could be introduced between the mixing chamber and the observation window A to allow a suitable time period for the steady state to be adjusted. Thus, the practical time range, after mixing, was approximately 6.5 to 33 msec with a 1-cm spacer, and approximately 32.5 to 165 msec with

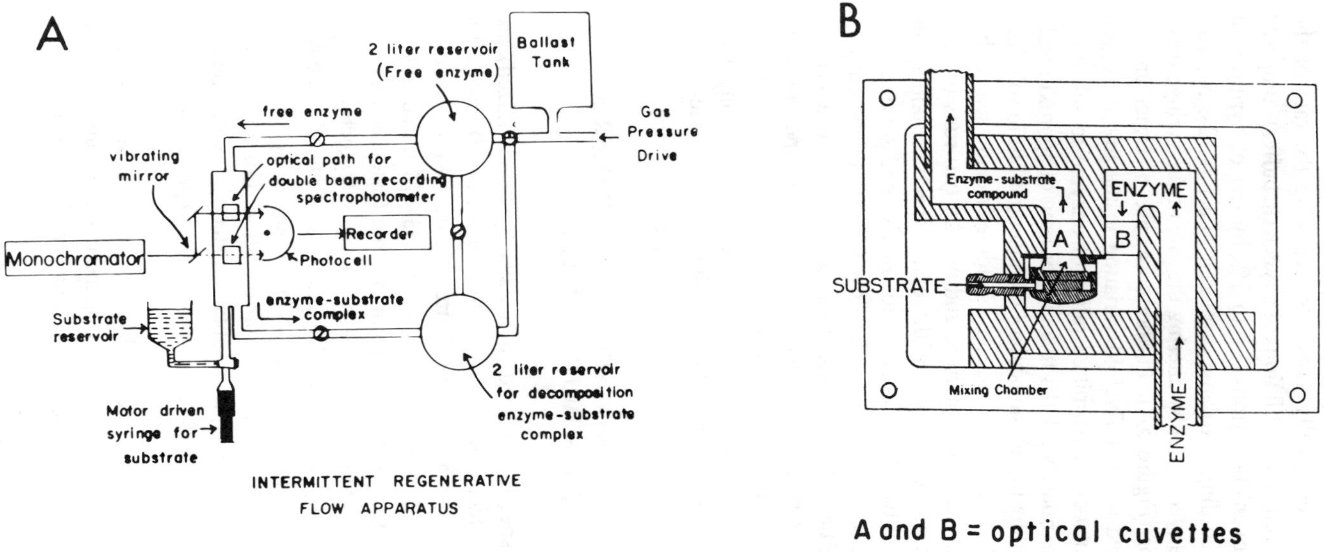

FIGURE 5. Schematic diagram of the intermittent regenerative continuous-flow apparatus. (A) The flow-system assembly is described as a general application to the difference spectrophotometric measurements of labile enzyme-substrate intermediates. In the application described herein, the substrate reservoir contains a specific concentration of O_2 dissolved in buffer, and it is injected by the motor-driven syringe at a constant rate into the main flow of the cell suspension (e.g., Baker's yeast), previously reduced with a suitable reductant substrate. After mixing, the split-beam wavelength scanning spectrophotometer records the difference spectrum between reduced and steady-state oxidized cytochrome. Finally, after the passage of the yeast suspension (with the gas-pressure drive), from its source to the collection reservoir, it is returned again to the source reservoir, where it is allowed to become anaerobic before initiating the next cycle. (B) Some of the details of the mixing and observation chamber, with its two optical cuvettes, "A" and "B", are shown here. For a description of the turbulent-jet mixers, refer to References 442, 443, and 445.[439,440]

the 5-cm spacer (the spacers were also of approximately 1 cm^2 in cross section). For the studies with the yeast system, approximately 100 msec (about 50 cm^3/sec, flow velocity with approximately 5-cm spacer) proved to be a more than adequate time for the steady state to be adjusted. In those cases where the flow velocity of the enzyme (or cell suspension) was maintained constant, i.e., the time between mixing and observation was fixed, the initial concentration of oxygen injected was systematically varied by adjusting the rate of the syringe drive or the syringe volume displaced. Thus, dilutions of oxygen (in buffer) compared to the main stream could be varied conveniently from 0.1 to 2% by use of syringes of appropriate cross-sectional area (and accordingly volume displaced; e.g., 1-, 2-, 5-, 10-, 20-, 30-, and 50-cm^3 syringes were employed), and by altering the rate of the syringe drive through the use of a variable DC motor (Figure 5A). Furthermore, use of either an air-saturated or O_2-saturated buffer permitted another fivefold variation in the initial oxygen concentrations. This overall flexibility of design permitted ratios of enzyme to substrate-volume flows from a lower limit of approximately 2000/1 to an upper limit of approximately 50/1, or to allow a systematic variation in the initial molecular O_2 concentration, in solution, some 200-fold, from about 0.1 to 20 μM. The sensitive spectrophotometer could scan either the Soret region or visible spectrum, each in less than 20 s, and usually two runs through the regenerative-flow system more than sufficed to cover the entire spectrum under one particular set of conditions (e.g., at one initial O_2 concentration). In a second set of experiments, designed to penetrate the pre-steady-state region, the flow velocity of the enzyme (or cell suspension) was varied, but the initial substrate concentration (i.e., O_2 concentration) was fixed by a suitable and concomitant variation in the syringe drive. Thus, the time after mixing could be varied down to approximately 6.5 msec at a fixed initial O_2 concentration as low as 1.3 or 0.6 μM, and permitted estimation of pseudo first-order rate constants from 150 to 50 s^{-1}, respectively, or calculated second-order rate constants of the order of 10^8 M^{-1} s^{-1}.

A. MEASUREMENTS OF THE STEADY-STATE KINETICS (IN BAKER'S YEAST) OF OXIDATION OF THE CYTOCHROME CHAIN BY MOLECULAR OXYGEN

In Figure 6 two sets of difference spectra (reduced minus steady-state oxidized) are presented of the respiratory pigments of intact Baker's yeast cells, obtained 100 msec after the addition of a specific amount of oxygen. The upper figure (Figure 6A) shows three spectra obtained from a sample of "starved" yeast, at 1-, 13-, and 23-μM initial oxygen concentrations, respectively. It may be noted, that the peaks at 605 nm (cytochromes a + a_3), 550 nm (cytochrome c + c_1), 520 nm (cytochromes c + b), 445 nm (cytochrome a_3), and 420 nm (mainly cytochrome c) are clearly delineated. On the lower figure (Figure 6B) the difference spectrum is shown of the respiratory pigments of a sample of "unstarved yeast" obtained 100 msec after the addition of 13-μM O_2 concentration. For comparison, a difference spectrum is presented of the same preparation obtained statically with conventional 1-cm cuvettes, at a relatively large initial oxygen concentration of approximately 400 μM and at about 30 s after mixing.

One notes that the pigments appear maximally oxidized in the steady state at 400-μM O_2 concentration, but that at 13-μM O_2 concentration, most of the pigments appear to be *near* maximally oxidized. The most conspicuous difference between these two cases in Figure 6B is that the "flavoproteins" appear to be largely oxidized at 400 μM O_2 as shown by the deep trough at 460 nm, in contrast to that observed at 100 msec and 13 μM O_2, where very little "flavoprotein" oxidation appears to have occurred. Finally, although Figures 6A and B are for two separate yeast preparations, "starved" and "unstarved", there appear to be only very little "significant" differences between them when both curves are normalized.

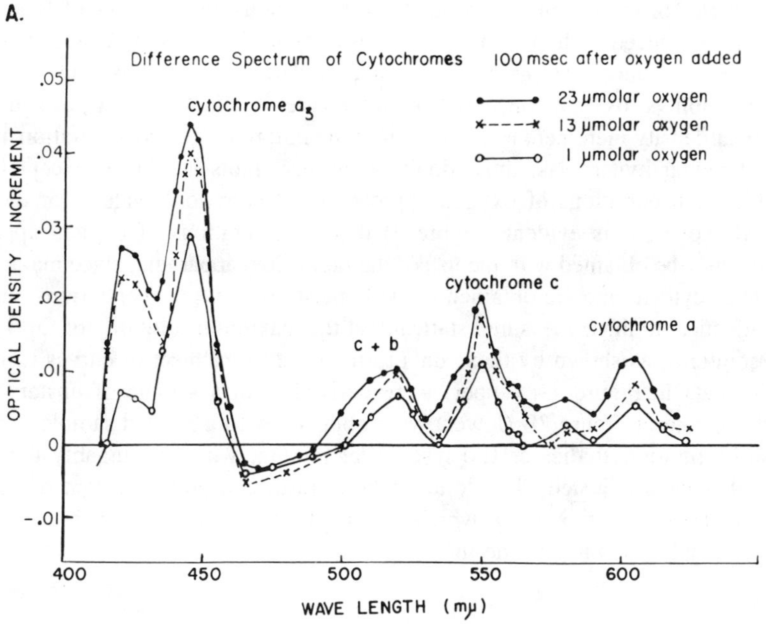

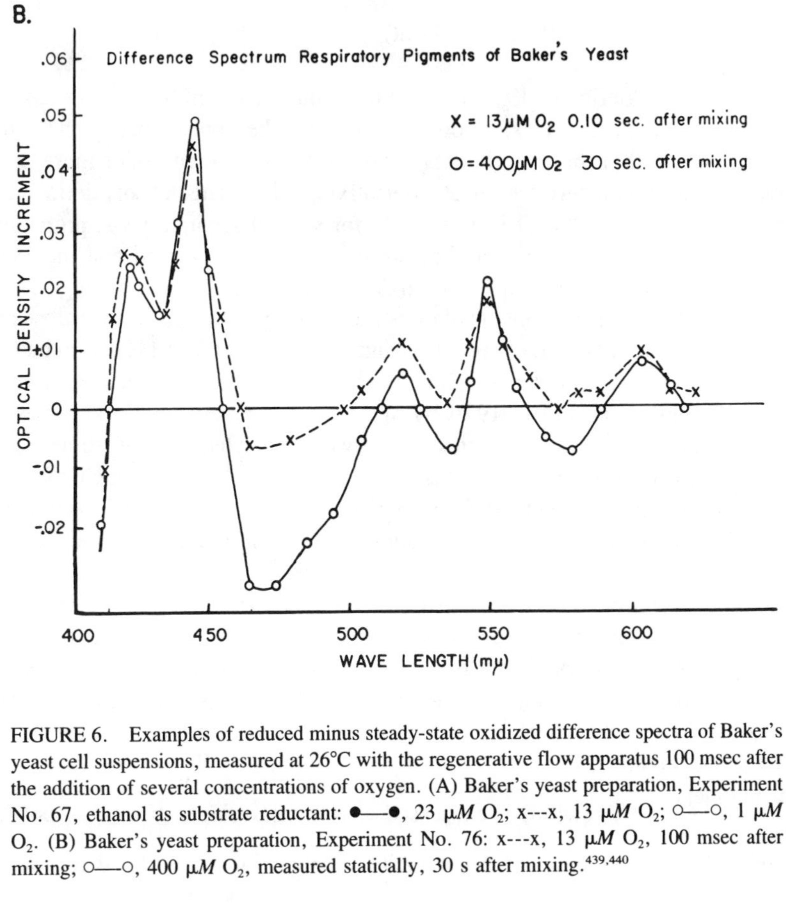

FIGURE 6. Examples of reduced minus steady-state oxidized difference spectra of Baker's yeast cell suspensions, measured at 26°C with the regenerative flow apparatus 100 msec after the addition of several concentrations of oxygen. (A) Baker's yeast preparation, Experiment No. 67, ethanol as substrate reductant: ●—●, 23 μM O_2; x---x, 13 μM O_2; ○—○, 1 μM O_2. (B) Baker's yeast preparation, Experiment No. 76: x---x, 13 μM O_2, 100 msec after mixing; ○—○, 400 μM O_2, measured statically, 30 s after mixing.[439,440]

Measurements of the extents of oxidation of the various cytochromes of Baker's yeast cells (previously reduced with ethanol as the substrate) at 25 to 100 msec after mixing with a number of initial concentrations of O_2 yielded similarly shaped plots (Figure 7) for the extent of oxidation vs. oxygen concentration in the case of cytochromes a_3, a, and c. These plots of optical-density increment as a function of the initial oxygen concentration all appear to yield rectangular hyperbolas; thus, double-reciprocal plots are linear except for those relatively high concentrations of oxygen approaching "saturation" values for the various pigments. Moreover, it is evident (Figure 7) that concentrations of oxygen approaching "saturation" may be obtained with the use of the rapid-flow apparatus, since maximal ΔOD values for each cytochrome are obtained at the highest initial O_2 concentrations, and which are almost identical to those measured statically at the maximum solubility of O_2 (i.e., 1200 μM O_2 in solution), as shown by those data to the right of each set of curves (Figures 7A and B). The data in Figure 7A (upper) were derived from a sample of unstarved yeast; whereas, those data in Figure 7B (lower) were obtained with a starved sample of yeast. In all of these O_2-titration studies at 100 msec after mixing, when presumably the complex steady state has been adjusted, the plotted data seem to conform to a type of Michaelis-Menton[14] expression, a derivation of which, and applicable to this case, is presented below and where it may be shown that one may define a

$$K_m \cong \left[\frac{k_{-1} + k_{+2}(a^{+2})}{k_{+1}}\right] \tag{22}$$

and a so-called "turnover number"

$$v_{max}^{respiration}/(E_t) = (dx/dt)_{max}/(a_3^{total}) = k_{+2}(a^{+2}) \tag{23}$$

For this system, the "defined" K_m value, or the value of the initial oxygen concentration corresponding to half-maximal oxidation, was found to be approximately the same for all three cytochromes within the experimental errors when measured under these steady-state conditions (i.e., approximately 100 msec after mixing). Estimates of this defined K_m for O_2 ranged between 0.6×10^{-6} to 0.8×10^{-6} M for several separate yeast preparations (see Table 3) and independent of whether they were starved or unstarved and indicative of an unusually high affinity for oxygen by the oxidase. Surprisingly, the average K_m value of 0.7 μM (estimated from measurements of cytochrome a_3 oxidation) is in very close agreement with the determination of Winzler[448] in 1941, and by Longmuir[449] in 1954 who had measured the respiration of yeast cells polarographically, and is in good agreement with the tentative value obtained manometrically in 1929 by Warburg and Kubowitz.[450] The measurements made by these three groups of early investigators were restricted to measurements of oxygen concentration as a function of time, and it is all the more remarkable that such good agreement is observed with the present methodology, which, in essence, could be considered a spectrophotometric titration of the oxygenating pigment, cytochrome a_3 (however, see Reference 451).

The "turnover numbers" for four of these yeast preparations were estimated from the maximal respiration (oxygen uptake) measured polarographically (converted into units of μ equivalents per liter per second) and the approximate a_3 concentration estimated from the ΔOD_{max} measured at 445 nm and the $\Delta\epsilon_m M$ assumed for cytochrome a_3.[452,453] These data are summarized in Table 3 where, curiously, the unstarved preparations of yeast seemed to yield somewhat higher values for the turnover number. If $k_{-1} \ll k_{+2}$ (a^{+2}), then $\frac{\text{(turnover number)}}{K_m} \simeq k_{+1}$ (with K_m expressed also in equivalents of oxygen per liter), and a *minimum* value for the second-order velocity constant for a postulated reaction between

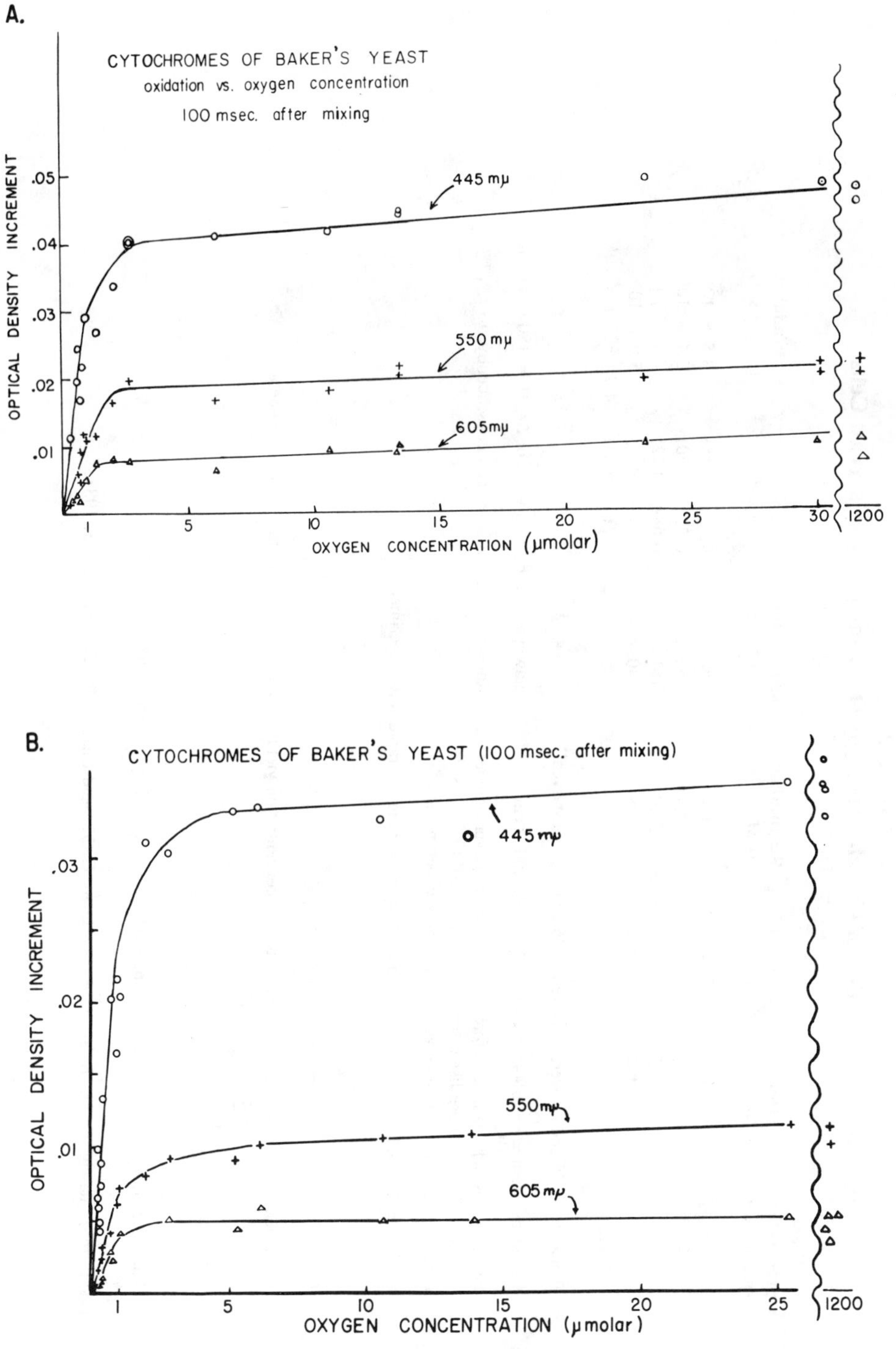

FIGURE 7. Difference spectrophotometric measurements at 26°C of the extent of oxidation of the various cytochromes of Baker's yeast cells, 100 msec after mixing with various concentrations of O_2. (A) Yeast preparation, Experiment No. 76, initially reduced with ethanol as substrate: o—o, Δ OD measurements at 445 nm minus baseline; + --- +, Δ OD at 550 nm minus baseline; △—△, Δ OD at 605 nm minus baseline. Results to the right of the wavy line were determined statically, before and after the series of determinations. (B) Yeast preparation, Experiment No. 69, same conditions and description as in A above.[439,440]

TABLE 3
Steady-State Oxidation of Cytochrome a_3 by O_2 in Intact Baker's Yeast Cells

Yeast preparation[a] (experiment no.)	Total a_3[b] (μM) conc	Dry wt.[c] (mg/ml)	Respiration[d] ($\mu M \cdot s^{-1}$)	Turnover no.[e] (s^{-1})	K_m[f] (μM)	k_{+1}[g] (minimal $l \cdot eq^{-1}\, s^{-1}$)
68	0.358	10.1	8.9	99.3	0.72 (±0.15)	3.5×10^7
69	0.398	11.3	9.8	98.2	0.73 (±0.20)	3.4×10^7
76	0.575	14.9	26.0	181	0.68 (±0.12)	6.7×10^7
84	0.424	10.9	14.8	140	—	5.0×10^7[h]
Average					0.71 (±0.03)	$4.7 (\pm 1.6) \times 10^7$

[a] Preparations 68 and 69 were "starved", 76 and 84, "unstarved".

[b] Calculated from $\frac{\Delta OD}{\Delta\epsilon_{mM}}$ at 445 nm where $\Delta\epsilon = 87\ mM^{-1}\ cm^{-1}$.[452,453] In Ludwig and Kuby,[439] a value of $\Delta\epsilon_m M = 180$ had been employed, which leads therefore to calculated turnover numbers 2.07 times those given here, and consequently, to $k_{+1}^{(minimal)}$ values 2.07 times those listed (or to about $1 \times 10^8\ l \cdot Eq^{-1}\ s^{-1}$).

[c] Concentration of yeast employed in rapid-flow experiments.

[d] $-\frac{dx}{dt}$ for O_2 uptake, calculated for the dilution (c) used in the rapid-flow experiments.

[e] Calculated from $\frac{-\frac{dx}{dt} \times 4}{[a_3]^{(\mu M)}}$ (in equivalents s^{-1})

[f] Estimated from measurements of the initial O_2 concentration yielding half-maximal oxidation of cytochrome a_3.

[g] Calculated from $-\left[\frac{\frac{dx}{dt} \times 4/[a_3]}{K_m \times 4}\right] \overline{(\text{in Eq. } 0_2 \cdot l^{-1})}$

[h] For calculations of $k_1^{(minimal)}$, K_m assumed to be 0.7 μM, as in preparations 68, 69, and 76.

Taken from Ludwig, G. D., Kuby, S. A., Edelman, G. M., and Chance, B., *Enzyme*, 29, 73, 1983.

reduced cytochrome c oxidase and oxygen may be calculated. These data are given in the last column to Table 3, yielding an average value of very approximately 5×10^7 l · equivalent^{-1} · s^{-1}, and independent as to whether the preparations were starved.

B. PRE-STEADY-STATE KINETICS OF OXIDATION OF CYTOCHROME C OXIDASE BY MOLECULAR OXYGEN IN INTACT YEAST CELLS

Kinetic experiments were conducted in which the reaction time was varied by systematically altering the flow velocity of the yeast cell suspension with the oxygen concentration fixed for each run by the appropriate adjustment of the substrate injection rate. Under these conditions, the maximum flow velocity attainable gave a time after mixing of approximately 6 msec. It was observed that as the time for reaction was progressively shortened, e.g., at an initial oxygen concentration of 1.3 μ*M*, an invariant steady-state value was maintained from approximately 25 down to 10 msec, at which point the system appeared to enter into a transient state. This steady-state value for the extent of oxidation of cytochrome a_3, i.e., about 71 to 72%, at 1.3 μ*M* initial O_2 concentration is in excellent agreement with three other estimates measured under steady-state conditions (averaging approximately 70% oxidation at 1.3 μ*M* O_2, for preparation numbers 68, 69, and 76), lending credence to these measurements. With the assumption of a bimolecular reaction between cytochrome a_3^{2+} and molecular oxygen,[454,455] one may calculate a pseudo first-order velocity coefficient; and pseudofirst-order constants of 148 and 45 s^{-1}, are obtained at 1.3 μ*M* and 0.6 μ*M* initial oxygen concentrations, respectively. If one adopts the concentration scale of equivalents per liter for the oxygen concentration, then a second order velocity constant of about 3×10^7 and 2×10^7 l · equivalent^{-1} s^{-1}, respectively, may be calculated. These values are in good agreement with the minimum estimate of k_{+1} given in Table 3, that is, approximately 5×10^7 l · equiv^{-1} · s^{-1}, and estimated simply from the turnover numbers and K_m values, and which points to only a small contribution by the back reaction after the injection of an oxygen pulse into an initially reduced cytochrome system.

A better estimate of the second-order velocity constant, k_{+1}, for the reaction

$$a_3^{+2} + \frac{1}{4} O_2 \xrightarrow{k_{+1}} a_3^{+3} + H_2O \tag{24}$$

is given by the integrated second-order plot (Figure 8), derived on the basis of negligible contributions of either the back reaction (k_{-1}) or by successive unimolecular irreversible reactions (k_{+2} to k_{+7}) at these relatively short reaction times in the pre-steady-state, leading to the integrated expression

$$k_{+1}t = \frac{2.303}{(B_o - A_o)} \log_{10}\left[\frac{A_o\,(B_o - x)}{B_o\,(A_o - x)}\right] \tag{25}$$

where B_0 is the initial oxygen concentration expressed in equivalents per liter, A_0 is the initial a_3^{+2} concentration in equivalents per liter, and x is the decrease in concentration (in equivalents per liter) of either reactant in a given time t. Thus, from the slope of Figure 8, which incorporates three separate sets of data, k_{+1} is estimated to be 2.7 (±0.9) $\times 10^7$ l · equiv^{-1} · s^{-1}.

This value for k_{+1} is in remarkable agreement with the estimates of Gibson and coworkers,[454,455] which were made on an isolated preparation of cytochrome c oxidase in solution, if one converts their data to the concentration scale of equivalents per liter for oxygen.

In Ludwig and Kuby,[439] a value of K_{+1} is given as approximately 1×10^8 l · mol^{-1} · s^{-1} and is identical to Gibson's[454,455] estimate; however, K_{+1} had been calculated from the

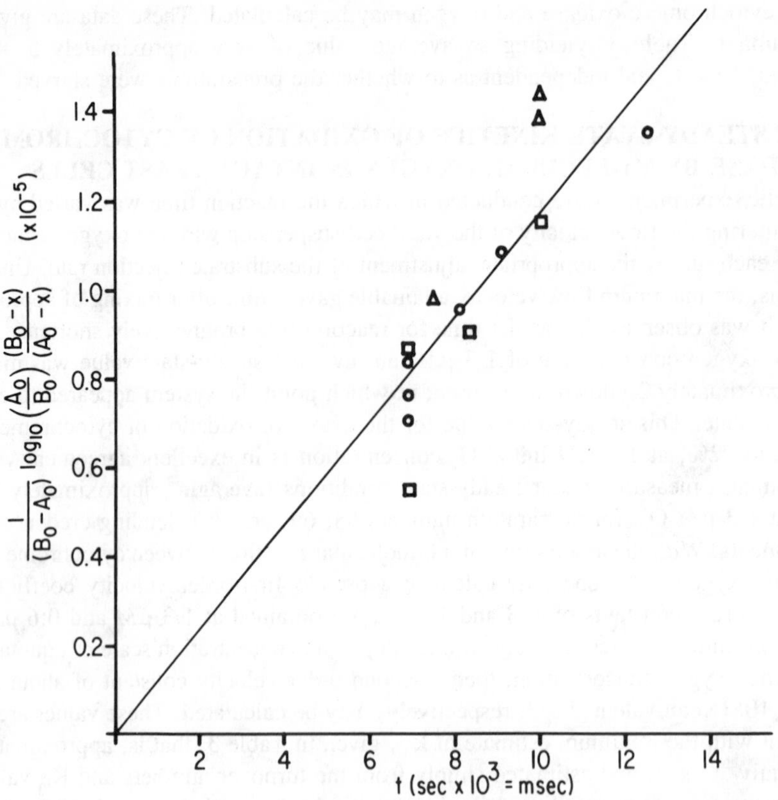

FIGURE 8. Kinetics of reaction at 26°C of cytochrome a_3 in Baker's yeast cell suspensions with O_2. Second-order plot of three separate sets of data at several fixed O_2 concentrations and at various reaction times. △—△, Experiment No. 84 A; ○—○, Experiment No. 84 B; □—□, Experiment No. 85. (From Ludwig, G. D., Kuby, S. A., Edelman, G. M., and Chance, B., *Enzyme*, 29, 73, 1983. With permission.)

pseudofirst-order velocity constant of approximately 150 s^{-1} obtained at an initial concentration of 1.3 μM O_2, where the concentration scale for O_2 employed for calculation was in moles per liter; thus, a recalculation in terms of an oxygen concentration of 5.2×10^{-6} equiv/l results in $150/(5.2 \times 10^{-6}) \simeq 3 \times 10^7$ l · equiv^{-1} s^{-1}, or the same value as listed above, and certainly one of the most rapid enzyme-substrate reactions reported. Furthermore, the agreement between the two estimates for k_{+1} for the reaction between cytochrome c oxidase and oxygen in either the isolated system[454,455] in solution or in the intact yeast cell leads to the surprising conclusion that there appears to be little, if any, barrier offered by the yeast cell to the diffusion of molecular oxygen into the cell.

A derivation of a defined K_m and turnover number for the steady-state reaction of cytochrome oxidase and O_2 in the intact yeast cell is as follows.

Assume the following mechanism to hold, after the introduction of an O_2 pulse,

$$a_3^{+2} + \underset{(x)}{\frac{1}{4}O_2} \underset{k_{-1}}{\overset{k_{+1}}{\rightleftharpoons}} a_3^{+3}$$

$$a_3^{+3} + a^{+2} \xrightarrow{k_{+2}} a_3^{+2} + a^{+3} \dashrightarrow$$

SCHEME I.

$$\frac{d(a_3^{+3})}{dt} = k_{+1}(a_3^{+2})(x) - (a_3^{+3})[k_{-1} + k_{+2}(a^{+2})] \tag{26}$$

$$\frac{d(x)}{dt} = -k_{+1}(a_3^{+2})(x) + k_{-1}(a_3^{+3}) \tag{27}$$

Provided a steady state with respect to (a_3^{+3}) may be adjusted, i.e.,

$$\frac{d(a_3^{+3})}{dt} = 0 \tag{28}$$

and assume the conservation expression for

$$a_3^{total} = E_t = a_3^{+3} + a_3^{+2} \tag{29}$$

where K_m is ''defined'' as

$$\left[\frac{k_{-1} + k_{+2}a^{+2}}{k_{+1}}\right] \tag{30}$$

$$\therefore \quad (a_3^{+3}) = \frac{E_t(x)}{\left[\frac{k_{-1} + k_{+2}a^{+2}}{k_{+1}} + (x)\right]} = \frac{E_t(x)}{K_m + (x)} \tag{31}$$

From

$$\frac{d(a_3^{+3})}{dt} + \frac{dx}{dt} = -k_{+2}(a_3^{+3})(a^{+2}) \tag{32}$$

the steady state with respect to a_3^{+3}, $\frac{d(a_3^{+3})}{dt} = 0$, and the above expression for (a_3^{+3}), it follows that

$$-\frac{dx}{dt} = \frac{k_{+2}a^{+2}E_t(x)}{K_m + (x)} \tag{33}$$

Since the maximum respiration or maximum velocity,

$$V_{max}^{resp} = -\left(\frac{dx}{dt}\right)_{max} = \lim_{(a^{+3} \to E_t)} k_{+2}(a_3^{+3})(a^{+2}) = k_{+2}E_t(a^{+2});$$

$$-\left(\frac{dx}{dt}\right) = \frac{V_{max}^{resp.}(x)}{K_m + (x)} = v_o^{resp.} \tag{34}$$

Also, the turnover number may be defined as

$$\frac{-\left(\frac{dx}{dt}\right)_{max}}{E_t} = \frac{-\left(\frac{dx}{dt}\right)_{max}}{a_3^{total}} = \frac{V_{max}^{resp.}}{E_t} = \frac{V_{max}^{resp.}}{a_3^{total}} = k_{+2}(a^{+2}) \tag{35}$$

A derivation of the integrated second-order plot (Figure 8) for estimation of the velocity constant for reaction of cytochrome a_3 with O_2 in intact yeast cells is as follows.

Consider the following series of bimolecular reactions (see also Reference 453).*

$$a_3^{2+} + \frac{1}{4} O_2 \xrightarrow{k_{+1}} a_3^{3+} + H_2O$$

$$a_3^{3+} + a^{2+} \xrightarrow{k_{+2}} a_3^{2+} + a^{3+}$$

$$a^{3+} + c^{2+} \xrightarrow{k_{+3}} a^{2+} + c^{3+}$$

$$c^{3+} + b^{2+} \xrightarrow{k_{+4}} c^{2+} + b^{3+}$$

$$b^{3+} + \frac{1}{2} Fl \cdot H_2 \xrightarrow{k_{+5}} b^{2+} + \frac{1}{2} Fl$$

$$Fl + DPNH + H^+ \xrightarrow{k_{+6}} Fl \cdot H_2 + DPN^+$$

$$DPN^+ + EtOH \xrightarrow{k_{+7}} DPNH + CH_3CHO + H^+$$

SCHEME II.

$$\frac{-d(O_2)}{dt} = k_{+1}(a_3^{+2}) \frac{(O_2)}{4} \tag{36}$$

$$\frac{-d(a_3^{2+})}{dt} = k_{+1}(a_3^{+2}) \frac{(O_2)}{4} - k_{+2}(a_3^{+3})(a^{+2}) \tag{37}$$

Since the sixth and seventh reactions are the slowest, they may be safely neglected at these fast reaction times, and after suitable substitutions,

$$\frac{-d(a_3^{2+})}{dt} = k_{+1}(a_3^{+2}) \frac{(O_2)}{4} + [\Sigma(\dot{a}^{+2}, \dot{c}^{+2}, \dot{b}^{+2}, \dot{F\ell} \cdot H_2) - \{k_{+4}(c^{3+})(b^{+2}) + k_{+6}(F\ell)(DPNH)\}] \tag{38}$$

It may be shown by calculation that under conditions of a pre-steady-state experiment (described above) at the concentrations in Baker's yeast and for the approximate turnover numbers for those reacting species given in the term in brackets (e.g., as estimated by Chance[453]), that the total correction (the algebraic sum of the terms in brackets) to the net rate of reaction of a_3^{2+} with oxygen will be quite small, and that

$$\frac{-d(a_3^{2+})}{dt} \simeq k_{+1}\ (a_3^{2+})\frac{(0_2)}{4} \tag{39}$$

appears to be a good approximation for the bimolecular kinetics, provided, however, that

* Not all of the components of the electron transport system, as it is known today, have been incorporated in this abbreviated series of bimolecular reactions,[453] e.g., the copper of cytochrome aa_3, cytochrome c_1, and ubiquinone reductase have been omitted. See Lehninger (Reference 926, Figure 5 in Chapter 17, and discussion) for a more up-to-date description of all the electron carriers known to participate in the electron-transport chain.

the primary assumption is valid, i.e., that oxygen does react in a second-order fashion directly with cytochrome a_3.[454,455] Integration of the expression follows by standard techniques (see above, Table 1) to yield

$$k_{+1}t = \frac{2.303}{B_o - A_o}\left[\log \frac{A_o\,(B_o - x)}{B_o\,(A_o - x)}\right] \quad (40)$$

The experiments described above were on intact *yeast cells*; studies on the *isolated* beef heart mitochondrial cytochrome oxidase[912-914] have shown it to be a large (M_r = 170,000) multisubunit protein.[915] The functional monomer presumably contains four metal ions: two coppers, designated Cu_A and Cu_B, and two irons contained in heme *a* prosthetic groups, designated Fe_a and F_{a_3}. Two of the metal ions, F_{a_3} and Cu_B, are situated within 5 Å of each other and constitute the dioxygen reduction site. The other two metal ions, Fe_a and Cu_A, rapidly accept electrons from cytochrome *c* and transfer them to the binuclear site.[916] The mechanism of dioxygen reduction of the enzyme has been the subject of several investigations.[454,455,917-921,923,924] Because the second-order rate constant (k_{+1}, see above) for the combination of dioxygen with the reduced enzyme is so rapid, $\sim 3 \times 10^7\ l \cdot equiv^{-1}\ s^{-1}$ [440,454,455] at 25°C, under physiological conditions, it is exceedingly difficult to detect and study any intermediates formed during the reaction and where the ensuing intramolecular electron transfers take place within a millisecond.[454,922] However, the reaction may be slowed by going to low temperatures and Chance et al.[921] developed the so-called triple-trapping technique to follow the reaction between dioxygen and cytochrome *c* oxidase at these low temperatures. Recently, Blair et al.[925] studied the low temperature reactions by EPR spectroscopy over a broader temperature range than previous studies and reported evidence for two intermediates at the "three-electron level" and entropic promotion of the bond-breaking step. Finally, a critical theoretical examination of the steady-state kinetic mechanism(s) of cytochrome oxidase was reported by Myers and Palmer recently.[1004]

One variation of the "continuous-flow technique" is the "chemical-stop" method which, for example, Bray and colleagues applied in an elegant series of experiments in their detailed study of the xanthine oxidase-catalyzed reaction and its intermediates.[457-459,467] In their method they allowed the reaction mixture to squirt in the form of a fine spray from the exit of a flow tube directly into a hydrocarbon liquid, such as isopentane, at a low temperature, e.g., −145°C so that the reaction mixture is frozen in a matter of milliseconds. This frozen reaction mixture and its composition fixed after a given extent or time of reaction could then be analyzed, e.g., by ESR spectroscopy for intermediates in the reactions catalyzed by xanthine oxidase.

Ruby[460] developed a mixing machine which yielded results of high precision when used as a "chemical-stop" device. It incorporated three separate mixing chambers with or without spacer units which enabled one to premix sensitive or unstable reagents immediately before initiating the primary reaction and to stop the reaction after a sufficient steady-state-of-flow had been attained. Reactions with half-lives as short as 6 msec (or as long as 1 to 6 s, with suitable spacers) could be followed with good precision. Kuby[461] later had combined the best features of the Ruby chemical-stop machine with a rapid- or stopped-flow feature, to be described next.

The major advantage of the stopped-flow technique is that it may be adapted to the use of comparatively small amounts of enzyme (or reactants) as developed by Chance and Gibson.[434,436,468]

A schematic diagram of a Gibson stopped-flow apparatus[435] is given in Figure 9.[419]

In the Gibson design,[435] the reactants are forced from two syringes through a mixing chamber to a specially designed observation chamber of a relatively long optical path length. The resulting mixture then is collected in the third or stopping syringe which is provided

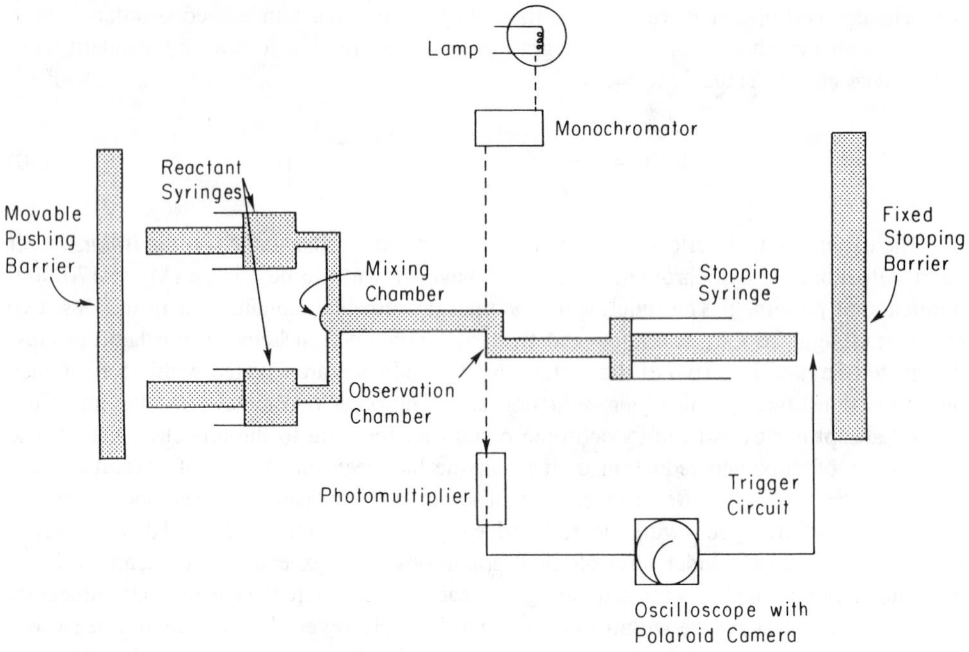

FIGURE 9. Schematic diagram of a Gibson-type of stopped-flow apparatus. (Taken from Halford, S. E., *Companion in Biochemistry,* Bull, A. T., Lagnado, J. R., Thomas, J. O., and Tipton, K. F., Eds., Longman Group, London, 1974, 197. With permission.)

with a rigid stop to suddenly arrest the flow. Observations are then conducted on the stationary mixture at one observation point and over a period of time. The stopping syringe also activates the trigger circuit of the previously calibrated cathode-ray oscilloscope which provides a tracing of the changes in the reaction mixture; this tracing, in turn, may be recorded by means of a polaroid camera. In this method, it is important to keep the time between mixing and the point of observation extremely short compared to the reaction time under investigation, otherwise, the initial phase of the reaction curve will be lost; also, this technique requires detection methods with very short response times, and spectrophotometric or fluorometric methods are commonly employed.

The combined problems of mixing time and detector response time in the Gibson-type of apparatus often limits the apparatus to dead-times of at least 5 msec before useful information may be gathered.

To eliminate some of the complexities of the rapid-flow methods in extremely fast reactions, Gibson, for example, attempted to initiate the reaction photochemically; thus, Gibson,[462] in a study of the association of myoglobin with carbon monoxide, first photodecomposed the complex with a light pulse and then followed its reassociation spectrophotometrically. Gibson and colleagues[454,455] also utilized the same technique in a study of the reaction of CO and cytochrome a_3 and in a competitive reaction with O_2 (as mentioned above).

Recently, in a remarkable series of experiments, Terner et al.[469] measured the picosecond resonance Raman spectrum of the oxyhemoglobin ($Hb \cdot O_2$) photoproduct which had been obtained with 30-psec, 532-nm pulses from a high-energy pulsed laser source. The spectrum obtained by Terner et al. differed from those reported previously for $Hb \cdot CO$, and for $Hb \cdot O_2$, with longer, weaker pulses. These differences were attributed to a ''prompt'' photoproduct, which could be associated with a previously observed $Hb \cdot O_2$ spectral intermediate which decays within 90 psec. They inferred, on the basis of the Raman resonance band frequencies, that this prompt intermediate was an electronically excited state of deoxy Hb with substantial Π-Π* character and that it might be a triplet Π-Π* state formed by dissociation of triplet O_2 from Hb · O_2.

In an interesting application of a "stopped-flow light scattering" technique and a "chemical quench-flow" method, Porter and Johnson[927,928] and Johnson[929,930,1002] studied the kinetics of the *Tetrahymena thermophila* dynein ATPase and evaluated the rate constants according to the following postulated pathway (i.e., in the absence of microtubules).

$$E + ATP \underset{k_{-1}}{\overset{k_1}{\rightleftharpoons}} E - ATP \underset{k_{-2}}{\overset{k_2}{\rightleftharpoons}} E - ADP - Pi \underset{k_{-3}}{\overset{k_3}{\rightleftharpoons}}$$

$$E - ADP + Pi \underset{k_{-4}}{\overset{k_4}{\rightleftharpoons}} E + ADP + Pi$$

Curiously, product release appeared to be the rate-limiting step in the steady state,[1005] but the dynein ATPase was apparently activated by the presence of tubulin or microtubules[931] which enhanced the rate of product release.

The kinetic constants they measured[930] were of a similar order of magnitude as those found for myosin-ATPase and its activation by actin.[932] Moreover, although ATP binding is approximately 100-fold tighter for myosin than for dynein, the thermodynamics of nucleotide binding appeared similar in both cases.[933]

Later, Johnson also applied a "stopped-flow fluorescent" technique to a study of the 5-enolpyruvoylshikimate-3-phosphate synthase reactions.[998-1000]

Before leaving the topic of flow techniques, it is appropriate to discuss very briefly the stirred-flow reactor technique of Denbigh and Page.[463] In principle, if one decreases the distance between the mixing chamber and the point of observation, so that the mixing chamber and observation point become one, then the stirred reactor results. If one controls the rate of addition of reactants and the outflow of products, eventually, after a given number of reactor volumes have traversed the system, a true steady state is attained in the reactor. This steady state is invariant in both time and space with the concentrations of all components, including any intermediates held fixed in the stirred reactor, at some given extent of the total reaction, which in turn is dependent on the flow rates and the chemical-reaction rates. The procedure, unfortunately, has not been widely used for the study of enzyme reactions, but it has many attractive features since it can simulate the behavior of biological systems. Denbigh et al.[464] have discussed the theory of this method and its relatively simple application to complex kinetic situations. Kuby[465] made some observations on the oxidation by molecular oxygen of the cytochrome chain of intact yeast cells, making use of the applications of Hammett and colleagues[466,641,642] to this technique in their study of certain very complex chemical reactions.

V. RELAXATION METHODS

The time for mixing components in solution, by rapid-flow methods, will probably never exceed 100 μsec for technical reasons. Reactions initiated by photo illumination techniques such as described earlier can penetrate the picosecond time scale; however, these methods are applicable to a limited number of cases. Eigen[470] developed a more general method which is capable of time resolution ranging from minutes to fractions of a nanosecond. In principle, the method involves measurement of the adjustment of a system following a relatively small perturbation; relaxation methods are capable of measuring reaction velocities almost seven orders of magnitude faster than can be measured conveniently by rapid-flow methods.[471] In principle, the perturbation may be applied as a periodic oscillation or as a single rapid change. At present, only the latter method is commonly employed and the following discussion will be limited to relaxation methods employing a single perturbation.

A number of physical perturbations are useful in chemical relaxation studies. In order of increasing rapidity, these are the pressure jump, temperature jump, ultrasonic methods, and electric field changes.[472-475] In the technique of temperature jump,[472] which has been most widely applied to enzyme systems, a high-voltage current is discharged through a reaction mixture with a high electrolyte conductance (i.e., with ionic strengths greater than 10^{-2} *M*). Within microseconds, this electric discharge causes a jump of several degrees in temperature as a result of the heat of friction due to the alignment of charged particles in the electrical field. Such a temperature-jump apparatus[472] can raise the temperature by 5°C within 10 μsec; following the perturbation, optical means (recorded with the aid of an oscilloscope) are usually used to observe the relaxation responses in 10^{-4} to 10^{-5} s. Because of the comparatively short duration of the entire experiment (usually less than 1 s), one may neglect the subsequent cooling. With solutions of low conductivity, other methods of heating have to be employed, e.g., use of pulses of high-energy laser beams.

The next most commonly employed relaxation technique, the pressure-jump method, involved the sudden release of pressure from about 50 to 60 atm to 1 atm[473] within about 60 μsec. The method is usually limited to the study of relaxation times of greater than 5×10^{-5} s and may be used for reactions that are too slow to be studied by the sound-absorption technique. In general, small volumes of reactants can be employed, and the simplicity of the technique makes it a convenient method for studying aqueous solution reactions whose positions can be perturbed by changes in pressure. Like sound-absorption methods, a volume change in the overall reaction being studied is required; the approach to the new equilibrium concentrations of the chemical species is usually followed by sensitive conductimetric measurements. With a pressure drop of 50 to 60 atm to equilibrium conditions at 1 atm, this technique has the advantage over some other perturbation methods in that the equilibrium data and rate constants obtained are applicable to the conditions employed in other kinetic studies.

If we consider a system (i.e., a solution initially at thermal equilibrium and containing substances coupled by chemical equilibria, which is perturbed by a chemical equilibria) which is perturbed by a rapid increase in temperature (over a few microseconds in rise time), the concentrations of the chemical species will change to new equilibrium values at the higher temperature. The magnitude of the concentration changes is determined by the laws of thermodynamics; in particular, van't Hoff's equation provides the relationship between the equilibrium constant, at constant pressure of each reaction, and the temperature, i.e.,

$$\left(\frac{\partial \ell nK}{\partial T}\right)_P = \frac{\Delta H^\circ}{RT^2} \tag{41}$$

where K is the equilibrium constant of each of the chemical reactions, at constant pressure, ΔH° is the standard enthalpy change of that chemical reaction, R is the gas constant, and T is the absolute temperature. It is evident that if a sequence of coupled reactions occurs, the system will be perturbed by a temperature jump if any one reaction is characterized by a nonzero enthalpy change; or, even if all the equilibria are insensitive to temperature, a perturbation of the system can be achieved by coupling those reactions of interest to a temperature-dependent reaction. An example of this case is the so-called "pH-jump", which can be achieved by use of a buffer with temperature-dependent ionization constants; with this method, pH-dependent reactions can be readily perturbed.

The rates of decay of the concentrations to their new equilibrium values are characterized by linear first-order differential equations, of which the solutions are linear combinations of exponentials. Each exponential term is associated with a relaxation time, τ, which may be thought of as a reciprocal first-order rate constant. Therefore, a complex reaction mechanism is associated with a spectrum of relaxation times. The explicit evaluation of the relaxation spectrum has been described in detail by Eigen and De Maeyer,[471] however, a few relatively simple cases will be given below. Also, no attempt will be made here at a comprehensive survey of the application of the temperature-jump method to biochemical mechanisms; refer, for example, to Hammes[869] for such a review.

In general, the relaxation times will have a unique dependence on the equilibrium concentrations and determination of the relaxation times of various concentrations allows one to postulate a reaction mechanism. At present, the time resolution of most temperature-jump equipment is approximately 10^{-6} s; however, in principle, the method should be capable of being extended to shorter times. Also, in our discussion, we have considered only the perturbation of a system at equilibrium. Relaxation methods, in principle, can be applied to any system characterized by a stationary state in net concentrations. Thus, steady-state systems can be perturbed and their relaxation spectra determined for those cases where the steady state persists for a sufficiently long time. French and Hammes[472] have described a temperature-jump apparatus directly coupled to a rapid-mixing flow system, a so-called "stopped-flow — temperature jump" apparatus. With their apparatus, perturbation of steady states with half-lives as short as milliseconds can be accomplished. Therefore, the temperature-jump method can be applied to both reversible and essentially irreversible reactions.

Before considering the relaxation equations characteristic of a few reactions, it is important to discuss the thermodynamics underlying the effect of pressure on chemical equilibrium, i.e., the pressure-jump method. This method depends upon a volume change in the reaction(s) under study where such a volume change permits a displacement of reactant concentrations from their equilibrium concentrations at 1 atm according to Equation 42,

$$\left(\frac{\partial \ell n K}{\partial P}\right)_T = -\frac{\Delta V^\circ}{RT} \tag{42}$$

where ΔV° = standard volume change for the reaction at constant temperature and K is the molal equilibrium constant for the reaction at constant temperature.

From Equation 42,

$$\frac{\partial(K)}{K\partial(P)} = \frac{-\Delta V^\circ}{RT} \simeq \frac{\Delta K}{K\Delta P} \tag{43}$$

or

$$\frac{\Delta K}{K} = \frac{-\Delta V^\circ(\Delta P)}{RT} \tag{44}$$

If $\Delta V^\circ = -10$ ml/mol, which is a reasonable value for a charge neutralization reaction occurring in aqueous solution, and if $\Delta P = 65$ atm at 25°C,

$$\therefore \quad \frac{\Delta K}{K} = \frac{[-10 \text{ ml/mol}][65 \text{ atm}]}{[82.05 \text{ cm}^3\text{atm/(°K mol)}][298 \text{ K}]} = 2.7 \times 10^{-2} \tag{45}$$

Therefore, a 2.7% change in the equilibrium constant results from a 65-atm pressure change. This small change, however, is well within the sensitivity of modern conductivity-detection circuits as described by Takahashi and Alberty.[473]

A. ANALYSES OF RELAXATION TIMES

The following discussion will consider changes measured after a rapid perturbation (e.g., by temperature jump or by pressure jump).

$$A \underset{k_{-1}}{\overset{k_{+1}}{\rightleftarrows}} B$$

SCHEME III.

1. First Order Reaction

Consider the chemical equilibrium described by a first-order process in either direction; after the rapid perturbation which causes a small disturbance to the chemical equilibrium, the change in concentration of any one of the reactants will follow a simple exponential curve (in fact, even regardless of the order of the reaction). If this equilibria described by Scheme III were actually the equilibrium between the two conformations of enzyme protein R and T, according to the allosteric model of Monod et al.[116]

$$E_R \underset{k\,-\,1}{\overset{k\,+\,1}{\rightleftarrows}} E_T$$

SCHEME IIIA.

Initially, the system is at an equilibrium determined by temperature, T, after the temperature is jumped to T′, the concentrations of the two enzyme species at time t are denoted as follows.

During the time of relaxation, the enzyme is redistributed between the two conformations until x, the distance from the new equilibrium position, is finally reduced to zero. At any time during the relaxation period, from the law of mass action,

$$\frac{d(E_T' - x)}{dt} = k_{+1}(E_R' + x) - k_{-1}(E_T' - x) \qquad (46)$$

or, after rearrangement,

$$\frac{dE_T'}{dt} - \frac{dx}{dt} = k_{+1}E_R' - k_{-1}E_T' + (k_{+1} + k_{-1})x \qquad (47)$$

Eventually, after the new equilibrium has been reached at the new temperature and the reaction system is once again stationary, one can write

$$\frac{dE_T'}{dt} = k_{+1}E_R' - k_{-1}E_T' = 0 \qquad (48)$$

Therefore, substracting these terms in Equation 48 from Equation 47 leaves

$$\frac{-dx}{dt} = (k_{+1} + k_{-1})x \qquad (49)$$

which, after integration, yields Equation 50.

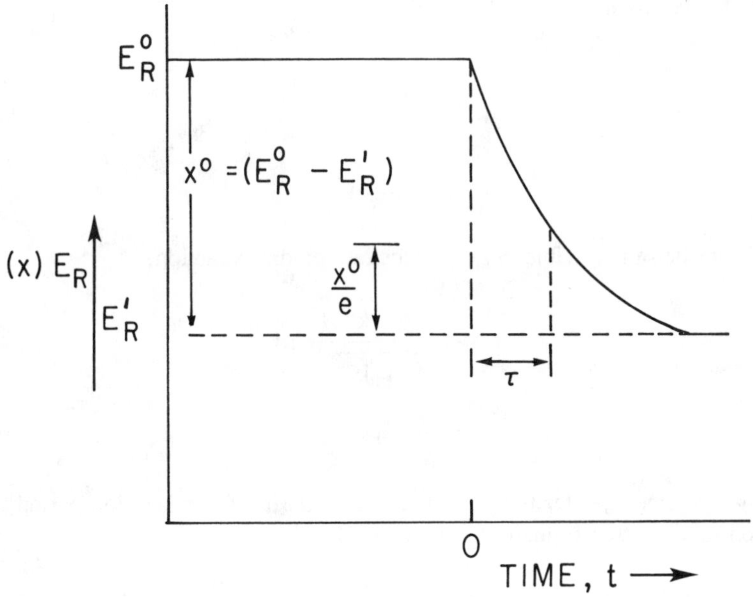

FIGURE 10. The process of relaxation for the system, $E_R \underset{k_{-1}}{\overset{k_{+1}}{\rightleftarrows}} E_T$.

$$x = x^o e^{-(k_{+1}+k_{-1})t} \tag{50}$$

If

$$x^o = E_R^o - E_R' \tag{51}$$

or the amplitude x at t = 0 at the start of the relaxation response; setting

$$\frac{1}{\tau} = k_{+1} + k_{-1} \tag{52}$$

where τ is the relaxation time required for x to be reduced to 1/e of its original amplitude (see Figure 10).

Figure 10 shows the process of relaxation for the system

$$\therefore \quad x = (E_R^o - E_R')e^{-t/\tau} \tag{53}$$

The reciprocal of τ, the relaxation time, therefore yields the sum of Equation 52, whereas, the ratio of $(\frac{k_{+1}}{k_{-1}})$ is obtained from the equilibrium constant at temperature T'.

$$K_{eq} = \frac{E_T'}{E_R'} = \frac{k_{+1}}{k_{-1}} \tag{54}$$

With a knowledge of the sum and the ratio of k_{+1} and k_{-1}, one may estimate the individual rate constants.

2. Second-Order Reaction

$$A + B \underset{k_{-1}}{\overset{k_{+1}}{\rightleftharpoons}} C$$

SCHEME IV.

Next, consider a reversible enzyme-substrate binding reaction.

$$E + S \underset{k_{-1}}{\overset{k_{+1}}{\rightleftharpoons}} ES$$

SCHEME V.

During any time, t, after the perturbation and during the relaxation period, the law of mass action requires that Equation 55 be obeyed,

$$\frac{d[(ES)' - X]}{dt} = k_{+1}(E' + X)(S' + X) - k_{-1}[(ES)' - X] \quad (55)$$

which, after rearrangements, yields Equation 56,

$$\frac{d(ES)'}{dt} - \frac{dx}{dt} = k_{+1}(E')(S') - k_{-1}(ES') + k_{+1}(E' + S')X + k_{-1}X + k_{+1}X^2 \quad (56)$$

At the new stationary equilibrium state corresponding to the new temperature, T′, one can write

$$\frac{d(ES)'}{dt} = k_{+1}(E')(S') - k_{-1}(ES)' = 0 \quad (57)$$

Subtraction of Equation 57 from Equation 56 leaves

$$\frac{-dx}{dt} = [k_{+1}(E' + S') + k_{-1}]X + k_1X^2 \quad (58)$$

At small values of X, one may neglect the X^2 term and integration of the remainder of Equation 58 yields

$$X = X^o e^{-[k_{+1}(E'+S')+k_{-1}]t} \quad (59)$$

which after defining $X^0 = (E^0 - E')$ and

$$\frac{1}{\tau} = k_{+1}(E' + S') + k_{-1} \quad (60)$$

yields

$$X = (E^o - E')e^{-t/\tau} \quad (61)$$

TABLE 4

Enzyme conformation	Initial concentration at temp, T	Final concentration at temp, T′	Concentration during time, t, of relaxation
R	E_R^0	E'_R	$E'_R + x$
T	E_T^0	E'_T	$E'_T - x$

TABLE 5

Reaction	$\left(\frac{1}{\tau}\right)$ Reciprocal of the relaxation time
$A \rightleftharpoons B$	$k_{+1} + k_{-1}$
$A + C \rightleftharpoons B + C$	$k_{+1} C_{eq} + k_{-1} C_{eq}$
$2A \rightleftharpoons A_2$	$4k_{+1} A_{eq} + k_{-1}$
$A + B \rightleftharpoons C$	$k_{+1} (A_{eq} + B_{eq}) + k_{-1}$
$A + B \rightleftharpoons C + D$	$k_{+1} (A_{eq} + B_{eq}) + k_{-1} (C_{eq} + D_{eq})$
$A + B + C \rightleftharpoons D$	$k_{+1} (A_{eq} B_{eq} + A_{eq} C_{eq} + B_{eq} C_{eq}) + k_{-1}$

From Havsteen, B., *Biochem. Soc. Symp.*, 26, 53, 1966.

From Equation 60, the reciprocal of the relaxation time is a linear function of the sum of the free enzyme and free substrate, with k_{+1} obtained by the slope of a plot of $\frac{1}{\tau}$ vs. $(E' + S')$, and k_{-1} from its intercept on the $\frac{1}{\tau}$ axis. Again a knowledge of $K_{eq} = \frac{k_{+1}}{k_{-1}}$ at T′, provides a check on the measurements of k_{+1} and k_{-1}.

We note that the derivation of Equation 61 required the assumption that x is small and consequently X^2 is negligible in Equation 58. In general, a valid application of chemical relaxation methods is dependent on working with concentration perturbations which are small relative to their total concentrations. Mechanisms with multiple reversible steps will generate multiple relaxation times. With the various reaction steps being coupled to one another, the spectrum of relaxation times are, in effect, the Eigen values of the system of rate equations. Hammes and Schimmel[476] and Czerlinski[477] have treated, in depth, the general problem of extracting rate constants from relaxation times. Experimentally, it should be noted, that the resolution of any two relaxation times will depend on how different they are in magnitude. Since there is always the possibility of poor resolutions, the apparent detection of a single relaxation time is not conclusive proof that a mechanism has only a single reversible equilibrium step, whereas detection of more than one relaxation time would provide evidence for multiple equilibria steps.

Havsteen[478] has summarized the expressions for the relaxation times of a number of reversible reactions, as given in Table 5.

All of the above reactions are, in essence, single-step reactions. For more complex systems, with a number of steps, the analysis of the experimental relaxation data in terms of relaxation times can become complicated. The relaxation of a system containing a number of steps will not necessarily proceed as a series of independent relaxations (unless each of the relaxation times can be separated by an order of magnitude), each corresponding to a specific step, since the concentration changes in any one step will affect the concentrations of species in succeeding steps.

3. Expressions for Relaxation Times with Two or More Steps

The derivation of relaxation times for such systems can be quite complex; Hammes and Schimmel[476] have summarized a method for such derivations which make use of determinants and leads to a method, for some conditions, whereby the relaxation times can be written by examination.

The equations for a number of systems likely to be encountered in enzyme systems have been tabulated by Yapel and Lumry[479] and by Dixon et al.[480] and several of these are given in Table 6.

As an illustration of the technique of measurement of relaxation spectra, consider the kinetic study (largely by the temperature-jump method) of Hammes and Hurst[338] of the reaction catalyzed by ATP-creatine transphosphorylase. For all their temperature-jump experiments, the pH of the reaction solutions was adjusted to 7.6 at approximately 25°C, the solutions were thermostated at 3°C, the equilibrium temperature after perturbation (with a 30-kV pulse which gave a temperature rise of approximately 8°C in less than 20 μsec) being approximately 11°C, and the ionic strength was maintained at 0.1 M with KCl. The approach to equilibrium after perturbation was monitored at 558 nm with phenol red as the indicator dye at a concentration of approximately 2×10^{-5} *M*. Consequently, only those effects producing a pH change could be observed. Consider the reaction of uncomplexed ADP with creatine kinase: a single rapid relaxation process (τ = 60 to 150 μsec) occurred when solutions of the enzyme and ADP were perturbed and the effect was studied over a fourfold variation in enzyme and 100-fold variation in ADP concentrations. Their data are shown in Figures 11 and 12. The relaxation time decreased rapidly with increasing concentrations of reactants at low reactant concentrations, but became independent of concentration at high ADP levels, which they felt was characteristic of a reaction mechanism of the type[481] given in Scheme VI.

$$\mathrm{E + ADP} \underset{k_{-1}}{\overset{k_{+1}}{\rightleftharpoons}} (\mathrm{E \cdot ADP})_1 \underset{k_{-2}}{\overset{k_{+2}}{\rightleftharpoons}} (\mathrm{E \cdot ADP})_2$$

SCHEME VI.

which, in essence, involves an isomerization of the enzyme through some conformational change in the protein. Refer to Table 6, System 1.

If the first step equilibrates rapidly relative to the second, the relaxation time for the slower process is

$$\frac{1}{\tau_2} \simeq k_2 + \frac{k_{+2}}{1 + K_{diss,1}\left[\dfrac{1}{(E)_{eq} + (S)_{eq}}\right]} \tag{62}$$

(Note, in Figures 11 and 12, Hammes and Hurst[338] use the symbol $(\bar{E})$ and $(\overline{ADP})$ for $(E)_{eq}$ and $(ADP)_{eq}$, respectively.)

Equation 62 can be rearranged:

$$\left[\frac{1}{\tau_2} - k_{-2}\right]^{-1} = \left(\frac{1}{k_{+2}}\right)\left[1 + K_{diss,1}\left(\frac{1}{(E)_{eq} + (S)_{eq}}\right)\right] \tag{63}$$

TABLE 6
Expressions for Relaxation Times with Two or More Steps

System 1

$$E + S \underset{k_{-1}}{\overset{k_{+1}}{\rightleftarrows}} ES \underset{k_{-2}}{\overset{k_{+2}}{\rightleftarrows}} ES'$$

$$\frac{1}{\tau_1} + \frac{1}{\tau_2} = k_{-1} + k_{+1}[(E)_{eq} + (S)_{eq}] + k_{+2} + k_{-2}$$

$$\frac{1}{\tau_1} \cdot \frac{1}{\tau_2} = k_{-1}k_{-2} + k_{+1}(k_{+2} + k_{-2})[(E)_{eq} + (S)_{eq}]$$

(where $(E)_{eq}$, $(S)_{eq}$, and $(P)_{eq}$ are the concentrations of free enzyme, free substrate, and free product reached at the second equilibrium, i.e., after the perturbation).

$$\text{For } \tau_1 << \tau_2\left(\text{i.e., } \frac{1}{\tau_1} >> \frac{1}{\tau_2}\right),$$

$$\frac{1}{\tau_2} \simeq k_{-2} + \frac{k_{+2}}{1 + \left(\frac{k_{-1}}{k_{+1}}\right)\left[\frac{1}{(E)_{eq} + (S)_{eq}}\right]}$$

$$\frac{1}{\tau_2} \simeq k_{-2} + \frac{k_{+2}}{1 + K_{diss,1}\left[\frac{1}{(E)_{eq} + (S)_{eq}}\right]}$$

System 2

$$E + S \underset{k_{-1}}{\overset{k_{+1}}{\rightleftarrows}} ES \underset{k_{-2}}{\overset{k_{+2}}{\rightleftarrows}} E + P$$

$$\frac{1}{\tau_1} + \frac{1}{\tau_2} = k_{+1}[(E)_{eq} + (S)_{eq}] + k_{-2}[(E)_{eq} + (P)_{eq}] + k_{-1} + k_{+2}$$

$$\frac{1}{\tau_1} \cdot \frac{1}{\tau_2} = (E)_{eq}[k_{+1}k_{-2}[(E)_{eq} + (S)_{eq} + (P)_{eq}) + k_{+1}k_{+2} + k_{-1}k_{-2}]$$

$$\text{For } \frac{1}{\tau_1} >> \frac{1}{\tau_2},\ \frac{1}{\tau_1} \simeq k_{+1}[(E)_{eq} + (S)_{eq}] + k_{-2}[(E)_{eq} + (P)_{eq}] + k_{-1} + k_{-2}$$

$$\frac{1}{\tau_2} \simeq \frac{(E)_{eq}[k_{+1}k_{-2}[(E)_{eq} + (S)_{eq} + (P)_{eq}] + k_{+1}k_{+2} + k_{-1}k_{-2}]}{k_{+1}\,[(E)_{eq} + (S)_{eq}] + k_{-2}[(E)_{eq} + (P)_{eq}] + k_{-1} + k_{+2}}$$

System 3

$$E + S \underset{\substack{k_{-1}\\ \text{fast}}}{\overset{k_{+1}}{\rightleftarrows}} ES \underset{\substack{k_{-2}\\ \text{slow}}}{\overset{k_{+2}}{\rightleftarrows}} EP \underset{\substack{k_{-3}\\ \text{fast}}}{\overset{k_{+3}}{\rightleftarrows}} E + P$$

$$\frac{1}{\tau_1} + \frac{1}{\tau_2} = (k_{+1} + k_{-3})(E)_{eq} + k_{+1}(S)_{eq} + k_{-3}(P)_{eq} + k_{-1} + k_{+3}$$

$$= (k_{+1} + k_{-3})(E)_{eq} + \left(k_{+1} + k_{-3}\frac{(P)_{eq}}{(S)_{eq}}\right)(S)_{eq} + k_{-1} + k_{+3}$$

$$\frac{1}{\tau_1} \cdot \frac{1}{\tau_2} = k_{+1}k_{-3}(E)^2_{eq} + k_{+1}k_{-3}\left(1 + \frac{(P)_{eq}}{(S)_{eq}}\right)(E)_{eq}(S)_{eq} +$$

$$(k_{+1}k_{+3} + k_{-1}k_{-2})(E)_{eq} + \left(k_{+1}k_{+3} + k_{-1}k_{-3}\frac{(P)_{eq}}{(S)_{eq}}\right)(S)_{eq} + k_{-1}k_{+3}$$

TABLE 6 (continued)
Expressions for Relaxation Times with Two or More Steps

$$\frac{1}{\tau_3} = \frac{k_{+2}}{1 + \left(\frac{k_{+1}}{k_{-1}}\right)\frac{F_{x2}}{(E)_{eq}}} + \frac{k_{-2}}{1 + \left(\frac{k_{+2}}{k_{-2}}\right)\frac{F_{x1}}{(E)_{eq}}}$$

where

$$F_{x1} = \frac{\left[\frac{k_{+1}}{k_{-1}} + (E)_{eq} + (S)_{eq} + (P)_{eq}\left(\frac{k_{+1}k_{-2}}{k_{-1}k_{+2}}\right)\right]}{\left[\frac{k_{+1}}{k_{-1}} + (E)_{eq} + (P)_{eq} + (S)_{eq}\right]}$$

$$F_{x2} = \frac{\left[\frac{k_{+2}}{k_{-2}} + (E)_{eq} + (P)_{eq} + (S)_{eq}\left(\frac{k_{-1}k_{+2}}{k_{+1}k_{-2}}\right)\right]}{\left[\frac{k_{+2}}{k_{-2}} + (E)_{eq} + (P)_{eq} + (S)_{eq}\right]}$$

System 4

$$E + S \underset{k_{-1}}{\overset{k_{+1}}{\rightleftarrows}} X_1 \underset{k_{-2}}{\overset{k_{+2}}{\rightleftarrows}} X_2 \underset{k_{-3}}{\overset{k_{+3}}{\rightleftarrows}} X_3 \underset{k_{-5}}{\overset{k_{+4}}{\rightleftarrows}} X_4$$

fast   slow   very slow   fast

$$\frac{1}{\tau_1} = k_{+1}((E)_{eq} + (S)_{eq}) + k_{-1}$$

$$\frac{1}{\tau_2} = k_{-2} + \frac{k_{+2}}{1 + \left(\frac{k_{-1}}{k_{+1}}\right)\left(\frac{1}{(E)_{eq} + (S)_{eq}}\right)}$$

$$\frac{1}{\tau_3} = \frac{k_{+3}}{1 + \left(\frac{k_{-2}}{k_{+2}}\right)\left[\frac{1}{1 + k_{-1}/k_{-2}[1/(E)_{eq} + (S)_{eq}]}\right]} + \frac{k_{-3}}{\left[1 + \frac{k_{+4}}{k_{-4}}\right]}$$

$$\frac{1}{\tau_4} = k_{+4} + k_{-4}$$

and a plot of

$$\left[\frac{1}{\tau_2} - k_{-2}\right]^{-1} \text{ vs. } \left[\frac{1}{(E)_{eq} + (S)_{eq}}\right] \text{ yields as Y-intercept } = \frac{1}{k_{+2}}$$

$$\text{and slope } = \left(\frac{K_{diss,1}}{k_{+2}}\right) \tag{64}$$

Various values of k_{-2} can be assumed in a trial-and-error process until a linear fit of the data is obtained. Application of this plot produced a fairly good straight line (Figure 12) with k_{-2} taken as $2.4 \times 10^3\ s^{-1}$ and yielded the calculated parameters of $k_{+2} = 1.67\ (\pm 0.35) \times 10^4\ s^{-1}$, and $K_{diss,1} = 6.0\ (\pm 1.5) \times 10^{-4}\ M$. Since $\left(\frac{k_{+2}}{k_{-2}}\right) = K_{diss,2}$, $K_{diss,2} = \frac{1.67 \times 10^4}{2.4 \times 10^3} = 7.0$. The calculated overall binding constant for ADP

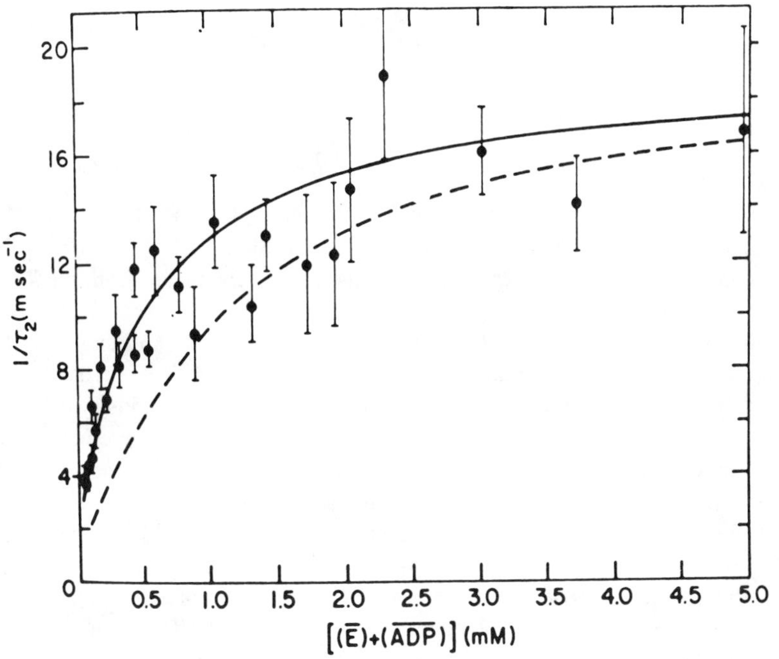

FIGURE 11. The reciprocal relaxation time for creatine kinase-ADP interactions, $1/\tau_2$, as a function of the sum of the equilibrium concentrations of enzyme and ADP. Each data point is the average of four to eight determinations; error limits are the standard deviations of the determinations. The solid curve represents the theoretical line for the uncoupled system calculated from the kinetic constants determined as in Figure 12. The dashed curve is the corresponding theoretical line for the coupled system. (From Hammes, G. G. and Hurst, J. K., *Biochemistry,* 8, 1083, 1969. With permission.)

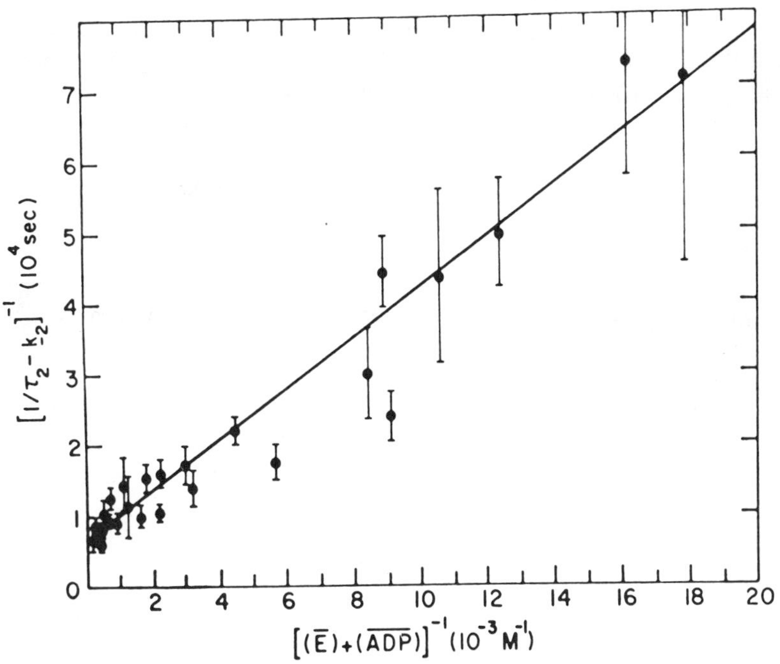

FIGURE 12. A plot of $[(1/\tau_2) - k_{-2}]^{-1}$ vs. $[(\bar{E}) + (\overline{ADP})]^{-1}$ with $k_{-2} = 2.4 \times 10^3\ s^{-1}$. Data points are the same as those plotted in Figure 11; the line was determined by a least-squares analysis of the data. (From Hammes, G. G. and Hurst, J. K., *Biochemistry,* 8, 1083, 1969. With permission.)

$$\overline{K}_{ADP} = \frac{(E)(ADP)}{[(E \cdot ADP)_1 + (E \cdot ADP)_2]} \tag{65}$$

$$\overline{K}_{ADP} = \frac{K_{diss,1}}{(1 + K_{diss,2})} = \frac{6.0 \times 10^{-4}}{1 + 7.0} = 7.5 \times 10^{-5}\,M \tag{66}$$

which agrees favorably with the thermodynamically measured value of 1.0×10^{-4} obtained by Kuby et al.[59] at 3°C, 0.15 (Γ/2), and pH 7.9 (see Table 4 in Chapter 8).

The theoretical curve, for this mechanism Scheme VI, calculated from the kinetic constants with the aid of Equation 62, is plotted as the solid line in Figure 11. The theoretical curve for the entire coupled system, i.e., with the use of the full expressions for System 1 in Table 6 (with values for $k_{+1} = 2.3 \times 10^7\,M^{-1}\,s^{-1}$ and $k_{-1} = 1.8 \times 10^4\,s^{-1}$, evaluated from the analysis of relaxation effects in solutions containing metal ions (see Hammes and Hurst[338]) is shown in Figure 11 as the dashed line. The reasonable agreement between the theoretical curves lends credence to the approximations made by Hammes and Hurst[338] in this analysis. (Compare also the more recent kinetic studies of Travers et al.[849-851]).

In a different application of temperature-jump measurements, Fisher and Sligar[967] examined the ferric spin-state equilibrium and relaxation rate of *Pseudomonas putida* Cytochrome P-450_{cam} with the use of several camphor substrate analogues known to induce different mixed spin stages in the substrate-bound complexes.[968] They found that all of the temperature-induced spectral changes were monophasic and that the spin-state relaxation rate reached a limiting value at high substrate concentrations. K_{spin}, a ferric spin-equilibrium constant, was defined as $K_{spin} = k_1/k_{-1}$ = P-450 (HS)/P-450 (LS) and $k_{obs} = k_1 + k_{-1}$. (HS) and (LS) represented the high-spin (S = 5/2) and low-spin (S = $^1/_2$) ferric iron, respectively; k_{obs} was the spectrally observed spin-state relaxation rate. They observed a good correlation between the fraction of high-spin species and a decrease in the rate constant, k_{-1}. With several substrate analogues, an increase in the fraction of high-spin species at equilibrium was concluded to be due to a restricted access and to the recombination rate of the axial ligand water molecule with the ferric iron. Thus, their data suggested that the substrate molecules are intimately involved in controlling the macroscopic "on" rate of the water molecule which serves as the sixth axial ligand to the heme iron in the low-spin form of cytochrome P-450_{cam}.[967]

VI. CONCLUDING REMARKS

With the advent of genetic engineering techniques[969,970] or, specifically, site-directed mutagenesis methods where individual amino acid residues of an enzyme molecule can be replaced by essentially any other naturally occurring amino acid residue in locations chosen by the experimenter,[971,987,1011] there has been a resurgence of activity-structure studies and of computer-aided modeling and simulation methods in enzyme proteins of known structure.[965,972] Add to this modern chemical technology[348,861,973,987,1010] and all this has required a further and deeper background in the fundamentals of enzyme catalysis, kinetics, and substrate binding (as presented here in Volume I) and an intimate knowledge of enzyme mechanisms (to be presented in Volume II).

APPENDIX

NOTES ADDED IN RETROSPECT

The characteristic and paramount property of enzymes is their ability to catalyze certain discreet chemical reactions. Moreover, and in contradistinction to inorganic catalysts, enzymes display an unusual degree of specificity towards their substrates; as such, enzymes have served as important characterization aids to chemists since their advent by Emil Fischer in his pioneering studies on the chemistry of carbohydrates[1] or as important tools in the covalent structural evaluation of macromolecules, including the enzymes themselves.[2]

A number of chemical models and explanations have been introduced to explain their most unusual property, that is, structural and optical specificity, beginning also with the "lock and key hypothesis" advanced by Emil Fischer.[1] But, probably, nowhere in the course of the development of modern biochemistry has one aspect occupied the attention of so many researchers, namely, the mechanism of action of these extraordinary protein catalysts (e.g., see Volume II[966]).

Towards this goal, a survey of modern biochemistry reveals that only in a few notable instances have very detailed studies been carried out on the relations between the structure of the complex enzyme protein molecule (at all of its levels of complexity) and its enzymatic function. The chemist must seek along all avenues for an explanation of the secret of the enzyme's extraordinary catalytic powers and its remarkable display of specificity toward its substrates — an explanation which must lie somewhere within the covalent nature of the protein, or within its three-dimensional conformation — or even as a result of its ability sometimes to alter or change its three-dimensional structure after binding its substrate, i.e., somewhere within its "conformational ability"[3] or within its conformational fluctuations.[955] Fundamental to all these studies is the hope, by the chemist, that a description of the enzyme's kinetic behavior will ultimately lend itself to a complete mathematical and physicochemical evaluation.

In recent years, a great deal of advance has been made along one kinetic approach, in particular, with the one simplifying assumption as to the existence of the hypothetical "steady state" of Briggs and Haldane,[4] a great deal of sophistication has been introduced into the unraveling of many enzymatic reaction mechanisms with their reaction rate expressions given in explicit differential form, so that the time appears ripe for a presentation, if possible, of a unification and recodification of these current ideas and kinetic approaches used in the study of the steady-state kinetic behavior of the enzymes. With the advent of many highly specialized tools for the study of their pre-steady-state behavior in recent years, an additional thesis could accordingly deal with this more complex technique, and the final chapter of Volume I (Part B) is concerned with transient-state kinetics. Finally, with efforts being devoted to the extraordinary complex physiological state of the "open system" and the coordinated sequence of catalytic enzymatic reactions which go to make up life's processes, a future undertaking would logically encompass the kinetic behavior of coordinated enzyme systems, and the interactions between catalytic enzymatic components.[5] This last task must, unfortunately, await more precise data gathering, evaluation, and processing on complex systems, an overall task which still seems too complex even for the single individual chemist to tackle by himself (alas, the days of Von Humboldt are unfortunately past where one man could hope to master all of man's knowledge) but of a nature so intriguing as to undoubtedly attract the attention of future teams of chemists and the employment of banks of high-speed computers. But a beginning is being made with the recent developments in computer science and the set of techniques known as ''artificial intelligence'' which is now rendering the complex multienzyme (metabolic) modeling process easier.[957] Also, the dynamics of complex metabolic systems seem to be more effectively analyzed by the judicious use of intrinsic time constants and dynamic modes of motion, both of which may be defined within the context of linear algebra as described very recently by Palsson et al.[958]

It is, therefore, the more immediate and modest task to present in a clear and concise fashion present-day information relating to the specific problem of the study of the steady-state properties of enzymes and then some discussion of their pre-steady-state behavior. However, from the past, the future is born, and from the past history on enzyme kinetics a greal deal of information still lies unrevealed, and for a firm grasp of present ideas a firm knowledge is necessary of the foundations derived from an orderly procession of ideas from the past. So, accordingly, this text will begin with fundamentals illustrating firm chemical principles, and which will include a review of the history of the development of the steady-state approach to the study of enzyme kinetics, to their modes of catalytic behavior, and to their mechanisms of action. Then, in outline fashion and in more sophisticated terms, the current state of the art will be delineated. It must be clearly indicated at the outset that there is, in principle, nothing mysterious about enzyme kinetics; all the known laws of reaction rate chemistry and catalytic behavior are obeyed. The problem lies in their complexity, in the numbers and kinds of reaction rate steps involved, and that, in essence, even in the simplest of cases, for a single-substrate, single-product reaction, catalysis by enzymes invariably leads to overall velocity expressions resting upon a set of nonlinear differential equations which do not lend themselves readily to analytic integral solutions. The enzyme chemist resorted, accordingly, to expressions of velocity data in terms of ''initial velocities'' or to initial time derivatives relating disappearance of substrate with appearance of product, as a function of time. It should also be remarked that in many cases those intermediates, postulated or real, in the reaction schemes, also, do not lend themselves to experimental or to measurable rate expressions. Thus, certain simplifying assumptions are common, i.e., that during the initial velocity where the substrate concentration is not far removed from its initial value that all the time derivatives for all the intermediates simultaneously approach zero in this restricted application by the enzymologist of the steady-state hypothesis; an hypothesis which, however, is not a true steady state, since not all reactants nor products themselves are held invariant with respect to time, as in, for example, the steady-state machines invented by Denbigh and Page[6] where a time invariance is maintained with respect to all reactants, all intermediates, and all products, for each given extent of the overall reaction. Moreover, it is, of course, unrealistic to assume that every time derivative for each intermediate must approach zero simultaneously (and a minimum of two enzyme-substrate intermediates is required for even the simplest of the single-substrate, single-product cases so as not to violate one of the basic tenets of the kinetics of chemical reactions, i.e., that of ''microscopic reversibility''). Fortunately, however, the assumption as to a restricted steady state often leads to a negligible error as pointed out in several situations;[7] nevertheless, one should always bear in mind the distinct possibility for certain cases that the assumption as to the steady state is not without risk — nor always valid. But the simplicity of this approach with its approximations usually leads to velocity expressions which even in the fairly complicated cases of two- or three-substrate reactions may shed some light on the kinetic mechanism. One should also bear in mind that which had been pointed out earlier,[8] that there are distinct possibilities that two different mechanisms may lead to rate equations that are of the same form, i.e., the two mechanisms being termed ''homeomorphic''. Thus, kinetic studies alone cannot distinguish them. In fact, it is a truism to state that kinetic studies alone cannot establish, with absolute certainty, that any mechanism is correct, for one is never in a position to unequivocally state that any particular mechanism does not possess its identical homeomorphic counterpart. On the other hand, it is equally *true*, that often *only* kinetic studies may be used to determine the *incorrect* nature of a particular pathway. Thus, the kinetic approach possesses the power to reject among the several possible plausible mechanisms. Whereas, thermodynamics often yields little or no information about the pathway, yet it is the latter, in fact (as demonstrated in Chapter 8), which can provide support for the selection as to the most likely mechanism (or mechanisms) and the self-consistent nature of the overall kinetic evaluation. Thus, it is the very combination of the two approaches, the kinetic and the thermodynamic, which has, in fact, provided a powerful combination and yielded much information on the ''mechanism of action'' of certain enzymatic reactions.[966]

One final comment, whereas, there have been a number of classical studies on ligand-induced conformational changes, the concept that protein enzymes are not rigid, but on the contrary dynamic, has led to a new era of thinking and to an active set of current research in a number of laboratories with a wide range of techniques to relate the mobility of specific parts of protein molecules to reaction mechanisms[955,976] (see a number of chapters in Volume II[966]). In fact, there is a vigorous search for and characterization of intermediate states in the process of ligand binding to proteins.[977-979] Whereas, many excellent studies have dealt with conformational changes on ligand binding and ligand dissociation on the time scale of milliseconds, ideas on molecular dynamics are now considering the picosecond fluctuations which are thought to be the basis for these relatively slower overall changes (see Chapter 1,[955] Section 1 in Volume II).

ACKNOWLEDGMENT

Much of the work by S. K. and coworkers cited in this volume was supported, in part, by grants-in-aid from the National Institutes of Health, U.S. Public Health Service.

References

REFERENCES

1. **Fischer, E.,** Einfluss dur configuration auf die wirkung dur enzyme, *Chem. Ber.*, 27, 2985, 1894.
2. *Methods in Enzymology,* Vol. 25—27 and 47, Hirs, C. H. W. and Timasheff, S. N., Eds., Academic Press, New York, 1972, 1972, 1973, 1977.
3. **Theorell, H.,** Function and structure of liver alcohol dehydrogenase, in *The Harvey Lectures,* Ser. 61, Academic Press, New York, 1965, 17.
4. **Briggs, G. E. and Haldane, J. B. S.,** I., A note on the kinetics of enzyme action, *Biochem. J.*, 19, 338, 1925.
5. **Eschrich, K., Schellenberger, W., and Hofmann, E.,** *In vitro* demonstration of alternate stationary states in an open enzyme system containing phosphofructokinase, *Arch. Biochem. Biophys.*, 205, 114, 1980.
6. **Denbign, K.,** Continuous reactions. Part II. The kinetics of steady state polymerisation, *Trans. Faraday Soc.*, 43, 648, 1947; **Denbign, K., Hicks, M., and Page,** The kinetics of open reaction systems, *Trans. Faraday Soc.*, 44, 479, 1948.
7. **Wong, J. T.,** *Kinetics of Enzyme Mechanisms,* Academic Press, New York, 1975.
8. **Wong, J. T. F. and Hanes, C. S.,** Kinetic formulations for enzymic reactions involving two substrates, *Can. J. Biochem. Physiol.*, 40, 763, 1962.
9. **Dixon, M. and Webb, E. C.,** *Enzymes,* 1st ed., Academic Press, New York, 1958.
10. **Dixon, M. and Webb, E. C.,** *Enzymes,* 2nd ed., Academic Press, New York, 1964.
11. **Henri, V.,** Théorie générale de l'action de quelques diastases, *Academie des Sciences, Paris,* 135, 916, 1902; Lois Générales de l'action des diastases, Hermann, Paris, 1903.
12. **Brown, A. J.,** Enzyme action, *Trans. J. Chem. Soc.*, 81, 373, 1902.
13. **Sörenson, S. P. L.,** Über die nessung und die bedeutung der wasserstoffionenkonzentration bei enzymatischen prozessen, *Biochem. Z.*, 21, 131, 1909.
14. **Michaelis, L. and Menten, M. L.,** Die Kinetik der invertinwirkung, *Biochem. Z.*, 49, 333, 1913.
15. **King, E. L. and Altman, C.,** A schematic method of deriving the rate laws for enzyme-catalyzed reactions, *J. Phys. Chem.*, 60, 1375, 1956.
16. **King, E. L.,** Unusual kinetic consequences of certain enzyme catalysis mechanisms, *J. Phys. Chem.*, 60, 1378, 1956.
17. **Monod, J., Wyman, J., and Changeux, J.-P.,** On the nature of the allosteric transitions: a plausible model, *J. Mol. Biol.*, 12, 88, 1965.
18. **Rubin, M. M. and Changeux, J.-P.,** On the nature of allosteric transitions: implications of non-exclusive ligand binding, *J. Mol. Biol.*, 21, 265, 1966.
19. **Blangy, K., Buc, H., and Monod, J.,** Kinetics of the allosteric interactions of phosphofructokinase from *Escherichia coli, J. Mol. Biol.*, 31, 13, 1968.
20. **Koshland, D. E., Nemethy, G., and Filmer, D.,** Comparison of experimental binding data and theoretical models in proteins containing subunits, *Biochemistry,* 5, 365, 1966.
21. **Koshland, D. E.,** Enzyme flexibility and enzyme action, *J Cell. Comp. Physiol.*, 54, (Suppl. 1) 245, 1959.
22. **Cornish-Bowden, A. J.,** *Principles of Enzyme Kinetics,* Butterworths, London, 1976.
23. **Orsi, B. A. and Tipton, K. F.,** Kinetic analysis of progress curves, in *Methods in Enzymology,* Vol. 63, Purich, D. L., Ed., Academic Press, New York, 1979, chap. 8.
24. **Walker, A. C. and Schmidt, C. L.,** Studies on histidase, *Arch. Biochem. Biophys.*, 5, 445, 1944.
25. **Kuby, S. A.,** Studies on β-D-Galactosidase from *Escherichia coli,* Part A, Studies on ATP-Creatine Transphosphorylase from Rabbit Skeletal Muscle, Part B, Ph.D. thesis, University of Wisconsin, 1953.
26. **Lilly, M. D., Hornby, W. E., and Crook, E. M.,** Kinetics of carboxymethylcellulose-ficin in packed beds, *Biochem. J.*, 100, 718, 1966.
27. **O'Neill, S. P., Lilly, M. D., and Rowe, P. N.,** Multiple steady states in continuous flow stirred tank enzyme reactors, *Chem. Eng. Sci.*, 26, 173, 1971.
28. **Dixon, M. and Webb, E. C.,** Inhibitors with very high affinities, in *Enzymes,* 2nd ed., Academic Press, New York, 1964, 331.
29. **Kuby, S. A., Hamada, M., Gerber, D., Tsai, W. C., Jacobs, H. K., Cress, M. C., Chua, G. K., Fleming, G., Wu, L. H., Fischer, A. H., Frischat, A., and Maland, L.,** Studies on adenosine triphosphate transphosphorylases. XII. Isolation and several properties of the crystalline calf ATP-AMP transphosphorylases (adenylate kinases) from muscle and liver and some observations on the rabbit muscle adenylate kinase, *Arch. Biochem. Biophys.*, 187, 34, 1978.
30. **Van Slyke, D. D. and Cullen, G. E.,** The mode of action of urease and of enzymes in general, *J. Biol. Chem.*, 19, 141, 1914.
31. **Lineweaver, H. and Burk, D.,** The determination of enzyme dissociation constants, *J. Am. Chem. Soc.*, 56, 658, 1934.
32. **Wilkinson, G. N.,** Statistical estimations in enzyme kinetics, *Biochem. J.*, 80, 324, 1961.

33. **Johansen, G. and Lumry, R.,** *C.R. Trav. Lab. Carlsberg,* 32, 185, 1961.
34. **Cleland, W. W.,** The statistical analysis of enzyme kinetic data, in *Advances in Enzymology,* Vol. 29, Nord, F. F., Ed., Wiley-Interscience, New York, 1967, 1.
35. **Bliss, C. I. and James, A. T.,** Fitting the rectangular hyperbola, *Biometrics,* 22, 573, 1966.
36. **Hanson, K. R., Ling, R., and Havir, E.,** A computer program for fitting data to the Michaelis-Menton equation, *Biochem. Biophys. Res. Commun.,* 29, 194, 1967.
37. **Eadie, G. S.,** The inhibition of cholinesterase by physostigmine and prostigmine, *J. Biol. Chem.,* 146, 85, 1942.
38. **Hofstee, B. H.,** Non-inverted versus inverted plots in enzyme kinetics, *Nature,* 184, 1296, 1959.
39. **Hultin, E.,** Statistical calculations on the accuracy of the Michaelis constant from viscosimetric determinations of polymetaphosphatase and dextranase activity, *Acta Chem. Scand.,* 21, 1575, 1967.
40. **Haldane, J. B. S.,** *Enzymes,* Longmans, Green and Co., London, 1930; reprinted by M. I. T. Press, Cambridge, MA, 1965.
41. **Lumry, R., Smith, E. L., and Glantz, R. R.,** Kinetics of carboxypeptidase action. I. Effect of various extrinsic factors on kinetic parameters, *J. Am. Chem. Soc.,* 73, 4330, 1951.
42. **Kistiakowsky, G. B. and Shaw, W. H. R.,** On the mechanism of the inhibition of urease, *J. Am. Chem. Soc.,* 75, 866, 1953.
43. **Cleland, W. W.,** The kinetics of enzyme-catalyzed reactions with two or more substrates or products, *Biochim. Biophys. Acta,* 67, 104, 1963; **Cleland, W. W.,** The kinetics of enzyme catalyzed reactions with two or more substrates or products. II. Inhibition: nomenclature and theory, 173; ibid., The kinetics of enzyme catalyzed reactions with two or more substrates or products. III. Prediction of initial velocity and inhibition patterns by inspection, 188.
44. **Dalziel, K.,** Initial steady state velocities in the evaluation of enzyme-coenzyme-substrate reaction mechanisms, *Acta Chem. Scand.,* 11, 1706, 1957.
45. **Dalziel, K.,** The determination of constants in a general coenzyme reaction mechanism by initial rate measurements in the steady state, *Trans. Faraday Soc.,* 54, 1247, 1958.
46. **Dalziel, K.,** Kinetic studies of liver alcohol dehydrogenase, *Biochem. J.,* 84, 244, 1962.
47. **Dalziel, K.,** The interpretation of kinetic data for enzyme-catalysed reactions involving three substrates, *Biochem. J.,* 114, 547, 1969.
48. **Kuby, S. A. and Noltmann, E. A.,** ATP-creatine transphosphorylase, in *The Enzymes,* 2nd ed., Vol. 6, Boyer, P. D., Lardy, H. A., and Myrbäck, K., Eds., Academic Press, New York, 1962, 515.
49. **Jacobs, H. K. and Kuby, S. A.,** Studies on adenosine triphosphate transphosphorylases. X. Kinetic properties of the crystalline adenosine triphosphate-creatine transphosphorylase from calf brain, *J. Biol. Chem.,* 245, 3305, 1970.
50. **Hamada, M. and Kuby, S. A.,** Studies on adenosine triphosphate transphosphorylases. XIII. Kinetic properties of the crystalline rabbit muscle ATP-AMP transphosphorylase (adenylate kinase) and a comparison with the crystalline calf muscle and liver adenylate kinases, *Arch. Biochem. Biophys.,* 190, 772, 1978.
51. **Alberty, R. A.,** Enzyme kinetics, in *Advances in Enzymology,* Vol. 17, Nord, F. F., Ed., Interscience, New York, 1978, 1.
52. **Alberty, R. A.,** The rate equation for an enzymic reaction, in *The Enzymes,* Vol. 1, 2nd ed., Boyer, P. D., Lardy, H., Myrbäck, K., Eds., Academic Press, New York, 1959, 143.
53. **Florini, J. R. and Vestling, C. S.,** Graphical determination of dissociation constants for two-substrate enzyme systems, *Biochim. Biophys. Acta,* 25, 575, 1957.
54. **Theorell, H. and Chance, B.,** Studies on liver alcohol dehydrogenase. II. The kinetics of the compound of horse liver alcohol dehydrogenase and reduced diphosphopyridine nucleotide, *Acta Chem. Scand.,* 5, 1127, 1951.
55. **Theorell, H. and Bonnichsen, R.,** Studies on liver alcohol dehydrogenase. I. Equilibria and initial reaction velocities, *Acta Chem. Scand.,* 5, 1105, 1951.
56. **Nygaard, A. P. and Theorell, H.,** The reaction mechanism of yeast alcohol dehydrogenase (ADH). Studies by overall reaction velocities, *Acta Chem. Scand.,* 9, 1300, 1955; Studies on liver alcohol dehydrogenase. III. The influence of pH and some anions on the reaction velocity constant, 9, 1502, 1955.
57. **Alberty, R. A.,** The relationship between Michaelis constants, maximum velocities and the equilibrum constant for an enzyme-catalyzed reaction, *J. Am. Chem. Soc.,* 75, 1928, 1953.
58. **Kuby, S. A., Wu, J. T., and Roy, R. N.,** Glucose 6-phosphate dehydrogenase from Brewers' yeast (Zwischenferment). IV. Further observations on the ligand induced macromolecular association phenomenon; kinetic properties of the two-chain protein species; and studies on the enzyme-substrate interactions, *Arch. Biochem. Biophys.,* 165, 153, 1974.
59. **Kuby, S. A., Mahowald, T. A., and Noltmann, E. A.,** Studies on adenosine triphosphate transphosphorylases. IV. Enzyme-substrate interactions, *Biochemistry,* 1, 748, 1962.
60. **Theorell, H. and McKinley-McKee, J. S.,** Mechanism of action of liver alcohol dehydrogenase, *Nature,* 192, 47, 1961.

61. **Thorell, H. and McKinley-McKee, J. S.,** Liver alcohol dehydrogenase. I. Kinetics and equilibria without inhibitors, *Acta Chem. Scand.*, 15, 1797, 1961.
62. **Frieden, C.,** The calculation of an enzyme-substrate dissociation constant from the over-all initial velocity for reactions involving two substrates, *J. Am. Chem. Soc.*, 79, 1894, 1957.
63. Recommendations of the Commission on Biochemical Nomenclature on the Nomenclature and Classification of Enzymes Together with their Units and the Symbols of Enzyme Kinetics (1972) Approved by the International Union of Biochemistry, Elsevier, Amsterdam, 1973.
64. **Plapp, B. V.,** On calculations of rate and dissociation constants from kinetic constants for the ordered bi bi mechanism of liver alcohol dehydrogenase, *Arch. Biochem. Biophys.*, 156, 112, 1973.
65. **Ainslie, G. R., Jr. and Cleland, W. W.,** Isotope exchange studies on liver alcohol dehydrogenase with cyclohexanol and cyclohexanone as reactants, *J. Biol. Chem.*, 247, 946, 1972.
66. **Schimerlik, M. I. and Cleland, W. W.,** Inhibition of creatine kinase by chromium nucleotides, *J. Biol. Chem.*, 248, 8418, 1973.
67. **Koshland, D. E.,** Stereochemistry and the mechanism of enzymatic reactions, *Biol. Rev.*, 28, 416, 1953.
68. **Frieden, C.,** Glutamic dehydrogenase. III. The order of substrate addition in the enzymatic reaction, *J. Biol. Chem.*, 234, 2891, 1959.
69. **Bloomfield, V., Peller, L., and Alberty, R. A.,** Multiple intermediates in steady-state enzyme kinetics. II. Systems involving two reactants and two products, *J. Am. Chem. Soc.*, 84, 4367, 1962.
70. **Alberty, R. A.,** On the determination of rate constants for coenzyme mechanisms, *J. Am. Chem. Soc.*, 80, 1777, 1958.
71. **Kuby, S. A. and Roy, R. N.,** Glucose-6-phosphate dehydrogenase from brewer's yeast. V. The effects of pH and temperature on the steady-state kinetic parameters of the two-chain protein species, *Biochemistry*, 15, 1975, 1976.
72. **Ottolenghi, P.,** The effects of hydrogen ion concentration on the simplest steady-state enzyme systems, *Biochem. J.*, 123, 445, 1971.
73. **Cohn, E. J. and Edsall, J. T.,** *Proteins, Amino Acids and Peptides*, Van Nostrand-Reinhold, New York, 1943.
74. **Edsall, J. T. and Wyman, J.,** Some general aspects of molecular interactions, in *Biophysical Chemistry*, Vol. 1, Academic Press, New York, chap. 11.
75. **Benesch, R. E. and Benesch, R.,** The acid strength of the -SH group in cysteine and related compounds, *J. Am. Chem. Soc.*, 77, 5877, 1955.
76. **Stern, A. E.,** Kinetics of biological reactions with special reference to enzymic processes, in *Advances in Enzymology*, Vol. 9, Nord, F. F., Ed., Interscience, New York, 1949, 25.
77. **Pauling, L.,** The hydrogen bond. A summary of the conditions for the formation and the properties of the hydrogen bond, in *The Nature of the Chemical Bond*, 2nd ed., Cornell University Press, Ithaca, NY, 1948, 333.
78. **Kuby, S. A. and Lardy, H. A.,** Purification and kinetics of β-D galactosidase from *Escherichia coli*, strain K-12, *J. Am. Chem. Soc.*, 75, 890, 1953.
79. **Bright, H. J.,** On the mechanism of divalent metal activation of β-methylaspartase, *J. Biol. Chem.*, 240, 1198, 1965.
80. **Kuby, S. A., Noda, L., and Lardy, H. A.,** Adenosinetriphosphate-creatine transphosphorylase. I. Isolation of the crystalline enzyme from rabbit muscle, *J. Biol. Chem.*, 209, 191, 1954; Adenosinetriphosphate-creatine transphosphorylase. II. Homogeneity and physicochemical properties, *J. Biol. Chem.*, 209, 203, 1954.
81. **Noda, L., Nihei, T., and Morales, M. F.,** The enzymic activity and inhibition of adenosine 5′-triphosphate-creatine transphosphorylase, *J. Biol. Chem.*, 235, 2830, 1960.
82. **Cleland, W. W.,** Enzyme kinetics, *Annu. Rev. Biochem.*, 36, 77, 1967.
83. **Morrison, J. F. and James, E.,** The mechanism of the reaction catalysed by adenosine triphosphate-creatine phosphotransferase, *Biochem. J.*, 97, 37, 1965.
84. **Morrison, J. F. and Cleland, W. W.,** Isotope exchange studies catalyzed by adenosine triphosphate: creatine phosphotransferase, *J. Biol. Chem.*, 241, 673, 1966.
85. **Mohan, M. S. and Rechnitz, G. A.,** Ion-electrode study of magnesium (II)-ATP and manganese (II)-ATP association, *Arch. Biochem. Biophys.*, 162, 194, 1974.
86. **Monk, C. B.,** Electrolytes in solutions of amino acids. Part IV. Dissociation constants of metal complexes of glycine, alanine and glycyl-glycine from pH titrations, *Trans. Faraday Soc.*, 47, 297, 1951.
87. **Kuby, S. A., Noda, L., and Lardy, H. A.,** Adenosinetriphosphate-creatine transphosphorylase. III. Kinetic studies, *J. Biol. Chem.*, 210, 65, 1954; Adenosinetriphosphate-creatine transphosphorylase. IV. Equilibrium studies, *J. Biol. Chem.*, 210, 83, 1954.
88. **Benzinger, T., Kitzinger, C., Hems, R., and Burton, K.,** Free-energy changes of the glutaminase reaction and the hydrolysis of the terminal pyrophosphate bond of adenosine triphosphate, *Biochem. J.*, 71, 400, 1959.
89. **Robbins, E. A. and Boyer, P. D.,** Determination of the equilibrium of the hexokinase reaction and the free energy of hydrolysis of adenosine triphosphate, *J. Biol. Chem.*, 224, 121, 1957.

90. **Alberty, R. A., Smith, R. M., and Bock, R. M.,** The apparent ionization constants of the adenosine-phosphates and related compounds, *J. Biol. Chem.*, 193, 425, 1951.
91. **Manchester, K. L.,** Resolution with a protein synthesising system of conflicting estimates for the stability constant for $KATP^{3-}$ formation, *Biochim. Biophys. Acta,* 630, 232, 1980.
92. **Mahowald, T., Noltmann, E., and Kuby, S. A.,** Studies on adenosine triphosphate transphosphorylases. III. Inhibition reactions, *J. Biol. Chem.*, 237, 1535, 1962.
93. **Gellert, M. and Sturtevant, J. M.,** The enthalpy change in the hydrolysis of creatine phosphate, *J. Am. Chem. Soc.*, 82, 1497, 1960.
94. **Boyer, P. D.,** Uses and limitations of measurements of rates of isotopic exchange and incorporation in catalyzed reaction, *Arch. Biochem. Biophys.*, 82, 387, 1960.
95. **Schacter, H.,** The use of the steady-state assumption to derive kinetic formulations for the transport of a solute across a membrane, in *Metabolic Pathways,* Vol. 6, Hoken, L. E., Ed., Academic Press, New York, 1972, chap. 1.
96. **Fromm, H. J.,** *Initial Rate Enzyme Kinetics,* Springer-Verlag, Heidelberg, 1975.
97. **Switzer, R. L.,** Regulation and mechanism of phosphoribosylpyrophosphate synthetase. II. Exchange reactions catalyzed by the enzyme, *J. Biol. Chem.*, 245, 483, 1970.
98. **Doudoroff, M., Barker, H. A., and Hassid, W. F.,** Studies with bacterial sucrose phosphorylase. I. The mechanism of action of sucrose phosphorylase as a glucose-transferring enzyme (transglucosidase), *J. Biol. Chem.*, 168, 725, 1947.
99. **Koshland, D. E., Jr.,** Isotopic exchange criteria for enzyme mechanisms, *Disc. Faraday Soc.*, 20, 143, 1955.
100. **Cedar, H. and Schwartz, N. H.,** The asparagine synthetase of *Escherichia coli.* II. Studies on mechanism, *J. Biol. Chem.*, 244, 4122, 1969.
101. **Ferdinand, W.,** The interpretation of non-hyperbolic rate curves for two-substrate enzymes. A possible mechanism for phosphofructokinase, *Biochem. J.*, 98, 278, 1966.
102. **Hill, A. V.,** A new mathematical treatment of changes of ionic concentration in muscle and nerve under the action of electric currents, with a theory as to their mode of excitation, *J. Physiol.*, 40, 190, 1910.
103. **Westley, J.,** *Enzymic Catalysis,* Harper and Row, New York, 1969.
104. **Rabin, B. R.,** The chemical nature of the products obtained by the action of cabbage-leaf phospholipase D on lysolecithin; the structure of lysolecithin, *Biochem. J.*, 102, 220, 1967.
105. **Cennamo, C. J.,** Steady-state kinetics of one-substrate enzymic mechanisms involving two enzyme conformations. I. Effects of modifiers on a mechanism postulating a single enzyme-substrate complex, *J. Theoret. Biol.*, 21, 260, 1968.
106. **Frieden, C.,** Kinetic aspects of regulation of metabolic processes: the hysteretic enzyme concept, *J. Biol. Chem.*, 245, 5788, 1970.
107. **Ainslie, G. R., Jr., Shill, P. J., and Neet, K. E.,** Transients and cooperativity. A slow transition model for relating transients and cooperative kinetics of enzymes, *J. Biol. Chem.*, 247, 7088, 1972.
108. **Worcel, A., Goldman, D. S., and Cleland, W. W.,** An allosteric reduced nicotinamide adenine dinucleotide oxidase from *Mycobacterium tuperculosis, J. Biol. Chem.*, 240, 3399, 1965.
109. **Atkinson, D. E., Hathaway, J. A., and Smith, E. C.,** Kinetics of regulatory enzymes. Kinetic order of the yeast diphosphopyridine nucleotide isocitrate dehydrogenase reaction and a model for the reaction, *J. Biol. Chem.*, 240, 2682, 1965.
110. **Taketu, K. and Pogell, B. M.,** Allosteric inhibition of rat liver fructose 1,6-diphosphatase by adenosine 5′-monophosphate, *J. Biol. Chem.*, 240, 651, 1965.
111. **Frieden, C.,** Treatment of enzyme kinetic data. I. The effect of modifiers on the kinetic parameters of single substrate enzymes, *J. Biol. Chem.*, 239, 3522, 1964.
112. **Atkinson, D. E. and Walton, G. M.,** Kinetics of regulatory enzymes. *Escherichia coli* phosphofructokinase, *J. Biol. Chem.*, 240, 757, 1965.
113. **Sweeny, J. R. and Fisher, J. R.,** An alternative to allosterism and cooperativity in the interpretation of enzyme kinetics data, *Biochemistry,* 7, 561, 1968.
114. **Fischer, J. R. and Hoagland, V. D.,** A systematic approach to kinetic studies of multisubstrate enzyme system, *Adv. Biol. Med. Phys.*, 12, 163, 1968.
115. **Griffin, C. C. and Brand, L.,** Kinetic implications of enzyme-effector complexes, *Arch. Biochem. Biophys.*, 126, 856, 1968.
116. **Monod, J., Wyman, J., and Changeux, J.-P.,** On the nature of allosteric transitions: a plausible model, *J. Mol. Biol.*, 12, 88, 1965.
117. **Frieden, C.,** Treatment of enzyme kinetic data. II. The multisite case: comparison of allosteric models and a possible new mechanism, *J. Biol. Chem.*, 242, 4045, 1967.
118. **Wong, J. F. T. and Endrenyi, L.,** Interpretation of nonhyperbolic behavior in enzymic systems. I. Differentiation of model mechanisms, *Can. J. Biochem.*, 49, 568, 1971.
119. **Chan, M. S. and Wong, J. F. T.,** Interpretation of nonhyperbolic behavior in enzymic systems. II. Quantitative characteristics of rate and binding functions, *Can. J. Biochem.*, 49, 581, 1971.

120. **Changeux, J.-P., Gerhart, J. C., and Schachman, H. K.,** Allosteric interactions in aspartate transcarbamylase. I. Binding of specific ligands to the native enzyme and its isolated subunits, *Biochemistry,* 7, 531, 1968.
121. **Changeux, J.-P. and Rubin, M. M.,** Allosteric interactions in aspartate transcarbamylase. III. Interpretation of experimental data in terms of the model of Monod, Wyman, and Changeux, *Biochemistry,* 7, 553, 1968.
122. **Koshland, D. E., Nemethy, G., and Filmer, D.,** Comparison of experimental binding data and theoretical models in proteins containing subunits, *Biochemistry,* 5, 365, 1966.
123. **Haber, J. E. and Koshland, D. E.,** Relation of protein subunit interactions to the molecular species observed during cooperative binding of ligands, *Proc. Natl. Acad. Sci. U.S.A.,* 58, 2087, 1967.
124. **Koshland, D. E., Conway, A., and Kirtley, M. E.,** Conformational changes and the mechanism of action of rabbit muscle glyceraldehyde-3-phosphate dehydrogenase, in FEBS *Symp. Regul. Enzyme Activity and Allosteric Interactions,* Kvamme, E. and Pihl, A., Eds., Associated Press, London, 1968, 131.
125. **Cornish-Bowden, A. and Koshland, D. E., Jr.,** A general method for the quantitative determination of saturation curves for multisubunit proteins, *Biochemistry,* 9, 3325, 1970.
126. **Gerhart, J. D. and Pardee, A. B.,** The enzymology of control by feedback inhibition, *J Biol. Chem.,* 237, 891, 1962; The effect of the feedback inhibitor, CTP, on subunit in aspartate transcarbamylase, in *Cold Spring Harbor Symposia Quant. Biol.,* Vol. 28, Cold Spring Harbor of Quantitative Biology, New York, 1963, 491.
127. **Gerhart, J. C. and Schachman, H. K.,** Allosteric interactions in aspartate transcarbamylase. II. Evidence for different conformational states of the protein in the presence and absence of specific ligands, *Biochemistry,* 7, 538, 1968.
128. **Bethell, M. R., Smith, K. E., White, J. S., and Jones, M. E.,** Carbamyl phosphate: an allosteric substrate for aspartate transcarbamylase of *Escherichia coli, Proc. Natl. Acad. Sci. U.S.A.,* 60, 1442, 1961.
129. **McClintock, D. K. and Markus, G.,** Conformational changes in aspartate transcarbamylase. I. Proteolysis of the intact enzyme, *J. Biol. Chem.,* 243, 3855, 1968.
130. **Weber, K.,** Aspartate transcarbamylase from *Escherichia coli*. Characterization of the polypeptide chains by molecular weight, amino acid composition, and amino-terminal residues, *J. Biol. Chem.,* 243, 543, 1968; New structural model of *E. coli* aspartate transcarbamylase and the amino-acid sequence of the regulatory polypeptide chain, *Nature,* 218, 1116, 1968.
131. **Wiley, D. C., Evans, D. R., Warren, S. G., McMurray, C. H., Edwards, B. F. P., Franks, W. A., and Lipscomb, W. N.,** The effect of NAD^+ on the catalytic efficiency of glyceraldehyde-3-phosphate dehydrogenase from rabbit muscle, in *Cold Spring Harbor Symp. Quant. Biol.,* 36, 285, 1972.
132. **Hammes, G. G. and Wu, C. W.,** Relaxation spectra of aspartate transcarbamylase. Interaction of the native enzyme with aspartate analogues, *Biochemistry,* 10, 1051, 1971.
133. **Teipel, J. and Koshland, D. E.,** The effect of NAD^+ on the catalytic efficiency of glyceraldehyde-3-phosphate dehydrogenase from rabbit muscle, *Biochim. Biophys. Acta,* 198, 183, 1970.
134. **Kirschner, K., Eigen, M., Bittman, R., and Voigt, B.,** The binding of nicotinamide-adenine dinucleotide to yeast D-glyceraldehyde-3-phosphate dehydrogenase: temperature-jump relaxation studies on the mechanism of an allosteric enzyme, *Proc. Natl. Acad. Sci. U.S.A.,* 56, 1661, 1966.
135. **Kirschner, K.,** Cooperative binding of nicotinamide-adenine dinucleotide to yeast glyceraldehyde-3-phosphate dehydrogenase. II. Stopped-flow studies at pH 8.5 and 40°C, *J. Mol. Biol.,* 58, 51, 1971.
136. **Kirschner, K., Gallego, E., Shuster, I., and Goodall, D.,** Cooperative binding of nicotinamide-adenine dinucleotide to yeast glyceraldehyde-3-phosphate dehydrogenase. I. Equilibrium and temperature-jump studies at pH 8.5 and 40°C, *J. Mol. Biol.,* 58, 29, 1971.
137. **Jaenicke, R. and Gratzer, W. B.,** Cooperative and noncooperative conformational effects of the coenzyme on yeast glyceraldehyde-3-phosphate dehydrogenase, *Eur. J. Biochem.,* 10, 158, 1969.
138. **Cook, R. A. and Koshland, D. E.,** Positive and negative cooperativity in yeast glyceraldehyde 3-phosphate dehydrogenase, *Biochemistry,* 9, 3337, 1970.
139. **Conway, A. and Koshland, D. E.,** Negative cooperativity in enzyme action. The binding of diphosphopyridine nucleotide to glyceraldehyde 3-phosphate dehydrogenase, *Biochemistry,* 7, 4011, 1968.
140. **Velick, S. F.,** Fluorescence spectra and polarization of glyceraldehyde-3-phosphate and lactic dehydrogenase coenzyme complexes, *J. Biol. Chem.,* 233, 1455, 1958.
141. **De Vijlder, J. J. M. and Hansen, B. J. M.,** Binding of NAD^+ to rabbit muscle glyceraldehydephosphate dehydrogenase as studied by optical rotary dispersion and circular dichroism, *Biochim. Biophys. Acta,* 178, 434, 1969.
142. **Price, N. C. and Radda, G. R.,** The binding of NAD^+ to rabbit muscle glyceraldehyde-3-phosphate dehydrogenase studied by protein fluorescence quenching, *Biochim. Biophys. Acta,* 235, 27, 1971.
143. **Klotz, I. M.,** The application of the law of mass action to binding by proteins. Interactions with calcium, *Arch. Biochem. Biophys.,* 9, 109, 1946.
144. **Scatchard, G.,** The attractions of proteins for small molecules and ions, *Ann. N.Y. Acad. Sci.,* 51, 660, 1949.
145. **Klotz, I. M.,** Protein interactions, in *The Proteins,* Vol. 1, 1st ed., Part B, Neurath, H. and Bailey, K., Eds., Academic Press, New York, 1953, chap. 8, 727.

146. **Klotz, I. M., Triwush, H., and Walker, F. M.,** The binding of organic ions by proteins. Competition phenomena and denaturation effects, *J. Am. Chem. Soc.*, 70, 2935, 1948.
147. **Scatchard, G., Scheinberg, I. H., and Armstrong, S. H.,** Physical chemistry of protein solutions. IV. The combination of human serum albumin with chloride ion, *J. Am. Chem. Soc.*, 72, 535, 1950; Physical chemistry of protein solutions, V. The combination of human serum albumin with thiocyanate ion, 22, 540, 1950.
148. **Klotz, I. M. and Hunston, D. L.,** Properties of graphical representations of multiple classes of binding sites, *Biochemistry,* 10, 3065, 1971.
149. **Adams, E. O.,** Relations between the constants of dibasic acids and of amphoteric electrolytes, *J. Am. Chem. Soc.*, 38, 1503, 1916.
150. **Edsall, J. T. and Wyman, J.,** Polybasic acids, bases, and ampholytes, including proteins, in *Biophysical Chemistry,* Academic Press, New York, 1958, chap. 9.
151. **Fletcher, J. E., Spector, A. A., and Ashbrook, J. D.,** Analysis of macromolecule-ligand binding by determination of stepwise equilibrium constants, *Biochemistry,* 9, 4580, 1970.
152. **Dalziel, K.,** The interpretation of kinetic data for enzyme-catalysed reactions involving three substrates, *Biochem. J.*, 114, 547, 1969.
153. **Elliott, K. R. F. and Tipton, K. F.,** Kinetic studies of bovine liver carbamoyl phosphate synthetase, *Biochem. J.*, 141, 807, 1974.
154. **Dixon, M., Webb, E. C., Thorne, C. J. R., and Tipton, K. F.,** Unspecific inhibition effects, in *Enzymes,* 3rd ed., Academic Press, New York, 1979, chap. 8, 361.
155. **Henderson, P. J. F.,** A linear equation that describes the steady-state kinetics of enzymes and subcellular particles interacting with tightly bound inhibitors, *Biochem. J.*, 127, 321, 1972.
156. **Chou, K.-C. and Forsén, S.,** Graphical rules for enzyme-catalysed rate laws, *Biochem. J.*, 187, 829, 1980.
157. **Volkenstein, M. W. and Goldstein, B. N.,** A new method for solving the problems of the stationary kinetics of enzymological reactions, *Biochim. Biophys. Acta,* 115, 471, 1966.
158. **Fromm, H. J.,** A simplified schematic method for deriving steady-state rate equations using a modification of the "theory of graphs" procedure, *Biochem. Biophys. Res. Commun.*, 40, 692, 1970.
159. **Reiner, J. M.,** *Behavior of Enzyme Systems,* Burgess, Minneapolis, 1959.
160. **Ainsworth, S. and Kinderlerer, J.,** A computer program to derive the rate equations of enzyme catalysed reactions with up to ten enzyme-containing intermediates in the reaction mechanism, *Int. J. Biomed. Comp.*, 7, 1, 1976.
161. **Fisher, D. D. and Schulz, A. R.,** Connection matrix representation of enzyme reaction sequences, *Math. Biosci.*, 4, 189, 1969.
162. **Hurst, R. O.,** A computer program for writing the steady-state rate equation for a multisubstrate enzymic reaction, *Can. J. Biochem.*, 47, 941, 1969.
163. **Inge, K. J. and Childs, R. E.,** A new method for deriving steady-state rate equations suitable for manual or computer use, *Biochem. J.*, 155, 567, 1967.
164. **Plowman, K. M.,** Complex mechanisms, in *Enzyme Kinetics,* McGraw-Hill, New York, 1972, 101.
165. **Fromm, H. J.,** The use of alternative substrates in studying enzymic mechanisms involving two substrates, *Biochim. Biophys. Acta,* 81, 413, 1964.
166. **Rudolph, F. B. and Fromm, H. J.,** Use of Isotope competition and alternative substrates for studying the kinetic mechanism of enzyme action. I. Experiments with hexokinase and alcohol dehydrogenase, *Biochemistry,* 9, 4660, 1970.
167. **Zewe, V., Fromm, H. J., and Fabiano, R. J.,** The effect of manganous ion on the kinetics and mechanism of the yeast hexokinase reaction, *J. Biol. Chem.*, 239, 1625, 1964.
168. **Cha, S.,** A simple method for derivation of rate equations for enzyme-catalyzed reactions under the rapid equilibrium assumption or combined assumptions of equilibrium and steady state, *J. Biol. Chem.*, 243, 820, 1968.
169. **Segel, I. H.,** Varieties of nonhyperbolic velocity curves. Hybrid ping pong-ordered and ping pong-random Bi Bi systems, in *Enzyme Kinetics,* John Wiley & Sons, New York, 1975, 662.
170. **Remington, R. D. and Schork, M. A.,** *Statistics with Applications to the Biological and Health Sciences,* Prentice-Hall, Englewood Cliffs, N.J., 1970.
171. **Lark, P. D. and Craven, B. R.,** *The Handling of Chemical Data,* Pergamon Press, Elmsford, N.Y., 1969.
172. **Scheflet, W. C.,** *Statistics for the Biological Sciences,* Addison-Wesley, Reading, MA, 1969.
173. **Bishop, O. N.,** *Statistics for Biology,* Houghton-Mifflin, Boston, 1966.
174. **Williams, E. J.,** *Regression Analysis,* John Wiley & Sons, New York, 1959.
175. **Bennett, C. A. and Franklin, H. L.,** *Statistical Analysis in Chemistry and the Chemical Industry,* John Wiley & Sons, New York, 1954.
176. **Levy, H. and Baggott, E. A.,** *Numerical Solutions of Differential Equations,* Dover, New York, 1950.

177. **Colquhoun, D.,** *Lectures on Biostatistics,* Oxford University Press, London, 1971.
178. **Mellor, J. W.,** *Higher Mathematics for Students of Chemistry and Physics,* 4th ed., Dover, New York, 1955.
179. **Aitken, A. C.,** *Determinants and Matrices,* Oliver and Boyd, Edinburgh, 1939.
180. **Hanes, C. S.,** Studies on plant amylases. I. The effect of starch concentration upon the velocity of hydrolysis by the amylase of germinated barley, *Biochem. J.,* 26, 1406, 1932.
181. **Dowd, J. E. and Riggs, D. S.,** A comparison of estimates of Michaelis-Menton kinetic constants from various linear transformations, *J. Biol. Chem.,* 240, 863, 1965.
182. **Cleland, W. W.,** Computer programs for processing kinetic data, *Nature,* 198, 463, 1963.
183. **Plowman, K. M. and Cleland, W. W.,** Purification and kinetic studies of the citrate cleavage enzyme, *J. Biol. Chem.,* 242, 4239, 1967.
184. **Segal, H. L.,** The development of enzyme kinetics, in *The Enzymes,* Vol. 1, 2nd ed., Boyer, P. D., Lardy, H., and Myrbäck, K., Eds., Academic Press, New York, 1959, 1.
185. **Hearon, J. Z., Bernhard, S. A., Friess, S. L., Botts, D. J., and Morales, M. F.,** Enzyme kinetics, in *The Enzymes,* Vol. 1, 2nd ed., Boyer, P. D., Lardy, H., and Myrbäck, K., Eds., Academic Press, New York, 1959, 49.
186. **Dixon, M., Webb, E. C., Thorne, C. J. R., and Tipton, K. F.,** *Enzymes,* 3rd ed., Academic Press, New York, 1979.
187. **Mahler, H. R. and Cordes, E. H.,** Enzyme kinetics, in *Biological Chemistry,* 2nd ed., Harper & Row, New York, 1971, 267.
188. **Reiner, J. M.,** *Behavior of Enzyme Systems,* 1st ed., Burgess Publishing, Minnesota, 1959; 2nd ed., Van Nostrand-Reinhold, 1969.
189. **Wilson, P. W.,** Kinetics and mechanisms of enzyme reactions, in *Respiratory Enzymes,* Lardy, H. A., Ed., Burgess Publishing, Minnesota, 1949, 16.
190. *The Enzymes,* 2nd ed., Boyer, P. D., Lardy, H., and Myrbäck, K., Eds., Vol. 1—8, Academic Press, New York, 1961 to 1963.
191. *The Enzymes,* 3rd ed., Vol. 1, 2, 4, 8, 9, and 11 to 13, Boyer, P. D., Ed., 1970 to 1976.
192. **Laidler, K.,** *The Chemical Kinetics of Enzyme Action,* 1st ed., Oxford University Press, London, 1958.
193. **Laidler, K. J. and Bunting, P. S.,** *The Chemical Kinetics of Enzyme Action,* 2nd ed., Clarendon Press, Oxford, 1973.
194. **Walter, C.,** *Steady-State Applications in Enzyme Kinetics,* Ronald Press, New York, 1965.
195. **Cleland, W. W.,** Steady state kinetics, in *The Enzymes,* Vol. 2, 3rd ed., Boyer, P. D., Ed., Academic Press, New York, 1970, 1.
196. **Dawes, E. A.,** Enzyme Kinetics, in *Comprehensive Biochemistry,* Vol. 12, Florkin, M. and Stotz, E. H., Elsevier, New York, 1964, 89.
197. **Engel, P. C.,** *Enzyme Kinetics — The Steady-State Approach,* John Wiley & Sons, New York, 1977.
198. **Cornish-Bowden, A. J.,** *Fundamentals of Enzyme Kinetics,* Butterworths, London, 1979.
199. *Methods in Enzymology,* Vol. 63, Enzyme kinetics and mechanism, Part A, Purich, D. L., Ed., chapters on steady-state kinetics, methodology, Colowick, S. P. and Kaplan, N. O., Eds., Academic Press, New York, 1979.
200. **Lumry, R.,** Some aspects of the thermodynamics and mechanism of enzymic catalysis, in *The Enzymes,* Vol. 1, 2nd ed., Boyer, P. D., Lardy, H., and Myrbäck, K., Eds., Academic Press, New York, 1959, 157.
201. **Walter, C.,** *Enzyme Kinetics, Open and Closed Systems,* Ronald Press, New York, 1966.
202. **Jencks, W. P.,** *Catalysis in Chemistry and Enzymology,* McGraw-Hill, New York, 1969.
203. **Yonetani, T. and Theorell, H.,** Studies on liver alcohol dehydrogenase complexes. III. Multiple inhibition kinetics in the presence of two competitive inhibitors, *Arch. Biochem. Biophys.,* 106, 243, 1964; Studies on liver alcohol dehydrogenase complexes. IV. Spectrophotometric observations on the enzyme complexes, 106, 253, 1964.
204. **Peller, L. and Alberty, R. A.,** Multiple intermediates in steady-state enzyme kinetics. I. The mechanism involving a single substrate and product, *J. Am. Chem. Soc.,* 81, 5907, 1959.
205. **Alberty, R. A., Bloomfield, V., and Peller, L.,** Multiple intermediates in steady-state enzyme kinetics. III. Analysis of the kinetics of some reactions catalyzed by dehydrogenases, *J. Am. Chem. Soc.,* 84 4375, 1962; **Alberty, R. A., Bloomfield, V., Peller, L., and King, E. L.,** Part IV. The steady-state kinetics of isotopic exchange in enzyme catalyzed reactions, 84, 4381, 1962.
206. **Frieden, C., Wolfe, R. G., and Alberty, R. A.,** Studies of the enzyme fumarase. IV. The dependence of the kinetic constants at 25° on buffer concentration, composition and pH, *J. Am. Chem. Soc.,* 79, 1523, 1957.
207. **Alberty, R. A. and Peirce, W. H.,** Studies of the enzyme fumarase. V. Calculation of minimum and maximum values of constants for the general fumarase mechanism, *J. Am. Chem. Soc.,* 79, 1526, 1957.
208. **Segal, H. L., Kachmar, J. F., and Boyer, P. D.,** Kinetic analysis of enzyme reactions. I. Further considerations of enzyme inhibition and analysis of enzyme activation, *Enzymologia,* 15, 187, 1952.

209. **Frieden, C. and Alberty, R. A.,** The effect of pH on fumarase activity in acetate buffer, *J. Biol. Chem.,* 212, 859, 1955.
210. **Cha, S.,** Kinetics of enzyme reactions with competing alternative substrates, *Mol. Pharmacol.,* 4, 621, 1968.
211. **Janson, C. A. and Cleland, W. W.,** The kinetic mechanism of glycerokinase, *J. Biol. Chem.,* 249, 2562, 1974.
212. **Morrison, J. F. and Ebner, K. E.,** Studies on galactosyltransferase. Kinetic investigations with N-acetylglucosamine as the galactosyl group acceptor, *J. Biol. Chem.,* 246, 3977, 1971.
213. **Rudolph, F. B. and Fromm, H. J.,** Kinetics of three substrate enzyme systems. Treatment of partially random mechanisms using equilibrium assumptions, *J. Theoret. Biol.,* 39, 363, 1973.
214. **Rudolph, F. B. and Fromm, H. J.,** Use of isotope competition and alternative substrates for studying the kinetic mechanism of enzymic action. II. Rate equations for three substrate enzymes systems, *Arch. Biochem. Biophys.,* 147, 515, 1971.
215. **Fromm, H. J.,** The use of competitive inhibitors in studying the mechanism of action of some enzyme systems utilizing three substrates, *Biochim. Biophys. Acta,* 139, 221, 1967.
216. **Hanson, T. L. and Fromm, H. J.,** Rat skeletal muscle hexokinase. I. Kinetics and reaction mechanism, *J. Biol. Chem.,* 240, 4133, 1965.
217. **Lucas, J. J., Burchiel, S. W., and Segel, I. H.,** Choline sulfatase of *Pseudomonas aeruginosa, Arch. Biochem. Biophys.,* 153, 664, 1972.
218. **Silverstein, E. and Boyer, P. D.,** Equilibrium reaction rates and the mechanisms of bovine heart and rabbit muscle lactate dehydrogenases, *J. Biol. Chem.,* 239, 3901, 1964; Equilibrium reaction rates and the mechanisms of liver and yeast alcohol dehydrogenase, *J. Biol. Chem.,* 239, 3908, 1964.
219. **Fromm, H. J., Silverstein, E., and Boyer, P. D.,** Equilibrium and net reaction rates in relation to the mechanism of yeast hexokinase, *J. Biol. Chem.,* 239, 3645, 1964.
220. **Reynard, A. M., Hass, L. F., Jacobsen, D. D., and Boyer, P. D.,** The correlation of reaction kinetics and substrate binding with the mechanism of pyruvate kinase, *J. Biol. Chem.,* 236, 2277, 1961.
221. **Umbarger, H. E.,** Evidence for a negative-feedback mechanism in the biosynthesis of isoleucine, *Science,* 123, 848, 1956.
222. **Umbarger, H. E. and Brown, B.,** Isoleucine and valine metabolism in *Escherichia coli.* VII. A negative feedback mechanism controlling isoleucine biosynthesis, *J. Biol. Chem.,* 233, 415, 1958.
223. **Yates, R. A. and Pardee, A. E.,** Control of pyrimidine biosynthesis in *Escherichia coli* by a feed-back mechanism, *J. Biol. Chem.,* 221, 757, 1956.
224. **Atkinson, D. E.,** Regulation of enzyme activity, *Annu. Rev. Biochem.,* 35, 85, 1960.
225. **Brown, W. E. L. and Hill, A. V.,** The oxygen-dissociation curve of blood, and its thermodynamic basis, *Proc. R. Soc. London,* B94, 297, 1922 and 1923.
226. **Levitski, A. and Koshland, D. E., Jr.,** Negative cooperativity in regulatory enzymes, *Proc. Natl. Acad. Sci., U.S.A.,* 62, 1121, 1969.
227. **Lam, C. F. and Priest, D. G.,** Enzyme kinetics. Systematic generation of valid King-Altman patterns, *Biophys. J.,* 12, 248, 1972.
228. **Gear, C. W.,** *Numerical Initial Value Problems in Ordinary Differential Equations,* Prentice-Hall, Englewood Cliffs, N.J., 1971.
229. **Bocher, M.,** *Introduction to Higher Algebra,* Macmillan, New York, 1907.
230. **Rosenbach, J. B., Whitman, E. A., Meserne, B. E., and Whitman, P. M.,** *College Algebra,* 4th ed., Ginn and Co., Boston, 1958.
231. **Frost, A. A. and Pearson, R. C.,** *Kinetics and Mechanisms,* 2nd ed., John Wiley & Sons, New York, 1961.
232. **Daniels, F. and Alberty, R. A.,** in *Physical Chemistry,* 2nd ed., John Wiley & Sons, New York, 1961, 294; *Physical Chemistry* 3rd ed., 1966, chap. 10.
233. **Laidler, K. J.,** *Chemical Kinetics,* 2nd ed., McGraw-Hill, New York, 1965.
234. **Capellos, C. and Bielski, H. J.,** *Kinetic Systems — The Mathematical Description of Chemical Kinetics in Solutions,* Wiley Interscience, New York, 1972.
235. **Amis, E. S.,** *Kinetics of Chemical Change in Solution,* Macmillan, New York, 1949.
236. **Glasstone, S., Laidler, K. J., and Eyring, H.,** *The Theory of Rate Processes,* McGraw-Hill, New York, 1941.
237. **Chanutin, A., Ludewig, S., and Masket, A. V.,** Further purification of catecholase (tyrosinase), *J. Biol. Chem.,* 147, 737, 1942.
238. **Hughes, T. R. and Klotz, I. M.,** Analysis of metal-protein complexes, in *Methods of Biochemical Analysis,* Vol. 3, Glick, D., Ed., Interscience, New York, 1956, 265.
239. **Velick, S. F., Hayes, I. E., and Harting, I.,** The binding of diphosphopyridine nucleotide by glyceraldehyde-3-phosphate dehydrogenase, *J. Biol. Chem.,* 203, 527, 1953.
240. **Hummel, J. P. and Dryer, W. J.,** Measurement of protein-binding phenomena by gel filtration, *Biochim. Biophys. Acta,* 63, 530, 1962.

241. **Fairclough, G. I., Jr. and Fruton, J. S.,** Peptide-protein interaction as studied by gel filtration, *Biochemistry,* 5, 673, 1966.
242. **Casman, M. and King, R. C.,** Subunit interactions and ligand binding in supernatant malic dehydrogenase. Cooperative binding of reduced nicotinamide adenine dinucleotide associated with a monomer-dimer equilibrium of the protein, *Biochemistry,* 11, 4937, 1972.
243. **Adair, G. S.,** The hemoglobin system. VI. The oxygen dissociation curve of hemoglobin, *J. Biol. Chem.,* 63, 529, 1925.
244. **Weber, G. and Anderson, S. R.,** Multiplicity of binding. Range of validity and practical test of Adair's equation, *Biochemistry,* 4, 1942, 1965.
245. **Pauling, L.,** The oxygen equilibrium of hemoglobin and its structural interpretation, *Proc. Natl. Acad. Sci. U.S.A.,* 21, 186, 1935.
246. **Eigen, M.,** Kinetics of reaction control and formation transfer in enzymes and nucleic acids, in *Nobel Symp.,* 5, 341, 1967.
247. **Kirtley, M. E. and Koshland, D. E., Jr.,** Models for cooperative effects in proteins containing subunits. Effects of the two interacting ligands, *J. Biol. Chem.,* 242, 4192, 1967.
248. **Nygaard, A. P. and Theorell, H.,** Dissociation constants in the yeast alcohol dehydrogenase system, calculated from overall reaction velocities, *Acta Chem. Scand.,* 9, 1551, 1955.
249. **Hamada, M., Palmieri, R. H., Russell, G. A., and Kuby, S. A.,** Studies on adenosine triphosphate transphosphorylases. XVI. Equilibrium binding properties of the crystalline rabbit and calf muscle ATP-AMP transphosphorylase (adenylate kinase) and derived peptide fragments, *Arch. Biochem. Biophys.,* 195, 155, 1979.
250. **Koshland, D. E., Jr. and Neet, D. E.,** The catalytic and regulatory properties of enzymes, in *Annual Review of Biochemistry,* Vol. 37, Snell, E. E., Boyer, P. D., Meister, A., and Sinsheimer, R. L., Eds., *Annual Reviews, Inc.,* California, 1968, 359.
251. **Gutfreund, H. and Knowles, J. R.,** The foundations of enzyme action, in *Essays in Biochemistry,* Vol. 3, Campbell, P. N. and Greville, G. D., Eds., Academic Press, London, 1967, 25.
252. **Rose, I. A.,** Enzyme reaction stereospecificity: a critical review, *Critical Review of Biochemistry,* CRC Press, Boca Raton, 1, 33, 1971.
253. **Dickerson, R. E. and Geiss, I.,** *The Structure and Action of Proteins,* Harper & Row, New York, 1969.
254. **Popjak, G.,** Stereospecificity of enzymic reactions, in *The Enzymes,* Vol. 2, 3rd ed., Boyer, P. D., Ed., Academic Press, New York, 1970, 115.
255. **Blow, D. M. and Steitz, T. A.,** X-Ray diffraction studies of enzymes, in *Annual Review of Biochemistry,* Vol. 39, Snell, E. E., Boyer, P. D., Meister, A., and Sinsheimer, R. L., Eds., Annual Reviews Inc., California, 1970, 63.
256. **Vennesland, B.,** Some applications of deuterium to the study of enzyme mechanisms, *Disc. Faraday Soc.,* 20, 240, 1955.
257. **Michaelis, L. and Schubert, M. D.,** The theory of reversible two-step oxidation involving free radicals, *Chem. Rev.,* 22, 437, 1938.
258. **Timasheff, S. N.,** Irreversible inhibition, in *The Enzymes,* Vol. 2, 3rd ed., Boyer, P. D., Ed., Academic Press, New York, 1970, 371.
259. **Richards, F. J. and Wyckoff, H. W.,** Bovine pancreatic ribonuclease, in *The Enzymes,* Vol. 4, 3rd ed., Boyer, P. D., Ed., Academic Press, New York, 1971, 647.
260. **Mildvan, A. S. and Cohn, M.,** Aspects of enzyme mechanisms studied by nuclear spin relaxation induced by paramagnetic probes, in *Advances in Enzymology,* Vol. 33, Nord, F. F., Ed., Interscience, New York, 1970, 1.
261. **Mildvan, A. S., Cohn, M., and Leigh, J. S., Jr.,** *Magnetic Resonance in Biological Systems,* Ehrenberg, A., Malstrom, B. G., and Vanngard, T., Eds., Pergamon Press, Elmsford, New York, 1967.
262. **Mildvan, A. S.,** Magnetic resonance studies of the conformations of enzyme-bound substrates, *Acc. Chem. Res.,* 10, 246, 1977.
263. **Rossman, M. G. et al.,** Structural constraints of possible mechanisms of lactate dehydrogenase as shown by high resolution studies of the apoenzyme and a variety of enzyme complexes, *Cold Spring Harbor Symp. Quant. Biol.,* 36, 179, 1972.
264. **Dwek, R. A.,** *NMR in Biochemistry, Applications to Enzyme Systems,* Oxford University Press, London, 1973.
265. **Zeffren, E. and Hall, P. L.,** *The Study of Enzyme Mechanisms,* John Wiley & Sons, New York, 1973.
266. **Bender, M. L.,** *Mechanism of Homogeneous Catalysis from Protons to Proteins,* John Wiley & Sons, New York, 1971.
267. **Gray, D. J.,** *Enzyme Catalyzed Reactions,* Van Nostrand-Reinhold, New York, 1971.
268. **Gutfreund, H.,** *Enzymes: Physical Principles,* John Wiley & Sons, New York, 1972.
269. **Glazer, A. N.,** Specific chemical modification of proteins, in *Annual Review of Biochemistry,* Vol. 39, Snell, E. E., Boyer, P. D., Meister, A., and Sinsheimer, R. L., Eds., Annual Reviews Inc., California, 1970, 101.

270. **Means, G. E. and Feeney, R. E.,** *Chemical Modification of Proteins,* Holden-Day, San Francisco, 1971.
271. **Soskel, N. T. and Kuby, S. A.,** A steady-state kinetic analysis of the prolyl-4-hydroxylase mechanism, *Connect. Tissue Res.,* 9, 121, 1981.
272. **Prockop, D. J., Kivirikko, K. I., Tuderman, L., and Guzman, N. A.,** I. The biosynthesis of collagen and its disorders, *N. Engl. J. Med.,* 301, 13, 1979; Part II. The biosynthesis of collagen and its disorders, *N. Engl. J. Med.,* 301, 77, 1979.
273. **Hayaishi, O., Nozaki, M., and Abbott, M. T.,** *Oxygenases: Dioxygenases,* in *The Enzymes,* Vol. 12, 3rd ed., Part B, Boyer, P. D., Ed., Academic Press, New York, 1975, 119.
274. **Cardinale, G. J. and Udenfriend, S.,** Prolyl hydroxylase, *Advances in Enzymology,* Vol. 41, Meister, A., Ed., Interscience, New York, 1974, 245.
275. **Puistola, V., Turpeennieme-Hujanen, T. M., Myllylä, R., and Kivirikko, K. I.,** Studies on the lysyl hydroxylase reaction. I. Initial velocity kinetics and related aspects, *Biochim. Biophys. Acta,* 611, 40, 1980.
276. **Puistola, V., Turpeennieme-Hujanen, T. M., Myllylä, R. and Kivirikko, K. I.,** Studies on the lysyl hydroxylase reaction. II. Inhibition kinetics and the reaction mechanism, *Biochim. Biophys. Acta,* 611, 51, 1980.
277. **Kondo, A., Blanchard, J. S., and England, S.,** Purification and properties of calf liver γ-butyrobetaine hydroxylase, *Arch. Biochem. Biophys.,* 212, 338, 1981.
278. **Tuderman, L., Myllylä, R., and Kivirikko, K. I.,** Mechanism of the prolyl hydroxylase reaction. I. Role of co-substrates, *Eur. J. Biochem.,* 80, 341, 1977.
279. **Myllylä, R., Tuderman, L., and Kivirikko, K. I.,** Mechanism of the prolyl hydroxylase reaction. II. Kinetic analysis of the reaction sequence, *Eur. J. Biochem.,* 80, 349, 1977.
280. **Myllylä, R., Kuutti-Savolainen, E.-R., and Kivirikko, K. I.,** The role of ascorbate in the prolyl hydroxylase reaction, *Biochem. Biophys. Res. Commun.,* 83, 441, 1978.
281. **Bright, H. J.,** Divalent metal activation of β-methylaspartase. The importance of ionic radius, *Biochemistry,* 6, 1191, 1967.
282. **Plowman, K. M.,** Appendix A: product inhibition patterns for termolecular enzymic reactions, in *Enzyme Kinetics,* McGraw-Hill, New York, 1972, 155.
283. **Tryon, E., Cress, M. C., Hamada, M., and Kuby, S. A.,** Studies on NADPH-cytochrome c Reductase. I. Isolation and several properties of the crystalline enzyme from ale yeast, *Arch. Biochem. Biophys.,* 197, 104, 1979.
284. **Noda, L., Kuby, S. A., and Lardy, H. A.,** ATP-creatine transphosphorylase. IV. Equilibrium studies, *J. Biol. Chem.,* 210, 83, 1954.
285. **Brown, T. R.,** Is creatine phosphokinase in equilibrium in skeletal muscle?, *Fed. Proc.,* 41, 174, 1982.
286. **Keutel, H. J., Okabe, K., Jacobs, H. K., Ziter, F., Maland, L., and Kuby, S. A.,** Studies on adenosine triphosphate transphosphorylases. XI. Isolation of the crystalline adenosine triphosphate-creatine transphosphorylases from the muscle and brain of man, calf and rabbit; and a preparation of their enzymatically active hybrids, *Arch. Biochem. Biophys.,* 150, 648, 1972.
287. **Yue, R. H., Jacobs, H. K., Okabe, K., Keutel, H. J., and Kuby, S. A.,** Studies on adenosine triphosphate transphosphorylases. VII. Homogeneity and physicochemical properties of the crystalline adenosine triphosphate-creatine transphosphorylase from calf brain, *Biochemistry,* 7, 4291, 1968.
288. **Kuby, S. A., Palmieri, R. H., Okabe, K., Frischat, A., and Cress, M. C.,** Studies on muscular dystrophy. II. A comparison by physical and chemical means of the normal human ATP-creatine transphosphorylases (creatine kinases) with those from tissues of Duchenne muscular dystrophy, *Arch. Biochem. Biophys.,* 194, 336, 1979.
289. **Okabe, K., Jacobs, H. K., and Kuby, S. A.,** Studies on adenosine triphosphate transphosphorylases. X. Reactivity and analysis of the sulfhydryl groups of the crystalline adenosine triphosphate-creatine transphosphorylase from calf brain, *J. Biol. Chem.,* 245, 6498, 1970.
290. **Scopes, R. K.,** Acid denaturation of creatine kinase, *Arch. Biochem. Biophys.,* 110, 320, 1965.
291. **Birktoft, J. J. and Ottesen, M.,** The effect of chemical modification on the acid denaturation of rabbit skeletal muscle creatine kinase, *Biochim. Biophys. Acta,* 175, 204, 1969.
292. **Beychok, S. and Warner, R. C.,** Denaturation and electrophoretic behavior of lysozyme, *J. Am. Chem. Soc.,* 81, 1892, 1959.
293. **Steinhardt, J.,** *K. Dan. Vidensk. Selsk. Mat.-Fys. Medd.,* 14(11), 1937.
294. **Yue, R. H., Palmieri, R. H., Olson, O. E., Maland, L., and Kuby, S. A.,** Studies on adenosine triphosphate transphosphorylases. VI. Studies on the polypeptide chains of the crystalline adenosine triphosphate-creatine transphosphorylase from rabbit skeletal muscle, *Biochemistry,* 6, 3204, 1967.
295. **Dawson, D. M., Eppenberger, H. M., and Kaplan, N. O.,** The comparative enzymology of creatine kinase, *J. Biol. Chem.,* 242, 210, 1967.
296. **Kistiakowski, G. B. and Lumry, R.,** Anomalous temperature effects in the hydrolysis of urea by urease, *J. Am. Chem. Soc.,* 71, 2006, 1949.

297. **Keutel, H. J., Jacobs, H. K., Okabe, K., Yue, R. H., and Kuby, S. A.,** Studies on adenosine triphosphate transphosphorylases. VII. Isolation of the crystalline adenosine triphosphate-creatine transphosphorylase from calf brain, *Biochemistry,* 7, 4283, 1968.
298. **Nihei, T., Noda, L., and Morales, M. F.,** Kinetic properties and equilibrium constant of the adenosine triphosphate-creatine transphosphorylase-catalyzed reaction, *J. Biol. Chem.,* 236, 3203, 1961.
299. **Jacobs, H. K. and Kuby, S. A.,** unpublished observations, 1970.
300. **Jacobs, H. K. and Kuby, S. A.,** Studies on muscular dystrophy. III. A comparison of the steady-state kinetics of the normal human ATP-creatine transphosphorylase isoenzymes (creatine kinases) with those from tissues of Duchenne muscular dystrophy, *J. Biol. Chem.,* 255, 8477, 1980.
301. **Kuby, S. A., Keutel, H. J., Okabe, K., Jacobs, H. K., Ziter, F., Gerber, D., and Tyler, F. H.,** Isolation of the human ATP-creatine transphosphorylases (creatine phosphokinases) from tissues of patients with Duchenne muscular dystrophy, *J. Biol. Chem.,* 252, 8382, 1977.
302. **Der Terrossian, E., Pradel, L. A., Kassab, R., and Thoai, N. V.,** Comparative structural studies of the active site of ATP-guanidine phosphotransferase. The essential cysteine tryptic peptide of arginine kinase from *Homarus vulgaris* muscle, *Eur. J. Biochem.,* 11, 482, 1969.
303. **Schapira, F.,** Variations pathologiques, en rapport avec l'onlogénèse, des formes moléculaires multiples de la lactico-deshydrogénase, de la créatine-kinase et de l'aldolase, *Bull. Soc. Chim. Biol.,* 49, 1647, 1967.
304. **Schapira, F., Dreyfus, J. C., and Allard, D.,** Les isozymes de la créatine kinase et de l'aldolase du muscle foetal et pathologique, *Clin. Chim. Acta,* 20, 439, 1968.
305. **Eppenberger, H. M., Eppenberger, M. E., Richterrich, R., and Aebi, H.,** The ontogeny of creatine kinase isozymes, *Dev. Biol.,* 10, 1, 1964.
306. **Cao, A., De Virginnis, S., and Falorni, A.,** The ontogeny of creatine-kinase isoenzymes, *Biol. Neonat.,* 13, 375, 1968.
307. **Dawson, D. M., Eppenberger, H. M., and Eppenberger, M. E.,** Multiple molecular forms of creatine kinases, *Ann. N.Y. Acad. Sci.,* 151, 616, 1968.
308. **Butterfield, D. A.,** Electron spin resonance studies of erythrocyte membranes in muscular dystrophy, *Acc. Chem. Res.,* 10, 111, 1977.
309. **Watts, D. C. and Rabin, B. R.,** A study of the "reactive" sulphydryl groups of adenosine 5′-triphosphate-creatine phosphotransferase, *Biochemical Journal,* 85, 507, 1962.
310. **Watts, D. C.,** Creatine Kinase (adenosine 5′-triphosphate-creatine phosphotransferase). VII. Chemical investigations of the enzyme mechanism, in *The Enzymes,* 3rd ed., Boyer, P. D., Ed., Academic Press, New York, 1973, 431.
311. **Mahowald, T. A., Noltmann, T. A., and Kuby, S. A.,** Studies on adenosine triphosphate transphosphorylases. III. Inhibition reactions, *J. Biol. Chem.,* 237, 1535, 1962.
312. **Smith, D. J. and Kenyon, G. L.,** Nonessentiality of the active sulfhydryl group of rabbit muscle creatine kinase, *J. Biol. Chem.,* 249, 3317, 1974.
313. **Borders, D. L., Jr. and Riordan, J. F.,** An essential arginyl residue at the nucleotide binding site of creatine kinase, *Biochemistry,* 14, 4699, 1975.
314. **James, T. L.,** Binding of adenosine 5′-diphosphate to creatine kinase. An investigation using intermolecular nuclear Overhauser effect measurements, *Biochemistry,* 15, 4724, 1976.
315. **Vasak, M., Nagoyama, K., Wüthrich, K., Mertens, M. L., and Kägi, J. H. R.,** Creatine kinase. Nuclear magnetic resonance and fluorescence evidence for interaction of adenosine 5′-diphosphate with an aromatic residue, *Biochemistry,* 18, 5050, 1979.
316. **Cook, P. F., Kenyon, G. L., and Cleland, W. W.,** Use of pH studies to elucidate the catalytic mechanism of rabbit muscle creatine kinase, *Biochemistry,* 20, 1204, 1981.
317. **Marletta, M. A. and Kenyon, G. L.,** Affinity labeling of creatine kinase by N-(2,3-epoxypropyl)-N-amidinoglycine, *J. Biol. Chem.,* 254, 1879, 1979.
318. **Mahowald, T. S.,** Identification of an epsilon amino group of lysine and a sulphydryl group of cysteine near the reactive cysteine residue in rabbit muscle creatine kinase, *Fed. Proc.,* 28, 601, 1969.
319. **James, T. L. and Cohn, M.,** The role of the lysyl residue at the active site of creatine kinase, *J. Biol. Chem.,* 249, 2599, 1974.
320. **Caldin, E. F.,** in *Fast Reactions in Solutions,* Blackwell, Oxford, 1964, 273.
321. **Clarke, D. E. and Price, N. C.,** The reaction of rabbit muscle creatine kinase with diethyl pyrocarbonate, *Biochem. J.,* 181, 467, 1979.
322. **Rabin, B. R. and Watts, D. C.,** A theory of the mechanism of action of creatine phosphokinase, *Nature,* 188, 1163, 1960.
323. **Rosevear, P. R., Desmeules, P., Kenyon, G. L., and Mildvan, A. S.,** Nuclear magnetic resonance studies of the role of histidine residues at the active site of rabbit muscle creatine kinase, *Biochemistry,* 20, 6155, 1981.
324. **Dunaway-Mariano, D. and Cleland, W. W.,** Preparation and properties of chromium (III) adenosine 5′-triphosphate, chromium (III) adenosine 5′-diphosphate, and related chromium complexes, *Biochemistry,* 19, 1496, 1980.

325. **McIlwain, H.,** *Biochemistry and the Central Nervous System,* 3rd ed., J. & A. Churchill, London, 1966.
326. **Nachmansohn, D. and Wilson, E. B.,** The enzymic hydrolysis and synthesis of acetylcholine, in *Advances in Enzymology,* Vol. 12, Nord, F. F., Ed., Interscience, New York, 1951, 259.
327. **Lui, N. S. T. and Cunningham, L. W.,** Cooperative effects of substrate analogs on the conformation of creatine phosphokinase, *Biochemistry,* 5, 144, 1966.
328. **Kägi, J. H. and Li, T. K.,** Extrinsic cotton effect in complexes of creatine phosphokinase with ADP and ATP, *Fed. Proc.,* 24, 285.
329. **Samuels, A. J., Nihei, T., and Noda, L.,** A change in optical rotation of creatine-ATP transphosphorylase during enzyme substrate interaction suggesting an alteration in conformation, *Proc. Natl. Acad. Sci. U.S.A.,* 47, 1992, 1961.
330. **Rouston, C., Kassab, R., Pradel, L. H., and Thoai, N. V.,** Interaction des ATP: guanidine phosphotransférases avec leurs substrats, étudiée par spectrophotometrie différentielle, *Biochim. Biophys. Acta,* 167, 326, 1968.
331. **Morrison, J. F. and Uhr, M. L.,** The function of bivalent metal ions in the reaction catalysed by ATP: creatine phosphotransferase, *Biochim. Biophys. Acta,* 122, 57, 1966.
332. **James, E. and Morrison, J. F.,** The reaction of nucleotide substrate analogues with adenosine triphosphate-creatine phosphotransferase, *J. Biol. Chem.,* 241, 4758, 1966.
333. **O'Sullivan, W. J., Diefenback, H., and Cohn, M.,** The effect of magnesium on the reactivity of the essential sulfhydryl groups in creatine kinase-substrate complexes, *Biochemistry,* 5, 2666, 1966.
334. **O'Sullivan, W. J. and Cohn, M.,** Nucleotide specificity and conformation of the active site of creatine kinase, *J. Biol. Chem.,* 241, 3116, 1966.
335. **Watts, D. C.,** Studies on the mechanism of action of adenosine 5′-triphosphate-creatine phosphotransferase. Inhibition by manganese ions and by *p*-nitrophenyl acetate, *Biochem. J.,* 89, 220, 1963.
336. **Samuels, A. J.,** Immunoenzymological evidence suggesting a change in conformation of adenylic acid deaminase and creatine kinase during substrate combination, *Biophys. J.,* 1, 437, 1961.
337. **Pradel, L. A. and Kassab, R.,** Site actif des ATP; guanidine phosphotransférases. II. Mise en évidence de résidues histidine essentiels an moyen du pyrocarbonate d'ethyle, *Biochim. Biophys. Acta,* 167, 317, 1968.
338. **Hammes, G. G. and Hurst, J. K.,** Relaxation spectra of adenosine triphosphate-creatine phosphotransferase, *Biochemistry,* 8, 1083, 1969.
339. **Jacobs, G. and Cunningham, L. W.,** Creatine kinase. The relationship of trypsin susceptibility to substrate binding, *Biochemistry,* 7, 143, 1968.
340. **Thomson, A. R., Eveleigh, J. W., and Miles, B. J.,** Amino acid sequence around the reactive thiol groups of adenosine triphosphate-creatine phosphotransferase, *Nature,* 203, 267, 1964.
341. **Taylor, J. S., Leigh, J. S., and Cohn, M.,** Magnetic resonance studies of spin-labeled creatine kinase system and interaction of two paramagnetic probes, in *Proc. Natl. Acad. Sci. U.S.A.,* 64, 219, 1969.
342. **Rao, B. D. N. and Cohn, M.,** ^{31}P NMR of enzyme-bound substrates of rabbit muscle creatine kinase. Equilibrium constants, interconversion rates, and NMR parameters of enzyme-bound complexes, *J. Biol. Chem.,* 256, 1716, 1981.
343. **Noda, L.,** Adenylate kinase, in *The Enzymes,* Vol. 8, 3rd ed., Boyer, P. D., Academic Press, New York, 1973, 279.
344. **Noda, L.,** Adenosine triphosphate-adenosine monophosphate. III. Kinetic studies, *J. Biol. Chem.,* 232, 237, 1958.
345. **Callaghan, O. H. and Weber, G.,** Kinetic studies on rabbit-muscle myokinase, *Biochem. J.,* 73, 473, 1959.
346. **Noda, L.,** Nucleoside triphosphate-nucleoside monophosphokinases, in *The Enzymes,* Vol. 6, 2nd ed., Boyer, P. D., Lardy, H., and Myrbäck, K., Eds., Academic Press, New York, 1962, 139.
347. **Rhoads, D. G. and Lowenstein, J. M.,** Initial velocity and equilibrium kinetics of myokinase, *J. Biol. Chem.,* 243, 3963, 1968.
348. **Hamada, M., Palmieri, R. H., Russell, G. A., and Kuby, S. A.,** Studies on adenosine triphosphate transphosphorylases. XVI. Equilibrium binding properties of the crystalline rabbit and calf muscle ATP-AMP transphosphorylase (adenylate kinase) and derived peptide fragments, *Arch. Biochem. Biophys.,* 195, 155, 1979.
349. **Su, S. and Russell, P. J.,** Adenylate kinase from baker's yeast. III. Equilibria: equilibrium exchange and mechanism, *J. Biol. Chem.,* 243, 3826, 1968.
350. **Rao, B. D. N., Cohn, M., and Noda, L.,** Differentiation of nucleotide binding sites and role of metal ion in the adenylate kinase reaction by ^{31}P NMR. Equilibria, interconversion rates, and NMR parameters of bound substrates, *J. Biol. Chem.,* 253, 1149, 1978.
351. **Markland, F. S. and Wadkins, C. L.,** Adenosine triphosphate-adenosine 5′-monophosphate phosphotransferase of bovine liver mitochondria. II. General kinetic and structural properties, *J. Biol. Chem.,* 241, 4136, 1966.
352. **Brownson, C. and Spencer, M.,** Kinetic studies on the two common inherited forms of hyman erythrocyte adenylate kinase, *Biochem. J.,* 130, 805, 1972.

353. **Bock, R. M.,** Adenine nucleotide and properties of pyrophosphate compounds, in *The Enzymes,* Vol. 2, 2nd ed., Boyer, P. D., Lardy, H., and Myrbäck, K., Eds., Academic Press, New York, 1960, 3.
354. **Hartridge, H. and Roughton, F. J. R.,** The kinetics of hoemoglobin. III. The velocity with which oxygen combines with reduced hoemoglobin, A107, 654, 1925.
355. **Linehard, G. E. and Secemski, I. I.,** P^1, P^5-di(adenosine-5′)pentaphosphate, a potent multisubstrate inhibitor of adenylate kinase, *J. Biol. Chem.,* 248, 1121, 1973.
356. **Schulz, G. E., Elzinga, M., Marx, F., and Schirmer, R. H.,** Three-dimensional structure of adenylate kinase, *Nature,* 250, 120, 1974.
357. **Sachsenheimer, W. and Schulz, G. E.,** Two conformations of crystalline adenylate kinase, *J. Mol. Biol.,* 114, 23, 1977.
358. **Pai. E. F., Sachsenhemier, W., Schirmer, R. H., and Schulz, G. E.,** Substrate positions and induced-fit in crystalline adenylate kinase, *J. Mol. Biol.,* 114, 37, 1977.
359. **Anderson, C. M., Zucker, F. H., and Steitz, T. A.,** Space-filling models of kinase clefts and conformation changes, *Science,* 204, 375, 1979.
360. **Adolfsen, R. and Moudrianakis, E. N.,** Control of complex metal ion equilibria in biochemical reaction systems. Intrinsic and apparent stability constants of metal adenine nucleotide complexes, *J. Biol. Chem.,* 253, 4378, 1978.
361. **Smith, R. M. and Martell, A. E.,** *Critical Stability Constants,* Vol. 2, Plenum Press, New York, 1975.
362. **Noda, L. and Kuby, S. A.,** ATP-AMP transphosphorylase (myokinase). II. Homogeneity measurements and physicochemical properties, *J. Biol. Chem.,* 226, 551, 1957.
363. **Kuby, S. A., Hamada, M., Palmieri, R. H., Tsai, W.-C., Jacobs, H. K., Maland, L., Wu, L., and Fischer, A.,** ATP-AMP transphosphorylase (myokinase) from rabbit and calf muscle-structure and equilibrium binding, *Fed. Proc. Abstr.,* 35, 1629, 1976.
364. **Noda, L., Kuby, S. A., and Lardy, H. A.,** ATP-creatine transphosphorylase. II. Homogeneity and physicochemical properties, *J. Biol. Chem.,* 209, 203, 1954.
365. **Noltmann, E. A., Mahowald, T. A., and Kuby, S. A.,** Studies on adenosine triphosphate transphosphorylases. II. Amino acid composition of adenosine triphosphate-creatine transphosphorylases, *J. Biol. Chem.,* 237, 1146, 1962.
366. **Noltmann, E. A., Mahowald, T. A., and Kuby, S. A.,** Studies on ATP-transphosphorylases. I. Amino acid composition of adenosine triphosphate-adenosine 5′-phosphate transphosphorylase (myokinase), *J. Biol. Chem.,* 237, 1138, 1962.
367. **Höber, R.,** Some properties of aqueous solutions, in *Physical Chemistry of Cells and Tissues,* The Blakiston Co., Philadelphia, 1945, chap. 5.
368. **Hughes, T. R. and Klotz, I. M.,** Analysis of metal-protein complexes, in *Methods of Biochemical Analysis,* Vol. 3, Glick, D., Ed., Interscience, New York, 1956, 265.
369. **Burton, K.,** Formation constants for the complexes of adenosine di- or triphosphates with magnesium or calcium ions, *Biochem. J.,* 71, 388, 1959.
370. **O'Sullivan, W. J. and Perrin, D. D.,** The stability constant of $MgATP^{2-}$, *Biochim. Biophys. Acta,* 52, 612, 1961.
371. **Melchior, N. C.,** Sodium and potassium complexes of adenosine-triphosphate: equilibrium studies, *J. Biol. Chem.,* 208, 615, 1954.
372. **Smith, R. M. and Alberty, R. A.,** Apparent stability constants of ionic complexes of various adenosine phosphates with divalent cations, *J. Am. Chem. Soc.,* 78, 2376, 1956.
373. **Smith, R. M. and Alberty, R. A.,** The apparent stability constants of ionic complexes of various adenosine phosphates with monovalent cations, *J. Phys. Chem.,* 60, 180, 1956.
374. **Smith, R. M.,** Stability constants of various adenosine phosphates with divalent and monovalent cations, thesis, University of Wisconsin, 1955.
375. **Alberty, R. A., Smith, R. M., and Bock, R. M.,** The apparent ionization constants of the adenosine phosphates and related compounds, *J. Biol. Chem.,* 193, 425, 1951.
376. **Morrison, J. F., O'Sullivan, W. J., and Ogston, A. G.,** Kinetic studies of the activation of creatine phosphoryltransferase by magnesium, *Biochim. Biophys. Acta,* 52, 82, 1961.
377. **Price, N. C., Reed, G. H., and Cohn, M.,** Magnetic resonance studies of substrate and inhibitor binding to porcine muscle adenylate kinase, *Biochemistry,* 12, 3322, 1973.
378. **Secrist, J. A., III, Barrio, J. R., Leonard, J. J., and Weber, G.,** Fluorescent modification of adenosine-containing coenzymes. Biological activities and spectroscopic properties, *Biochemistry,* 11, 3499, 1972.
379. **McDonald, G. D., Cohn, M., and Noda, L.,** Proton magnetic resonance spectra of porcine muscle adenylate kinase and substrate complexes, *J. Biol. Chem.,* 250, 6947, 1975.
380. **Heil, A., Müller, G., Noda, L., Pinder, T., Schirmer, H., Schirmer, I., and von Zabern, I.,** Location of chromophoric residues in proteins by solvent perturbation, *Eur. J. Biochem.,* 43, 131, 1974.
381. **Parker, C. A.,** *Photoluminescence of Solutions,* Elsevier, Amsterdam, 1968.
382. **Bennet, C. A. and Franklin, M. L., Eds.,** *Statistical Analysis in Chemistry and the Chemical Industry,* John Wiley & Sons, New York, 1954.

383. **Herskovits, T. T. and Laskowski, V., Jr.**, Location of chromophoric residues in proteins by solvent perturbation. I. Tyrosyls in serum albumins, *J. Biol. Chem.*, 237, 2481, 1962.
384. **Herskovits, T. T.**, Difference spectroscopy, in *Methods in Enzymology*, Vol. 11, Hirs, C. H. W., Ed., Academic Press, New York, 1967, 748.
385. **Anderson, C. M., Zucker, F. H., and Steitz, T. A.**, Space-filling models of kinase clefts and conformation changes, *Science*, 204, 375, 1979.
386. **Levedahl, B. H. and James, T. W.**, Correction of rotatory dispersion characteristics of adenosine triphosphate and related compounds, *Biochim. Biophys. Acta*, 23, 332, 1957.
387. **Hotta, K., Brahms, J., and Morales, M. F.**, Evidence for interaction between magnesium and purine or pyrimidine rings of 5′-ribonucleotides, *J. Am. Chem. Soc.*, 83, 997, 1961.
388. **Levedahl, B. H. and James, T. W.**, The configuration of adenosine triphosphate as deduced from rotatory dispersion, *Biochim. Biophys. Acta*, 21, 298, 1956.
389. **McCormick, W. G. and Levedahl, B. H.**, Rotatory dispersion of inosine triphosphate and the influence of metals on ITP and ATP, *Biochim. Biophys. Acta*, 34, 303, 1959.
390. **Martell, A. E. and Schwarzenbach, G.**, Adenosinphosphate und triphosphat als komplexbildner für calcium und magnesium, *Helv. Chim. Acta*, 39, 653, 1956.
391. **Hammes, G. G., Maciel, G. E., and Waugh, J. S.**, The conformation of metal-adenosine triphosphate complexes in solution, *J. Am. Chem., Soc.*, 83, 2394, 1961.
392. **Szent-Gyorgyi, A.**, *Bioenergetics*, Academic Press, New York, 1957.
393. **Epp, A., Ramasarma, T., and Wetter, L. R.**, Infrared studies on complexes of Mg^{++} with adenosine phosphates, *J. Am. Chem. Soc.*, 80, 724, 1958.
394. **Williams, R. J. P.**, Coordination, chelation, and catalysis, in *The Enzymes*, Vol. 1, 2nd ed., Boyer, P. D., Lardy, H., and Myrbäck, K., Eds., Academic Press, New York, 1959, 391.
395. **Cohn, M. and Hughes, T. R., Jr.**, Nuclear magnetic resonance spectra of adenosine di- and triphosphate. II. Effect of complexing with divalent metal ions, *J. Biol. Chem.*, 237, 176, 1962; Phosphorus magnetic resonance spectra of adenosine di- and triphosphate. I. Effect of pH, *J. Biol. Chem.*, 235, 3250, 1960.
396. **Cohn, M.**, Magnetic resonance studies of metal-enzyme-substrate interactions, *Science*, 136, 325, 1962.
397. **Mildvan, A. S.**, The role of metals in enzyme catalyzed substitutions at each of the phosphorus atoms of ATP, *Advances in Enzymology*, Vol. 49, Meister, A., Ed., Interscience, New York, 1979, 103.
398. **Brown, F. F., Campbell, I. D., Henson, R., Hirst, C. W. J., and Richards, R. E.**, A study of interaction of manganese ions with ATP by ^{31}P Fourier-transform nuclear magnetic resonance, *Eur. J. Biochem.*, 38, 54, 1973.
399. **Bailar, J. C., Ed.**, *The Chemistry of the Coordination Compounds*, Reinhold Publishing, New York, 1956.
400. **Cleland, W. W. and Mildvan, A. S.**, Chromium (III) and Cobalt (III) nucleotides as biological probes, *Advances in Inorganic Biochemistry*, Vol. 1, Eichkorn, G. L. and Marzilli, L. G., Eds., Elsevier/North-Holland, New York, 1979, 163.
401. **Cornelius, R. D., Hart, P. A., and Cleland, W. W.**, Phosphorus-31 NMR studies of complexes of adenosine triphosphate, adenosine diphosphate, tripolyphosphate, and pyrophosphate with cobalt (III) ammines[1a], *Inorg. Chem.*, 16, 2799, 1977.
402. **Merritt, E. A., Sundaralingam, M., Cornelius, R. D., and Cleland, W. W.**, X-ray crystal and molecular structure and absolute configuration of (dihydrogen tripolyphosphato) tetraamminecobalt (III) monohydrate, $Co(NH_3)_4H_2P_3O_{10} \cdot H_2O$. A model for a metal-nucleoside polyphosphate complex, *Biochemistry*, 17, 3274, 1978.
403. **Dunaway-Mariano, D. and Cleland, W. W.**, Preparation and properties of chromium (III) adenosine 5′-triphosphate, chromium (III) adenosine 5′-diphosphate, and related chromium (III) complexes, *Biochemistry*, 19, 1496, 1980.
404. **Dunaway-Mariano, D. and Cleland, W. W.**, Investigations of substrate specificity and reaction mechanism of several kinases using chromium (III) adenosine 5′-triphosphate and chromium (III) adenosine 5′-diphosphate, *Biochemistry*, 19, 1506, 1980.
405. **Jaffe, E. K. and Cohn, M.**, Diastereomers of the nucleoside phosphorothioates as probes of the structure of the metal nucleotide substrates and of the nucleotide binding site of yeast hexokinase, *J. Biol. Chem.*, 254, 10839, 1979.
406. **Huang, S. L. and Tsai, M. D.**, Does the magnesium (II) ion interact with the α-phosphate of adenosine triphosphate? An investigation by oxygen-17 nuclear magnetic resonance, *Biochemistry*, 21, 951, 1982.
407. **Britzinger, H.**, The structures of adenosine triphosphate metal ion complexes in aqueous solutions, *Biochim. Biophys. Acta*, 77, 343, 1963.
408. **Kuntz, G. P. P. and Swift, T. J.**, Contrasting structures of the magnesium and calcium adenosine triphosphate complexes as studied by nuclear reaction, *Fed. Proc.*, 32, 546, 1973.
409. **Tran-Dinh, S., Roux, M., and Ellenberger, M.**, Interaction of Mg^{2+} ions with nucleoside triphosphates by phosphorus magnetic resonance spectroscopy, *Nucleic Acids Res.*, 2, 1101, 1975.
410. **Gupta, R. K. and Mildvan, A. S.**, Structures of enzyme-bound metal complexes in the phosphoryl transfer reaction of muscle pyruvate kinase. ^{31}P NMR studies with magnesium and kinetic studies with chromium nucleotides, *J. Biol. Chem.*, 252, 5967, 1977.

411. **Ramirez, F. and Marecek, J. F.,** Coordination of magnesium with adenosine 5′-diphosphate and triphosphate, *Biochim. Biophys. Acta,* 589, 21, 1980.
412. **Bishop, E. O., Kimber, S. J., Orchard, D., and Smith, B. I.,** A ^{31}P-NMR study of mono- and dimagnesium complexes of adenosine 5′ triphosphate and model systems, *Biochim. Biophys. Acta,* 635, 63, 1981.
413. **Jaffe, E. K. and Cohn, M.,** ^{31}P nuclear magnetic resonance spectra of the thiophosphate analogues of adenine nucleotides; effects of pH and Mg^{2+} binding, *Biochemistry,* 17, 652, 1978.
414. **Myers, T. C.,** in *^{31}P Nuclear Magnetic Resonance,* Crutschfield, M. M., Dunga, C. H., Letcher, J. H., Mark, V., and Van Wazer, J. R., Eds., Interscience, New York, 1967, 314.
415. **Tran-Dinh, S. and Roux, M.,** A phosphorus-magnetic-resonance study of the interaction of Mg^{2+} with adenyl-5′-yl imidodiphosphate. Binding sites of Mg^{2+} ion on the phosphate chain, *Eur. J. Biochem.,* 76, 245, 1977.
416. **Tsai, M. D., Huang, S. L., Kozlowski, J. G., and Chang, C.-C.,** Applicability of the phosphorus-31 (oxygen-17) nuclear magnetic resonance method in the study of enzyme mechanism involving phosphorus, *Biochemistry,* 19, 3531, 1980.
417. **Gutfreund, H.,** Transients and relaxation kinetics of enzyme reactions, *Annu. Rev. Biochem.,* 40, 315, 1971.
418. **Gutfreund, H.,** Ligand binding, in *Enzymes: Physical Principles,* Wiley-Interscience, London, 1972, 69.
419. **Halford, S. E.,** in *Companion to Biochemistry,* Bull, A. T., Lagnado, J. R., Thomas, J. O., and Tipton, K. F., Eds., Longman Group, London, 1974, 197.
420. **Roughton, F. J. W.,** The study of fast reactions, *Disc. Faraday Soc.,* 17, 116, 1954.
421. **Livingston, R.,** Evaluation and interpretation of rate data, in *Technique of Organic Chemistry, Investigation of Rates and Mechanisms of Reactions,* Vol. 8, Freiss, S. L. and Weissberger, A., Eds., Interscience, New York, 1953, 169.
422. **Gutfreund, H.,** The physical chemistry of enzymes, *Disc. Faraday Soc.,* 10, 167, 1955.
423. **Laidler, K. J. and Bunting, P. S.,** The time course of enzyme reactions, in *The Chemical Kinetics of Enzyme Action,* 2nd ed., Clarendon Press, Oxford, 1973, 185.
424. **Gutfreund, H.,** The study of fast reactions, *Disc. Faraday Soc.,* 17, 220, 1954.
425. **Bernhard, S. A.,** A new method for the determination of the amidase activity of trypsin: kinetics of the hydrolysis of benzoyl-L-arginineamide, *Biochem. J.,* 59, 506, 1955.
426. **Gutfreund, H. and Hammond, B. R.,** Steps in the reactions of chymotrypsin with tyrosine derivatives, *Biochem. J.,* 73, 526, 1959.
427. **Hammond, B. R. and Gutfreund, H.,** The mechanism of ficin-catalyzed reactions, *Biochem. J.,* 72, 349, 1959.
428. **Darvey, I. G.,** Transient phase kinetics of enzyme reactions, *J. Theoret. Biol.,* 19, 215, 1968.
429. **Hijazi, N. H. and Laidler, K. J.,** Transient-phase and steady-state kinetics for enzyme activation, *Can. J. Biochem.,* 51, 806, 1973.
430. **Ouellet, L. and Stewart, J. A.,** Theory of the transient phase in an enzyme system involving two enzyme-substrate complexes, *Can. J. Chem.,* 37, 737, 1959.
431. **Hammes, G. G. and Schimmel, P. R.,** Rapid reactions and transient states, in *The Enzymes,* Vol. 2, 3rd ed., Boyer, P. D., Ed., Academic Press, New York, 1970, 67.
432. **Hartridge, H. and Roughton, F. J. R.,** A method of measuring the velocity of very rapid chemical reactions, *Proc. R. Soc. London,* A104, 376, 1923.
433. **Kupiecki, F. P. and Coon, M. J.,** "Hydroxylamine kinase" and pyruvic kinase, *J. Biol. Chem.,* 235, 1944, 1960.
434. **Chance, B.,** Reaction kinetics of enzyme-substrate compounds, in *Technique of Organic Chemistry, Investigation of Rates and Mechanisms of Reaction,* Vol. 8, Friess, S. L. and Weissberger, A., Eds., Interscience, New York, 1953, 627.
435. **Gibson, Q. H.,** Rapid mixing: stopped flow, in *Methods in Enzymology,* Vol. 16, 187, 1969.
436. **Chance, B.,** Detailed description of Chance's apparatus, in *Techniques in Organic Chemistry,* Vol. 8, 2nd ed., part 2, Friess, S. L., Lewis, E. S., and Weissberger, A., Eds., Wiley-Interscience, New York, 1963, 733.
437. **Gutfreund, H.,** Rapid mixing; continuous flow, in *Methods in Enzymology,* Vol. 16, Justin, K., Ed., Academic Press, New York, 1969, 229.
438. **Chance, B.,** The pulsed flow apparatus, in *Rapid Mixing and Sampling Techniques in Biochemistry,* Chance, B., Eisenhardt, R., Gibson, Q., and Lonberg-Holm, K., Eds., Academic Press, New York, 1964, 125.
439. **Ludwig, G. D. and Kuby, S. A.,** A kinetic study of the reaction of molecular oxygen with the cytochrome chain of yeast, *Fed. Proc. Abstr.,* 14, 247, 1955.
440. **Ludwig, G. D., Kuby, S. A., Edelman, G. M., and Chance, B.,** A kinetic study of the oxidation by molecular oxygen of the cytochrome chain of intact yeast cells, acetobacter suboxydans cells, and of particulate suspensions of heart muscle, *Enzyme,* 29, 73, 1983.
441. **Chance, B.,** Regeneration and recirculation of reactants in the rapid-flow apparatus, *Disc. Faraday Soc.,* 17, 120, 1954.

442. **Chance, B. and Legallais, V.,** Regeneration and recirculation of reactants in the rapid-flow apparatus. Part 2. Practical designs, *Disc. Faraday Soc.*, 17, 123, 1954.
443. **Chance, B., DeVault, D., Legallais, V., Mela, L., and Yonetani, T.,** Fast reactions and primary processes in chemical kinetics, in *Nobel Symp.*, Vol. 5, Claesson, S., Ed., Interscience, New York, 1967, 437.
444. **Chance, B. and Pring, M.,** in *Biochemie des Sauerstoffs,* Hess, B. and Staudinger, H. J., Eds., Springer-Verlag, Berlin, 1968, 102.
445. **Asakura, T., Kobayashi, L., and Chance, B.,** Kinetic studies on the reaction between apocytochrome *c* peroxidase and protoporphyrin. IX, *J. Biol. Chem.*, 249, 1799, 1974.
446. **Yang, C.-C. and Legallais, V.,** A rapid and sensitive recording spectrophotometer for the visible and ultraviolet region. I. Description and performance, *Rev. Sci. Instr.*, 25, 801, 1954.
447. **Yang, C.-C.,** A rapid and sensitive recording spectrophotometer for the visible and ultraviolet region. II. Electronic circuits, *Rev. Sci. Instr.*, 25, 807, 1954.
448. **Winzler, R. J.,** The respiration of Baker's yeast at low oxygen tension, *J. Cell. Comp. Physiol.*, 17, 263, 1941.
449. **Longmuir, I. S.,** Respiration rate of bacteria as a function of oxygen concentration, *Biochem. J.*, 57, 81, 1954.
450. **Warburg, O. and Kubowitz, F.,** Atmung bei sehr kleinen sauerstoffdrucken, *Biochem. Z.*, 214, 5, 1929.
451. **Ferguson-Miller, S., Brautigan, D. L., and Margoliash, E.,** Correlation of the kinetics of electron transfer activity of various eukaryotic cytochromes *c* with binding to mitochondrial cytochrome *c* oxidase, *J. Biol. Chem.*, 251, 1104, 1976.
452. **Chance, B.,** The carbon monoxide compounds of the cytochrome oxidases. III. Molecular extinction coefficients, *J. Biol. Chem.*, 202, 407, 1953.
453. **Chance, B.,** Spectra and reaction kinetics of respiratory pigments of homogenized and intact cells, *Nature,* 169, 215, 1952.
454. **Greenwood, C. and Gibson, Q. H.,** The reaction of reduced cytochrome *c* oxidase with oxygen, *J. Biol. Chem.*, 242, 1782, 1967.
455. **Wharton, D. C. and Gibson, Q. H.,** Studies of the oxygenated compound of cytochrome oxidase, *J. Biol. Chem.*, 243, 702, 1968.
456. **Frost, A. H. and Pearson, R. G.,** Complex reactions, in *Kinetics and Mechanism,* John Wiley & Sons, New York, 1953, 154.
457. **Bray, R. C.,** Sudden freezing as a technique for the study of rapid reactions, *Biochem. J.*, 81, 189, 1961.
458. **Bray, R. C., Palmer, G., and Beinert, H. C.,** Direct studies on the electron transfer sequence in xanthine oxidase by electron paramagnetic resonance spectroscopy. II. Kinetic studies employing rapid freezing, *J. Biol. Chem.*, 239, 2667, 1964.
459. **Palmer, G., Bray, R. C., and Beinert, H. C.,** Direct studies on the electron transfer sequence in xanthine oxidase by electron paramagnetic resonance spectroscopy. I. Techniques and description of spectra, *J. Biol. Chem.*, 239, 2657, 1964.
460. **Ruby, W. R.,** Mixing machine of high precision for study of rapid reactions, *Rev. Sci. Instr.*, 26, 460, 1955.
461. **Kuby, S. A.,** unpublished experiments, 1958—1963.
462. **Gibson, Q. H.,** An apparatus for flash photolysis and its application to the reactions of myoglobin with gases, *J. Physiol.*, 134, 112, 1956.
463. **Denbigh, K. G. and Page, F. M.,** The capacity flow method in chemical kinetics, *Disc. Faraday Soc.*, 17, 145, 1954.
464. **Denbign, K. B., Hicks, M., and Page, F. M.,** The kinetics of open reaction systems, *Trans. Faraday Soc.*, 44, 479, 1948.
465. **Kuby, S. A.,** unpublished observations, 1954.
466. **Young, H. H., Jr. and Hammett, L. P.,** Kinetic measurements in a stirred flow reactor; the alkaline bromination of acetone, *J. Am. Chem. Soc.*, 72, 280, 1950.
467. **Bray, R. C.,** The chemistry of Xanthine oxidase. 8. Electron-spin-resonance measurements during the enzymic reaction, *Biochem. J.*, 81, 196, 1961.
468. **Chance, B.,** Accelerated and stopped-flow methods using spectrophotometric measurements, in *Technique of Organic Chemistry, Investigation of Rates and Mechanisms of Reaction,* Vol. 8, Friess, S. L. and Weissberger, A., Eds., Interscience, New York, 1953, 690.
469. **Terner, J., Voss, D. F., Paddock, C., Miles, R. B., and Spiro, T. G.,** Picosecond resonance raman spectrum of the oxyhemoglobin photoproduct. Evidence for an electronically excited state, *J. Phys. Chem.*, 86, 859, 1982.
470. **Eigen, M.,** Methods for investigation of ionic reactions in aqueous solutions with half-times as short as 10^{-9} sec. Applications to neutralization and hydrolysis reactions, *Disc. Faraday Soc.*, 17, 194, 1954.
471. **Eigen, M. and De Maeyer, L.,** Relaxation methods, in *Techniques of Organic Chemistry,* Vol. VIII, part 2, Friess, S. L., Lewis, E. S., and Weissberger, A., Eds., Interscience, New York, 1963, 895.

472. **French, T. C. and Hammes, G. G.,** Fast reactions, in *Methods in Enzymology,* Vol. 16, Kustin, K., Ed., Academic Press, New York, 1969, 3.
473. **Takahashi, M. T. and Alberty, R. A.,** Fast reactions, in *Methods in Enzymology,* Vol. 16, Kustin, K., Ed., Academic Press, New York, 1969, 31.
474. **Eggers, F. and Kustin, K.,** Fast reactions, in *Methods in Enzymology,* Vol. 16, Kustin, K., Ed., Academic Press, New York, 1969, 55.
475. **De Maeyer, L. C. M.,** Fast reactions, in *Methods in Enzymology,* Vol. 16, Kustin, K., Ed., Academic Press, New York, 1969, 80.
476. **Hammes, G. G. and Schimmel, P. R.,** Relaxation spectra of enzymatic reactions, *J. Phys. Chem.,* 71, 917, 1967.
477. **Czerlinski, G. H.,** Chemical relaxation of cyclic enzyme reactions. III. Experimental implication of previous results, *J. Theoret. Biol.,* 21, 408, 1968.
478. **Havsteen, B.,** Studies of rapid reactions by flow and relaxation techniques, *Biochem. Soc. Symp.,* 26, 53, 1966.
479. **Yapel, A. F., Jr. and Lumry, R.,** A practical guide to the temperature-jump method for measuring the rate of fast reactions, *Methods Biochem. Anal.,* 20, 169, 1971.
480. **Dixon, M., Webb, E. C., Thorne, C. J. R., and Tipton, K. F.,** Transient phases of enzyme reactions, in *Enzymes,* 3rd ed., Academic Press, New York, 1979, 205.
481. **Amdur, I. and Hammes, G. G.,** Rate equations near equilibrium, in *Chemical Kinetics,* McGraw-Hill, New York, 1966, 138.
482. **Laidler, K. J. and Bunting, P. S.,** General kinetic principles, in *The Chemical Kinetics of Enzyme Action,* 2nd ed., Clarendon Press, Oxford, 1973, 44.
483. **Eyring, H.,** Activated complex in chemical reactions, *J. Chem. Phys.,* 3, 107, 1935; The absolute rate of reactions in condensed phases, *J. Chem. Phys.,* 3, 492, 1935.
484. **Glasstone, S.,** Calculation of energy and heat capacity. VI. Classical theory, in *Thermodynamics for Chemists,* D. Van Nostrand, New York, 1950, 95.
485. **Eyring, H.,** The activated complex and the absolute rate of chemical reactions, *Chem. Rev.,* 17, 65, 1935.
486. **Melander, L.,** *Isotope Effects on Reaction Rates,* Ronald Press, New York, 1960.
487. **Laidler, K. J.,** in *Theories of Chemical Reaction Rates,* McGraw-Hill, New York, 1969, 86.
488. **Wolsberg, M.,** Theoretical evaluation of experimentally observed isotope effects, *Acc. Chem. Res.,* 5, 225, 1972.
489. **Cleland, W. W., O'Leary, M. H., and Northrop, D. B., Eds.,** *Isotope Effects on Enzyme-Catalyzed Reactions,* University Park Press, Baltimore, 1977.
490. **Northrop, D. B.,** Determining the absolute magnitude of hydrogen isotope effects, in *Isotope Effects on Enzyme-Catalyzed Reactions,* Cleland, W. W., O'Leary, M. H., and Northrop, D. B., Ed., University Park Press, Baltimore, 1977, 122; Appendix D: table for use with Northrop's method, in *Isotope Effects on Enzyme-Catalyzed Reactions,* Cleland, W. W., O'Leary, M. H., and Northrop, D. B., Eds., University Park Press, Baltimore, 1977, 280.
491. **Bell, R. P.,** in *The Proton in Chemistry,* Cornell University Press, Ithaca, New York, 1959, 187.
492. **Bunton, C. A. and Shiner, V.,** Isotope effects in deuterium oxide solution. I. Acid-base equilibria, *J. Am. Chem. Society,* 83, 42, 1961.
493. **Bigeleisen, J. W. and Wolfsberg, M.,** Theoretical and experimental aspects of isotope effects in chemical kinetics, *Adv. Chem. Phys.,* 1, 15, 1958.
494. **Jencks, W. P.,** Isotope effects, in *Catalysis in Chemistry and Biochemistry,* McGraw-Hill, New York, 1969, chap. 4.
495. **Bell, R. P., Fendley, J. A., and Hulett, J. R.,** The hydrogen isotope effect in the bromination of 2-carbethoxy*cyclo*pentanone, *Proc. R. Soc. London,* A235, 453, 1956.
496. **Hulett, J. R.,** The bromination of 2-carbethoxy*cyclo*pentanone at low temperatures, *Proc. R. Soc. London,* A251, 274, 1959.
497. **Caldin, E. F. and Harbron, E.,** A non-linear Arrhenius plot for the proton-transfer reaction between acetic acid and the anion of 2, 4, 6-trinitrotoluene in ethanol at low temperatures, *J. Chem. Soc.,* 3454, 1962; **Caldin, E. F., Kasparian, M., and Tomalin, G.,** Kinetics of proton-transfer reaction between 4-nitrobenzyl-cyanide and ethoxide ion in ethanol + ether from −60 to −124°C, *Trans. Faraday Soc.,* 64, 2802, 1968; **Caldin, E. F. and Tomalin, G.,** Kinetic isotope effect in reaction between 4-nitrobenzyl-cyanide and ethoxide ion in ethanol + ether, *Trans. Faraday Soc.,* 64, 2814, 1968.
498. **Rossini, F. D., et al.,** Selected values of chemical thermodynamic properties, *Natl. Bur. Stand. (U.S.) Circ.,* 500, 1952.
499. **Malmberg, C. G.,** Journal of Research of the *Natl. Bur. Stand. (U.S.) Circ.,* 60, 609, 1958.
500. **Bender, M. L., Pollock, E. J., and Neveu, M. C.,** Deuterium oxide solvent isotope effects in the nucleophilic reactions of phenyl esters, *J. Am. Chem. Soc.,* 84, 595, 1962.
501. **Swain, C. G., Wiles, R. A., and Bader, R. F. W.,** Use of substituent effects of isotope effects to distinguish between proton and hydride transfers. Part 1. Mechanism of oxidation of alcohols by bromine in water, *J. Am. Chem. Soc.,* 83, 1945, 1961.

502. **Belleau, B. and Moran, J.,** Deuterium isotope effects in relation to the chemical mechanism of monoamine oxidase, *Ann. N.Y. Acad. Sci.,* 107, 822, 1963.
503. **O'Leary, M. H.,** Studies of enzyme reaction mechanisms by means of heavy-atom isotope effects, in *Isotope Effects on Enzyme-Catalyzed Reactions,* Cleland, W. W., O'Leary, M. H., and Northrop, D. B., Eds., University Park Press, Baltimore, 1977, 233.
504. **Schowen, R. L.,** Solvent isotope effects on enzymic reactions, in *Isotope Effects on Enzyme-Catalyzed Reactions,* Cleland, W. W., O'Leary, M. H., and Northrop, D. B., Eds., University Park Press, Baltimore, 1977, 64.
505. **Kirsch, J. F.,** Secondary-kinetic isotope effects, in *Isotope Effects on Enzyme-Catalyzed Reactions,* Cleland, W. W., O'Leary, M. H., and Northrop, D. B., Eds., University Park Press, Baltimore, 1977, 100.
506. **Streitwieser, A., Jr., Jagow, R. H., Fahey, R. C., and Suzuki, S.,** Kinetic isotope effects in the acetolyses of deuterated cyclopentyl tosylates, *J. Am. Chem. Soc.,* 80, 2326, 1958.
507. **Melander, L.,** Prediction of rate-constant ratios from molecular data, in *Isotope Effects on Reaction Rates,* Ronald Press, New York, 1960, chap. 2.
508. **Northrop, D. B.,** Steady-state analysis of kinetic isotope effects in enzymic reactions, *Biochemistry,* 14, 2644, 1975.
509. **Swain, C. G., Stivers, E. C., Reuwer, J. F., Jr., and Schaad, L. J.,** Use of hydrogen isotope effects to identify the attacking nucleophile in the enolization of ketones catalyzed by acetic acid, *J. Am. Chem. Soc.,* 80, 5885, 1958.
510. **Bigeleisen, J.,** *International Atomic Energy Agency,* 1, 161, 1962.
511. **Stern, M. J. and Vogel, P. C.,** Relative tritium-deuterium isotope effects in the absence of large tunneling factors, *J. Am. Chem. Soc.,* 93, 4664, 1971.
512. **Stern, M. J. and Weston, R. E., Jr.,** Phenomenological manifestations of quantum-mechanical tunneling. III. Effect on relative tritium-deuterium kinetic isotope effects, *J. Chem. Phys.,* 60, 2815, 1974.
513. **Carter, R. E. and Melander, L.,** Experiments on the nature of steric isotope effects, *Adv. Phys. Organ. Chem.,* 10, 1, 1973.
514. **O'Leary, M. H.,** Heavy-atom isotope effects in enzyme-catalyzed reactions, in *Transition States of Biochemical Processes,* Schower, R. L. and Gandour, R., Eds., Plenum Press, New York, 1977.
515. **Schimerlik, M. I., Grimshaw, C. E., and Cleland, W. W.,** Determination of the rate-limiting steps for malic enzyme by the use of isotope effects and other kinetic studies, *Biochemistry,* 16, 571, 1977.
516. **Bruice, T. C.,** Proximity effects and enzyme catalysis, in *The Enzymes,* Vol. 2, 3rd ed., Boyer, P. D., Ed., Academic Press, New York, 1970, 217.
517. **Rose, I. A.,** Hydrogen isotope effects in proton transfer to and from carbon, in *Isotope Effects on Enzyme-Catalyzed Reactions,* Cleland, W. W., O'Leary, M. H., and Northrop, D. B., Eds., University Park Press, Baltimore, 1977, 209.
518. **Rose, I. A. and Iyengar, R.,** The D_2O effect on catalysis by triose phosphate isomerase requires isotope exchange on the enzyme, *J. Am. Chem. Soc.,* 105, 295, 1983.
519. **Bender, M. L. and Hamilton, G. A.,** Kinetic isotope effects of deutrium oxide on several α-chymotrypsin-catalyzed reactions, *J. Am. Chem. Soc.,* 84, 2570, 1962.
520. **Brubacher, L. J. and Bender, M. L.,** The preparation and properties of *trans*-cinnamoyl-papain, *J. Am. Chem. Soc.,* 88, 5871, 1966.
521. **Green, J. R. and Westley, J.,** Mechanism of rhodanese action: polarographic studies, *J. Biol. Chem.,* 236, 3047, 1961.
522. **Westley, J. and Nakamoto, T.,** Mechanism of rhodanese action: isotopic tracer studies, *J. Biol. Chem.,* 237, 547, 1962.
523. **Page, M. I.,** The principles of enzymatic catalysis, *Int. J. Biochem.,* 10, 471, 1979.
524. **Koshland, D. E., Jr.,** Mechanisms of transfer enzymes, in *The Enzymes,* Vol. 1, 2nd ed., Boyer, P. D., Lardy, H., Myrbäck, K., Eds., Academic Press, New York, 1959, 305.
525. **Koshland, D. E., Jr.,** The comparison of non-enzymic and enzymic reaction velocities, *J. Theoret. Biol.,* 2, 75, 1962.
526. **Koshland, D. E., Jr.,** Enzyme flexibility and enzyme action, *J. Cell. Comp. Physiol.,* 54 (Suppl. 1), 245, 1959.
527. **Jencks, W. P.,** Part 1. Mechanisms for catalysis, in *Catalysis in Chemistry and Enzymology,* McGraw-Hill, New York, 1969, chap. 1.
528. **Page, M. I. and Jencks, W. P.,** Entropic contributions to rate accelerations in enzymic and intermolecular reactions and the chelate effect, in *Proc. Natl. Acad. Sci. U.S.A.,* 68, 1678, 1971.
529. **Storm, D. R. and Koshland, D. E., Jr.,** A source for the special catalytic power of enzymes: orbital steering, in *Proc. Natl. Acad. Sci. U.S.A.,* 66, 445, 1970.
530. **Bruice, T. C., Brown, A., and Harris, D. O.,** On the concept of orbital steering in catalytic reactions, in *Proc. Natl. Acad. Sci., U.S.A.,* 68, 658, 1971.
531. **Brønsted, J. N. and Wynne-Jones, W. F. K.,** Acid catalysis in hydrolytic reactions, *Trans. Faraday Soc.,* 25, 59, 1929.

532. **Koshland, D. E., Jr.,** The active site and enzyme action, *Advances Enzymology,* Vol. 22, Nord, F. F., Ed., Interscience, New York, 1960, 45.
533. **Laidler, K. J.,** *Chemical Kinetics,* 2nd ed., McGraw-Hill, New York, 1965.
534. **Jencks, W. P.,** *Catalysis in Chemistry and Enzymology,* McGraw-Hill, New York, 1969.
535. **Jencks, W. P. and Carriuolo, J.,** Imidazole catalysis. II. Acyl transfer and the reactions of acetyl imidazole with water and oxygen anions, *J. Biol. Chem.,* 234, 1272, 1959.
536. **Bender, M. L. and Kezdy, F. J.,** Mechanism of action of proteolytic enzymes, *Annu. Rev. Biochem.,* 34, 49, 1965.
537. **Wilson, I. B. and Bergman, F.,** Acetylcholinesterase. VIII. Dissociation constants of the active groups, *J. Biol. Chem.,* 186, 683, 1950.
538. **Lowry, T. M. and Smith, G. F.,** Studies of dynamic isomerism. Part XXIV. Neutral-salt action in mutarotation, *J. Chem. Soc.,* 130, 2539, 1927.
539. **Lowry, T. M.,** Studies of dynamic isomerism. XX. Amphoteric solvents as catalysis for the mutarotation of sugars, *J. Chem. Soc.,* 128, 2883, 1925.
540. **Swain, C. G. and Brown, J. F., Jr.,** Concerted displacement reactions. VII. The mechanism of acid-base catalysis in non-aqueous solvents, *J. Am. Chem. Soc.,* 74, 2534, 1952.
541. **Wang, J. H.,** Facilitated proton transfer in enzyme catalysis. It may have a crucial role in determining the efficiency and specificity of enzymes, *Science,* 161, 328, 1968.
542. **Blow, D. M., Birktoff, J. J., and Hartley, B. S.,** Role of a buried acid group in the mechanism of action of chymotrypsin, *Nature,* 221, 337, 1969.
543. **Kasserra, J. P. and Laidler, K. J.,** Mechanisms of action of trypsin and chymotrypsin, *Can. J. Chem.,* 47, 4031, 1969.
544. **Eigen, M. and Hammes, G. G.,** Elementary steps in enzyme reactions (as studied by relaxation spectrometry), in *Advances in Enzymology,* Vol. 25, Nord, F. F., Ed., Interscience, New York, 1963, 1.
545. **Eigen, M., Hammes, G. G., and Kustin, K.,** Fast reactions of imidazole studied with relaxation spectrometry, *J. Am. Chem. Soc.,* 82, 3482, 1960.
546. **Witzel, H.,** The function of the pyrimidine base in the ribonuclease reaction, in *Progress in Nucleic Acid Research,* Vol. 2, Davidson, J. N. and Cohn, W. E., Eds., Academic Press, New York, 1963, 221.
547. **Ramsden, E. N. and Laidler, K. J.,** Kinetics and mechanisms of reactions catalyzed by pancreatic ribonuclease, *Can. J. Chem.,* 44, 2597, 1966.
548. **Koshland, D. E., Jr.,** Application of a theory of enzyme specificity to protein synthesis, *Proc. Natl. Acad. Sci., U.S.A.,* 44, 98, 1958.
549. **Wang, J. H.,** Directional character of proton transfer in enzyme catalysis, *Proc. Natl. Acad. Sci. U.S.A.,* 66, 874, 1970.
550. **Koshland, D. E., Jr. and Neet, K. E.,** The catalytic and regulatory properties of enzymes, *Annu. Rev. Biochem.,* 37, 359, 1968.
551. **Laidler, K. J.,** The influence of pressure on the rates of biological reactions, *Arch. Biochem.,* 30, 226, 1951; Laidler, K. J., VIII. The influence of pressure on enzyme reactions, in *Chemical Kinetics of Enzyme Action,* 1st ed., 1958, 224.
552. **Bender, M. L., Kezdy, F. J., and Gunter, C. R.,** The anatomy of enzymic catalysis. α-chymotrypsin, *J. Am. Chem. Soc.,* 86, 3714, 1964.
553. **Trayser, K. A. and Colowick, S. P.,** Properties of crystalline hexokinase from yeast. II. Studies on ATP-enzyme interaction, *Arch. Biochem. Biophys.,* 94, 161, 1961.
554. **Haldane, J. B. S.,** *Enzymes,* Longmans, Green and Co., London, 1930.
555. **Jencks, W. P.,** Strain, distortion and conformational change, *Catalysis in Chemistry and Enzymology,* McGraw-Hill, New York, 1969, chap. 5.
556. **Jencks, W. P.,** Binding energy, specificity, and enzymic catalysis: the circe effect, in *Advances in Enzymology,* Vol. 43, Meister, A., Ed., Interscience, New York, 1975, 219.
557. **Jencks, W. P.,** What everyone wanted to know about tight-binding and enzyme catalysis, but never thought of asking, in *Molecular Biology, Biochemistry and Biophysics, Chemical Recognition in Biology,* Vol. 32, Chapeville, F. and Haennt, A., Eds., Springer-Verlag, New York, 1980, 3.
558. **Blake, C. C. F., Koenig, D. F., Meir, G. A., North, A. C. T., Phillips, D. C., and Sarna, V. R.,** Structure of hen egg-white lysozyme. A three-dimensional Fourier synthesis at 2 Å resolution, *Nature,* 206, 757, 1965.
559. **Blake, C. C. F., Meir, G. A., North, A. C. T., Phillips, D. C., and Sarna, V. R.,** On the conformation of the hen egg-white lysozyme molecule, *Proc. R. Soc. London,* B167, 365, 1967.
560. **Wishnia, A.,** Substrate specificity at the alkane binding sites of hemoglobin and myoglobin, *Biochemistry,* 8, 5064, 1969.
561. **Bernhard, S. A. and Gutfreund, H.,** *Proceedings of the International Symposium on Enzyme Chemistry,* Tokyo, 1958, 124.
562. **Bernhard, S. A.,** Abstract for enzyme flexibility and enzyme action, *J. Cell. Comp. Physiol.,* 54 (Suppl. 1), 256, 1959.

563. **Spencer, T. and Sturtevant, J. M.,** The mechanism of chymotrypsin-catalyzed reactions, *J. Am. Chem. Soc.,* 81, 1874, 1959.
564. **Hein, G. E. and Niemann, C.,** Steric course and specificity of α-chymotrypsin-catalyzed reactions, *J. Am. Chem. Soc.,* 84, 4495, 1962.
565. **Niemann, C.,** Alpha-chymotrypsin and the nature of enzyme catalysis. The problem of enzyme catalysis is considered in terms of the behavior of a single hydrolytic enzyme, *Science,* 143, 1287, 1964.
566. **Hamilton, C. L., Niemann, C., and Hammond, G. S.,** A quantitative analysis of the binding of *N*-acyl derivatives of α-aminoamides by α-chymotrypsin, *Proc. Natl. Acad. Sci. U.S.A.,* 55, 664, 1966.
567. **Kanamori, K. and Roberts, J. D.,** ^{15}N NMR studies of biological systems, *Acc. Chem. Res.,* 16, 35, 1983.
568. **Kustin, K.,** Fast reactions, in *Methods in Enzymology,* Vol. 16, Colowick, S. P. and Kaplan, N. O., Eds., Academic Press, New York, 1969, chap. 1.
569. **Purich, D. L.,** Enzyme kinetics and mechanism, Part B. Isotopic probes and complex enzyme systems, *Methods in Enzymology,* Vol. 64, Colowick, S. P. and Kaplan, N. O., Eds., Academic Press, New York, 1980.
570. **Raushel, F. M. and Seiglie, J. L.,** Kinetic mechanism of argininosuccinate synthetase, *Arch. Biochem. Biophys.,* 225, 979, 1983.
571. **Tsai, C. S., Wand, A. J., Templeton, D. M., and Weiss, P. M.,** Multifunctionality of lipoamide dehydrogenase: promotion of election transferase reaction, *Arch. Biochem. Biophys.,* 225, 554, 1983.
572. **Grazi, E., Trombetta, G., and Lanzara, V.,** Fructose-1,6-bisphosphate aldolase from rabbit muscle. Kinetic resolution of the enamine phosphate from the enaminealdehyde intermediate at low temperature, *Biochemistry,* 22, 4434, 1983.
573. **Branlant, G., Eiler, B., Biellmann, J. F., Lutz, H. P., and Luisi, P. L.,** Applicability of the induced fit model to glyceraldehyde-3-phosphate dehydrogenase from sturgeon muscle. Study of the binding of oxidized nicotinamide adenine dinucleotide and nicotinamide 8-bromoadenine dinucleotide, *Biochemistry,* 22, 4437, 1983.
574. **Smith, T. and Bell, J. E.,** Mechanism of hysteresis in bovine glutamate dehydrogenase: role of subunit interactions, *Biochemistry,* 21, 733, 1982.
575. **Chan, K. F., Hurst, M. D., and Graves, D. J.,** Phosphorylase kinase specificity. A comparative study with *c*AMP-dependent protein kinase on synthetic peptides and peptide analogs of glycogen synthase and phosphorylase, *J. Biol. Chem.,* 257, 3655, 1982.
576. **Kahler, S. G. and Kirkman, H. N.,** Intracellular glucose-6-phosphate dehydrogenase does not monomerize in human erythrocytes, *J. Biol. Chem.,* 258, 717, 1983.
577. **Keizer, J.,** Nonequilibrium statistical thermodynamics and the effect of diffusion on chemical reaction rates, *J. Phys. Chem.,* 86, 5052, 1982.
578. **Cohn, M.,** Some properties of the phosphorothioate analogues of adenosine triphosphate as substrates of enzymic reactions, *Acc. Chem. Res.,* 15, 326, 1982.
579. **Brenner, D. G. and Knowles, J. R.,** Penicillanic acid sulfone: an unexpected isotope effect in the interaction of 6α- and 6β-monodeuterio and of 6,6-dideuterio derivatives with RTEM β-lactamase from *Escherichia coli, Biochemistry,* 20, 3680, 1981.
580. **Kemal, C. and Knowles, J. R.,** Penicillanic acid sulfone: interaction with RTEM β-lactamase from *Escherichia coli, Biochemistry,* 20, 3688, 1981.
581. **Hermes, J. D., Roeska, C. A., O'Leary, M. H., and Cleland, W. W.,** Use of multiple isotope effects to determine enzyme mechanisms and intrinsic isotope effects. Malic enzyme and glucose-6-phosphate dehydrogenase, *Biochemistry,* 21, 5106, 1982.
582. **Hansen, D. E. and Knowles, J. R.,** The stereochemical course of the reaction catalyzed by creatine kinase, *J. Biol. Chem.,* 256, 5967, 1981.
583. **Campbell, P., Nashed, N. T., Lapinskas, B. A., and Gurrieri, J.,** Thionesters as a probe for electrophilic catalysis in the serine protease mechanism, *J. Biol. Chem.,* 258, 59, 1983.
584. **Beaty, N. B. and Ballou, D. P.,** The reductive half-reaction of liver microsomal FAD-containing monooxygenase, *J. Biol. Chem.,* 256, 4611, 1981.
585. **Beaty, N. P. and Ballou, D. P.,** Transient kinetic study of liver microsomal FAD-containing monooxygenase, *J. Biol. Chem.,* 255, 3817, 1980.
586. **Solheim, L. P. and Fromm, H. J.,** Effect of inorganic phosphate on the reverse reaction of bovine brain hexokinase, *Biochemistry,* 22, 2234, 1983.
587. **Chance, B., Fischetti, R., and Powers, L.,** Structure and kinetics of the photoproduct of carboxymyoglobin at low temperatures: an X-ray absorption study, *Biochemistry,* 22, 3820, 1983.
588. **Sammons, R. D., Frey, P. A., Bruzik, K., and Tsai, M. D.,** Effects of ^{17}O and ^{18}O on ^{31}P: further investigation and applications, *J. Am. Chem. Soc.,* 105, 5455, 1983.
589. **Trybus, K. M. and Taylor, E. W.,** Transient kinetics of adenosine 5′-diphosphate and adenosine 5′-(β,γ-imidotriphosphate) binding to subfragment 1 and actinsubfragment 1, *Biochemistry,* 21, 1284, 1982.
590. **Brouwer, A. C. and Kirsch, J. F.,** Investigation of diffused-limited rates of chymotrypsin reactions by viscosity variation, *Biochemistry,* 21, 1302, 1982.

591. **Viola, R. E., Raushel, F. M., Rendina, A. R., and Cleland, W. W.,** Substrate synergism and the kinetic mechanism of yeast hexokinase, *Biochemistry,* 21, 1295, 1982.
592. **Davies, P. B.,** Laser magnetic resonance spectroscopy, *J. Phys. Chem.,* 85, 2599, 1981.
593. **Hoffman, B. M., Roberts, J. E., Kang, C. H., and Margoliash, E.,** Electron paramagnetic and electron nuclear double resonance of the hydrogen peroxide compound of cytochrome *c* peroxidase, *J. Biol. Chem.,* 256, 6556, 1981.
594. **Yoshida, A., Impraim, C. C., and Huang, I.-Y.,** Enzymatic and structural differences betweeen usual and atypical human liver alcohol dehydrogenases, *J. Biol. Chem.,* 256, 12430, 1981.
595. **Edelson, D.,** Computer simulation in chemical kinetics, *Science,* 214, 981, 1981.
596. **Dower, S. K., De Lisi, C., Titus, J. A., and Segal, D. M.,** Mechanism of binding of multivalent immune complexes to Fc receptors. I. Equilibrium binding, *Biochemistry,* 20, 6326, 1981.
597. **Dower, S. K., De Lisi, C., Titus, J. A., and Segal, D. M.,** Mechanism of binding of multivalent immune complexes to Fc receptors. II. Kinetics of binding, *Biochemistry,* 20, 6335, 1981.
598. **Reutimann, H., Straub, B., Luisi, P. L., and Holmgren, A.,** A conformational study of thioredoxin and its tryptic fragments, *J. Biol. Chem.,* 256, 6796, 1981.
599. **Araiso, T. and Dunford, H. B.,** Horseradish peroxidase. Complex formation with anions and hydrocyanic acid, *J. Biol. Chem.,* 256, 10099, 1981.
600. **Araiso, T. and Dunford, H. B.,** Effect of modification of heme propionate groups on the reactivity of horseradish peroxidase, *Arch. Biochem. Biophys.,* 211, 346, 1981.
601. **Simmondsen, R. P. and Tollin, G.,** Transient kinetics of redox reactions of flavodoxin: effects of chemical modification of the flavin mononucleotide prosthetic group on the dynamics of intermediate complex formation and electron transfer, *Biochemistry,* 22, 3008, 1983.
602. **Massey, V. and Hemmerich, P.,** *Biochem. Rev.,* 8, 246, 1980.
603. **Schopfer, L. M., Haushalter, J. P., Smith, M., Milad, M., and Morris, M. D.,** Resonance raman spectra for flavin derivatives modified in the 8 position, *Biochemistry,* 20, 6734, 1981.
604. **Shelnutt, J. A., Rousseau, D. L., Dethmers, J. K., and Margoliash, E.,** Protein influences on porphyrin structure in cytochrome *c*: evidence from Raman difference spectroscopy, *Biochemistry,* 20, 6485, 1981.
605. **Wu, S. T., Pieper, G. M., Salhany, J. M., and Eliot, R. S.,** Measurement of free magnesium in perfused and ischemic arrested heart muscle. A quantitative phosphorus-31 nuclear magnetic resonance and multiequilibria analysis, *Biochemistry,* 20, 7399, 1981.
606. **Northrop, D. B.,** Minimal kinetic mechanism and general equation for deutrium isotope effects on enzymic reactions. Uncertainty in detecting a rate-limiting step, *Biochemistry,* 20, 4056, 1981.
607. **Northrop, D. B.,** Interpretations of isotope effects on V/K, *Fed. Proc.,* 41, 628, 1981.
608. **Ray, W. J., Jr.,** Rate-limiting step: a quantitative definition. Application to steady-state enzymic reactions, *Biochemistry,* 22, 4625, 1983.
609. **Truhlar, D. G., Hase, W. L., and Hynes, J. T.,** Current status of transition-state theory, *J. Phys. Chem.,* 87, 2664, 1983.
610. **Grace, S. and Dunaway-Mariano, D.,** Examination of the solvent perturbation technique as a method to identify enzyme catalytic groups, *Biochemistry,* 22, 4238, 1983.
611. **Harris, W. R.,** Thermodynamic binding constants of the zinc-human serum transferrin complex, *Biochemistry,* 22, 3920, 1983.
612. **Walsh, C. T.,** *Enzymatic Reaction Mechanisms,* W. H. Freeman, San Francisco, 1979.
613. **Warshel, A. and Levitt, M.,** Theoretical studies of enzymic reactions: dielectric, electrostatic and steric stabilization of the carbonium ion in the reaction of lysozyme, *J. Mol. Biol.,* 103, 227, 1976.
614. **Welch, G. R., Somogyi, B., and Damjanovich, S.,** The role of protein fluctuation in enzyme action: a review, *Prog. Biophys. Mol. Biol.,* 39, 109, 1982.
615. **Green, D. E. and Ji, S.,** Transductional and structural principles of the mitochondrial transducing unit, *Proc. Natl. Acad. Sci. U.S.A.,* 70, 904, 1973.
616. **Warshel, A.,** Energetics of enzyme catalysis. Isoenzyme/enzyme mechanism/dielectric effects in enzymes and in solutions/relationship between protein folding and catalysis, in *Proc. Natl. Acad. Sci. U.S.A.,* 75, 5250, 1978.
617. **Raushel, F. M. and Seiglie, J. L.,** Kinetic mechanism of arginineosuccinate synthetase, *Arch. Biochem. Biophys.,* 225, 979, 1983.
618. **Welsh, K. M., Jacobyansky, A., Springs, B., and Cooperman, B. S.,** Catalytic specificity of yeast inorganic pyrophosphatase for magnesium ion as cofactor. An analysis of divalent metal ion and solvent isotope effects on enzyme function, *Biochemistry,* 22, 2243, 1983.
619. **Webb, M. R. and Trentham, D. R.,** The mechanism of ATP hydrolysis catalyzed by myosin and actomyosin, using rapid reaction techniques to study oxygen exchange, *J. Biol. Chem.,* 256, 10910, 1981.
620. **Klotz, I. M.,** Numbers of receptor sites from Scatchard graphs: facts and fantasies, *Science,* 217, 1247, 1982.
621. **Feldman, H. A.,** Statistical limits in Scatchard analysis, *J. Biol. Chem.,* 258, 12865, 1983.

622. **Klotz, I. M.,** Number of receptor sites from Scatchard and Klotz graphs: a constructive critique, *Science,* 220, 981, 1983.
623. **Scatchard, G.,** The attractions of proteins for small molecules and ions, *Ann. N.Y. Acad. Sci.,* 51, 660, 1949.
624. **Munson, P. J. and Rodbard, D.,** Number of receptor sites from Scatchard and Klotz graphs: a constructive critique, *Science,* 220, 979, 1983.
625. **Page, J. D. and Wilson, I. B.,** The inhibition of acetylcholinesterase by arsenite and fluoride, *Arch. Biochem. Biophys.,* 226, 492, 1983.
626. **McCarthy, M. B. and White, R. E.,** Competing modes of peroxyacid flux through Cytochrome p-450, *J. Biol. Chem.,* 258, 11610, 1983.
627. **Jacobson, M. A. and Colman, R. F.,** Resonance energy transfer between the adenosine 5′-diphosphate site of glutamate dehydrogenase and a guanosine 5′-triphosphate site containing a tyrosine labeled with 5′-[*p*-(fluorosulfonyl) benzoyl]-1, N^6-ethenoadenosine, *Biochemistry,* 22, 4247, 1983.
628. **Dalbey, R. E., Weiel, J., and Yount, R. G.,** Förster energy transfer measurements of thiol 1 to thiol 2 distances in myosin subfragment 1, *Biochemistry,* 22, 4696, 1983.
629. **Brylawski, B. P. and Caplow, M.,** Rate for nucleotide release from tubulin, *J. Biol. Chem.,* 258, 760, 1983.
630. **Posner, I., Wang, C.-S., and McConathy, W. J.,** The comparative kinetics of soluble and heparin-sepharose-immobilized bovine lipoprotein lipase, *Arch. Biochem. Biophys.,* 226, 306, 1983.
631. **Snyder, G. H., Annerazzo, M. J., Karalis, A. J., and Field, D.,** Electrostatic influence of local cysteine environments on disulfide exchange kinetics, *Biochemistry,* 20, 6509, 1981.
632. **Watanabe, T., Lewis, D., Nakamoto, R., Kurzmack, M., Fronticelli, C., and Inesi, G.,** Modulation of calcium binding in sarcoplasmic reticulum adenosinetriphosphatase, *Biochemistry,* 20, 6617, 1982.
633. **Bruist, M. F. and Hammes, G. G.,** Further characterization of nucleotide binding sites on chloroplast coupling factor one, *Biochemistry,* 20, 6298, 1981.
634. **Salhany, J. M. and Gaines, E. D.,** Steady state kinetics of erythrocyte anion exchange. Evidence for site-site interactions, *J. Biol. Chem.,* 256, 11080, 1981.
635. **Reinhart, G. D. and Lardy, H. A.,** Rat liver phosphofructokinase: kinetic activity under near-physiological conditions, *Biochemistry,* 19, 1477, 1980.
636. **Reinhart, G. D. and Lardy, N. A.,** Rat liver phosphofructokinase: kinetic and physiological ramifications of the aggregation behavior, *Biochemistry,* 19, 1491, 1980.
637. **Uyeda, K., Furuya, E., and Luby, L. J.,** The effect of natural and synthetic D-fructose 2,6-bisphosphate on the regulatory kinetic properties of liver and muscle phosphofructokinases, *J. Biol. Chem.,* 256, 8394, 1981.
638. **Horrocks, W. DeW., Jr. and Sudnick, D. R.,** Lanthanide ion luminescence probes of the structure of biological macromolecules, *Acc. Chem. Res.,* 14, 384, 1981.
639. **Antonini, E. and Ascenzi, P.,** The mechanism of trypsin catalysis at low pH. Proposal for a structural model, *J. Biol. Chem.,* 256, 12449, 1981.
640. **Young, H. H., Jr. and Hammett, L. P.,** Kinetic measurements in a stirred flow reactor; the alkaline bromination of acetone, *J. Am. Chem. Soc.,* 72, 280, 1950.
641. **Saldick, J. and Hammett, L. P.,** Rate measurements by continuous titration in a stirred flow reactor, *J. Am. Chem. Soc.,* 72, 283, 1950.
642. **Rand, M. J. and Hammett, L. P.,** Reaction rates and heats by the temperature rise in a stirred flow reactor, *J. Am. Chem. Soc.,* 72, 287, 1950.
643. **Scanion, W. J. and Eisenberg, D.,** Solvation of crystalline proteins. Solvent bound in sperm whale metmyoglobin type A crystals at 6.1 and 23.5°C, *J. Phys. Chem.,* 85, 3251, 1981.
644. **Nagumo, M., Nicol, M., and El-Sayed, M. A.,** Polarized resonance Raman spectroscopy of the photointermediate of oxyhemoglobin on the picosecond time scale, *J. Phys. Chem.,* 85, 2435, 1981.
645. **Geren, L. M. and Millet, F.,** Fluorescence energy transfer studies of the interaction between adrenodoxin and cytochrome *c, J. Biol. Chem.,* 256, 10485, 1981.
646. **Maurer, P. J. and Nowak, T.,** Fluoride inhibition of yeast enolase. 1. Formation of the ligand complexes, *Biochemistry,* 20, 6894, 1981.
647. **Nowak, T. and Maurer, P. J.,** Fluoride inhibition of yeast enolase. 2. Structural and kinetic properties of the ligand complexes determined by nuclear relaxation rate studies, *Biochemistry,* 20, 6901, 1981.
648. **Holmes, M. A. and Mathews, B. W.,** Binding of hydroxamic acid inhibitors to crystalline thermolysin suggests a pentacoordinate zinc intermediate in catalysis, *Biochemistry,* 20, 6912, 1981.
649. **Jordan, F. and Polgar, L.,** Proton nuclear magnetic resonance evidence for the absence of a stable hydrogen bond between the active site aspartate and histidine residues of native subtilisins and for its presence in thiolsubtilisins, *Biochemistry,* 20, 6366, 1981.
650. **Cooperman, B. S., Panackal, A., Springs, B., and Hamm, D. L.,** Divalent metal ion, inorganic phosphate, and inorganic phosphate analogue binding to yeast inorganic pyrophosphatase, *Biochemistry,* 21, 6051, 1981.

651. **Wu, S. T., Pieper, G. M., Salhany, J. M., and Eliot, R. S.,** Measurement of free magnesium in perfused and ischemic arrested heart muscle. A quantitative phosphorus-31 nuclear magnetic resonance and multiequilibria analysis, *Biochemistry,* 20, 7399, 1981.
652. **Springs, B., Welsh, K. M., and Cooperman, B. S.,** Thermodynamics, kinetics, and mechanism in yeast inorganic pyrophosphatase catalysis of inorganic pyrophosphate: inorganic phosphate equilibration, *Biochemistry,* 20, 6384, 1981.
653. **Hutny, J.,** Kinetics of hog pancrease α-amylase; theoretical model of the dual site enzyme, *Acta Biochim. Polon.,* 28, 123, 1981.
654. **Kossiakoff, A. A. and Spencer, S. A.,** Direct determination of the protonation states of aspartic acid-102 and histidine-57 in the tetrahedral intermediate of the serine proteases: neutron structure of trypsin, *Biochemistry,* 20, 6462, 1981.
655. **Gorga, F. R. and Lienhard, G. E.,** Equilibria and kinetics of ligand binding to the human erythrocyte glucose transporter. Evidence for an alternating conformation model for transport, *Biochemistry,* 20, 5108, 1981.
656. **Minton, A. P. and Wilf, J.,** Effect of macromolecular crowding upon the structure and function of an enzyme: glyceraldehyde-3-phosphate dehydrogenase, *Biochemistry,* 20, 4821, 1981.
657. **Grimshaw, C. E. and Cleland, W. W.,** Kinetic mechanism of *Bacillus subtilis* L-alanine dehydrogenase, *Biochemistry,* 20, 5650, 1981.
658. **Wilkinson, K. D. and Rose, I. A.,** Study of crystalline hexokinase-glucose complexes by isotope trapping, *J. Biol. Chem.,* 256, 9890, 1981.
659. **Ahmad, I. and Tollin, G.,** Solvent effects on flavin electron transfer reactions, *Biochemistry,* 20, 5925, 1981.
660. **Hermann, R., Jaenicke, R., and Rudolph, R.,** Solvent effects on flavin electron transfer reactions, *Biochemistry,* 20, 5195, 1981.
661. **Skolnick, P., Moncada, V., Barker, J. L., and Paul, S. M.,** Pentobarbital: dual actions to increase brain benzodiazephine receptor affinity, *Science,* 211, 1448, 1981.
662. **Feldman, H. A.,** Mathematical theory of complex ligand-binding systems at equilibrium: some methods for parameter fitting, *Anal. Biochem.,* 48, 317, 1972.
663. **Munson, P. J. and Rodbard, D.,** Ligand: a versatile computerized approach for characterization of ligand-binding systems, *Anal. Biochem.,* 107, 220, 1980.
664. **Fletterick, R. J. and Madsen, N. B.,** The structures and related functions of phosphorylase *a, Annu. Rev. Biochem.,* 49, 31, 1980.
665. **Weber, I. T., Johnson, L. N., Wilson, K. S., Yeates, D. G. R., Wild, D. L., and Jenkins, J. A.,** Crystallographic studies on the activity of glycogen phosphorylase *b, Nature,* 274, 433, 1978.
666. **Fletterick, R. J., Sygusch, J., Murray, N., Madsen, N. B., and Johnson, L. N.,** Low resolution structure of the glycogen phosphorylase *a* monomer and comparison with phosphorylase *b, J. Mol. Biol.,* 103, 1, 1976.
667. **Fletterick, R. J., Sygusch, J., Semple, H., and Madsen, N. B.,** Structure of glycogen phosphorylase *a* at 3.0 Å resolution and its ligand binding sites at 6 Å, *J. Biol. Chem.,* 251, 6142, 1976.
668. **Black, W. J. and Wang, J. H.,** Studies on the allosteric activation of glycogen phosphorylase by nucleotides. I. Activation of phosphorylase *b* by inosine monophosphate, *J. Biol. Chem.,* 243, 5892, 1968.
669. **Rahim, Z. H. A., Perrett, D., and Griffiths, J. R.,** Skeletal muscle purine nucleotide levels in normal and phosphorylase kinase deficient mice, *FEBS Lett.,* 69, 203, 1976.
670. **Johnson, L. N., Stura, E. A., Wilson, K. S., Sansom, M. S. P., and Weber, I. T.,** Nucleotide binding to glycogen phosphorylase *b* in the crystal, *J. Mol. Biol.,* 134, 639, 1979.
671. **Lorek, A., Wilson, K. S., Stura, E. A., Jenkins, J. A., Zanotti, G., and Johnson, L. N.,** Proposals for the catalytic mechanism of glycogen phosphorylase *b* prompted by crystallographic studies on glucose 1-phosphate binding, *J. Mol. Biol.,* 140, 565, 1980.
672. **Sprang, S. R., Goldsmith, E. J., Fletterick, R. J., Withers, S. G., and Madsen, N. B.,** Catalytic site of glycogen phosphorylase: structural changes during activation and mechanistic implications, *Biochemistry,* 21, 5364, 1982.
673. **Withers, S. G., Madsen, N. B., Sprang, S. R., and Fletterick, R. J.,** Catalytic site of glycogen phosphorylase: structure of the T state and specificity for α-D-Glucose, *Biochemistry,* 21, 5372, 1982.
674. **Jenkins, J. A., Johnson, L. N., Stuart, D. I., Stura, E. A., Wilson, K. S., and Zanotti, G.,** Phosphorylase: control and activity, *Phil. Trans. R. Soc. London,* B293, 23, 1981.
675. **Ho, H. C. and Wang, J. H.,** A calorimetric study of the interactions between phosphorylase *b* and its nucleotide activators, *Biochemistry,* 12, 4750, 1973.
676. **Morange, M., Garcia Blanco, F., Vandenbunder, B., and Buc, H.,** AMP analogs: their function in the activation of glycogen phosphorylase *b, Eur. J. Biochem.,* 65, 553, 1976.
677. **Kasvinsky, P. J., Madsen, N. B., Fletterick, R. J., and Sygusch, J.,** X-ray crystallographic and kinetic studies of oligosaccharide binding to phosphorylase, *J. Biol. Chem.,* 253, 1290, 1978.

678. **Kasvinsky, P. J., Madsen, N. B., Sygusch, J., and Fletterick, R. J.,** The regulation of glycogen phosphorylase *a* by nucleotide derivatives, *J. Biol. Chem.*, 253, 3343, 1978.
679. **Sprange, S. R., Fletterick, R. J., Stern, M., Yang, D., Madsen, N. B., and Sturtevant, J. M.,** Analysis of an allosteric binding site: the nucleotide inhibitor site of phosphorylase *a, Biochemistry,* 21, 2036, 1982.
680. **Madsen, N. B., Shechosky, S. and Fletterick, R. J.,** Site-site interactions in glycogen phosphorylase *b* probed by ligands specific for each site, *Biochemistry,* 22, 4460, 1983.
681. **Madsen, N. B., Auramovic-Zikic, O., Lue, P. F., and Honkel, K. O.,** Studies on allosteric phenomena in glycogen phosphorylase *b, Mol. Cell. Biochem.*, 11, 35, 1976.
682. **Johnson, M., Cress, M., and Kuby, S. A.,** Studies on NADH (NADPH)-cytochrome *c* reductase from ale yeast-isolation and characterization, *Fed. Proc.*, 42, 2114, 1983.
683. **Alston, T. A., Porter, D. J. T., and Bright, H. J.,** Enzyme inhibition by nitro and nitroso compounds, *Acc. Chem. Res.*, 16, 418, 1983.
684. **Laidler, K. J. and King, M. C.,** The development of transition-state theory, *J. Phys. Chem.*, 87, 2657, 1983.
685. **Renisch, J., Rojas, C., and McFarland, J. T.,** Kinetic methods for the study of the enzyme systems of β-oxidation, *Arch. Biochem. Biophys.*, 227, 21, 1983.
686. **Shreve, D. S., Holloway, M. P., Haggerty, J. C., and Sable, H. Z.,** The catalytic mechanism of transketolase. Thiamin pyrophosphate-derived transition states for transketolase and pyruvate dehydrogenase are not identical, *J. Biol. Chem.*, 258, 12405, 1983.
687. **Appleby, C. A., Bradbury, J. H., Morris, R. J., Writtenberg, B. A., Writtenberg, J. B., and Wright, P. E.,** Leghemoglobin. Kinetic, nuclear magnetic resonance, and optical studies of pH dependence of oxygen and carbon monoxide binding, *J. Biol. Chem.*, 258, 2254, 1983.
688. **Birktoft, J. J. and Baniszak, L. J.,** The presence of a histidine-aspartic acid pair in the active site of 2-hydroxyacid dehydrogenases. X-ray refinement of cytoplasmic malate dehydrogenase, *J. Biol. Chem.*, 258, 472, 1983.
689. **Knoght, W. B., Ting, S.-J., Chuang, S., Dunaway-Mariano, D., Haromy, T., and Sundaralingam, M.,** Yeast inorganic pyrophosphatase substrate recognition, *Arch. Biochem. Biophys.*, 227, 302, 1983.
690. **Chang, Y. C., McCalmont, T., and Graves, D. J.,** Functions of the 5′ phosphate in phosphorylase: a study using pyridoxal-reconstituted enzyme as a model system, *Biochemistry,* 22, 4987, 1983.
691. **McClintock, D. K. and Markus, G.,** Conformational changes in aspartate transcarbyamylase. II. Concerted or sequential mechanism?, *J. Biol. Chem.*, 244, 36, 1969.
692. **Anderson, E. G. and Pratt, R. F.,** Pre-steady state β-lactamase kinetics and the interpretation of pH rate profiles, *J. Biol. Chem.*, 258, 13120, 1982.
693. **Hammes, G. G.,** *Enzyme Catalysis and Regulation,* Academic Press, New York, 1982.
694. **Perutz, M. F.,** Electrostatic effects in proteins, *Science,* 201, 1187, 1978.
695. **Kauzmann, W.,** Some factors in the interpretation of protein denaturation, *Adv. Protein Chem.*, 14, 1, 1959.
696. **Levitt, M. and Warshel, A.,** Computer simulation of protein folding, *Nature,* 253, 694, 1977.
697. **Pauling, L.,** quoted by Perutz as a "prediction [made] in a lecture at Nottingham University, England, in 1948." See also, **Pauling, L.,** Nature of forces between large molecules of biological interest, *Nature,* 161, 707, 1948.
698. **Phillips, D. C.,** The three-dimensional structure of an enzyme molecule, *Sci. Am.*, 215, 78, 1966.
699. **Blake, C. C. F., Johnson, L. N., Mair, G. A., North, A. C. T., Phillips, D. C., and Sarma, V. R.,** Crystallographic studies of the activity of hen egg-white lysozyme, *Proc. R. Soc. London,* B167, 378, 1967.
700. **Vernon, C. A.,** The mechanisms of hydrolysis of glycosides and their relevance to enzyme-catalysed reactions, *Proc. R. Soc. London,* B167, 389, 1967.
701. **Warshel, A.,** Energetics of enzyme catalysis, *Proc. Natl. Acad. Sci. U.S.A.*, 75, 5250, 1978.
702. **Page, M. I.,** Minireview. Transition states, standard states and enzymic catalysis, *Int. J. Biochem.*, 11, 331, 1980.
703. **Green, D. E.,** A framework of principles for the unification of bioenergetics, *Ann. N.Y. Acad. Sci.*, 227, 6, 1974.
704. **Careri, G.,** in *Quantum Statistical Mechanics in the Natural Sciences,* Kursunoglu, B., Mintz, S. L., and Widmayer, S. M., Eds., Plenum Press, New York, 1974, 15.
705. **Careri, G.,** in *New Trends in the Description of the General Mechanism and Regulation of Enzymes,* Vol. 21, Damjanovich, S., Elödi, P. and Somogyi, B., Eds., Academiai Kiado, Budapest, 1978, 151.
706. **Careri, G.,** *Molecular Hydration and Its Possible Role in Enzymes,* in Proc. Water Biophysics, Gerton College, Cambridge, June 29 to July 3, 1981.
707. **Careri, G. and Gratton, E.,** *Biol. Sys.*, 8, 185, 1977.
708. **Careri, G., Fasella, P., and Gratton, E.,** Statistical time events in enzymes: a physical assessment, *CRC Crit. Rev. Biochem.*, 3, 141, 1975.

709. **Careri, G., Fasella, P., and Gratton, E.,** Enzyme dynamics: the statistical physics approach, *Annu. Rev. Biophys. Bioeng.*, 8, 69, 1979.
710. **Gavish, B.,** The role of geometry and elastic strains in dynamic states of proteins, *Biophys. Struct. Mechanism,* 4, 37, 1978.
711. **Gavish, B.,** Position-dependent viscosity effects on rate coefficients, *Phys. Rev. Lett.*, 44, 1160, 1980.
712. **Gavish, B.,** A model for non-harmonic thermal behavior of proteins, *Fed. Proc.*, 39, 1760, 1980.
713. **Beece, D., Einstein, L., Frauenfelder, H., Good, D., Marden, M. C., Reinisch, L., Reynolds, A. H., Sorensen, L. B., and Yue, K. T.,** Solvent viscosity and protein dynamics, *Biochemistry,* 19, 5147, 1980.
714. **Welch, G. R.,** On the role of organized multienzyme systems in cellular metabolism: a general synthesis, *Prog. Biophys. Mol. Biol.*, 32, 103, 1977.
715. **Welch, G. R.,** On the free energy "cost of transition" in intermediary metabolic processes and the evolution of cellular infrastructure, *J. Theoret. Biol.*, 68, 267, 1977.
716. **Welch, G. R. and Keleti, T.,** On the "cystosociology" of enzyme action *in vivo*: a novel thermodynamic correlate of biological evolution, *J. Theoret. Biol.*, 93, 701, 1981.
717. **Kole, R. and Altman, S.,** Properties of purified ribonuclease P from *Escherichia coli, Biochemistry,* 20, 1902, 1981.
718. **Reed, R. E., Baer, M. F., Guerrier-Takada, C., Donis-Keller, H., and Altman, S.,** Nucleotide sequence of the gene encoding the RNA subunit (M1 RNA) of ribonuclease P from *Escherichia coli, Cell,* 30, 627, 1982.
719. **Reed, R. E. and Altman, S.,** Repeated sequences and open reading frames in the 3′ flanking region of the gene for the RNA subunit of *Escherichia coli* ribonuclease P, in *Proc. Natl. Acad. Sci. U.S.A.*, 80, 5359, 1983.
720. **Guerrier-Takada, C., Gardiner, K., Marsh, T., Pace, N., and Altman, S.,** The RNA moiety of ribonuclease P is the catalytic subunit of the enzyme, *Cell,* 35, 849, 1983.
721. **Guerrier-Takada, C. and Altman, S.,** Catalytic activity of an RNA molecule prepared by transcription *in vitro, Science,* 223, 285, 1984.
722. **Edsall, J. T. and Gutfreund, H.,** *Biothermodynamics, the Study of Biochemical Processes at Equilibrium,* John Wiley & Sons, New York, 1983.
723. **Wishnia, A.,** On the thermodynamic basis of induced fit. Specific alkane binding to proteins, *Biochemistry,* 8, 5070, 1969.
724. **Hammes, G. G. and Wu, C.-W.,** Regulation of enzyme activity. The activity of enzymes can be controlled by a multiplicity of conformational equilibria, *Science,* 172, 1205, 1971.
725. **Frieden, C.,** Protein-protein interaction and enzymatic activity, *Annu. Rev. Biochem.*, 40, 653, 1971.
726. **Frieden, C.,** Treatment of enzyme kinetic data. II. The multisite case: comparison of allosteric models and a possible new mechanism, *J. Biol. Chem.*, 242, 4045, 1967.
727. **Nichol, L. W., Jackson, W. J. H., and Winzor, D. L.,** A theoretical study of the binding of small molecules to a polymerizing protein system. A model for allosteric effects, *Biochemistry,* 6, 2449, 1966.
728. **Silman, I. and Katchalski, E.,** Water-insoluble derivatives of enzymes, antigens, and antibodies, *Annu. Rev. Biochem.*, 35, 873, 1966.
729. **Boguslaski, R. C., Smith, R. S., and Mhatre, N. S.,** Applications of bound biopolymers in enzymology and immunology, *Curr. Top. Microbiol. Immunol.*, 58, 1, 1972.
730. **Stark, G. R., Ed.,** Sequential degradation of peptides using solid supports, in *Biochemical Aspects of Reactions on Solid Supports,* Academic Press, New York, 1971, 171.
731. **Laidler, K. J. and Bunting, P. S.,** The kinetics of immobilized enzyme systems, in *Enzyme Kinetics and Mechanism, Part B, Methods in Enzymology,* Vol. 64, Purich, D. L., Colowick, S. P., and Kaplan, N. O., Eds., Academic Press, New York, 1980, 227.
732. **Gough, D. A. and Andrade, J. D.,** Enzyme electrodes. Electrodes containing immobilized enzymes could be used to monitor specific metabolites, *Science,* 180, 380, 1973.
733. **Guilbault, G. G. and Nagy, G.,** Improved urea electrode, *Anal. Chem.*, 45, 417, 1973.
734. **Papariello, G. J., Mukherji, A. K., and Shearer, C. M.,** A penicillin selective enzyme electrode, *Anal. Chem.*, 45, 790, 1973.
735. **Henahan, J. F.,** *Chem. Eng. News,* 49, 42, 1971.
736. **Goldman, R., Kedem, O., Silman, I. H., Caplan, S. R., and Katchalski, E.,** Papain-collodion membranes. I. Preparation and properties, *Biochemistry,* 7, 486, 1968.
737. **Wingard, L. B., Jr., Eds.,** Films bearing reticulated enzymes: applications to biological models and to membrane biotechnology, in *Enzyme Engineering: Biotechnology and Bioengineering Symposium,* No. 3, John Wiley & Sons, New York, 1972, 299.
738. **Solmona, M., Saronio, C., and Garattini, S., Eds.,** *Insolubilized Enzymes,* Raven Press, New York, 1974.
739. **Sharp, A. K., Kay, G., and Lilly, M. D.,** The kinetics of β-galactosidase attached to porous cellulose sheets, *Biotechnol. Bioeng.*, 11, 363, 1969.

740. **Goldman, R., Kedem, O., and Katchalski, E.,** Papain-collodion membranes. II. Analysis of the kinetic behavior of enzymes immobilized in artificial membranes, *Biochemistry,* 7, 4518, 1968.
741. **Goldman, R., Kedem, O., and Katchalski, E.,** Kinetic behavior of alkaline phosphatase-collodion membranes, *Biochemistry,* 10, 165, 1971.
742. **Goldman, R. and Katchalski, E.,** Kinetic behavior of a two-enzyme membrane carrying out a consecutive set of reactions, *J. Theoret. Biol.,* 32, 243, 1971.
743. **Sundaram, P. V., Tweedale, A., and Laidler, K. J.,** Kinetic laws for solid-supported enzymes, *Can. J. Chem.,* 48, 1498, 1970.
744. **Kobayashi, T. and Laidler, K. J.,** Kinetic analysis for solid-supported enzymes, *Biochim. Biophys. Acta,* 302, 1, 1973.
745. **Bunting, P. S. and Laidler, K. J.,** Kinetic studies on solid-supported β-galactosidase, *Biochemistry,* 11, 4477, 1972.
746. **Sélégney, E., Broun, G., Geffroy, J., and Thomas, D.,** *J. Chim. Phys. Physiochim. Biol.,* 66, 391, 1969.
747. **Kasche, V., Lundquist, H., Bergman, R., and Axen, R.,** A theoretical model describing steady-state catalysis by enzyme immobilized in spherical gel particles. Experimental study of α-chymotrypsin-sepharose, *Biochem. Biophys. Res. Commun.,* 45, 615, 1971.
748. **Kasche, V. and Bergwall, M.,** Intrinsic molecular properties and inhibition of immobilized enzymes. Theory and experimental observations on α-chymotrypsin: sepharose, in *Insolubilized Enzymes,* Salmona, M., Saronio, C., and Garattini, S., Eds., Raven Press, New York, 1974, 77.
749. **Boguslaski, R. C., Blaedel, W. J., and Kissel, T. R.,** Kinetic behavior of enzymes immobilized in artificial membranes, in *Insolubilized Enzymes,* Salmona, M., Saronio, C., and Garattini, S., Eds., Raven Press, New York, 1974, 87.
750. **Laidler, K. J. and Bunting, P. S.,** The denaturation of proteins, in *The Chemical Kinetics of Enzyme Action,* Oxford University Press, London, 1973, 382.
751. **Goldstein, L.,** Kinetic behavior of immobilized enzyme systems, in *Methods in Enzymology, Immobilized Enzymes, Vol. 44,* Mosback, K., Ed., Academic Press, New York, 1976, 397.
752. **Engasser, J. M. and Horvath, C.,** Diffusion and kinetics with immobilized enzymes, in *Applied Biochemistry and Bioengineering,* Vol. 1, Wingard, L. B., Katchalski-Katzir, E., and Goldstein, L., Eds., Academic Press, New York, 1976, 127.
753. **Hinberg, I., Korus, R., and O'Driscoll, K. F.,** Gel entrapped enzymes: kinetic studies of immobilized β-galactosidase, *Biotechnol. Bioeng.,* 16, 943, 1974.
754. **Kobayashi, T. and Laidler, K. J.,** Theory of kinetics of reactions catalysed by enzymes attached to the interior surfaces of tubes, *Biotechnol. Bioeng.,* 16, 99, 1974.
755. **Blaedel, W. J. and Boguslaski, R. C.,** Kinetic behavior of enzymes immobilized in artificial membranes, *Biochem. Biophys. Res. Commun.,* 47, 248, 1972.
756. **Daka, N. J. and Laidler, K. J.,** Flow kinetics of lactate dehydrogenase chemically attached to nylon tubing, *Can. J. Biochem.,* 56, 774, 1978.
757. **Allison, R. D. and Purich, D. L.,** Practical considerations in the design of initial velocity enzyme rate assays, in *Methods in Enzymology, Enzyme Kinetics and Mechanism,* Vol. 63, Purich, D. L., Colowick, S. P., and Kaplan, N. O., Eds., Academic Press, New York, 1979, 3.
758. **Fromm, H. J.,** *Initial Rate Enzyme Kinetics,* Springer-Verlag, Berlin, 1975.
759. **Caldin, E. F. and Tomalin, G.,** Computations on quantum-mechanical tunneling in proton-transfer reactions in solutions, *Trans. Faraday Soc.,* 64, 2823, 1968.
760. **Maugh, T. H., II,** A renewed interest in immobilized enzymes. A host of potential new applications for immobilized enzymes and cells presages what some call "a new industrial revolution," *Science,* 223, 474, 1984.
761. **Kuby, S. A., Noda, L., and Lardy, J. A.,** ATP-creatine transphosphorylase. III. Kinetic studies, *J. Biol. Chem.,* 210, 65, 1954.
762. **Green, A. A. and McElroy, W. D.,** Crystalline firefly luciferase, *Biochim. Biophys. Acta,* 20, 170, 1956.
763. **Lienhard, G. E. and Secemski, I. I.,** P^1, P^5-di(adenosine-5′)pentaphosphate, a potent multisubstrate inhibitor of adenylate kinase, *J. Biol. Chem.,* 248, 1121, 1973.
764. **Rudolph, F. B., Baugher, B. W., and Beissner, R. S.,** Techniques in coupled enzyme assays, in *Methods in Enzymology, Enzyme Kinetics and Mechanisms,* Vol. 63, Purich, D. L., Colowick, S. P. and Kaplan, N. O., Eds., Academic Press, New York, 1979, 22.
765. **Horecker, B. L. and Kornberg, A.,** The extinction coefficients of the reduced band of pyridine nucleotides, *J. Biol. Chem.,* 175, 385, 1948.
766. **Bergmeyer, H. U., Gloger, M., and Tischer, M.,** Determination of the catalytic activity of enzymes, in *Methods of Enzymatic Analysis,* 3rd ed., Vol. 1, Section 2.3, Bergmeyer, H. U., Ed., Deerfield Beach, FL, 1983.
767. **Bowen, W. J. and Kerwin, T. P.,** The purification of myokinase with ion exchange resin, *Arch. Biochem. Biophys.,* 57, 522, 1955.

768. International Union of Biochemistry, *Enzyme Nomenclature,* Elsevier, Amsterdam, 1965.
769. **Theorell, J.,** Über hemmung der reaktionsgeschwindigkeit durch phosphat in Warburgs und Christians system, *Biochem. Z.,* 275, 416, 1935.
770. **Cantor, C. R. and Schimmel, P. R.,** *Biophysical Chemistry Part III. The Behavior of Biological Macromolecules,* W. H. Freeman, San Francisco, 1980, chap. 15 to 17, 849—971.
771. **Eisenthal, R. and Cornish-Bowden, A.,** The direct linear plot. A new graphical procedure for estimating enzyme kinetic parameters, *Biochem. J.,* 139, 715, 1974.
772. **Cornish-Bowden, A.,** *Principles of Enzyme Kinetics,* Butterworth, London, 1975.
773. **Purich, D. L., Ed.,** *Contemporary Enzyme Kinetics and Mechanism,* Academic Press, New York, 1983.
774. **Purich, D., Ed.,** Intermediates, steriochemistry, and rate studies, in *Enzyme Kinetics and Mechanism, Part C, Methods in Enzymology,* Colowick, S. P. and Kaplan, N. O., Eds., Academic Press, New York, 1984.
775. **Cleland, W. W.,** Determining the chemical mechanisms of enzyme-catalyzed reactions by kinetic studies, *Advances in Enzymology and Related Areas of Molecular Biology,* Vol. 45, Meister, A., Ed., Interscience, New York, 1977, 273.
776. **Cleland, W. W.,** The use of pH studies to determine chemical mechanisms of enzyme-catalyzed reactions, in *Methods in Enzymology,* 87, 390, 1982.
777. **Tipton, K. F. and Dixon, H. B. F.,** Effects of pH on enzymes, in *Methods in Enzymology,* 63, 183, 1979.
778. **Findlay, D., Mathias, A. P., and Rabin, B. R.,** The active site and mechanism of action of bovine pancreatic ribonuclease. V. The charge types at the active centre, *Biochem. J.,* 85, 139, 1962.
779. **Bacarella, A. L., Grunwald, E., Marshall, H. P., and Purlee, E. L.,** The potentiometric measurement of acid dissociation constants and pH in the system methanol-water. pK_A values for carboxylic acids and anilinium ions, *J. Org. Chem.,* 20, 747, 1955.
780. **Danyluk, S. S., Taniguchi, H., and Janz, G. J.,** The thermodynamics of hydrochloric acid and the ionization constants of formic, acetic and propionic acids in 82 weight percent dioxane-water solutions, *J. Phys. Chem.,* 61, 1679, 1957.
781. **Donzou, P.,** Enzymology at sub-zero temperatures, *Advances in Enzymology Related Areas of Molecular Biology,* Vol. 45, Meister, A., Ed., Interscience, New York, 1977, 188.
782. **Harned, H. S.,** Experimental studies of the ionization of acetic acid, *J. Phys. Chem.,* 43, 275, 1939.
783. **Maurel, P., Hui Hon Hoa, G., and Douzou, P.,** The pH dependence of the tryptic hydrolysis of benzoyl-L-arginine ethyl ester in cooled mixed solvents, *J. Biol. Chem.,* 250, 1376, 1975.
784. **Shedlovsky, T.,** *Electrolytes,* Pesce, B., Ed., Pergamon Press, Elmsford, N.Y., 1959, 146.
785. **Spivey, H. O. and Shedlovsky, T.,** Studies of electrolytic conductance in alcohol-water mixtures. II. The ionization constant of acetic acid in ethanol-water mixtures at 0, 25, and 35°, *J. Phys. Chem.,* 71, 2171, 1967.
786. **Steel, B. J., Robinson, R. A., and Bates, R. G.,** *J. Res. Natl. Bur. Stand.,* 71, 11, 1967.
787. **Paabo, M., Bates, R. G., and Robinson, R. A.,** Dissociation of ammonium ion in methanol-water solvents, *J. Phys. Chem.,* 70, 247, 1966.
788. **Grace, S. and Dunaway-Mariano, D.,** Examination of the solvent perturbation technique as a method to identify enzyme catalytic groups, *Biochemistry,* 22, 4238, 1983.
789. **Krishnaswamy, M. and Kenkare, U.,** The effect of pH, temperature, and organic solvents on the kinetic parameters of *Escherichia coli* alkaline phosphatase, *J. Biol. Chem.,* 245, 3956, 1970.
790. **Rife, J. and Cleland, W. W.,** Determination of the chemical mechanism of glutamate dehydrogenase from pH studies, *Biochemistry,* 19, 2328, 1980.
791. **Knight, W. B., Fitts, S. W., and Dunaway-Mariano, D.,** Investigation of the catalytic mechanism of yeast inorganic pyrophosphate, *Biochemistry,* 20, 4079, 1981.
792. **Viola, R. E. and Cleland, W. W.,** Use of pH studies to elucidate the chemical mechanism of yeast hexokinase, *Biochemistry,* 17, 4111, 1978.
793. **Viljoen, C. and Bates, D. P.,** A steady state analysis of pH and temperature effects on the action of an arginine esterase from the venom of the gaboon adder, *Bitis gabonica,* on synthetic arginine substrates, *Hoppe-Seyler's Z. Physiol. Chem.,* 362, 95, 1981.
794. **Williams, J. and Morrison, J.,** Chemical mechanism of the reaction catalyzed by dihydrofolate reductase from streptococcus fascium: pH studies and chemical modification, *Biochemistry,* 20, 6024, 1981.
795. **Stone, S. R. and Morrison, J. F.,** Catalytic mechanism of the dihydrofolate reductase as determined by pH studies, *Biochemistry,* 23, 2753, 1984.
796. **Cook, P. F., Kenyon, G., and Cleland, W. W.,** Use of pH studies to elucidate the catalytic mechanism of rabbit muscle creatine kinase, *Biochemistry,* 20, 1204, 1981.
797. **Schimerlik, M. I. and Cleland, W. W.,** pH variation of the kinetic parameters and the catalytic mechanism of malic enzyme, *Biochemistry,* 16, 576, 1977.
798. **Veloso, D., Cleland, W. W., and Porter, J.,** pH properties and chemical mechanism of action of 3-hydroxy-3-methylglutaryl coenzyme A reductase, *Biochemistry,* 20, 887, 1981.

799. **Godinot, C. and Penin, F.,** *Biochem. Int.,* 2, 595, 1981.
800. **Nitta, U., Kunikata, T., and Wanatabe, T.,** Kinetic study of soybean β-amylase. The effect of pH, *J. Biochem. Tokyo,* 85, 41, 1979.
801. **Solheim, L. P. and Fromm, H. J.,** pH kinetic studies of bovine brain hexokinase, *Biochemistry,* 19, 6074, 1980.
802. **Grimshaw, C. E., Cook, P. F., and Cleland, W. W.,** Use of isotope effects and pH studies to determine the chemical mechanism of *Bacillus subtilis* L-alanine dehydrogenase, *Biochemistry,* 20, 5655, 1981.
803. **Hammes, G. G. and Wu, C.-W.,** Relaxation spectra of aspartate transcarbamylase; interaction of the native enzyme with carbamyl phosphate, *Biochemistry,* 10, 2150, 1971.
804. **Laidler, K. J. and Bunting, P. S.,** The influence of hydrogen ion concentration, in *The Chemical Kinetics of Enzyme Action,* 2nd ed., Clarendon Press, Oxford, 1973, 162.
805. **Knowles, J.,** The intrinsic pK_a-values of functional groups in enzymes: improper deductions from the pH dependence of steady-state parameters, *CRC Crit. Rev. Biochem.,* 4, 165, 1976.
806. **Fersht, A.,** Measurement and magnitude of enzymatic rate constants, in *Enzyme Structure and Mechanism,* W. H. Freeman, San Francisco, 1977, 153.
807. **Lipscomb, W. N.,** Acceleration of reactions by enzymes, *Acc. Chem. Res.,* 15, 232, 1982.
808. **Rebek, J., Jr.,** Binding forces, equilibria, and rates: new models for enzymic catalysis, *Acc. Chem. Res.,* 17, 258, 1984.
809. **Garfinkel, L. and Garfinkel, D.,** Calculation of free Mg^{2+} concentration in adenosine 5′-triphosphate containing solutions *in vitro* and *in vivo, Biochemistry,* 23, 3547, 1984.
810. **Gupta, R. K. and Moore, R. D.,** ^{31}P NMR studies of intracellular free Mg^{2+} in intact frog skeletal muscle, *J. Biol. Chem.,* 255, 3987, 1984.
811. **O'Sullivan, W. J. and Smithers, G. W.,** Stability constants for biologically important metal-ligand complexes, in *Methods Enzymol.,* 63, 294, 1979.
812. **Fromm, H. J.,** Computer-assisted derivation of steady-state rate equations, in *Methods Enzymol.,* 63, 84, 1979.
813. **Theorell, H. and McKinley-McKee, J. S.,** Liver alcohol dehydrogenase. II. Equilibria, *Acta Chim. Scand.,* 15, 1811, 1961.
814. **Duggleby, R. G.,** A nonlinear regression program for small computers, *Anal. Biochem.,* 110, 9, 1981.
815. **Mosteller, F. and Tukey, J. W.,** A class of mechanisms for fitting, in *Data Analysis and Regression,* Addison-Wesley, Reading, MA, 1977, 333.
816. **Pecoraro, B. L., Hermes, J. D. and Cleland, W. W.,** Stability constants of Mg^{2+} and Cd^{2+} complexes of adenine nucleotides and thionucleotides and rate constants for formation and dissociation of MgATP and MgADP, *Biochemistry,* 23, 5262, 1984.
817. **Vanoni, M. A. and Matthews, R. G.,** Kinetic isotope effects on the oxidation of reduced nicotinamide adenine dinucleotide phosphate by the flavoprotein methylenetetrahydrofolate reductase, *Biochemistry,* 23, 5272, 1984.
818. **Kutzback, C. and Stockstad, E. L. R.,** Mammalian methylenetetrahydrofolate reductase. Partial purification, properties and inhibition by *S*-adenosylmethionine, *Biochim. Biophys. Acta,* 250, 459, 1971.
819. **Vanoni, M. A., Daubner, S. C., Ballou, P., and Matthews, R. G.,** Studies on methylenetetrahydrofolate reductase from pig liver: catalytic mechanism and regulation by adenosylmethionine, in *The Chemistry and Biology of Pteridines,* Blair, J. A., Ed., de Gruyter, Berlin, 1983, 235.
820. **Daubner, S. C. and Matthews, R. G.,** Purification and properties of methylenetetrahydrofolate reductase from pig liver, *J. Biol. Chem.,* 257, 140, 1982.
821. **Vanoni, M. A., Ballou, D. P., and Matthews, R. G.,** Methylenetetrahydrofolate reductase. Steady state and rapid reaction studies on the NADPH-methylenetetrahydrofolate, NADPH-menadione, and methylenetetrahydrofolate-menadione oxidoreductase activities of the enzyme, *J. Biol. Chem.,* 258, 11510, 1983.
822. **Albery, W. J. and Knowles, J. R.,** Evolution of enzyme function and the development of catalytic efficiency, *Biochemistry,* 15, 5631, 1976.
823. **Rose, I. A.,** Failure to confirm previous observations on triosephosphate isomerase intermediate and bound substrate complexes, *Biochemistry,* 23, 5893, 1984.
824. **Melander, L.,** Secondary hydrogen isotope effects, in *Isotope Effects on Reaction Rates,* Ronald Press, New York, 1960, chap. 5.
825. **Hermes, J. S. and Cleland, W. W.,** Evidence from multiple isotope effect determinations for coupled hydrogen motion and tunneling in the reaction catalyzed by glucose-6-phosphate dehydrogenase, *J. Am. Chem. Soc.,* 106, 7263, 1984.
826. **Mares-Guia, M., Silva, E., and Tunes, H.,** Competitive inhibition by the D-isomer in racemic mixtures used as substrate in kinetic studies: a simple method for data treatment, *Arch. Biochem. Biophys.,* 228, 278, 1984.
827. **Cleland, W. W., Gross, M., and Folk, K. E.,** Inhibition patterns obtained where an inhibitor is present in constant proportion to variable substrate, *J. Biol. Chem.,* 248, 6541, 1973.

828. **Tryon, E., Cress, M. C., Hamada, M., and Kuby, S. A.,** Studies on NADPH-cytochrome c reductase. I. Isolation and several properties of the crystalline enzyme from ale yeast, *Arch. Biochem. Biophys.*, 197, 104, 1979.
829. **Tryon, E. and Kuby, S. A.,** Studies on NADPH-cytochrome c reductase. II. Steady-state kinetic properties of the crystalline enzyme from ale yeast, *Enzyme*, 31, 197, 1984.
830. **Graves, D. J. and Martensen, T. M.,** The use of alternative substrates to study enzyme-catalyzed chemical modification, in *Methods in Enzymology*, Vol. 64, Purich, D. L., Colowick, S. P. and Kaplan, N. O., Eds., Academic Press, New York, 1980, 325.
831. **Morrison, J. F. and White, A.,** Isotope exchange studies of the reaction catalyzed by ATP: creatine phosphotransferase, *Eur. J. Biochem.*, 3, 145, 1967.
832. **Maggio, E. T., Kenyon, G. L., Markham, G. D., and Reed, G. H.,** Properties of a CH_3S-blocked creatine kinase with altered catalytic activity. Kinetic consequences of the presence of the blocking group, *J. Biol. Chem.*, 252, 1202, 1977.
833. **Engelborghs, Y., Marsh, A., and Gutfreund, H.,** A quenched-flow study of the reaction catalysed by creatine kinase, *Biochem. J.*, 151, 47, 1975.
834. **Milner-White, E. J. and Watts, D. C.,** Inhibition of adenosine 5′-triphosphate-creatine phosphotransferase by substrate-anion complexes. Evidence for the transition-state organization of the catalytic site, *Biochem. J.*, 122, 727, 1971.
835. **Cook, P. F., Kenyon, G. L., and Cleland, W. W.,** Use of pH studies to elucidate the catalytic mechanism of rabbit muscle creatine kinase, *Biochemistry*, 20, 1204, 1981.
836. **James, T. L. and Cohn, M.,** The role of the lysyl residue at the active site of creatine kinase. Nuclear Overhauser effect studies, *J. Biol. Chem.*, 249, 2599, 1974.
837. **Rosevear, P. R., Desmeules, P., Kenyon, G. L., and Mildvan, A. S.,** Nuclear magnetic resonance studies of the role of histidine residues at the active site of rabbit muscle creatine kinase, *Biochemistry*, 20, 6155, 1981.
838. **Lui, N. S. T. and Cunningham, L.,** Cooperative effects of substrate and substrate analogues on the conformation of creatine phosphokinase, *Biochemistry*, 5, 144, 1966.
839. **Nageswara Rao, B. D. and Cohn, M.,** ^{31}P NMR of enzyme-bound substrates of rabbit muscle creatine kinase. Equilibrium constants, interconversion rates, and NMR parameters of enzyme-bound complexes, *J. Biol. Chem.*, 256, 1716, 1981.
840. **Dunaway-Mariano, D. and Cleland, W. W.,** Preparation and properties of chromium (III) adenosine 5′-triphosphate, chromium (III) adenosine 5′-diphosphate, and related chromium complexes, *Biochemistry*, 19, 1496, 1980.
841. **Theorell, H. and McKinley-McKee, J. S.,** Liver alcohol dehydrogenase. III. Kinetics in the presence of caprate, *iso*butyramide and imidazole, *Acta Chem. Scand.*, 15, 1834, 1961.
842. **Theorell, H. and McKinley-McKee, J. S.,** Liver alcohol dehydrogenase. IV. Kinetics in the presence of zinc binding agents, *Acta Chem. Scand.*, 15, 1836, 1961.
843. **Henderson, P. J. F.,** Steady-state enzyme kinetics with high-affinity substrates or inhibitors. A statistical treatment of dose-response curves, *Biochem. J.*, 135, 101, 1973.
844. **Viola, R. E. and Cleland, W. W.,** Initial velocity analysis for terreactant mechanisms, in *Methods in Enzymology*, Vol. 87, Purich, D. L., Colowick, S. P. and Kaplan, N. O., Eds., Academic Press, New York, 1982, 353.
845. **Hermes, J. D., Weiss, P. M., and Cleland, W. W.,** Use of nitrogen-15 and deuterium isotope effects to determine the chemical mechanism of phenylalanine ammonia-lyase, *Biochemistry*, 24, 2959, 1985.
846. **Viola, R. E.,** Kinetic studies of the reactions catalyzed by glucose-6-phosphate dehydrogenase from *Leuconostoc mesenteroides*: pH variation of kinetic parameters, *Arch. Biochem. Biophys.*, 228, 415, 1984.
847. **Levy, H. R., Christoff, M., Ingulli, J., and Ho, E. M. L.,** Glucose-6-phosphate dehydrogenase from *Leuconostoc mesenteroides*: revised kinetic mechanism and kinetics of ATP-inhibition, *Arch. Biochem. Biophys.*, 222, 473, 1983.
848. **Putney, S., Herlihy, W., Royal, N., Pang, H., Aposhian, H. V., Pickering, L., Belagaje, R., Biemann, K., Page, D., Kuby, S. A., and Schimmel, P. J.,** Rabbit muscle creatine phosphokinase. cDNA cloning, primary structure and detection of human homologues, *J. Biol. Chem.*, 259, 14317, 1984.
849. **Travers, F., Barman, T. E., and Bertrand, R.,** Transient phase studies on the creatine kinase reaction. The analysis of a reaction pathway with three intermediates, *Eur. J. Biochem.*, 100, 149, 1979.
850. **Barman, T. E., Brun, A., and Travers, F.,** A flow-quench apparatus for cryoenzymic studies. Application to the creatine kinase reaction, *Eur. J. Biochem.*, 110, 397, 1980.
851. **Travers, F. and Barman, T. E.,** Cryoenzymic studies on the transition-state analog complex creatine kinase · ADPMg · nitrate · creatine, *Eur. J. Biochem.*, 110, 405, 1980.
852. **Lawson, J. W. R. and Veech, R. L.,** Effects of pH and free Mg^{2+} on the K_{eq} of the creatine kinase reaction and other phosphate hydrolyses and phosphate transfer reactions, *J. Biol. Chem.*, 254, 6528, 1979.

853. **Lerman, C. L. and Cohn, M.,** ^{31}P NMR quantitation of the displacement of equilibria of arginine, creatine, pyruvate, and 3-P-glycerate kinase reactions by substitution of sulfur for oxygen in the β-phosphate of ATP, *J. Biol. Chem.*, 255, 8756, 1980.
854. **Kuby, S. A., Palmieri, R. H., Frischat, A., Fischer, A. H., Wu, L. H., Maland, L., and Manship, M.,** Studies on adenosine triphosphate transphosphorylases. XVI. Amino acid sequence of rabbit muscle ATP-AMP transphosphorylase, *Biochemistry*, 23, 2393, 1984.
855. **Kuby, S. A., Palmieri, R. H., Okabe, K., Cress, M. C., and Yue, R. H.,** Studies on adenosine triphosphate transphosphorylases. XVII. Physico-chemical comparison of the ATP-creatine transphosphorylase (creatine kinase) isoenzymes from man, calf, and rabbit, *J. Protein Chem.*, 2, 469, 1984.
856. **Kenyon, G. L. and Reed, G. H.,** Creatine kinase: structure-activity relationships, *Advances in Enzymology and Related Areas of Molecular Biology*, Vol. 54, Meister, A., Ed., John Wiley & Sons, New York, 1983, 367.
857. **Kalbitzer, H. R., Marquetant, R., Rösch, P., and Schirmer, R. H.,** The structural isomerisation of human-muscle adenylate kinase as studied by ^{1}H-nuclear magnetic resonance, *Eur. J. Biochem.*, 126, 531, 1982.
858. **Fry, D. C., Kuby, S. A., and Mildvan, A. S.,** NMR studies of the MgATP binding site of adenylate kinase and of a 45 amino acid fragment of the enzyme, *Fed. Proc.*, A43, 1837, 1984.
859. **Fry, D. C., Kuby, S. A., and Mildvan, A. S.,** Measurements of absolute interproton distances by frequency dependent nuclear Overhauser effects: conformation of bound MgATP on adenylate kinase and a peptide of the enzyme, *Am. Chem. Soc. Div. Biol. Chem. Abstr., Biochemistry*, 23, 3357, 1984.
860. **Fry, D. C., Kuby, S. A., and Mildvan, A. S.,** The ATP-binding site of adenylate kinase: homology to F1 ATPase, RAS P21, myosin, and other nucleotide binding proteins, *Fed. Proc. Abstr.*, 44, 672, 1985.
861. **Fry, D. C., Kuby, S. A., and Mildvan, A. S.,** NMR studies of the MgATP binding site of adenylate kinase and of a 45 residue peptide fragment of the enzyme, *Biochemistry*, 24, 4680, 1985.
862. **Fry, D. C., Kuby, S. A., and Mildvan, A. S.,** The ATP-binding site of adenylate kinase: mechanistic implications for its homology with other nucleotide-binding proteins, *Proc. Natl. Acad. Sci.*, 83, 907, 1986.
863. **Crivellone, M. D., Hermodson, M., and Axelrod, B.,** Inactivation of muscle adenylate kinase by site-specific destruction of tyrosine 95 using potassium ferrate, *J. Biol. Chem.*, 260, 2657, 1985.
864. **Walker, J. E., Saraste, M., Runswick, M. J., and Gay, N. J.,** Distantly related sequences in the α- and β-subunits of ATP synthase, myosin, kinases and other ATP-requiring enzymes and a common nucleotide binding fold, *EMBO J.*, 1, 945, 1982.
865. **Stewart, J. M. and Young, J. D.,** *Solid Phase Peptide Synthesis*, W. H. Freeman, San Francisco, 1969.
866. **Stewart, J. M. and Young, J. D.,** *Solid Phase Peptide Synthesis*, 2nd ed., Pierce Chemical, Rockford, IL, 1984.
867. **Jarori, G. K., Ray, B. D., and Nageswara Rao, B. D.,** Structure of metal-nucleotide complexes bound to creatine kinase: ^{31}P NMR measurements using Mn (II) and Co (II), *Biochemistry*, 24, 3487, 1985.
868. **Leyh, T. S., Goodhart, P. J., Nguyen, A. G., Kenyon, G. L., and Reed, G. H.,** Structures of manganese (II) complexes with ATP, ADP, and phosphocreatine in the reactive central complexes with creatine kinase: electron paramagnetic resonance studies with oxygen-17-labeled ligands, *Biochemistry*, 24, 308, 1985.
869. **Hammes, G. G.,** Relaxation spectrometry of biological systems, in *Advances in Protein Chemistry*, Vol. 23, Anfinsen, C. B., Jr., Anson, M. L., Edsall, J. T., and Richards, F. M., Eds., Academic Press, New York, 1968, 1.
870. **Degani, H., Laughlin, M., Campbell, S., and Shulman, R. G.,** Kinetics of creatine kinase in heart: a ^{31}P NMR saturation- and inversion-transfer study, *Biochemistry*, 24, 5510, 1985.
871. **Johnson, M. S. and Kuby, S. A.,** Studies on NADH(NADPH)-cytochrome *c* reductase (FMN-containing) from yeast. Isolation and physicochemical properties of the enzyme from top-fermenting ale yeast, *J. Biol. Chem.*, 260, 12341, 1985.
872. **Johnson, M. S. and Kuby, S. A.,** Studies on NADH(NADPH)-cytochrome *c* reductase from yeast. II. Further characterization and steady-state kinetic properties of the flavoenzyme from top-fermenting ale yeast, *Arch. Biochem. Biophys.*, 245, 271, 1986.
873. **Tietz, A. and Ochoa, S.,** "Fluorokinase" and pyruvic kinase, *Arch. Biochem. Biophys.*, 78, 477, 1958.
874. **Kupiecki, F. P. and Coon, M. J.,** Bicarbonate- and hydroxylamine-dependent degradation of adenosine triphosphate, *J. Biol. Chem.*, 234, 2428, 1959.
875. **Weiss, P. M., Hermes, J. D., Dougherty, T. M., and Cleland, W. W.,** *N*-hydroxycarbamate is the substrate for the pyruvate kinase catalyzed phosphorylation of hydroxylamine, *Biochemistry*, 23, 4346, 1984.
876. **Kayne, F. J.,** Pyruvate kinase catalyzed phosphorylation of glycolate, *Biochem. Biophys. Res. Commun.*, 59, 8, 1974.
877. **Creighton, D. J. and Rose, I. A.,** Studies on the mechanism and stereochemical properties of the oxalacetate decarboxylase activity of pyruvate kinase, *J. Biol. Chem.*, 251, 61, 1976.
878. **Creighton, D. J. and Rose, I. A.,** Oxalacetate decarboxylase activity in muscle is due to pyruvate kinase, *J. Biol. Chem.*, 251, 69, 1976.

879. **Robinson, J. L. and Rose, I. A.,** The proton transfer reactions of muscle pyruvate kinase, *J. Biol. Chem.*, 247, 1096, 1972.
880. **Gupta, R. K., Oesterling, R. M., and Mildvan, A. S.,** Dual divalent cation requirement for activation of pyruvate kinase: essential roles of both enzyme- and nucleotide-bound metal ions, *Biochemistry,* 15, 2881, 1976.
881. **Nageswara Rao, B. D., Kayne, F. J., and Cohn, M.,** ^{31}P NMR studies of enzyme-bound substrates of rabbit muscle pyruvate kinase. Equilibrium constants, exchange rates, and NMR parameters, *J. Biol. Chem.*, 254, 2689, 1979.
882. **Baek, Y. H. and Nowak, T.,** Kinetic evidence for a dual cation role for muscle pyruvate kinase, *Arch. Biochem. Biophys.*, 217, 491, 1982.
883. **Dougherty, T. M. and Cleland, W. W.,** pH studies on the chemical mechanism of rabbit muscle pyruvate kinase. I. Alternate substrates oxalacetate, glycolate, hydroxylamine, and fluoride, *Biochemistry,* 24, 5870, 1985.
884. **Dougherty, T. M. and Cleland, W. W.,** pH studies on the chemical mechanism of rabbit muscle pyruvate kinase. II. Physiological substrates and phosphoenol-α-ketobutyrate, *Biochemistry,* 24, 5875, 1985.
885. **Mildvan, A. S., Sloan, D. L., Fung, C. H., Gupta, R. K., and Melamud, E.,** Arrangement and conformations of substrates at the active site of pyruvate kinase from model building studies based on magnetic resonance data, *J. Biol. Chem.*, 251, 5431, 1976.
886. **Blättler, W. A. and Knowles, J. R.,** Stereochemical course of glycerol kinase, pyruvate kinase, and hexokinase: phosphoryl transfer from chiral [γ(*S*)-^{16}O, ^{17}O, ^{18}O]ATP, *J. Am. Chem. Soc.*, 101, 510, 1979.
887. **Kayne, F. J.,** Pyruvate kinase, in *The Enzymes,* 3rd ed., Vol. 8, Boyer, P. D., Ed., Academic Press, New York, 1973, 353.
888. **Mildvan, A. S. and Nowak, T.,** Nuclear magnetic resonance studies of the function of potassium in the mechanism of pyruvate kinase, *Biochemistry,* 11, 2819, 1972.
889. **Messmer, C. H. and Kägi, J. H. R.,** Tryptophan residues of creatine kinase: a fluorescence study, *Biochemistry,* 24, 7172, 1985.
890. **Kägi, J. H. R., Li, T. K., and Vallee, B. L.,** Extrinsic cotton effects in complexes of creatine phosphokinase with adenine coenzymes, *Biochemistry,* 10, 1007, 1971.
891. **Price, N. C.,** The interaction of nucleotides with kinases, monitored by changes in protein fluorescence, *FEBS Lett.*, 24, 21, 1972.
892. **Vasák, M., Nagayama, K., Wüthrich, K., Mertens, M. L., and Kägi, J. H. R.,** Creatine kinase. Nuclear magnetic resonance and fluorescence evidence for interaction of adenosine 5′-diphosphate with aromatic residue(s), *Biochemistry,* 18, 5050, 1979.
893. **Lehrer, S. S. and Leavis, P. C.,** Solute quenching of protein fluorescence, in *Methods in Enzymology,* Vol. 49, Academic Press, New York, 1978, 222.
894. **Ratliff, R. L., Weaver, R. H., Lardy, H. A., and Kuby, S. A.,** Nucleoside diphosphokinase. I. Isolation of the crystalline enzyme from brewers' yeast, *J. Biol. Chem.*, 239, 301, 1964.
895. **Kuby, S. A. and Fleming, G.,** unpublished, 1980.
896. **Kuby, S. A., Fleming, G., Frischat, A., Cress, M. C., and Hamada, M.,** Studies on adenosine triphosphate transphosphorylases. XV. Human isoenzymes of adenylate kinase: isolation and physicochemical comparison of the crystalline human ATP-AMP transphosphorylases from muscle and liver, *J. Biol. Chem.*, 258, 1901, 1983.
897. **Kuby, S. A., Fleming, G., Gubler, C., and Tryon, E.,** unpublished observations.
898. **Sabat, M., Cini, R., Haromy, T., and Sundaralingam, M.,** Crystal structure of the α, β, γ-tridentate manganese complex of adenosine 5′-triphosphate cocrystallized with 2, 2′-dipyridylamine, *Biochemistry,* 24, 7827, 1985.
899. **Cini, R., Cinquantini, A., Burla, M. C., Nunzi, A., Polidori, G., and Zanazzi, P. F.,** *Chim. Ind. (Milan),* 64, 826, 1982.
900. **Cini, R., Sabat, M., Sundaralingam, M., Burla, M. C., Nunzi, A., Polidori, G., and Zanazzi, P. F.,** Interaction of adenosine 5′-triphosphate with metal ions. X-ray structure of ternary complexes containing Mg (II), Ca (II), Mn (II), Co (II), ATP and 2,2′-dipyridylamine, *J. Biomol. Struct. Dyn.*, 1, 633, 1983.
901. **Cini, R., Burla, M. C., Nunzi, A., Polidori, G. P., and Zanazzi, P. F.,** Preparation and physicochemical properties of the ternary complexes formed between adenosine 5′-triphosphoric acid, bis(2-pyridyl)amine, and divalent metal ions. Crystal and molecular structures of the compounds containing Mg^{II} and Ca^{II}, *J. Chem. Soc., Dalton Trans.*, 2467, 1984.
902. **Mildvan, A. S.,** Conformation and arrangement of substrates at active sites of ATP-utilizing enzymes, *Phil. Trans. R. Soc., London,* B293, 65, 1981.
903. **Rosevear, P. R., Bramson, H. N., O'Brian, C., Kaiser, E. T., and Mildvan, A. S.,** Nuclear overhauser effect studies of the conformations of tetraamminecobalt (III)-adenosine 5′-triphosphate free and bound to bovine heart protein kinase, *Biochemistry,* 22, 3439, 1983.

904. **Krenitsky, T. A.,** Pentosyl transfer mechanisms of the mammalian nucleoside phosphorylases, *J. Biol. Chem.,* 243, 2871, 1968.
905. **Gallo, R. C., Perry, S., and Breitman, T. R.,** The enzymic mechanisms for deoxythymidine synthesis in human leukocytes. I. Substrate inhibition by thymine and activation by phosphate or arsenate, *J. Biol. Chem.,* 242, 5059, 1967.
906. **Gallo, R. C. and Breitman, T. R.,** The enzymatic mechanisms for deoxythymidine synthesis in human leukocytes. I. Comparison of deoxyribosyl donors, *J. Biol. Chem.,* 243, 4936, 1968.
907. **Schwartz, M.,** Thymidine phosphorylase from *Escherichia coli.* Properties and kinetics, *Eur. J. Biochem.,* 21, 191, 1971.
908. **Iltzsch, M. H., Kouni, J. H., and Cha, S.,** Kinetic studies of thymidine phosphorylase from mouse liver, *Biochemistry,* 24, 6799, 1985.
909. **Lijk, L. J., Kalk, K. H., Brandenburger, N. P., and Hol, W. G. J.,** Binding of metal cyanide complexes to bovine liver rhodanese in the crystalline state, *Biochemistry,* 22, 2952, 1983.
910. **Chow, S. F. and Horowitz, P. M.,** Tetracyanonickelate probes of the activity site of sulfur-free rhodanese, *J. Biol. Chem.,* 260, 15516, 1985.
911. **Somogyi, B., Papp, S., Rosenberg, A., Seres, E., Matkó, J., Welch, G. R., and Nagy, P.,** A double quenching method for studying protein dynamics: separation of the fluorescence quenching parameters characteristic of solvent-exposed and solvent-masked fluorophors, *Biochemistry,* 24, 6674, 1985.
912. **Hartzell, C. R. and Beinert, H.,** Components of cytochrome *c* oxidase detectable by EPR spectroscopy, *Biochim. Biophys. Acta,* 368, 318, 1974.
913. **Wikström, M.,** Proton translocation by cytochrome oxidase, *Curr. Top. Membr. Transp.,* 16, 303, 1982.
914. **Vik, S. B. and Capaldi, R. A.,** Conditions for optimal electron transfer activity of cytochrome *c* oxidase isolated from beef heart mitochondria, *Biochem. Biophys. Res. Commun.,* 94, 348, 1980.
915. **Merle, P. and Kadenback, B.,** The subunit composition of mammalian cytochrome *c* oxidase, *Eur. J. Biochem.,* 105, 499, 1980.
916. **Wikström, M., Krab, K., and Saraste, M.,** *Cytochrome Oxidase — A Synthesis,* Academic Press, New York, 1981.
917. **Chance, B. and Leigh, J. S., Jr.,** Oxygen intermediates and mixed valence states of cytochrome oxidase: infrared absorption difference spectra of compounds A, B, and C of cytochrome oxidase and oxygen, *Proc. Natl. Acad. Sci. U.S.A.,* 74, 4777, 1977.
918. **Clore, G. M. and Chance, E. M.,** The mechanism of reaction of ferricyanide-pretreated mixed-valence-state membrane-bound cytochrome oxidase with oxygen at 173° K, *Biochem. J.,* 173, 811, 1978.
919. **Clore, G. M. and Chance, E. M.,** The kinetics and thermodynamics of the reaction of solid-state fully reduced membrane bound cytochrome oxidase with carbon monoxide as studied by dual-wavelength multichannel spectroscopy and flash photolysis, *Biochem. J.,* 175, 709, 1978.
920. **Clore, G. M., Andréasson, L. E., Karlsson, B., Aasa, R., and Malmström, B. G.,** Characterization of the low temperature intermediates of the reaction of fully reduced soluble cytochrome oxidase with oxygen by electron-paramagnetic-resonance with optical spectroscopy, *Biochem. J.,* 185, 139, 1980.
921. **Chance, B., Saronio, C., and Leigh, J. S., Jr.,** Functional intermediates in the reaction of membrane bound cytochrome oxidase with oxygen, *J. Biol. Chem.,* 250, 9226, 1975.
922. **Gibson, Q. H. and Greenwood, C.,** Reactions of cytochrome oxidase with oxygen and carbon monoxide, *Biochem. J.,* 86, 541, 1963.
923. **Babcock, G. T., Jean, J. M., Johnston, L. N., Palmer, G., and Woodruf, W. H.,** Time-resolved resonance Raman spectroscopy of transient species formed during the oxidation of cytochrome oxidase by dioxygen, *J. Am. Chem. Soc.,* 106, 8305, 1984.
924. **Karlsson, B., Aasa, R., Vánngård, T., and Malmström, B. G.,** An EPR-detectable intermediate in the cytochrome oxidase-dioxygen reaction, *FEBS Lett.,* 131, 186, 1981.
925. **Blair, D. F., Witt, S. N., and Chan, S. I.,** Mechanism of cytochrome *c* oxidase-catalyzed dioxygen reduction at low temperatures. Evidence for two intermediates at the three electron level and entropic promotion of the bond-breaking step, *J. Am. Chem. Soc.,* 107, 7389, 1985.
926. **Lehninger, A. L.,** *Principles of Biochemistry,* Worth Publishers, New York, 1982.
927. **Porter, M. E. and Johnson, K. A.,** Transient state kinetic analysis of the ATP-induced dissociation of the dynein-microtubule complex, *J. Biol. Chem.,* 258, 6582, 1983.
928. **Porter, M. E. and Johnson, K. A.,** Characterization of the ATP-sensitive binding of *Tetrahymena* 30 S dynein to bovine brain microtubules, *J. Biol. Chem.,* 258, 6575, 1983.
929. **Johnson, K. A.,** The pathway of ATP hydrolysis by dynein. Kinetics of a presteady state phosphate burst, *J. Biol. Chem.,* 258, 13825, 1983.
930. **Johnson, K. A.,** Pathway of the microtubule-dynein ATPase and the structure of dynein: a comparison with actomyosin, *Annu. Rev. Biophys. Biophys. Chem.,* 14, 161, 1985.
931. **Omoto, C. K. and Johnson, K. A.,** Activation of the dynein adenosinetriphosphatase by microtubules, *Biochemistry,* 25, 419, 1986.
932. **Taylor, E. W.,** Mechanism of actomyosin ATPase and the problem of muscle contraction, *CRC Crit. Rev. Biochem.,* 6, 103, 1979.

933. **Holzbaur, E. L. F. and Johnson, K. A.,** Rate of ATP synthesis by dynein, *Biochemistry,* 25, 428, 1986.
934. **Cornish-Bowden, A. and Endrenyi, L.,** Robust regression of enzyme kinetic data, *Biochem. J.,* 234, 21, 1986.
935. **Cornish-Bowden, A.,** Estimation of kinetic constants, in *Fundamentals of Enzyme Kinetics,* Butterworths, London, 1979, 205.
936. **Cornish-Bowden, A. and Endrenyi, L.,** Fitting of enzyme kinetic data without prior knowledge of weights, *Biochem. J.,* 193, 1005, 1981.
937. **Cornish-Bowden, A.,** in *Techniques in Protein and Enzyme Biochemistry,* Part II supplement, BS 115, Tipton, K. F., Ed., Elsevier, Limerick, Ireland, 1985, 1.
938. **Perlmutter-Hayman, B.,** Cooperative binding to macromolecules. A formal approach, *Acc. Chem. Res.,* 19, 90, 1986.
939. **Weetall, H. H. and Pitcher, W. H., Jr.,** Scaling up an immobilized enzyme system, *Science,* 232, 1396, 1986.
940. **Krebs, E. G. and Beavo, J. A.,** Phosphorylation-dephosphorylation of enzymes, *Annu. Rev. Biochem.,* 48, 923, 1979.
941. **Chock, P. B., Rhee, S. G., and Stadtman, E. R.,** Interconvertible enzyme cascades in cellular regulation, *Annu. Rev. Biochem.,* 49, 813, 1980.
942. **Rosen, O. M. and Krebs, E. G., Eds.,** *Protein Phosphorylation,* Cold Spring Harbor Laboratory, Cold Spring Harbor, N.Y., 1981.
943. **Cohen, P.,** The role of protein phosphorylation in neural and hormonal control of cellular activity, *Nature,* 296, 613, 1982.
944. **Nestler, E. J. and Greengard, P.,** *Protein Phosphorylation in the Nervous System,* Wiley-Interscience, New York, 1984.
945. **Cohen, P., Ed.,** *Enzyme Regulation by Reversible Phosphorylation: Further Advances,* Elsevier, Amsterdam, 1984.
946. **Schacter, E., Chock, P. B., and Stadtman, E. R.,** Regulation through phosphorylation/dephosphorylation cascade systems, *J. Biol. Chem.,* 259, 12252, 1984.
947. **Schacter, E., Chock, P. B., and Stadtman, E. R.,** Energy consumption in a cyclic phosphorylation/dephosphorylation cascade, *J. Biol. Chem.,* 259, 12260, 1984; correction, *J. Biol. Chem.,* 260, 6501, 1985.
948. **Goldbeter, A. and Koshland, D. E.,** Energy expenditure in the control of biochemical systems by covalent modification, *J. Biol. Chem.,* 262, 4460, 1987.
949. **Hwang, J. K. and Warshel, A.,** Semiquantitative calculations of catalytic free energies in genetically modified enzymes, *Biochemistry,* 26, 2669, 1987.
950. **Wells, J. A., Cunningham, B. C., Craycar, T. P., and Estell, D. A.,** Importance of hydrogen-bond formation in stabilizing the transition state of subtilism, *Phil. Trans. R. Soc. London,* A317, 415, 1986.
951. **Bryan, P., Pantoliano, M. W., Quill, S. G., Hsiao, H.-Y., and Poulos, T.,** Site-directed mutagenesis and the role of the oxyanion hole in subtilisin, *Proc. Natl. Acad. Sci. U.S.A.,* 83, 3743, 1986.
952. **Warshel, A. and Russell, S. T.,** Theoretical correlation of structure and energetics in the catalytic reaction of trypsin, *J. Am. Chem. Soc.,* 108, 6569, 1986.
953. **Hwang, J.-K. and Warshel, A.,** Microscopic examination of free-energy relationships for electron transfer in polar solvents, *J. Am. Chem. Soc.,* 109, 715, 1987.
954. **Fersht, A.,** *Enzyme Structure and Mechanism,* 2nd ed., W. H. Freeman, San Francisco, 1984.
955. **Lumry, R.,** Mechanical force, hydration and conformational fluctuations in enzymic catalysis, in *A Study of Enzymes,* Vol. II, Kuby, S. A., Ed., *CRC Press,* Boca Raton, FL, in press.
956. **Williams, R. J. P.,** Complexation and catalysis in biology, in *A Study of Enzymes,* Vol. 2, Kuby, S. A., Ed., CRC Press, Boca Raton, FL, in press.
957. **Garfinkel, D., Kulikowski, C. A., Soo, V.-W., Maclay, J., and Achs, M. J.,** Modeling and artificial intelligence approaches to enzyme systems, *Fed. Proc.,* 46, 2481, 1987.
958. **Palsson, B. O., Joshi, A., and Ozturk, S. A.,** Reducing complexity in metabolic networks: making metabolic meshes manageable, *Fed. Proc.,* 46, 2485, 1987.
959. **Kurz, L. C., Weitkamp, E., and Frieden, C.,** Adenosine deaminase: viscosity and the mechanism of substrate and of ground- and transition-state analogue inhibitors, *Biochemistry,* 26, 3027, 1987.
960. **Aflalo, C. and DeLuca, M.,** Continuous monitoring of adenosine 5′-triphosphate in the microenvironment of immobilized enzymes by firefly luciferase, *Biochemistry,* 26, 3913, 1987.
961. **Srivastava, D. K. and Bernhard, S. A.,** Metabolite transfer via enzyme-enzyme complexes, *Science,* 234, 1081, 1986.
962. **Srere, P. A.,** Why are enzymes so big?, *Trends Biochem. Sci.,* 9, 387, 1984.
963. **Dixon, M., Webb, E. C., Thorne, C. J. R., and Tipton, K. F.,** Enzyme inhibition and activation, in *Enzymes,* 3rd ed., Academic Press, New York, 1979, 332.
964. **Hamada, M., Takenaka, H., Sumida, M., and Kuby, S. A.,** Adenylate Kinase, in *A Study of Enzymes,* Vol. 2, Kuby, S. A., Ed., CRC Press, Boca Raton, FL, in press.

965. **Fry, D. C., Kuby, S. A., and Mildvan, A. S.,** NMR studies of the AMP-binding site and mechanism of adenylate kinase, *Biochemistry,* 26, 1645, 1987.
966. **Kuby, S. A., Ed.,** *A Study of Enzymes,* Vol. 2, CRC Press, Boca Raton, in press.
967. **Fisher, M. T. and Sligar, S. G.,** Temperature jump relaxation kinetics of the P-440_{cam} spin equilibrium, *Biochemistry,* 26, 4797, 1987.
968. **Gould, P., Gelb, M., and Sligar, S. G.,** Interaction of 5-bromocamphor with cytochrome P-450_{cam}. Production of 5-ketocamphor from a mixed spin state hemoprotein, *J. Biol. Chem.,* 256, 6686, 1981.
969. **Wells, J. M. C. and Fersht, A. R.,** Hydrogen bonding in enzymatic catalysis analyzed by protein engineering, *Nature,* 316, 656, 1985.
970. **Thomas, P. G., Russell, A. J., and Fersht, A. R.,** Tailoring the pH dependence of enzyme catalysis using protein engineering, *Nature,* 318, 375, 1985.
971. **Craik, C. S., Largman, C., Fletcher, T., Roczniak, S., Barr, P. J., Fletterick, R., and Rutter, W. J.,** Redesigning trypsin: alteration of substrate specificity, *Science,* 228, 291, 1985.
972. **Wodak, S. J.,** Computer aided design in protein engineering, in *Annals of the New York Academy of Sciences, Enzyme Engineering 8,* Vol. 508, Laskin, a. I., Mosback, K., Thomas, D., and Wingard, L. B., Eds., New York Academy of Sciences, New York, 1987, 1.
973. **Kaiser, E. T.,** Preparation of flavopapain and other semisynthetic enzymes, in *Annals of the New York Academy of Sciences, Enzyme Engineering 8,* Vol. 508, Laskin, A. I., Mosback, K., Thomas, D., and Wingard, L. B., Eds., New York Academy of Sciences, New York, 1987, 14.
974. **Spring, S., Standing, T., Fletterick, R. J., Stroud, R. M., Finer-Moore, J., Xuong, N.-H., Hamlin, R., Rutter, W. J., and Craik, C. S.,** The three-dimensional structure of asn^{102} mutant of trypsin: role of asp^{102} in serine protease catalysis, *Science,* 237, 905, 1987.
975. **Craik, C. S., Roczniak, S., Largman, C., and Rutter, W. J.,** The catalytic role of the active site aspartic acid in serine proteases, *Science,* 237, 909, 1987.
976. **Welch, G. R., Ed.,** *The Fluctuating Enzyme,* John Wiley & Sons, New York, 1986.
977. **Chance, B., Korszun, K., Khalid, S., Alter, C., Sorge, J., and Gabbidon, E.,** Time resolved studies of active site structural changes in solution, in *Structural Biological Applications of X-Ray Absorption, Scattering, and Diffraction,* Bartunik, H. D. and Chance, B., Eds., Academic Press, New York, 1986, 49.
978. **Frauenfelder, H.,** Ligand binding and protein dynamics, in *Structure and Motion: Membranes, Nucleic Acids, and Proteins,* Clementi, E., Corongiu, G., Sarma, M. H., and Sarma, R. H., Eds., Academic Press, New York, 1985, 205.
979. **Powers, L., Chance, B., Chance, M., Campbell, B., Friedman, J., Khalid, S., Kumar, C., Naqui, A., Reddy, K. S., and Zhou, Y.,** Kinetic, structural, and spectroscopic identification of geminate states of myoglobin: a ligand binding site on the reaction pathway, *Biochemistry,* 26, 4785, 1987.
980. **Morrison, J. F.,** Dihydrofolate reductase, in *A Study of Enzymes,* Vol. 2, Kuby, S. A., Ed., CRC Press, Boca Raton, FL, in press.
981. **Kuby, S. A., Fleming, G., and Wooden, S.,** ATP-creatine transphosphorylase (creatine Kinase) from rabbit muscle. A peptide fragment with nucleotide binding properties, *FASEB J.,* 2, A996, 1988.
982. **Mildvan, A. S. and Fry, D. C.,** NMR studies of the mechanism of enzyme action, in *Advances in Enzymology and Related Areas of Molecular Biology,* Meister, A., Ed., John Wiley & Sons, New York, 1987, 241.
983. **Madsen, N. B.,** Glycogen phosphorylase and glycogen synthetase, in *A Study of Enzymes,* Vol. 2, Kuby, S. A., Ed., CRC Press, Boca Raton, FL, in press.
984. **Yüksel, K. U. and Gracy, R. W.,** Triose- and hexose-phosphate isomerases, in *A Study of Enzymes,* Vol. 2, Kuby, S. A., Ed., CRC Press, Boca Raton, FL, in press.
985. **Rosevear, P. R., Powers, V. M., Dowan, D., Mildvan, A. S., and Kenyon, G. L.,** Nuclear Overhauser effect studies on the conformation of magnesium adenosine 5′-triphosphate bound to rabbit muscle creatine kinase, *Biochemistry,* 26, 5338, 1987.
986. **Wada, H., Manganiello, V. C., and Osborne, J. C., Jr.,** Analysis of the kinetics of cyclic AMP hydrolysis by the cyclic GMP-stimulated cyclic nucleotide phosphodiesterase, *J. Biol. Chem.,* 262, 13938, 1987.
987. **Kim, H. J., Nishikawa, S., Tanaka, T., Uesugi, S., Takenaka, H., Hamada, M., and Kuby, S. A.,** Synthetic genes for human muscle-type adenylate kinase in *Escherichia coli, Protein Eng.,* 2, 379, 1989.
988. **Johnson, M. S. and Kuby, S. A.,** Cytochrome *c*, quinone, and P-450, reductases, in *A Study of Enzymes,* Vol. 2, Kuby, S. A., Ed., CRC Press, Boca Raton, FL, in press.
989. **Morrison, J. F.,** The slow-binding and slow, tight-binding inhibition of enzyme catalysed reactions, *Trends Biochem. Sci.,* 7, 102, 1982.
990. **Morrison, J. F. and Walsh, C. T.,** The behavior and significance of slow-binding enzyme inhibitors, *Adv. Enzymol. Rel. Areas Mol. Biol.,* Vol. 61, Meister, A., Ed., John Wiley & Sons, New York, 1987, 201.
991. **Schloss, J. V.,** Significance of slow-binding enzyme inhibition and its relationship to reaction-intermediate analogues, *Acc. Chem. Res.,* 21, 348, 1988.

992. **Jacobs, J. and Schultz, P. G.,** Catalytic antibodies, *J. Am. Chem. Soc.*, 109, 2174, 1987.

993. **Benkovic, S. J., Napper, A. D., Janda, K., and Lerner, R. A.,** Catalytic antibodies, *Biochemistry,* 28, 1934, 1989.

994. **Gates, C. A. and Northrop, D. B.,** Alternative substrate and inhibition kinetics of aminoglycoside nucleotidyltransferase 2''-I in support of a Theorell-Chance kinetic mechanism, *Biochemistry,* 27, 3826, 1988.

995. **Gates, C. A. and Northrop, D. B.,** Substrate specificities and structure-activity relationships for the nucleotidylation of antibiotics catalyzed by aminoglycoside nucleotidyltransferase 2''-I, *Biochemistry,* 27, 3820, 1988.

996. **Gates, C. A. and Northrop, D. B.,** Determination of the rate-limiting segment of aminoglycoside nucleotidyltransferase 2''-I by pH- and viscosity-dependent kinetics, *Biochemistry,* 27, 3834, 1988.

997. **Iverson, B. L. and Lerner, R. A.,** Sequence-specific peptide cleavage catalyzed by an antibody, *Science,* 243, 1184, 1989.

998. **Anderson, K. S., Sikorski, J. A., and Johnson, K. A.,** Evaluation of 5-enolpyruvoylshikimate-3-phosphate synthase substrate and inhibitor binding by stopped-flow and equilibrium fluorescence measurements, *Biochemistry,* 27, 1604, 1988.

999. **Anderson, K. S., Sikorski, J. A., and Johnson, K. A.,** A tetrahedral intermediate in the EPSP synthase reaction observed by rapid quench kinetics, *Biochemistry,* 27, 7395, 1988.

1000. **Anderson, K. S., Sikorski, J. A., Benesi, A. J., and Johnson, K. A.,** Isolation and structural elucidation of the tetrahedral intermediate in the EPSP synthase enzymatic pathway, *J. Am. Chem. Soc.*, 110, 6577, 1988.

1001. **Cha, Y., Murray, C. J., and Klinman, J. P.,** Hydrogen tunneling in enzyme reactions, *Science,* 243, 1325, 1989.

1002. **Johnson, K. A.,** Rapid kinetic analysis of mechanochemical adenosine triphosphatases, in *Methods in Enzymology,* Vol. 134, Vallee, R. B., Colowick, S. P., and Kaplan, N. O., Eds., Academic Press, New York, 1986, 677.

1003. **Kivirikko, K. I., Myllylä, R., and Pihlajaniemi, T.,** Protein hydroxylation: prolyl 4-hydroxylase, an enzyme with four cosubstrates and a multifunctional subunit, *FASEB J.*, 3, 1609, 1989.

1004. **Myers, D. and Palmer, G.,** The kinetic mechanism(s) of cytochrome oxidase. Techniques for their analysis and criteria for their validation, in *Cytochrome Oxidase: Structure, Function and Physiopathology,* Brunori, M. and Chance, B., Eds., *Ann. N.Y. Acad. Sci.*, 550, 85, 1988.

1005. **Holzbaur, E. L. F. and Johnson, K. A.,** ADP release is rate limiting in steady-state turnover by the dynein adenosinetriphosphatase, *Biochemistry,* 28, 5577, 1989.

1006. **Leatherbarrow, R. J.,** Enzfitter. *A Non-Linear Regression Data Analysis Program for the IBM PC (and True Compatibles),* Elsevier, Amsterdam, 1987.

1007. **Chou, K.-C.,** Graphic rules in steady and non-steady state enzyme kinetics, *J. Biol. Chem.*, 264, 12074, 1989.

1008. **Schultz, P. G.,** Catalytic antibodies, *Acc. Chem. Res.*, 22, 287, 1989.

1009. **Kwiatkowski, A. P., Huang, C. Y., and King, M. M.,** Kinetic mechanism of the type II calmodulin-dependent protein kinase: Studies of the forward and reverse reactions and observation of apparent rapid-equilibrium ordered binding, *Biochemistry,* 29, 153, 1990. Appendix: Huang, C. Y., Various cases of rapid-equilibrium ordered bireactant mechanisms--their bases and differentation, *Biochemistry,* 29, 158, 1990.

1010. **Kuby, S. A., Hamada, M., Johnson, M. S., Russell, G. A., Manship, M., Palmieri, R., H., Fleming, G., Bredt, D. S., and Mildvan, A. S.,** Studies on adenosine triphosphate transphosphorylases. XVIII. Synthesis and preparation of peptides and peptide fragments of rabbit muscle ATP-AMP transphosphorylase (adenylate kinase) and their nucleotide-binding properties, *J. Protein Chem.*, 8, 549, 1989.

1011. **Kim, H. J., Nishikawa, S., Tokutomi, Y., Takenaka, H., Hamada, M., Kuby, S. A., and Uesugi, S.,** In vitro mutagenesis studies at the arginine residues of adenylate kinase. A revised binding site for AMP in the X-ray-deduced model, *Biochemistry,* 29, 1107, 1990.

Index

INDEX

A

B

C

D

E

F

G

H

I

K

L

M

N

O

P

Q

R

T

U

V

W

X

Y

Z